METHODS IN ENZYMOLOGY

VOLUME I

METHODS IN ENZYMOLOGY

I: Preparation and Assay of Enzymes

II: Preparation and Assay of Enzymes

III: Preparation and Assay of Substrates

IV: Special Techniques for the Enzymologist

V: Preparation and Assay of Enzymes

VI: Preparation and Assay of Enzymes (*Continued*)
Preparation and Assay of Substrates
Special Techniques

VII: Cumulative Subject Index

METHODS IN ENZYMOLOGY

Edited by

SIDNEY P. COLOWICK and NATHAN O. KAPLAN

McCollum-Pratt Institute
The Johns Hopkins University, Baltimore, Maryland

VOLUME I

ACADEMIC PRESS INC., PUBLISHERS
NEW YORK, 1955

ACADEMIC PRESS INC.
111 FIFTH AVENUE
NEW YORK, NEW YORK 10003

United Kingdom Edition
Published by
ACADEMIC PRESS INC. (LONDON) LTD.
BERKELEY SQUARE HOUSE, LONDON W. 1

Library of Congress Catalog Card Number: 54-9110

Fifth Printing, 1968

Preface

In this four-volume work is presented, for the first time in the English language, a comprehensive compilation of the methods used in the study of enzymes. In certain respects, this work should serve as a companion piece for Sumner and Myrbäck's "The Enzymes," in which methodology has not been emphasized. The primary aim has been to provide workers in the field with laboratory directions for the preparation and assay of enzymes (Volumes I and II), for the preparation and determination of substrates (Volume III), and for certain special techniques of particular use to the enzymologist (Volume IV).

In order to accomplish this aim, it has been necessary to enlist the aid of some 300 investigators, each a specialist on the particular subject which he has contributed, since it is only in this way that one can hope to obtain an authoritative description of the most modern and reliable procedure in each case. The editors have been fortunate in obtaining contributions from leading scientists in all accessible parts of the world. The number of contributions from American laboratories is relatively large; this is in part a reflection of the growth of many new centers of enzyme research in this country during the past decade.

The need for a treatise of this kind is well attested to by the enthusiastic response of the great majority of those investigators who were asked to contribute. It is hoped that such a work will prove useful not only to specialists in the field of enzymology but even more so to those in bordering fields, who often have difficulty in finding or evaluating critically the enzyme methods needed in their work.

The field of enzymology, strictly defined, concerns the chemistry and mechanism of enzyme-catalyzed reactions, as well as the chemistry of the enzymes themselves. In recent years, however, the interest in enzymology has spread into practically all branches of biological and medical sciences. The embryologist is interested in enzyme changes during development. The geneticist seeks relationships between the enzymatic make-up and the gene pattern of the cell. The pathologist is interested in changes in the enzyme pattern of tissues in disease states, and the clinician makes use of changes in the enzyme composition of the blood-plasma as a diagnostic tool. The cell physiologist seeks the enzymatic basis for the performance of work in living systems, as in motility, bioluminescence, absorption and excretion, and electrical activity. The endocrinologist seeks an explanation for hormone action in the control of active enzyme concentration in

v

the cell. The pharmacologist seeks a rational approach to chemotherapy through studies on the enzymatic basis of drug action. The nutritionist seeks a basis for vitamin and mineral requirements in terms of their function as integral parts of enzyme systems. The cytologist is engaged in determining the locus of various enzymatic activities in different portions of the cell. The radiologist seeks an explanation for the biological effects of radiation in terms of their effects on enzyme systems. Finally, of course, the biochemist, when he seeks to determine the chemical events occurring in living cells by means of isotopes, optical methods, nutritional studies, or the like, is in reality studying an integrated series of enzyme-catalyzed reactions, and often resorts to enzymological studies in cell-free systems in order to confirm and extend his findings.

In all these examples, the interest in the enzymes is based on their role in cellular function. There is another great field of interest in enzymes, based on their use as reagents. Space does not permit enumeration of the many applications of enzymes as reagents in industrial technology. In the past, the use of enzymes as reagents has been largely limited to crude preparations which serve some degradative function in digestive proc-esses. A relatively new application of enzymes, made possible by their isolation in pure form, lies in their use as analytical reagents. There is no reagent more specific than a pure enzyme, and enzymes as analytical reagents, thus far exploited mainly by the enzymologists themselves, will certainly find wide use in the years ahead in the hands of analytical chemists.

With a view toward eliminating some of the artificial barriers which have sprung up between workers using different approaches to problems in enzymology, the editors have included in these volumes, particularly in Volume IV, a number of subjects not ordinarily regarded as within the scope of enzymology. For example, techniques are described for the study of metabolism of intact cells and tissues by optical and histochemical methods, and for the study of tissues in varying states of disorganization (slices, homogenates, mitochondria, chloroplasts, etc.). At the other end of the scale, methods are included for the physicochemical characteri-zation of enzymes by a variety of techniques designed to give information on the purity, size, shape, essential groups, and amino acid composition and sequence. A large section is also devoted to the use of radioactive and stable isotopes, including methods of measurement as well as application in metabolic studies. Although labeled compounds have been used thus far primarily for tracing metabolic pathways in living cells, their use in elucidating reaction mechanisms in cell-free systems is becoming more and more prevalent, and the inclusion of such procedures therefore serves a twofold purpose.

Although the scope of this work is necessarily broad, one of its main functions is to provide methods for the purification and assay of enzymes, and the bulk of Volumes I and II is devoted to this subject. After some general considerations of the methodology involved in the extraction of enzymes from various types of cells and in the fractionation of enzymes by means of salts, organic solvents, adsorbents, and chromatography, there follows a detailed description of the assay, purification, and properties of each of the individual enzymes covered in this treatise.

An attempt has been made at complete coverage of all known enzymes, but this is manifestly impossible for a number of reasons. The most obvious of these is the continuing growth of the field, with the discovery of newly found enzymes being announced in practically every issue of the current journals. The past five years have witnessed remarkable developments, particularly with respect to the enzyme systems concerned in fat and nucleic acid metabolism, in alternate pathways of carbohydrate metabolism, and in synthetic (energy-requiring) processes. In the face of continuing progress, the time nevertheless appears ripe for gathering together what is known of these systems, since we are now in a position where we can at least put many of them on an equal footing with the enzymes of the glycolytic scheme, which for so long served as the major source of problems and achievements for the enzymologist.

Progress has been evident not only in the discovery of a multitude of new enzyme systems but also in the development of a variety of new techniques. Most prominent has been the development of column and paper chromatographic techniques for the separation of enzymes and for the identification of reaction products. To a large degree, the availability of improved physical equipment has been responsible for major advances in enzyme technology. For example, equipment for low-temperature, high-speed centrifugation has made possible protein separations not previously feasible. The use of spectrophotometric methods for following the course of enzyme-catalyzed reactions has become more widespread and has supplanted to a considerable extent the use of manometric techniques.

In general, it has not been considered within the scope of this treatise to provide the reader with operating instructions for various types of apparatus. This kind of information is readily available from other sources.

A number of editorial problems have arisen which seem worthy of mention. One question was whether or not two enzymes which catalyze the same reaction but are obtained from different sources should both be described fully in the present work. The policy decided upon was that both should be covered only when distinct differences in properties have been demonstrated (e.g., differences in metal or coenzyme requirement, or, when pure, differences in amino acid composition, etc.). Another question was whether enzymes deserve description when they have been detected in crude extracts of tissues but have not been purified extensively. It was decided that such preparations should be included, since the method of assay should prove useful to those interested in studying the distribution or further purification of the enzyme.

With respect to the arrangement of subject matter, an attempt has been made to organize the material on the basis of metabolic relationships. For example, the enzymes of carbohydrate metabolism are presented more or less in the order of their appearance in metabolic schemes, and this plan has been followed throughout wherever possible. An alphabetic arrangement would theoretically be more desirable, but with the present state of multiple nomenclature in enzymology, such an arrangement is not feasible.

Concerning the problem of nomenclature, we have taken the view that the author writing on a particular subject is in the best position to evaluate the nomenclature in that area. For this reason, no attempt has been made to introduce any revisions in nomenclature. Instead, wherever an enzyme or substrate has common names other than those used by the author, all alternative names appear in the subject index.

A major point of organization concerned the question of whether the preparation and properties of substrates and coenzymes should be described along with the related enzymes. In view of the fact that a particular substrate or coenzyme might be involved in a number of different enzyme-catalyzed reactions, the only way to avoid overlapping was to describe substrates and coenzymes in a separate volume (Volume III). In only a few cases, in which it was quite obvious that the substrate concerned was involved in only one enzyme system (e.g., the luciferin-luciferase system) do the descriptions of enzyme and substrate appear in the same section. Although it is realized that this kind of organization may cause some inconvenience to the user of these volumes, no other arrangement appeared suitable. In order to facilitate use, copious cross references are included, not only between volumes, but between articles in the same volume.

Another point with respect to the description of substrates was the question of whether the preparation of substances which are at present commercially available need be included. Here the policy has been to include a description of the preparation of many commercially available substrates and enzymes. The reasons for this were the following. In many cases, commercially available materials must be subjected to further purification, for which directions are desirable. Second, in the case of those materials which are isolated from biological sources, there is often a need, as, for example, in isotope experiments, for performing the isolation procedure in the laboratory. Third, in some cases commercial materials are not readily obtainable, for reasons of cost or accessibility. Finally, it seems desirable to include descriptions of this kind, if only for the training of students, who should become familiar with the preparations of certain materials, even when they are commercially available.

In order to keep the size of this treatise within reasonable bounds, it was decided that figures and photographs should be kept to the minimum. It appeared sufficient in most cases to give values for Michaelis' constants, pH optima, molecular weights, R_f values, etc., without including Lineweaver-Burk plots, pH-activity curves, sedimentation diagrams, chromatograms, etc. Photographs of crystalline enzymes were not included, since it seemed sufficient for the purposes of this treatise to include simply a description of crystal form.

We wish to express our gratitude to all the contributors who made the publication of this treatise possible. We wish also to thank the members of the Advisory Board, who helped in the choice of subject matter, as well as in the selection of appropriate authors. We wish particularly to acknowledge the secretarial assistance of Miss Helen Yates, whose efforts in this work went far beyond the call of duty. We are also grateful to Dr. John B. Wolff for his painstaking translation of several contributions, including articles 40, 41, 57, 58, 63, 65, 66 and 94 in Volume I. Finally, we wish to pay tribute to Mr. Kurt Jacoby of Academic Press, who was responsible for initiating the negotiations for the preparation of this work, and whose enthusiastic and energetic administration served as an effective stimulus to the activities of the somewhat less energetic editors.

SIDNEY P. COLOWICK
NATHAN O. KAPLAN

Baltimore, Maryland
Feb. 15, 1955

Contributors to Volume I

Article numbers are shown in parentheses following the names of contributors.

CHRISTIAN B. ANFINSEN, JR. (115), *National Institutes of Health, Bethesda, Maryland*

BERNARD AXELROD (4, 51), *Western Regional Laboratory, Albany, California*

J. S. D. BACON (30), *The University, Sheffield, England*

H. A. BARKER (91, 99, 103), *University of California, Berkeley, California*

GERWIN BEISENHERZ (57, 58), *University of Marburg, Marburg, Germany*

PETER BERNFELD (17, 31), *Tufts University, Boston, Massachusetts*

RONALD BENTLEY (45), *University of Pittsburgh, Pittsburgh, Pennsylvania*

M. BIER (106), *Fordham University, New York, N. Y.*

SIMON BLACK (81), *National Institutes of Health, Bethesda, Maryland*

WALTER D. BONNER, JR. (121), *Smithsonian Institute, Washington, D. C.*

ROGER BONNICHSEN (78), *Medical Nobel Institute, Stockholm, Sweden*

MAX BOVARNICK (88), *New York State Department of Health, Albany, New York*

NORMAN G. BRINK (78), *Department of Organic and Biochemical Research, Merck and Company, Rahway, New Jersey*

THEODOR BÜCHER (58, 63, 65, 66), *University of Marburg, Marburg, Germany*

ROBERT M. BURTON (59, 84), *Johns Hopkins University, Baltimore, Maryland*

WILLIAM L. BYRNE (38), *Duke University School of Medicine, Durham, North Carolina*

WALTER CHRISTIAN (40), *C. F. Boehringer and Söhne, Mannheim-Waldhof, Germany*

SEYMOUR S. COHEN (47, 52), *University of Pennsylvania, Philadelphia, Pennsylvania*

SIDNEY P. COLOWICK (11), *The Johns Hopkins University, Baltimore, Maryland*

GERTY T. CORI (23, 24, 25), *Washington University, St. Louis, Missouri*

ROBERT K. CRANE (33), *Washington University, St. Louis, Missouri*

R. D. DEMOSS (43, 80), *Johns Hopkins University, Baltimore, Maryland*

MALCOLM DIXON (68), *Cambridge University, Cambridge, England*

ALBERT DORFMAN (19), *University of Chicago, Chicago, Illinois*

M. DOUDOROFF (28), *University of California, Berkeley, California*

K. A. C. ELLIOTT (1), *Montreal Neurological Institute, Montreal, Canada*

DAVID S. FEINGOLD (29), *Hebrew University, Jerusalem, Israel*

WILLIAM H. FISHMAN (31), *Tufts University, Boston, Massachusetts*

C. S. FRENCH (9), *Carnegie Institution, Stanford, California*

KARL-HEINZ GARBADE (58), *University of Marburg, Marburg, Germany*

JOHN GERGELY (100), *Massachusetts General Hospital, Boston, Massachusetts*

MARTIN GIBBS (62), *Brookhaven National Laboratory, Brookhaven, New York*

G. GOMORI (16), *University of Chicago, Chicago, Illinois*

PAUL R. GORHAM (5), *National Research Council, Ottawa, Ontario, Canada*

ARDA A. GREEN (10), *The Johns Hopkins University, Baltimore, Maryland*

I. C. GUNSALUS (7), *University of Illinois, Urbana, Illinois*

G. DE LA HABA (54), *Yale University School of Medicine, New Haven, Connecticut*

L. P. HAGER (75), *Massachusetts General Hospital, Boston, Massachusetts*

OSAMU HAYAISHI (14, 110), *Washington University, St. Louis, Missouri*

EDWARD J. HEHRE (21), *Cornell University School of Medicine, New York, N. Y.*

LEON A. HEPPEL (15), *National Institutes of Health, Bethesda, Maryland*

DENIS HERBERT (125), *Experimental Station, Porton, England*

H. G. HERS (34), *Université de Louvain, Louvain, Belgium*

SHLOMO HESTRIN (29), *Hebrew University, Jerusalem, Israel*

C. H. W. HIRS (13), *Rockfeller Institute, New York, N. Y.*

GEORGE H. HOGEBOOM (3), *National Cancer Institute, Bethesda, Maryland*

B. L. HORECKER (42, 53, 55), *National Institutes of Health, Bethesda, Maryland*

WALTER L. HUGHES (10), *The Johns Hopkins University, Baltimore, Maryland*

BARBARA ILLINGWORTH (23), *Washington University, St. Louis, Missouri*

MARY ELLEN JONES (96), *Massachusetts General Hospital, Boston, Massachusetts*

ELLIOT JUNI (72), *University of Illinois, Urbana, Illinois*

NATHAN O. KAPLAN (46), *The Johns Hopkins University, Baltimore, Maryland*

SEYMOUR KAUFMAN (120), *New York University, New York, N. Y.*

PATRICIA J. KELLER (23, 24), *Washington University, St. Louis, Missouri*

Z. I. KERTESZ (18), *Agricultural Experiment Station, Cornell University, Ithaca, New York*

HANS KLENOW (50), *The University, Copenhagen, Denmark*

H. J. KOEPSELL (74), *The Upjohn Company, Kalamazoo, Michigan*

S. KORKES (76, 77), *Duke University, Durham, North Carolina*

ARTHUR KORNBERG (42, 67, 99, 111, 117, 118), *Washington University, St. Louis, Missouri*

EDWIN G. KREBS (61), *University of Washington, Seattle, Washington*

KIYOSHI KURAHASHI (126), *Western Reserve University, Cleveland, Ohio*

HENRY A. LARDY (38), *University of Wisconsin, Madison, Wisconsin*

JOSEPH LARNER (27), *University of Illinois, Urbana, Illinois*

ALBERT L. LEHNINGER (89), *The Johns Hopkins University, Baltimore, Maryland*

LUIS F. LELOIR (35, 48), *Instituto de Investigaciones Bioquimicas, Buenos Aires, Argentina*

FRANZ LEUTHARDT (41), *University of Zurich, Zurich, Switzerland*

I. LIEBERMAN (103), *Washington University, St. Louis, Missouri*

KUO-HUANG LING (38), *University of Wisconsin, Madison, Wisconsin*

FRITZ LIPMANN (75, 96), *Massachusetts General Hospital, Boston, Massachusetts*

HENRY N. LITTLE (87), *University of Massachusetts, Amherst, Massachusetts*

FEODOR LYNEN (94), *University of Munich, Munich, Germany*

HENRY R. MAHLER (85, 92), *University of Wisconsin, Madison, Wisconsin*

VINCENT MASSEY (122), *Cambridge University, Cambridge, England*

A. MAZUR (108), *College of the City of New York, New York, N. Y.*

MARGARET R. McDONALD (32), *Carnegie Institution, Cold Spring Harbor, New York*

H. W. MILNER (9), *Carnegie Institution, Stanford, California*

ROBERT K. MORTON (6), *University of Melbourne, Melbourne, Australia*

DAVID NACHMANSOHN (104, 107), *Columbia University, New York, N. Y.*

VICTOR A. NAJJAR (36), *The Johns Hopkins University, Baltimore, Maryland*

ALVIN NASON (8, 87), *The Johns Hopkins University, Baltimore, Maryland*

J. B. NEILANDS (69), *University of California, Berkeley, California*

SEVERO OCHOA (114, 116, 123, 124), *New York University, New York, N. Y.*

PETER OESPER (64), *Hahneman Medical College, Philadelphia, Pennsylvania*

G. PFLEIDERER (66), *University of Marburg, Marburg, Germany*

G. W. E. PLAUT (119), *New York University College of Medicine, New York, N. Y.*

R. R. PORTER (12), *National Institute, Mill Hill, London, England*

VAN R. POTTER (2), *McCardle Laboratories, University of Wisconsin, Madison, Wisconsin*

J. H. QUASTEL (112), *Montreal General Hospital, Montreal, Canada*

E. RACKER (54, 56, 70, 79, 83), *Yale University, New Haven, Connecticut*

IRWIN A. ROSE (97), *Yale University, School of Medicine, New Haven, Connecticut*

HENRY Z. SABLE (49), *Tufts University, Boston, Massachusetts*

MICHAEL SCHRAMM (29), *Hebrew University, Jerusalem, Israel*

J. EDWIN SEEGMILLER (82), *Boston City Hospital, Boston, Massachusetts*

H. W. SEELEY (105), *Cornell University, Ithaca, New York*

THOMAS P. SINGER (71), *University of Wisconsin, Madison, Wisconsin*

MILTON W. SLEIN (37), *Camp Detrick, Frederick, Maryland*

P. Z. SMYRNIOTIS (42, 53, 55), *National Institutes of Health, Bethesda, Maryland*

ALBERTO SOLS (33), *Washington University, St. Louis, Missouri*

MORRIS SOODAK (102), *Massachusetts General Hospital, Boston, Massachusetts*

E. R. STADTMAN (84, 98, 99), *National Institutes of Health, Bethesda, Maryland*

THRESSA C. STADTMAN (113), *National Institutes of Health, Bethesda, Maryland*

JOSEPH R. STERN (93, 95), *New York University, New York, N. Y.*

GEORGE DE STEVENS (20), *Remington Rand, Inc., Middletown, Connecticut*

ELMER STOTZ (109), *Rochester Medical School, Rochester, New York*

HAROLD J. STRECKER (44, 73), *Psychiatric Institute, Columbia University, College of Physicians and Surgeons, New York, N. Y.*

P. K. STUMPF (90), *University of California, Berkeley, California*

S.-C. SUNG (119), *University of Wisconsin, Madison, Wisconsin*

E. W. SUTHERLAND (26), *Western Reserve University, Cleveland, Ohio*

HERBERT TABOR (101), *National Institutes of Health, Bethesda, Maryland*

J. F. TAYLOR (39), *Louisville School of Medicine, Louisville, Kentucky*

RAÚL E. TRUCCO (35, 48), *Instituto de Investigaciones Bioquimicas, Buenos Aires, Argentina*

MILTON F. UTTER (126), *Western Reserve University, Cleveland, Ohio*

SIDNEY F. VELICK (60), *Washington University, St. Louis, Missouri*

W. J. WHELAN (22), *University College of North Wales, Bangor, Caernarvonshire, Great Britain*

OTTO WIELAND (94), *University of Munich, Munich, Germany*

IRWIN B. WILSON (104, 107), *Columbia University, New York, N. Y.*

H. P. WOLF (41), *University of Zurich, Zurich, Switzerland*

JOHN B. WOLFF (46), *The Johns Hopkins University, Baltimore, Maryland*

ISRAEL ZELITCH (86), *Connecticut Agricultural Experiment Station, New Haven, Connecticut*

Outline of Organization

VOLUME I
PREPARATION AND ASSAY OF ENZYMES

	Article Numbers	Pages
Section I. General Preparative Procedures		
A. Tissue Slice Technique	1	3–9
B. Tissue Homogenates	2	10–15
C. Fractionation of Cellular Components	3–5	16–25
D. Methods of Extraction of Enzymes	6–9	25–67
E. Protein Fractionation	10–15	67–138
F. Preparation of Buffers	16	138–146
Section II. Enzymes of Carbohydrate Metabolism		
A. Polysaccharide Cleavage, and Synthesis	17–27	149–225
B. Disaccharide Hexoside and Glucuronoside Metabolism	28–31	225–269
C. Metabolism of Hexoses	32–48	269–356
D. Metabolism of Pentoses	49–56	357–386
E. Metabolism of Three-Carbon Compounds	57–77	387–495
F. Reactions of Two-Carbon Compounds	78–86	495–535
G. Reactions of Formate	87,88	536–541
Section III. Enzymes of Lipid Metabolism		
A. Fatty Acid Oxidation	89–59	545–585
B. Acyl Activation and Transfer	96–105	585–627
C. Lipases and Esterases	106–109	627–660
D. Phospholipid and Steroid Enzymes	110–113	660–681
Section IV. Enzymes of Citric Acid Cycle	114–126	685–763

VOLUME I

TABLE OF CONTENTS

	Page
PREFACE	v
CONTRIBUTORS TO VOLUME I	xi
OUTLINE OF ORGANIZATION, VOLUME I	xv
OUTLINE OF VOLUMES I–VI	xxiii

Section I. General Preparative Procedures

1. Tissue Slice Technique	K. A. C. ELLIOTT	3
2. Tissue Homogenates	VAN R. POTTER	10
3. Fractionation of Cell Components of Animal Tissues	GEORGE H. HOGEBOOM	16
4. Preparation of Mitochondria from Plants	BERNARD AXELROD	19
5. Preparation of Chloroplasts and Disintegrated Chloroplasts	PAUL R. GORHAM	22
6. Methods of Extraction of Enzymes from Animal Tissues	ROBERT K. MORTON	25
7. Extraction of Enzymes from Microorganisms (Bacteria and Yeast)	I. C. GUNSALUS	51
8. Extraction of Soluble Enzymes from Higher Plants	ALVIN NASON	62
9. Disintegration of Bacteria and Small Particles by High-Pressure Extrusion	C. S. FRENCH and H. W. MILNER	64
10. Protein Fractionation on the Basis of Solubility in Aqueous Solutions of Salts and Organic Solvents	ARDA ALDEN GREEN and WALTER L. HUGHES	67
11. Separation of Proteins by Use of Adsorbents	SIDNEY P. COLOWICK	90
12. The Partition Chromatography of Enzymes	R. R. PORTER	98
13. Chromatography of Enzymes on Ion Exchange Resins	C. H. W. HIRS	113
14. Special Techniques for Bacterial Enzymes. Enrichment Culture and Adaptive Enzymes	OSAMU HAYAISHI	126
15. Separation of Proteins from Nucleic Acids	LEON HEPPEL	137
16. Preparation of Buffers for Use in Enzyme Studies	G. GOMORI	138

Page

Section II. Enzymes of Carbohydrate Metabolism

17. Amylases, α and β PETER BERNFELD   149

18. Pectic Enzymes   Z. I. KERTESZ   158

19. Mucopolysaccharidases   ALBERT DORFMAN   166

20. Cellulase Preparation from *Helix pomatia* (Snails)   GEORGE DE STEVENS   173

21. Polysaccharide Synthesis from Disaccharides   EDWARD J. HEHRE   178

22. Phosphorylases from Plants   W. J. WHELAN   192

23. Muscle Phosphorylase   GERTY T. CORI, BARBARA ILLINGWORTH, and PATRICIA J. KELLER   200

24. The PR Enzyme of Muscle   PATRICIA J. KELLER and GERTY T. CORI   206

25. Amylo-1,6-glucosidase   GERTY T. CORI   211

26. Polysaccharide Phosphorylase, Liver   EARL W. SUTHERLAND   215

27. Branching Enzyme from Liver   JOSEPH LARNER   222

28. Disaccharide Phosphorylases   M. DOUDOROFF   225

29. Hexoside Hydrolases   SHLOMO HESTRIN, DAVID S. FEINGOLD, and MICHAEL SCHRAMM   231

30. Methods for Measuring Transglycosylase Activity of Invertases   J. S. D. BACON   258

31. Glucuronidases   WILLIAM H. FISHMAN and PETER BERNFELD   262

32. Yeast Hexokinase   MARGARET R. McDONALD   269

33. Animal Tissue Hexokinases   ROBERT K. CRANE and ALBERTO SOLS   277

34. Fructokinase (Ketohexokinase)   H. G. HERS   286

35. Galactokinase and Galactowaldenase   LUIS F. LELOIR and RAÚL E. TRUCCO   290

36. Phosphoglucomutase from Muscle   VICTOR A. NAJJAR   294

37. Phosphohexoisomerases from Muscle   MILTON W. SLEIN   299

38. Phosphohexokinase   KUO-HUANG LING, WILLIAM L. BYRNE, and HENRY A. LARDY   306

		Page
39. Aldolase from Muscle	JOHN FULLER TAYLOR	310
40. Aldolase from Yeast	WALTER CHRISTIAN	315
41. Fructose-1-Phosphate Aldolase from Liver	FRANZ LEUTHARDT and H. P. WOLF	320
42a. Glucose-6-phosphate Dehydrogenase	A. KORNBERG and B. L. HORECKER	323
b. 6-Phosphogluconic Dehydrogenase	B. L. HORECKER and P. Z. SMYRNIOTIS	323
43. Glucose-6-phosphate and 6-phosphogluconic Dehydrogenases from *Leuconostoc mesenteroides*	R. D. DeMoss	328
44. Glucose Dehydrogenase from Liver	HAROLD J. STRECKER	335
45. Glucose Aerodehydrogenase (Glucose Oxidase)	RONALD BENTLEY	340
46. Hexose Phosphate and Hexose Reductase	JOHN B. WOLFF and NATHAN O. KAPLAN	346
47. Gluconokinase	SEYMOUR S. COHEN	350
48. Synthesis of Glucose-1,6-diphosphate	LUIS F. LELOIR and RAÚL E. TRUCCO	354
49. Yeast Pentokinase	HENRY Z. SABLE	357
50. Phosphoribomutase from Muscle	HANS KLENOW	361
51. Pentose Phosphate Isomerase	BERNARD AXELROD	363
52. Pentose Isomerases	SEYMOUR S. COHEN	366
53. Transketolase from Liver and Spinach	B. L. HORECKER and P. Z. SMYRNIOTIS	371
54. Crystalline Transketolase from Baker's Yeast	G. DE LA HABA and E. RACKER	375
55. Transaldolase	B. L. HORECKER and P. Z. SMYRNIOTIS	381
56. Deoxyribose Phosphate Aldolase (DR-Aldolase)	E. RACKER	384
57. Triosephosphate Isomerase from Calf Muscle	GERWIN BEISENHERZ	387
58. α-Glycerophosphate Dehydrogenase from Rabbit Muscle	GERWIN BEISENHERZ, THEODOR BÜCHER, and KARL-HEINZ GARBADE	391
59. Glycerol Dehydrogenase from *Aerobactor aerogenes*	ROBERT MAIN BURTON	397

Page

60. Glyceraldehyde-3-phosphate Dehydrogenase from Muscle SIDNEY F. VELICK 401

61. Glyceraldehyde-3-phosphate Dehydrogenase from Yeast EDWIN G. KREBS 407

62. TPN Triosephosphate Dehydrogenase from Plant Tissue MARTIN GIBBS 411

63. Phosphoglycerate Kinase from Brewer's Yeast THEODOR BÜCHER 415

64. Phosphoglyceric Acid Mutases PETER OESPER 423

65. Enolase from Brewer's Yeast THEODOR BÜCHER 427

66. Pyruvate Kinase from Muscle THEODOR BÜCHER and G. PFLEIDERER 435

67. Lactic Dehydrogenase of Muscle ARTHUR KORNBERG 441

68. Lactic Dehydrogenase from Yeast MALCOLM DIXON 444

69. Lactic Dehydrogenase of Heart Muscle J. B. NEILANDS 449

70. Glyoxalases E. RACKER 454

71. Plant Carboxylases THOMAS P. SINGER 460

72. Acetoin Formation in Bacteria ELLIOT JUNI 471

73. Phosphoroclastic Split of Pyruvate, Yielding Formate (*E. coli*) HAROLD J. STRECKER 476

74. Phosphoroclastic Split of Pyruvate, Yielding Hydrogen (*Clostridium butylicum*) H. J. KOEPSELL 479

75. Phosphate-Linked Pyruvic Acid Oxidase from *Lactobacillus delbrückii* L. P. HAGER and FRITZ LIPMANN 482

76. Coenzyme A-Linked Pyruvic Oxidase (Animal) S. KORKES 486

77. Coenzyme A-Linked Pyruvic Oxidase (Bacterial) S. KORKES 490

78. Liver Alcohol Dehydrogenase ROGER BONNICHSEN and NORMAN G. BRINK 495

79. Alcohol Dehydrogenase from Baker's Yeast E. RACKER 500

80. TPN-Alcohol Dehydrogenase from *Leuconostoc mesenteroides* R. D. DeMOSS 504

81. Potassium-Activated Yeast Aldehyde Dehydrogenase SIMON BLACK 508

82. TPN-Linked Aldehyde Dehydrogenase from Yeast J. EDWIN SEEGMILLER 511

Page

83. Liver Aldehyde Dehydrogenase   E. RACKER   514

84. Aldehyde. Dehydrogenase from *Clostridium kluyveri*   E. R. STADTMAN and ROBERT M. BURTON   518

85. Flavin-Linked Aldehyde Oxidase   HENRY R. MAHLER   523

86. Glycolic Acid Oxidase and Glyoxylic Acid Reductase   ISRAEL ZELITCH   528

87. Formic Dehydrogenase from Peas   ALVIN NASON and HENRY N. LITTLE   536

88. Formic Hydrogenlyase from *Escherichia coli*   MAX BOVARNICK   539

Section III. Enzymes of Lipid Metabolism

89. Fatty Acid Oxidation in Mitochondria   ALBERT L. LEHNINGER   545

90. Fatty Acid Oxidation in Higher Plants   P. K. STUMPF   549

91. Butyrate Enzymes of *Clostridium kluyveri*   H. A. BARKER   551

92. Butyryl Coenzyme A Dehydrogenase   HENRY R. MAHLER   553

93. Crystalline Crotonase from Ox Liver   JOSEPH R. STERN   559

94. β-Ketoreductase   FEODOR LYNEN and OTTO WIELAND   566

95. Enzymes of Acetoacetate Formation and Breakdown   JOSEPH R. STERN   573

96. Aceto-CoA-Kinase   MARY ELLEN JONES and FRITZ LIPMANN   585

97. Acetate Kinase of Bacteria (Acetokinase)   IRWIN A. ROSE   591

98. Phosphotransacetylase from *Clostridium kluyveri*   E. R. STADTMAN   596

99. Coenzyme A Transphorase from *Clostridium kluyveri*   H. A. BARKER, E. R. STADTMAN, and ARTHUR KORNBERG   599

100. Deacylases (Thiol Esterase)   JOHN GERGELY   602

101. Acetylation of Amines with Pigeon Liver Enzyme   HERBERT TABOR   608

102. Acetylation of D-Glucosamine by Pigeon Liver Extracts   MORRIS SOODAK   612

103. Amino Acid Acetylase of *Clostridium kluyveri*   I. LIEBERMAN and H. A. BARKER   616

104. Choline Acetylase   DAVID NACHMANSOHN and IRWIN B. WILSON   619

Page

105. Acetoacetate Decarboxylase — H. W. SEELEY — 624

106. Lipases — M. BIER — 627

107. Acetylcholinesterase — DAVID NACHMANSOHN and IRWIN B. WILSON — 642

108. Dialkylphosphofluoridase — A. MAZUR — 651

109. Liver Esterase — ELMER STOTZ — 657

110. Phospholipases — OSAMU HAYAISHI — 660

111. Enzymatic Phospholipid Synthesis: Phosphatidic Acid — ARTHUR KORNBERG — 673

112. Choline Oxidase — J. H. QUASTEL — 674

113. Cholesterol Dehydrogenase from a *Mycobacterium* — THRESSA C. STADTMAN — 678

Section IV. Enzymes of Citric Acid Cycle

114. Crystalline Condensing Enzyme from Pig Heart — SEVERO OCHOA — 685

115. Aconitase from Pig Heart Muscle — CHRISTIAN B. ANFINSEN, JR. — 695

116. Isocitric Dehydrogenase System (TPN) from Pig Heart — SEVERO OCHOA — 699

117. Isocitric Dehydrogenase of Yeast (TPN) — ARTHUR KORNBERG — 705

118. Isocitric Dehydrogenase of Yeast (DPN) — ARTHUR KORNBERG — 707

119. Diphosphopyridine Nucleotide Isocitric Dehydrogenase from Animal Tissues — G. W. E. PLAUT and S.-C. SUNG — 710

120. α-Ketoglutaric Dehydrogenase System and Phosphorylating Enzyme from Heart Muscle — SEYMOUR KAUFMAN — 714

121. Succinic Dehydrogenase — WALTER D. BONNER, JR. — 722

122. Fumarase — VINCENT MASSEY — 729

123. Malic Dehydrogenase from Pig Heart — SEVERO OCHOA — 735

124. "Malic" Enzyme — SEVERO OCHOA — 739

125. Oxalacetic Carboxylase of *Micrococcus lysodeikticus* — DENIS HERBERT — 753

126. Oxalacetate Synthesizing Enzyme — MERTON F. UTTER and KIYOSHI KURAHASHI — 758

AUTHOR INDEX 765

SUBJECT INDEX 785

Outline of Volumes I through VI

VOLUME I

PREPARATION AND ASSAY OF ENZYMES

Section I. General Preparative Procedures

A. Tissue Slice Technique. **B.** Tissue Homogenates. **C.** Fractionation of Cellular Components. **D.** Methods of Extraction of Enzymes. **E.** Protein Fractionation. **F.** Preparation of Buffers.

Section II. Enzymes of Carbohydrate Metabolism

A. Polysaccharide Cleavage and Synthesis. **B.** Disaccharide, Hexoside, and Glucuronide Metabolism. **C.** Metabolism of Hexoses. **D.** Metabolism of Pentoses. **E.** Metabolism of Three-Carbon Compounds. **F.** Reactions of Two-Carbon Compounds. **G.** Reactions of Formate.

Section III. Enzymes of Lipid Metabolism

A. Fatty Acid Oxidation. **B.** Acyl Activation and Transfer. **C.** Lipases and Esterases. **D.** Phospholipid and Steroid Enzymes.

Section IV. Enzymes of Citric Acid Cycle

VOLUME II

PREPARATION AND ASSAY OF ENZYMES

Section I. Enzymes of Protein Metabolism

A. Protein Hydrolyzing Enzymes. **B.** Enzymes in Amino Acid Metabolism (General). **C.** Specific Amino Acid Enzymes. **D.** Peptide Bond Synthesis. **E.** Enzymes in Urea Synthesis. **F** Ammonia Liberating Enzymes. **G.** Nitrate Metabolism.

Section II. Enzymes of Nucleic Acid Metabolism

A. Nucleases. **B.** Nucleosidases. **C.** Deaminases. **D.** Oxidases. **E.** Nucleotide Synthesis.

Section III. Enzymes in Phosphate Metabolism

A. Phosphomonoesterases. **B.** Phosphodiesterases. **C.** Inorganic Pyro- and Poly-phosphatases. **D.** ATPases. **E.** Phosphate-Transferring Systems.

Section IV. Enzymes in Coenzyme and Vitamin Metabolism

A. Synthesis and Degradation of Vitamins. **B.** Phosphorylation of Vitamins. **C.** Coenzyme Synthesis and Breakdown.

Section V. Respiratory Enzymes

A. Pyridine Nucleotide-Linked, Including Flavoproteins. **B.** Iron-Porphyrins. **C.** Copper Enzymes. **D.** Unclassified.

VOLUME III

PREPARATION AND ASSAY OF SUBSTRATES

Section I. Carbohydrates

A. Polysaccharides. **B.** Monosaccharides. **C.** Sugar Phosphates and Related Compounds. **D.** Unphosphorylated Intermediates and Products of Fermentation and Respiration.

Section II. Lipids and Steroids

A. Isolation and Determination of Lipids and Higher Fatty Acids. **B.** Preparation and Analysis of Phospholipids and Derivatives. **C.** Fractionation Procedures for Higher and Lower Fatty Acids. **D.** Preparation and Assay of Cholesterol and Ergosterol.

Section III. Citric Acid Cycle Components

A. Chromatographic Analyses of Organic Acids. **B.** Specific Procedures for Individual Compounds.

Section IV. Proteins and Derivatives

A. General Procedures for Determination of Proteins and Amino Acids. **B.** General Procedures for Preparation of Peptides and Amino Acids. **C.** Specific Procedures for Isolation and Determination of Individual Amino Acids.

Section V. Nucleic Acids and Derivatives

A. Determination, Isolation, and Characterization of Nucleic Acids. **B.** Determination, Isolation, Characterization, and Synthesis of Nucleotides and Nucleosides.

Section VI. Coenzymes and Related Phosphate Compounds

A. General Procedures for Isolation, Determination, and Characterization of Phosphorus Compounds. **B.** Specific Procedures for N-Phosphates and Individual Coenzymes.

Section VII. Determination of Inorganic Compounds

VOLUME IV

SPECIAL TECHNIQUES FOR THE ENZYMOLOGIST

Section I. Techniques for Characterization of Proteins (Procedures and Interpretations)

A. Electrophoresis; Macro and Micro. **B.** Ultracentrifugation and Related Techniques (Diffusion, Viscosity) for Molecular Size and Shape. **C.** Infra-red Spectrophotometry. **D.** X-ray Diffraction. **E.** Light Scattering Measurements. **F.** Flow Birefringence. **G.** Fluorescence Polarization and Other Fluorescence Techniques. **H.** The Solubility Method for Protein Purity. **I.** Determination of Amino Acid Sequence in Proteins. **J.** Determination of Essential Groups for Enzyme Activity.

Section II. Techniques for Metabolic Studies

A. Measurement of Rapid Reaction Rates; Techniques and Applications, Including Determination of Spectra of Cytochromes and Other Electron Carriers in Respiring Cells. **B.** Use of Artificial Electron Acceptors in the Study of Dehydrogenases. **C.** Use of Percolation Technique for the Study of the Metabolism of Soil Microorganisms. **D.** Methods for Study of the Hill Reaction. **E.** Methods for Measurement of Nitrogen Fixation. **F.** Cytochemistry.

Section III. Techniques for Isotope Studies

A. The Measurement of Isotopes. **B.** The Synthesis and Degradation of Labeled Compounds (Including Application to Metabolic Studies): Monosaccharides and Polysaccharides; Citric Acid Cycle Intermediates; Glycolic, Glyoxylic and Oxalic Acids; Purines and Pyrimidines; Porphyrins; Amino Acids and Proteins; Steroids; Methylated Compounds and Derivatives; Sulfur Compounds; Fatty Acids; Phospholipids; Coenzymes; Iodinated Compounds; Intermediates of Photosynthesis; O^{18}-Labeled Phosphorus Compounds.

VOLUME V

PREPARATION AND ASSAY OF ENZYMES

Section I. General Preparative Procedures

A. Column Chromatography of Proteins. **B.** Preparative Electrophoresis. **C.** Preparation and Solubilization of Particles (Bacterial, Mammalian, and Higher Plant). **D.** Mammalian Cell Culture. **E.** Protoplasts.

Section II. Enzymes of Carbohydrate Metabolism

A. Polysaccharide Cleavage and Synthesis. **B.** Disaccharide, Hexoside and Glucuronide Metabolism. **C.** Metabolism of Hexoses, Pentoses, and 3-Carbon Compounds. **D.** Hexosamine and Sialic Acid Metabolism. **E.** Aromatic Ring Synthesis.

Section III. Enzymes of Lipid Metabolism

A. Fatty Acid Synthesis and Breakdown. **B.** Acid Activating Enzymes. **C.** Phospholipid Synthesis and Breakdown. **D.** Steroid Metabolism.

Section IV. Enzymes of Citric Acid Cycle

A. Krebs Cycle. **B.** Krebs-Kornberg Cycle. **C.** Related Enzymes.

Section V. Enzymes of Protein Metabolism

A. Proteolytic Enzymes. **B.** Amino Acid Dehydrogenases and Transaminases. **C.** Amino Acid Activating Enzymes. **D.** Other Enzymes of Amino Acid Breakdown and Synthesis. **E.** Enzymes of Sulfur Metabolism.

VOLUME VI

PREPARATION AND ASSAY OF ENZYMES

Section I. Enzymes of Nucleic Acid Metabolism

A. RNA, DNA, and Protein Synthesis. **B.** Nucleases. **C.** Nucleotide Synthesis and Breakdown.

Section II. Enzymes of Phosphate Metabolism

A. Cleavage and Synthesis of Phosphate Bonds. **B.** Oxidative Phosphorylation. **C.** Photosynthetic Phosphorylation.

Section III. Enzymes of Coenzyme and Vitamin Metabolism

A. Breakdown and Synthesis of Various Vitamins and Coenzymes. **B.** Transformations of Folic Acid Coenzymes.

Section IV. Respiratory Enzymes

A. Cytochromes. **B.** Electron-Transferring Flavoproteins. **C.** Pyridine Nucleotide Enzymes. **D.** Other Enzymes.

PREPARATION AND ASSAY OF SUBSTRATES

Section I. Carbohydrates

A. Sialic Acids and Derivatives. **B.** Sugar Phosphates. **C.** Shikimic Acid Precursors. **D.** Diacetylmethylcarbinol.

Section II. Lipids and Steroids

A. Intermediates in Cholesterol Synthesis. **B.** Gas Chromatography of Fatty Acids. **C.** Thiolesters of Coenzyme A. **D.** β-Hydroxy and Keto acids.

Section III. Proteins and Derivatives

A. Amino Acids and Derivatives. **B.** Intermediates in Metabolism of Methionine, Histidine, Tryptophan, and Aspartic Acid. **C.** Spermine and Related Compounds. **D.** Soluble Collagens.

Section IV. Nucleic Acids, Coenzymes and Derivatives

A. Nucleotides, Acyl Nucleotides, and Purines. **B.** Synthesis and Isolation of RNA, Aminoacyl-sRNA, and DNA. **C.** "Nearest Neighbor" Sequences. **D.** UDPG and Related Compounds. **E.** Assay of Pyridine Nucleotides. **F.** Folic Acid and Derivatives.

SPECIAL TECHNIQUES

A. Amino Acid Analysis and Sequence. **B.** Immuno-diffusion. **C.** Density Gradient Sedimentation. **D.** Tritium-Labeling. **E.** Neutron Activation of Phosphorus. **F.** Histochemical Methods and Starch Gel Electrophoresis for Dehydrogenase Studies. **G.** Nuclear Magnetic Resonance and Electron Spin Resonance. **H.** Infra-red Measurements in Aqueous Media. **I.** Optical Rotatory Dispersion.

Errata for Volume III

P. 165, line 4 should read "$C_6H_{10}O_4(PO_4H)_2Ba$" instead of "$C_6H_{10}O_4(PO_3H)_2Ba$."

P. 210, line 28 and P. 213, line 7 should read "$C_3H_5O_7PBa$" instead of "$C_3H_5O_8PBa$."

Errata for Volume VI

P. viii: Paul F. Knowles should read Peter F. Knowles.

The affiliation for E. Haslam, R. D. Haworth, and Peter F. Knowles should read "The Department of Chemistry, University of Sheffield, Sheffield, England" instead of "The University of California, Davis, California".

Section I

General Preparative Procedures

[1] Tissue Slice Technique

By K. A. C. ELLIOTT

Principle. The use of tissue slices was introduced by Otto Warburg[1] as a method for studying, by simple *in vitro* methods, the metabolism of tissue which has not been extensively disrupted and is still capable of metabolic activity similar to that which it would perform *in vivo*. The method consists in the preparation of slices thin enough to allow oxygen from the surrounding medium to reach the innermost layers of the slice by diffusion, in spite of consumption of the oxygen as it diffuses through the outer layers, but thick enough that the proportion of cells disrupted by the slicing may be small. The slices are suspended in an appropriate medium in a vessel containing a suitable gas mixture and incubated with shaking, usually at 38°.

Warburg calculated the maximum permissible slice thickness, using a diffusion constant for oxygen in tissue arrived at by Krogh and assuming that maximum rate of oxygen consumption by cells could occur at very low oxygen concentrations. He had observed that the latter was true for unicellular organisms, and it has since been shown to be true for mammalian tissue particles.[2,3] For a rate of oxygen consumption of about 3 ml./g./hr., the calculated maximum thickness is about 0.5 mm. when the oxygen tension over the suspending medium is about 1 atm. The calculated limiting thickness is inversely proportional to the square root of the respiration rate. Experimental studies[4] have more or less confirmed Warburg's calculations.

For a more extensive discussion of this subject, the article by Field[5] is recommended.

Preparation of Slices. The organ should be excised immediately after the animal is killed, and preparation of slices should start at once. If delay is unavoidable, the tissue is usually kept cold on ice.

The Stadie-Riggs microtome[6] is a very satisfactory instrument for

---

[1] O. Warburg, *Biochem. Z.* **142**, 317 (1923).

[2] C. O. Warren, *J. Cellular Comp. Physiol.* **19**, 193 (1942).

[3] K. A. C. Elliott and M. Henry, *J. Biol. Chem.* **163**, 351 (1946).

[4] See, e.g., F. A. Fuhrman and J. Field, 2nd, *Arch. Biochem.* **6**, 337 (1945).

[5] J. Field, 2nd, *in* "Methods in Medical Research" (Potter, ed.), Vol. 1, p. 289. Yearbook Publishers, Chicago, 1948.

[6] W. C. Stadie and B. C. Riggs, *J. Biol. Chem.* **154**, 687 (1944). The authors give complete details of construction and use of this instrument. A slightly modified version of the microtome, and blades, can be obtained from Arthur H. Thomas Co. This microtome is satisfactory, although the writer prefers the instrument as originally described, since the latter has a longer working space which makes it easier to free the slice from between the blade and the plastic. (The commercial instrument tested by us cut slices appreciably thinner than specified.)

preparing slices of uniform thickness. A block of tissue of convenient size is placed on a piece of moist filter paper on the microtome pedestal. (The paper prevents the tissue from slipping.) The microtome is held over the tissue with enough pressure to cause a little flattening of the tissue block, and the slice is cut by a to and fro movement of the blade. The cut slice lies between the blade and the top section of the microtome. Sliding the blade backward or forward usually leaves the slice adhering to the top section, free of the blade, so that it can be lifted off. Sometimes it is necessary to unscrew the apparatus in order to remove the slice. This often happens, with brain, when the tissue has not been wetted, and occurrence of this trouble may be minimized by using a fresh section of the blade for each slice—by pushing the blade further in after each slice—and by wiping the inside of the instrument free of grease. The blade and microtome should be washed with soap and water after each use and sometimes in the course of cutting a series of slices.

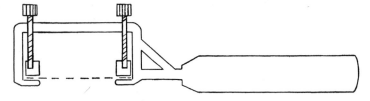

Fig. 1. A holder for safety razor blades.

Slices can be cut satisfactorily free-hand with a good "cutthroat" razor or with a safety razor blade held in a bow-shaped holder similar to that illustrated in Fig. 1. For this instrument half of a thin double-edged blade which has been broken lengthwise is used. (The holes in unbroken blades catch into the tissue block. Break the blade *before* removing the wrapping paper.) The writer prefers to lay the tissue on filter paper and slice horizontally. The contour of the tissue surface has to be followed, although the tissue can be slightly flattened by the blade. Other slicing techniques have been used successfully, such as flattening the tissue with a frosted microscope slide and cutting underneath it.[7] With any instrument, the movement of the blade should be more to and fro than straight ahead through the slice. Any blade should be resharpened or discarded when at all dull. The edges of slices may be trimmed to eliminate ragged pieces, but this is not usually necessary.

Cutting slices is easier if the tissue and instrument have been moistened with a saline medium. However, by working in a humid chamber good slices can be cut without wetting the tissue.

[7] W. Deutsch, *J. Physiol.* **87**, 56P (1936).

The Stadie-Riggs microtome gives slices of fairly constant thickness; the thickness can be varied by using top sections with depressions of different depths. Fairly constant thickness can be achieved by the free-hand methods with practice. The thickness of a slice can be determined by weighing it to obtain its volume, assuming a density of 1, and spreading it on, or floating it over, graph paper to find its area. Slices should not be cut very much thinner than the limiting thickness; this increases the proportion of damaged cells and the proportion of tissue subjected to high oxygen tension.

The above techniques apply to tissues like liver, brain, or kidney. Retina can be stripped easily from the eyeball and used directly. Testis cannot be sliced, but an equivalent preparation can be obtained by teasing out bunches of tubules. For many muscle studies, pieces of mouse or young rat diaphragm are used. Slices of heart muscle have been used frequently. Slicing skeletal muscle involves damaging most of the cells. Satisfactory preparations of skeletal muscle have been obtained by dissecting out fiber bundles.[8] Smooth muscle can be obtained by dissection from arteries or small intestine.

Handling Slices. Slices may be handled with flat-bladed coverslip forceps or curved fine forceps. Delicate slices should be pushed into a bundle before lifting. Slices may be stored for a short time in a humid atmosphere, or they may be stored in a suitable saline medium—if possible the same medium as that in which they are to be incubated and containing glucose or other suitable substrate. When using a humid chamber (see below), slices are best kept on watch glasses covered with petri dishes to prevent spray from falling on them and to ensure 100% humidity.

Slices are rapidly weighed on a small aluminum or wire pan with a torsion balance. Slices which have been cut or stored wet must be drained first. With a firm tissue like kidney cortex, draining on filter paper is no problem. With a tissue like brain, draining is never satisfactory. Such slices can be drained on semihardened paper with moderate damage, or they may be lightly drained by placing them on a small perforated glazed disk which is pressed onto filter paper.

The slice is best introduced into the reaction vessel by pushing it from the weighing pan onto the edge of the vessel and then pushing it down by means of any thin, slightly curved instrument. The slice should be wetted with a droplet of saline medium so that it will slide easily without damage. If slices have stuck together or become folded, it is advisable to swirl the vessel until the slices are open and separate.

8 H. B. Richardson, E. Shorr, and R. O. Loebel, *J. Biol. Chem.* **86**, 551 (1930).

Humid Chamber. Figure 2 shows the simple chamber used by the writer. It is a modification[9] of the chamber described by Sperry and Brand.[10] Humidity is ensured by covering the entire base of the chamber with several layers of filter paper and keeping them thoroughly wet and by a stream of bubbles from an aquarium aerator in a large beaker full of

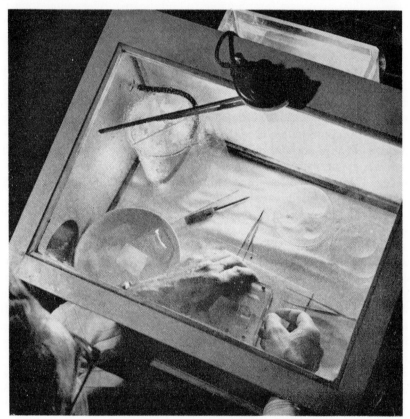

Fig. 2. Humid chamber with slicing equipment.

water. A sheet of glass, leaning against the beaker, minimizes the spread of spray. An automobile windshield wiper clears the glass top of fog which tends to collect as soon as warm hands work in the chamber. The whole chamber is illuminated through the frosted glass back; a trough of water intercepts heat from the lamp. A similar chamber has been described which includes a refrigeration coil,[11] but a cool humid chamber without

[9] K. A. C. Elliott, *Proc. Soc. Exptl. Biol. Med.* **63,** 234 (1946).
[10] W. M. Sperry and F. C. Brand, *Proc. Soc. Exptl. Biol. Med.* **42,** 147 (1939).
[11] F. A. Fuhrman and J. Field, 2nd, *J. Biol. Chem.* **153,** 515 (1944).

actual refrigeration is satisfactory provided that slices are not stored for long periods. The chamber can be cooled appreciably by placing ice in the beaker and dishes, but this is not usually necessary.

Media. The solution to be used as suspending medium depends upon the problem under study. Rates of metabolism are affected considerably by the inorganic, as well as the organic, constitution of the medium. The media used commonly consist of a saline mixture of osmotic pressure and inorganic constitution similar to that of serum. The writer prefers a composition similar to that of cerebrospinal fluid in which the concentration of calcium corresponds to the ionized and not to the total calcium of serum. Bicarbonate ion is the third most prevalent ion in body fluids, and bicarbonate buffered medium should be used if closely physiological conditions are required. With such a solution, the overlying gas mixture must contain 5% carbon dioxide to ensure a physiological pH. The use of bicarbonate-carbon dioxide involves complexities in aerobic manometric and other studies. For convenience, phosphate or other buffer is commonly substituted for bicarbonate. When using such a solution it must be recognized that the medium is not very physiological; particularly it should be remembered that a constant and physiological calcium concentration cannot be maintained in the presence of a high phosphate concentration at neutral pH. For aerobic studies with slices, 95 to 100% oxygen must be used. For anaerobic manometric studies on acid formation, the bicarbonate-carbon dioxide buffered medium is the most convenient.

The composition of the media usually used by the writer are as follows, in millimoles per liter:

Bicarbonate buffered medium: 122 NaCl, 3 KCl, 1.2 $MgSO_4$, 1.3 $CaCl_2$, 0.4 KH_2PO_4, 10 glucose, 25 $NaHCO_3$.

Phosphate buffered medium. The same as above, but with 17.5mM. sodium phosphate buffer, pH about 7.8, substituted for bicarbonate, and calcium omitted because it precipitates to a variable extent in the presence of phosphate. The pH of this medium usually falls rapidly during preliminary manipulations after introducing slices and reaches about 7.2 at the end of the experiment.

These solutions are most conveniently prepared by mixing appropriate volumes of isotonic solutions of the salts (0.154 M for NaCl, KCl, $MgSO_4$, or $NaHCO_3$, and 0.11 M for $CaCl_2$, $MgCl_2$, or phosphate).

Krebs[12] has designed a series of media which contain pyruvate or lactate, fumarate, and glutamate in addition to the usual salts and glucose, and which gives unusually high oxygen uptake rates. Other media are used for various specific purposes. For instance, maximum synthesis

[12] H. A. Krebs, *Biochim. et Biophys. Acta* **4,** 249 (1950).

of acetylcholine by brain requires 27 mM. potassium, and maximum synthesis of glycogen from pyruvate by liver requires a still higher potassium concentration. When additions are made to the medium, an equiosmolar amount of sodium chloride should be omitted. Serum, or serum which has been freed of bicarbonate by acidifying to about pH 5.5, evacuating, and reneutralizing, is occasionally used. The protein provides good buffering, but the presence of lactate and other substances in serum in variable concentrations introduces complications.

Mixtures of oxygen or nitrogen containing 5% carbon dioxide can be obtained commercially. Most commercial nitrogen or nitrogen-carbon dioxide mixtures contain significant amounts of oxygen, and, for anaerobic studies, it is advisable to remove this oxygen by passing the gas over hot copper. A convenient apparatus for this purpose has been described.[13]

The Basis of Calculation. Warburg and many others have expressed results in terms of the weight found when the tissue is removed at the end of the experimental period, drained or rinsed in water, and dried at 100 to 110° to constant weight. This is not a satisfactory basis, because fragments of the slice, and materials leeched out of the slice, are not accounted for, and results per unit weight are therefore falsely high.

Trichloroacetic acid may be added to the medium. The slice, fragments, and precipitate are then collected and washed with water on a sintered glass filter. On drying the filter the trichloroacetic acid is decomposed and volatilized, and the complete dry weight is obtained.[14] The filters should be cleaned out and dried to constant weight before every use because accumulated dried tissue is hydroscopic and its weight is affected by repeated heating.

A figure for the "initial dry weight" of slices which have been prepared wet may be obtained from their drained weight, and the wet weight/dry weight ratio determined on other slices which are weighed after similar draining and again after drying. This ratio is variable, however, since the draining is never constant.

The writer prefers to express results in terms of the actual fresh weight of the tissue used. This can be determined accurately only on slices which have been prepared in a humid chamber without wetting and without being allowed to dry out. Adherent fluid cannot be easily drained away from fragile tissue without damaging the slice, and tissue which has been wetted absorbs 10 or 20% or more of its weight of fluid—usually irregularly.

Other bases for calculation of activity, such as total nitrogen, deoxyribonucleoprotein, or fat-free dry weight, are legitimately applied according to the particular problem.

[13] K. A. C. Elliott, *Can. J. Research,* **F27,** 299 (1949).
[14] W. C. Stadie, B. C. Riggs, and N. Haugaard, *J. Biol. Chem.* **160,** 191 (1945).

Results. Results of measurements of the metabolism of various tissues by many workers have been assembled by Krebs and Johnson.[15] The following table summarizes a few examples of the oxygen uptake rates. Figures vary from species to species, and they vary with the media used and the weight basis used in the calculations (see, e.g., ref. 16).

RESPIRATION RATES OF VARIOUS RAT TISSUES[a]

Tissue	$-Q_{O_2}$	Tissue	$-Q_{O_2}$
Cerebral cortex slices	10–16	Intestinal mucosa, intact	12
Retina, intact	22–33	Diaphragm, intact	7–8
Kidney cortex slices	21–24	Lung slices	8
Liver slices	8–17	Pancreas slices	4
Spleen slices	11–12	Tumors, various, slices	6–14
Testis, teased out	11–12	Whole resting animal	4.8–6.0

[a] In bicarbonate-saline-glucose medium. Figures from Krebs and Johnson.[15] $-Q_{O_2}$ means microliters of oxygen consumed per milligram dry weight (dried at the end of the experiment) per hour.

Limitations. The tissue slice technique gives over-all pictures of metabolism rather than information concerning single reactions and enzymes. Although the slice technique is used chiefly to provide information concerning metabolism under simple but "physiological" conditions, it is at present impossible to provide a completely physiological environment containing all humoral and other physiological agents.

The slice has no capillary circulation, and all exchanges between the tissue and the suspending medium must occur by diffusion through distances up to half the thickness of the slice. To maintain an adequate oxygen tension in the innermost layers of an actively respiring slice, an abnormally high and actually toxic concentration of oxygen has to be in contact with the outer layers. Similarly, the concentration of any rapidly utilized substrate provided in the medium has to be higher than the likely physiological level. The concentration of a substrate added to the medium does not represent the concentration within the slice; this will vary with the depth in the slice and the rate of consumption. The carbon dioxide tension, the hydrogen ion concentration, and the concentrations of lactic acid and other metabolically produced substances tend to be significantly higher within the slice than in the surrounding medium.[1,17]

[15] H. A. Krebs and W. A. Johnson, *Tabulae biologicae* **19**, Part 3, 100 (1948).
[16] K. A. C. Elliott and N. Henderson, *J. Neurophysiol.* **11**, 473 (1945); K. A. C. Elliott, *in* "Neurochemistry" (Elliott, Page, and Quastel, eds.). Charles C Thomas, Springfield, Ill., 1954.
[17] K. A. C. Elliott and M. K. Birmingham, *J. Biol. Chem.* **177**, 51 (1949).

[2] Tissue Homogenates

By VAN R. POTTER

Definition

Tissue homogenates are now widely used for the study of enzymatic reactions in biological material, either directly in the form of whole homogenates, or as the starting material for the preparation of the particulate components of cells. The definition of a homogenate is at best an arbitrary matter, and it may be either operational or conceptual. In the original paper by Potter and Elvehjem[1] the term "tissue suspension" was used, and it was not until 1941 that the term "homogenate" was employed.[2] Since that time the term has been used in literally hundreds of reports with some variation in the sense intended, but usually in connection with the use of the device originally described or some modification of it, and thus usually in an operational sense. Since even the operations that involve the same apparatus can yield widely varying results, it is desirable to consider a conceptual definition. Prior to 1947[3] the tissue suspension was considered to be an adequate definition of a homogenate. It was thought that a greater or smaller number of intact cells could be obtained, depending more on the suspension medium than on the grinding technique, and efforts were made to evaluate the extent of cell breakage by enzymatic criteria.[4] However, as early as 1941[2] the concept of a suspension of respiratory particles probably identical with mitochondria had been advanced, and, with the development of the improved techniques for the isolation of mitochondria from sucrose homogenates in Claude's laboratory by Hogeboom *et al.*,[5] the basis for the present conceptual definition was established. Thus, it may be stated that since 1947 a homogenate as employed in this laboratory has meant a whole tissue preparation in which the cell disruption is as complete as possible, while at the same time the destruction of nuclei, mitochondria, and other cell particulates is held as close to the minimum as present knowledge permits. Operationally, the pestle homogenizer in our opinion remains the apparatus of choice. Although various types of blendors have been employed elsewhere to prepare homogenates, the literature on their advantages and disadvantages cannot be reviewed here. However, the clearest indication for their use is

---

[1] V. R. Potter and C. A. Elvehjem, *J. Biol. Chem.* **114**, 495 (1936).

[2] V. R. Potter, *J. Biol. Chem.* **141**, 775 (1941).

[3] V. R. Potter, *J. Biol. Chem.* **169**, 17 (1947).

[4] V. R. Potter, *J. Biol. Chem.* **163**, 437 (1946).

[5] G. H. Hogeboom, W. C. Schneider, and G. E. Palade, *Proc. Soc. Exptl. Biol. Med.* **65**, 320 (1947).

in the large-scale preparation of cell fractions to be used as the starting material for the isolation of individual enzymes that are localized in a particular fraction, as has been pointed out by Green.[6] For certain purposes water or isotonic KCl homogenates may still be very useful, but it seems clear that the cell components are less "physiological" under these conditions than when they are suspended in isotonic sucrose. In summary, then, every user of a homogenate should define his preparation in both operational and conceptual terms, the former with respect to apparatus and media, the latter with respect to the criteria for judging cell disruption and the condition of the cell particulates.

Apparatus

Construction. The apparatus consists of a test tube and a close-fitting power-driven pestle.[1] Similar devices for hand operation had been described for other purposes earlier by Hagan and by TenBroeck, as noted in 1946,[4] and the TenBroeck model is available from the Scientific Glass Apparatus Co., which also supplies glass and plastic models as described herein. The Erway Co. (Oregon, Wisconsin) supplies glass units, and the Arthur H. Thomas Co. supplies plastic models. For many purposes the tubes can be ordinary 16 × 150-mm. Pyrex test tubes. The pestles can be made from thick-walled glass tubing that approximates the inside diameter of the test tubes. The stem of the pestle can be made from a 220-mm. length of 6-mm. o.d. capillary tubing which will fit the chuck of a Cenco or Sargent cone-driven stirring motor. A cylindrical bulb with parallel sides about 20 mm. long is made from the thick-walled tubing. One end is sealed off to fit the inside of the test tubes, and the other end is sealed to the capillary tubing that forms the stem. The bulb is then molded while hot after it has been inserted into a test tube that represents the larger inside diameter of the range of tubes to be used. The capillary tube permits the operator to blow the glass. The shaft can best be lined up with the pestle on a glass-working lathe by a professional glass blower. The top of the shaft should be sealed off without enlarging the outside diameter. The bottom of the pestle should have six or seven small glass beads sealed to it. These should be ground down to 0.3 to 0.5 mm. in height with emery cloth, and they should then fit the inside of the test tubes. The slight irregularities in the side walls of the pestles should be removed by grinding inside a test tube at moderate speed, using a very dilute suspension of fine Carborundum powder in water. There should be a series of test tubes for each pestle, and the test tubes should be lightly ground to remove irregularities on their inside walls.

Frequently there are misconceptions about the closeness of the fit

[6] D. E. Green, *Biol. Revs.* in press (1954).

that should be attained, and some expert glassworkers have turned out units that fit like good hypodermic syringes. Such units are too tight. In the original report, a clearance of 0.23 mm. in good units was reported.[1] Recently, the writer found that the optimum clearance was different for each tissue in a series that included liver, kidney, heart, diaphragm, and skeletal muscle. The latter three tissues could be homogenized easily after being finely minced with scissors, but the fit of the homogenizer was critical and had to be determined by experience. The easiest way to standardize and select appropriate test tubes is to test them with water. A tube that can be lifted by raising the pestle is about right for liver and kidney, whereas for minced muscle the fit should be so loose that the tube cannot be lifted in this manner.

The ground-glass surfaces appear to speed up the cell disruption, but unskilled operators frequently fail to keep the tubes in line with the pestles and therefore erode the pestle until it is barrel-shaped instead of cylindrical. When operated properly, glass-to-glass contacts should be negligible, and glass powder should not enter the homogenate. Under these circumstances glass breakage will be negligible.

In many situations, more rugged homogenizers can be built to similar specifications using plastic pestles (Lucite, Teflon, or Kel-F), often in conjunction with stainless steel shafts. Machined Lucite cylinders with flat inside bottoms have been used in place of test tubes by Miller and co-workers.[7] In the construction of more rugged devices machined to close fits, it must be borne in mind that the device is most effective when the material above and below the rotating pestle is rapidly transferred back and forth, and that whenever the fit is so close that this is impossible there is a danger of local frictional heating.

The apparatus is powered by a cone-driven stirring motor with a friction drive, as produced by the Sargent Co. or the Central Scientific Co. The shaft of the pestle is held in the chuck on the variable speed shaft of the motor, and the speed is set near the maximum or at about 1000 r.p.m. The tube is always held by hand.

Operation. After tissues have been removed from the animal and dissected to yield a representative sample that is as homogeneous as possible, they are usually dropped into a cold isotonic solution of sodium chloride (0.15 M). After a few minutes the tissue is blotted, and about 1 g., accurately weighed, is placed in a homogenizer tube containing a measured volume (2 to 5 ml.) of the cold homogenization medium. The final amount of medium is usually calculated to give a total of 9 vol. of medium per gram of tissue, and the remainder is added prior to homogenization to

[7] J. M. Price, E. C. Miller, J. A. Miller, and G. M. Weber, *Cancer Research* **10,** 18 (1950).

give a 10% homogenate which is a useful dilution. If less tissue is needed, a 10% homogenate is prepared and diluted just before it is used. More concentrated suspensions can be prepared, but in general they have not exceeded a concentration of 20%. After adding the medium, the shaft of the pestle is fixed to the drive shaft of the motor, and the tube is moved up and down while the pestle is rotated by the motor. The tube is moved up and down as rapidly as possible, with complete transfers of the homogenate to the spaces above and below the pestle as soon as this can be accomplished. The tissue is always kept below the pestle until it is sufficiently reduced to pass to the upper space. Incorporation of air into the homogenate is undesirable. The operation is usually carried out in a refrigerated room.

Media

The choice of medium is still undergoing evolution and depends on the uses intended, but in general there is much to be said for 0.25 M sucrose which is considered isotonic. Studies carried out at various sucrose concentrations suggest that the final concentration of reactants and sucrose should combine to give a tonicity equivalent to 0.25 M sucrose or a value slightly below this.[8]

Water homogenates have been used successfully for systems in which the necessary cofactors could be added to the reaction mixture, as, for example, in the succinic[9] and malic[10] oxidizing systems. Oxidative phosphorylation has been observed in water homogenates of kidney,[3,11] but after 20 or 30 minutes the system decays rapidly,[12] although parallel systems in isotonic KCl were maintained somewhat longer.[12] Marked tissue differences were noted, and water homogenates of liver were inactivated much more rapidly in the reaction system.[12] Although isotonic KCl is superior to water for maintenance of the coordinated systems of the mitochondria, it appears that KCl has no advantage over sucrose and that the latter is superior for such systems. However, even in sucrose the ability of rat liver mitochondria to carry out oxidative phosphorylation is rapidly lost at 30 or 38° in the absence of substrate, and ATP-ase activity increases concomitantly.[13] These changes are correlated with the loss of substances that have ultraviolet absorption characteristic of nucleo-

[8] V. R. Potter and R. O. Recknagel, in "Phosphorus Metabolism," (McElroy and Glass, eds.) Vol. 1, p. 377. The Johns Hopkins Press, Baltimore, 1951.

[9] W. C. Schneider and V. R. Potter, J. Biol. Chem. **149**, 217 (1943).

[10] V. R. Potter, J. Biol. Chem. **165**, 311 (1946).

[11] V. R. Potter, Arch. Biochem. **6**, 439 (1945).

[12] V. R. Potter, G. A. LePage, and H. L. Klug, J. Biol. Chem. **175**, 619 (1948).

[13] V. R. Potter, P. Siekevitz, and H. C. Simonson, J. Biol. Chem. **205**, 893 (1953).

tides.[14] The finding of a number of new acid-soluble nucleotides that are analogous to the AMP, ADP, ATP series[15] suggests that some of these compounds may be of importance in the reactions that involve ATP,[16] and it is clear that further advances in the development of reaction media for homogenates may be expected. However, the original concept[1] of the homogenate has not been modified in so far as it called for the restoration of the cofactors for a given reaction, with side reactions minimized by the omission of the cofactors specific for them.

Uses

Between the whole animal and the individual isolated enzyme, the homogenate and the isolated cell fractions represent a half-way station that was formerly occupied solely by the tissue slice, and, although there are still many reactions that occur in slices and fail in homogenates, the further understanding of even these reactions may depend on the results obtained by means of homogenates. The studies with homogenates are so closely allied to the studies with individual cell fractions that investigations with whole homogenates tend to evolve into studies with individual and combined cell fractions.

Space does not permit the description of all of the uses to which tissue homogenates have been put, but in general they are used for the study of multiple enzyme systems that stress sequences of metabolic reactions with the intention of (1) assaying the tissue for a specific individual enzyme, or (2) studying the regulatory factors that determine which metabolic sequences occur when many alternatives are possible[17] or which reactions are rate-controlling.[18] Assays for individual enzymes are at all times dependent upon the knowledge of the rate-controlling factors and can proceed no further than the collateral knowledge permits. Nevertheless, such assays still occupy a basic role in establishing a base line for our description of the metabolic properties of a tissue. The present tendency is to determine *what* metabolic reactions occur in the tissues of the whole animal, by means of isotopic tracer techniques, and then to study *how* these reactions are controlled by studying them with the homogenate techniques.[19] These techniques include separation into cell fractions, recombination of cell fractions, comparison with the whole homogenate,

[14] P. Siekevitz and V. R. Potter, *Federation Proc.* **12,** 267 (1953).

[15] H. Schmitz, R. B. Hurlbert, and V. R. Potter, *J. Biol. Chem.* **209,** 41 (1954).

[16] V. R. Potter, E. Herbert, Y. Takagi, P. Siekevitz, and A. Brumm, *Federation Proc.* **13,** 276 (1954).

[17] R. O. Recknagel and V. R. Potter, *J. Biol. Chem.* **191,** 263 (1951).

[18] P. Siekevitz and V. R. Potter, *J. Biol. Chem.* **201,** 1 (1953).

[19] V. R. Potter, R. O. Recknagel, and R. B. Hurlbert, *Federation Proc.* **10,** 646 (1951).

development of *intracellular* reaction media,[19] and development of techniques for the measurement of the extent of changes involving multiple reaction systems.

Evaluation

Studies with homogenates should include tests of the effect of variation in homogenization time and in the closeness of the fit between pestle and tube. In most studies, the optimum duration of the homogenization may be doubled without affecting the results, but in a recent study by Bucher *et al.*[20] it was reported that the time was very critical in the case of a rat liver homogenate incorporating labeled acetate into cholesterol.

One of the questions that has recently been considered is the extent of cell disruption. Price *et al.*[7] counted the nuclei in homogenates with the help of a hemocytometer. Data on the extent of cell disruption have been provided by Hogeboom *et al.*,[21] who showed that in rat liver they were able to find 7×10^6 intact cells per gram of liver tissue in unfiltered homogenates, as compared with 268×10^6 free nuclei per gram in the same preparation. The resulting ratio is, of course, not strictly a measure of the extent of cell disruption because of the occurrence of binucleate cells in rat liver, but it shows that the number of intact cells can be reduced to a very low level in this tissue. When the extent of cell disruption is being brought to a maximum, it is necessary to guard against nuclear breakdown. This can be determined to some extent by centrifugal separation of the nuclear fraction and determination of whether the DNA is localized satisfactorily. In the study referred to,[21] this fraction accounted for 93% of the original DNA.

The evaluation of the state of the mitochondria, microsomes, and other constituents of homogenates is still not a matter of routine, and there is considerable reason to believe that the physical state of these particulates is not the same as it is in the cell. Although under certain conditions the Janus green staining[5,19] of the mitochondria is similar to the properties of these particles in whole cells, the staining properties are affected by a number of factors. Biochemical properties may be expected to be given increasing emphasis as criteria of the condition of the cellular particulates,[20] and visual examination of photographs taken at the highest magnifications, as in the studies by Palade,[22] will serve as a guide to the biochemical studies.

[20] N. L. R. Bucher, N. H. McGovern, R. Kingston, and M. H. Kennedy, *Federation Proc.* **12**, 184 (1952).
[21] G. H. Hogeboom, W. C. Schneider, and M. J. Striebich, *J. Biol. Chem.* **196**, 111 (1952).
[22] G. E. Palade, *J. Histochem. and Cytochem.* **1**, 188 (1953).

[3] Fractionation of Cell Components of Animal Tissues

By George H. Hogeboom

General Method for the Isolation of Liver Cell Components[1]

Principle. A broken cell suspension, or homogenate, is prepared with isotonic sucrose as the medium. The structural elements of the cell are then isolated by means of differential centrifugation. The procedure has been used, either essentially as described here or with some modifications, for the fractionation of kidney,[2] certain tumors,[3] cardiac[4] and skeletal[5] muscle, lymphoid tissues,[6] and brain.[7]

Reagents and Equipment

> 0.25 M and 0.34 M sucrose.
>
> International refrigerated centrifuge,[8] model PR-1, with horizontal head No. 269, multispeed attachment, and angle rotor No. 291.
>
> Spinco model L ultracentrifuge[9] with angle rotor No. 30.
>
> Homogenizers of the Potter-Elvehjem[10] type. The pestles are constructed from a chemically resistant plastic (Kel-F,[11] a monochlorotrifluoroethylene polymer) that can be cleaned in chromic acid cleaning solution. The heads of the pestles have a smooth surface and are hemispherical at each end; their length is approximately twice as great as their diameter, and they are machined to fit graduated test tubes of various capacities, ranging from 10 to 50 ml. The stem of each pestle consists of a Kel-F rod, $\frac{1}{4}$ or $\frac{5}{16}$ inch in diameter, that fits in a threaded hole in one end of the head.

[1] The present method is a modification of procedures described previously [G. H. Hogeboom, W. C. Schneider, and G. E. Palade, *J. Biol. Chem.* **172,** 619 (1948); W. C. Schneider, *J. Biol. Chem.* **176,** 259 (1948); W. C. Schneider and G. H. Hogeboom, *J. Biol. Chem.* **183,** 123 (1950)].

[2] W. C. Schneider and V. R. Potter, *J. Biol. Chem.* **177,** 893 (1949).

[3] W. C. Schneider and G. H. Hogeboom, *J. Natl. Cancer Inst.* **10,** 969 (1950).

[4] G. W. E. Plaut and K. A. Plaut, *J. Biol. Chem.* **199,** 142 (1952); J. W. Harmon and M. Feigelson, *Exptl. Cell Research* **3,** 509 (1952); K. W. Cleland and E. C. Slater, *Biochem. J.* **53,** 547 (1953).

[5] S. V. Perry, *Biochim. et Biophys. Acta* **8,** 499 (1952); A. Kitiyakara and J. W. Harmon, *J. Exptl. Med.* **97,** 553 (1953).

[6] M. L. Petermann, R. B. Alfin-Slater, and A. M. Larack, *Proc. Soc. Exptl. Biol. Med.* **69,** 542 (1948).

[7] T. M. Brody and J. A. Bain, *J. Biol. Chem.* **195,** 685 (1952).

[8] International Equipment Co., Boston, Mass.

[9] Specialized Instruments Corp., Belmont, Calif.

[10] V. R. Potter and C. A. Elvehjem, *J. Biol. Chem.* **114,** 495 (1936).

[11] Obtained in the form of rods from the Plax Corp., Hartford, Conn.

Preparation of Homogenates. The livers are chilled immediately after removal by immersion in 0.25 M sucrose at 0° and are then blotted, minced with scissors into pieces about 0.5 cm. in thickness, and weighed. 9.0 ml. of cold 0.25 M sucrose is added to each 1.0 g. of liver, and the tissue is homogenized for 2 minutes. In practice, the homogenization is interrupted after 1 minute to permit rechilling of the suspension. The pestle of the homogenizer is rotated at 600 to 1000 r.p.m. by means of a drill press or fairly powerful stirring motor, and the test tube is moved rapidly up and down over the pestle. Experience in the author's laboratory has indicated that a higher percentage of cells are disrupted if there is an appreciable clearance rather than a tight fit between the pestle and test tube. Apparently, the major factor in cell disruption by this type of homogenizer is the rapidity with which the tissue suspension can be forced between the wall of the tube and the rotating pestle. All subsequent steps in the fractionation procedure are carried out at 0 to 4°.

Isolation of Nuclear Fraction. 10 ml. of the homogenate is layered carefully over 10 ml. of 0.34 M sucrose in a 30-ml. graduated test tube.[12] The mixture is centrifuged for 10 minutes at 2000 r.p.m. (average centrifugal force, 700 $\times$ g) in head No. 269 of the International centrifuge. The centrifuge is accelerated and decelerated slowly to avoid mixing of the two layers. The entire supernatant is removed with a capillary pipet,[13] and the sediment is resuspended in a known volume of 0.25 M sucrose by brief homogenization.

The nuclear fraction isolated by this procedure is cytologically inhomogeneous. Although an essentially quantitative recovery of cell nuclei is achieved, the fraction also contains erythrocytes, connective tissue, residual intact cells, and approximately 10%[14] of the free mitochondria of the homogenate.

Isolation of Mitochondria. The supernatant from the first centrifugation is transferred to a Lusteroid tube,[15] 1⅛ inches in diameter, and centrifuged for 10 minutes at 9200 r.p.m. (5000 $\times$ g) in the International rotor No. 291. The opalescent supernatant is removed, together with a pink, partially sedimented layer of particles above the firmly packed pellet of mitochondria. The pellet is transferred quantitatively with 0.25 M sucrose to a 10-ml. graduated test tube and is redispersed by homogenization in 5 ml. of the sucrose solution. The homogenizing pestle and test tube are kept cold and used again when the mitochondria are

---

[12] Kimble Glass, Toledo, Ohio; Catalogue No. 46355.

[13] A convenient type of capillary pipet for the removal of the supernatant without disturbance of the sediment has been described by J. M. Price, E. C. Miller, and J. A. Miller, *J. Biol. Chem.* **173,** 345 (1948). The tip of the capillary is sealed, and a hole is made in the side of the capillary just above the tip.

[14] G. H. Hogeboom and W. C. Schneider, *J. Biol. Chem.* **204,** 233 (1953).

[15] Lusteroid Container Co., Maplewood, N. J.

subsequently resedimented and resuspended. The suspension is transferred to a Lusteroid tube and centrifuged for 10 minutes at 20,000 r.p.m. (24,000 $\times$ g) in the No. 291 rotor.[16] The procedure of resuspension of the mitochondria, centrifugation at 20,000 r.p.m., and removal of the supernatant and slowly sedimenting particles is repeated.

The final suspension of thrice-sedimented mitochondria is reasonably pure on the basis of *cytological* criteria of homogeneity and represents a yield of about 80%. The biochemical homogeneity of the preparation is open to some question.[17]

Isolation of Microsomes. The combined supernatants obtained in the isolation of mitochondria are made up to a volume of 35 ml. with 0.25 M sucrose and centrifuged for 60 minutes at 25,000 r.p.m. (54,000 $\times$ g) in the No. 30 rotor of the Spinco ultracentrifuge. The firmly packed, reddish, transparent pellet is resuspended in a known volume of 0.25 M sucrose, and the supernatant, which contains the soluble material of the homogenate, is saved. If the Spinco centrifuge is not available, microsomes can be sedimented at maximum speed in the multispeed attachment of the International centrifuge. In this case, the time of centrifugation must be prolonged (2 hours), and the tubes should have a small diameter ($\frac{1}{2}$ or $\frac{5}{8}$ inch) to shorten the path of sedimentation.

The microsomal fraction consists of cytoplasmic particulate material that is too small to be resolved clearly in the optical microscope. The results of several investigations[17,18] indicate that the fraction is both cytologically and biochemically heterogeneous.

Isolation of Liver Cell Nuclei in an Aqueous Medium[19]

Principle. Livers are homogenized in a sucrose solution containing a low concentration of $CaCl_2$.[20] In this medium, the cell nuclei are not

[16] Adaptors machined from Teflon (polytetrafluorethylene, U. S. Gasket Co.) make it possible to use $\frac{5}{8}$-inch tubes in the No. 291 rotor. Occasionally, part of the pellet will leak through the base of defective tubes. It is therefore desirable to use clean adaptors, so that a complete recovery of the sediment can always be obtained.

[17] A. B. Novikoff, E. Podber, J. Ryan, and E. Noe, *J. Histochem. and Cytochem.* **1**, 27 (1953); E. L. Kuff and W. C. Schneider, *J. Biol. Chem.* **206**, 677 (1954).

[18] C. P. Barnum and R. D. Huseby, *Arch. Biochem. and Biophys.* **19**, 17 (1950).

[19] The present method has been reported previously [G. H. Hogeboom, W. C. Schneider, and M. J. Striebich, *J. Biol. Chem.* **196**, 111 (1952)]. On the assumption that the nuclear membrane is permeable to proteins, cell nuclei have also been isolated by laborious procedures in organic solvents of low polarity [M. Behrens, *Z. physiol. Chem.* **27**, 258 (1939); A. L. Dounce, G. H. Tishkoff, S. R. Barnett, and R. M. Freer, *J. Gen. Physiol.* **33**, 629 (1950); V. G. Allfrey, H. Stern, A. E. Mirsky, and H. Saetren, *J. Gen. Physiol.* **35**, 529 (1952)]. In the opinion of the author [G. H. Hogeboom and W. C. Schneider, *J. Biol. Chem.* **197**, 611 (1952); *Nature* **170**, 374 (1952); *Science* **118**, 419 (1953)], the necessity for resorting to the use of nonaqueous media remains to be proved.

[20] R. M. Schneider and M. L. Petermann, *Cancer Research* **10**, 751 (1950).

aggregated (as they are in 0.25 M sucrose), retain their normal morphological characteristics, and are resistant to mechanical injury. The nuclei are separated from mitochondria by means of a layering technique and repeated centrifugation. The extent of contamination of the isolated nuclei by intact liver cells is reduced to the minimum by repeated homogenization.

Procedure. The entire procedure is carried out at 0 to 4°. Livers of rats or mice are perfused first with 0.145 M NaCl and then with 0.25 M sucrose containing 0.0018 M $CaCl_2$. The livers are then minced with scissors and homogenized as described above in 0.25 M sucrose—0.0018 M $CaCl_2$. In order to remove connective tissue and large clumps of liver cells, the homogenate is filtered through one layer of single napped flannelet (nap side up). A 10-ml. aliquot of the filtered homogenate is layered carefully over 20 ml. of 0.34 M sucrose—0.00018 M $CaCl_2$[21] in a 30-ml. test tube,[12] and the mixture is centrifuged for 10 minutes at 2000 r.p.m. in the International head No. 269. The entire supernatant is removed,[13] and the sediment is homogenized for 15 seconds in 5 ml. of 0.25 M sucrose —0.00018 M $CaCl_2$. 10 ml. of 0.34 M sucrose—0.00018 M $CaCl_2$ are then introduced slowly beneath the suspension of nuclei (by means of a pipet with a long capillary tip), and the mixture is centrifuged again for 10 minutes at 2000 r.p.m. The procedure of homogenization of the pellet, layering, and centrifugation is repeated twice more.

The preparations of cell nuclei obtained by this procedure represent a yield of 70 to 80% and contain less than 1% of the intact cells of the original whole tissue and less than 0.5% of the free mitochondria of the original homogenate. In addition, the preparations are contaminated by an unknown but apparently relatively small number of structures resembling cell membranes.

[21] When the nuclei are removed from the homogenate, it is necessary to reduce the $CaCl_2$ concentration; otherwise, the nuclei undergo serious morphological alteration.

[4] Preparation of Mitochondria from Plants

By BERNARD AXELROD

Preparation

General Principles. The methods by which metabolically active mitochondria are isolated from plant tissues and studied derive directly from the procedures which have been developed for the investigation of animal mitochondria.[1] Basically, the winning of mitochondria from any tissue depends on its disintegration to a subcellular level and a separation of the desired particles by differential centrifugation under suitably mild phys-

[1] For discussion of mitochondria from animal sources, see Vol. I [3].

ical conditions. In actual practice there is a considerable variation in these conditions, depending on the particular plant tissue involved, and each material must be considered individually. Two procedures are therefore given, one for mung bean mitochondria, an example of relatively sensitive particles, and one for avocado mitochondria, which appear to be much hardier.

Preparation of Mung Bean Mitochondria.[2] Seeds of *Phaseolus aureus* which have been soaked for 10 minutes in 0.5% NaOCl are placed in distilled H_2O for 60 minutes and then allowed to germinate in vermiculite at 26° for 90 hours in red light of low intensity, and with the humidity at 90%. Next, 30 g. of the hypocotyls is ground for 2 minutes in a mortar with 10 g. of sand and 40 ml. of 0.01 M PO_4 buffer (KH_2PO_4 and Na_2-HPO_4), pH 7.1, in 0.4 M sucrose. The brei is strained through muslin and centrifuged at 500 $\times$ g for 5 minutes. The supernatant is then centrifuged for 15 minutes at 10,000 $\times$ g. After the sedimented material is resuspended in 20 ml. of fresh sucrose buffer, it is again centrifuged at high speed. The resulting precipitate is dispersed in 3.5 ml. of 0.5 M sucrose in 0.02 M PO_4 buffer, pH 7.2, and is then ready to be assayed. It contains approximately 1.3 mg. protein N per milliliter. All operations must be performed at 0 to 4°. Such a preparation, unlike those from other plant sources, is fairly homogeneous morphologically and stains with Janus Green B in the manner characteristic of intact mitochondria.

Preparation of Avocado Mitochondria.[3] A ripe Fuerte avocado is peeled and grated, and 150 g. is blended for 60 seconds with 300 ml. of 0.5 M sucrose in a Waring blendor whose speed is controlled by a variable transformer set to deliver approximately 35 to 47 volts, depending on the state of firmness of the material. The speed of the blending is highly critical and should be ascertained for the particular equipment used. The brei is forced through four layers of cheesecloth with a potato masher to yield a fairly uniform homogenate which is then centrifuged for 5 minutes at 500 $\times$ g. After the layer of fat which accumulates on the surface is removed with the aid of suction, the supernatant is distributed among six 50-ml. centrifuge tubes and spun at 17,000 $\times$ g for 15 minutes. The pellets in each tube are resuspended in 20 ml. of 0.5 M sucrose with the aid of a Lucite homogenizer, and the high-speed centrifugation is repeated. The precipitate from all six tubes is taken up in 4 ml. of 0.5 M sucrose and is then ready to be assayed. All the above steps are carried out at approximately 0°.

[2] A. Millerd, J. Bonner, B. Axelrod, and R. S. Bandurski, *Proc. Natl. Acad. Sci. U.S.* **37**, 855 (1951).

[3] J. B. Biale and R. E. Young, *Western Sect., Am. Soc. Plant Physiologists, Abstr.* (June 1953).

Measurement of Activity

Theory. The quality of a mitochondrial preparation is not easily defined, inasmuch as not all the reactions which mitochondria can catalyze are known, nor is it feasible to test for all that are known. The particular reactions of interest to the investigator should be the basis of the criteria. However, since the deviations from mild conditions of preparation always seem to interfere with the oxidation of the Krebs cycle acids and even more readily with the accompanying phosphorylation, the "health" of mitochondria is ascertained by following these reactions. Ketoglutarate oxidation is measured manometrically in the Warburg apparatus. Adenylate in catalytic amounts, glucose, and PO_4 are included in the reaction mixture, to permit phosphorylation which is followed analytically by the decrease in PO_4^{---}. In the cases of mung bean and avocado the necessary hexokinase is associated with the particles. NaF is added to inhibit ATP-ase.

Procedure. This procedure for mung bean mitochondria[4] is also suitable for the avocado particles. First, 0.5 ml. of the enzyme is added to a chilled 15-ml. Warburg flask containing the following substances in such amounts that their final concentrations are as given: sucrose ($0.2\ M$), glucose ($0.1\ M$), ATP ($10^{-3}\ M$), $MgSO_4$ ($10^{-3}\ M$), phosphate ($0.01\ M$), NaF ($0.01\ M$), and α-ketoglutarate ($0.02\ M$). The volume of the reaction mixture should be 1.5 ml., and the pH, 7.1. The flask is then attached to the manometer and equilibrated for 5 to 10 minutes before the "zero" reading is made. The reaction is carried out at 20°. PO_4^{---} uptake is measured in a parallel reaction using double the quantities of reactants. At 0 and 30 minutes 1.0-ml. aliquots are removed to tubes at 0° containing 1.0 ml. of 2 N HCl. The PO_4^{---} content of the samples is then determined, after the precipitate has been removed by centrifugation. Oxidation rates and phosphorylation rates are expressed as microatoms of O_2 or P per milligram protein N per hour, respectively. The phosphorylative efficiency is expressed as the P:O ratio. Mung bean preparations usually give an oxidation rate of about 16 microatoms per hour per milligram protein N and a P:O ratio near unity. Comparable results are to be expected with avocados except that the P:O ratio tends to go significantly above 1.0.

General Applicability of the Methods

In preparing mitochondria from various sources, each case must be treated individually. Thus, whereas PO_4^{---} must be present in the extracting medium within certain narrow limits of concentration when mung beans are the source, for optimum results, it is not essential when avocado

---

[4] J. Bonner and A. Millerd, *Arch. Biochem. and Biophys.* **42,** 135 (1953).

is used. The tonicity of the extracting medium is not as critical with mung beans as with avocado. On the other hand, Laties[5] finds that cauliflower mitochondria are extremely sensitive to the tonicity of the extracting medium but are indifferent to the presence of PO_4^{---}. The optimum procedure for comminuting the tissue will, of course, vary with the tissue. Among the other sources treated in the literature (potato, Millerd;[6] pea, Stafford;[7] various flowering plants, Bhagvat and Hill[8]) the cauliflower is excellent because of its ready availability and uniformity. It has been extensively studied by Laties,[9] and much information is available on the technique of preparation of these mitochondria, their stability, metabolic behavior, and response to various cofactors.

[5] G. G. Laties, *J. Exptl. Botany* **5,** 49 (1954).
[6] A. Millerd, *Proc. Linnean Soc. N. S. Wales* **76,** 123 (1951).
[7] H. A. Stafford, *Physiol. Plantarum* **4,** 696 (1951).
[8] K. Bhagvat and R. Hill, *New Phytologist,* **50,** 112 (1951).
[9] G G. Laties, *Plant Physiol.* **28,** 555 (1953); *Physiol. Plantarum* **6,** 199 (1953); **6,** 215 (1953).

[5] Preparation of Chloroplasts and Disintegrated Chloroplasts

By PAUL R. GORHAM

Preparative Methods

Principle. Chloroplasts and disintegrated chloroplasts are prepared by macerating leaves or other plant material in water or hypertonic solution, removing coarse cellular debris by filtration, and other protoplasmic constituents by differential centrifugation. The isolation procedures described below are based on the facts that intact chloroplasts vacuolate and burst in water or hypotonic solution[1] and that vigorous grinding with little or no fluid promotes fragmentation. Weier and Stocking[2] have critically reviewed the numerous preparative procedures that have been described.

Applicability. The methods described are satisfactory for preparing chloroplast suspensions from leaves of spinach, Swiss chard, tobacco, and several other species, and also from certain algae. The methods are not applicable to all types of plant material, however, since ease of maceration and centrifugation vary considerably with species.[1,3]

General Instructions. Carry out all operations at temperatures close to 0°.[4] Rinse dirt from freshly harvested or cold-stored leaves. Remove

[1] S. Granick, *Am. J. Botany* **25,** 558 (1938); *in* "Photosynthesis in Plants" (Franck and Loomis, eds.), p. 113. Iowa State College Press, Ames, 1949.
[2] T. E. Weier and C. R. Stocking, *Botan. Rev.* **18,** 14 (1952); *Am. J. Botany* **39,** 720 (1952).
[3] K. A. Glendenning and P. R. Gorham, *Can. J. Research* **C28,** 114 (1950).
[4] Unless the enzyme system being studied is thermostable.

petioles, midribs, and larger veins, soak leaf-blade material in distilled water until fully turgid, and blot between filter paper. Wash algal cells by filtration or centrifugation. Examine chloroplast suspensions under the microscope (400 to 1000 ×) before use.

Whole Chloroplasts from Leaves. Grind 25 g. of leaf-blade material in 140 ml. of 0.5 M sucrose[5] in a Waring blendor operated at full speed for no more than 3 minutes. Filter the macerate through parachute Nylon or muslin to remove cell walls and fibers. Centrifuge the filtrate 1 minute at 200 × g[6] to remove extraneous particles and intact cells. Centrifuge the supernatant at 800 to 1800 × g for 5 minutes to sediment intact chloroplasts. Discard the supernatant. Suspend[7] the sediment in sucrose solution, and recentrifuge. Discard the supernatant, and resuspend the washed whole chloroplasts in sucrose or other hypertonic medium. Yields are generally low.

Disintegrated Chloroplasts from Leaves. Grind 50 g. of leaf-blade material in a mortar with quartz sand, adding little or no water. Alternatively, blend in 60 ml.[8] of distilled water at full speed for 3 to 5 minutes. Filter the macerate through parachute Nylon or muslin. Centrifuge the filtrate at 800 × g for 5 minutes, and discard the sediment of extraneous particles, whole cells, and whole chloroplasts. Centrifuge the supernatant at 15,000 × g for 15 minutes. Discard the supernatant, suspend the sediment in water, recentrifuge, and resuspend the washed, disintegrated chloroplasts in water or other solution. Good yields are generally obtained.

Disintegrated Chloroplasts from Unicellular Algae. Some filamentous algae are readily macerated by the methods used for leaves, but the unicellular green algae *Chlorella* and *Scenedesmus* have very resistant cell walls and are only partially disintegrated after 5 minutes in a Potter homogenizer.[3] Cells of *Chlorella pyrenoidosa* or *C. vulgaris* (but not of *Scenedesmus obliquus*, strain D3, which remain intact) can be 99.5 to 100% disintegrated by the following procedure: Force 35 ml. of a suspension containing 0.8 to 1.0 ml. of washed cells through the steel needle-valve of a French pressure disintegrator[9] at a pressure of 56,000 p.s.i. and a rate of 5 to 7 ml./min. Centrifuge the macerates at 600 × g for 1 minute

[5] Hypertonic. Dilute buffer, 0.05 M, or 0.01 M KCl for full photochemical activity, may be added to macerating and suspending fluids.

[6] For 200 to 800 × g an International clinical centrifuge and for 1000 to 15,000 × g a Servall Type SS-1 centrifuge are convenient.

[7] By gentle use of a rubber policeman in a small volume of fluid. Break up any lumps by filtering and rubbing through a thin pad of glass wool.

[8] Just enough to cover the blendor blades in operation. Addition of crushed ice while blending promotes fragmentation. Dilute buffer may be added, but KCl for full photochemical activity should be added only to suspending fluid.

[9] H. W. Milner, N. S. Lawrence, and C. S. French, *Science* **111**, 633 (1950).

to remove intact cells and debris. Centrifuge the supernatants at 15,000 × g, wash, and resuspend disintegrated chloroplasts in water or other solution as for preparations from leaves.

Properties

Particle Size. The diameters of disintegrated and whole chloroplasts range from approximately 0.1 to 2 μ and 3 to 6 μ, respectively.[10] Although sometimes termed "grana," disintegrated chloroplasts do not correspond either in size or absence of nonpigmented stroma to *grana* as defined by cytologists.[1,2] Disintegrated or whole chloroplasts seem to differ little in enzymatic activity, but colloidal dispersion into particles smaller than 10 mμ in diameter causes a reduction in photochemical activity of 75% or more.[11]

Purity. The methods described do not ensure complete freedom from enzyme-containing nuclear, mitochondrial, or proteinaceous material that may be trapped, adsorbed, or precipitated on or about the chloroplast particles. Purity should be assessed by microscopic, histochemical, and other control tests.[2] For minimum contamination, whole chloroplasts should be isolated, and then fragmented, if desired, by means of a supersonic or pressure disintegrator.

Cofactors. Certain enzyme systems of chloroplasts require cofactors from the cytoplasmic fraction for full activity. The photochemical system, for example, requires anions,[12,13] such as chloride, which are present in the cytoplasm. Laties[14] has also reported a heat-stable cytoplasmic factor needed to activate a chloroplast oxidase system.

Stability, Preservation. The stability of different chloroplast enzyme systems varies greatly. For example, photochemical activity has a half-life of only a few hours at 1°, whereas lecithinase activity shows a high degree of thermostability[15] and remains unchanged for 2 to 3 days at 5°. Deterioration of photochemical activity may be retarded by macerating and suspending chloroplasts in 0.05% KCl plus 0.05 M phosphate buffer, pH 6.2, 10% propylene glycol, or 15 to 20% methanol.[11,13,16] By freezing chloroplasts in 0.5 M sucrose[17] with dry ice, storing at a temperature of

---

[10] Swelling or bursting in distilled water provides a test for whole chloroplasts.

[11] H. W. Milner, C. S. French, M. L. G. Koenig, and N. S. Lawrence, *Arch. Biochem.* **28**, 193 (1950).

[12] O. Warburg, "Heavy Metal Prosthetic Groups and Enzyme Action" (transl. by A. Lawson). Clarendon Press, Oxford, 1949.

[13] P. R. Gorham and K. A. Clendenning, *Arch. Biochem. and Biophys.* **37**, 199 (1952).

[14] G. G. Laties, *Arch. Biochem.* **27**, 404 (1950).

[15] M. Kates, *Nature* **172**, 814 (1953).

[16] P. R. Gorham and K. A. Clendenning, *Can. J. Research* **C28**, 513 (1950).

[17] Freezing in the presence of 0.01 M KCl damages the photochemical system, however.

−40° or lower, and rapidly thawing portions as needed, photochemical activity can be preserved for long periods with little change.[16] Lyophilization, although unsatisfactory for retaining photochemical activity,[16] is useful for preserving other enzyme systems.

Chlorophyll Assay. It is frequently desired to express chloroplast concentration in terms of total chlorophyll content. Run 1 to 2 ml. of suspension into 80% acetone,[18] filter through Whatman No. 5 paper coated with Johns-Manville analytical filter aid, make to suitable volume, and determine the total concentration of chlorophylls a and b with a spectrophotometer having adequate resolution and sensitivity to give spectral slit widths of 3 mμ or less at 660 mμ.[19] Use this standardized extract to calibrate a test-tube colorimeter (660 filter)[20] for use in all subsequent determinations.

[18] Taking care to rinse the pipet in the acetone.
[19] D. I. Arnon, *Plant Physiol.* **24**, 1 (1949); J. J. Wolken and F. A. Schwertz, *J. Gen. Physiol.* **37**, 111 (1953); F. P. Zscheile, *J. Phys. & Colloid Chem.* **51**, 903 (1947). Spectral slit widths of 1.5 mμ are readily obtained with the Cary recording spectrophotometer. Difficulties may be experienced with individual Beckman DU instruments.
[20] The Evelyn colorimeter or Coleman Junior spectrophotometer (isolating 35-mμ bands) used at 652 mμ are convenient.

[6] Methods of Extraction of Enzymes from Animal Tissues

By ROBERT K. MORTON

I. General Procedures

Principles

With the exception of the plant prolamines, which contain a high proportion of nonpolar side chains, most purified nonconjugated proteins are soluble in dilute salt solutions at pH values removed from their isoelectric points. This is especially true of enzymes, which are typically albumins or globulins and of relatively small molecular size.

However, enzymes generally occur in the cell as complexes with other enzymes and inert proteins, nucleic acids, polysaccharides, or lipids. Even those which occur in true solution in the cytoplasm are probably associated in molecular aggregates, whereas those of cytoplasmic particles belong to complexes composed of several groups of compounds and containing relatively large amounts of lipid. These multimolecular aggregates are thus generally insoluble in distilled water or dilute salt solutions. Therefore, in order to extract the bound enzymes into solution, it is often necessary to dissociate them from other molecules as well as physically to disrupt the particles. In other cases, it may be necessary to obtain the enzyme complex (such as nucleoprotein or mucoprotein) in solution and subsequently to dissociate the enzyme during purification. It may also be necessary to prevent adsorption of enzymes onto cellular structures, such as microsomes and mitochondria. Hence the conditions for extracting certain enzymes into solution may be quite critical and require investigation of many variables.

Although the dispersion of molecular aggregates can often be achieved mechanically, dissociation of many enzyme complexes requires specific chemical forces. Undoubtedly, chemical methods must be related to the type of bonding between the enzyme and associated material. In few cases can this be described with certainty. However, electrostatic (salt) linkages, hydrogen bonds, and van der Waals forces are probably of considerable importance, and coordinate and covalent bonds much less frequent.

In view of this complexity, it is not surprising that most methods for extraction of enzymes from animal cells have been developed empirically. However, increasing knowledge of the intracellular localization of enzymes, of their chemical nature, and particularly of requirements for activity now makes possible a more rational approach to the selection of methods for extraction of any enzyme than was previously possible. Methods found useful with particular enzymes are described in detail in Vols. I and II. Theoretical aspects of problems of protein isolation have been discussed by Taylor.[1] In this chapter, consideration is given to those factors which directly influence the extraction of enzymes, and to methods which have general application to animal tissues.

Selection and Preparation of Tissue

There is considerable variability in enzymic activity between different animals, between different organs of the one animal, and according to conditions of growth and development. Extraction and subsequent purification may often be simplified by appropriate selection of the source of enzyme. For example, human prostate gland (since it virtually lacks diesterase) is preferable as a source of acid monophosphoesterase to pancreas, spleen, or liver, although the latter organs are usually more readily available. Specific nucleotidase is readily prepared from bull semen but only with difficulty from bull testes.[1a]

Where the nutritive status of the animal can be controlled, consideration should be given to simplifying extraction and purification by adjusting the diet. Starvation of rats for 24 hours is generally desirable to reduce glycogen when liver enzymes are to be prepared.

[1] J. F. Taylor, in "The Proteins" (Neurath and Bailey, eds.), Vol. 1, Part A, p. 1. Academic Press, New York, 1953.
[1a] L. A. Heppel and R. J. Hilmoe, J. Biol. Chem. 188, 665 (1951).

Perfusion of organs such as the liver is often a valuable preliminary procedure for removal of unwanted protein. As far as possible, depot fat and connective tissue and other inactive regions of the organ should be removed manually.

A priori, the organ should be collected and used immediately after death. Postmortem changes, such as fall in ATP concentration, may considerably influence the form and extractability of the enzyme, as in the case of muscle myosin ATP-ase (Szent-Györgyi[2]). Fresh hearts are essential for preparation of succinic dehydrogenase free of contaminant products derived from hemoglobin.

It is usually desirable to cool the tissue to 0° in crushed ice, in order to prevent autolysis and breakdown of cofactors. Enzymic changes in the tissue may be prevented by rapid freezing. Where storage is essential, prior packing in solid CO_2 is preferable to slower cooling in a cabinet, since rapid cooling forms smaller ice crystals. In many cases, however, it is desirable to prevent freezing, since this causes disruption of cells and cellular components. Lovelock[3] has reported that storage in 40% glycerol prevents lysis of erythrocytes stored at −15°. The author has also found that storage of tissues at −2° in molar sucrose or at −15° in 50% glycerol (containing 0.05 M phosphate buffer at pH 7.4) is useful for preventing loss of enzymic activity during storage. Porter *et al.*[3a] have reported similar observations.

Importance of Intracellular Localization of Enzymes

Knowledge of the intracellular localization of the enzyme being studied largely determines the method of extraction to be used. Methods for determining localization are discussed in Vol I [3]. Enzymes may be considered to fall within the following groups.

1. Enzymes in true solution in the cytoplasm, viz., remaining in the "final supernatant" after complete removal of *all* particulate components from an homogenate in sucrose (0.25 M) or physiological saline. Disruption of cells in suitable buffer usually releases such enzymes into the dispersion medium.

2. Enzymes associated with cellular structures, viz., with nuclei, mitochondria, microsomes, and "Golgi material." These enzymes may occur (a) in solution within a limiting semipermeable membrane (mitochondrial acid phosphatase, fumarase, etc.; see Schneider[4]) or (b) in-

[2] A. Szent-Györgyi, "The Chemistry of Muscular Contraction," 2nd ed. Academic Press, New York, 1951.

[3] J. E. Lovelock, *Biochim. et Biophys. Acta* **10**, 414 (1953); **11**, 28 (1953).

[3a] V. S. Porter, N. P. Deming, R. C. Wright, and E. M. Scott, *J. Biol. Chem.* **205**, 883 (1953).

[4] W. C. Schneider, *J. Histochem. and Cytochem.* **1**, 212 (1953).

timately associated with insoluble lipoprotein material. Among the latter group are microsomal enzymes such as alkaline phosphatases[5,6] and those of mitochondrial membranes and cristae such as the succinic oxidase system.[7,8] Whereas enzymes of the former group (a) may be extracted after disruption of the membrane, those of the latter (b) can be obtained in solution only by procedures which dissociate the lipoprotein complex.

Differential extraction procedures, based on the knowledge of the intracellular localization and relative solubility, greatly facilitate the purification of enzymes. This principle was used by Morton[6] for purification of alkaline phosphatases.

Methods for Disintegration of Tissues and Cellular Particles

Mechanical Treatment. The degree of subdivision markedly influences the extraction of enzymes. In general, animal tissues should be first cut into cubes of about 3-mm. section in a standard meat chopper or Latapie mincer (see Vol. I [2]). A machine which cuts cleanly, rather than tears the tissues, is preferable. Tough tissues, such as mammary gland, should be cut into thin slices with a sharp knife or a rotating blade (as in commercial bacon cutters) before mincing. Frozen tissues may be packed in crushed solid CO_2 and disintegrated in a precooled mincer. This is especially useful for organs such as pituitary glands. The machine developed by McIlwain and Buddle[9] for cutting very small and thin tissue slices may be found of value for small-scale preparative work. Single-cell suspensions may be prepared by dissolving away the intercellular "cementing substance" with pyrophosphate buffer.[10] However, in view of the ability of this buffer to extract many enzymes into solution (see p. 31), caution is necessary where quantitative recovery of enzymes is to be achieved. Enzymically-active cell suspensions may also be prepared by shaking (in a Warburg bath) small pieces of liver or brain tissue suspended in neutral buffer containing papain and cysteine (personal communication from Dr. H. McIlwain).

Many enzymes may be extracted into solution from the coarse mince, and this procedure sometimes has considerable advantages in facilitating manipulation (see Hopkins and Morgan[11]). However, further disintegration is usually necessary to disrupt the tissue cells. Methods of preparing tissue homogenates are discussed in Vol. I [2].

[5] R. K. Morton, *Biochem. J.* **55**, 786 (1953); **57**, 231 (1954); *Nature* **171**, 734 (1953).

[6] R. K. Morton, *Nature* **166**, 1092 (1950); *Biochem. J.* **55**, 795 (1953); **57**, 595 (1954).

[7] K. W. Cleland and E. C. Slater, *Biochem. J.* **53**, 547 (1953).

[8] G. E. Palade, *J. Histochem. and Cytochem.* **1**, 188 (1953).

[9] H. McIlwain and H. L. Buddle, *Biochem. J.* **53**, 412 (1953).

[10] N. G. Anderson, *Science* **117**, 627 (1953).

[11] F. G. Hopkins and E. J. Morgan, *Biochem. J.* **32**, 611 (1938).

Morphological structures may also be disrupted by repeated freezing and thawing of tissues. Disintegration is associated with physical damage by ice-crystals combined with plasmolysis due to the high salt concentration of the remaining fluid.[3,3a] Repetition of the procedure disrupts mitochondria and other cellular structures as well as whole cells. The method has been used for preparation of pyruvic[12] and α-ketoglutaric[13] oxidases from heart muscle.

High-Frequency Oscillations. Disruption of cellular particles by high-frequency oscillations was used by Haas[14] to prepare an optically clear suspension containing an active cytochrome oxidase of heart muscle. Hogeboom and Schneider[15] adopted a similar technique to disrupt liver mitochondria. After disruption, the enzymes associated with the mitochondrial membranes and cristae (such as the succinic dehydrogenase system of heart muscle) can be sedimented on high-speed centrifuging, but other enzymes, such as cytochrome oxidase, remain in the supernatant.

Mitochondria were isolated[15] from an homogenate of rat liver in 0.25 M sucrose (see Vol. I [3]). The mitochondria were suspended in 0.25 M sucrose, or in distilled water or dilute phosphate buffer (0.002 M KH_2PO_4–0.016 M K_2HPO_4, pH 7.7). The suspension was exposed to 9 kc./sec., in a commercial magnetostriction oscillator (Type R-22-3, Ratheon Co., Waltham, Mass.) at 2 to 3°. After 35 to 55 minutes, the opalescent suspension was centrifuged (5000 × g, 10 minutes) and the supernatant then recentrifuged (148,000 × g, 30 minutes). The clear yellow supernatant so obtained contained almost 60% of the total nitrogen, 30% of the DPN-cytochrome c reductase, 25% of the cytochrome oxidase, but only 5% of the succinic oxidase system. The latter system was largely sedimented in the two precipitates. The specific activities of the enzymes occurring in the final supernatant were considerably less than those of the original mitochondrial suspension.

Drying tissues *in vacuo* causes somewhat similar disruption to freezing and thawing. The dried tissue may be further disintegrated by grinding in a dry mortar or in a ball mill. Drying with acetone is a special method of particular value and is discussed below (p. 34).

Extraction of Enzymes with Aqueous Solutions

The factors discussed in this section relate to the extraction of enzymes with dilute salts or buffers. Whole minces, tissue homogenates, cellular particles, or material previously dried, either *in vacuo* or with organic solvents, may be used as a source of enzyme.

[12] V. Jagannathan and R. S. Schweet, *J. Biol. Chem.* **196**, 551 (1952).

[13] D. R. Sanadi, J. W. Littlefield, and R. M. Bock, *J. Biol. Chem.* **197**, 851 (1952).

[14] E. Haas, *J. Biol. Chem.* **148**, 481 (1943).

[15] G. H. Hogeboom and W. C. Schneider, *Nature* **166**, 302 (1950); *Science* **113**, 355 (1951).

Extraction Method. Between 2 and 5 vol. of extractant is normally used, and the suspension is stirred mechanically, with care to avoid frothing, for periods up to 24 hours. Normally, 1 to 2 hours suffices. Prior to extraction, dried powders should be dispersed into a uniform paste, with a Potter-Elvehjem homogenizer, or a mortar and pestle.

Insoluble material may be removed by appropriate centrifuging or by vacuum filtration through a thick bed of suitable filter aid, such as Hyflo Super-Cel, over a No. 1 Whatman paper on a Büchner funnel. It is often advantageous to squeeze a mince, such as that of muscle, through a double layer of cheesecloth or coarse muslin. Re-extraction of the insoluble material, using about half the initial volume of solvent, is desirable to increase recoveries.

Temperature of Extraction. The temperature should normally be held at about 0° to prevent enzymic changes during extraction. However, higher temperatures, such as 35 to 40°, for short periods (15 to 30 minutes) may aid extraction in some cases. Significant autolysis usually occurs at such temperatures, however.

Influence of pH. The choice of pH of extraction is determined by the stability of the enzyme, the nature of the enzyme complex, and the solubility of the enzyme. Electrostatic (salt) linkages may frequently be disrupted by adjustment of pH, particularly to acid (pH 3 to 6) values. However, since the isoelectric points of many enzymes are in the acid range (pH· 4 to 6), these proteins are generally more soluble at alkaline pH values. Hence it may be advantageous to vary the pH of extraction within the range of enzyme stability. Cytochrome c is extracted from heart muscle mitochondria with dilute TCA,[16] and ribonuclease from pancreas with 0.25 N H_2SO_4.[17] These are exceptional cases, however, and the pH of extraction is usually limited to the range 4 to 9.

Influence of Salt Concentration. Most proteins exhibit a marked "salting-in" effect. The amount of protein dissolved is thus a function of salt concentration up to levels at which "salting-out" effects are evident. Therefore, a range of salt concentrations should be investigated.

There is some interaction between pH and salt concentration. At neutral pH values higher salt concentrations are generally required for extraction of protein. In general, an ionic strength equal to that of 0.15 M NaCl is most suitable for enzyme extraction. Enzyme complexes, such as nucleoproteins, however, may be dissociated only by higher salt concentrations (0.5 to 2 M NaCl), and these concentrations may therefore be necessary in particular cases.

[16] D. Keilin and E. F. Hartree, *Proc. Roy. Soc. (London)* **B122,** 298 (1937); *Biochem. J.* **39,** 289 (1945).
[17] M. Kunitz, *J. Gen. Physiol.* **24,** 15 (1940).

Distilled water may be more effective than salt solutions in certain cases, probably owing to the disruptive effect of the markedly hypotonic conditions on cellular structures.

Specific Salt Effects. Both salt linkages and hydrogen bonds may be largely dissociated by suitable salts. Salts with univalent cations and multivalent anions are most effective. Considerable selectivity of enzyme extraction is obtainable by proper choice of anion and cation. The solvent power of cations diminishes in the order $Li > K > Na$; and of anions in the order $P_2O_7 > B_4O_7 > PO_4 > CNS > HCO_3 > I > Cl$ (at neutral or alkaline pH) and citrate > acetate (at acid pH). This classification is based on the author's experience but should be considered only as a general guide. It may not always apply.

In the author's experience, the most generally useful solutions for extraction of enzymes, arranged in increasing order of effectiveness, are: $0.15\ M$ NaCl; 0.02 or $0.05\ M$ potassium phosphate buffer, 0.02 or $0.05\ M$ sodium pyrophosphate buffer, all from pH 7.4 to 7.8. For the acid range, $0.02\ M$ sodium citrate buffer at pH 5.5 is employed. With the solutions in this sequence, the author has obtained selective extraction of enzymes from heart muscle tissue mince (see below). Pyrophosphate buffer has been particularly useful for enzyme extraction, since (a) it buffers over a wide pH range ($pK_{a_3} = 6.54$, $pK_{a_4} = 8.44$); (b) it has a marked dissociating effect on hydrogen bonds and salt linkages; (c) it chelates many divalent cations; and (d) it has a specific protective effect on certain enzymes, such as succinic dehydrogenase. At pH values below 6, the author prefers citrate buffer, which has some of the advantages shown by pyrophosphate buffer and which strongly chelates calcium ions.

The procedure for differential extraction of enzymes from heart muscle is at present under investigation by the author. Preliminary details of the method are given below.

Freshly collected horse heart is cooled in crushed ice during transport to the laboratory. Fat and unwanted tissue is cut away, and the muscle minced in a mechanical mincer previously cooled to 0°.

The mince is immediately extracted with various buffer solutions as indicated. Extraction is carried out using between 3 and 10 vol. of buffer, at 0°, with continuous mechanical stirring. The mince is allowed to settle, the supernatant solution decanted, the mince gently squeezed through a double layer of cheesecloth, and the residue re-extracted. The extracts may be clarified by high-speed centrifuging ($20,000 \times g$, 1 hour, at 0°) or by slow filtration under vacuum through Hyflo Super-Cel at 0°.

Fraction I. Dilute buffer mixture: $0.10\ M$ NaCl, $0.05\ M$ sodium succinate, $0.001\ M$ KCN, $0.02\ M$ Versene, adjusted to pH 7.4. Care should be exercised when large volumes of this dilute solution of cyanide are used.

Re-extraction is continued until the final extract is colorless.

Hemoglobin, myoglobin, and cytochrome c are removed, together with many soluble enzymes such as creatine phosphokinase and some aconitase.

Fraction II. The colorless residue is re-extracted with pyrophosphate buffer mixture: $0.02\ M$ sodium pyrophosphate, $0.05\ M$ sodium succinate, $0.001\ M$ KCN; pH 7.8.

Actomyosin, traces of cytochrome c, and considerable amounts of flavoproteins, including diaphorase are thus extracted. The detailed enzymic activity of this extract is still under investigation.

Fraction III. The residue is extracted with sodium citrate buffer (0.02 M, pH 5.5).

Fraction IV. The residue is disintegrated in 0.05 M phosphate buffer, pH 7.4, in a Waring blendor for 5 minutes at 0°. Frothing should be prevented by completely filling the vessel or covering the liquid surface with a solid paraffin plug or other suitable antifrothing agent. The supernatant obtained by centrifuging (20,000 $\times$ g, 1 hour) contains enzymes which are now under investigation.

It has recently been reported[18] that treatment of a pig heart preparation with M potassium thiocyanate enabled extraction of a fibrinokinase which had not previously been obtained in solution. Since this salt has a pronounced dissociating effect on salt linkages, it would appear to merit wider application for enzyme extraction.

In addition to dissociation of electrostatic forces and hydrogen bonds, salts may specifically influence extraction of enzymes by chelation of metals. Many metal-protein complexes are insoluble, and glycine and glycylglycine have been used to obtain these in solution.[19] Attention is drawn to the remarkable ability of ATP to chelate metals and to assist solution of insoluble compounds such as heavy metal phosphates.[20] It is quite possible that ATP may assist solution of metal-protein complexes and so aid enzyme extraction, in addition to its more specific ability to dissociate protein complexes such as that of actomyosin.[2] Other metal-chelating agents, such as pyrophosphate and citrate as mentioned above, and ethylenediaminetetraacetic acid (Versene), may be of value for bringing proteins into solution. Calcium may be removed from casein by both citrate and Versene, for example, and the decalcified protein is then soluble at neutral pH values.

Specific Substrate Effects. The solubility of enzymes may be influenced by association with substrates. Apoenzymes may be expected to have different solubility characteristics from those of holoenzymes, since combination with coenzymes may frequently alter the charge distribution on the protein surface. Hence it is quite possible that the addition of substrates or coenzymes may often assist dissociation of enzyme complexes or otherwise aid extraction of enzymes. For example, the addition of β-glycerophosphate prevents alkaline phosphatase adsorption on animal

[18] T. Astrup and A. Stage, *Nature* **170**, 929 (1952).

[19] E. J. Cohn, F. R. N. Gurd, D. M. Surgenor, B. A. Barnes, R. K. Brown, G. Derouaux, J. M. Gillespie, F. W. Kahnt, W. F. Lever, C. H. Liu, D. Mittelman, R. F. Mouton, K. Schmid, and E. Uroma, *J. Am. Chem. Soc.* **72**, 465 (1950).

[20] C. Neuberg and I. Mandl, *Arch. Biochem.* **23**, 499 (1949); I. Mandl, A. Grauer, and C. Neuberg, *Biochim. et Biophys. Acta* **8**, 654 (1952); **10**, 540 (1953).

charcoal,[21] and cytochrome c is much more readily eluted in the reduced than in the oxidized state.[22]

Differential Extraction of Tissue

The value of different buffer solutions for obtaining selective extraction of enzymes has been indicated above. The principle has been extensively developed by Cohn et al.[23] for purification of proteins from liver (see below). Formation of metal-protein complexes of differing solubilities in ethanol and in glycine solutions has also been used by the same group for purification of plasma proteins.[19] Dilute glycerol solutions may also be of advantage for the same purpose and may often have a stabilizing effect on enzyme systems.

In the procedure of Cohn et al.[23] bovine liver is perfused, frozen, and then finely divided mechanically. Differential extraction of proteins is achieved with ethanol-water and buffer mixtures as follows:

Fraction C_1. Extracted with 30% ethanol at $-8°$, pH 5.8, ionic strength 0.03. Fraction contains albumin and hemoglobin with nonprotein nitrogenous material and certain pigments.

Fraction C_2. Residue re-extracted with 19% ethanol at $-5°$, pH 5.8, and ionic strength 0.02. This fraction contains glutamic dehydrogenase, arginase, xanthine oxidase, esterase, and acid and alkaline phosphatases.

Fraction C_3. Residue extracted with salt solution at $0°$, pH 5.8, and ionic strength 0.15. Fraction contains catalase, peroxidase, and D-amino acid oxidase.

Fraction B_1. Residue extracted with buffer at $0°$, pH 7.2, and ionic strength 0.04. Fraction contains proteinases, deoxyribonucleoproteins, and ribonucleoproteins.

Fraction B_2. Residue extracted with buffer at $0°$, pH 7.4, and ionic strength 0.15. Fraction contains some proteinases.

Special Methods for Extraction

Correct use of the above methods should extract enzymes which are in solution in cytoplasm or within the mitochondrial membranes. Some enzymes associated with carbohydrates (such as mucoproteins) and with nucelic acid (nucleoproteins) may also be extracted. However, those associated with microsomes and with mitochondrial membranes and cristae are usually bound to lipid, and special methods have been developed for obtaining such enzymes in solution. Dissociation of lipoproteins in such a way that protein is undenatured is often one of the major problems in obtaining enzymes in true solution (see Ball and Cooper[24]). The following methods have been used.

[21] M. A. M. Abul-Fadl and E. J. King, Biochem. J. 44, 431 (1949).
[22] K. Zeile and F. Reuter, Z. physiol. Chem. 221, 101 (1933).
[23] E. J. Cohn, D. M. Surgenor, and M. J. Hunter, in "Enzymes and Enzyme Systems" (Edsall, ed.). Harvard University Press, Cambridge, Mass., 1951.
[24] E. G. Ball and O. Cooper, J. Biol. Chem. 180, 113 (1949).

Freezing and Thawing. The formation of ice crystals, in which water molecules are strongly bonded to one another, undoubtedly weakens hydrogen bonds and van der Waals forces between lipids and proteins. The material is usually frozen slowly at $-15°$ and allowed to thaw at room temperature. The α-keto acid oxidases have been extracted from heart muscle by this method.[12,13] Extraction may be assisted by the further addition of ethanol (and other solvents) as in the case of DPN-cytochrome c reductase of breast muscle.[25] Addition of ether may dissociate some lipoproteins (such as those of plasma) if the temperature is lowered to $-25°$ in order to remove water as ice.[26]

Butyl Alcohol Method. The unique ability of butyl alcohol to dissociate lipoprotein complexes without causing denaturation (Morton[6]) enables many enzymes to be extracted from animal tissues and obtained in true solution. The various procedures are described later.

Acetone Powder Method. Drying with acetone facilitates extraction of enzymes from dispersions of whole tissue or of cytoplasmic particles. The method depends on rapid removal of water. This reduces denaturation of protein by the solvent and causes concomitant disintegration of cellular structures and disruption of certain lipid-protein bonds. After complete drying and mechanical grinding to a powder, the enzymes are extracted with dilute buffer or salt solutions as previously described (p. 29). The author has used the following methods.

Finely disintegrated whole tissue or cytoplasmic particles are suspended in 0.01 M phosphate or 0.005 M pyrophosphate buffer and brought to pH 6.5 and cooled to 0°. The dispersion is added very slowly to 10 vol. of dry acetone at $-15°$ with rapid mechanical stirring. The suspension is held at $-15°$ for about 10 minutes, and the supernatant aqueous acetone is decanted. The precipitate is collected either by centrifuging at $-15°$, or by vacuum filtration through a No. 1 Whatman paper on a Büchner funnel in a cold room at 0°. The precipitate is washed twice by suspending on each occasion in about 3 vol. (calculated from the original volume of dispersion) of acetone at $-15°$. The acetone is finally removed from the precipitate first by using a stream of nitrogen and then by drying the powder *in vacuo* over H_2SO_4. Sometimes the last acetone treatment is followed by washing with dry peroxide-free diethyl ether (at $-15°$), which greatly facilitates rapid drying. Since the author has observed this to cause loss of enzymic activity in some cases, however, it is suggested that this further treatment should be investigated or used with caution. The dried material is stored *in vacuo* over $CaCl_2$ or NaOH at 0° or $-15°$. It may retain activity for long periods.

[25] H. Edelhoch, O. Hayaishi, and L. J. Teply, *J. Biol. Chem.* **197**, 97 (1952).
[26] A. S. McFarlane, *Nature* **149**, 439 (1942).

Alternatively, the whole tissue may be disintegrated directly in 10 vol. of acetone at $-15°$ in a Waring blendor (for 3 minutes), and the precipitate retreated with acetone as described above. The former method requires somewhat larger volumes of acetone but is generally preferred by the author.

It has been found that washing the first acetone precipitate with n-butanol (at $-15°$) removes much lipid material and greatly improves the extraction of enzymes (see below).

The value of acetone powders for obtaining enzymes in solution is illustrated by the successful extraction of liver succinic dehydrogenase,[27] TPN-cytochrome c reductase,[28] and a fatty acid oxidase system[29] by this method. Considerable purification of some mitochondrial-bound enzymes has been achieved from acetone-powders of these particles.[29a]

Extraction of Dry Proteins with Lipid Solvents. Enzymes which are denatured by organic solvents added to the aqueous phase may sometimes be obtained in solution by extracting the frozen-dried material with lipid solvents. Dixon and Thurlow[30] extracted dried milk proteins with ether at low temperature ($0°$) and obtained xanthine oxidase in solution on subsequent dispersion in aqueous media.

The extraction of dried brain tissue and red blood cells with n-butanol enables extraction of "true" cholinesterase in solution from these tissues (see p. 47).

Use of Detergents and Related Compounds

Natural detergents, such as the cholic acids and phospholipids, as wel as the synthetic cationic, anionic, and "nonionic" compounds, may all combine with lipoproteins. Under conditions of low ionic strength and at appropriate pH, such detergents can form stable micelles and hence cause physical dispersion of water-insoluble material. Theoretical aspects of the use of detergents for "solubilization" are discussed by Klevens.[31]

Electrostatic (salt) linkages and van der Waals forces appear to be involved in the linkage of proteins and detergents,[32] and hence the formation of complexes is very dependent on pH conditions. Anionic detergents, for example, combine with the basic groups of the protein and hence

[27] G. H. Hogeboom, *J. Biol. Chem.* **162**, 739 (1946).

[28] B. L. Horecker, *J. Biol. Chem.* **183**, 593 (1950).

[29] G. R. Drysdale and H. A. Lardy, *J. Biol. Chem.* **202**, 119 (1953).

[29a] H. R. Mahler, S. J. Wakil, and R. M. Bock, *J. Biol. Chem.* **204**, 453 (1953); S. J. Wakil, D. E. Green, S. Mii, and H. R. Mahler, *J. Biol. Chem.* **207**, 631 (1954).

[30] M. Dixon and S. Thurlow, *Biochem. J.* **18**, 971 (1924).

[31] H. B. Klevens, *Chem. Revs.* **47**, 1 (1950).

[32] F. W. Putnam, *Advances in Protein Chem.* **4**, 80 (1948).

should be used at neutral or alkaline pH conditions. In many cases there is stoichiometric relationship between the detergent and protein, and very small concentrations (0.001 to 0.01 M) of synthetic detergents are needed to form stable micelles.

The naturally occurring detergents, such as sodium deoxycholate, are apparently not sufficiently active to dissociate lipoproteins, but dissociation may sometimes be induced by synthetic cationic and anionic agents. However, these compounds further react with the liberated protein, usually causing denaturation, precipitation, formation of a new protein complex, or even catalyzed hydrolysis of peptide bonds.[32] Hence detergents cannot be considered desirable for extraction of enzymes.

Sodium deoxycholate[33] has been used to obtain water-soluble complexes of the residue material of heart muscle mitochondria (see below). Such complexes may retain some enzymic activity. Tryptic digestion of a deoxycholate dispersion has been used by Smith and Stotz[34] to obtain considerable purification of cytochrome oxidase, but some deoxycholate remains firmly bound to the enzyme. Such preparations are very valuable for spectrophotometric studies. Sodium deoxycholate also forms complexes with the lipoprotein of microsomal particles, but the firmly bound enzymes, such as alkaline phosphatase, are not extracted or obtained in true solution by this treatment.[5,35]

Wainio et al.[33] prepared a suspension of washed lamb heart muscle as described by Keilin and Hartree[35a] (see Vol. I [121]). The precipitate at pH 4.6 was suspended in an equal weight of 0.1 M phosphate buffer, pH 7.4. Sodium deoxycholate (80 mg.) was added to 2 ml. of this suspension (at 4°), and the material ground (in a Potter-Elvehjem type of homogenizer) for 10 minutes at 4°. The suspension was centrifuged (20,000 × g, 1 hour), and the clear, red-brown supernatant containing dispersed particles was diluted in 25 vol. of cold distilled water. The supernatant contained cytochrome oxidase, purified about twice as compared with the original heart muscle suspension.

Ord and Thompson[36] investigated a range of naturally occurring and synthetic detergents for extraction of brain cholinesterase into solution. This enzyme is apparently associated with a lipoprotein complex. Enzymically active dispersions of the lipoprotein were obtained with sodium cholate and deoxycholate and with the nonionic agent, Lubrol W (a cetyl alcohol-polyoxyethylene condensate). There was no evidence of dissocia-

[33] D. Keilin and E. F. Hartree, Proc. Roy. Soc. (London) B127, 167 (1939); W. W. Wainio, S. J. Cooperstein, S. Kollen, and B. Eichel, J. Biol. Chem. 173, 145 (1948).
[34] L. Smith and E. Stotz, Federation Proc. 9, 230 (1950); L. Smith, Federation Proc. 10, 249 (1951).
[35] J. C. Mathies, Biochim. et Biophys. Acta 7, 387 (1951).
[35a] D. Keilin and E. F. Hartree, Biochem. J. 41, 500 (1947).
[36] M. G. Ord and R. H. S. Thompson, Biochem. J. 49, 191 (1951).

tion of the enzyme from the lipid material. However, brain cholinesterase may be obtained in true solution by dissociation of the lipoprotein complex with butyl alcohol (see p. 47).

Ord and Thompson[36] purified brain cholinesterase as follows. An homogenate (10%, w/v) of rat brain in glass-distilled water was centrifuged for 10 minutes at 4000 r.p.m. An active, cloudy supernatant (at pH 6.0) was obtained. Lubrol W was added to this supernatant with continuous stirring to a final concentration of 1% (w/v), thus yielding a translucent suspension. This was adjusted to pH 4.5, a precipitate removed by centrifuging, the supernatant brought to pH 8.0, and ammonium sulfate added to 24% saturation. The precipitate so obtained was removed by filtration through No. 42 Whatman paper, and the filtrate brought to 50% saturation with ammonium sulfate. The precipitate rose to the surface on centrifuging. It was collected and dissolved in water, giving a yellow solution containing 25% of the original cholinesterase material.

Tween 20 (polyoxyethylene sorbitan monolaurate) has also been used to extract acetylcholinesterase from red cell stroma.[36a] Both the detergent and lipid could ultimately be removed from the dried preparation by successive extraction with acetone followed by n-butanol or ethanol.

Williams and coworkers[36b] have applied a somewhat similar procedure to acetone powders and have obtained clear solutions containing liver choline and succinic dehydrogenases.

Mitochondria were isolated from an homogenate of rat liver in 0.25 M sucrose. The mitochondria were washed twice with water and resedimented by centrifuging. The precipitate was dispersed in 20 vol. of cold acetone and centrifuged, and this procedure repeated. The sedimented material was taken up in 20 vol. of cold diethyl ether and again centrifuged, the precipitate air-dried on filter paper (2 to 3 minutes), and the ether finally removed in vacuo.

The acetone powder (20 mg.) and sodium choleate (12 mg. in concentrated solution at pH 6.8) were suspended in 1 ml. of 0.1 M sodium phosphate buffer (pH 6.8) and dispersed in a Potter-Elvehjem glass homogenizer for 1 to 2 minutes. The material was centrifuged at 25,000 × g for 30 minutes. The clear supernatant contained all the choline dehydrogenase activity of the acetone powder. When the precipitate obtained between 26 and 33% acetone was treated with pancreatic lipase, a portion of the activity remained in the supernatant after centrifuging at 144,000 × g for 30 minutes.

The anionic detergent Teepol XL (0.05%) or the nonionic Triton X-100 (0.12%) added to a water homogenate of mouse liver releases β-glucuronidase from the cytoplasmic particles.[37] However, the enzyme may also be obtained in solution by mechanical disruption of the particles, by

---

[36a] C. A. Zittle, E. S. Della Monica, and J. H. Custer, *Arch. Biochem. and Biophys.* **48**, 43 (1954).

[36b] J. N. Williams, Jr., and A. Sreenivasan, *J. Biol. Chem.* **203**, 899 (1953); K. Ebisuzaki and J. N. Williams, Jr., *Federation Proc.* **13**, 202 (1954).

[37] P. G. Walker and G. A. Levvy, *Biochem. J.* **49**, 620 (1951).

freezing and thawing, by treatment with acetone, or by incubation in 0.1 M acetate buffer, pH 5.2, for 4 hours at 37°.[37] This suggests that release of the enzyme is here dependent on destruction of a particulate membrane surrounding the particles, rather than on dissociation of a lipid-enzyme complex.

The pyridine nucleotide transhydrogenase of animal tissues has been found to be associated with particulate components of animal tissues (Kaplan et al.[37a]). These workers obtained an active extract from the particles as follows.

A digitonin solution was prepared by suspending 2 g. of digitonin in a few milliliters of distilled water, adding 20 ml. of 5 N NaOH and stirring until dissolved, adjusting to pH 7.0 with N HCl, and diluting to 100 ml. with water. A suspension of the enzymically active particulate material in 0.1 M phosphate buffer, pH 7.0, was treated with an equal volume of the 2 % digitonin solution and stirred for 20 minutes. The supernatant obtained by centrifuging contained about 75 % of the enzymic activity of the suspension. The enzyme preparation was found to be very labile at this stage of purification.

Urea and guanidine HCl cause dissociation of many protein complexes, especially of nucleoproteins, but they are also very active protein-denaturing agents, and it is unlikely that such compounds are of any value for extraction of enzymes in solution.

Heparin, a natural polysaccharide, occurs in association with lipoproteins in mast cells and liver.[38] It has been used to solubilize lipoproteins, such as the lung thromboplastic factor, but it does not cause dissociation of the lipoprotein of particulate cellular components.[39] Hence heparin is unlikely to be of any greater value than sodium deoxycholate for extraction of enzymes.

Enzymic Extraction Methods

Autolysis. Most animal tissues undergo rapid enzymic changes on storage. Liver, kidney, pancreas, and intestinal mucosa, for example, must be treated very rapidly to prevent modification of the proteins, even when held at 0°. Autolysis involves both proteolysis and lipolysis, as well as many other enzymic changes (see Haehn[40]). This method has been widely used to extract alkaline phosphatase from microsomes. Albers and Albers[41] introduced autolysis of kidney mince, in the presence of 25 % acetone containing ethyl acetate and toluene, to extract kidney alkaline phosphatase. However, preliminary removal of lipid by extraction of the dried tissue

[7a] N. O. Kaplan, S. P. Colowick, and E. F. Neufeld, *J. Biol. Chem.* **205**, 1 (1953).
[38] O. Snellman, B. Sylvén, and C. Julén, *Biochim. et Biophys. Acta* **7**, 98 (1951).
[39] E. Chargaff, *Advances in Protein Chem.* **1**, 1 (1944).
[40] H. Haehn, *Ergeb. Enzymforsch.* **5**, 117 (1936).
[41] H. Albers and E. Albers, *Z. physiol. Chem.* **232**, 165, 189 (1935).

with organic solvents or addition of "pancreatin" is necessary to extract the enzyme from hog kidneys by this method.[35]

Since the changes induced by autolysis cannot be predicted and must be very complex, this method cannot be recommended for enzyme extraction. The butyl alcohol procedures (see below) were originally developed by Morton[6] in order to avoid such undesirable procedures for extraction of alkaline phosphatase.

Use of Purified Enzymes. Addition of enzymes of known specificity or inhibition of undesirable enzymic activities in a complex system should have many advantages over autolysis for extraction of enzymes.

Since trypsin acts specifically on peptide linkages involving basic amino acids, certain proteins are resistant to tryptic digestion, whereas others are rapidly attacked. Hence enzymes such as xanthine oxidase,[42] alkaline phosphatase,[43] and TPN-cytochrome c reductase[28] have been obtained in solution by addition of trypsin at alkaline pH. However, subsequent purification of the enzyme may be complicated by the presence of peptides of varying chain length. These may become persistent contaminants, although it is possible that ion exchange columns may prove useful for their removal. Furthermore, it is always uncertain as to the extent of modification of the enzyme induced by tryptic action. The trypsin, of course, must be subsequently removed or inhibited, as, for example, with soybean trypsin inhibitor.

The use of lipases to split lipid-protein bonds is fundamentally preferable to the use of proteases. Crude pancreatic lipase and lecithinase have been used to obtain xanthine oxidase[42] and alkaline phosphatase[44] in solution. Some disruption of lipid-protein bonds by prior preparation of an acetone powder is apparently necessary before steapsin can disrupt kidney microsomes sufficiently to release DPN-cytochrome c reductase.[45] It is possible that the lecithinases are important in disruption of lipoprotein complexes during tissue autolysis. These enzymes may prove useful for aiding lipid removal with butanol (see below) and thus assist extraction and purification of enzymes of lipoprotein complexes.

It is probable that digestion of nucleic acids, as well as of mucopolysaccharides, contributes to the efficiency of autolysis for enzyme extraction. Ribonucleases and hyaluronidases (relatively free of proteases) have not so far been widely employed for obtaining enzymes in solution. However,

[42] E. G. Ball, *J. Biol. Chem.* **128**, 51 (1939).
[43] G. Ehrensvärd, *Z. physiol. Chem.* **217**, 274 (1933); G. Schmidt and S. J. Thannhauser, *J. Biol. Chem.* **149**, 369 (1943).
[44] C. A. Zittle and E. S. Della Monica, *Proc. Soc. Exptl. Biol. Med.* **76**, 193 (1951); *Arch. Biochem. and Biophys.* **35**, 321 (1952).
[45] L. A. Heppel, *Federation Proc.* **8**, 205 (1949).

this method would appear to be worthy of investigation in cases in which other methods are unsatisfactory. Deoxyribonuclease has been used to purify the proteins of certain viruses.[46] Combined with lipid removal by organic solvents (see below), the use of such purified enzymes may be expected to aid extraction of many refractory enzymes.

II. Butyl Alcohol Method

Enzymes of microsomes and of mitochondrial membranes and cristae are more or less firmly bound to lipid material. In order to separate proteins from these and other lipoprotein complexes (such as those of blood plasma) it is necessary to remove the lipid without denaturation of the associated protein. The use of butyl alcohol was introduced by Morton[47] in 1950 for this purpose, after an investigation of methods for obtaining alkaline phosphatases in true solution from microsomal particles. The original procedures[47] have been extended by Morton[48] and other workers (see table, p. 50) to enable separation of a number of enzymes and other proteins from lipoprotein complexes. The method has also been employed for the purification of a number of soluble proteins.[49,50] The use of isobutanol for obtaining DPN-cytochrome c reductase in solution from liver microsomes was independently found by Kun.[51] Hurst[52] had earlier noted the specific ability of butanol to remove lipid from lipoprotein complexes of insect integument.

Principle

It has been determined[48] that the relative effectiveness of organic solvents for dissociating lipoprotein complexes in aqueous media and for releasing undenatured protein in general may be expressed as follows: (1) most effective: n- and isobutanols; (2) partially effective: sec-butanol, cyclohexanol and tert-amyl alcohol; (3) ineffective: all other solvents tested, including chloroform, carbon tetrachloride, toluene, ether, acetone et al. Hence butanol has a unique effect, which may be attributed to:

1. A very marked lipophilic property and especial affinity for phospholipids. This results in rapid penetration of the complex.

2. Concomitant hydrophilic property, partly expressed by the solubility in water of 10.5% at 0°.

[46] R. M. Herriott and J. L. Barlow, *J. Gen. Physiol.* **36**, 17 (1952).
[47] R. K. Morton, *Nature* **166**, 1092 (1950).
[48] R. K. Morton, Thesis, University of Cambridge, 1952.
[49] B. A. Askonas, Thesis, University of Cambridge, 1951.
[50] B. D. Polis and H. W. Shmukler, *J. Biol. Chem.* **201**, 475 (1953).
[51] E. Kun, *Proc. Soc. Exptl. Biol. Med.* **77**, 441 (1951).
[52] H. Hurst, *Discussions Faraday Soc.* **3**, 193 (1948).

3. Orientation of the butanol molecules between the lipid and water molecules, thus producing a detergent-like action. n-Butanol and iso-butanol are the only alcohols of the homologous aliphatic series possessing *both* hydrophilic and lipophilic properties to an appreciable extent.

4. Displacement of lipid from the protein by the butanol. Effective competitive action prevents the displaced lipid from re-forming a complex with the protein. The butanol possibly competes with the phospholipid for polar side chains of the protein (see Hurst[52]).

Although butanol is a valuable lipid solvent when used with dried materials, it is even more effective in dissociating lipoprotein complexes when these are in aqueous media. By contrast, for effective removal of protein-bound lipid by other organic solvents, it is necessary to remove associated water by drying the material, or by freezing to $-25°$ (McFarlane[53]), or by addition of TCA, methanol, or other protein-denaturing agents.

The procedure to be used is determined largely by the relative stability of the protein or enzyme when obtained in solution and separated from the lipid material. One of the valuable features of the butyl alcohol method is the possibility of variation of the procedure according to the requirements of each individual problem. The following procedures have been developed by the author for use with both *aqueous materials* and *dried materials*.

Methods Used with Aqueous Material

The lipoprotein complex is isolated as free as possible of other proteins. n-Butanol is added slowly to the aqueous material, either to a limited concentration (procedure B, one-phase) or to saturate the aqueous phase (procedure A, two-phase). The protein is recovered from the aqueous phase after removal of insoluble material by centrifuging.

The following general considerations apply to all methods with aqueous materials.

Isolation of Lipoproteins. Although the method may be used with whole-tissue dispersions, in order to facilitate later purification the water-insoluble complexes should be separated from the tissue dispersion wherever possible, and exhaustively extracted to remove soluble proteins and other components. Water-soluble proteins should be purified as far as possible before butanol treatment. Mitochondrial material or microsomes may be obtained by differential sedimentation or by precipitation, extracted with 0.15 M NaCl (pH 7.4) or 0.05 M pyrophosphate buffer (pH 7.8) and then with distilled water (Morton[47,54]).

[53] A. S. McFarlane, *Nature* **149**, 439 (1942).
[54] R. K. Morton, *Biochem. J.* **55**, 795 (1953); and *Biochem. J.* in press.

Influence of Ionic Strength and pH. In general the material should be suspended in distilled water or buffer of low ionic strength (about 0.03) before addition of butanol. However, the ionic strength may be varied somewhat according to the stability of the protein to be isolated.

Dissociation of lipoproteins by butanol is generally more rapid at either acid (3 to 6) or alkaline (8 to 10) pH values than at pH 7. This suggests that electrostatic bonds may be involved in some lipid-protein complexes and thus butanol action may be facilitated by appropriate pH adjustment. However, the pH during butanol treatment is largely determined by the stability of the enzyme or protein to be purified. Alkaline phosphatase[47,54,55] and amino peptidase[56] have been dissociated from lipoprotein complexes at pH 5 to 6.0, γ-glutamyl transpeptidase[47,57] at neutral pH, and uricase[58] at pH 10. Heme a was dissociated from heart muscle lipoprotein at pH 3.5 and obtained in the butanol phase.[48] The ability of butanol to dissociate lipoprotein complexes at pH values close to neutrality makes this solvent of particular value for protein purification.

Effect of Temperature. The temperature may be varied from about -2 to 40°, according to the stability of the dissociated protein. It should be noted that the solubility of n-butanol in water varies inversely with temperature, viz., solubility 10.5% at 0°; 8.9% at 10°; 7.8% at 20°; 7.1% at 30°; 6.6% at 40° (Durrans[59]). The increased concentration of butanol at lower temperatures does not appear to increase the rate of denaturation of the protein. Butanol has been used at about 35° for purification of alkaline phosphatase[47,54] and uricase[58] at 5° for amino peptidase,[56] and at 0° and below for many other enzymes.[47,48,57,59a]

Mechanical Treatment

The whole tissue, or isolated insoluble material, should first be disintegrated as finely as possible, using methods as described on p. 28 in order to aid rapid penetration of butanol into the particles. The dispersion is then transferred to a suitable beaker and stirred rapidly but so as to avoid frothing. A mechanical stirrer with a small four-blade paddle, arranged so as to prevent any whip, is most suitable. Butanol is then added slowly down the side of the container from a reservoir (such as a graduated separating funnel) with a capillary orifice. Since rubber is soluble in butanol, rubber connections should not be used in contact with butanol.

[55] C. A. Zittle and E. S. Della Monica, *Arch. Biochem. and Biophys.* **35**, 321 (1952).
[56] D. S. Robinson, S. M. Birnbaum, and J. P. Greenstein, *J. Biol. Chem.* **202**, 1 (1953).
[57] P. J. Fodor, A. Miller, and H. Waelsch, *J. Biol. Chem.* **202**, 551 (1953).
[58] E. Leone, *Biochem. J.* **54**, 393 (1953).
[59] I. H. Durrans, "Solvents," 2nd ed. Chapman and Hall, London, 1950.
[59a] O. Hayaishi and A. Kornberg, *J. Biol. Chem.* **206**, 647 (1954).

The equipment described by Askonas[60] for protein precipitation with organic solvents is very suitable for the butanol addition and has the advantage that temperature may be accurately controlled. With attention to these details, it is possible to reduce surface denaturation to the minimum during butanol treatment.

There appears to be a critical concentration of butanol for adequate disruption of lipoproteins, at which the associated protein is released into solution in the aqueous medium. The concentration varies according to the stability of the complex and the experimental conditions (pH, ionic strength, temperature, etc.) but is generally between 3 and 6% (v/v) butanol (see procedure B, p. 45). The addition of butanol may be continued beyond this concentration, and when the aqueous phase is saturated, lipids pass into the butanol phase (see procedure A, p. 45).

In some cases, selective denaturation of unwanted protein may be combined with butanol treatment by shaking the material after addition of butanol, in order to improve purification.[48,56] The disruption of lipoproteins, however, depends on the specific effects of butanol and not on protein denaturation. By contrast, the action of chloroform (see Schwimmer and Pardee[61]) depends essentially on protein denaturation induced by shaking.

Period of Butanol Treatment. The disruption of lipoprotein appears to be almost instantaneous in some cases, but the necessary period of exposure to butanol action depends on experimental conditions. Normally, the addition of butanol takes about 15 to 30 minutes, and the material is stirred for another 30 minutes to 1 hour. However, longer periods of butanol treatment may be necessary, particularly where procedure B (one-phase procedure) is used. Leone,[58] for example, obtained both increasing amounts of uricase in solution and greater purification by holding the material for periods up to 18 hours after addition of butanol.

Methods of Protein Protection. Although the use of butanol enables removal of lipid with minimal denaturation of proteins, many enzymes, such as succinic dehydrogenase, have been found to be extremely labile when obtained in solution and free of lipid.[48] Other enzymes, such as brain cholinesterase, are extremely labile in the presence of aqueous organic solvents. For such enzymes, the butanol procedures with dried materials (see p. 46) are often applicable.

However, specific protective agents may sometimes be used to prevent loss of enzymic activity when butanol is employed with aqueous materials. Morton and Tsao[62] found that addition of ATP and actin was

[60] B. A. Askonas, *Biochem. J.* **48**, 42 (1951).
[61] S. Schwimmer and A. B. Pardee, *Advances in Enzymol.* **14**, 375 (1953).
[62] T.-C. Tsao, *Biochim. et Biophys. Acta* **11**, 368 (1953).

necessary to protect myosin ATP-ase from denaturation by n-butanol. The author has used succinate and pyrophosphate to protect succinic dehydrogenase, and addition of reducing agents (such as glutathione), together with treatment under nitrogen, to protect this and other enzymes with sensitive —SH groups from denaturation by n-butanol. The addition of enzyme substrates may not only protect active centers (see Bach *et al.*[63]) but may also change the general properties of the protein, as shown by the difference in solubility of oxidized and reduced cytochrome.[64]

The Butanol-Soluble Material

Where procedure A (two-phase method) is used with mitochondria or microsomes, the excess butanol contains considerable amounts of material. This is mostly phospholipid, but other compounds, such as the ketogenic factor of horse muscle and calf intestine (Stewart and Young[65]), are also extracted. The butanol therefore may sometimes contain essential coenzymes, such as lipothiamide. During purification of alkaline phosphatase, Morton[54] found that holding the residual butanol layer at -15 to $-20°$ caused precipitation of much of the butanol-soluble material, from which was obtained the crystalline ketogenic factor of calf intestinal mucosa.[65] The precipitate is collected by centrifuging at $-15°$. The final residue in the butanol may be recovered by vacuum distillation of the butanol.

Recovery of Protein

Enzymes and other proteins stable to acetone may be directly precipitated from the aqueous phase with acetone at low temperature, using the procedure described by Askonas.[60] The presence of butanol reduces the concentration of acetone necessary for precipitation of a given protein. When proteins are labile to such solvent treatment, the butanol may be removed preferably by dialysis against buffers of low ionic strength at $0°$ or by freezing and drying *in vacuo*.

Choice of Solvents

Both n- and isobutanols appear to be effective.[47,54] n-Butanol has been preferred by the author because, in general, it is less toxic to enzymes, but isobutanol may be advantageous in certain cases (see Kun[51]). Admixture of other solvents, such as ethanol, ether, chloroform, etc., has been found of no advantage and often causes complete denaturation of

[63] S. J. Bach, M. Dixon, and L. G. Zerfas, *Biochem. J.* **40**, 229 (1946).
[64] D. Keilin, *Proc. Roy. Soc. (London)* **B106**, 418 (1930); K. Zeile and F. Reuter, *Z. physiol. Chem.* **221**, 101 (1933).
[65] H. B. Stewart and F. G. Young, *Nature* **170**, 976 (1952).

enzymes. However, the author has observed that a mixture of 1 part of benzene to 2 parts of *n*- or isobutanol may remove some lipid material in addition to that extracted by butanol alone. This solvent mixture may be found to be of some advantage for very stable complexes, but the stability of enzymes to this treatment has not as yet been determined. Cyclo-hexanol has been found partly to disrupt some enzyme complexes. Al-though it offers no advantage over butanol in the cases studied, it may be found of value in particular circumstances.

Various Butanol Procedures for Extraction of Aqueous Material

The following procedures have been used by the author and other workers to meet the needs of particular problems.

Procedure A (Two-Phase Procedure). The procedure is illustrated in the purification of alkaline phosphatases (Morton[47,54]). Butanol is added to saturate the aqueous phase and to give some excess (about 20% v/v *n*-butanol). The dispersion is centrifuged (2000 × *g* or greater for 30 min-utes), the emulsion separating into excess butanol overlying butanol-saturated residue material and a clear aqueous phase containing dissolved proteins. At higher forces and with longer centrifuging the residue material is sedimented and the excess butanol directly overlies the aqueous phase. The excess butanol should be removed by suction, and the aqueous layer collected and filtered, if necessary, to obtain a water-clear extract.

Procedure B (One-Phase Procedure). As indicated above, most lipo-proteins are dissociated at concentrations of butanol well below that re-quired for saturation of the aqueous phase. The critical concentration for dissociation of the lipoprotein varies with experimental conditions. Morton and Tsao[62] used butanol at a final concentration of about 5% (v/v) to obtain myosin ATP-ase in solution at low ionic strength. Leone[58] used a final concentration of about 3% (v/v) for obtaining uricase in solution.

It is therefore suggested that, in investigating this procedure, the pH of the material should be close to either the acid or the alkaline limit of enzyme stability, and butanol should be added slowly to give a final con-centration of between 3 and 5% (v/v). The material should be centrifuged at high speed (20,000 × *g* or higher) for 1 hour to remove insoluble mate-rial, and the aqueous extract filtered if necessary.

Procedure C (Butanol-Acetone Powder). The aqueous material is held at 0 to −2°, and cold butanol (−15°) is added as described above, either to saturate the aqueous phase (cf. procedure A) or to a limited concentra-tion (cf. procedure B). The whole suspension (at about −2°) is then added slowly to acetone held at −15° and continuously stirred. The precipitate is collected and dried as described for preparation of acetone powders (see above, p. 34). In this way a stable, dry preparation is obtained. Subsequent extraction with dilute buffers (0.15 M NaCl or 0.05 M phosphate buffer, pH 7.4, for example) is used to obtain enzymes or other proteins in true solution. The combined action of the butanol and acetone drying causes complete disruption of many lipoprotein complexes, whereas the protein is recovered in a native state. A highly purified preparation of actin has been obtained from rabbit skeletal muscle by Tsao and Bailey[66] with a butanol-acetone powder.

Procedure D (Butanol-Enzymic Procedure). In order to obtain certain proteins in solution from insoluble complexes, it may be necessary to dissociate other water-insoluble material, such as structural proteins, carbohydrates, and nucleic acids, as well as lipids. Enzymic digestion of nucleic acids by crystalline ribonuclease has been used by the author in combination with procedure B. This method is at present being investigated in detail. Digestion of nucleoprotein in powders obtained by procedure C and procedures E and F (pp. 47 and 48) may also be useful in obtaining refractory enzymes in solution. Preliminary removal of lipid with butanol appears to be essential for effective enzymic digestion of many protein complexes.

Butanol Procedures for Extraction of Dried Material

A number of enzymes and other proteins are denatured by organic solvents, including butanol, when these are added to the aqueous material. However, such proteins are frequently quite stable when exposed to dry organic solvents. Preliminary removal of water by drying the material *in vacuo* or with acetone, followed by treatment with dry butanol, therefore may permit removal of lipid without denaturation of the associated enzyme.

The following general considerations apply to the two procedures which employ dry butanol for extraction of lipid.

Preparation of Material. The material should be thoroughly disintegrated (see above, p. 28) before drying. The dried preparation should then be ground to a fine powder, preferably in a ball mill (about 18 hours) or with a mortar and pestle, prior to butanol extraction.

The preparation of the dispersion before drying may have considerable influence on subsequent extraction of lipids. The effect of varying the pH, salt concentration, nature of salt, addition of enzyme substrates, and degree of dispersion of material has not, as yet, been properly investigated. In general, preparations have been adjusted to about pH 6.5 and ionic strength 0.05 before drying.

[66] T.-C. Tsao and K. Bailey, *Biochim. et Biophys. Acta* **11**, 102 (1953).

Extraction with Butanol. Water should be excluded from the system as far as possible during solvent extraction; otherwise protein denaturation may occur. The effect of temperature during extraction with butanol has not yet been systematically studied. Extractions have been carried out both at 0° and at room temperature (about 15°) by the author with equal success.

Extraction of Proteins. The specific differences between salts as discussed previously should be remembered when extracting the lipid-free material. In general, extraction of the dry powder with 0.05 M phosphate or 0.02 M pyrophosphate buffer at pH 7.4 to 7.8 is preferred by the author.

The following procedures have been used by the author and other workers.

Procedure E. Butanol Extraction of Vacuum-Dried Material. The isolated lipoprotein, or the whole tissue dispersion, is frozen and dried *in vacuo.* Alternatively, pressed yeasts or bacteria may be dried in an air stream at about 30°. The dried material is ground to a fine powder which is then thoroughly dispersed in dry *n*-butanol, using about 50 ml. of butanol per gram of dry material. Dispersion is preferably carried out in a Potter-Elvehjem homogenizer or a Waring blendor. Alternatively, the material may be ground into a thick paste with butanol in a mortar and pestle. The material is extracted with the solvent for about 20 minutes (or overnight at 0°) with continuous stirring and then collected by centrifuging. The material should be re-extracted with dry butanol as before, and then with dry acetone (at −15°). Dry peroxide-free diethyl ether (at −15°) and light petroleum ether (40–60° b.p.) have also been used by the author for removal of butanol. After complete evaporation of the organic solvents *in vacuo* over $CaCl_2$, the precipitate is rubbed into a thick paste with dilute buffer (such as 0.05 M phosphate or 0.02 M pyrophosphate buffer, pH 7.8), further buffer is added, and the material is extracted for about 1 to 2 hours at 0°. The extract is finally centrifuged at high speed (40,000 × g or greater, for 1 to 2 hours) to remove insoluble material.

The cholinesterases of brain[48] and erythrocyte stroma[67] have been obtained in true solution by this procedure. The erythrocyte enzyme was considerably purified.[67] Other organic solvents apparently are not successful in releasing brain[68] or erythrocyte[67] cholinesterases into true solution by this procedure. Mitochondrial-bound glutaminase I has recently been obtained in solution by a similar procedure.[68a]

[67] J. A. Cohen and M. G. P. J. Warringa, *Biochim. et Biophys. Acta* **10**, 195 (1953).
[68] K. Bullock, *Biochem. J.* **49**, vii (1951).
[68a] M. C. Otey, S. M. Birnbaum, and J. P. Greenstein, *Arch. Biochem. and Biophys.* **49**, 245 (1954).

Procedure F (*Butanol Extraction of Acetone-Dried Material*). Acetone powders prepared as described on p. 34 contain considerable amounts of acetone-insoluble phospholipid. Treatment of the acetone powder with butanol removes this phospholipid and thus greatly facilitates obtaining the protein into true solution on subsequent extraction with dilute salt solutions.

The aqueous dispersion of tissue or protein complex is added slowly to 10 vol. of acetone held at $-15°$, and the precipitate is collected as described on p. 34. The precipitate is then dispersed in *n*-butanol at $-15°$ (about 3 vol., as calculated from the original suspension), in a Waring blendor (1 minute) or a Potter-Elvehjem homogenizer, and thoroughly extracted by stirring for 15 minutes. The precipitate is collected either by centrifuging (at $-15°$) or by vacuum filtration through a No. 1 Whatman paper on a Büchner funnel, washed twice with acetone at $-15°$, and dried *in vacuo*. Peroxide-free diethyl ether (at $-15°$) used after acetone may assist drying but should be employed with caution (see p. 34).

The dried material is ground to a fine powder in a ball mill (about 18 hours) and then extracted for 1 to 2 hours at $0°$ with 0.05 M phosphate or 0.02 M pyrophosphate buffer at pH 7.8 and centrifuged (40,000 $\times g$; 1 hour) to remove insoluble material. The author has found that a number of enzymes, such as liver esterase, may be obtained in true solution by this procedure.[48]

Procedure for Extracting Mitochondria

Particular interest is attached to the extraction of enzymes from mitochondrial membranes and cristae. The various principles as discussed in the preceding sections are all applicable to this problem, and no detailed directions can be given which are suitable for all mitochondrial enzymes, since no procedure so far investigated permits of extraction of all the protein of mitochondria in water-soluble form. However, the procedure developed by the author for the preparation of succinic dehydrogenase of heart muscle[47] illustrates the principles discussed above and is therefore given in detail here.

Freshly collected horse heart is transported to the laboratory in crushed ice. Fat and unwanted tissue is cut away, and the muscle minced in a mechanical meatgrinder previously cooled to $0°$. The mince is washed with tap water, with continuous stirring, the mince being squeezed in muslin at regular intervals. When the extracting fluid is colorless (after 4 to 6 hours of extraction), the mince is disintegrated in a Waring blendor for 5 minutes, using 200-g. lots of mince and 300 ml. of 0.05 M potassium phosphate buffer, pH 7.4.

The dispersion is centrifuged (2000 × g, 15 minutes) at room temperature, and the precipitate is discarded. The cloudy supernatant is recentrifuged (20,000 × g, 30 minutes, at 0°). Alternatively, the supernatant may be cooled to 0°, adjusted to pH 5.6 with 0.5 N acetic acid, and centrifuged (2000 × g, 30 minutes) at 0°. The precipitate, which consists mainly of mitochondrial membranes and cristae, is dispersed in 0.85% NaCl and adjusted to pH 7.5 with 0.02 M NaHCO$_3$. The suspension is recentrifuged (20,000 × g, 1 hour at 0°) and the precipitate rubbed into 0.02 M NaHCO$_3$ and adjusted to pH 7.8. Sufficient bicarbonate solution is used to obtain a thin paste.

The material is cooled to −2°, and n-butanol (to 40%, v/v, at −2°) is slowly added with continuous mechanical stirring. After holding for 30 minutes, the material is centrifuged (5000 × g, 30 minutes, at −2°), and the supernatant excess butanol layer removed. The underlying clear yellow aqueous layer is transferred to a dialysis sac (at 0°) and dialyzed with continuous stirring against several changes of 0.01 M NaHCO$_3$ solution (at 0°) until all butanol is removed.

This extract contains diaphorase and succinic dehydrogenase in true solution. The latter enzyme, however, is highly unstable. Activity may be completely lost during dialysis. The succinic dehydrogenase activity may be precipitated with ammonium sulfate at pH 6.8, between 0.45 and 0.7 saturation. Activity may be retained for about 24 to 48 hours by keeping the precipitate under 0.7 saturated ammonium sulfate solution at pH 6.8 and at 0°. However, considerable difficulty has been experienced in obtaining reproducible preparations with succinic dehydrogenase activity although the diaphorase activity is consistently high. It should be noted that an intact lipid group may be essential for activity of certain enzymes.[56,69]

Application of Butanol to Assay Procedures

Assay of Alkaline Phosphatase. This enzyme is intimately associated with insoluble cellular particles.[54] It is desirable to separate the enzyme from the particles in order to remove natural inhibitors and to facilitate preparation of accurate dilutions. Treatment of the tissue dispersion with butanol permits separation of the enzyme from the particles, and dialysis of the clear solution then removes dialyzable inhibitors. The preparation may be frozen and dried *in vacuo* and assayed at a convenient period. The procedure is described in Vol. II [**81**].

Summary of Procedures

The proteins obtained in solution or purified by the butanol method during the period 1950–1953 are shown in the following table. Although

[69] W. W. Kielley and O. Meyerhof, *J. Biol. Chem.* **183**, 391 (1950).

PROTEINS PURIFIED BY THE BUTYL ALCOHOL METHOD

Protein	Reference
Actin	Tsao and Bailey[a]
Alkaline phosphatase of milk	Morton[b,c]
	Zittle and Della Monica[d]
Alkaline phosphatases of intestine, kidney, liver, and mammary gland	Morton[b,e]
Amino peptidase of hog kidney	Robinson et al.[f]
ATP-ase (myosin)	Morton and Tsao; see Morton,[g] Tsao[h]
Casein	Morton[g]
Cholinesterase (intestinal)	Morton[b]
Cholinesterase (brain)	Morton[g]
Cholinesterase (erythrocyte)	Cohen and Warringa[i]
Creatine phosphokinase	Askonas[j]
Cytochrome b_2 (crystallized from yeast)	Appleby and Morton[k]
Diaphorase (heart muscle)	Morton[g]
DPN-cytochrome c reductase (liver)	Kun[l]
Esterase (insect)	Morton[b]
Glutaminase I (kidney)	Otey et al.[m]
γ-Glutamyl transpeptidase (glutathionase)	Morton[b]
	Fodor et al.[n]
Lactic dehydrogenase (bacterial flavoprotein)	Szulmajster et al.[o]
Lactic dehydrogenase of yeast (cytochrome b_2)	Appleby and Morton[k]
Lecithinase (bacterial)	Hayaishi and Kornberg[p]
Red protein of milk	Polis and Shmukler[q]
Succinic dehydrogenase	Morton[b]
Transaminase (mitochondrial)	Morton[b]
Uricase	Leone[r]
Xanthine oxidase (milk)	Morton[b]

[a] T.-C. Tsao and K. Bailey, Biochim. et Biophys. Acta 11, 102 (1953).

[b] R. K. Morton, Nature 166, 1092 (1950).

[c] R. K. Morton, Biochem. J. 55, 795 (1953).

[d] C. A. Zittle and E. S. Della Monica, Arch. Biochem. and Biophys. 35, 321 (1952).

[e] R. K. Morton, Biochem. J. 57, 595 (1954).

[f] D. S. Robinson, S. M. Birnbaum, and J. P. Greenstein, J. Biol. Chem. 202, 1 (1953).

[g] R. K. Morton, Thesis, University of Cambridge, 1952.

[h] T.-C. Tsao, Biochim. et Biophys. Acta 11, 368 (1953).

[i] J. A. Cohen and M. G. P. J. Warringa, Biochim. et Biophys. Acta 10, 195 (1953).

[j] B. A. Askonas, Thesis, University of Cambridge, 1951.

[k] C. A. Appleby and R. K. Morton, Nature 173, 749 (1954).

[l] E. Kun, Proc. Soc. Exptl. Biol. Med. 77, 441 (1951).

[m] M. C. Otey, S. M. Birnbaum, and J. P. Greenstein, Arch. Biochem. and Biophys. 49, 247 (1954).

[n] P. J. Fodor, A. Miller, and H. Waelsch, J. Biol. Chem. 202, 551 (1953).

[o] J. Szulmajster, M. Grunberg-Manago, and C. Delavier-Klutchko, Bull. soc. chim. biol. 35, 1381 (1953).

[p] O. Hayaishi and A. Kornberg, J. Biol. Chem. 206, 647 (1954).

[q] B. D. Polis and H. W. Shmukler, J. Biol. Chem. 201, 475 (1953).

[r] E. Leone, Biochem. J. 54, 393 (1953).

the method has so far been applied mostly with animal tissues, the procedures described above should be just as suitable for obtaining active enzymes from plant tissues, yeasts, and bacteria.

[7] Extraction of Enzymes from Microorganisms
(Bacteria and Yeast)

By I. C. GUNSALUS

Principle

The methods which have proved effective in liberating enzymes from microbial cells have been largely mechanical rupture of the cell wall and membrane, frequently with fragmentation of the latter. In specific instances enzymatic,[1] including autolysis, and chemical[2] treatments have proved useful.

The choice of procedure will depend on the particular species or strain of microorganism used, including the ease with which the cell is ruptured, the quantity of enzyme required, and the type of preparation from which the extract is prepared. The choice will also vary with the sensitivity and the localization of the enzyme within the cell. Most methods in use were empirically derived; thus, consulting the original literature in specific cases is useful—for example, a recent review by Hugo[3] carries more than a hundred references to specific cases.

Bacterial and yeast cells for the preparation of enzyme extracts may be obtained as products of the fermentation industries; that is, brewer's yeast as sludge from a spent fermentation tank, baker's yeast in pound cakes or in bulk—usually starch-free—from the baking or food industries. A few bacterial strains are available from pharmaceutical procedures which do not impair enzyme activity. More frequently, however, bacterial cells are grown by the investigator, usually under conditions which support optimum growth or a maximum enzyme yield.[4] In the latter case, the age and enzymatic content of the cells, the means of harvest—usually Sharples continuous centrifuge—and the type of cell preparation from which the extract is prepared can be controlled. For small quantities, batch centrifugation may be employed, in which case aeration during harvest can also be controlled, should this be important to the investiga-

---

[1] M. Penrose and J. H. Quastel, *Proc. Roy. Soc. (London)* **B107**, 168 (1930); M. F. Utter, L. O. Krampitz, and C. H. Werkman, *Arch. Biochem.* **9**, 285 (1946).

[2] P. V. B. Cowles, *Yale J. Biol. and Med.* **19**, 835 (1947).

[3] W. B. Hugo, *Bacteriol. Revs.* **18**, 87 (1954).

[4] A. J. Wood and I. C. Gunsalus, *J. Bacteriol.* **44**, 333 (1942); E. F. Gale, *Bacteriol. Revs.* **7**, 139 (1943).

tion. In such cases freshly harvested cell paste with a minimum of autolysis is obtained.

In the use of freshly harvested or commercially prepared cell paste, cakes, or sludges, washing may be employed if the experiments so indicate, using caution with aged cells lest the enzymes be extracted by the washing procedure.

The cells may be preserved by storage at reduced temperatures, for example, in the deep-freeze or dry ice chest at -10 to $-40°$; or they may be reduced immediately to extracts, in which state the enzymes are usually stable at deep-freeze temperatures. If neither of these proves convenient, stable dried cell preparations can usually be obtained by air, vacuum, or solvent drying procedures.

In summary: The extraction and preservation of the activity of any given enzyme is an empirical process based on the investigator's own or his colleagues' experience. In the absence of example, the original literature offers a wide variety of methods and of organisms used successfully for various enzymes.[3]

Dried Cell Preparations

The ease of release of enzymes from cells depends to a great extent on their previous treatment—drying may serve to preserve enzyme activity or may be used as a means of liberating or extracting enzymes. Dried cell preparations frequently have sufficiently lost the "permeability" properties of living cells to serve as enzyme preparations—that is, they will metabolize phosphate esters and polybasic acids and can be activated by coenzymes[5]—and the coenzymes and metabolic intermediates are often leeched from the cells during the drying procedure or after resuspending in buffers. In many cases the drying procedure involves enzymatic autolysis, which enhances enzyme liberation.

Procedure

Dried cell preparations can be obtained by:

1. Air drying—frequently accompanied by autolysis;
2. Slow vacuum drying—usually from cell pastes, also accompanied by autolysis and yielding glass-like preparations which can be later powdered readily by mortar and pestle;
3. Lyophylization—desiccation from the frozen state, usually of a 10 to 20% by wet weight cell suspension;
4. Dehydrating with water-miscible solvents.

[5] I. C. Gunsalus and W. W. Umbreit, *J. Bacteriol.* **49**, 347 (1945); I. C. Gunsalus, W. D. Bellamy, and W. W. Umbreit, *J. Biol. Chem.* **155**, 685 (1944).

Air Drying (of Yeast). The finding by Lebedev[6] that air-dried yeast could be extracted by mild grinding or by stirring with buffer greatly accelerated studies with soluble enzymes. Since Lebedev's finding, many methods have been employed for drying[7] and for extracting[8] air-dried yeast and bacteria. The important factors considered in the early experiments are summarized by Harden.[9] For more recent data specific manuscripts should be consulted. In general, yeast at the consistency of yeast cake, or cells solidly packed by centrifugation after washing, are crumbled by hand or by forcing through a 10-mesh sieve or potato dicer onto trays in layers less than an inch deep and allowed to dry 2 to 3 days at temperatures ranging from 25 to 30°, occasionally as high as 35 to 40°. The layer of yeast may be turned at half-day intervals to promote drying; a fan may be employed to accelerate the process. Air-dried yeast has undergone partial autolysis and will usually yield active enzyme extracts if suspended in 3 to 5 vol. of water or suitable buffer and stirred 2 to 3 hours at room temperature.[8,9] (Occasionally yeast is pretreated—for example, mild bicarbonate hydrolysis before extraction.[8]) The cell debris can then be removed by centrifugation, and the enzymes subjected to further purification if desired. If a particular enzyme is not solubilized by the drying procedure, one of the more rigorous methods of extraction may be resorted to.

Slow Drying in Vacuum. This procedure has been advantageously applied to enzyme extraction of several types of bacteria.[10–12] The packed cells obtained by centrifugation are transferred by a spatula to a beaker or evaporating dish, placed in a desiccator over calcium chloride, P_2O_5 or anhydrous calcium sulfate (Drierite), and the desiccator evacuated with an aspirator or vacuum pump, sealed, and allowed to stand, usually overnight. Occasionally a thick, 20% wet weight, cell suspension is dried by a similar procedure. The dried preparations are usually of hard, glassy consistency and have undergone some autolysis. The cells may be extracted by suspending directly in buffer but usually more successfully by grinding the dried cell mass to a powder with a mortar and pestle before extraction. If the paste has been deposited in layers more than 0.5 to 1 cm. thick, gentle grinding followed by further drying may be required. More labile enzymes are destroyed by the slow drying procedure;

[6] A. von. Lebedev, *Compt. rend.* **152**, 49 (1911).
[7] C. Neuberg and H. Lustig, *Arch. Biochem.* **1**, 191 (1943).
[8] B. L. Horecker and P. Z. Smyrniotis, *J. Biol. Chem.* **193**, 371 (1951).
[9] A. Harden, "Alcoholic Fermentation," Chapter 2. Longmans, Green, & Co., London, 1923.
[10] F. Lipmann, *Cold Spring Harbor Symposia Quant. Biol.* **7**, 248 (1939).
[11] E. R. Stadtman and H. A. Barker, *J. Biol. Chem.* **180**, 1085 (1949).
[12] B. P. Sleeper, M. Tsuchida, and R. Y. Stanier, *J. Bacteriol.* **59**, 129 (1950).

thus partial reaction sequences may be studied successfully with such preparations.[12]

As with air-dried yeast, the soluble enzymes can usually be extracted from the slow vacuum-dried bacterial or yeast preparations by gentle agitation with buffer or water. The incorporation of reducing substances—that is, cysteine, glutathione, or sodium sulfide[11]—before drying, especially with metabolically anaerobic bacteria, has been helpful in preventing the loss of otherwise labile enzymes. These additions have been particularly beneficial in preserving those enzymes which appear to possess mercapto groups or to be sulfhydryl activated.[11]

Lyophylization. Drying quick-frozen, thick cell suspensions in vacuum has proved especially beneficial in preserving labile enzymes. In fact, a similar procedure on a smaller scale is in common use to preserve stock cultures in a viable state.[13] For enzyme work, vacuum-dried cells are usually stored in the presence of air in a sufficiently tight container to prevent rehydration. The cells soon lose their viability and in many cases are more permeable to substrates, including cofactors, than are viable cells.[5] Usually, more drastic means are required to extract enzymes from lyophylized than from air or slow vacuum-dried cells, very probably because autolysis does not occur. Lyophylization of masses of cells can be accomplished most conveniently by suspending 10 to 40% wet weight of cells in distilled water. With some cells a washing step to remove salts and residual media ingredients is desirable to reduce autorespiration (endogenous substrate oxidation); with large batches of cells a quick freeze before drying is also desirable. For preparations up to 10 to 15 g. of cells, a single 100-mm. Petri dish containing 10 to 40 ml. of suspension can be placed over 4- to 8-mesh Drierite in a 250-mm. desiccator and evacuated to less than 0.1 mm. mercury. Under these conditions the rate of evaporation will be rapid enough to cause the suspension to freeze. If the vacuum is preserved—usually it is well to keep the pump running—the preparation will dry without thawing in a period of 6 to 8 hours (usually well to leave overnight) to yield a porous, friable mass which may be reduced to a powder by working gently with a spatula on a clean paper. The enzyme activities in lyophylized cells may remain nearly constant for several years if stored at $-20°$ and protected from the re-entry of water. The moisture content of lyophylized cells is frequently less than 0.1%, usually lower than in air-dried preparations.

To prepare large batches of lyophylized cells, shell freezing in plasma bottles or round-bottom distillation flasks, up to 2- to 3-l. volume, is useful. The desiccation is usually accomplished by connecting the flask

[13] L. A. Rogers, *J. Bacteriol.* **57**, 137 (1949); L. Atkin, W. Moses, and P. P. Gray, *ibid.* **57**, 575 (1949).

to a large-diameter, short-path vacuum system containing a mechanically refrigerated, or dry ice-cooled, receiver to remove the water by sublimation. Convenient units for this purpose are available commercially or may be fabricated from glass or from metal.[14,15]

Acetone (Solvent) Dried Preparations. Large quantities of cells may be conveniently dried by adding an aqueous suspension to a large volume of water-miscible solvent at a sufficiently low temperature to prevent denaturation of the cell proteins. The most common procedure is to add an aqueous, or buffered, cell suspension slowly with vigorous stirring to not less than 10 volumes of acetone previously cooled to $-20°$ or to the temperature of dry ice. After a brief stirring the cells are allowed to settle, the supernatant, which frequently contains gummy material, is decanted, and the residual solvent is removed on a Buchner filter. The filter cake of cells is washed on the Buchner with 2 to 5 vol. of $-20°$ acetone—or suspended in 2 to 5 vol. of chilled acetone and again collected on a Buchner— and washed once with 2 to 3 vol. of $-20°$ peroxide-free ether, sucked dry on the filter, and transferred to a large sheet of wrapping paper and worked gently with a spatula until the solvent has evaporated, leaving a dry powder. If the atmosphere is humid, the dried cells should not remain in contact with the air for more than a few minutes, and the removal of solvent can be completed, although less satisfactorily, in a desiccator in the presence of paraffin to absorb the solvent. Acetone-dried preparations usually contain no living cells—bacterial endospores being an exception—and retain fewer of the permeability properties of the viable cell than do lyophylized preparations. If freshly prepared cells are acetone-dried, little, if any, autolysis will have occurred. Acetone-dried cells may yield active enzyme extracts on stirring with suitable buffer or with water,[16] or more rigorous extraction may be required.[17] Other water-miscible solvents including dioxane[18] have also been employed, although less frequently, to prepare dried cells.

Toluene, chloroform, or similar solvents[19] may be used to alter the permeability of cells, permitting enzyme assays without respect to cell permeability. This type of preparation is seldom subjected to further

[14] E. W. Flosdorf, "Freeze-Drying," Reinhold Publishing Co., New York, 1949; Virtis Co., Inc., Yonkers, New York (metal).

[15] W. J. Elser, R. A. Thomas, and E. I. Steffen, *J. Immunol.* **28**, 433 (1935); D. H. Campbell and D. Pressman, *Science* **99**, 285 (1944) (glass).

[16] E. F. Gale, *Advances in Enzymol.* **6**, 1 (1946); G. W. E. Plaut and H. A. Lardy, *J. Biol. Chem.* **180**, 13 (1949).

[17] M. I. Dolin, Thesis, Indiana University, 1950.

[18] Procedures similar to J. B. Sumner and A. L. Dounce, *J. Biol. Chem.* **127**, 439 (1939).

[19] E. Buchner and R. Rapp, *Ber.* **30**, 2668 (1897); see also ref. 9, p. 36.

purification, even to the extent of removing the enzymes from cellular debris.

Toluene may, however, be used in conjunction with grinding procedures to enhance enzyme liberation.[20] Butanol,[21] among other solvents, has been used to solubilize particle-bound enzymes. Methods for the solubilization and analysis of the enzymatic components of the cytochrome-containing bacterial particles have, however, not been devised.

Mechanical Rupture of Cells

Mechanical rupture of the rigid cell wall is employed as a means of liberating particularly the soluble enzymes, which presumably occur in the cell cytoplasm. Such procedures also fragment the more friable cell membrane but do not usually remove the particle-bound enzymes from their cellular components. These more rigorous methods are used for cells which do not yield extracts after autolysis or drying and are not subject to known specific enzyme treatment.[1,22] Most mechanical methods of cell rupture combine techniques, such as the addition of abrasives, freezing the cell mass—with the ice crystals serving as abrasive--or the combination of crushing and shearing forces.

The mechanical rupture of microbial cells to yield active enzyme preparations will be discussed under the following headings:

1. Mechanical pressure; hydraulic or fly press.
2. Pressure release; rapid release of compressed gas.
3. Sonic or ultrasonic waves; tuned magnetostriction or piezoelectric oscillator.
4. Mechanical shaking with abrasives; tuning fork or blendor-homogenizer.
5. Grinding; mechanical or manual.

Among the abrasives most commonly used with these procedures are fine quartz sand, powdered Pyrex glass, micro-size ballotini or reflector beads, Carborundum, jeweler's rouge, powdered dry ice, and alumina. The most effective particle size is 500-mesh or finer; uniform particle size affords more effective cell fragmentation. In the use of all the abrasives tested, except alumina, mechanical forces seem to predominate. With alumina a chemical process seems also to be involved.[23,24] The use of abrasives, particularly alumina, will be discussed in greater detail below.

[20] J. Berger, M. J. Johnson, and W. H. Peterson, *Enzymologia* 4, 31–35 (1937).
[21] R. K. Morton, *Nature* 166, 1092 (1950).
[22] L. H. Stickland, *Biochem. J.* 23, 1187 (1929).
[23] H. McIlwain, *J. Gen. Microbiol.* 2, 288 (1948).
[24] O. Hayaishi and R. Y. Stanier, *J. Bacteriol.* 62, 691 (1951).

Mechanical Pressure. Two means of preparing microbial "juices" or cell extracts employing hydraulic pressure have been employed. Buchner and Hahn's (1903)[25] preparation of yeast zymase, by applying hydraulic pressure to a mixture of yeast and kieselguhr, constituted an important step forward in establishing on a firm basis the enzyme concept. An excellent review of these early studies, including grinding with quartz sand and preparation of extracts from dried cells and from fresh yeast by hydraulic pressure, appears in Chapter 2 of Harden's monograph, "Alcoholic Fermentation."[9] Buchner and Hahn mixed kieselguhr with 3 to 4 parts of yeast paste, subjected the mass to hydraulic pressure, and were able to collect 30 to 50% of the cell volume as an extract—zymase. Such concentrated extracts were used in early experiments, before diffusible cofactor requirements were recognized as one of the prime causes for loss of activity on dilution. The hydraulic press, although useful, has not been employed extensively with bacteria and seems not to be readily applicable to larger quantities of cells.

More recently, Hughes[26] has introduced the use of a fly press to apply instantaneously pressures up to 10 to 15 tons per square inch. The cells may be mixed with an abrasive or cooled to -20 to $-30°$, in which case the ice crystals serve as abrasive. Rupture of a very high percentage of the cells has been reported[3,26]—up to 99% of the cells in *Escherichia coli* pastes. Bacterial endospores have yielded enzyme preparations by repeated treatment in the Hughes press, interspersed with alternate freezing and thawing.[27]

Pressure Release. The rupture of microorganisms by the sudden release of pressure after compressing a water-soluble gas has recently been reinvestigated by Fraser[28] for the release of bacteriaphage and of enzymes from *E. coli*. Fraser's procedure can be applied to small volumes, 5 to 10 ml., and to dilute cell suspensions, 10^8 cells per milliliter. A single decompression passage using nitrous oxide at pressures up to 900 psi ruptured more than 75% of the cells in an *E. coli* suspension. The method can be applied to heavier cell suspensions, but the percentage of cells ruptured decreases. Some organisms are refractory even to several decompression passages. This procedure has been especially useful in releasing particulate cellular components, including bacteriaphage.

Sonic Waves. Treatment of cell suspensions at sonic frequencies in the Raytheon 50-watt 9-kc. or 200-watt 10-kc. magnetostriction oscillator,

[25] E. Buchner and H. Haehn, "Die Zymase Garung," p. 58. R. Oldenburg, Munich, 1903; see also ref. 9, p. 23.
[26] D. E. Hughes, *Brit. J. Exptl. Pathol.* **32,** 97 (1951).
[27] M. R. Pollock, *J. Gen. Microbiol.* **8,** 186 (1953).
[28] D. Fraser, *Nature* **167,** 33 (1951).

either alone or with finely powdered abrasives, has simplified and accelerated studies with bacterial extracts. The two Raytheon instruments are similar except for power output and size of the sample cup. The suspension is treated in a cylindrical stainless steel, water-cooled cup, to which the power is applied through a bottom diaphragm activated by a laminated nickel rod held in an oscillating magnetic field. The instrument is "tuned" for maximum energy input, but to date no means of measuring the power actually delivered to the liquid in the treatment cup has been devised—thus each treatment is more or less empirical. The treatment cup of the 9-kc. machine has a total volume of 60 ml., in which 25 to 35 ml. of cell suspension can be treated satisfactorily; the 10-kc. cup volume is 150 ml., in which 60 to 100 ml. can be treated. The best disintegration has been obtained with cell suspensions containing 50 to 100 mg. dry weight per milliliter; thus the 10-kc. machine will allow treatment of 2 to 10 g. of dry cells. Gram-negative rods (pseudomonads or coli-aerogenes types) are readily ruptured—a 10- to 15-minute treatment in the 10-kc. machine will break more than 90% of the cells—thus allowing treatment of up to 25 to 50 g. of cells per hour. The Gram-positive organisms of the *Bacillus* and *Clostridium* genera are also readily ruptured, but the lactic acid bacteria, the staphylococci and the corynebacteria, are more refractory, frequently requiring periods of 30 to 60 minutes in the 10-kc. machine, and as long as 3 hours in the 9-kc. machine. In some cases even this extended treatment releases less than half of the enzyme content in soluble form, the remainder being recoverable in the cell residue.[17] Yeast cells and bacterial endospores are more refractory to rupture in the Raytheon equipment, only limited success having been obtained to date.

The use of a quartz (piezoelectric) crystal employing supersonic frequencies up to 600 kc. previously employed[29] has proved less effective than the magnetostriction method because of the small volumes which may be employed, the critical nature of the tuning, the relatively dilute suspensions which may be successfully treated, and the extended period of treatment—up to 4 hours—required.

Mechanical Shaking with Abrasives. Another method applied to cell suspensions is rapid shaking—300 to 3000 oscillations a minute—with regular small particles (50 to 500 μ). Important variables are the size and uniformity of the particles, the speed of shaking, and the nature and density of the cell suspension.[30] Mechanical shaking with a Mickle[31] or

[29] P. K. Stumpf, D. E. Green, and F. W. Smith, *J. Bacteriol.* **51**, 487 (1946).
[30] G. Furness, *J. Gen. Microbiol.* **7**, 335 (1952); P. D. Cooper, *J. Gen. Microbiol.* **9**, 199 (1953).
[31] H. Mickle, *J. Roy. Microscop. Soc.* **68**, 10 (1948).

Microid[30] mechanical shaker, using ballotini beads, grade 9 to 16 (average diameter 300 to 100 μ), at cell densities up to 2 to 5 mg. dry weight per milliliter, has been used to liberate enzymes and to prepare cell walls for examination.[30,32] The Mickle disintegrator (H. Mickle, Hampton, Middlesex) carries two 15-ml. cups, each mounted on the end of an electrically driven tuning fork, and the Microid shaker (Griffin & Tatlock, London) carries four 3 × 14-cm. (about 100-ml. volume) cups at speeds up to about 2500 oscillations per minute, or two cups at speeds to 3000 oscillations per minute. There are few data available on enzyme extraction by either of these procedures, but those available indicate promise.

Lamanna and Mallette[33] have recently reported the rupture of dense cell suspensions, particularly yeast, by disintegration through mechanical agitation in the presence of grade 10 glow beads (60 to 80 mesh, 200-mμ average diameter), obtainable from Minnesota Mining & Manufacturing Company, Minneapolis. For this purpose they employed a Waring blendor or a Virtis homogenizer. As typical examples, 99% of the cells in a pound of yeast were ruptured in the presence of 800 g. of beads in a total volume of approximately 1 l. by 90-minute treatment in the Waring blendor. To avoid surface denaturation by foaming, a paraffin block carved to fit the top of the Waring cup was pressed tightly against the liquid surface. With the Virtis homogenizer, which operates at higher speeds, 99% of the cells of a 1.5-ml. *E. coli* suspension, containing 4 × 10^{10} organisms per milliliter and 2 g. of beads, were ruptured after 5 minutes' treatment. Heat generated by either of these instruments can be dissipated by water cooling—the micro cup of the Waring blendor has been ice-cooled by filling with crushed ice an 8- to 10-inch cone soldered to its base. Lamanna and Mallette give no data on enzyme extraction—presumably the enzymes would be liberated and not destroyed by these procedures.

Grinding. Cell suspensions have been milled mechanically by feeding continuously through a Booth-Green mill,[34] which employs a motor-driven cone and close-fitting roller bearing which turns a machined race. The heat generated is dissipated by circulating the cell suspension with a small metering pump through a heat exchanger submerged in ice water. This procedure, useful in early experiments, has now been largely replaced by more convenient methods which may be applied to larger quantities of cells. As a typical example of the Booth-Green procedure, 2 hours were required to extract *E. coli* or *Bacillus subtilis* enzymes and as much as 4 hours for refractory organisms such as *Sarcina lutea*.

[32] M. R. J. Salton and R. W. Horne, *Biochim. et Biophys. Acta* **7**, 177 (1951).
[33] C. Lamanna and M. F. Mallette, *J. Bacteriol.* **67**, 503 (1954).
[34] B. H. Booth and D. E. Green, *Biochem. J.* **32**, 855 (1938).

The majority of the grinding procedures now in use are applied to (1) cell pastes, (2) frozen cell pastes, or (3) dried cells. With both hand and mechanical grinding, abrasives are usually added. As with the hydraulic press, the most effective abrasives are of small, uniform particle size and appear to exert a mechanical effect in the rupture of the cells. Freezing procedures, and the addition of dry ice, appear also to involve abrasive action of the ice, or the dry ice crystals; a single exception, the use of powdered alumina, introduced by McIlwain,[23] appears also to exert a chemical action.[24]

Cell pastes mixed with powdered Pyrex or, more recently, with other abrasives, have been ground by forcing the mixture between a motor-driven, ice-filled cone and a close-fitting funnel.[35] In early studies this grinding method was used to prepare concentrated extracts which were of considerable value in establishing pathways in microbial systems. More recently, alumina grinding and the use of sonic oscillators have somewhat replaced this method because of the rather laborious nature of the method, the difficulty in collecting the ground cells without loss, and the necessity for recycling if a high percentage of the cells is to be ruptured. It is possible that the use of uniform-size, small glass beads or other fine-mesh abrasives, including alumina, would enhance the effectiveness and thereby the usefulness of this procedure. McIlwain suggests that one criterion for effective grinding is uniform particle size and particles of a size approximating that of the cells to be ruptured.

With the introduction of alumina, it was found that cell pastes to which one or two weights of alumina, 500 mesh or finer, had been added could be ruptured by hand grinding for 2 to 5 minutes with a chilled mortar and pestle,[23] in many cases liberating more than 80% of an enzymatic activity measurable in the cell suspension or dried cells. An empirical but convenient method of determining when the paste has been ground sufficiently is to observe a change to a darker, more viscous (tacky), somewhat moister preparation. With further experience, the grade and exact size of alumina particles seem not to be highly important, so long as the alumina used does not contain excess alkalinity. McIlwain suggests Griffin and Tatlock's "microid polishing alumina" which has been washed and dried at 100°. More recently in this country, "levigated alumina," obtainable from Buehler, Ltd., Metallurgical Apparatus, Evanston, Illinois,[24] or Alcoa A-301, −325 mesh, or A-303 alumina, obtainable from Aluminum Company of America, Chemicals Division, Pittsburgh, Pennsylvania, have been used with equal success.

After grinding cell pastes, or dried cells, to which buffer or water has

[35] G. Kalnitsky, M. J. Utter, and C. H. Werkman, *J. Bacteriol.* **49**, 595 (1945).

been added along with alumina, the thick paste can be diluted gradually to 50 to 100 mg. dry weight of cells per milliliter, and the alumina, unruptured cells, and cell walls removed by centrifugation at 10,000 to 12,000 r.p.m. in an SS-1 Servall centrifuge or other centrifuge giving approximately the same gravitational field—that is, 15,000 to 20,000 $\times$ g. Differential centrifugation has been applied to separate unbroken cells and cell walls from the fragmented amber fraction, presumably the cell membrane, which contains the hydrogen transport and some of the substrate dehydrogenase enzymes.[36] Centrifugation of ground cells to separate debris is best accomplished by dilution of the extracts to 50 to 100 mg. dry cell weight per milliliter to reduce the density, and particularly the viscosity, sufficiently to allow effective separation. The fractionation of bacterial extracts proceeds best after removal of the nucleic acids—otherwise sharp separations are not possible.[36,37] The problems encountered in fractionation of microbial extracts will be dealt with elsewhere.

Dried cells can also be ruptured by alumina grinding with a mortar and pestle, followed by an addition of a suitable buffer and continued grinding. By this procedure active extracts of both easily rupture cells, i.e., *E. coli*,[37] and more difficultly ruptured cells, such as *Streptococcus faecalis*,[17] have been prepared. From 100 mg. dry weight of *E. coli*, 15 to 20 mg. of protein is extractable; from *S. faecalis*, 8 to 10 mg. These extracts are subject to fractionation after removal of the nucleic acids. Both organisms, as dry powders, can be similarly ruptured by grinding with powdered dry ice with analogous yields of protein, but so far these extracts have resisted purification.

Other Procedures. For special cases a vacuum ball mill[38] has yielded active cell extracts, particularly of those enzymes subject to air oxidation. Similar procedures have been employed to obtain labile antigens, frequently by submerging the ball mill in a dry ice-solvent mixture to render the cells more brittle.

Enzymatic methods have been employed in very few cases, owing to the limited variety of cells which can be ruptured with specific enzymes, as for example, lysozyme.[1] It may well prove that cell wall hydrolyzing enzymes, or even proteolytic enzymes, will come into wider use as these methods are explored.[22]

Chemical lysis has so far been relegated to agents such as glycine,

[36] I. C. Gunsalus, R. Y. Stanier, and C. F. Gunsalus, *J. Bacteriol.* **66**, 535, 548 (1953).
[37] S. Korkes, A. del Campillo, I. C. Gunsalus, and S. Ochoa, *J. Biol. Chem.* **193, 721** (1951).
[38] W. W. Umbreit, R. H. Burris, and J. F. Stauffer, "Manometric Techniques," p. 129. Burgess Press, Minneapolis, 1949.

which exerts generalized action on proteins, often with denaturation. Recently, polyelectrolytes (i.e., a dimethyl aminoethyl acryllate polymer) active in low concentration have been introduced by Puck[39] for the liberation of bacteriaphage and promise also to be useful in the liberation of enzymes. These methods have not, however, been explored sufficiently to allow general application.

[39] T. T. Puck, *Cold Spring Harbor Symposia Quant. Biol.* **18**, 153 (1953).

[8] Extraction of Soluble Enzymes from Higher Plants

By ALVIN NASON

Preparation

Principle. The procedures outlined below are intended to extract soluble protein containing maximum enzymatic activity from plant cells under conditions least favorable for protein denaturation. The use of a buffered extracting solution, generally ranging on the alkaline side, is usually desirable. The addition of sulfhydryl compounds, other reducing substances, or metal-binding agents may enhance the extraction, activity, and stability of some enzymes, depending on their properties.[1] High-speed centrifugation is necessary to free the extract of chloroplast fragments, grana, and mitochondria.

Reagents

0.1 M potassium phosphate buffer, pH 7.5, with or without 10^{-3} M cysteine or Versene (ethylenediaminetetraacetic acid).
Chilled acetone ($-15°$).

Procedures. The variety of procedures which have been used for obtaining soluble cell-free enzyme extracts from higher plants can be classified into the following three main groups. In most cases either fresh or frozen tissue may be used as starting material, depending on the activity obtained in the final preparation. In the case of seed materials, the seed may be soaked overnight by covering with tap water and then extracting essentially by the methods below; or the dried seeds may first be ground or milled to a powder and then extracted essentially by method 2.

METHOD 1. Extracts of pulpy tissues such as storage organs (e.g., the carrot or potato) can be prepared by first passing the tissue through a meat grinder. This is followed by pressing out the juice by hand or mechanical press through muslin or three to four layers of cheesecloth. An active extract of starch phosphorylase has been prepared from potato

[1] See M. Gibbs, Vol. I [62].

in this manner.[2] Overnight freezing and subsequent thawing of etiolated barley shoots followed by pressing by hand through cheesecloth yielded a sap for the study of the ascorbic acid system.[3] Centrifuging for 10 minutes at 20,000 $\times$ g further frees the liquid of suspended particles.

METHOD 2. For the preparation of extracts from leaves and meristems the tissue is first ground with a cold mortar and pestle, then homogenized with a Ten Brock glass homogenizer in two to five times its weight of cold buffer, pressed through cheesecloth, and finally centrifuged at 20,000 $\times$ g for 10 minutes at 4°.[4] The Waring blendor may be substituted for the mortar and pestle, provided that the buffer is added with the leaves at this first stage. The use of alumina powder in the homogenization process was highly effective in solubilizing nitrate reductase from soybean leaves.[5] In the latter procedure the extract is made by grinding one weight of fresh leaves, three weights of cold 0.1 M K_2HPO_4 buffer (pH 9.0), and two weights of alumina powder (Alcoa A-301) in a Waring blendor for 2 minutes at 4°. The mixture is further ground for 3 minutes in a Ten Brock homogenizer at 0 to 4°, then centrifuged in a Servall centrifuge at 20,000 $\times$ g for 10 minutes at 4°, yielding the clear supernatant enzyme solution. The use of sand or ground glass with the mortar and pestle excludes the subsequent use of the glass homogenizer. Such a procedure is desirable for preparing extracts from fibrous roots, since the latter tends to crack the glass homogenizer.

METHOD 3. Acetone-dried powders are prepared by placing the tissue, usually leaves, in a Waring blendor, covering with at least 5 vol. of chilled acetone, and blending vigorously for 30 seconds to 1 minute. The resulting slurry is filtered through a Büchner funnel, washed with an excess of chilled acetone, and the residue spread out on filter paper and allowed to dry at room temperature. The acetone powder, which is stored in a desiccator at 4°, may be extracted with approximately 5 to 10 vol. of phosphate buffer essentially as described in method 2. Acetone powders have the advantage of being chlorophyll-free and tend to have a minimum of gum- and resin-like material which might otherwise interfere in subsequent fractionation of the extract.

[2] R. M. McCready and W. Z. Hassid, *J. Am. Chem. Soc.* **66**, 560 (1944).

[3] W. O. James and J. M. Cragg, *New Phytologist* **42**, 28 (1943).

[4] A. Nason, H. A. Oldewurtel, and L. M. Propst, *Arch. Biochem. and Biophys.* **38**, 1 (1952).

[5] H. J. Evans and A. Nason, *Plant Physiol.* **28**, 233 (1953).

[9] Disintegration of Bacteria and Small Particles by High-Pressure Extrusion

By C. S. FRENCH and H. W. MILNER

Introduction

The high-pressure extrusion method is used to bring particles, already small, such as bacteria or fragments of plant or animal cells, into a highly dispersed state. The method consists in forcing a liquid suspension at high pressure through a very fine aperture, which causes the disruption of the particles to much smaller size. One fortunate aspect of this procedure is that it does not require the addition of abrasives to the suspension. This avoids the dangers of adsorption or contamination.

For the ultrafine disintegration of cellular components, such as chloroplasts, it has been thoroughly tested by the authors. In collaboration with other individuals, they have successfully used it for the extraction of glutamic acid oxidase from *E. coli*, for the disintegration of fungus spores, and for preparation of a colloidal dispersion of the pigment system of *Rhodospirillum rubrum*.

The results with yeast have been less successful.[1] Strangely enough, the cells of yeast may not show disintegration immediately after passing through the valve, but after standing a few hours after such treatment the disintegration of the cell walls may take place. This phenomenon has not yet been studied in detail.

Moderate success has been achieved with the disintegration of *Chlorella* cells. Several passages through the apparatus have been found necessary for a high degree of dispersion.

This apparatus, when used at lower pressure and with larger apertures, might work adequately as a tissue homogenizer to break down tissues into individual cells. This application has not yet been tested.

We have found the device to be vastly superior to the use of supersonic irradiation in regard to both effectiveness and reproducibility of treatment.

Construction

The apparatus consists of a steel cylinder, bored with a central hole which empties to the outside through a needle valve. A plunger, which fits the hole snugly, puts the pressure on the liquid. The first model was made from a cold-rolled steel bar 3 inches in diameter and 5 inches long. It had a 1-inch hole bored to a depth of 4 inches in its center. A needle

[1] H. W. Milner, N. S. Lawrence, and C. S. French, *Science* **111**, 633 (1950).

valve, made of steel (for use with ammonia), was screwed into the other end. The central hole was made by drilling and reaming, and the plunger consisted of stock cold-rolled steel bar about 0.003 inch under size. Below the plunger was placed a leather washer, and below this a section of rubber stopper. Although the external needle valve may be dangerous, most of our work has been done with this simple device, which is illustrated in ref. 1.

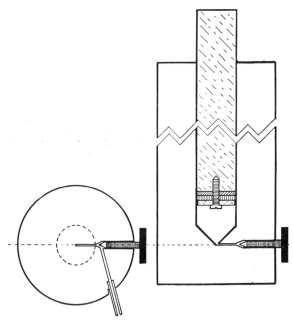

Fig. 1. Apparatus for the disintegration of biological materials in water suspension by high pressure extrusion.

An improved design, which unfortunately was made in too large a size for convenience, has been described in ref. 2, from which Fig. 1 has been taken. The recommended size for the hole is $\frac{3}{4}$ inch in diameter, for use with a common laboratory press. The upper limit on the size of the hole is determined by the ability of the press to maintain a high pressure while giving a usable flow rate through the orifice without exceeding the possible horsepower output of the operator. In this model the needle valve is an integral part of the main cylinder and consists of a $\frac{1}{16}$-inch hole from the center hole to the valve seat which is a short section of a 40° cone, and the rest of the hole is threaded for a $\frac{1}{4}$-28 screw. The point of

[2] C. S. French and H. W. Milner, *Symposia Soc. Exptl. Biol.* **5**, 232 (1951).

the needle valve is turned to the same angle as the cone seat and hardened. Since the seat wears rapidly, an appropriate reamer and also a brass lapping tool are on hand for frequent touching up of the valve seat.

Several interesting variations in design have been made by the American Instrument Company with the thought of putting this device on the market commercially. Their modifications consist in the use of an 0-ring seal backed up by a Teflon washer and the use of a removable base plug with a seal similar to that of the plunger. This removable base plug would greatly facilitate the internal grinding of the cylinder to give a very smooth wall. Although these modifications seem to be excellent in principle, we have not had an opportunity to test this model.

In the construction and use of this device it should be realized that the high pressure can be dangerous. The operator must make sure that the needle valve is on the side away from the operator and that no possibility of squirting streams of high-velocity liquid exists. It is necessary to make sure by computation that the strength of the steel is not exceeded and that the ends of the cylinder as they taper toward the bottom are rounded rather than left sharp.

The rubber seal backed by a leather washer behaves quite well until the leather becomes worn, after which the rubber is extruded up into the space between the plunger and the cylinder. When this happens, it is difficult to remove the plunger. To facilitate its removal by force, the protruding end of the plunger is threaded. The washers are conveniently turned by hand on the lathe, and their frequent replacement is advisable.

Operating Procedure

The entire unit is cooled to 0° immediately before use. The suspension is then introduced, the needle valve closed, the plunger inserted, the unit turned over so as to rest on the plunger, and, for safety, the needle valve opened until all the air has been displaced and the liquid starts to come out. The assembly is then placed in the press with the needle valve closed and the pressure built up to 20,000 p.s.i. as computed from the known diameter of the hole and the reading of the gage on the press. The needle valve is then opened slightly, and the pressure maintained at 20,000 p.s.i. while the material flows out through the needle valve. A rise in temperature of about 15 degrees occurs during the passage through the needle valve so that the extruded homogenate is caught in a cooled vessel. A used can that has an inert lacquer on its inside is very convenient for this purpose when packed in ice. It may be that frozen material can be extruded as well as liquid suspensions, but this attempt has not yet been made. Although the rate of flow seems to have little effect upon chloroplast disintegration, at least within the limits of 2 to 16 ml./min., the pressure used has a direct effect upon the percentage of the material which is dispersed with a given fineness. Further details as to the effects of pressure and flow rates are given in refs. 1 and 2.

[10] Protein Fractionation on the Basis of Solubility in Aqueous Solutions of Salts and Organic Solvents

By Arda Alden Green *and* Walter L. Hughes

I. General Considerations

Proteins are large molecules with definite size, shape, and charge. Such entities have reproducible solubility in a given medium which can be characterized as well as can that of any smaller molecule. The successful separation of proteins on the basis of their solubility depends on the control of certain variables. These include pH and temperature as well as the composition of the solvent and the character of the proteins present. The solvent may be varied by the addition of varying amounts of salts including those of the heavy metals or by changing the dielectric constant by using aqueous mixtures of organic solvents instead of water.

Since in general separations are made after the solid phase and solution have come into equilibrium with each other, certain thermodynamic principles are useful in describing and formulating the observed results. Thus, at equilibrium, the activity of any substance in both phases must be the same. Then, if the state and composition of the solid phase containing the component under consideration is unchanged upon changing the composition of the solvent, the activity, a, of the dissolved component at equilibrium must be unchanged, and any simultaneous change in the concentration, c, of the dissolved component must be due to a change in its activity coefficient, γ. By definition, $\gamma = a/c$. The ratio of the solubility of a substance in water, S_0, to its solubility in a solvent of given composition, S, is the activity coefficient, γ, of the substance in solution; $\gamma = S_0/S$.

If the solid phase remains unchanged, solubility phenomena may be described completely by a consideration of the solvent composition. In the case of proteins, where the solid phase contains additional components, such as water, salts, and other solvent molecules, solubility changes may be due to changes in the composition of the solid phase. Nevertheless, considerable success in explaining solubility phenomena has been obtained by assuming that the activity of the protein in the solid phase is constant, unless the solid phase contains more than one protein component. Consequently, we will begin by discussing as the simplest case the factors affecting the solubility of "pure" proteins. If, in a mixture, the behavior of each protein component with respect to these parameters is known, a method for the separation of such a mixture may be designed to yield the pure solid components.

The most important single characteristic of a protein is the number and arrangement of the charges on the molecule. This is dependent on the amino acid composition—in particular, on the number of free carboxyl groups of aspartyl and glutamyl residues on the one hand and of the basic groups of histidyl, arginyl, and lysyl residues on the other. Free SH and phenolic OH groups play a negligible role, since they dissociate above pH 10. Charged groups not derived from amino acids, such as phosphate esters in lipids or coenzyme prosthetic groups, must also have an effect.

Thus, the protein molecule, containing many groupings which are reacting with the solvent as acids or bases, results in the presence of many species in equilibrium with each other. Of course, any such set of species may be considered as a component for thermodynamic purposes. When the composition of the solvent changes, the change in distribution of these species must be treated statistically. However, in principle it may readily be seen that any change in the solvent which produces new protein species in the solution, without markedly affecting the chemical potential (activity) of those already present, must result in an increased solubility of the protein. This explains the remarkable sensitivity of protein solubility to variation of the pH. Thus since the solid phase in protein equilibrium usually corresponds to an approximately isoelectric protein, protein solubility usually rises sharply on either side of the isoelectric point. It would be reasonable, other things being equal, that the most insoluble protein phase (i.e., lowest activity) would occur when the protein carried zero net charge (i.e., isoelectric), since under these conditions repulsive forces between protein molecules would be at a minimum. Of course, ions which are tightly bound to a protein molecule also contribute to its charge, and this contribution must be included in calculating the net charge on the protein.

In addition to altering protein solubility by changing the protein species present, solubility may be affected by changes in the solvent's composition. These effects seem largely explicable in terms of the Debye-Hückel theory. Qualitatively the protein in the solid phase may be pictured as held together by the coulombic forces between opposite charges on adjacent molecules. These are relatively long-range forces, since they fall off only as the square of the distance (r): $F = ee'/Dr^2$ (e and e' represent the opposing charges, and D the dielectric constant of the solvent). This explains the precipitating action of organic solvents which, by lowering the dielectric constant of the solvent, increase the attractive forces between the molecules.

Ions also affect protein solubility. In small amounts they act to shield the protein molecules from each other by coming between the opposing charges and interacting with them in the Debye-Hückel sense. Thus they increase protein solubility, a phenomenon which is called "salting-in."

However, in high concentrations salts decrease protein solubility. This behavior, called "salting-out," is common to all nonelectrolyte solutes and appears to be due to "dehydration" of the protein molecule. Thus the solvent becomes organized about the salt ions to such a degree that its normal organization around the protein is decreased and the protein molecules are able to associate in a solid phase. Solvation effects (combination of the solvent with the solute) are also undoubtedly important in explaining the effects of organic precipitants.

In spite of numerous efforts to explain the solubility behavior of proteins, it must be remembered that solubility is poorly understood for even simple substances and is still formulated in empirical terms.[1] Unknown factors frequently play a predominant role. Excellent examples of this may be found among the hemoglobins, where slight changes in the groups attached to the iron atoms markedly affect the solubility. Thus, the addition of oxygen to human hemoglobin increases the solubility manyfold, whereas in horse hemoglobin exactly the opposite effect is observed. Furthermore, even changing the valence state of the iron atoms from ferrous to ferric also has a marked effect. Pauling explains this in terms of changes in the "complementariness" of adjoining molecules in the solid phase. This is an attractive concept which, unfortunately, cannot be subjected to experimental verification at the present time.

II. Effect of Electrolytes

1. Solubility in Solutions of Low Ionic Strength. Globulins are defined as proteins relatively insoluble at their isoelectric points with solubility increasing upon the addition of salt. This solubility increase was first described in the middle of the last century. In 1905 appeared the classical papers of Hardy[2] and of Mellanby[3] on serum globulin and of Osborne and Harris[4] on edestin.

Differences in the effect of different salts, especially those of different valence types, on protein behavior were studied by Mellanby, who described in the following terms the solubility of globulins in salt solutions.

---

[1] For a thorough discussion of solubility the reader is referred to J. H. Hildebrand and R. L. Scott, "The Solubility of Nonelectrolytes," 3rd ed. Reinhold, New York, 1950.

[2] W. B. Hardy, *J. Physiol.* **33**, 251 (1905–06).

[3] J. Mellanby, *J. Physiol.* **33**, 338 (1905–06).

[4] T. B. Osborne and I. Harris, *Am. J. Physiol.* **14**, 151 (1905).

"Solution of globulin by a neutral salt is due to forces exerted by its free ions. Ions with equal valencies, whether positive or negative, are equally efficient, and the efficiencies of ions of different valencies are directly proportional to the squares of their valencies."[3]

This statement of Mellanby is now recognized as a description of the principle of the ionic strength which was later formulated by Lewis.[5]

However, when solubility is plotted against a function of the ionic strength, the curves approach each other in very dilute solution but fan out at higher concentrations. It is these differences that can be described by the Debye[6] theory which takes into account, besides the valence of the ions of the electrolyte, the effective diameter of the ions and their distance apart, as well as the apparent valence type of the protein.

As pointed out above, solubility can be treated as a variation in activity coefficients in solutions of differing electrolyte content and dielectric constant. These activity coefficients have been described by the Debye theory, based on electrostatic forces, developed for small charged particles in dilute solution, and by Hückel's theory of the solubility of nonelectrolytes in concentrated electrolyte solutions. The theory has been modified by Scatchard and Kirkwood[7] and Kirkwood[8] to apply to proteins. This is not the place to describe the theory in detail, but the simplified equations will be used in the following discussions.

The simplified form of the Debye equation including the "salting out" term added by Hückel[9] is

$$- \log \gamma = \log S - \log S_0 = \frac{0.5 Z_1 Z_2 \sqrt{\mu}}{1 + A \sqrt{\mu}} - K_s \mu \tag{1}$$

in which γ is the activity coefficient, S is the solubility, S_0 is the solubility in the absence of electrolytes, μ is the ionic strength, $Z_1 Z_2$ is the valence type, K_s is the "salting-out" constant, 0.5 is a theoretical constant having this value at 25°, and A is a constant which depends on the mean effective diameter, b, of all ions in a solution and on κ, a "reciprocal distance."

The application of this theory to oxyhemoglobin in phosphate solutions at 0° was first discussed by Cohn and Prentiss.[10] The solubility of horse carboxyhemoglobin in various electrolytes is presented in Fig. 1. The curves are drawn according to equation 1.

[5] G. N. Lewis and M. Randall, "Thermodynamics and the Free Energy of Chemical Substances." McGraw Hill, New York and London, 1923.
[6] P. Debye and E. Hückel, *Physik. Z.* **24**, 185 (1923).
[7] G. Scatchard and J. G. Kirkwood, *Physik. Z.* **33**, 297 (1932).
[8] J. G. Kirkwood, *J. Chem. Phys.* **2**, 351 (1934).
[9] E. Hückel, *Physik. Z.* **26**, 93 (1925).
[10] E. J. Cohn and A. M. Prentiss, *J. Gen. Physiol.* **8**, 619 (1927).

The Debye theory as modified by Scatchard and Kirkwood[7] takes into account the number and position of the charges on the protein molecule and describes solubility as a function of ionic strength rather than of the square root of the ionic strength. The data on the solubility of carboxyhemoglobin in chlorides and sulfates at constant pH[11] can be accurately described by this equation.[12] It can also be satisfactorily described by a purely empirical simple linear equation with two constants.[11] Practically speaking, regardless of the theoretical description, chlorides are more effective in dissolving proteins than either sulfate or phosphates.

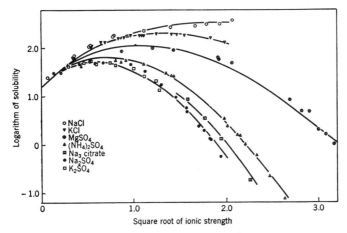

FIG. 1. The solubility of carboxyhemoglobin in various electrolytes at 25°. The curves are drawn according to equation 1 (from Green[11]).

The isoelectric precipitation of proteins depends on two factors. (1) All proteins exhibit minimum solubility, in solutions of constant ionic strength, at their isoelectric points. The effect of pH on solubility is described in detail in a later section. (2) At low concentrations of electrolyte, or in the absence of electrolyte, some proteins are sufficiently insoluble to form a precipitate. This property has been useful as a means of crystallizing proteins, e.g., edestin by Osborne in 1892[13] and muscle phosphorylase.[14]

Some proteins become insoluble in the complete absence of electrolyte and stay that way; they are denatured. However, it may be safe to precipitate the same globulin in a dilute solution of electrolyte.

[11] A. A. Green, J. Biol. Chem. **95**, 47 (1932).

[12] E. J. Cohn, Chem. Revs. **19**, 241 (1936).

[13] T. B. Osborne, Am. Chem. J. **14**, 662 (1892).

[14] A. A. Green and G. T. Cori, J. Biol. Chem. **151**, 21 (1943).

The term "relatively insoluble" has been used advisedly, since iso-electric precipitation may prove a profitable method of separation even when a measurable amount of protein remains in the supernatant. If two proteins have the same isoelectric point and one is more soluble than the other, it is possible to remove the less soluble at an electrolyte concentra-tion just low enough to precipitate most of it and then lower the elec-trolyte concentration by dialysis or by dilution to precipitate the second protein.

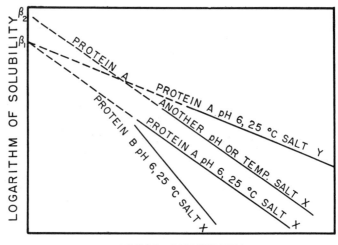

FIG. 2

The addition of heavy metal salts, e.g., zinc, or of protamine, nucleic acid, or trichloroacetic acid may precipitate proteins ordinarily com-pletely water-soluble. These precipitating agents will be dealt with in later sections.

2. Solubility in Concentrated Salt Solutions. In concentrated elec-trolyte solutions the decrease in the logarithm of the solubility of a protein is a linear function of the increasing salt concentration or, better, of the ionic strength which is defined in terms of the square of the valence of the ions of the electrolyte.

$$\log S = \beta - K_s \mu \tag{2}$$

where S is the solubility of the protein, β is the intercept constant, K_s' (the slope of the line) is the "salting-out" constant, and μ is the ionic strength expressed in terms of salt per 1000 g. of water. A similar straight line with slightly different constants is obtained if the ionic strength is defined as $\Gamma/2$ (per liter of solution). The ionic strength is one-half the sum of the concentration (in moles per liter for $\Gamma/2$ and in moles per 1000 g. of H_2O for μ) of each of the ions in solution multiplied by the square of its valence.

Thus, the solubility of a protein in concentrated salt solutions can be defined in terms of two constants β and K_s. The most important variables

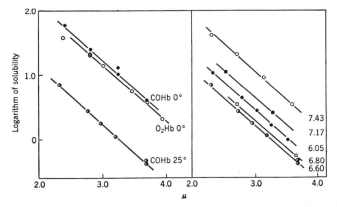

FIG. 3. The solubility of hemoglobin in concentrated phosphate buffers of varying temperature and pH (from Green[15]).

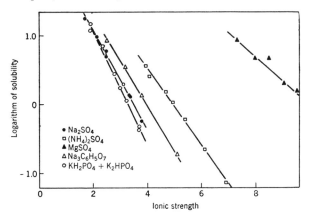

FIG. 4. The solubility of carboxyhemoglobin at 25° and pH 6.6 in concentrated solutions of various electrolytes (from Green[15]).

to be considered are temperature, pH, and kind of salt. In studying the effect of any one variable the others must be kept constant. Figure 2 is a diagrammatic sketch of the solubility of proteins at increasing concentrations of electrolytes at high ionic strength. Protein A at a given pH and temperature in salt X can be described in terms of a straight line whose

[15] A. A. Green, J. Biol. Chem. 93, 495 (1931).

intercept is β_1. β is the hypothetical solubility in the absence of salt, and the curves have been extended as dotted lines to the intercept. The same protein in the same salt at a different pH or temperature is described as a parallel line; K_s is constant, but β is different, β_2. Protein A at the original

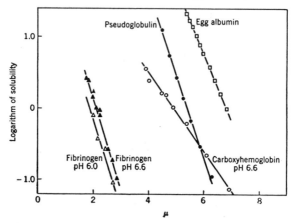

Fig. 5. The solubility of various proteins in concentrated ammonium sulfate solutions (from Green[16]).

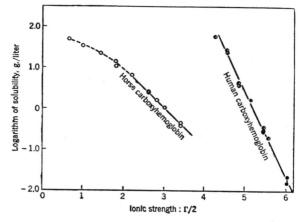

Fig. 6. Solubility of the carboxyhemoglobin of the horse and of man in phosphate buffers at 25°. Carboxyhemoglobin of horse, ○; of man, preparation 1 ◑, preparation 2 ◐, preparation 3 ● (from Green et al.[16]).

pH and temperature but in a different salt, Y, describes a curve with a different slope but the same β_1. Protein B at the original pH, temperature, and salt has a completely different curve. Figures 3, 4, 5, and 6 show some of the original data from which the above relationships were derived.

[16] A. A. Green, E. J. Cohn, and M. H. Blanchard, *J. Biol. Chem.* **109**, 631 (1935).

The effect of varying the parameters may be briefly described as follows.

a. pH. At constant ionic strength the solubility increases at pH values either acid or alkaline to the isoelectric point. This is the most important variable and will be dealt with in a separate section.

b. Temperature. In dilute aqueous solutions of electrolyte most proteins are more soluble at higher temperatures than at $0°$. In concentrated salt solutions this effect may be reversed, for example, carboxyhemoglobin in phosphate buffers. In ammonium sulfate egg albumin has a minimum solubility at $20°$. Advantage may be taken of the negative temperature coefficient in the crystallization of certain proteins by allowing a cold solution to warm up very gradually.

c. Type of Electrolyte. Uni-univalent salts are relatively ineffective in precipitating proteins. Only the more insoluble proteins such as fibrinogen are precipitated by these salts even when they are used in high concentrations. Salts of higher valence produce much higher ionic strengths and are generally more effective as precipitating agents.

Horse carboxyhemoglobin is a globulin. Its solubility is increased upon the addition of low concentrations of salt, and it is "salted-out" at higher concentrations of salt. Figure 4 represents the solubility of crystalline horse carboxyhemoglobin at constant pH and temperature in concentrated solutions of various electrolytes.

The slopes of the lines, and therefore the K_s' values, vary with the kind of salt used. For this protein, K_s' decreases in the order $KH_2PO_4 + K_2HPO_4$, Na_2SO_4, $Na_3C_6H_5O_7$, $(NH_4)_2SO_4$, and $MgSO_4$. This is essentially the same order as the series given by Hofmeister in 1888.

Ammonium sulfate is the salt most commonly used for protein precipitation because it is very soluble and also effective. The following practical considerations may be useful. There are three methods of addition of this or any other electrolyte; (1) add the solid salt; (2) add a saturated solution of the salt; (3) dialyze the salt through a cellulose membrane. All these procedures are more effective if carried out with mechanical stirring to prevent local excess of electrolyte. The last of these methods is the most gradual and therefore the most effective although often impractical. The protein is usually placed inside the dialyzing sac, and the concentration of the outside solution is increased as slowly as desired, usually determining the concentration of the equilibrated salt solution by means of its specific gravity. The traditional method is the addition of a solution of saturated ammonium sulfate, and the concentration of salt is described in terms of "per cent saturation"; e.g., equal volumes of protein solution and salt solution result in "50% saturation." The volume change of mixing is ignored; it isn't very great or very impor-

TABLE I

	Final concentration of ammonium sulfate, % saturation																
Initial concentration of ammonium sulfate, % saturation	10	20	25	30	33	35	40	45	50	55	60	65	70	75	80	90	100
	Grams solid ammonium sulfate to be added to 1 l. of solution																
0	56	114	144	176	196	209	243	277	313	351	390	430	472	516	561	662	767
10		57	86	118	137	150	183	216	251	288	326	365	406	449	494	592	694
20			29	59	78	91	123	155	189	225	262	300	340	382	424	520	619
25				30	49	61	93	125	158	193	230	267	307	348	390	485	583
30					19	30	62	94	127	162	198	235	273	314	356	449	546
33						12	43	74	107	142	177	214	252	292	333	426	522
35							31	63	94	129	164	200	238	278	319	411	506
40								31	63	97	132	168	205	245	285	375	469
45									32	65	99	134	171	210	250	339	431
50										33	66	101	137	176	214	302	392
55											33	67	103	141	179	264	353
60												34	69	105	143	227	314
65													34	70	107	190	275
70														35	72	153	237
75															36	115	198
80																77	157
90																	79

tant, since the essential thing in any procedure is to be able to repeat it exactly. However, the difference between the concentration of a saturated solution of ammonium sulfate at 25°, 4.1 M, and the concentration at 0°, 3.9 M, is significant. Calculation of the amount of saturated solution necessary to go from one concentration to another is simplified by first calculating the amount of water and the amount of saturated solution necessary to make the initial solution and then calculating the total amount of saturated solution required to bring that amount of water to the desired concentration.

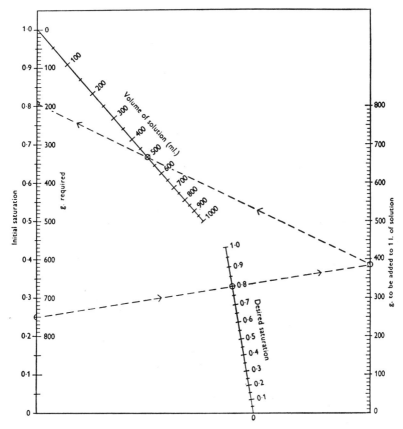

Nomogram for Saturation of Ammonium Sulfate at Room Temperature (from Dixon[17]).

A straight line through the initial saturation and the desired saturation gives the amount of solid ammonium sulfate to be added to 1 liter of solution. A line from this point passing through the volume of the solution gives the amount required.

Many investigators also choose to ignore second-order volume changes when adding the solid salt, assuming that the addition of 1 g. of salt increases the volume by the same amount regardless of the concentration of the solution. It is thus possible to use a simple equation[18] to calculate

[17] M. Dixon, *Biochem. J.* **54**, 457 (1953).
[18] M. Kunitz, *J. Gen. Physiol.* **35**, 423 (1952).

the amount of salt necessary to produce any required change in concentration. It is, however, preferable to know the exact concentration in terms of moles per liter, and Table I has been prepared accordingly. A saturated solution of ammonium sulfate at 25° is 4.1 M and requires 767 g. of salt for each liter of water. The table presents grams of ammonium sulfate necessary to take 1 l. of solution from one "per cent saturation" to another, meaning by "per cent saturation" per cent of 4.1 M. The values have been calculated from tables of per cent salt, specific gravity, and grams per liter at the various concentrations. Dixon[17] has published a convenient nomogram (see p. 77) covering the material given in Table I.

III. Effect of pH on Solubility

Proteins become positively or negatively charged on either side of the isoelectric point, and these forms are more soluble than the electrically neutral but highly charged molecule. As acid is added to the molecule the COO^- groups accept hydrogen ions, and as alkali is added the NH_3^+ groups become NH_2, and the protein becomes positively and negatively charged, respectively. One would thus expect solubility to be directly related to the titration curve. Linderstrøm-Lang[19] has developed an equation describing the solubility of a pure protein as a function of pH, provided that the rate of change of charge with pH is known for the molecule, and has successfully applied it to some solubility data for β-lactoglobulin, a relatively insoluble protein. However, for other data on this same protein it proved inexact. This may have been due to changes in the solid phase, which must remain constant in composition if his equation is to be applied.

The solubility of some proteins (hemoglobin, egg albumin, and casein[20] varies with pH as though they had two divalent positive and two divalent negative charges which were being titrated on either side of the isoelectric point in the range where solubility can be measured. This is obviously only an apparent phenomenon, but it makes it possible to describe the solubility of a protein at varying hydrogen ion concentration by the empirical equation

$$\frac{S}{^{++}\text{Protein}^{--}} = \frac{S}{S_n} = 1 + \frac{a_{H^+}^2}{K_1'K_2'} + \frac{K_3'K_4'}{a_{H^+}^2} \tag{3}$$

S is the solubility, S_n the solubility of the neutral molecule, a_{H^+} the activity of the hydrogen ions, $K_1'K_2'$ the dissociation constants in acid solution, and $K_3'K_4'$ the dissociation constants in alkaline solution. If these

[19] K. Linderstrøm-Lang, *Arch. Biochem.* **11**, 191 (1946).
[20] A. A. Green, *J. Biol. Chem.* **93**, 517 (1931).

dissociation constants are sufficiently far apart in value, solubility in alkaline solution may be described by the equation

$$\frac{\text{Protein}^{--}}{^{++}\text{Protein}^{--}} = \frac{S}{S_n} = 1 + \frac{K_3'K_4'}{a_{H^+}^2} \tag{4}$$

and solubility in acid solution by the equation

$$\frac{\text{Protein}^{++}}{^{++}\text{Protein}^{--}} = \frac{S}{S_n} = 1 + \frac{a_{H^+}^2}{K_1'K_2'} \tag{5}$$

Since the solubility curves at constant pH with varying high salt concentration are parallel for different pH values, the description of the

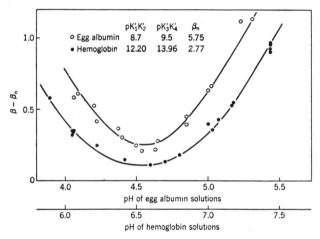

FIG. 7. The solubility of hemoglobin and of egg albumin in concentrated salt solutions of varying pH (from Green[20]).

variation of solubility with pH in this region can be described with one equation. If all the results are plotted as the ratio of the increased solubility of the protein in acid or in alkaline solution to that of the neutral molecule, all the points will fall on one curve.

This has been done in Fig. 7, which describes the variation of solubility, expressed as β, with pH of both hemoglobin in phosphate buffers and egg albumin in ammonium sulfate (20). β_n is the logarithm of the hypothetical solubility of the neutral molecule of the protein at zero salt concentration. Thus,

$$\frac{S}{^{++}\text{Protein}^{--}} = \frac{S}{S_n} = \frac{\text{antilog } \beta}{\text{antilog } \beta_n} \tag{6}$$

The curves drawn through the points are theoretical according to equation 3. The values for S_n or β_n are not the values for minimum solubility as actually determined but are corrected for slight amounts of both negatively and positively charged molecules present with the neutral molecule even at the pH of minimum solubility.

The solubility of proteins in constant low electrolyte concentration can also be described by equation 3. The values for the apparent dissociation constants differ from the constant values found at higher salt concentrations. As the salt concentration decreases, $pK_1'K_2'$ and $pK_3'K_4'$ increase; $pK_1'K_2'/2$ approaches the isoelectric point and $pK_3'K_4'/2$ goes away from the isoelectric point. This means that two curves for S/S_n,

TABLE II
INTERPOLATED VALUES FOR THE MOLE FRACTION OF TOTAL ACETATE AS CH_3COONa
OF SOLUTIONS OF GIVEN CONCENTRATION AND pH
Calculated by means of the equation

$$pH + \log \frac{(CH_3COO^-)}{(CH_3COOH)} = pK - \frac{0.5Z^2 \sqrt{\mu}}{1 + \kappa b} + K_s\mu = 4.77 - \frac{0.5 \sqrt{\mu}}{1 + \kappa b} - 0.16\,\mu$$

The value of pK, 4.77, results from the measurements of Cohn made with the hydrogen electrode at room temperature, uncorrected for liquid junction potential, calculated by means of the appropriate values for the $N/10$ calomel half-cell between 0.3357 volt at 18° and 0.3353 volt at 25°. The activity coefficients calculated by Cohn's equation are assumed to be independent of temperature over this range.

	Total concentration of acetate in moles per liter											
	0.0	0.05	0.1	0.2	0.4	0.6	0.8	1.0	1.2	1.4	1.6	1.8	2.0
pH	Mole fraction of total acetate as CH_3COONa											
3.8	0.097	0.105	0.109	0.115	0.121	0.125	0.128	0.131	0.133	0.138	0.139	0.140	0.139
3.9	.124	.132	.136	.141	.149	.153	.156	.159	.161	.166	.167	.168	.167
4.0	.152	.160	.166	.171	.180	.184	.188	.190	.192	.195	.196	.198	.197
4.1	.179	.193	.197	.205	.215	.220	.223	.228	.228	.229	.230	.231	.230
4.2	.211	.230	.239	.248	.259	.264	.268	.270	.270	.270	.270	.270	.271
4.3	.252	.276	.284	.295	.306	.313	.317	.320	.320	.320	.319	.317	.316
4.4	.299	.328	.335	.347	.358	.365	.369	.370	.370	.370	.368	.365	.362
4.5	.351	.382	.390	.403	.414	.420	.423	.421	.420	.419	.417	.413	.411
4.6	.404	.436	.449	.462	.473	.477	.480	.478	.476	.471	.468	.464	.461
4.7	.459	.496	.508	.519	.530	.534	.533	.530	.527	.522	.520	.515	.509
4.8	.518	.556	.566	.578	.588	.589	.589	.587	.584	.580	.575	.571	.565
4.9	.566	.613	.624	.634	.641	.642	.639	.636	.632	.626	.618	.613	.607
5.0	.629	.666	.678	.686	.691	.692	.689	.684	.679	.672	.664	.657	.652
5.1	.681	.716	.724	.733	.738	.734	.732	.727	.722	.716	.710	.703	.695
5.2	.730	.762	.768	.775	.780	.776	.773	.768	.762	.756	.750	.744	.736
5.3	.772	.801	.807	.812	.815	.813	.810	.807	.799	.793	.787	.780	.774
5.4	.809	.835	.840	.844	.846	.846	.841	.838	.832	.826	.820	.814	.807
5.5	.840	.862	.868	.873	.873	.874	.869	.866	.860	.855	.848	.843	.837
5.6	.867	.888	.893	.895	.895	.895	.891	.889	.881	.877	.870	.867	.862

TABLE III

INTERPOLATED VALUES FOR THE MOLE FRACTION OF TOTAL PHOSPHATE AS K_2HPO_4 OF SOLUTIONS OF GIVEN CONCENTRATION AND pH

Calculated by means of the equation

$$\mathrm{pH} + \log\frac{(\mathrm{KH_2PO_4})}{(\mathrm{K_2HPO_4})} = pK_2 - \frac{0.5(Z^{-2}-Z^{-2})}{1+\kappa b}\sqrt{\mu} + K_s\mu = 7.20 - \frac{1.5\sqrt{\mu}}{1+1.5\sqrt{\mu}} + K_s\mu$$

The value of pK_2, 7.20, results from the measurements of Cohn made with the hydrogen electrode at room temperature, uncorrected for liquid junction potentials, and recalculated using the appropriate values for the $N/10$ calomel half-cell between 0.3357 volt at 18° and 0.3353 volt at 25°. The activity coefficients calculated by Cohn's equation are assumed to be independent of temperature over this range.

Total concentration of phosphate in moles per liter — Mole fraction of total phosphate as K_2HPO_4

pH	0	0.01	0.04	0.10	0.20	0.30	0.40	0.50	0.60	0.80	1.0	1.2	1.4	1.6	1.8	2.0
5.3												0.110	0.122	0.133	0.146	0.158
5.4											0.115	.129	.142	.154	.168	.180
5.5									0.099	0.119	.134	.150	.164	.177	.191	.202
5.6								0.109	.121	.141	.157	.172	.188	.203	.216	.228
5.7						0.104	0.121	.132	.145	.165	.182	.198	.216	.230	.242	.254
5.8				0.085	0.110	.129	.146	.158	.171	.192	.212	.227	.244	.258	.269	.280
5.9	0.048	0.065	0.083	.106	.135	.155	.173	.186	.200	.224	.244	.259	.274	.286	.295	.306
6.0	.059	.081	.103	.132	.163	.185	.203	.219	.236	.259	.277	.292	.305	.316	.323	.330
6.1	.074	.100	.126	.160	.195	.220	.239	.256	.273	.295	.312	.325	.336	.345	.351	.358
6.2	.091	.122	.155	.192	.232	.261	.281	.298	.312	.333	.349	.360	.369	.376	.381	.386
6.3	.112	.150	.190	.232	.276	.305	.326	.341	.354	.372	.386	.395	.401	.407	.412	.416
6.4	.137	.183	.230	.278	.325	.353	.373	.385	.398	.414	.424	.432	.436	.440	.443	.446
6.5	.166	.222	.274	.328	.376	.403	.421	.435	.444	.458	.466	.471	.474	.476	.477	.478
6.6	.201	.266	.325	.381	.429	.457	.473	.484	.493	.503	.508	.510	.510	.510	.510	.510
6.7	.240	.315	.380	.438	.486	.511	.526	.535	.543	.549	.551	.550	.548	.546	.544	.541
6.8	.285	.369	.440	.497	.543	.565	.578	.586	.590	.594	.594	.590	.586	.582	.578	.574
6.9	.334	.425	.498	.557	.598	.617	.629	.634	.637	.638	.636	.631	.624	.618	.612	.607
7.0	.387	.484	.556	.615	.651	.669	.677	.681	.683	.681	.676	.671	.661	.655	.648	.641
7.1	.443	.544	.614	.668	.701	.716	.722	.724	.725	.721	.715	.707	.699	.690	.681	.673
7.2	.500	.604	.670	.717	.747	.758	.764	.763	.762	.758	.751	.742	.732	.723	.713	.704
7.3	.557	.659	.720	.762	.785	.796	.801	.800	.797	.790	.784	.774	.764	.754	.743	.733
7.4	.613	.710	.763	.802	.822	.829	.832	.832	.828	.821	.814	.803	.793	.782	.771	.761
7.5	.666	.756	.805	.837	.854	.860	.860	.859	.855	.848	.840	.830	.820	.809	.798	.788
7.6	.715	.796	.840	.866	.880	.883	.884	.883	.879	.872	.864	.855	.844	.835	.824	.813
7.7	.759	.831	.869	.890	.902	.905	.905	.904	.901	.894	.885	.876	.867	.858	.849	.837

one at high and one at low constant electrolyte content, cross each other at the pH of minimum solubility. Thus the pH of minimum solubility usually decreases with increasing salt concentration. This was formerly explained in terms of the activity coefficients of the dissociating groups but is probably due to the combination of ions with the protein, since anions are usually bound more tightly than cations. This effect is greatly magnified in the case of some proteins. On the acid side of the isoelectric point small amounts of added electrolyte may render the protein insoluble until relatively high acidities are reached.

Adequate pH control during protein purification necessitates the presence of suitable buffers. This is true even in cases where sufficient protein is present to exert some buffering action, since the separation of a solid protein phase may be accompanied by the absorption or release of protons by this component. The reader is referred to Tables II and III for acetate and phosphate buffers[21] over a wide range of concentrations and to another article for a presentation of various buffers.[22]

IV. Organic Solvents

The addition of organic solvents to aqueous solutions of proteins produces marked effects on the solubility. Much of the effect is undoubtedly due to changes in the dielectric constant of the medium (Table IV).

TABLE IV

CONTROL OF THE DIELECTRIC CONSTANT BY ORGANIC SOLVENTS

Dielectric constant	Weight per cent of solvent in water producing dielectric constant indicated at 20°[a]				
	CH_3OH	C_2H_5OH	$n\text{-}C_3H_7OH$	$(CH_3)_2CO$	1,4-dioxane
80	0	0	0	0	0
70	22	18	15	18	12
60	42	34	29	34	22
50	62	51	42	49	33
40	82	68	56	64	45
30	—	88	73	81	58
32	100%	—	—	—	—
25	—	100%	—	—	—
21	—	—	100%	—	—
20	—	—	—	100%	—
2	—	—	—	—	100%

[a] The values have been interpolated from G. Åkerlof, *J. Am. Chem. Soc.* **54**, 4125 (1932) and G. Åkerlof and A. Short, *ibid.* **58**, 1241 (1936).

[21] A. A. Green, *J. Am. Chem. Soc.* **55**, 2331 (1933).
[22] G. Gomori, Vol. I [16].

Thus by decreasing the dielectric constant, one increases the coulombic attractive forces between the unlike charges of protein molecules and solubility is lowered. At low ionic strengths the addition of increasing amounts of organic solvents continuously depresses the solubility of proteins at constant pH and ionic strength. In this way very soluble proteins, such as albumin, may be made sufficiently insoluble for purposes of purification. It should be noted that the addition of small amounts of solvent and of salt have opposite effects on the solubility of proteins. This frequently makes it possible to adjust the solubility of a protein to a convenient value at a variety of hydrogen ion concentrations by balancing the solvent action of salt with the precipitating action of the organic solvent.

Of course, the action of organic solvents is not limited to their effect on the dielectric constant, but direct solvation of the protein must also occur and part of this solvation may represent exchange with protein-bound water. Also, since useful organic solvents must be very soluble in water, their interaction with water may cause dehydration of the protein. Although solvation should increase protein solubility and dehydration should decrease solubility, the net effect of these factors is unpredictable. However, in the case of certain proteins such as zein (the prolamines) which contain relatively few charged groups, organic solvents become excellent solvents, presumably because of their solvating effects. In this connection it is perhaps worth mentioning that mixed solvents frequently dissolve proteins better than pure solvents. Thus 90% ethanol dissolves zein better than absolute ethanol, and 90% acetic acid dissolves serum albumin better than glacial acetic acid. Some organic solvents, such as the glycols, show no precipitating action, presumably because of their close resemblance to water.

In general, organic solvents show marked tendencies to denature proteins. However, these effects may frequently be minimized by operating at low temperatures. These temperatures may of course be appreciably below 0°, owing to the depression of the freezing point by the organic solvent (see Table V). Proteins with loosely bound prosthetic groups such as lipoproteins are particularly sensitive to organic solvents. However, below 0° even plasma lipoproteins appear stable to concentrations of 15% or more of ethanol.

Protein solubility in the presence of organic solvents usually decreases markedly with decreasing temperature, and this, of course, also decreases the amount of solvent required for precipitation. This marked sensitivity to temperature also necessitates adequate temperature control during fractionation, since frequently one degree of variation may markedly change solubilities. Provided that the components are sufficiently stable, variation of temperature may thus prove a very useful parameter in protein purification.

Of the various organic solvents which might be useful precipitants, ethanol has probably been studied in most detail, because of its acceptability in pharmaceuticals. However, methanol and acetone have also proved valuable and might seem inherently better than ethanol. Thus, methanol has been claimed to have a lesser denaturing action on the plasma proteins than ethanol,[23] and acetone is widely used for making dried protein preparations (acetone powders). Acetone appears to be a much milder precipitant of the very soluble human carboxyhemoglobin than alcohols (methyl, ethyl, or propyl). Ethyl ether has also been used

TABLE V

PHYSICAL CHEMICAL CONSTANTS OF ETHANOL-WATER MIXTURES USED IN PLASMA FRACTIONATION

Volume % ethanol at 25°	Mole fraction, ethanol	Density at 25° referred to water at 4°[a]	Refractive index, 25°[b]	Freezing point, °C.[c]
8	0.0266	0.9860	1.3366	− 2.6
10	0.0328	0.9835	1.3377	− 3.4
15	0.0509	0.9775	1.3405	− 5.4
18	0.0624	0.9741	1.3422	− 6.8
25	0.0907	0.9659	1.3463	−10.7
40	0.1630	0.9448	1.3537	−23.0
53.3	0.2462	0.9198	1.3583	−33.5

[a] Data at 25° taken from H. Landolt and R. Börnstein, "Physikalisch-chemische Tabellen," p. 448, J. Springer, Berlin, 1923. Densities at other temperatures estimated by extrapolation.

[b] Taken from "Official and Tentative Methods of Analysis," 3rd ed., published by the Association of Official Agricultural Chemists, Washington, D. C., 1930.

[c] Landolt and Börnstein, loc. cit. p. 1457.

as a protein precipitant,[24] although its small solubility limits its effectiveness to the more insoluble proteins. Dioxane and tetrahydrofurane have adequate solubilities in water, but like ether they are usually contaminated with peroxides, which would have very deleterious effects on proteins. A study of the relative precipitability of several proteins by these various reagents as a function of the dielectric constant would seem very desirable. Some study of their relative denaturing action should also be included.

[23] L. Pillemer and M. C. Hutchinson, J. Biol. Chem. **158**, 299 (1945).
[24] R. A. Kekwick, B. R. Record, and M. E. MacKay, Nature **157**, 629 (1946).

Thus, organic solvents provide an additional variable in the precipitation of proteins. A detailed description of the use of alcohol will be found in Section VI.

V. Ionic Precipitants

Heavy metal ions and many complex anions have long been known as excellent protein precipitants, although usually the products were denatured. However, denaturation is not an inevitable sequel to their action, and careful control of the conditions for precipitation frequently makes them quite innocuous. As in ethanol precipitation, low temperatures appear desirable, and pH control is essential. (This is particularly true in the case of metal ions which displace protons.) The amounts of reagent effective are frequently considerably smaller than the reagent's maximum stoichiometric combination with the protein.

These reagents act primarily by changing the charge on the protein and hence altering the isoelectric point. Consequently, optimal precipitation will occur at considerably different pH values in the presence of such ions. Thus, serum albumin, normally precipitated at pH 5, precipitates at pH 7 in the presence of zinc ions. Of course, the zinc complex separates as a new solid phase and such new phases have very different solubilities from that of the original protein. One explanation of this would be in the ability of multivalent ions to form cross-links between protein molecules, thus stabilizing the solid phase. If large amounts of ions are bound to the protein the resultant electrostatic forces may markedly distort the molecular geometry. This would also affect solubility and also may explain the denaturing effects of an excess of reagent.

Of all the ions studied in the fractionation of plasma proteins, zinc appears to be particularly useful, being effective at 0.5 to 20 mM (at higher concentrations salting-in occurs). The pH must be above 6 for combination of zinc with the protein. Other examples of the precipitation of proteins by cations are casein by calcium ions and the classical crystallization of insulin with zinc[25] and of enolase with mercury.[26]

When proteins combine with negative ions such as acids of high molecular weight or anions containing groupings which interact strongly with proteins, the expected effect on the isoelectric point is observed, and the protein becomes decreasingly soluble with increasing acidity. Classical examples of acid precipitants are trichloroacetic, phosphotungstic, and sulfosalicylic acids. The fact that such reagents may have a denaturing action, while accentuating the phenomena cannot be solely responsible for their precipitating action, since denatured proteins are also soluble

---

[25] D. A. Scott, *Biochem. J.* **28**, 1592 (1934).
[26] O. Warburg and W. Christian, *Biochem. Z.* **310**, 384 (1922).

when they carry sufficient charge. Furthermore, certain enzymes of relatively low molecular weight and some polypeptides may be precipitated by these reagents without denaturation, especially in the presence of electrolyte. Their use in separating serum proteins has been briefly described.[27] Again, in using these ionic precipitants one must be assured that the enzyme is not inactivated, and a convenient method for their subsequent removal should also exist.

Among ionic substances used in protein fractionation might be included detergents. However, although they have frequently proved useful in "solubilization" of proteins, their value in fractionation seems limited. The reason for this would appear related to their mode of action which involves strong bonding between the lipophilic portions of the protein and the detergent. Consequently, they may react in an "all or none" sense with the protein placing a very high charge on the protein molecule.[28] Their action thus resembles acid denaturation and frequently they do, in fact, denature the protein. Moreover, even in those cases where denaturation does not occur, the large amounts of detergent bound to the protein components present similar surfaces to the solution, largely hiding the inherent differences in the protein components and causing them to behave similarly in any fractionation procedure. Their "all or none" behavior, like selective heat denaturation, may prove useful if one carefully adjusts the detergent concentration to a value at which selective reaction with only certain of the components present occurs.

Polyelectrolytes, in contrast to detergents, interact primarily through their charged groups, forming salt-like linkages, and they appear to afford unusually mild reagents for protein precipitation. Thus nucleic acid has been used as a polyanion by the Warburg school[29,30] and others,[31,32] and protamine as a polycation.[33] However, synthetic polyelectrolytes appear superior for these purposes.[34] Since, in reaction, they neutralize the charge on the protein, they too tend to differentiate proteins on the basis of the proteins' isoelectric points. The complexes formed may be very insoluble, permitting separation from dilute solutions. In their use advantage may be taken of all the variables discussed above (pH, ionic strength, temperature, dielectric constant) plus variation in the nature and amount of the polyelectrolyte. Systems of such complexity would

[27] T. Astrup and A. Birch-Anderson, *Nature* **160**, 637 (1947).

[28] H. Neurath, G. R. Cooper, and J. O. Erickson, *J. Biol. Chem.* **142**, 249 (1942).

[29] O. Warburg and W. Christian, *Biochem. Z.* **303**, 40 (1939).

[30] F. Kubowitz and P. Ott, *Biochem. Z.* **314**, 94 (1943).

[31] A. Kleczkowski, *Biochem. Z.* **40**, 677 (1946).

[32] K. B. Björnesjö and T. Teorell, *Arkiv Kemi, Mineral. Geol.* **A19**, No. 34 (1945).

[33] F. Haurowitz, *Kolloid-Z.* **74**, 208 (1936).

[34] H. Morawetz and W. L. Hughes, *J. Phys. Chem.* **56**, 64 (1952).

appear to offer unusual opportunities in protein fractionation, although evidence that this is so has not been accumulated to date. At least in certain cases the protein may readily be recovered from the complex by precipitation of the polyelectrolyte with a suitable counter ion (e.g., polymethacrylate is precipitated by barium ions).

VI. Separation of Proteins from Mixtures

It would seem obvious that, by taking advantage of the several parameters discussed above, any proteins could be separated if their solubility properties were known. However, in practice, proteins must first be purified so that their solubility characteristics may be learned, and hence the foregoing can be applied only as general rules to guide the investigator.

The multiplicity of separation techniques and variables in each method make it impossible to map out a general procedure for the separation of a given protein. It is strictly a matter of trial and error, but there are certain aspects that are worth consideration here.

The most obvious yet often neglected requirement for the isolation of a protein from a mixture is a method for identification of the protein. Enzyme chemists are fortunate in that it is usually possible to devise an assay system reproducible *in vitro*. Other things being equal, the degree of success in any isolation procedure depends on the selectivity of the identification system.

Next it would seem advisable to map out the stability limits of the enzyme, especially with respect to pH, temperature, and organic solvents, so that one knows the breadth of conditions within which he must operate. It should be borne in mind that these limits may change in either direction as purification progresses.

Having found conditions for the initial precipitation, the most potent variable is pH. The importance of the isoelectric point in protein fractionation lies in its wide variation from protein to protein. A protein completely soluble at a given pH may be precipitated almost completely by a change of 1 pH unit, whereas a second protein soluble at the second pH may be insoluble at the original pH. At a pH in between both proteins might be partially precipitated. Of course at the pH intermediate between the two isoelectric points the proteins may combine with each other to form complexes which are occasionally more but usually less soluble than the individual proteins.

For this reason it is advisable to start any fractionation procedure at a pH either more alkaline than the most alkaline isoelectric point or more acid than the most acid isoelectric point (if the enzymes will remain active) and approach the other pH range as fractionation progresses.

Occasionally alteration of acid and alkaline precipitation is particularly effective. In any event, the pH must be controlled if one expects to have a reproducible procedure.

Since purification of a protein requires many steps, a good yield should be obtained at each step. Even if one obtains an 80% yield in each of only five steps, the final yield is 33%.

Choice of the best concentration of protein for effective fractionation is usually a compromise based on several factors. First, in general, the more dilute the system, the less the interaction between the several protein components and the better the separation. However, at too great dilutions, the large volumes become difficult to process; mechanical losses from incomplete recovery of precipitates become serious; and some proteins show greater lability (presumably due to surface denaturation or to the increased ratio of solvent impurities such as metals or peroxides which have a denaturing action). A good "round" figure for protein concentration would then be 1%. However, this figure may vary at least tenfold either way in particular instances.

Some investigators are surprised when an enzyme exhibits different solubility limits as it becomes purified. This may be due to protein-protein interaction. However, one frequently forgets that proteins have finite solubilities, and therefore when one is working with purified enzymes in higher concentration precipitation will commence with smaller amounts of the precipitant than were necessary in the original (and dilute) extracts.

Although most procedures for isolating proteins are carried out with but a single component in mind, there is obvious merit in inclusive systems which will simultaneously isolate several components. In addition to the economy in material effected, such procedures, stressing yields (or conservation of specific activity) and degrees of separation achieved, are particularly suited for the elucidation of proper conditions for separations. The study of the protein-protein interactions, inevitably observed, may also shed light on the nature of these processes *in vivo*.

An example of such a comprehensive fractionation procedure and, at the same time, an example of the use of organic solvents is to be found in fractionation of plasma with ethanol as done in Cohn's laboratory. The original plan of the Cohn ethanol scheme was to separate a series of protein fractions of successively lower isoelectric points by adjustment of the pH acid to neutrality. The ethanol concentration and ionic strength would also be suitably varied to achieve the best fractionation at each step. Fractionation would be followed by electrophoretic analysis (which describes the protein system in terms of net charge on the different components and consequently tends to identify components in terms of their isoelectric point). This scheme of fractionation should minimize

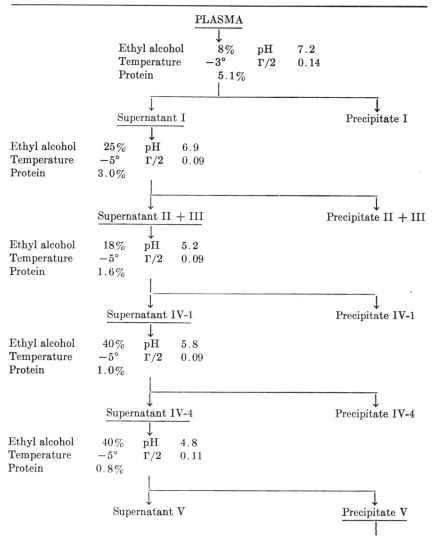

The fractionation of human plasma by ethanol according to method 6 (from Hughes[35]).

protein-protein interaction, which is strongest between proteins bearing opposite electrical charges, since all the proteins would be negatively charged except the protein being precipitated, which would have zero

[35] W. L. Hughes, in "The Proteins" (H. Neurath and K. Bailey, eds.), Vol. 2, Part B, Academic Press, New York, 1954.

charge. The effectiveness of this scheme will, of course, depend on the sharpness of the solubility minima as a function of pH for the various components present.

One of the most successful of the fractionation schemes for plasma is Cohn's method 6.[36] The procedure is given in the diagram on page 89.

Although such elaborate schemes are economical in relation to starting material, they are sometimes impracticable. In fact it may be advisable to purposely denature certain impurities by heat or acid in order to facilitate their removal.

Finally, it is worth noting that purity can be defined only in terms of *lack of specified impurities.* Consequently, the "pure" protein must be tested for impurities in all possible ways rather than relying on a single criterion such as electrophoretic homogeneity. The modern analytical tools of protein chemistry are very useful in testing purity, but they should not supersede the older methods based on solubility; constant solubility with excess saturating body, and testing whether the "pure" protein with constant properties is precipitated under the widest possible range of precipitation conditions. If this latter test is not fulfilled, a method for further fractionation becomes automatically available.

There are two articles on the solubility of proteins that the reader may find useful.[37,38]

[36] E. J. Cohn, L. E. Strong, W. L. Hughes, D. J. Mulford, J. N. Ashworth, M. Melin, and H. L. Taylor, *J. Am. Chem. Soc.* **68,** 459 (1946).
[37] J. T. Edsall, *in* "Chemistry of the Amino Acids and Proteins" (Schmidt, ed.). Charles C Thomas, Springfield, Ill., 1938.
[38] J. F. Taylor, *in* "The Proteins" (H. Neurath and K. Bailey, eds.), Vol. I, Part A. Academic Press, New York, 1953.

[11] Separation of Proteins by Use of Adsorbents

By Sidney P. Colowick

A. Introductory Remarks

The use of adsorbents for enzyme purification, a procedure introduced by Willstätter more than thirty years ago,[1] remains today one of the principle tools of the enzymologist. A survey of the contributions to Volumes I and II of this treatise reveals that the two adsorbents which are most widely used today in batch adsorption processes are Willstätter's Cγ modification of aluminum hydroxide gel[2] and Keilin and Hartree's

[1] R. Willstätter, *Ber.* **55,** 3601 (1922).
[2] R. Willstätter and H. Kraut, *Ber.* **56,** 1117 (1923).

tricalcium phosphate gel.[3] The remarks which follow will therefore be made principally with reference to these two reagents. It should be emphasized that no general principles can be laid down for the use of these reagents, since their very usefulness lies in the fact that different proteins behave quite differently with respect to both adsorption and elution characteristics. The development of an appropriate procedure for purification of a given enzyme is therefore a highly empirical process which cannot be based on known chemical or physical properties of the particular protein but must depend on exhaustive preliminary experiments covering a wide range of conditions. Although many investigators have devised schemes suitable for the purification of particular enzymes, not one of them would lay claim to the title of expert in the field of adsorption, and for this reason the task of writing this section has devolved upon one of the editors.

B. Selective Adsorption

Negative Adsorption. Selective adsorption can be carried out in two ways, sometimes referred to as "negative adsorption" and "positive adsorption." Negative adsorption is applied when the enzyme under investigation is not readily adsorbed and consists in adding the maximal amount of adsorbent which can be tolerated without extensive removal of the desired enzyme from solution. Proteins and other impurities having a high affinity for the adsorbent are thereby removed. When large volumes of adsorbent are required, it is advisable to submit the gel suspension itself to preliminary centrifugation and to use the packed gel in order to avoid dilution of the resulting enzyme solution. To cite two of many specific examples of the application of negative adsorption, one may mention the use of alumina $C\gamma$ for removal of cytochrome c in the preparation of cytochrome b_2[4] or the use of zinc hydroxide gel to remove red and brown pigments in the preparation of the "green enzyme," butyryl CoA dehydrogenase.[5] In most cases, the protein impurities removed do not have distinguishable spectral characteristics, and the success of the procedure must be gaged by measurements of specific activity in the supernatant fluid. In general, negative adsorption does not lead to extensive purification in terms of specific activity, but serves (a) for the removal of interfering proteins and (b) for the preparation of the solution for subsequent positive adsorption.

Positive Adsorption. Positive adsorption is applied directly when the bulk of the enzyme under investigation is readily taken up by an amount

---

[3] D. Keilin and E. F. Hartree, *Proc. Roy. Soc. (London)* **B124**, 397 (1938).

[4] M. Dixon, Vol. I [68].

[5] H. R. Mahler, Vol. I [92].

of adsorbent which removes only a small fraction of the total protein. When the enzyme is not readily adsorbed, it is preferable to remove impurities first by negative adsorption and then proceed with positive adsorption. It should be emphasized that the ease of adsorption of a given enzyme depends greatly on the nature of the impurities present. An enzyme which is readily adsorbed from pure solutions will not be readily adsorbed when impurities are present in sufficient quantity to saturate the adsorbent. In carrying out positive adsorption, one uses the minimum amount of adsorbent which serves to remove the bulk (80% or more) of the desired enzyme.

Protein Concentration and Gel-Protein Ratio. The conditions which must be considered in carrying out selective adsorption include the following: protein concentration, gel-protein ratio, salt concentration, and pH. It is common practice to dilute protein solutions to concentrations of 1% or less before carrying out adsorption. Presumably, the purpose of this is to minimize protein-protein interactions and thereby favor selective adsorption. However, a much more important factor appears to be the gel-protein ratio. The amount of gel added is best expressed in terms of dry weight, determined by drying a known volume of gel to constant weight at 100°. In practice the gel-protein ratio is ordinarily in the range 0.1 to 2.0, depending on the ease of adsorption of the desired enzyme. Within reasonable limits, the gel-protein ratio required for a given degree of adsorption is generally assumed to be fairly independent of the protein concentration, and for this reason the specification of a procedure in terms of the gel-protein ratio has become accepted as the best way to approach the ideal of reproducibility.

Salt Concentration. Salt concentration is one of the major factors requiring control in selective adsorption. In general, a low salt concentration favors adsorption, and for this reason many investigators subject their material to dialysis against water before attempting positive adsorption. However, it should be emphasized that this precaution is in many cases unnecessary. Satisfactory adsorption can often be carried out on undialyzed aqueous extracts of tissues or on undialyzed aqueous solutions of fractions obtained by precipitation with salts or organic solvents. The presence of high concentrations of polyvalent anions interfere with adsorption (see section on selective elution).

Hydrogen Ion Concentration. The other factor of major importance is pH. It is almost universal practice to carry out positive adsorption at low pH values. Most commonly, the pH is maintained in the range 5.0 to 5.5 by the addition of dilute acetate buffer or acetic acid (final concentration around 0.01 M). For the purposes of negative adsorption, higher pH values and higher salt concentrations may, of course, be desirable.

Temperature and Time of Adsorption. Other factors to be considered in adsorption are the temperature and the time of contact with the adsorbent. The temperature is selected mainly on the basis of the stability of the enzyme, and for this reason adsorption is ordinarily carried out at or near 0°. However, there is no reason why adsorption may not be carried out at room temperature when one is dealing with stable enzymes. Systematic studies on the effectiveness of adsorption of enzymes as a function of time or temperature are, to this writer's knowledge, non-existent. The time selected for contact with the adsorbent is commonly specified as 10 to 15 minutes, and this length of time appears, from general experience, to be more than ample for reaching adsorption equilibrium.

Age of Gel and Reproducibility. Another factor which has not been a subject of systematic study in most enzyme purification studies is the age of the gel at the time of use. It is customary to store alumina or calcium phosphate gels for several months prior to use, the idea being that the gels undergo subtle changes in physical state during aging, involving changes in degree of hydration. The purpose of aging is to permit these changes to occur before initial use, in order that subsequent use will be more likely to reproduce the original results. However, quantitative studies of the effect of aging on the efficacy of a gel for adsorption of a given enzyme are certainly lacking in most cases.

When one reads that a certain enzyme has been purified by the use of a calcium phosphate gel which was at least nine months old, this usually means that a gel of this age happened to be sitting on the investigator's shelf at the time he began his studies. It most certainly does not mean that one who is interested in repeating this procedure must wait nine months after preparing the fresh gel. If this were a means of insuring reproducibility, it would be worth while. However, there are other intangible factors which make it difficult to reproduce an adsorption procedure even when two gels of the same age are used under apparently identical conditions. In attempting to repeat a procedure described in the literature, the investigator must in no case proceed blindly according to the written directions but must make careful preliminary experiments to determine whether his own gel preparation behaves in the manner described. Discrepancies observed will usually be of a quantitative rather than qualitative nature, and satisfactory results can usually be obtained by adjusting the quantity of gel used. It is this writer's opinion that, since preliminary tests are in any case essential, not too much attention need be paid in most cases to specifications concerning the age of the gel.

That aged gels are not to be regarded as more efficient adsorbents than freshly prepared gels is supported by the fact that in some cases the use of the freshly prepared gel is actually recommended.[6] Furthermore, there are other cases in which the adsorbent is actually formed in the enzyme solution, for example by the addition of calcium chloride to phosphate-containing enzyme solutions.[7]

Fractional Adsorption. This name is applied to the procedure in which a number of successive small amounts of adsorbent are added and removed by centrifugation until the enzyme has been almost completely removed from solution. Each of the gel fractions is eluted individually, and the fractions of highest specific activity are collected. This procedure, which has been employed extensively in the Ochoa laboratory,[8] takes fullest advantage of the selective action of the adsorbent, and it is surprising that adsorption procedures are not more often carried out in this way. Such a method has the additional advantage that it can be carried out directly on a large batch of enzyme without preliminary trials, since one is assured of eventual recovery of the enzyme, even when the total quantity of adsorbent required differs markedly from that prescribed.

C. Selective Elution

Once an enzyme has been adsorbed, the problem of elution becomes the real test of the artistry, ingenuity, imagination, and, above all, the patience of the investigator. There is no special trick to eluting almost all the adsorbed proteins by using a high concentration of a polyvalent anion at a high pH. Classically, phosphate at a concentration of 0.1 to 1.0 M and in the pH range 7 to 9 is used for this purpose. In most, but not all, cases, this treatment will remove the desired enzyme, but its very effectiveness makes it unsuitable as a selective eluting procedure. It is only by employing special tricks for elution of the desired enzyme that extensive purification can be expected.

The principle commonly employed is to make use whenever possible of what might be called "negative elution," that is the use of solutions of low salt concentration and/or low pH to extract protein impurities while leaving the desired enzyme adsorbed to the gel. The first step usually consists in "washing" the gel with water or dilute acetate buffer at pH 5.0 to 5.5. This is often followed by successive extractions with phosphate buffers of gradually increasing concentration[9] and/or pH.[10] Each eluate is tested for activity and protein content, and those with highest specific activity are collected.

[6] B. L. Horecker, A. Kornberg, and P. Z. Smyrniotis, Vol. I [42].
[7] S. Ochoa, Vol. I [123].
[8] S. Ochoa, Vol. I [124].
[9] L. A. Heppel, Vol. II [57].
[10] A. Meister, Vol. II [52].

If one is especially fortunate, even the highest concentration of phosphate at the highest pH tested will fail to elute the enzyme. This is most likely to occur, of course, when the enzyme was readily adsorbed by a small amount of adsorbent in the first place. In such a case, other eluants which may be effective include the following: ammonium sulfate,[11] pyrophosphate,[12] or a mixture of phosphate and ammonium sulfate.[4] It is interesting to note in this connection that a very high concentration of ammonium sulfate (e.g., 0.6 saturated) may be a less effective eluant than a lower concentration (e.g., 0.3 saturated) by virtue of the salting out effect obtained with the former.[11] Other agents which have found occasional use for elution are borate buffers in the pH range 8.5 to 10.2,[13–15] sodium bicarbonate,[16,17] trisodium citrate,[18] and 80% glycerol.[19]

One finding worthy of special mention is that of Heppel,[20] namely that the enzyme inorganic pyrophosphatase can be eluted by extremely low concentrations (0.001 M) of its substrate, inorganic pyrophosphate. If this represents a case of elution made possible by specific enzyme-substrate binding, it certainly offers a fruitful approach for possible specific elution of other enzymes based on this principle.

The rate of elution appears in general to be slower than the rate of adsorption. It is quite common practice to elute three or four times with the same eluting solvent, allowing 10 or 15 minutes of contact each time. This is probably a reflection of the long time required to reach elution equilibrium, which in turn is probably due to the difficulty in getting adequate dispersion of a tightly packed gel in the eluting solution. The method described by Ochoa,[21] which involves shaking gel plus eluting fluid for 1 hour with glass beads, with the addition of capryl alcohol to avoid foaming, appears to provide a satisfactory answer to this problem.

D. Choice of Adsorbent

It is quite clear that the various adsorbents are not interchangeable in their action. For example, alumina and calcium phosphate gel have been found to differ quite markedly in their ability to adsorb 3'-nucleoti-

[11] G. W. E. Plaut and S.-C. Sung, Vol. I [119].
[12] A. Nason and H. J. Evans, Vol. II [59].
[13] E. Haas, Vol. II [125].
[14] E. Volkin, Vol. II [82].
[15] R. W. McGilvery, Vol. II [84].
[16] L. Shuster and N. O. Kaplan, Vol. II [86].
[17] M. Slein, Vol. I [37].
[18] G. Schmidt, Vol. II [79].
[19] C. V. Smythe, Vol. II [41].
[20] L. A. Heppel, Vol. II [91].
[21] S. Ochoa, Vol. I [114].

dase.[16] Another example is the superiority of alumina over zinc hydroxide gel for the adsorption of butyryl-CoA dehydrogenase.[5] A third example is the use of fuller's earth or kaolin for the adsorption of prostatic phosphatase, which is not adsorbed by alumina gel.[18] For this reason, more than one kind of adsorbent is often used during the course of a purification procedure. Most commonly, calcium phosphate gel and alumina C_γ are used in combination,[22] but further exploitation of other adsorbents would appear desirable. In addition to fuller's earth, kaolin, and zinc hydroxide gel, one may mention bentonite,[23] cellulose,[24] charcoal,[25,14] and cornstarch[26] as adsorbents which have thus far seen only limited use in enzyme purification. Charcoal has been used to advantage for removing nucleotides and nucleic acids from enzyme preparations,[25] but it has not seen wider use because of the apparent difficulty in eluting proteins once they have been adsorbed on charcoal. However, Volkin[14] has successfully separated bone monoesterase from diesterase by means of charcoal, taking advantage of the fact that the former, but not the latter, may be eluted with borate.

It is worth while to note that certain substances not ordinarily regarded as adsorbents may function as such when used for other purposes. For example, Celite (Johns-Manville), which is commonly used as a filter aid, sometimes for the purpose of aiding in the removal of gels by filtration,[27] has been reported to cause appreciable loss of enzyme in at least one case.[28] In another case, Celite is actually recommended as the adsorbent of choice.[29]

Another substance in this category is anhydrous aluminum oxide (e.g., Alcoa A-303), which is commonly used for grinding bacterial cells prior to extraction.[30] In several cases,[31,13] this material has proved to be an effective adsorbent for certain enzymes.

Use of Adsorbents on Columns. The separation of proteins on columns is described elsewhere in this volume[32,33] and will not be discussed here. Most separations of proteins on columns are not based on the principles

[22] S. Ratner, Vol. II [48].
[23] L. F. Leloir and R. E. Trucco, Vol. I [35].
[24] G. C. Butler, Vol. II [89].
[25] P. J. Keller and G. T. Cori, Vol. I [24].
[26] J. Larner, Vol. I [27].
[27] C. R. Dawson and R. J. Magee, Vol. II [145].
[28] A. Kornberg, Vol. II [112].
[29] J. B. Sumner, Vol. II [137].
[30] I. C. Gunsalus, Vol. I [7].
[31] E. Juni, Vol. I [72].
[32] C. H. W. Hirs, Vol. I [13].
[33] R. R. Porter, Vol. I [12].

of adsorption and elution but rather on either ion exchange chromatography[32] or partition chromatography.[33] The gelatinous adsorbents are not suitable for column work because of the resistance which they offer to the flow of aqueous solutions. However, certain nongelatinous forms of the adsorbents have been used in columns with some success.[29,34]

E. Preparation of Adsorbents

Described below are methods for the preparation of alumina C_γ and calcium phosphate gel. The writer has chosen to describe the standard preparations[2,3] rather than more recent modifications for the preparation of both alumina[27] and calcium phosphate.[35-38,20] He feels that this is more desirable in the interests of reproducibility, since most enzyme purifications now in the literature have been carried out with these so-called standard preparations.

Alumina C_γ ($Al_2O_3 \cdot 3H_2O$). The Willstätter school developed a series of aluminum hydroxide gels, differing in physical and chemical properties, which they designated A, B, C_α, C_β, and C_γ. The latter is the only form of any importance now, since it is the most stable and therefore the most readily reproducible. The directions given below are translated almost literally from E. Bauer,[39] who presents essentially the original procedure of Willstätter and Kraut,[2] simplified somewhat according to a personal communication to him from Kraut.

"A hot solution of 340 grams of aluminum ammonium sulfate in 500 ml. of water is poured all at once into 3.25 liters of ammonium sulfate-ammonia water at 60°. The latter reagent contains 100 g. of ammonium sulfate and 215 ml. of 20% ammonia, that is, a slight excess. During the precipitation and for an additional ¼ hour, the mixture is stirred vigorously and the temperature should not fall below 60°. The first voluminous precipitate gradually becomes flocculent. Then the mixture is diluted to 20 liters and the supernatant fluid decanted as soon as the precipitate has settled out. The washing with water by decantation is repeated two more times. To destroy any remaining basic aluminum sulfate 40 ml. of 20% ammonia are added during the fourth washing; then the washing with pure water is resumed.

"Between the 12th and 20th washings, the supernatant fluid will no longer become clear. Then continue the washing two more times, for

[34] B. D. Polis and H. W. Shmukler, Vol. II [144].

[35] J. B. Sumner and D. J. O'Kane, *Enzymologia* **12**, 251 (1948).

[36] A. C. Maehly, Vol. II [143].

[37] A. Meister, *Biochem. Preparations* **2**, 18 (1952).

[38] S. Swingle and A. Tiselius, *Biochem. J.* **48**, 171 (1951).

[39] E. Bauer, *in* "Die Methoden der Fermentforschung" (E. Bamann and K. Myrbäck, eds.), Vol. 2, p. 1426. Academic Press, New York, 1945.

which at least several days will be necessary. The complete conversion into the γ-modification after several weeks can be confirmed by the heating test with ammonia. In general the alumina C_γ should not be used for a period of three months after preparation.

"The heating test allows the distinction of the C-modifications. Hot 10% ammonia converts the alumina C_α and C_γ into a heavy white powder of composition $Al_2O_3 \cdot 3H_2O$. The C_β form will not change in outward appearance, although its water content falls to 27%."

Calcium Phosphate Gel. The following description is quoted directly from Keilin and Hartree.[3]

"150 c.c. calcium chloride solution (132 g. $CaCl_2 \cdot 6H_2O$ per litre) is diluted to about 1600 c.c. with tap water and shaken with 150 c.c. trisodium phosphate solution (152 g. $Na_3PO_4 \cdot 12H_2O$ per litre). The mixture is brought to pH 7.4 with dilute acetic acid and the precipitate washed three or four times by decantation with large volumes of water (15–20 l.). The precipitate is finally washed with distilled water in a centrifuge. 9.1 g. calcium phosphate is obtained."

[12] The Partition Chromatography of Enzymes

By R. R. Porter

Introduction

Many enzymes have been purified during the last twenty years chiefly by fractional precipitation, and, although other steps have been introduced, precipitation by salt or organic solvents remains a universal procedure. Its resolving power is limited so that in the later stages where proteins of similar solubility are being separated purification may be accompanied by heavy loss of the desired material. There are examples of extremely similar proteins with clearly distinguished biological properties such as antibodies, where no resolution has ever been achieved by any method. There is clearly a need for more effective methods of protein purification, and it is not surprising, in view of the striking success of chromatographic and countercurrent methods in the fractionation of small molecules, that attempts should be made to apply similar principles to proteins. It is probable that such techniques will have their greatest use in testing the homogeneity of apparently pure proteins and in completing the purification of material from which much of the accompanying impurities have been removed by the classical methods.

Principle

The principle of partition chromatography, whether for small or for large molecules, is the same as was first demonstrated by Martin and

Synge[1] in 1941. The material to be fractionated is subject to continuous partitioning between two immiscible liquids. This is achieved by immobilizing one liquid as a film or small droplets on an inert carrier such as kieselguhr while the second liquid is allowed to flow slowly past. If other factors are not superimposed, the rate of movement of the solute relative to the rate of flow of the moving liquid is determined by the partition coefficient of the solute between the two liquids. This relation has been expressed by Martin and Synge[1] in the equation

$$R = \frac{A}{A_L + \alpha A_s}$$

where $R = \dfrac{\text{Volume of the column}}{\substack{\text{Volume of the eluate when maximum} \\ \text{concentration of solute appears}}}$.

A = Area of cross section of the column.

A_L = Area of cross section of the mobile phase.

A_s = Area of cross section of the stationary phase.

α = Partition coefficient.

In the extension of this method to proteins, two important difficulties are met. First, it is not easy to find immiscible liquids in which proteins will partition with a coefficient of the required order (say 5/1 to 15/1) and yet at the same time retain sufficient solubility in the phase in which it is least soluble. Second, many proteins are easily denatured by organic solvents and also after adsorption, such as may occur either at the liquid-solid or at the liquid-liquid interfaces present in a partition chromatographic column.

Generally, proteins are soluble only in aqueous mixtures which have at least 50% water, and hence the choice of immiscible liquids is limited. Two notable exceptions to this have been reported: insulin will partition between butanol and aqueous dichloroacetic acid with sufficient solubility in both phases,[2] and casein behaves similarly in water-phenol mixtures.[3] It seems probable, however, that such examples will remain exceptional.

The systems to be described here are the two liquid phases produced when a solute is added in sufficient concentration to a mixture of water and a water-miscible organic solvent. The solute must have a high solubility in water and very low solubility in the organic solvent for two liquid phases to be produced. Inorganic salts such as K_2HPO_4, NaH_2PO_4, and $(NH)_2SO_4$ have been used as solute, but nonelectrolytes such as sucrose may also be effective and have the advantage of possessing little precipi-

[1] A. J. P. Martin and R. L. M. Synge, *Biochem. J.* **35**, 1558 (1941).

[2] E. J. Harfenist and L. C. Craig, *J. Am. Chem. Soc.* **74**, 3083 (1953).

[3] I. O. Walter, Thesis, University of Berne, 1952.

tating power for proteins. Of the many water-miscible organic solvents, the glycol ethers (Cellosolves and Carbitols) appear to be the most generally useful in that they give rise to immiscible phases both of which have a high water content. Like other ethers, they have the disadvantage of readily forming peroxides, and hence they have to be freshly distilled before use.

Denaturation caused by organic solvents is usually avoidable by working at temperatures below 0°, but the denaturation which follows adsorption is less easily overcome. Adsorption on the solid (kieselguhr) which holds the stationary phase has been found in only one instance. Thus, rabbit γ-globulin is adsorbed irreversibly by Hyflo Super-Cel. Celite 545, a larger particle silica, has less adsorptive capacity, and, with both, this is reduced further in more-alkaline solution so that at pH 8.8 there is no adsorption on Celite 545.

Adsorption at the liquid-liquid interface during partition chromatography is frequently met, as judged by the low rate of movement of a protein on a column compared to the rate calculated from the partition coefficient. In many cases this adsorption is not followed by denaturation, and it may increase the resolving power of the system. If denaturation does occur, as shown by poor recovery of soluble protein, it may be avoidable with other phase systems.

Method

Materials. The kieselguhrs used have included several grades supplied by Johns-Manville, such as Hyflo Super-Cel and Celite 545. Before use they are stood in 2 *N* HCl for 24 hours with occasional stirring, washed with distilled water until free of acid, and dried at 105°.

Silane-treated kieselguhr is prepared as described by Howard and Martin.[4] About 500 g. of kieselguhr is put into a closed vessel together with a beaker containing 10 ml. of dimethyldichlorosilane. The vapor reacts with the kieselguhr with which it is allowed to stand for 24 hours or longer. The container is opened (considerable fuming occurs, owing to the HCl which has formed), and the kieselguhr is tested for complete reaction by stirring a sample with water; none should sink. It is washed with methanol until the washings are neutral and then dried at 105°.

All solvents are redistilled before use and should be free of material adsorbing in the ultraviolet light and free of peroxides.

Estimations. Protein concentration is estimated by the adsorption at 280 mμ, as this offers a rapid method with sufficient sensitivity for most purposes.

---

[4] G. A. Howard and A. J. P. Martin, *Biochem. J.* **46**, 532 (1948).

When working with impure material, partition coefficients and elution from the column are determined by assay of the biological activity of the protein which it is desired to fractionate.

Preparation of Phase Diagrams. The success of the chromatography depends on the choice of a suitable phase system, and in order to increase the range of the method many systems have been investigated. Phase

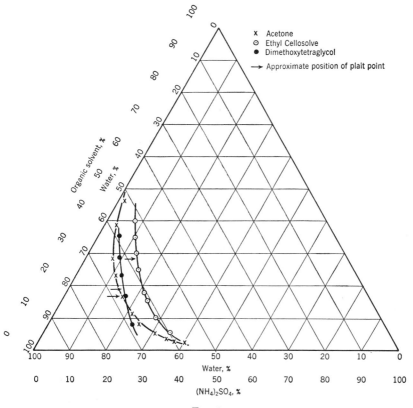

Fig. 1

diagrams with intermediate characteristics can be obtained by mixing solvents or solutes. A phase diagram of sufficient accuracy may be prepared as follows: 1 g. of solute, e.g. $(NH_4)_2SO_4$, is dissolved in 2 ml. of water, and the organic solvent is added from a buret until opalescence just appears. The volume added is noted, a further 1 ml. of water added, and the titration repeated. At the beginning the mixture will be largely bottom phase, but later it will become almost entirely top phase. At intermediate positions there will be more nearly equal amounts of both phases, and at one point it will be observed that the mixture has just changed from having more bottom phase to more top phase. The composition at this point is approximately that of the plait point. The series of points thus obtained give the phase diagram. Examples of phase diagrams are shown in Figs. 1 to 6, and it can be seen that systems with widely varying characteristics are obtainable.

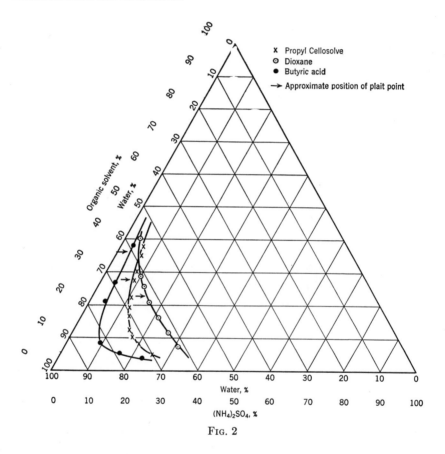

Fig. 2

Choice of Phase System. The choice of phase system most suitable for the chromatography of a given protein depends on a variety of factors. In the diagrams the plait point is shown. This is a theoretical point at which the concentration of components is such that two phases are just produced and these phases have almost identical composition. In any mixture which produces two phases, one (possibly both) will contain less water and therefore more salt or organic solvent than at the plait point. Therefore, if the protein is unlikely to be soluble in a mixture whose composition is similar to that at the plait point, it will not have sufficient solubility in both phases for use. Hence inspection of the phase diagrams and knowledge of the solubility characteristics of the protein such as the concentration of $(NH_4)_2SO_4$ or ethanol required for precipitation will give guidance as to the systems most likely to be useful. As solvent and solute influence the

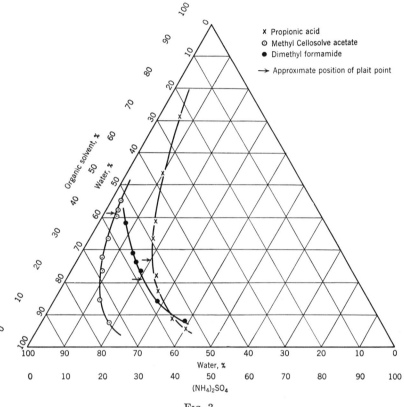

FIG. 3

effect of each other, this estimation can be only approximate, and trial of solubility in the phases in test-tube experiments is necessary. At the same time it will be observed if denaturation occurs at room temperature, and consequently if the work will have to be carried out at lower temperatures. Insolubility and denaturation may be quickly distinguished by addition of water. In the first case the mixture will become clear, but it will remain cloudy if denaturation has occurred. The partition coefficient can also be measured by partitioning a few milligrams of material between 2 ml. of the top and bottom phases. It is usually preferable first to try a mixture near to the plait point and then to move further away if higher coefficients are required.

By these means a phase system may be found which will give a partition coefficient of, say, 5/1 and a solubility in the least soluble phase of 1 mg./ml. If significant denaturation does not occur when the protein is

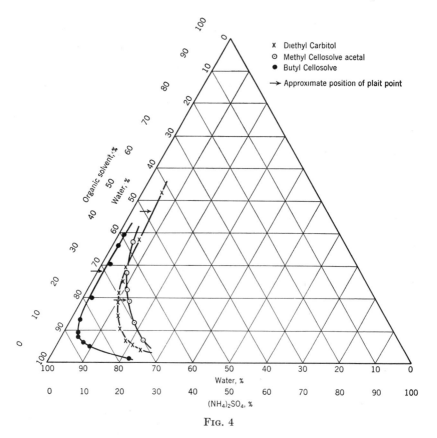

x   Diethyl Carbitol
⊙   Methyl Cellosolve acetal
●   Butyl Cellosolve

→ Approximate position of plait point

FIG. 4

dissolved in either phase over, say, 24 hours, then chromatography may be commenced.

Preparation of Column. The stationary phase is that in which the protein is most soluble. If this is the aqueous phase, the supporting solid will be kieselguhr, and, if the organic phase is to be stationary, silane-treated kieselguhr is used. The only exception to this is with the system ammonium sulfate, water, and ethyl Cellosolve, where, for reasons not understood, the organic phase is held by acid-washed Hyflo.

To prepare a column, the kieselguhr, say 6 g., is added 3 ml. of stationary phase and mixed well in a beaker. Moving phase is added to produce a smooth cream, and the mixture is stirred until free of lumps and air bubbles. It is then poured into a glass tube (diameter about 1 cm.), the bottom of which has been turned in slightly to support a perforated silver disk and a filter paper disk which fits the tube. The slurry

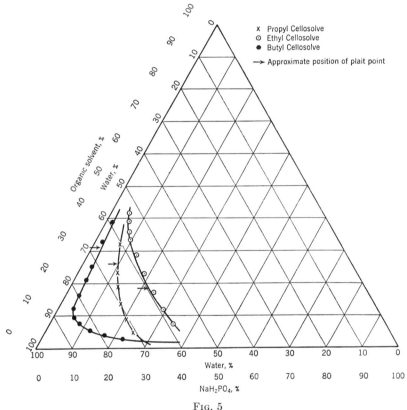

Fig. 5

is stirred with a similar perforated disk on a thin stainless steel rod until free of air bubbles and then is allowed to settle under slight pressure. More moving phase may be added if there is any evidence of impurities being washed from the kieselguhr, and the column is then allowed to drain until the supernatant liquid has almost disappeared. It is never allowed to go dry, for the column will crack and have to be repacked. The protein, say 10 mg., is dissolved in 2 ml. of the moving phase and is run onto the top of the column with a bent-tipped pipet in order to avoid disturbing the surface. When it has sunk into the column, washings of the containing vessel are added and allowed to drain similarly. More moving phase is now added to give sufficient head to maintain a rate of flow of about 5 ml./hr. The rate of flow is not critical. Aliquots of effluent are collected, and the protein content or biological activity estimated. The whole procedure may be scaled up as required if the small columns prove satisfactory.

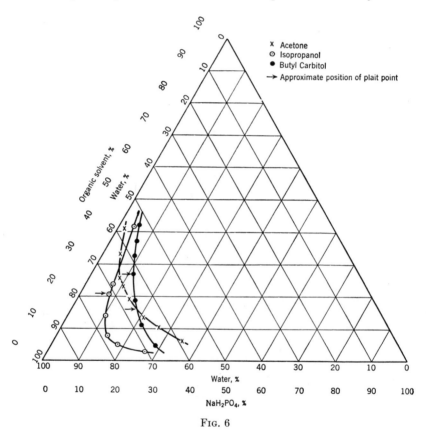

FIG. 6

The weight of protein which can be chromatographed will be limited by its solubility, by the spreading of the peaks which accompanies higher loading, and also by the influence of the proteins on the moving phase. At higher protein concentration it will produce opalescence, as would an addition of smaller amounts of the phase system solute. A slight effect on the column can be tolerated, but if excessive it will disrupt the phase system and cause stripping of the stationary phase so that all the material comes through at the liquid front without fractionation. Losses on the column will result from adsorption and denaturation on the kieselguhr or at the liquid-liquid interface. The latter will probably be observed during estimation of the partition coefficient, and the former may be confirmed by direct estimation after stirring 1 g. of kieselguhr with 5 ml. of either phase in which the protein has been dissolved. If denaturation occurs, it is necessary to change the phase system or supporting solid.

Notes on Method. In addition to the general procedure outlined above there are several practical points which may be recorded.

1. TEMPERATURE LABILITY OF PHASE SYSTEMS. All the phase systems investigated are altered by changes in temperature of several degrees, but those containing butyl Cellosolve are much more labile, and control within $\pm 0.5°$ is essential. It follows that phase mixtures must be equilibrated at the working temperature.

2. SOLUBILITY. Freeze-dried proteins which are dissolved only with difficulty in the organic phase may become more amenable if a small drop of water (say 0.01 ml. to 10 mg. of protein) is added to reduce them to a paste. If subsequently dissolved in several milliliters of the organic phase, this has no effect on the chromatography.

3. CHOICE OF PHASE MIXTURE. In addition to the main points made earlier, it is obvious that in the choice of phase mixture attention must be paid to the pH of optimal stability of the protein being investigated, or of minimal activity of contaminants such as proteolytic enzymes. Advantage may also be taken of the stabilization of enzymes by metal ions; thus $MnSO_4$ would be a suitable solute in attempts to chromatograph aminopeptidase. If sulfhydril enzymes were being investigated, it would probably be necessary to add a little thioglycollate or other reducing agent, as traces of peroxide may reappear rapidly in glycol ethers.

Examples of the Partition Chromatography of Proteins

As an illustration of the application of the method, examples of the chromatograms obtained, together with details of the experimental condition used, are given below for several enzymes and other proteins.

Ribonuclease.[5]

Source of enzyme: Beef pancreas.
Approximate purity of starting material: 30 to 100%.
Experimental conditions: Room temperature; acid-washed Hyflo used throughout.
System 1 (Fig. 7): 15 g. of $(NH_4)_2SO_4$; 30 g. of ethyl Cellosolve; 55 g. of water; 6 g. of Hyflo in a tube of 1.2-cm. internal diameter; 5 mg. of ribonuclease. Moving phase-aqueous phase.

[5] A. J. P. Martin and R. R. Porter, *Biochem. J.* **49**, 215 (1951).

System 2 (Fig. 7): 20 g. of $(NH_4)_2SO_4$; 24 g. of ethyl Cellosolve; 56 g. of water; 6 g. of Hyflo in a tube of 1.2-cm. internal diameter; 5 mg. or ribonuclease. Moving phase-aqueous phase.

System 3 (Fig. 7): 16.5 g. of $(NH_4)_2SO_4$; 36.5 g. of ethyl Cellosolve; 47 g. of water; 3 g. of Hyflo in tube of 1.2-cm. internal diameter; 5 mg. of ribonuclease. Moving phase-aqueous phase.

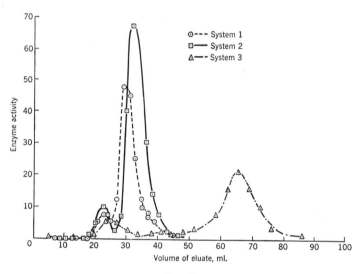

FIG. 7

It will be observed that, with phase mixtures increasingly different from the plait point of the system, the partition coefficient increases, the rate of movement of the protein falls, and the separation of the two enzymically active components is greater. Complete separation from inactive contaminants was achieved if such were present.

Insulin[6]

Source of enzyme: Beef pancreas.

Approximate purity of starting material: 2 to 100%.

Experimental conditions (Fig. 8): Room temperature.

Phase system: 15 ml. of water; 6.67 ml. of ethyl Cellosolve; 3.3 ml. of butyl Cellosolve; 9 ml. of $5M$ NaH_2PO_4 ($+ H_3PO_4$ to pH 3). Moving phase-aqueous phase.

50 g. of silane-treated Hyflo in a tube of 2.3-cm. internal diameter. Weight of crude insulin, 100 mg.

[6] R. R. Portèr, *Biochem. J.* **53,** 320 (1953).

The small peak running ahead of the insulin had hyperglycemic, but no hypoglycemic, activity. In this and many other chromatograms the impurities moved very slowly if at all and hence are not seen.

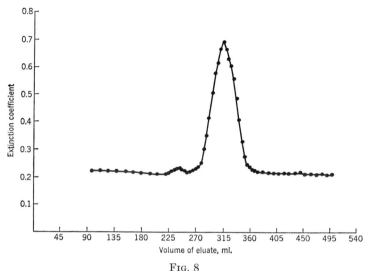

FIG. 8

Other systems used successfully for the chromatography of insulin are as follows:

1. 83 ml. of water; 45.5 ml. of butyl Cellosolve; 10 ml. of 2.5 M phosphate buffer, pH 7.6.
2. 13 ml. of water; 5.33 ml. of ethyl Cellosolve; 2.67 ml. of butyl Cellosolve; 15 ml. of 2.5 M phosphate buffer, pH 7.6.
3. 40 ml. of water; 21 ml. of isopropanol; 25 ml. of 5 M NaH_2PO_4 adjusted to pH 3 by addition of H_5PO_4.

Chymotrypsinogen and Chymotrypsin[7]

Source: Beef pancreas.
Approximate purity: 100 %.
Experimental conditions (Fig. 9): Room temperature.
Phase system: 40 ml. of water; 32 ml. of 5 M NaH_2PO_4 ($+$ H_3PO_4 to pH 3); 17.33 ml. of ethyl Cellosolve; 8.67 ml. of diethyl Carbitol. Moving phase-aqueous phase. 6 g. of silane-treated Hyflo in a tube of 1.2-cm. internal diameter; 10 mg. of protein chromatographed.

[7] R. R. Porter, "The Chemical Structure of Proteins," Ciba Symposium. J. & A. Churchill, London, 1953.

In this chromatogram the zymogen and the enzyme are clearly resolved, with possibly a smaller third component also present. Other system used successfully: 6 g. of KH_2PO_4 + 0.5 g. of KOH; 61 ml. of water;

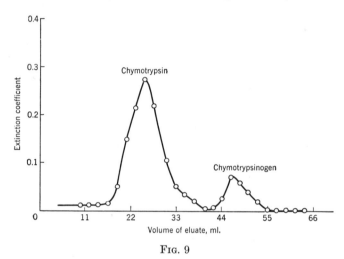

FIG. 9

33 ml. of methyl Cellosolve acetate. Moving phase-organic phase. Acid-washed Hyflo at temperature $-5°$.

Trypsin

Source: Beef pancreas.

Approximate purity: 100%.

Experimental conditions (Fig. 10): Room temperature.

Phase system: 130 g. of $NaH_2PO_42H_2O$; 206 ml. of water; 64 ml. of ethyl Cellosolve. Moving phase-aqueous phase. 8 g. of silane-treated Hyflo in a tube of 1.2-cm. internal diameter. 8 mg. of trypsin chromatographed.

Penicillinase

Source: Filtrate from adapted *B. cereus* culture.

Approximate purity: Very impure.

Experimental conditions (Fig. 11): Room temperature.

Phase system: 90 ml. of 2.5 M potassium phosphate, pH 7; 150 ml. of water; 100 ml. of propyl Cellosolve. Moving phase-organic phase. 5 g. of acid-washed Hyflo. 75 mg. of solid put on the column.

It can be seen that the enzyme was entirely free of material adsorbing in the ultraviolet light and was insufficient to weigh. The purification

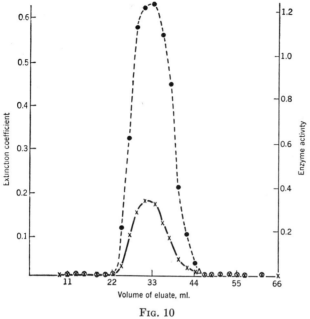

FIG. 10

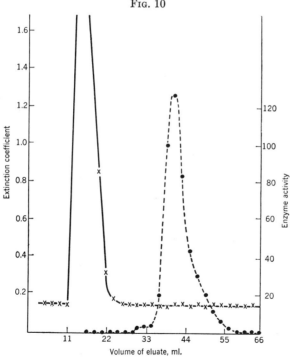

FIG. 11

achieved was therefore high. However, the enzyme was slowly denatured at room temperature, and hence a system at a low temperature was required.

Alternative System.

Experimental conditions: $-2°$ temperature.

Phase system: 85 ml. of 2.5 M potassium phosphate, pH 7; 150 ml. of water; 57 ml. of propyl Cellosolve; 38 ml. of butyl Cellosolve. Moving phase-organic phase.

Conclusion

The resolving power of the method is determined both by the efficiency of the column and by the selectivity of the phase system used. The column efficiency is usually as high as or higher than that attained in the partition chromatography of small molecules, but the selectivity, that is the difference in partition coefficient between the proteins which it is desired to fractionate, is impossible to forecast. With crude insulin and penicillinase all the impurities behaved very differently from the protein to be isolated —so much so that they either traveled with the solvent front or remained stationary on the column. This has been the general experience, and so it prevents one chromatogram giving a panoramic view of a protein mixture such as is obtained by electrophoresis. The effectiveness of a chromatogram in separating extremely similar proteins can be determined only by trial.

With this, and indeed with any method of fractionation, there is always the possibility that a protein will associate with another and that the complex will behave under the conditions used as a single component. Evidence of such behavior was observed in a chromatographic study[5] of the digestion of ribonuclease by pepsin and carboxypeptidase where the proteolytic enzyme and ribonuclease formed an enzyme-substrate complex which gave only a single peak, moving at a rate distinct from that of either component.

The value of a method such as this lies largely in its range of applicability. Insufficient proteins have so far been examined to judge of this, but it is already clear that, although it should be possible to fractionate many proteins by partition chromatography, there are others whose susceptibility to surface denaturation will be too great, and again others will be too easily thrown out of solution by salts and organic solvents. An important defect of the method is the difficulty of stating definite conditions by which to judge whether any particular protein can be handled by this technique.

[13] Chromatography of Enzymes on Ion Exchange Resins

By C. H. W. Hirs

Chromatographic Method

Scope. Mixtures of amino acids and peptides have been fractionated effectively by elution analysis on columns of ion exchange resins.[1] Successful chromatography of enzymatically active proteins by means of the same type of procedure has been limited to stable, basic substances of low molecular weight such as cytochrome c,[2] lysozyme from the white of hens' eggs,[3] two low molecular weight proteins with lysozyme activity present in extracts of rabbit spleen,[4] and ribonuclease[5] and chymotrypsinogen[6] from bovine pancreas. In each case, the carboxylic acid cation exchange resin IRC-50 (XE-64) has been used.

A description of the general procedure will be given, together with detailed directions for the application of the method to the chromatography of ribonuclease, chymotrypsinogen, and lysozyme.

General Procedure

Preparation of Resin. Amberlite IRC-50[7] has thus far been found suitable for chromatographic work with proteins only when in the form of a powder obtained by crushing the resin. The pulverized material is available commercially under the designation Amberlite XE-64, a preparation which contains particles ranging in size from approximately 50 to 500 mesh. Before use, the commercial material must be further purified. To each 1.5 kg. of the resin[8] 3.5 l of water is added, and the mixture is stirred mechanically for 20 minutes. After a settling period of 2 hours, the supernatant suspension and some foam are removed. The settling process is repeated four or five times with 2-l. portions of water until the supernatant liquid is clear after about 1 hour of settling. With the finest particles removed in this manner, it becomes possible to filter off the resin on

[1] S. Moore and W. H. Stein, *Ann. Rev. Biochem.* **21**, 521 (1952).

[2] S. Paleus and J. B. Neilands, *Acta Chem. Scand.* **4**, 1024 (1950).

[3] H. H. Tallan and W. H. Stein, *J. Biol. Chem.* **200**, 507 (1953).

[4] G. Jollès and C. Fromageot, *Biochim. et Biophys. Acta* **11**, 95 (1953).

[5] C. H. W. Hirs, S. Moore, and W. H. Stein, *J. Biol. Chem.* **200**, 493 (1953).

[6] C. H. W. Hirs, *J. Biol. Chem.* **205**, 93 (1953).

[7] Some of the most important characteristics of Amberlite IRC-50 have been summarized by R. Kunin and R. J. Myers in 'Ion Exchange Resins,'' John Wiley and Sons, New York, 1950.

[8] The powdered resin should be handled with care when it is in a dry state. Precautions should be taken to avoid inhaling the dust that is inevitably swirled up during manipulations involving fine powders.

a Büchner funnel and to dry it by drawing air through it. The dry resin is stirred in 4 l. of acetone for 3 hours, the suspension filtered, and the resin washed with acetone (about 8 l.) until the filtrate is clear and colorless. The air-dried resin is suspended in water, stirred until all bubbles are eliminated, and the last traces of acetone removed by washing copiously with water (24 l.) on a Büchner funnel.

The resin, suspended in 5 l. of water, is converted to the sodium salt by the addition, over a period of 30 minutes, of 560 g. of NaOH (in the form of a 40% aqueous solution). The pH of the suspension should rise at least to 11. Stirring is continued until the evolution of heat has subsided (3 hours), and the sodium salt of the resin is then washed by decantation with five 2-l. portions of water. A 2-hour settling period is allowed each time. The XE-64 is transferred to a filter and washed with water (about 12 l.) until the filtrate has a pH of about 10. Resin prepared in this way may be employed for work with crude tissue extracts, as illustrated in the following experiments with ribonuclease and lysozyme. A column of finer particles might become clogged at the top.

For the resolving power desirable in analytical work, such as that involving the determination of the chromatographic homogeneity of an enzyme preparation, the resin should be screened through a 200-mesh sieve. From 5 to 25% of the product obtained by the procedure described above can be driven through a 200-mesh screen (74-μ openings) by a strong jet of water. After the passage of each few liters of water, the screen must be inverted and brushed with a stiff brush under a stream of water to free the pores before screening the next 100 g. or so of resin. The through-200-mesh resin is filtered off and should be a fairly uniform product which settles cleanly from water in 2 hours. The resin thus prepared can be used over and over again, indefinitely, with appropriate washing with HCl and NaOH between uses.

The XE-64 must next be equilibrated with the buffer to be employed in the subsequent chromatography. The volume of the resin per gram varies with its degree of neutralization, and therefore less resin is required to produce a column of given dimensions at pH 7 than will be required for an identical column at pH 6. As a typical example, the process will be illustrated for a column at pH 6.47 (0.2 M phosphate buffer), suitable for preparative work with ribonuclease. The buffer used for this purpose is prepared by dissolving 165.6 g. of $NaH_2PO_4 \cdot H_2O$ and 113.6 g. of Na_2HPO_4 (anhydrous) in water and diluting the solution to a volume of 10 l. If all the resin obtained from the original 1.5 kg. of powder is used in the following process, sufficient material will be produced for two columns 7.5 cm. in diameter and 60 cm. high. For analytical experiments the equilibration is done on a smaller scale.

The sodium salt of the XE-64 is suspended in water, transferred to a Büchner funnel, and converted to the acid form by passing 10 l. of 3 N HCl through the filter over a 4-hour period. The product is washed with 6 l. of water, or until the filtrate is neutral. The acid form of the resin is stirred in 5 l. of the buffer at pH 6.47. The pH of the suspension falls immediately, and approximately 30-ml. portions of 40% NaOH are added every few minutes to maintain the suspension at about pH 5.5. As neutralization of the carboxyl groups continues, the suspension is brought to pH 6.4 by additions of alkali at 15-minute intervals. When the pH remains at about 6.55 after 15 minutes of stirring, the suspension is stirred overnight. By this time equilibrium will have been reached, and the pH will have fallen to between 6.45 and 6.50. The resin is now transferred to a Büchner funnel and washed with 30 l. of buffer over a period of 8 hours. The pH of the final filtrate should be the same as that of the inflowing buffer.

Preparation of Columns. Analytical scale columns are conveniently prepared in chromatograph tubes 0.9 cm. in diameter, equipped with a sintered glass plate. The wet resin, equilibrated with the appropriate buffer, is suspended in an equal volume of buffer. The suspension is introduced into the tube in portions, designed to give 10-cm. segments of settled resin, and each addition is permitted to settle completely before the next is made.[9] Upon completion of a column of the desired length, a reservoir of the buffer is attached to the tube, and approximately 25 to 30 holdup volumes of buffer are permitted to pass through the column. At pH 6.47, for instance, the holdup volume is approximately 0.31 ml./ cm.[3] of resin bed. This volume does not change much with the pH. The pH of the effluent should always be checked before the column is finally used for chromatography. When employed for analytical purposes with purified enzyme preparations, such columns may be used indefinitely. Between runs it is customary to wash the columns with about 20 holdup volumes of buffer. If the column is not required immediately, it should be closed off at both the top and the bottom in order to prevent drying out.

For work on a preparative scale, tubes up to 7.5 cm. in diameter have been used without any loss in resolving power. On this scale, the pouring procedure requires slight modification. As each portion of resin suspension is introduced, flow from the delivery tube of the column is stopped (pinchclamp). The added suspension is stirred up carefully in the tube and allowed to come to rest before flow through the column is permitted to start once more. In this way, each segment of the column settles with a

---

[9] It is best to have approximately 7 to 10 cm. of liquid above the resin surface at the time the next addition is made to the column. If less is present, the existing surface may be disturbed and an uneven column will then result.

perfectly horizontal surface. The final (upper) surface of the column is protected with a circle of filter paper of almost the same diameter as the tube, and the column is washed with 25 to 30 holdup volumes of buffer.

Operation of Columns. The effluent from an XE-64 column has a tendency not to wet glass and, therefore, to give irregular flow from the tips of narrow chromatograph tubes. The difficulty may be avoided by incorporating a detergent in the eluting buffers. The nonionic detergent BRIJ 35 (Atlas Powder Co.) has given excellent results with chromatograms of chymotrypsinogen preparations, but its use with other enzymes may be undesirable. When the use of the detergent is objectionable, regular flow may be obtained by inserting a roll of silver gauze into the tip of the chromatograph tubes. The upper end of the gauze should be in contact with the sintered plate, and the lower end compressed to a point projecting about 1 cm. below the glass tip. With preparative scale columns this modification is not necessary.

Samples to be added to the columns should preferably be of minimum volume and of a pH adjusted to be slightly lower than the pH of the column. Thus, for analytical work on a column 0.9 cm. in diameter, a load of 5 to 10 mg. of enzyme dissolved in 1 ml. of solution is a very satisfactory scale of operation, although very much smaller quantities may be effectively analyzed when the enzymatic activity determination to be used on the effluent fractions is sensitive enough. When the quantity applied to a column 0.9 cm. in diameter exceeds about 20 mg. of enzyme, the peaks on the effluent concentration curve start to spread, indicating that overloading is taking place. The load may be increased in proportion to the area of the column surface when experiments on a larger scale are performed.

The samples are introduced without disturbance of the column surface by means of pipets with bent tips. The rate of sample addition should preferably be the same as, or very little greater than, the rate at which elution of the column is to proceed thereafter. The transfer of sample is made quantitative by three consecutive washes with buffer. Thereupon a layer of approximately 5 cm. of buffer is introduced above the column, and a solvent reservoir is attached. The most convenient procedure for analytical work is to use a 300-ml. separatory funnel equipped with a constant head air-inlet tube for the reservoir. The funnel is attached by a length of Tygon tubing to the column and may therefore be used as a leveling bulb to control the rate of flow in the column. In the operation of larger columns, it is more convenient to store the buffer in a reservoir bottle at bench level and to pump the eluent to the column surface by means of suitably controlled constant air pressure.[10]

[10] A mechanical regulator, such as the Model 40-15 Nullmatic regulator produced by the Moore Products Co., Philadelphia, provides extremely constant pressure in

The column is mounted on a fraction collector, and elution allowed to proceed at a rate of 2.5 to 3.0 ml./cm.²/hr. in a room at constant or near constant temperature. When the flow rate is increased beyond about 4 ml./cm.²/hr., the zones begin to broaden, and resolving power is reduced. The incorporation of 0.5% BRIJ 35 into the eluent (cf. above) permits the rate of solvent flow to be doubled without sacrifice in the resolving power of the columns.

The effluent is collected in fractions of 0.5 to 1.0 ml. when columns 0.9 cm. in diameter are used, i.e., of about $V/15$, where V is the holdup volume of the column. In preparative scale experiments, the fraction size may be limited by the size of the collecting receivers, in which event it is usually necessary to analyze only every second or third tube.

In contrast to experiments involving purified preparations, when tissue extracts are analyzed it is possible to use a column for only about four chromatograms before the surface becomes clogged. When this occurs, the upper portion of the column may be stirred up and the suspended resin replaced by fresh, equilibrated resin. Contaminated resin may be purified by converting it to the acid form and warming it on a steam bath for several hours with 4 N HCl, followed by a wash with 1 N NaOH.

Analysis of Effluent Fractions. A measure of the protein concentration in the effluent may be obtained spectrophotometrically, as in the chromatography of cytochrome c.[2] The blue color obtained in the reaction with ninhydrin has been employed with chromatograms of ribonuclease,[5] lysozyme,[3] and chymotrypsinogen.[6] A suitable aliquot, usually not more than 0.2 ml., is withdrawn from the fractions with a micropipet and mixed with 1 ml. of ninhydrin reagent prepared with hydrindantin.[11] After heating for 15 minutes in a water bath at 100°, the solution is diluted with aqueous propanol, cooled, and the optical density determined at 570 mμ. The blank value under these conditions usually lies between 0.10 and 0.15 optical density unit. On the average, a given protein yields approximately 10% as much color on a weight basis as does leucine under identical conditions.

---

conjunction with a reducing valve and filter assembly; see S. Moore and W. H. Stein, *J. Biol. Chem.* **192**, 663 (1951).

[11] The ninhydrin reagent is the same as that employed in the method of Moore and Stein for the colorimetric determination of amino acids, except that hydrindantin (0.5 g./l.) is added to the reagent in place of stannous chloride. The presence of the tin in the reagent normally employed causes the formation of a turbidity when samples containing phosphate ion are analyzed. With the reagent containing preformed hydrindantin, this difficulty is not encountered and completely clear solutions are obtained. See Vol. III [76]; see also S. Moore and W. H. Stein, *J. Biol. Chem.* **176**, 367 (1948).

Enzymatic assay may also be used to analyze the effluent fractions. Provided that the procedures employed are sufficiently sensitive and precise, valuable information concerning the nature and homogeneity of a given peak on an effluent concentration curve may thus be obtained.

Recovery of Enzymes from the Effluent Fractions. The solutions obtained by combining the fractions corresponding to a peak on the effluent concentration curve may be worked up in several ways. The simplest procedure for obtaining the desired material in a salt-free condition is to remove the buffering salts by dialysis, as exemplified in the experiments with lysozyme. Tallan and Stein[3] found that an essentially salt-free preparation of the enzyme could be obtained in a yield of approximately 70% by means of two successive dialyses in collodion sacs for 20 hours at 4° against distilled water.

A considerable concentration of a protein solution may be attained by adsorbing the protein on IRC-50 at a pH of 5.4. Under these conditions, the resin is present in large part as the free acid in all but the strongest salt solutions (cf. ref. 7). Accordingly, by adjusting the pH of the solution obtained after chromatography to approximately 5.4, and then passing it through a small filter of the resin previously equilibrated to the same pH, the enzyme is quantitatively adsorbed onto the resin. (Cf. experiments of Boardman and Partridge[12] on the chromatography of hemoglobins on IRC-50.) Thereafter it may be completely eluted from the filter into a much smaller volume of buffer by gradually increasing the pH of the resin in suspension to a point where the protein is no longer held appreciably. The process has been worked out quantitatively for ribonuclease and will be described later.

Solutions of ribonuclease have been completely desalted by simultaneous treatment with the coarse bead forms of the resins Amberlite IRA-400 and Amberlite IR-120, mixed in proportion to their equivalent capacities to bind anions and cations, respectively. This approach, which utilizes what has been called the "molecular sieve" principle by Partridge,[13] unfortunately results in the loss of as much as 40% of the enzyme when it is applied to solutions of ribonuclease in phosphate buffer. Apparently the surface capacity of the IR-120 beads is sufficiently great to account for such losses. Another and more serious difficulty that has been encountered in this method is that it causes transformation of lysozyme when attempts are made to obtain salt-free preparations of the enzyme. Thus Tallan and Stein[3] found that a chromatographically homogeneous sample of lysozyme after treatment with mixed resins to remove buffer

[12] N. K. Boardman and S. M. Partridge, *Nature* **171**, 208 (1953).
[13] S. M. Partridge, *Nature* **169**, 496 (1952).

salts contained up to 18% of material moving faster than lysozyme on the XE-64 chromatogram.

Ribonuclease A

Direct Isolation from Pancreas—Analytical Methods. The effluent from the chromatogram is surveyed both by ninhydrin analyses and by determinations for ribonuclease activity. Ribonuclease activity is determined by the spectrophotometric procedure of Kunitz[14] on 0.3-ml. aliquots of the effluent. For the ninhydrin analyses, 0.2-ml. aliquots are taken.

Starting Material. Comminuted, lyophilized beef pancreas stored at −15° maintains its ribonuclease activity undiminished for over a year and is suitable for the preparation of the enzyme. The powder should be made from pancreas secured as soon as possible after the death of the animal.

Preparation of Extract. 20 g. of the pancreas powder is stirred at 4° with 150 ml. of 0.25 N sulfuric acid for 15 minutes. Mucous cloths and fatty material adhering to the stirrer are discarded, and the suspension is centrifuged for 10 minutes at low speed. The middle layer is withdrawn without disturbing the upper layer of mucoid and fat droplets. The extract is filtered through soft paper (Schleicher and Schuell, No. 588). By careful addition of 1 N NaOH the pH of the filtrate is brought to between 5.5 and 6.0, and the solution is permitted to stand for 10 minutes. The flocculent precipitate that forms is removed by centrifugation, and the extract is used for chromatography immediately.

Chromatography. Before an attempt is made to isolate the enzyme from the extract on a preparative scale, an analytical column should be run in order to make sure that the sample of pancreas which has been taken was fresh, as well as to check on the adequacy of the simple extraction procedure. For this purpose, 2 ml. of the extract is applied to a column of XE-64, 0.9 cm. in diameter and 30 cm. high, which has been equilibrated with 0.2 M sodium phosphate buffer at pH 6.47 (as described previously). After rinsing the sample in with buffer, buffer is passed through the column at a rate of 1.5 to 2.0 ml./hr. Fractions of 0.5-ml. volume are collected and analyzed. An elution curve of the type shown on the left of Fig. 1 will be obtained. If the extract is from fresh pancreas, the separation of the A and B components will be clean, and the activity plot will come to zero between the two peaks. The presence of active material with an

---

[14] See Vol. II [62]. The original procedure is modified in that a buffer of 0.1 M concentration is employed at pH 5. For analyses of fractions corresponding to a peak on the effluent curve, dilutions are made with this buffer to give an enzyme concentration optimal for the spectrophotometric method.

intermediate rate of travel indicates that the extract is unsatisfactory for preparative work.

When a satisfactory extract is obtained upon extraction of the pancreas powder, the entire quantity indicated above, freshly prepared, is applied to a column at pH 6.47 which is 7.5 cm. in diameter and 60 cm. long. It is rinsed in with 20-ml. portions of buffer, and the column is eluted at a rate of 120 ml./hr. Fractions of 50 ml. are collected and analyzed. The elution curve obtained in such a run will be similar to that

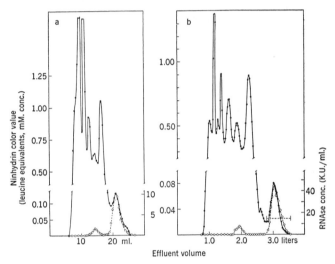

Fig. 1. Chromatography of 0.25 N sulfuric acid extracts of comminuted beef pancreas. Elution was performed with 0.2 M sodium phosphate buffer at pH 6.47 from a column of Amberlite IRC-50 (XE-64) equilibrated with the same buffer. a, curve obtained with a 0.9 × 30 cm. column used to test the extract. b, curve obtained with an extract prepared from the same sample of pancreas on a preparative scale column which was 7.5 × 60 cm. The closed circles indicate the ninhydrin color value; the open circles, ribonuclease activity in Kunitz units. Load: a, 0.27 gm.; b, 20.0 gm. (Reproduced by courtesy of the *Journal of Biological Chemistry*.)

shown on the right of Fig. 1. The fractions corresponding to ribonuclease A are combined, and the solution rendered salt-free in the manner described below.

Concentration. The solution of ribonuclease A in phosphate buffer is adjusted to pH 5.35 to 5.40 by addition of phosphoric acid. The solution is passed at a rate of 1 ml./min. through a short column of XE-64, 2 cm. in diameter, which has previously been equilibrated with sodium phosphate buffer at pH 5.38 (0.2 M in phosphate, 0.21 N in sodium ion). The capacity of the resin for ribonuclease under these conditions is about 30 mg./ml. of resin bed. The volume of resin taken for the column may be deduced from the analytical data given by the preparative column curve (cf. Fig. 1), allowing for a 10% excess of resin over the amount calculated. The enzyme will be completely removed from the solution. For elution, the resin is suspended in an equal volume of phosphate buffer at pH 7.96 (0.2 M in phosphate, 0.38 N in sodium ion). The suspension is stirred gently and brought to pH 8 by careful addition of 5 N NaOH, added in small droplets over a period of 30 minutes. The suspension is filtered, and the resin is washed with 1 vol. of buffer. The elution is made practically quantitative by resuspending the resin in an equal volume of the buffer at pH 8 for 30 minutes and repeating the filtering and washing procedure. The combined filtrates represent a solution of ribonuclease A concentrated 10- to 15-fold with respect to the buffer salts.

Desalting. Amberlite IR-120 (acid form) and Amberlite IRA-400 (free base) in the form of coarse beads (analytical grade), and of known titer, are mixed in equivalent quantities, and the mixture is suspended in water. For each milliliter of 0.2 M sodium phosphate buffer, 0.8 ml. of the settled resin mixture is required. The mixed resins are stirred gently for 20 to 30 minutes with the enzyme solution. After this time the benzidine-ammonium molybdate test (Feigl[15]) for phosphate ion should be negative. The supernatant is withdrawn, and the mixed resins are washed free of protein solution by decanting ten times with equal volumes of water. The last traces of salt are removed by passing the combined supernatant fluids through a column of the mixed resins which has a bed volume of 0.15 ml./ml. of the buffer originally taken. The column is rinsed with 10 vol. of water, after which the final solution is taken to dryness in a high vacuum from the frozen state. The yield is 20 mg. of ribonuclease A, having a nitrogen content of 17.8% on a moisture-free basis. The preparation is essentially ash-free and may be crystallized from aqueous ethanol. The activity of the enzyme is approximately 110 units (as defined by Kunitz[14]) per milligram. On chromatography at pH 6.47 it will give a single, symmetrical peak on the effluent curve.

From Crystalline Ribonuclease Prepared from Beef Pancreas by the Method of Kunitz. The most convenient procedure for the preparation of ribonuclease A is to chromatograph a crystalline preparation obtained from pancreas by the method of Kunitz and McDonald.[16] Such preparations contain material moving ahead of ribonuclease A on XE-64 chromatograms. In the best preparations this may amount to some 15% of the total material, whereas in poorer preparations of the crystalline enzyme the content of fast-moving protein may amount to as much as 40 to 50%. The procedure to be followed is identical to that which has just been de-

---

[15] F. Feigl, "Laboratory Manual of Spot Tests." Academic Press, New York, 1943.
[16] See Vol. II [62]; see also M. Kunitz, *Biochem. Preparations* **3,** 9 (1953).

scribed. However, in the absence of the considerable quantity of tissue contaminants present in crude extracts, a much larger load may be applied to the column. Thus, when 750 mg. of a good preparation of the crystalline enzyme is used, ribonuclease A may be secured from it in a yield of 400 mg. The preparation obtained in this manner is indistinguishable from that secured by the "direct" method.

Chymotrypsinogen α

Examination of a number of crystalline preparations of this zymogen on analytical scale chromatograms with the resin XE-64 at pH 6.02 has demonstrated that such preparations are not always homogeneous.[6] The columns may therefore be used to advantage to determine the homogeneity of chymotrypsinogen α obtained by isolation from pancreas.

Analytical Methods. It is not possible to attain complete activation of chymotrypsinogen α by treatment with trypsin. Enzymatic assays are therefore carried out after an arbitrary time of activation with trypsin. Aliquots of 0.05 to 0.10 ml. are pipetted from the appropriate effluent fractions and diluted to a volume of 2.5 ml. by addition of a freshly prepared solution of trypsin (0.002% in 0.10 M sodium phosphate buffer at pH 7.6). The solutions are placed in a bath at 35° for 3 hours, and the proteolytic activity released is determined by the spectrophotometric procedure of Kunitz[17] with Hammarsten's casein as substrate. The increase in optical density at 280 mμ is related to the chymotrypsinogen α concentration by means of a standard curve, prepared under the same conditions of activation and hydrolysis, and based on a sample of the zymogen which has been demonstrated to be homogeneous. A "zero time" tube is made up for each solution that is analyzed, and each assay tube is read against its own "zero time" tube as a blank. Ninhydrin determinations are carried out as mentioned previously.

Chromatography. The XE-64 resin is equilibrated at pH 6.02 with a 0.2 M phosphate buffer which is 0.24 N in sodium ion. In other respects the experimental procedure is the same as that described for ribonuclease. In Fig. 2 are shown two chromatograms of crystalline chymotrypsinogen α preparations. The first curve is characteristic of the result seen when a satisfactory preparation is analyzed, whereas the second curve, obtained from a sample ostensibly prepared by the same method as was used to obtain the first, demonstrates that there are aspects in the isolation procedure that are as yet not adequately controlled.

When a crude extract of pancreas, as prepared for the isolation of ribonuclease A, is to be chromatographed, a preliminary dialysis at 4° of

---

[17] See Vol. II [2]; see also J. H. Northrop, M. Kunitz, and R. M. Herriott, "Crystalline Enzymes," 2nd ed. Columbia University Press, New York, 1948.

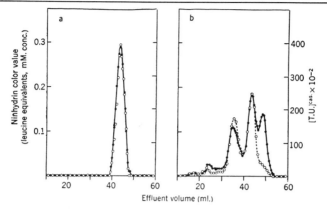

FIG. 2. Chromatography of preparations of crystalline chymotrypsinogen α on a 0.9 × 56-cm. column of IRC-50 (XE-64), with sodium phosphate buffer at pH 6.02 as eluent. *a* and *b*, curves obtained from two commercial preparations of the zymogen. The closed circles indicate the ninhydrin color value; the open circles, proteolytic activity measured after activation with trypsin. (Reproduced by courtesy of the *Journal of Biological Chemistry*.)

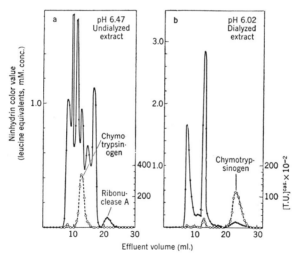

FIG. 3. Chromatography of 0.25 N sulfuric acid extracts of comminuted beef pancreas on a 0.9 × 30-cm. column of IRC-50 (XE-64). Elution was with 0.2 M sodium phosphate buffer in each instance. *a*, curve obtained at pH 6.47 with an undialyzed extract. *b*, curve obtained after dialysis of the extract and chromatography at pH 6.02. The ribonuclease A peak, which appears at 60 effluent ml. under these conditions, is not shown. The closed circles indicate the ninhydrin color value; the open circles, proteolytic activity measured after activation with trypsin. (Reproduced by courtesy of the *Journal of Biological Chemistry*.)

the extract against 0.2 N sulfuric acid is effective in removing impurities which prevent a complete resolution of the chymotrypsinogen α peak on the chromatograms. The relation of the positions of the peaks for chymotrypsinogen α and ribonuclease A and B will be apparent from a comparison of the chromatograms shown in the first parts of Figs. 1 and 3. The second half of Fig. 3 illustrates the result obtained when a dialyzed extract of the tissue is analyzed on XE-64 at pH 6.02. A chromatogram of this type may be used to assay the chymotrypsinogen α content of the extract. Trypsinogen is strongly held on the resin and does not appear in the effluent under these conditions.

Egg-White Lysozyme

Lysozyme may be obtained from egg white as the isoelectrically precipitated protein, as the chloride, and as the carbonate.[18] These preparations may be chromatographed on XE-64 at pH 7.18 in 0.2 M sodium phosphate buffer. With this technique Tallan and Stein[3] have demonstrated that the isolectric protein is the most nearly homogeneous preparation of the enzyme that may be obtained. A chromatographically homogeneous enzyme may be prepared from the isoelectric protein with the aid of a preparative column.

Analytical Methods. Lysozyme activity is determined by a modification of the method of Smolelis and Hartsell.[19] A suspension of the dried cells of *Micrococcus lysodeikticus* is prepared in 1/15 M phosphate buffer (Sørensen) of pH 6.24 such that, on a Coleman Junior spectrophotometer, the percentage transmittance of a portion of the suspension diluted with an equal volume of buffer has the arbitrary value of 25.0. Readings are taken at a wavelength of 570 mμ. An aliquot of the effluent fractions (generally less than 0.05 ml.) containing about 10 γ of enzyme is diluted to 3 ml. with the buffer of pH 6.24, followed by addition of 3 ml. of cell suspension. The mixture is kept at 25° and shaken at frequent intervals for 40 minutes. The transmittance is read, and the amount of enzyme present determined from a standard curve relating transmittance to enzyme concentration.[20] The curve is obtained from a sample of enzyme shown to be chromatographically homogeneous.

Ninhydrin analyses are performed in the manner previously mentioned under general procedures.

[18] H. L. Fevold and G. Alderton, *Biochem. Preparations* **1**, 67 (1949).
[19] A. N. Smolelis and S. E. Hartsell, *J. Bacteriol.* **58**, 731 (1949).
[20] Dilutions are made with the same buffer at pH 6.24 for those tubes corresponding to the peak on the curve. The assay is approximately linear below 10 γ of enzyme, and the dilutions should therefore be made accordingly.

Chromatography. Chromatograms are run on XE-64 equilibrated with 0.2 M sodium phosphate buffer at pH 7.18. The enzyme may be cleanly separated from other proteins in egg white by chromatography of the solution obtained by diluting the white of eggs with 4 vol. of 0.2 M phosphate buffer of the same pH as the column. For preparative purposes, it is most convenient to employ as a starting material lysozyme isolated from egg white by isoelectric precipitation. Tallan and Stein chromatographed 280 mg. of such a preparation on a column 4 cm. in diameter and

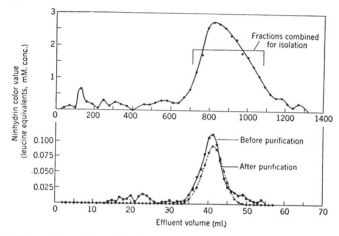

FIG. 4. Chromatographic purification of isoelectric lysozyme on columns of IRC-50 (XE-64) with 0.2 M sodium phosphate buffer at pH 7.18 as eluent. The enzymatic activity per unit of ninhydrin color of the major peak is the same before and after fractionation. *Top:* preparative scale column: 4 × 30 cm., load: 280 mg. lysozyme; *bottom:* analytical scale column: 0.9 × 30 cm., load: 5 mg. lysozyme. (Reproduced by courtesy of the *Journal of Biological Chemistry.*)

30 cm. high and obtained 114 mg. of a salt-free protein which was chromatographically homogeneous. The elution curve from an experiment of this kind is shown in Fig. 4, which also shows the corresponding curve obtained on an analytical scale column. For isolation, the fractions indicated in Fig. 4 were combined and freed of salts by dialysis in collodion sacs, followed by a second dialysis through fresh sacs. The final solution was lyophilized to obtain a powder which contained 0.35% ash and which possessed a nitrogen content of 18.7%.

[14] Special Techniques for Bacterial Enzymes. Enrichment Culture and Adaptive Enzymes

By OSAMU HAYAISHI

Introduction

Microorganisms in general, and bacteria in particular, offer numerous advantages over almost any kind of biological material for the study of comparative biochemical problems. Thus, the number of studies of microbial enzymes has increased to a large degree and has resulted in great and important contributions to a better understanding of a number of biochemical reactions involved in mammalian metabolism. In this way the concept of "unity in biochemistry" has been demonstrated time and again. But, aside from revelations concerning mammalian metabolism, these studies are important with regard to the microorganisms themselves because a great many chemical reactions which take place in nature are carried out by their action or, more precisely, by their enzymes. The general properties of microbial enzymes and the essential principles of metabolism of microorganisms conform to the same law as do those of animals and plants. But microbes appear to be much more versatile, to such an extent that one is able to find organisms capable of performing almost any type of biochemical reaction, even of a sort not known to take place or to be significantly active in animal and plant metabolism.

The first, and perhaps the most important, problem confronting an enzymologist when he starts to purify an enzyme is the choice of starting material. Usually he makes a survey of a number of different materials, such as various organs and tissues of animals and plants, or, as the case may be, microorganisms. If he chooses the latter, he will find that his preparation will generally be less expensive than animal tissues and also more uniform and reproducible in quality. In order to detect and isolate a suitable microorganism, the methodological principle of "enrichment culture" developed by Winogradsky in 1890, and later elaborated by Beijerinck, is commonly used. For a summary of Beijerinck's extensive work, see Stockhausen.[1] By this method the desired organism can be found and, when grown under appropriate conditions, can be generally relied upon to produce uniform and potent enzymatic activities.

Enrichment Culture

Principle. The basic principle of the enrichment culture technique can best be expressed by the well-known remark of Baas Becking: "Everything is everywhere; the environment selects." In order to select a desired

[1] F. Stockhausen, Ökologie, "Anhäufungen" nach Beijerinck, Berlin, 1907.

type of microorganism, a sample of soil, animal feces, or sometimes river water, containing a mixture of various kinds of microorganisms, is obtained and incubated with a specific substrate under certain physicochemical conditions (pH, oxygen tension, temperature, etc.). In this way one obtains only those organisms capable of metabolizing the particular substrate under the particular conditions in which they will grow and multiply. However, the inoculum usually contains an appreciable amount of other nutrient material, and other microorganisms present may grow to an appreciable extent in the first incubation mixture. In order to eliminate the effect of unknown material in the inoculum, the culture is transferred to another batch of the same medium as soon as visible growth appears. After several successive transfers, incubated under the same conditions, one type of organism will develop progressively and eventually become the predominant population. The principle of enrichment culture is, therefore, based upon a natural selection of a species or a mutant which can best fit the given environment and has nothing to do with the adaptive enzymes which will be discussed later. However, the organisms obtained by the enrichment culture technique are often of adaptive nature, and a number of their enzymes are adaptive enzymes, probably because these microorganisms are of the free-living type and must possess adaptability toward ever-changing ecological factors in order to survive and multiply.

Aerobic Enrichment. A simple and most common procedure for aerobic enrichment is illustrated by the following example, in which an organism which decomposes uracil and thymine was isolated.[2] Samples of soil were collected from several different sources including swamp mud, fertilized soil, sewage, humus, etc. In each case 1 part by weight of soil sample was suspended in 9 parts by volume of 0.8% NaCl solution. 1 ml. of this soil suspension was mixed with 9 ml. of the following medium in test tubes or small flasks and incubated at 26°.

Substrate:	Uracil or thymine	0.1 %
Inorganic salt mixture:	K_2HPO_4	0.15 %
	KH_2PO_4	0.05 %
	$MgSO_4 \cdot 7H_2O$	0.02 %

(Dissolved in a 1:1 mixture of distilled water and tap water)

After 2 days, growth of microorganisms was indicated by the appearance of turbidity, and microscopic examination of the culture showed proliferation of several kinds of bacteria. 0.5 ml. of the incubation mixture was immediately transferred to 9.5 ml. of fresh medium of the same composition. Several successive transfers were made in this way at daily intervals, and finally the culture was plated on a medium of the same composition,

---

[2] O. Hayaishi and A. Kornberg, *J. Biol. Chem.* **197**, 717 (1952).

but containing 2% agar in addition. Several colonies which appeared to be slightly different were isolated and plated on the solid medium, and the cells were examined under the microscope. If the appearance of colonies on the second plates was uniform and the microscopic examination established homogeneity of the flora, the strain was considered to be pure and was ready for large-scale cultivation.

The method described above is probably the most convenient for isolating an organism which utilizes and grows on a given substrate. However, it must be remembered that an organism which grows fastest under given conditions may not be the best source material for the desired enzymes. Moreover, there may be some other organisms which might slowly utilize the same compound by a different pathway and which might not be picked up by this method. For example, when DL-tryptophan was used as a substrate, only those strains of *Pseudomonas* which utilize both the D- and L-isomers of tryptophan were isolated by this procedure, because this type of organism (a quinoline-pathway organism)[3] overgrows the other type of *Pseudomonas* (an aromatic-pathway organism), which can utilize the L-isomer but not the D-isomer. These two different strains differ not only in their ability to utilize D-tryptophan but also in their entire metabolic pathway below kynurenine.

It is also known that in a mixed culture the presence of certain microorganisms can suppress the growth of other organisms (microbial antagonism). It is, therefore, not advisable to make many transfers in liquid media since, as successive transfers are made and the culture becomes more and more pure, there is less opportunity to pick up more than one kind of organism, even though there may be others which can utilize the particular substrate also, but perhaps through different pathways.

To surmount this difficulty, direct plating of an original 10% soil suspension is sometimes preferred. By direct plating only, one can detect the slowly growing organisms which may provide more interesting and useful information than the faster growing ones. For example, during a survey for a tryptophan-utilizing soil organism, several slow-growing strains which excreted some fluorescent pigment around the colonies were discovered by this method. Upon isolation and further investigation of these strains, it was revealed that they were natural mutants which had a defective ability of oxidizing anthranilic acid and that the fluorescent pigment excreted from the cells was in fact anthranilic acid.[4,5] The case is analogous to that of rare human specimens characterized by inborn errors of metabolism, which have provided useful tools for the study of intermediate

[3] R. Y. Stanier and O. Hayaishi, *Science* **114**, 326 (1951).

[4] O. Hayaishi and R. Y. Stanier, *Bacteriol. Proc.*, p. 131 (1951).

[5] R. Y. Stanier and O. Hayaishi, *J. Bacteriol.* **62**, 367 (1951).

metabolism.[6] The isolation of such strains could not have been done unless the original suspension was directly plated on the solid medium, because their growth is much slower than normal strains. The mutants apparently derive their energy mainly from the alanine moiety of kynurenine, and anthranilic acid moiety is only slowly metabolized or excreted into the medium without being utilized at all.

Another disadvantage of using liquid culture is that, as the organisms grow, the composition of the entire medium changes. Consequently, a number of organisms which cannot utilize the primary substrate but can grow on the degradation products thereof will start to grow after a while (symbiosis), unless subculture is made soon enough to prevent the proliferation of such secondary and tertiary flora. This situation is sometimes observed on plates as a "satellite" phenomenon, where a colony of one type of organism is surrounded by a number of small colonies of a different species which depend upon nutritional material excreted by the former.

Organisms appearing on secondary enrichment, instead of always being undesirable, are sometimes useful. As will be discussed later, enrichment as well as adaptation is not necessarily confined to a primary substrate but also can be observed with the secondary and tertiary products, which were derived from the primary substrate by the action of growing organisms. For example, in order to study metabolism of imidazolacetic acid, one can use histamine as a substrate in an enrichment medium, since the former is not commercially available whereas the latter is readily available and has been known to be oxidized to imidazoleacetic acid by mammalian tissues and bacterial preparations.

Composition of Enrichment Media. In the case of uracil and thymine, no carbon and nitrogen sources other than the substrate were added in the medium. If, however, the substrate served only as a carbon source, and since a nitrogen source has to be present in the medium in order to support the growth of microorganisms, inorganic nitrogenous compounds such as NH_4Cl or KNO_3 (the former being the more commonly used) are included at a concentration of 0.5 to 1.0%. Organic nitrogen sources such as amino acids and proteins are not commonly used. Occasionally only the carbon skeleton of the substrate molecule is utilized, even though nitrogen is present. For example, when a glycerophosphorylcholine (GPC)-utilizing organism was isolated from soil,[7] the addition of yeast extract was found to be indispensable. It was later found that the yeast extract was serving not only as a source of growth factors but also as a nitrogen source because the isolated organism was not able to metabolize the choline moiety of the

---

[6] A. E. Garrod, "Inborn Errors of Metabolism," 2nd ed., Oxford University Press, London, 1923.

[7] O. Hayaishi and A. Kornberg, *J. Biol. Chem.* **206,** 647 (1954).

GPC. The addition of extra carbon sources is not usually necessary; actually they may even be undesirable because the presence of fermentable sugars may alter the reaction of the medium and change the metabolic pathways or may even completely inhibit utilization of the particular substrate. For example, formation of most amino acid decarboxylases is greatly enhanced when the pH of the medium is below 6.0, and formation of deaminases is accelerated in an alkaline reaction. Therefore, an amino acid will be deaminated at high pH's and decarboxylated if a fermentable sugar is present in the medium and the pH of the medium is consequently brought down below 6.0.[8]

Although a number of free-living bacteria are capable of growing in a simple medium which contains only a single substrate and some inorganic salts, the presence of growth factors is sometimes required in addition to nitrogen and carbon sources. For this purpose yeast extract is commonly employed. Since a high concentration of yeast extract alone permits good growth of a number of bacteria, it is advisable to use less than a 0.1% concentration. In many instances, two series of media are prepared, one containing only substrate and a salt mixture, and the other containing 0.1% yeast extract in addition.

Minerals are usually supplied from the tap water, but addition of small amounts of Zn, Fe, and Ca appears to be helpful. For a detailed discussion on the mineral components of enrichment media, the reader is referred to a detailed treatment of this subject by Hutner et al.[9]

As has been indicated already, the metabolic pathways and the formation of enzymes in microbial cells are greatly influenced by the physicochemical factors of the environment. Although it is impossible to test all the possible conditions, it is useful to make several cultures and incubate at different temperatures, or alter the pH of the medium. Many free-living microbes seem to prefer approximately 30° or even slightly lower temperatures for growth. It is not infrequent that a soil organism will grow well at 26° and very poorly at 37°.

Inasmuch as agar has been most commonly used as a solidifying agent, gelatin is becoming obsolete for this purpose. On rare occasions it may interfere with isolation of certain organisms. As shown by Winogradsky in 1891, the presence of organic compounds may be inhibitory for certain organisms, especially some autotrophic organisms, and agar actually can be utilized by some microbes as a carbon source, although the rate appears

---

[8] E. F. Gale, "The Chemical Activities of Bacteria," 3rd ed., University Tutorial Press, London, 1952.

[9] S. H. Hutner, L. Provasoli, A. Schatz, and C. P. Haskins, *Proc. Am. Phil. Soc.* **94,** 152 (1950).

to be rather slow. In order to avoid the use of agar, Winogradsky devised a solid medium in which silicic acid, or so-called "silica jelly" was used as a solidifying agent.[10]

Anaerobic Enrichment. In contrast to aerobic microorganisms, anaerobes are not only incapable of growing but are actually killed in the presence of air. In essence, the principle of anaerobic culture is to eliminate oxygen from the medium and also to prevent diffusion of atmospheric oxygen into the medium. In order to achieve this, a small amount of reducing agent such as Na_2S (0.03%, w/v) is incorporated into the medium. A liquid medium is placed in a deep container, such as a measuring flask or a reagent bottle, almost up to the top, and sealed airtight with a rubber stopper or paraffin. When a solid medium is used, a deep culture tube method is usually preferred to the plating method, especially when the special apparatus needed for anaerobic conditions is not at hand. A deep culture tube (soft glass) of a suitable agar medium containing 0.03% Na_2S is heated until the agar is completely molten. The molten agar medium is kept in a water bath at 45°, and a sample of soil properly diluted is inoculated. After the contents are well mixed, the test tube is allowed to solidify in an upright position and is incubated at a suitable temperature. It is desirable to adjust the dilution of the seeding in order to obtain relatively few (ten to twenty) isolated colonies per tube. After growth becomes visible, the glass is cut at a short distance below the desired colony. The agar is deposited aseptically in a sterile Petri dish, and the colony is isolated by cutting the agar block with a sterile platinum needle.

It is often difficult to isolate anaerobic bacteria from an enrichment culture which also contains aerobic bacteria, since the latter might survive under anaerobic conditions. When anaerobic plating is done with the aid of anaerobic jars, some of the anaerobes tend to spread rapidly over the surface of the agar; therefore, the "pour plate" method is much preferred to surface streaked plates. Addition of a small amount of crystal violet (0.001%) to the medium inhibits aerobic spore formers which often interfere with isolation of anaerobic organisms. The plates should be placed in an anaerobic environment as soon as possible after the medium is inoculated.

For a general discussion on handling anaerobic bacteria, the reader is referred to the chapter on "The Study of Obligatory Anaerobic Bacteria" in the "Manual of Methods for Pure Culture Study of Bacteria."[11]

---

[10] S. Winogradsky, *Ann. inst. Pasteur* **5**, 92 (1891).

[11] The Committee on Bacteriological Technic of the Society of American Bacteriologists, "Manual of Methods for Pure Culture Study of Bacteria," Geneva, N. Y., 1946.

Enzymatic Adaptation and Other Factors Involved in the Formation of Microbial Enzymes

General Principle. The chemical constitution of media, as well as the optimal physicochemical conditions required for the best growth of bacteria, varies considerably among microorganisms. When they are grown as a source material for enzyme purification, it is particularly important to know the proper conditions for obtaining maximum yields of enzymes in a reproducible way. The yield of certain enzymes and the crop of cells, often expressed by weight or nitrogen, are two different things, each independent of the other. Among the many factors which influence the formation of bacterial enzymes, the presence of particular substrates in media has been known to be the most important. Some twenty years ago, Karström[12] found that a group of microbial enzymes is produced only when the cells are grown in the presence of a particular substrate, or an amount of enzymes is greatly increased when the cells are grown under such conditions, whereas the cell grown without addition of the substrate contains only a negligible amount of enzymes. The terms "adaptive enzymes" and "enzyme adaptation" were introduced by Karström and have been widely used since then. More recently "induced enzymes" and "enzymatic induction" have been preferred by some investigators.[13] The group of enzymes other than adaptive type is called the "constitutive enzymes," and their amount in the cell is not influenced by the supply of external substrates. Studies of adaptive enzymes have been conducted along two lines; although extensive efforts were made in order to study the basic mechanism of enzyme formation in microbial cells, this phenomenon has also been applied for tracing metabolic pathways in microorganisms and for furnishing excellent source material for enzymologists.

The adaptive response of microbial cells to a substrate is not confined to the primary substrate itself, because, as discussed in the previous chapter, the substrate undergoes transformation, owing to the metabolic activity of the cells. Thus, it was found independently by several workers[14-16] that a microorganism may respond to a single substrate by producing a series of enzymes involved in the whole series of reactions along the entire metabolic pathway. For example, when a substrate A is furnished to a bacterial cell, and this A substance may induce formation of an enzyme Ea, Ea will metabolize the A substance and produce a product, substance

12 H. Karström, *Ergeb. Enzymforsch.* **7**, 350 (1938).

13 M. Cohn, J. Monod, M. R. Pollock, S. Spiegelman, and R. Y. Stanier, *Nature* **172**, 1096 (1953).

14 R. Y. Stanier, *J. Bacteriol.* **54**, 339 (1947).

15 J. L. Karlsson and H. A. Barker, *J. Biol. Chem.* **175**, 913 (1948).

16 M. Suda, O. Hayashi, and Y. Oda, *J. Biochem. (Japan)* **37**, 355 (1950).

B. Substance B thus formed from A by the action of an adaptive enzyme Ea may then give rise to adaptive production of Eb, as if B were furnished to the cell from the outside. Therefore, if A is metabolized by way of B, C, D, etc., and these individual steps are catalyzed by a number of enzymes Ea, Eb, Ec, etc., and if these enzymes are of adaptive nature, there is no production of Ea, Eb, Ec, etc., unless the organism was exposed to one of these substrates. Now if the organism is grown in the presence of A, and the cells are found to possess not only Ea but also a series of enzymes Eb, Ec, Ed, etc., it can be presumed that A may be metabolized in this organism by way of B, C, D., etc. The method of analysis of metabolic pathways based upon this principle has been termed "simultaneous adaptation"[14] or "successive adaptation"[16] or, more recently, "sequential induction"[13] and has been found to be useful for the study of metabolism of aromatic substances, amino acids, purines, etc.[17]

It became apparent that the validity of the conclusions drawn from such experiments rested on two assumptions, namely: (1) the enzyme induction and enzymatic activity are rigidly specific, and (2) the cell membrane is freely permeable to all the compounds tested. These assumptions are not always correct, however. In the first case, certain enzymes are known to attack several different substrates closely resembling each other or having common structures or radicals. Moreover, enzymatic induction was shown to be evoked by unmetabolizable substrate analogs.[18] In the second case, a number of compounds have been shown not to be freely permeable to the bacterial cell membrane. The assumption of free permeability of microbial cells, which is an inherent danger with all the work conducted with intact cells, is more serious. Therefore, data obtained under these assumptions have to be interpreted with utmost care and, in any event, no longer should be considered to be ultimate proof. It is always desirable to provide more direct evidence such as is obtained with a purified enzyme preparation. It has to be emphasized, however, that, despite its drawbacks, the method of adaptive enzymes provides not only a convenient guide for tracing metabolic pathways but also an excellent source material for enzyme purification, especially when the substrate is not readily available or there is some difficulty in obtaining direct adaptation to the substrate. For example, *Pseudomonas kynureninase*[19] was found to be adaptive, and its specific activity in the crude extract was found to be 7.0 to 8.0 units/mg. of protein when the cells were grown in the presence of kynurenine, whereas that of the crude extract prepared from the cells grown on ordinary broth was only 0.05 to 0.08, which is

[17] R. Y. Stanier, *Ann. Rev. Microbiol.* **5**, 35 (1951).
[18] S. Spiegelman, M. Sussman, and D. B. Taylor, *Federation Proc.* **9**, 120 (1950).
[19] O. Hayaishi and R. Y. Stanier, *J. Biol. Chem.* **195**, 735 (1952).

comparable to the specific activity of crude rat liver extract (0.08 to 0.15). Considering losses sustained during purification procedures, a difference of 100-fold purity in starting materials is enormous. However, it was impractical to grow a large amount of cells in a medium containing kynurenine, since the latter substrate was not easily obtainable. This difficulty was solved by growing the cells on a medium containing tryptophan, previous work[20] having shown that the organism oxidizes tryptophan to kynurenine. The same principle was employed when a large amount of catechol-adapted cells was needed for the purification of pyrocatecase.[21] Since catechol is rather unstable and does not stand autoclaving, it had to be sterilized separately by filtration. In order to avoid this complication, cells were grown on an anthranilic acid-containing medium, catechol being generated from anthranilic acid while the cells were growing on it and metabolizing it.

Adaptive Technique. The simplest and most commonly employed method for obtaining adapted cells is to grow the organisms in the presence of a specific substrate and harvest them before the substrate is all used up. For example, *Pseudomonas* strain Tr. 24 was grown in a medium containing 0.1% L-tryptophan, 0.1% Difco yeast extract, 0.2% K_2HPO_4, 0.1% KH_2PO_4, and 0.02% $MgSO_4 \cdot 7H_2O$ for 18 hours at 30°, with constant mechanical shaking. When crude cell-free extracts were made from these adapted cells, the specific activities of the tryptophan peroxidase-oxidase system, kynureninase, and pyrocatecase were found to be 271, 610, and 107 arbitrary units, whereas those of crude extracts prepared from cells grown on asparagine were 7, 6, and 0 (undetectable by the method employed) units, respectively. The difference between adapted and nonadapted cells varies considerably, depending on the organisms, the enzymes, and the growth conditions. They may differ by several to more than 1000-fold.

The quantity of the substrate needed for the best yield of an enzyme has to be determined experimentally. In practice, 0.1 to 0.2% concentration has been employed in many instances.[2,20] As has been observed by Mirick[22] and others, the amount of substrate is often critical, since adaptive enzymes will disappear when the substrate is used up and the cells are still growing or metabolizing some other components of the medium. It is desirable to have an excess of substrate if it is inexpensive and readily available in a large quantity. However, since this is not always the case and when the substrate is available only in small quantities, the following methods are usually employed.

[20] O. Hayaishi and R. Y. Stanier, *J. Bacteriol.* **62**, 691 (1951).
[21] O. Hayaishi and Z. Hashimoto, *J. Biochem. (Japan)* **37**, 371 (1950).
[22] G. S. Mirick, *J. Exptl. Med.* **78**, 255 (1943).

1. INDIRECT ADAPTATION. As has been stated above, the adaptive response of the organism is not confined to a primary substrate but results in a sequential induction of a series of enzymes. Thus the amount of *Neurospora kynureninase* was reported to be increased as much as 600-fold by the inclusion of L-tryptophan (0.012%) in the medium.[23] In another instance, L-glycerophosphorylcholinediesterase was purified from *Serratia* grown in a medium containing 0.1% crude lecithin, since specific activity of lecithin-adapted cell extracts is about five times as high as that of unadapted cell extracts.[7]

2. ADAPTATION WITH RESTING CELLS. Enzyme induction occurs frequently with resting cell suspensions in the presence of a specific substrate. The rate and time required for adaptation as well as the final activity are variable. For example, the resting cells of *Pseudomonas* can adapt to catechol within 1 hour, whereas it may take more than 3 hours to adapt to tryptophan under the same conditions. Cells are grown in ordinary medium which supports the best growth of the organism. They are harvested, washed, and resuspended in saline or phosphate buffer containing the specific substrate in a concentration of about 0.01 to 0.005 M. The incubation is carried out until the cells show maximum activity (see below). Addition of a small amount of yeast extract (0.05%), a nitrogen source (0.1% NH_4Cl) or an energy source (0.05% glucose) usually accelerates the rate of adaptation. This method of adaptation with resting cells is sometimes employed to *readapt* cells which have lost some activity during the process of harvesting and washing. For example, an anaerobic organism, *Zymobacterium oroticum*, was grown in a medium containing 2% tryptone, 0.05% Difco yeast extract, 0.2% orotic acid, and 0.05% sodium thioglycolate. After centrifugation, the cells were resuspended in 0.01 M sodium orotate containing approximately 0.05 M potassium phosphate buffer, pH 7.0, and 0.005 M cysteine, and incubated *in vacuo* at 26° for 20 minutes. More active extracts were obtained when this readaption step was included.[24]

Influence of Substances Other Than the Substrate. The addition of other substances which increase the crop of cells may increase or decrease the total yield of enzymes. In many instances, inclusion of a small amount of yeast extract increases the yield of cells without decreasing the specific activity. However, inclusion of other metabolizable substances, such as glucose, might inhibit adaptation almost completely. For example, when a strain of *Mycobacterium* was grown in the presence of uracil, addition of 0.2% glucose increased the total cell crop about 1.5- to 2-fold, but strongly

[23] W. B. Jakoby and D. M. Bonner, *J. Biol. Chem.* **205**, 699 (1953).
[24] I. Lieberman and A. Kornberg, *Biochim. et Biophys. Acta* **12**, 223 (1953).

inhibited formation of uracil-thymine oxidase; on the other hand barbiturase activity was increased 3- to 4-fold.[2]

As indicated in the foregoing discussion, the time of harvest is critical, particularly when the amount of substrate is not excessive. In this respect also one has to consider that the enzyme for the primary substrate usually reaches its peak of activity faster than those of secondary and tertiary substrates, and also that, as soon as the substrate is consumed, the enzymatic activity starts to disappear. The best time for harvesting cells can be determined by taking aliquots and quickly assaying the activity. If the assay is time consuming, it is necessary to run small-scale experiments beforehand in order to determine the relation between either time or growth (weight or turbidity) and the activity of a particular enzyme. Growth is usually followed by turbidimetry with a photoelectric colorimeter using a wavelength of 600 or 660 mμ. Parallel experiments on a small scale, such as one done in a Warburg apparatus, are often employed when adaptation is being carried out with a resting cell suspension. Another method for estimating the time of harvest is to determine the amount of substrate remaining. The culture keeps growing as long as there is more than 0.001 M substrate remaining. As soon as the growth curve reaches its plateau (stationary phase), the cells should be harvested before the substrate is used up. In either case, it is best to grow cells in a liquid medium in order to follow enzymatic activity. When a small quantity of aerobic organisms is needed, a 4- or 6-l. Erlenmeyer flask containing about 1 l. of a desired medium is used and shaken on a mechanical shaker. The yield of cells ranges from 1.0 to 10.0 g./l. of wet cells. Larger quantities of aerobic cells can be obtained in an aerated liquid culture. For anaerobic organisms, stagnant carboys are generally used. Precautions must be taken frequently to maintain reducing conditions during harvest in order to minimize inactivation of the enzyme. The Sharples supercentrifuge is commonly used for harvesting a large quantity of culture, and other centrifuges can be used for small quantities. Salt solutions, such as 0.8% KCl or 0.5% NaCl–0.5% KCl solution, are preferred to distilled water for washing cells because cells pack down more tightly in a salt solution.

Physicochemical Factors Involved in Enzyme Formation. (1) AEROBIOSIS AND ANAEROBIOSIS. Gale reported that the formation of the oxidative L-alanine deaminase of *E. coli* is five to six times as great under aerobic conditions as under anaerobic conditions, and the formation of the anaerobic serine dehydrase is two to three times as great when growth is anaerobic as when it is aerobic.[8] Cytochrome peroxidase, catalase, and a pigment of *Pseudomonas* appear to be increased when the cells are grown without aeration, whereas under high oxygen tension only a trace of pigment was observed, no peroxidase activity was found, and the catalase activity was reduced, although diaphorase activity increased.[25]

2. HYDROGEN ION CONCENTRATION. The pH of the medium is another important factor which determines quantity and quality of the enzymes produced in the cell. For further information on this subject, see Gale.[8]

3. TEMPERATURE. For many free living organisms isolated by enrichment culture technique from soil or air, $20° \sim 30°$ appears to be the optimum temperature for growth. Sometimes a lower temperature is preferred. Not much is known about the effect of growth temperature on enzyme constitution. Gale[8] has shown that the amino acid decarboxylase of *E. coli* is formed to a greater extent when growth occurs at $20°$ than when it occurs at $37°$.

4. AGE OF CULTURE. Some enzymes are known to be produced chiefly at an early stage of the growth cycle, and some enzymes are produced chiefly at a later period. This may probably be explained by the combined effect of substrate change, pH, and other physicochemical factors.

[25] H. M. Lenhoff and N. O. Kaplan, *Nature* **172**, 730 (1953).

[15] Separation of Proteins from Nucleic Acids

By LEON HEPPEL

A number of procedures are available for separation of proteins in the denatured state from nucleic acids, e.g., shaking with octyl alcohol-chloroform or treatment with detergents. Separation of native protein from nucleic acids is more difficult. Treatment with neutral salts, or a combination of citric acid and strong salt, has been employed to separate histones from water-extracted nucleoprotein.

Extracts of bacterial cells contain much nucleic acid which interferes greatly with salt fractionation so that one obtains a series of fractions which differ very little in enzymatic specific activity. Separation of much of this nucleic acid by $MnCl_2$ has recently been reported.[1] Extracts prepared by alumina grinding are mixed with 0.05 their volume of 1 M $MnCl_2$, and the precipitate is removed by centrifugation. The supernatant solution can now be successfully fractionated.

Sometimes nucleic acid is added to an enzymatically active protein solution because subsequent fractionation is made more successful. An example is the purification of the enzyme which transfers P from D-1,3-diphosphoglyceric acid to ADP.[2] The protein–nucleic acid complex is

[1] S. Korkes, A. del Campillo, I. C. Gunsalus, and S. Ochoa, *J. Biol. Chem.* **193**, 721 (1951).

[2] T. Bücher, *Biochim. et Biophys. Acta* **1**, 292 (1947); see also Vol. I [62].

dissociated in this case by first lowering the salt content. This is done by ethanol precipitation. The precipitate is then dissolved, brought to pH 7, and to it is added a neutralized solution of protamine sulfate until the mixture remains clear after short centrifugation in the cold. More protamine sulfate is then added until drop tests on the supernatant solution are negative for both excess protamine and excess nucleic acid.

[16] Preparation of Buffers for Use in Enzyme Studies

By G. GOMORI

The buffers described in this section are suitable for use either in enzymatic or histochemical studies. The accuracy of the tables is within ±0.05 pH at 23°. In most cases the pH values will not be off by more than ±0.12 pH even at 37° and at molarities slightly different from those given (usually 0.05 M).

The methods of preparation described are not necessarily identical with those of the original authors. The titration curves of the majority of the buffers recommended have been redetermined by the writer. The buffers are arranged in the order of ascending pH range. For more complete data on phosphate and acetate buffers over a wide range of concentrations, see Vol. I [10].

1. Hydrochloric Acid–Potassium Chloride Buffer[1]

Stock Solutions

A: 0.2 M solution of KCl (14.91 g. in 1000 ml.).
B: 0.2 M HCl.
   50 ml. of A + x ml. of B, diluted to a total of 200 ml.

x	pH
97.0	1.0
78.0	1.1
64.5	1.2
51.0	1.3
41.5	1.4
33.3	1.5
26.3	1.6
20.6	1.7
16.6	1.8
13.2	1.9
10.6	2.0
8.4	2.1
6.7	2.2

[1] W. M. Clark and H. A. Lubs, *J. Bacteriol.* **2**, 1 (1917).

2. Glycine–HCl Buffer[2]

Stock Solutions

A: 0.2 M solution of glycine (15.01 g. in 1000 ml.).
B: 0.2 M HCl.
    50 ml. of A + x ml. of B, diluted to a total of 200 ml.

x	pH	x	pH
5.0	3.6	16.8	2.8
6.4	3.4	24.2	2.6
8.2	3.2	32.4	2.4
11.4	3.0	44.0	2.2

3. Phthalate–Hydrochloric Acid Buffer[1]

Stock Solutions

A: 0.2 M solution of potassium acid phthalate (40.84 g. in 1000 ml.).
B: 0.2 M HCl.
    50 ml. of A + x ml. of B, diluted to a total of 200 ml.

x	pH	x	pH
46.7	2.2	14.7	3.2
39.6	2.4	9.9	3.4
33.0	2.6	6.0	3.6
26.4	2.8	2.63	3.8
20.3	3.0		

4. Aconitate Buffer[3]

Stock Solutions

A: 0.5 M solution of aconitic acid (87.05 g. in 1000 ml.).
B: 0.2 M NaOH.
    20 ml. of A + x ml. of B, diluted to a total of 200 ml.

x	pH	x	pH
15.0	2.5	83.0	4.3
21.0	2.7	90.0	4.5
28.0	2.9	97.0	4.7
36.0	3.1	103.0	4.9
44.0	3.3	108.0	5.1
52.0	3.5	113.0	5.3
60.0	3.7	119.0	5.5
68.0	3.9	126.0	5.7
76.0	4.1		

[2] S. P. L. Sørensen, *Biochem. Z.* **21**, 131 (1909); **22**, 352 (1909).
[3] G. Gomori, unpublished.

5. Citrate Buffer[4]

Stock Solutions

A: 0.1 M solution of citric acid (21.01 g. in 1000 ml.).
B: 0.1 M solution of sodium citrate (29.41 g. $C_6H_5O_7Na_3\cdot2H_2O$ in 1000 ml.; the use of the salt with $5\frac{1}{2}$ H_2O is not recommended).
x ml. of A + y ml. of B, diluted to a total of 100 ml.

x	y	pH
46.5	3.5	3.0
43.7	6.3	3.2
40.0	10.0	3.4
37.0	13.0	3.6
35.0	15.0	3.8
33.0	17.0	4.0
31.5	18.5	4.2
28.0	22.0	4.4
25.5	24.5	4.6
23.0	27.0	4.8
20.5	29.5	5.0
18.0	32.0	5.2
16.0	34.0	5.4
13.7	36.3	5.6
11.8	38.2	5.8
9.5	41.5	6.0
7.2	42.8	6.2

6. Acetate Buffer[5]

Stock Solutions

A: 0.2 M solution of acetic acid (11.55 ml. in 1000 ml.).
B: 0.2 M solution of sodium acetate (16.4 g. of $C_2H_3O_2Na$ or 27.2 g. of $C_2H_3O_2Na\cdot3H_2O$ in 1000 ml.).
x ml. of A + y ml. of B, diluted to a total of 100 ml.

x	y	pH
46.3	3.7	3.6
44.0	6.0	3.8
41.0	9.0	4.0
36.8	13.2	4.2
30.5	19.5	4.4
25.5	24.5	4.6
20.0	30.0	4.8
14.8	35.2	5.0
10.5	39.5	5.2
8.8	41.2	5.4
4.8	45.2	5.6

[4] R. D. Lillie, "Histopathologic Technique," Blakiston, Philadelphia and Toronto, 1948.
[5] G. S. Walpole, *J. Chem. Soc.* **105**, 2501 (1914).

7. Citrate–Phosphate Buffer[6]

Stock Solutions

A: 0.1 M solution of citric acid (19.21 g. in 1000 ml.).
B: 0.2 M solution of dibasic sodium phosphate (53.65 g. of Na_2-$HPO_4 \cdot 7H_2O$ or 71.7 g. of $Na_2HPO_4 \cdot 12H_2O$ in 1000 ml.).
x ml. of A + y ml. of B, diluted to a total of 100 ml.

x	y	pH
44.6	5.4	2.6
42.2	7.8	2.8
39.8	10.2	3.0
37.7	12.3	3.2
35.9	14.1	3.4
33.9	16.1	3.6
32.3	17.7	3.8
30.7	19.3	4.0
29.4	20.6	4.2
27.8	22.2	4.4
26.7	23.3	4.6
25.2	24.8	4.8
24.3	25.7	5.0
23.3	26.7	5.2
22.2	27.8	5.4
21.0	29.0	5.6
19.7	30.3	5.8
17.9	32.1	6.0
16.9	33.1	6.2
15.4	34.6	6.4
13.6	36.4	6.6
9.1	40.9	6.8
6.5	43.6	7.0

8. Succinate Buffer[3]

Stock Solutions

A: 0.2 M solution of succinic acid (23.6 g. in 1000 ml.).
B: 0.2 M NaOH.
25 ml. of A + x ml. of B, diluted to a total of 100 ml.

x	pH	x	pH
7.5	3.8	26.7	5.0
10.0	4.0	30.3	5.2
13.3	4.2	34.2	5.4
16.7	4.4	37.5	5.6
20.0	4.6	40.7	5.8
23.5	4.8	43.5	6.0

[6] T. C. McIlvaine, *J. Biol. Chem.* **49**, 183 (1921).

9. Phthalate–Sodium Hydroxide Buffer[1]

Stock Solutions

A: 0.2 M solution of potassium acid phthalate (40.84 g. in 100 ml.).
B: 0.2 M NaOH.
    50 ml. of A + x ml. of B, diluted to a total of 200 ml.

x	pH	x	pH
3.7	4.2	30.0	5.2
7.5	4.4	35.5	5.4
12.2	4.6	39.8	5.6
17.7	4.8	43.0	5.8
23.9	5.0	45.5	6.0

10. Maleate Buffer[7]

Stock Solutions

A: 0.2 M solution of acid sodium maleate (8 g. of NaOH + 23.2 g.
of maleic acid or 19.6 g. of maleic anhydride in 1000 ml.).
B: 0.2 M NaOH.
    50 ml. of A + x ml. of B, diluted to a total of 200 ml.

x	pH	x	pH
7.2	5.2	33.0	6.2
10.5	5.4	38.0	6.4
15.3	5.6	41.6	6.6
20.8	5.8	44.4	6.8
26.9	6.0		

11. Cacodylate Buffer[8]

Stock Solutions

A: 0.2 M solution of sodium cacodylate (42.8 g. of $Na(CH_3)_2AsO_2\cdot$
$3H_2O$ in 1000 ml.).
B: 0.2 M HCl.
    50 ml. of A + x ml. of B, diluted to a total of 200 ml.

x	pH	x	pH
2.7	7.4	29.6	6.0
4.2	7.2	34.8	5.8
6.3	7.0	39.2	5.6
9.3	6.8	43.0	5.4
13.3	6.6	45.0	5.2
18.3	6.4	47.0	5.0
23.8	6.2		

[7] J. W. Temple, *J. Am. Chem. Soc.* **51**, 1754 (1929).
[8] M. Plumel, *Bull. soc. chim. biol.* **30**, 129 (1949).

12. Phosphate Buffer[2]

Stock Solutions

A: 0.2 M solution of monobasic sodium phosphate (27.8 g. in 1000 ml.).

B: 0.2 M solution of dibasic sodium phosphate (53.65 g. of $Na_2HPO_4 \cdot 7H_2O$ or 71.7 g. of $Na_2HPO_4 \cdot 12H_2O$ in 1000 ml.).

x ml. of A + y ml. of B, diluted to a total of 200 ml.

x	y	pH	x	y	pH
93.5	6.5	5.7	45.0	55.0	6.9
92.0	8.0	5.8	39.0	61.0	7.0
90.0	10.0	5.9	33.0	67.0	7.1
87.7	12.3	6.0	28.0	72.0	7.2
85.0	15.0	6.1	23.0	77.0	7.3
81.5	18.5	6.2	19.0	81.0	7.4
77.5	22.5	6.3	16.0	84.0	7.5
73.5	26.5	6.4	13.0	87.0	7.6
68.5	31.5	6.5	10.5	90.5	7.7
62.5	37.5	6.6	8.5	91.5	7.8
56.5	43.5	6.7	7.0	93.0	7.9
51.0	49.0	6.8	5.3	94.7	8.0

13. Tris(hydroxymethyl)aminomethane-maleate (Tris-maleate) Buffer[9,10]

Stock Solutions

A: 0.2 M solution of Tris acid maleate (24.2 g. of tris(hydroxymethyl)aminomethane + 23.2 g. of maleic acid or 19.6 g. of maleic anhydride in 1000 ml.).

B: 0.2 M NaOH.

50 ml. of A + x ml. of B, diluted to a total of 200 ml.

x	pH	x	pH
7.0	5.2	48.0	7.0
10.8	5.4	51.0	7.2
15.5	5.6	54.0	7.4
20.5	5.8	58.0	7.6
26.0	6.0	63.5	7.8
31.5	6.2	69.0	8.0
37.0	6.4	75.0	8.2
42.5	6.6	81.0	8.4
45.0	6.8	86.5	8.6

[9] A buffer-grade Tris can be obtained from the Sigma Chemical Co., St. Louis 3, Mo., or from Matheson Coleman & Bell, East Rutherford, N. J.

[10] G. Gomori, *Proc. Soc. Exptl. Biol. Med.* **68,** 354 (1948).

14. Barbital Buffer[11]

Stock Solutions

A: 0.2 M solution of sodium barbital (veronal) (41.2 g. in 1000 ml.).
B: 0.2 M HCl.
    50 ml. of A + x ml. of B, diluted to a total of 200 ml.

x	pH
1.5	9.2
2.5	9.0
4.0	8.8
6.0	8.6
9.0	8.4
12.7	8.2
17.5	8.0
22.5	7.8
27.5	7.6
32.5	7.4
39.0	7.2
43.0	7.0
45.0	6.8

Solutions more concentrated than 0.05 M may crystallize on standing, especially in the cold.

15. Tris(hydroxymethyl)aminomethane (Tris) Buffer[9]

Stock Solutions

A: 0.2 M solution of tris(hydroxymethyl)aminomethane (24.2 g. in 1000 ml.).
B: 0.2 M HCl.
    50 ml. of A + x ml. of B, diluted to a total of 200 ml.

x	pH
5.0	9.0
8.1	8.8
12.2	8.6
16.5	8.4
21.9	8.2
26.8	8.0
32.5	7.8
38.4	7.6
41.4	7.4
44.2	7.2

[11] L. Michaelis, *J. Biol. Chem.* **87**, 33 (1930).

16. Boric Acid–Borax Buffer[12]

Stock Solutions

A: 0.2 M solution of boric acid (12.4 g. in 1000 ml.).

B: 0.05 M solution of borax (19.05 g. in 1000 ml.; 0.2 M in terms of sodium borate).

50 ml. of A + x ml. of B, diluted to a total of 200 ml.

x	pH	x	pH
2.0	7.6	22.5	8.7
3.1	7.8	30.0	8.8
4.9	8.0	42.5	8.9
7.3	8.2	59.0	9.0
11.5	8.4	83.0	9.1
17.5	8.6	115.0	9.2

17. 2-Amino-2-methyl-1,3-propanediol (Ammediol) Buffer[13]

Stock Solutions

A: 0.2 M solution of 2-amino-2-methyl-1,3-propanediol (21.03 g. in 1000 ml.).

B: 0.2 M HCl.

50 ml. of A + x ml. of B, diluted to a total of 200 ml.

x	pH	x	pH
2.0	10.0	22.0	8.8
3.7	9.8	29.5	8.6
5.7	9.6	34.0	8.4
8.5	9.4	37.7	8.2
12.5	9.2	41.0	8.0
16.7	9.0	43.5	7.8

18. Glycine-NaOH Buffer[2]

Stock Solutions

A: 0.2 M solution of glycine (15.01 g. in 1000 ml.).

B: 0.2 M NaOH.

50 ml. of A + x ml. of B, diluted to a total of 200 ml.

x	pH	x	pH
4.0	8.6	22.4	9.6
6.0	8.8	27.2	9.8
8.8	9.0	32.0	10.0
12.0	9.2	38.6	10.4
16.8	9.4	45.5	10.6

[12] W. Holmes, *Anat. Record* **86**, 163 (1943).
[13] G. Gomori, *Proc. Soc. Exptl. Biol. Med.* **62**, 33 (1946).

19. Borax-NaOH Buffer[1]

Stock Solutions

A: 0.05 M solution of borax (19.05 g. in 1000 ml.; 0.02 M in terms of sodium borate).

B: 0.2 M NaOH.

　　50 ml. of A + x ml. of B, diluted to a total of 200 ml.

x	pH
0.0	9.28
7.0	9.35
11.0	9.4
17.6	9.5
23.0	9.6
29.0	9.7
34.0	9.8
38.6	9.9
43.0	10.0
46.0	10.1

20. Carbonate–Bicarbonate Buffer[14]

Stock Solutions

A: 0.2 M solution of anhydrous sodium carbonate (21.2 g. in 1000 ml.).

B: 0.2 M solution of sodium bicarbonate (16.8 g. in 1000 ml.).

　　x ml. of A + y ml. of B, diluted to a total of 200.

x	y	pH
4.0	46.0	9.2
7.5	42.5	9.3
9.5	40.5	9.4
13.0	37.0	9.5
16.0	34.0	9.6
19.5	30.5	9.7
22.0	28.0	9.8
25.0	25.0	9.9
27.5	22.5	10.0
30.0	20.0	10.1
33.0	17.0	10.2
35.5	14.5	10.3
38.5	11.5	10.4
40.5	9.5	10.5
42.5	7.5	10.6
45.0	5.0	10.7

[14] G. E. Delory and E. J. King, *Biochem. J.* **39**, 245 (1945).

Section II

Enzymes of Carbohydrate Metabolism

[17] Amylases, α and β

By Peter Bernfeld

Assay Method

A large number of valuable methods have been described for the assay of amylase. They are based on one or another of the following phenomena observed in the enzyme digests: (1) increase in reducing power of a solution of amylopectin or soluble starch; (2) change of the iodine-staining properties of the substrate; (3) decrease of the viscosity of a starch paste. All three phenomena are characteristic for the action of α-amylases; only the first one, however, can be used for the assay of β-amylase. The assay method described below[1] is based on the increase in reducing power and is applicable for both α- and β-amylases. Although any method for the determination of reducing sugars may be used, the one described here has proved to be simple, reliable, and rapid. It was first employed by Sumner[2] for the assay of saccharase.

Reagents

Dissolve 1 g. of potato amylopectin, or 1 g. of Schoch's B fraction from corn,[3] or, if neither is available, 1 g. of soluble starch, Merck, in 100 ml. of 0.02 M phosphate buffer, pH 6.9, containing 0.0067 M NaCl (for assay of human salivary α-amylase). For assay of sweet potato β-amylase, dissolve 1 g. of the substrate in 0.016 M acetate buffer, pH 4.8.

Dissolve at room temperature 1 g. of 3,5-dinitrosalicylic acid in 20 ml. of 2 N NaOH and 50 ml. H_2O, add 30 g. of Rochelle salt, and make up to 100 ml. with distilled H_2O. Protect this solution from CO_2.

Procedure. 1 ml. of properly diluted enzyme is incubated for 3 minutes at 20° with 1 ml. of the substrate solution. The enzyme reaction is interrupted by the addition of 2 ml. of dinitrosalicylic acid reagent. The tube containing this mixture is heated for 5 minutes in boiling water and then cooled in running tap water. After addition of 20 ml. of H_2O, the optical density of the solution containing the brown reduction product is determined photometrically, by means of a green filter,[4] and a blank is prepared

---

[1] G. Noelting and P. Bernfeld, *Helv. Chim. Acta* **31**, 286 (1948).

[2] J. B. Sumner and S. F. Howell, *J. Biol. Chem.* **108**, 51 (1935); J. B. Sumner, *J. Biol. Chem.* **62**, 287 (1924–25).

[3] T. J. Schoch, *Advances in Carbohydrate Chem.* **1**, 247 (1945); see also Vol. III [2].

[4] Filter No. 540 in an Evelyn photoelectric colorimeter, or filter No. 54 in a Klett-Summerson photoelectric colorimeter.

in the same manner without enzyme. A calibration curve established with maltose (0.2 to 2 mg. in 2 ml. of H_2O) is used to convert the colorimeter readings into milligrams of maltose.

Amylase Activity. Amylase activity is expressed in terms of milligrams of maltose ($C_{12}H_{22}O_{11} \cdot H_2O$) liberated in 3 minutes at 20° by 1 ml. of the enzyme solution, even though, in the case of α-amylatic action, the actual reaction products are dextrins rather than maltose. *Specific activity* is expressed as amylase activity per milligram of protein. Protein is determined by either micro-Kjeldahl or according to Lowry.[5] When the Kjeldahl method is used with solutions containing ammonium sulfate, NH_3 is previously eliminated by distillation in the presence of MgO.

Purification Procedure of Human Salivary α-Amylase

Human salivary amylase has been chosen as an example to describe the purification and the properties of an α-amylase, because the purification procedure of this enzyme is readily reproducible and the starting material is easily obtained.[6]

General Remarks. All operations are carried out between 0 and +4°. Precipitants are added to the enzyme solution under slow but efficient mechanical stirring of the latter. The time required for the addition is indicated in the text in parentheses. When the addition of precipitant is completed, the stirring is always continued for 15 minutes, before the centrifugation is begun. The centrifugations are done at about 3000 r.p.m. for 15 minutes, unless otherwise stated. All acetone concentrations indicated below are expressed as per cent by volume; all ammonium sulfate concentrations as per cent saturation.

Starting Material. Human saliva is collected in the presence of a few drops of 2-octanol to prevent foaming. On the average, one individual can produce 50 ml. of saliva in 1 hour by chewing large pieces of paraffin. The saliva is cleared by centrifugation for 1 hour. It can be stored for several days in the cold in the presence of a few drops of toluene.

Step 1. First Acetone Fractionation. Acetone is slowly added (45 minutes) to 1500 ml. of centrifuged saliva to give a final concentration of 48%. The precipitate is discarded after centrifugation, the solution is brought to 70% acetone (45 minutes), and the precipitate, after centrifugation and discarding of the supernatant solution, is dissolved in 350 ml. of 0.07 M sodium acetate.

[5] O. H. Lowry, N. J. Rosebrough, A. L. Farr, and R. J. Randall, *J. Biol. Chem.* **193**, 265 (1951).

[6] K. H. Meyer, E. H. Fischer, A. Staub, and P. Bernfeld, *Helv. Chim Acta* **31**, 2158 (1948).

Step 2. Second Acetone Fractionation. The precipitate obtained by adding acetone to the above solution to a final concentration of 50% (in 20 minutes) is discarded after centrifugation, and a second precipitate obtained at 70% acetone (20 minutes) is collected and dissolved in 250 ml. of 0.07 M sodium acetate.

Step 3. First Ammonium Sulfate Precipitation. 10 ml. of a solution of ammonium sulfate, saturated at 0°, is added rapidly to the enzyme solution from the previous step. The clear solution is then brought to pH 8.0 by the addition of a few drops of 0.1 N NH_4OH. Then a solution of ammonium sulfate, saturated at 0° and previously adjusted to pH 8.0 with NH_4OH, is added rapidly to raise the concentration of ammonium sulfate to 45.5% saturation. After 30 minutes of slow stirring, the precipitate is removed by centrifugation for 30 minutes and is dissolved in 80 ml. of 0.07 M sodium acetate.

Step 4. Second Ammonium Sulfate Precipitation. The operation outlined in the last paragraph is repeated with the enzyme solution of step 3, starting with the addition of 3 ml. of saturated ammonium sulfate to the enzyme solution, followed by adjustment of the clear solution to pH 8.0 and precipitation of the enzyme at 40% saturation of ammonium sulfate. The precipitate is dissolved in 40 ml. of 0.2 M sodium acetate.

Step 5. Third Acetone Precipitation. The solution of step 4 is adjusted to pH 7.0 with a few drops of 0.1 N NH_4OH. Then acetone is added (15 minutes) to a final concentration of 70%, and the precipitate is collected by centrifugation and dissolved in 15 ml. of H_2O. The purpose of this step is to remove from the enzyme solution the bulk of ammonium sulfate by discarding it with the supernatant fluid.

Step 6. Ion Exchange. The remaining sulfate ions are replaced with acetate ions by pouring the enzyme solution through a column (6 inches long, $\frac{5}{8}$ inch inside diameter) containing 20 ml. of anion exchange resin Amberlite IRA-400 which has been previously charged with acetate and washed with H_2O. The resin is maintained in place between two cotton plugs. The flow rate is adjusted to 3 ml./min. When the enzyme solution has disappeared below the upper cotton plug, 1 ml. of H_2O is added. In order to recover the enzyme from the column, this operation is repeated with four 1-ml. portions of H_2O and then with larger amounts, each time the liquid level reaches the upper cotton plug. Fractions of 5 ml. each are collected, and all those fractions having an optical density of 0.5 or more at 280 mμ are combined (usually fractions 2 through 7).

Step 7. Fourth Acetone Precipitation. Ammonium acetate is now the only electrolyte remaining in the enzyme solution; it does not interfere with the protein precipitation of step 7. In order to concentrate the enzyme solution and simultaneously to remove the salt, acetone is added (within 10 minutes) to make a final concentration of 70%. After centrifugation, the supernatant fluid is decanted and discarded, and the centrifuge tube containing the precipitate is inverted for 5 minutes, in order to drain the adhering solution. After the inside walls of the tube have been wiped off with a filter paper, the precipitate is triturated with 2 ml. of 0.1 N NaOH, added gradually, until a clear solution is obtained (final pH 8.0 to 8.4.

Crystallization. The above solution is kept in the icebox. About 60% of the enzyme deposits in crystalline form overnight, 80% in 2 days; 1500 ml. of saliva yields approximately 150 mg. of crystalline amylase.

Recrystallization. The centrifuged crystals are washed three times by rapidly triturating them with cold 30% acetone, immediately followed by centrifugation. The crystals are suspended in H_2O (1 ml. for 70 mg. of protein), and 0.04 N NaOH is added to raise the pH to 11.0. The crystals dissolve slowly; gentle continuous shaking hastens the dissolution. The pH decreases as the enzyme dissolves, and it should be maintained at 11.0 by occasional addition of small amounts of 0.04 N NaOH. After all the protein is dissolved, the solution is neutralized with 0.1 N acetic acid and immediately centrifuged to remove small amounts of insoluble material. The enzyme crystallizes upon standing in the ice chest; it may be necessary to add seeding crystals.

Properties of Human Salivary α-Amylase

Action. The enzyme hydrolyzes α-1,4-glucosidic bonds in polyglucosans (amylose, amylopectin, glycogen, and dextrins). The location in the molecule of the bond to be hydrolyzed is selected at random, but the terminal bonds are split much more slowly. The initial stage of action of this enzyme is characterized by a rapid decrease of the molecular weight of the substrate, resulting in a rapid change of its iodine-staining properties (dextrinizing activity) and, when starch paste is the substrate, in an extremely rapid loss of the viscosity; these phenomena are produced by the cleavage of a very small number of glucosidic bonds; i.e., they are accompanied by only a slight increase in the reducing power (saccharifying activity). The final products of action are maltose, limit dextrins of low degree of polymerization (3 to 7), and small amounts of glucose. The α-1,6-glucosidic bonds (branching points) are not hydrolyzed by this enzyme. These all appear in the limit dextrins.

Purity of Crystalline Salivary Amylase. After several recrystallizations of the enzyme, the specific amylase activity remaining in the mother liquor becomes equal to the specific amylase activity of the crystals. The crystalline enzyme migrates as a single protein on electrophoresis. It does not contain any maltase, nor could the presence of any other starch-

TABLE I

SUMMARY OF PURIFICATION PROCEDURE OF SALIVARY α-AMYLASE

Step of purification	Amylase activity per ml.	Protein,[a] mg./ml.	Specific amylase activity	Over-all enzyme recovery, %
Centrifuged saliva	410	3.3	125	—
1. First acetone	1,620	4.2	385	90
2. Second acetone	1,920	2.9	663	77
3. First $(NH_4)_2SO_4$	3,540	4.1	865	45
4. Second $(NH_4)_2SO_4$	5,720	6.1	940	36
5. Third acetone	12,800	13.4	955	30
6. Ion exchange	6,980	7.3	955	30
7. Fourth acetone	63,000	66.0	955	30
Crystallization	—	—	980	24
Recrystallization	—	—	1000	19

[a] From Kjeldahl N determination, assuming in all the fractions the same nitrogen content as found in pure amylase (15.8%).

digesting enzymes, such as β-amylase or α-1,6-glucosidase, be detected. Crystalline human salivary α-amylase, therefore, appears to be a pure protein.

Chemical Composition. The crystalline enzyme contains 15.8% N, less than 0.01% P, and only traces of S. There is no indication for the presence of a prosthetic group in this enzyme.

The solubility at 2° is 0.2 to 0.5 g. of crystalline protein in 100 ml. of H_2O buffered to pH 8.5.

The light absorption spectrum exhibits a maximum of the optical density at 280 mμ, and a slight irregularity at 292 mμ. There is practically no light absorption above 310 mμ.

The electrophoretic mobility of the enzyme in glycine buffer, pH 10.14, of ionic strength $\mu = 0.1$, is $u = 3.75 \times 10^{-5}$ cm.^{2} volt^{-1} sec.$^{-1}$. The isoelectric point is near 5.3.

Effect of pH on the Stability. An aqueous solution of the crystalline enzyme does not lose any activity during storage for 20 hours at 20° when the pH is between 4.5 and 11.0.

Effect of pH on the Enzyme Activity. The enzyme is active between pH 3.8 and 9.4 with a distinct optimum at pH 6.9.

Effect of Temperature on the Enzyme Activity. The activity of the crystallized enzyme increases with the temperature to about 40° and decreases at higher temperatures. The temperature coefficient is 2.3 between 10 and 20°, and 2.0 between 20 and 30°. The energy of activation, calculated from these data, is 13,350 cal./mole.

Activators. The activity of pure human salivary α-amylase decreases to about 15% of its original value when the enzyme has previously been dialyzed for 2 days against dilute NH_4OH (pH 9 to 10) and when the substrate has been dialyzed against 0.01 M phosphate buffer. The full enzyme activity is restored upon addition of Cl^- to a final concentration of 0.01 M, and about 80% of the original activity is recovered in the presence of 0.001 M Cl^-. Other anions, having a similar but weaker effect, are, in the order of decreasing activation power: Br^-, NO_3^-, and I^-. No activation has been detected with SO_4^{--}, CO_3H^-, PO_4H^{--}, CNS^-, or $CH_3CO_2^-$. This activation is entirely independent of the type of cation present.

Comparison with Other Amylases.[7] Crystalline human salivary α-amylase is identical with crystalline *human pancreatic* α-amylase.[8] Human salivary amylase hydrolyzes the substrate in the same way as does crystalline *pig pancreatic* α-amylase.[9] These two latter enzymes also have an identical optimum pH of action, and both require Cl^- for their full activity. They differ markedly from each other, however, in their specific amylase activity, in their electrophoretic mobility, in their solubility, and in their pH tolerance. Crystalline human salivary α-amylase exhibits the same substrate specificity as the crystalline α-amylases from malt,[10] from *B. subtilis*,[11] and from *A. oryzae*.[12] The end products obtained by extensive action of these three latter enzymes on the substrate, however, are of somewhat higher molecular weight than those obtained with salivary amylase. Furthermore, their optimum pH of action is more acid, and most of their physical and chemical properties are different from those of salivary amylase. The ratio of saccharifying to dextrinizing activity is the same for all these α-amylases. This ratio is about one-sixth of that observed with β-amylases.

Purification Procedure of Sweet Potato β-Amylase

This enzyme was the first amylase obtained in crystalline form. The original purification procedure of Balls *et al.*[13] is given here with some minor modifications.

General Remarks. When not otherwise stated, all operations are carried out at room temperature. Overnight storage of enzyme material, however, is done at 4°.

[7] P. Bernfeld, *Advances in Enzymol.* **12**, 379 (1951).
[8] P. Bernfeld, F. Duckert, and E. H. Fischer, *Helv. Chim. Acta* **33**, 1064 (1950).
[9] E. H. Fischer and P. Bernfeld, *Helv. Chim. Acta* **31**, 1831 (1948); K. H. Meyer, E. H. Fischer, and P. Bernfeld, *Arch Biochem.* **14**, 149 (1947); *Experientia* **3**, 106 (1947).
[10] S. Schwimmer and A. K. Balls, *J. Biol. Chem.* **179**, 1063 (1949).
[11] K. H. Meyer, M. Fuld, and P. Bernfeld, *Experientia* **3**, 411 (1947).
[12] E. H. Fischer and R. de Montmollin, *Helv. Chim. Acta* **34**, 1987 (1951).
[13] A. K. Balls, M. K. Walden, and R. R. Thompson, *J. Biol. Chem.* **173**, 9 (1948).

Step 1. Extraction of Sweet Potatoes. 10 lb. of sweet potatoes is washed, ground to a coarse mixture in a meat grinder, and then homogenized in small portions for 5 minutes in a Waring blendor with the addition of 2500 ml. of H_2O in the process. The solid is separated by centrifugation and discarded.

Step 2. Heating at 60°. About 1900 ml. of the turbid extract and a few drops of 2-octanol are placed in a 3000-ml. Erlenmeyer flask which is immersed in water of 85°. The flask is continuously shaken by hand until the temperature of the enzyme solution reaches 60° (8 to 10 minutes). The flask is immersed in water of 60°, kept there for 7 minutes without shaking, and then placed in cold running water, so that the temperature of the enzyme solution drops to 40° in about 5 minutes, and to 28° in 5 more minutes. The heated juice is stored under toluene for 4 days at 4° and then centrifuged. The precipitate is discarded.

Step 3. Precipitation with Basic Lead Acetate. The turbid enzyme solution is thoroughly mixed with a quantity of a lead subacetate suspension (200 g. $Pb(CH_3CO_2)_2 \cdot Pb(OH)_2$, Merck, per liter) that causes a precipitation of 25 to 33% of the amylase present. This quantity, which has to be determined in preliminary experiments, amounts to about 30 ml. of lead subacetate suspension per 1000 ml. of amylase solution. The precipitate formed is separated by centrifugation at 3000 r.p.m. and discarded.

Step 4. First Ammonium Sulfate Precipitation. Crystalline ammonium sulfate is added to the turbid enzyme solution to 70% saturation, the total volume of the mixture is measured, and the precipitate is collected by spinning the mixture for 60 minutes at 20,000 r.p.m. in a Spinco Model L ultracentrifuge (45,000 × g.). The clear supernatant liquid is decanted and discarded after its volume is measured. The precipitate is diluted with 380 ml. of water to make it 25% saturated with ammonium sulfate. The amount of water to be added is calculated from the volume of the precipitate (210 ml.) estimated by difference, and from the assumption that the precipitate is 70% saturated with ammonium sulfate as if it were a solution. The mixture, which has a pH of 5 to 5.2, is centrifuged at 3000 r.p.m., the precipitate is further diluted with 500 ml. of water, which dissolves it partially, then made 25% saturated with ammonium sulfate, and centrifuged again. The precipitate is discarded, and the turbid supernatant fluids are combined and clarified by high-speed centrifugation at 20,000 r.p.m.

Step 5. Second Ammonium Sulfate Precipitation. Crystalline ammonium sulfate is added to 70% saturation. After centrifugation at 3000 r.p.m., the supernatant fluid is discarded and the precipitate diluted with water to 38 ml.

Step 6. Dialysis. The turbid amylase solution is dialyzed for 2 days against four 2-l. portions of distilled water. During this time, the enzyme dissolves entirely, and the volume of the enzyme solution increases to about 100 ml. The amylase solution is then clarified by 10 minutes' centrifugation at 18,000 r.p.m.

Crystallization. The enzyme solution is diluted to 150 ml. and made 20% saturated with ammonium sulfate, the pH being 5 to 5.2: After cooling to 0°, the pH is lowered to 3.2 by the addition of N hydrochloric acid, and the precipitate is removed by centrifugation at 18,000 r.p.m. for 10 minutes at 0°. Saturated ammonium sulfate solution is then added to the supernatant fluid to 25% saturation, which raises the pH to 3.7, and seed crystals are added. The enzyme solution is placed in a dialysis bag (Visking seamless regenerated cellulose tubing) and dialyzed at 0° against twenty times its volume of 30% saturated ammonium sulfate, pH 3.8. After several hours, the concentration of ammonium sulfate is gradually raised to 45% saturation over a period of 12 hours, by adding saturated ammonium sulfate solution of pH 3.8 to the solution outside the dialysis tubing. Microscopic examination of the mixture reveals the presence of crystals along with the amorphous matter.

TABLE II

SUMMARY OF PURIFICATION PROCEDURE OF SWEET POTATO β-AMYLASE

Step of purification	Amylase activity per ml.[a]	Protein,[b] mg./ml.	Volume, ml.	Specific amylase activity	Over-all enzyme recovery, %
1. Crude extract	146	10.2	3850	14.3	—
2. Heating at 60°	165	8.6	3850	19.2	113
3. Lead subacetate	114	2.3	3080	49.5	63
4. First $(NH_4)_2SO_4$	222	2.15	1160	104	46
5. Second $(NH_4)_2SO_4$	5750	40.8	38	140	39
6. Dialysis	685	3.4	130	201	16
Crystallization	—	—	—	480	13
Recrystallization	—	—	—	560	11

[a] One amylase activity unit corresponds to 0.25 of Balls'[13] units.

[b] From measurements according to Lowry,[5] using crystallized salt-free chymotrypsin (Worthington Biochemical Lab.) as the reference protein.

Recrystallization. This procedure is adequately described by Englard and Singer.[14] Six to eight recrystallizations are necessary to obtain a product which is free from amorphous material.

Properties of Sweet Potato β-Amylase

Action. β-Amylase hydrolyzes α-1,4-glucosidic bonds in polyglucosans (amylose, amylopectin, glycogen, and dextrins). In contrast to the action

[14] S. Englard and T. P. Singer, *J. Biol. Chem.* **187,** 213 (1950).

of α-amylase, only the penultimate bond from the nonreducing end group of the substrate molecule is selected for cleavage by β-amylase. Thus, one molecule of maltose after the other (as β-maltose after Walden inversion) is detached from the substrate until the enzyme encounters an obstacle, i.e., a branching point. The enzyme action is stopped at this point. Accordingly, amylose is completely converted into maltose when the reaction is carried out properly.[15] Amylopectin yields about 60% of maltose and 40% of a high molecular weight limit dextrin, giving a purple iodine color.

All following properties, unless otherwise stated, have been reported by Balls et al.[13]

Purity of Crystalline Sweet Potato β-Amylase. Enzyme recrystallized eight times behaves like a homogeneous protein in a phase-solubility test and on electrophoresis; by ultracentrifugation, an impurity was detected, constituting about 3% of the total protein.[14] After several recrystallizations, the enzyme is free from α-amylase and maltose and contains no more traces of an acid phosphatase.

Chemical Composition. The crystallized sweet potato β-amylase contains 15.1% N (Kjeldahl), 0.83% amino N, 1.16% amide N, 6.0% arginine, 7.0% tyrosine, 0.79% cystine + cysteine, 4.32% methionine, and 0.66% ash. There is no indication that the enzyme contains a prosthetic group or is a heavy metal complex.

The diffusion constant has been measured to be $D_{20} = 5.77 \times 10^{-7}$ cm.^{2} sec.$^{-1}$.[14]

The molecular weight calculated from sedimentation and diffusion data is $152{,}000 \pm 15{,}000$.[14]

The isolectric point, as determined by electrophoresis, is pH 4.74 to 4.79.[14]

The turnover number has been calculated to be 250,000 glucosidic linkages hydrolyzed by 1 molecule of enzyme per minute at 30° and pH 4.80.[14,16]

Effect of pH on the Enzyme Activity. In acetate buffer, the enzyme is most active between pH 4 and 5; in citrate buffer the optimum is between pH 5 and 6.

Activators and Inhibitors. Unlike the animal α-amylases, sweet potato β-amylase does not require Cl^- or any other anions for its activity. In contrast to malt α-amylase, the β-amylases are not activated by Ca^{++}. Reagents recognized as selective for —SH groups are powerful inhibitors of sweet potato β-amylase,[16] i.e., 5×10^{-7} M p-chloromercuribenzoate, 10^{-5} M $AgNO_3$, 10^{-5} M $HgCl_2$, 10^{-3} M o-iodosobenzoate, and 10^{-3} M $CuSO_4$. The inhibition by 1.18×10^{-5} M p-chloromercuribenzoate can

[15] P. Bernfeld and P. Gürtler, *Helv. Chim. Acta* **31,** 106 (1948).
[16] S. Englard, S. Sorof, and T. P. Singer, *J. Biol. Chem.* **189,** 217 (1951).

be partially reversed by 2×10^{-5} to 2×10^4 M—SH glutathione or by 1.18×10^{-4} M 1,2-dithiopropanol.[16]

Comparison with Malt β-Amylase. The crystalline β-amylases from sweet potato and from malt[17] exhibit the same ratio of saccharifying to dextrinizing activity, indicating an identity in the enzyme specificity between these two amylases. They differ from each other, however, in many of their physical and chemical properties.

[17] K. H. Meyer, E. H. Fischer, and A. Piguet, *Helv. Chim. Acta* **34**, 316 (1951).

[18] Pectic Enzymes

By Z. I. KERTESZ

Introduction and Nomenclature

In this discussion the nomenclature adopted by the American Chemical Society[1] is used. Thus "pectins" or "pectinic acids" designate those water-soluble polygalacturonic acids of varying methyl ester contents and degree of neutralization which show colloidal properties and are capable of forming, under certain conditions, gels with sugar and acid.[2] "Pectic acids" are polygalacturonic acids of colloidal nature but essentially free of methyl ester groups.

"Protopectin" is the water-insoluble parent pectic substance which occurs in plants and which, upon limited acid hydrolysis, yields pectins of various sorts. The enzyme "protopectinase" supposedly hydrolyzes protopectin into pectin and perhaps cellulose. There is very little exact knowledge concerning protopectin, and even the existence of a separate chemical entity of this sort is doubtful. For this reason protopectinase will not be discussed here except to state that it supposedly occurs in some higher plants and in many microorganisms. Plant tissue slices rich in water-insoluble pectic constituents are used as a substrate of protopectinase.[3]

[1] G. L. Baker, G. H. Joseph, Z. I. Kertesz, H. H. Mottern, and A. G. Olsen, *Chem. Eng. News* **22**, 105 (1944).

[2] Z. I. Kertesz, "The Pectic Substances," Interscience Publishers, New York, 1951. For other recent reviews of the field of pectic enzymes, see H. J. Phaff and M. A. Joslyn, *Wallerstein Labs. Communs.* **10**, 133 (1947); Z. I. Kertesz and R. J. McColloch, *Advances in Carbohydrate Chem.* **5**, 79 (1949); H. Lineweaver and E. F. Jansen, *Advances in Enzymol.* **11**, 267 (1951); Z. I. Kertesz *in* "The Enzymes" (Sumner, and Myrbäck, eds.), Vol. I, Part 2, p. 745, Academic Press, New York, 1951; H. Deuel, J. Solms, and H. Altermatt, *Vierteljahrsschr. naturforsch. Ges. Zürich* **98**, 49 (1953).

[3] F. R. Davison and J. J. Willaman, *Botan. Gaz.* **83**, 329 (1927).

Disregarding protopectinase, there are two major groups of pectic enzymes which will be discussed. The first one is "pectinesterase"[4] (PE) (pectin-methylesterase,[5] pectase[6]), which catalyzes the fissure of the natural or synthetic methyl ester of polygalacturonic acids.[7] The second group consists of the pectolytic enzymes designated as "polygalacturo-nidases"[8] (PG)(pectinases), which cleave hydrolitically the glycosidic bonds of polygalacturonic acids.

A. Pectinesterase (PE)

$$Pectin + H_2O = Pectic\ acid + CH_3OH$$

Assay Method

Principle. The action of PE on pectin (or synthetic polymethyl esters of polygalacturonic acids) offers two practical possibilities for measurement. First, the CH_3OH split off might be measured.[7] This method is roundabout and time consuming. Second, the increase in free carboxyl groups may be followed titrimetrically while a constant pH in the reaction mixture is maintained.[5] The latter method, which has passed through several modifications,[4,9,10] is used now by most investigators and will be given below. Until about 1937 PE activity was mostly measured by the formation of the pectinate gel from the pectin in the presence of calcium salts ("pectase reaction").[6] This method is now regarded as unsuitable, since the appearance of the gel is influenced both by enzyme activity and by the suitability of the prevailing conditions for gel formation. The quantitative expression of such results is also difficult.

Reagents

Pectin in 1% solution and containing 0.1 M NaCl. Pectinum N.F. or any good-grade purified pectin may be used.[11] The mixture is best made up from stock solutions of 2% pectin and 1.5 M NaCl.

0.1 or 0.02 N NaOH. The more dilute alkali is used with low enzyme activities.

Enzyme solution. The roots, stems, leaves, and fruits of most higher plants contain PE.[2] Some of the extensively investigated sources of highly active PE are alfalfa, red clover, tomato fruit, tobacco

[4] H. Lineweaver and G. A. Ballou, *Arch. Biochem* **6,** 389 (1945).

[5] Z. I. Kertesz, *J. Biol. Chem.* **121,** 589 (1937).

[6] J. J. Willaman, *Minn. Stud. Biol. Sci.* **6,** 333 (1927).

[7] T. Fellenberg, *Biochem. Z.* **85,** 45 (1918).

[8] Z. I. Kertesz, *Ergeb. Enzymforsch.* **5,** 233 (1936).

[9] R. J. McColloch and Z. I. Kertesz, *Arch. Biochem.* **13,** 217 (1947).

[10] V. B. Fish and R. B. Dustman, *J. Am. Chem. Soc.* **67,** 1155 (1945).

[11] For the preparation of pectin and pectic acid, see Vol. III [5].

plants, and the flavedo and albedo of citrus fruit. PE is produced by a great many microorganisms (bacteria, molds, yeasts), but usually plant sources are used for purification, since such sources of large quantities of raw material are easily available. In the case of higher plants PE is usually strongly absorbed on the plant tissue and therefore extracts rather than plant sap or tissue particles are customarily used (see Purification Procedures).

Procedure. 20 ml. of 1% pectin solution containing 0.1 *M* NaCl is adjusted to pH 7.5 and placed in a constant temperature bath maintained at 30°. The enzyme solution (0.1 to 5.0 ml.) is then added, the pH is immediately readjusted to 7.5, and the time is noted. The alkali is added at the rate required to keep the mixture at pH 7.5 for 10 minutes or, with weak enzymes, for 30 minutes. Either a pH meter with glass electrode or an indicator solution such as Hinton's mixture[12] may be used. The quantity of enzyme is adjusted so that the pH does not drift below 7.0 before it can be readjusted to 7.5.

Definition of Unit. The activity is expressed[5] either as the milligrams of CH_3O liberated in 30 minutes per milliliter or gram of enzyme ($PMU_{ml.}$ or $PMU_g.$) or as the milliequivalent of ester hydrolyzed per minute per gram of enzyme[4] (PE. u./g.).

Purification Procedures

As noted, PE usually is strongly adsorbed on the water-insoluble cellular constituents of plants, and therefore plant sap, water extracts, or the juice expressed directly or after freezing contains only a fraction of the enzyme present in the tissue. For this reason either a 10% solution of NaCl or an extracting medium adjusted to pH 6.0 to 7.0 or a combination of both is used to obtain the enzyme from the ground tissue.[4,13] The fact that PE is adsorbed on the tissues may be utilized for a preliminary purification of the enzyme by removing the plant sap or juice containing only a little of the enzyme by filtering or centrifuging and quickly washing the water-insoluble tissue with a little water adjusted to pH 3 to 4. After this preliminary purification the tissue is extracted as noted above. It must be borne in mind, however, that PE becomes inactivated when standing in solutions having a pH below 3 or above 8.0.

A simple method of purifying the enzyme depends on the fact that, as the salt is removed from a solution containing PE, the enzyme precipitates.[14] The tissue extract may be best dialyzed in a rocking dialysis

[12] C. L. Hinton, "Fruit Pectins, Their Behaviour, and Jellying Properties," Chemical Publishing Co., New York, 1940.

[13] J. J. Willaman and C. H. Hills, U. S. Patent 2,358,429 (1944).

[14] R. J. McColloch, J. C. Moyer, and Z. I. Kertesz, *Arch. Biochem.* **10**, 479 (1946).

machine against flowing distilled water with a marble inserted to stir the solution within the viscose casing. The dialysis is continued until the water flowing out of the tubes proves to be free of Cl^- ions. During dialysis the enzyme precipitates and can be then separated from the contents of the viscose tube by centrifuging. The precipitates may be redissolved in a small quantity of salt solution and again dialyzed.

TABLE I
EXTRACTION AND PURIFICATION OF PE FROM DRAINED AND WASHED TOMATO PULP[a]

	Volume, ml.	PMU/ml.	Total PMU	Recovery, %
First 10% NaCl extract	100	64.8	6480	70
Second 10% NaCl extract	100	12.6	1260	83[b]
Third 10% NaCl extract	100	4.3	428	88[b]
First dialysis precipitate in 10% NaCl	15	206	3085	40
Second dialysis precipitate in 10% NaCl	1	3060	3060	99

[a] The 100 g. of wet pulp used was prepared by freezing the tomatoes, thawing, draining the serum, washing with 0.05 N HCl, and filtering off the free liquid.[14]
[b] Accumulative percentage of recovery.

In the absence of salt the enzyme is readily adsorbed from a dialyzed solution on diatomaceous earth (Celite).[4,15] The PE can be eluted with dilute salt solution. Precipitation with ammonium sulfate has been also applied to the purification of PE.

Properties

Specificity. PE is a very specific enzyme. Glycol and glycerine esters of pectic acids are not attacked,[16] and the de-esterification of fifty various nongalacturonide esters proceed at least 1000 times as slowly as that of natural pectin.[17] The synthetic polygalacturonic acid polymethyl ester methyl glycoside and α-D-galacturonic acid methyl ester and its methyl glycoside are also hydrolyzed much more slowly than the ester groups in natural pectin.[5] However, apparently PE is capable of hydrolyzing the half-calcium salt of methyl-D-tartaric acid.[18]

Activators and Inhibitors. The effects of salts on PE has been extensively studied. Many salts increase the activity of the PE of higher plants

[15] L. R. MacDonnell, E. F. Jansen, and H. Lineweaver, *Arch. Biochem.* **6,** 389 (1945).
[16] H. Deuel, *Helv. Chim. Acta* **30,** 1523 (1947).
[17] L. R. MacDonnell, R. Jang, E. F. Jansen, and H. Lineweaver, *Arch. Biochem.* **28,** 260 (1950).
[18] C. Neuberg and C. Ostendorf, *Biochem. Z.* **229,** 464 (1930).

several-fold, whereas in salt-free solutions PE is almost inactive.[4] With divalent cations and at pH 6.0, maximum activation is reached in the neighborhood of 0.03 M concentration.[15] Monovalent cations usually produce maximum activation at pH 6.0 in 0.10 M concentration and do not suppress activity below molarities of 1.0. Lineweaver and Ballou proposed a hypothesis to explain the mechanism of activation.[4] The salt activation of the PE of molds is much less pronounced.[9]

PE is unusually resistant to the effect of chemical agents, but synthetic detergents of the sodium lauryl sulfate and alkyl aryl sulfate type inactivate PE in low concentrations. Fungal PE is much more resistant to such agents than the PE of higher plants.[9]

Temperature Effects. The PE of higher plants is comparatively heat resistant; above 60° the enzyme is gradually inactivated, but at pH 4.0 the enzyme solution has to be heated to about 80° in order to inactivate the PE in a few seconds.[19] Mold PE is more sensitive to heat.

Effect of pH. The pH optimum is influenced by the amounts and kinds of cations present. In general, the effect of added salts is to lower the pH at which maximum activity is attained and to extend the activity range to lower pH regions. In the pH range 7.0 to 8.0 salts have little or no effect.[4,9,20] The activity of plant PE increases as the pH is raised gradually from 4.0 to 7.0; above pH 7.0 the situation is not clear.

Contrary to the PE of higher plants, a fungal PE is active in salt-free medium and shows a well-defined pH optimum at about 4.5 to 5.0. Addition of salts will not cause a drift of the optimum pH range as in the case of the PE of higher plants.[9]

B. Polygalacturonases (PG)

The PG cleave hydrolytically the glycoside bonds of pectic substances. Although until a few years ago only one type of PG was recognized,[2] recent investigations indicate variations both with respect to substrate specificity as related to extent of methylation of the polygalacturonic acid,[21] and concerning the extent of hydrolysis which PG are able to achieve (size specificity?).[22-26] At the present time and for the purpose of discussing

[19] Z. I. Kertesz, *Food Research* **4**, 113 (1939).
[20] M. Holden, *Biochem. J.* **40**, 103 (1946).
[21] C. G. Seegmiller and E. F. Jansen, *J. Biol. Chem.* **195**, 327 (1952).
[22] R. J. McColloch and Z. I. Kertesz, *Arch. Biochem.* **17**, 192 (1948).
[23] B. S. Luh and H. J. Phaff, *Arch. Biochem. and Biophys.* **33**, 212 (1951).
[24] A. Ayres, J. Dingle, A. Phipps, W. W. Reid, and G. L. Solomons, *Nature* **170**, 834 (1952).
[25] E. Roboz, R. W. Barratt, and E. L. Tatum, *J. Biol. Chem.* **195**, 459 (1952).
[26] J. Ozawa, A. Takeda, and K. Okamoto, *Nôgaku Kenkyū* **41**, 21 (1953).

enzyme methods the estimation of PG might be considered irrespectively of these complications, since the methods used are similar for these various types of PG. Often more than one method has to be applied to establish the type of PG dealt with.

Principles. PG is hardly ever estimated by the determination of the amount of substrate left unchanged, an approach not very suitable in case of polymers. Rather, the physical changes which occur or the increase in the reducing end groups (reducing power) are used.[2] There are still difficulties in the interpretation of the meaning of physical measurements, as of viscosity changes, for instance, yet this method is very convenient for comparative measurements. Ostwald or Ostwald-Cannon-Fenske viscosimeters at 30° are usually used with about 0.1% solutions containing 0.8% NaCl and 0.2% Calgon.[27] Only the end-group method will be described here. The recently proposed "cup-plate"[28] and nepholometric methods[29] are as yet insufficiently tested to obtain a clear picture of their usefulness.

Reagents

> Pectic acid solution, 0.5%. (In certain studies pectin or the synthetic polymethyl ester of pectic acid may be used to ascertain substrate specificity.)
> 1 M Na_2CO_3 solution.
> 0.1 N iodine solution.
> 2 M H_2SO_4.
> 0.05 N $Na_2S_2O_3$ solution.
> Galacturonic acid monohydrate solution (up to 1.0 ml.).

Procedure.[27] To 99 ml. of 0.5% pectic acid, previously adjusted to pH 4.0 and 25°, 1 ml. of enzyme solution is added. The time is noted, and 5-ml. aliquots are removed at various times (depending on enzyme activity) and added to 0.9 ml. of the Na_2CO_3 solution in a glass-stoppered flask. Now 5 ml. of iodine solution is added, and after thorough mixing the mixture is allowed to stand for exactly 20 minutes; then 2 ml. H_2SO_4 is added and the excess of iodine titrated with the $Na_2S_2O_3$ solution. A calibration curve is prepared with the galacturonic acid monohydrate solution. Under these conditions 1 meq. of reduced iodine corresponds to

[27] H. S. Owens, R. M. McCready, A. D. Shepherd, T. H. Schultz, E. L. Pippen, H. A. Swenson, J. C. Miers, R. F. Erlandsen, and W. D. Maclay, Methods used at Western Regional Research Laboratory for extraction and analysis of pectic materials. U.S. Department of Agriculture, AIC-340, June, 1952.

[28] J. Dingle, W. W. Reid, and G. L. Solomons, *J. Sci. Food Agr.* **4,** 149 (1953).

[29] K. Vas, *Acta Chim. Acad. Sci. Hung* **3,** 165 (1953).

0.513 mM. of aldose liberated. Controls are run without enzyme as well as with heated enzyme solution.

Definition of Unit. The activity may be expressed as (PG. u.)/ml., indicating the millimoles of reducing groups liberated per minute per milliliter of enzyme.

Occurrence. Studies of PG usually deal with PG of fungal origin. In addition to most fungi, bacteria and yeasts also produce pectolytic enzymes. Whereas many microorganisms capable of forming PG (or, more correctly, some types of PG) have been listed by several authors,[2] it is clear that the conditions of growth will to a great extent govern the PG activity. With our limited understanding of the influence of the various factors involved, it would seem of little help or even misleading to list organisms in which PG activity has been demonstrated. The pectolytic enzymes produced by microorganisms differ in characteristics (substrate and size specificity), but too little is known about this subject to warrant any detailed discussion on classification. Suffice it to say here that the PG of tomatoes[22] seems to be different from that produced by mold and used in commercial pectinase preparations. Some of the differences will be noted below under "Specificity."

Purification Procedures

PG is the major active principle in many commercial enzyme preparations used for fruit juice clarification and other purposes.[2] Such preparations contain, of course, many other enzymes in addition to PG. The laboratory methods for the purification of commercial PG or of PG obtained from microorganisms are not well developed. The following method is one which has been applied recently:[30] A cold (5°) 5% extract of the crude enzyme is adjusted to pH 3.0 with 4 N HCl and stirred in an ice bath with freshly washed alginic acid for about 5 minutes, using 0.1 g. of alginic acid per unit of PG. The alginic acid on which the enzyme is adsorbed is centrifuged off in the cold and washed with half the original volume of cold 1 M NaCl. The first eluate containing impurities and a little PG may be discarded; later similar extracts are quickly adjusted to pH 5 to prevent inactivation of the PG by the low pH. The elution is repeated two to three times. If inactivation of the PE in the eluate is desired, the pH is adjusted to 3.2 and the solution is allowed to stand with toluene at room temperature for 18 hours, and then the pH is adjusted to 5. The PG solution may be dialyzed and lyophilized (freeze-dried). The PG yield obtained by this method was about 42%.

[30] H. Lineweaver, R. Jang, and E. F. Jansen, *Arch. Biochem.* **20,** 137 (1949).

TABLE II
PURIFICATION OF PG OF THE FUNGAL PREPARATION "PECTINOL 45AP" WITH ALGINIC
ACID ADSORPTION AT 5°[a]

	Activity yield,[b] %	Activity, units/mg. total N
Original	100	0.24
Supernatant from adsorption on alginic acid	20	0.07
0.1 M NaCl eluate	<0.2	—
First 1 M NaCl eluate	26	0.83
Second 1 M NaCl eluate	18	0.76
Third 1 M NaCl eluate	8	0.71

[a] Adapted from H. Lineweaver, R. Jang, and E. F. Jansen, *Arch. Biochem.* **20,** 137 (1949).

[b] The yields are based on the original activity.

Properties

Specificity. The situation concerning the specificity of PG or the existence of various PG is far from clear. Pectic acid used to be considered the true substrate of mold PG, and indeed the increasing activity as the methyl ester is progressively removed from pectins has been demonstrated by several authors.[30-33] Purified PG seems unable to attack methyl-D-galacturonic acid. Glucuronic acid compounds such as mesquite gum, oxidized cellulose, carboxymethylcellulose, hyaluronic acid, starch, sucrose, maltose, inulin, *Pneumococcus* polysaccharide Type I (which contains some 60% anhydro-D-galacturonic acid), and several other compounds are not hydrolyzed.

The pectolytic enzyme of tomatoes was found to be able to hydrolyze the substrate (pectic acid) to a limited extent only and without the formation of (mono)galacturonic acid.[22] Several similar enzymes have been reported in recent years. These enzymes, designated as "depolymerases," may be regarded as size-specific PG, for the present at least. On the other hand, pectolytic enzymes found in *Neurospora crassa*[25] and in a commercial enzyme preparation[21] seem to show a preference for pectins (pectinic acids) as substrates. These enzymes have been called "polymethylgalacturonases." Some PG seem to attack preferentially the bonds by which terminal units of polygalacturonic acids are attached whereas other PG act in a random manner.[34,35]

[31] E. F. Jansen and C. R. MacDonnell, *Arch. Biochem.* **8,** 97 (1945).

[32] F. Weber and H. Deuel, *Mitt. Lebensm. Hyg.* **36,** 368 (1945).

[33] S. A. Waksman and M. C. Allen, *J. Am. Chem. Soc.* **55,** 3408 (1933).

[34] J. Matus, *Ber. Schweitz. Botan. Ges.* **58,** 319 (1948).

[35] J. Solms, H. Deuel, and L. Anyas-Weisz, *Helv. Chim. Acta* **35,** 2363 (1952).

The whole field of PG specificity is not too clear at the present time, and it is likely that a whole series of pectolytic enzymes of various specificity, activity and stability exist.

Activators and Inhibitors. PG seems to be activated according to the Hofmeister and Schulze-Hardy ionic series by the addition of various electrolytes.[34] For the present such reports should be regarded with some reservation, since there are several complicating factors not entirely eliminated in activation studies.

PG can be specifically inactivated by urea,[36] and glycine[37] and formaldehyde[38] also have been claimed to inactivate PG.

Temperature Effects. In the form of a dry powder the PG of molds is quite resistant to heat. In solution, PG is protected against thermal inactivation by the addition of alginate, glycerol, or even sucrose. PG is quite unstable even at room temperature, and when kept at 55° for 30 minutes or at pH 4.0 at 50° for 3 hours complete loss of activity may result.[2] As is common with enzymes, the thermal behavior of PG depends a great deal on the pH of the solution.

Effect of pH. Older experiments found the pH optimum of PG around 3.5, but more recent work indicates in many cases optima around 4.0 to 4.2.[2] Since it is known that there are different types of pectolytic enzymes and that these at times occur in mixtures, such variations are not surprising. PG is inactive at pH 7. There is still some uncertainty concerning the true pH optima of some of the more recently discovered pectolytic enzymes, most of which seem to have pH optima higher than that of PG.

All these observations concerning pH and temperature effects will have to be critically re-examined when the different types of pectolytic enzymes are clearly characterized and when one can have some assurance of the presence of only one type of enzyme in the material investigated.

[36] C. V. Smythe, B. B. Drake, and J. A. Miller, U. S. Patent 2,599,531 (1952).
[37] R. Otto and G. Winkler, German Patent 729,667 (1942).
[38] A. Mehlitz and H. Maass, *Z. Untersuch. Lebensm.* **70**, 180 (1935).

[19] Mucopolysaccharidases

By ALBERT DORFMAN

Although several mucopolysaccharides have now been well character-ized, little is known regarding mucopolysaccharidases other than Hyase.[1] This section is accordingly confined to a consideration of the assay of Hyase.

[1] Hyase is used to denote hyaluronidase, HA is used to denote hyaluronic acid and CSA to denote chondroitinsulfuric acid.

General Considerations. The assay of Hyase poses many difficulties from both a theoretical and a practical point of view for the following reasons:[2] (1) the enzyme has never been isolated as a homogeneous protein, although recently Malmgren[3] has claimed that a preparation of Hogberg appears to be homogeneous on the basis of sedimentation and diffusion studies; (2) the reaction catalyzed by this enzyme is not yet clearly defined; (3) there is evidence that Hyases from different sources are not uniform in their mechanism of action; (4) pure HA [1] is not readily available; (5) preparations of HA from certain natural sources are apt to be contaminated by inhibitory substances which are chemically related to HA; and (6) the kinetics of the reaction of Hyase on HA appear to be complex.

Although it is beyond the scope of this discussion to elucidate the present state of knowledge regarding these factors, some of these must be considered for an intelligent approach to the problem of determination of Hyase activity.

The term Hyase has usually been applied to the activity responsible for the enzymatically induced change in certain physical properties of the mucopolysaccharide, HA. Such activity is widely distributed in nature, although it has conclusively been demonstrated only in testicular tissue in the mammal. There is evidence that the action of Hyase on HA results in the liberation of reducing groups of the N-acetylglucosamine portion of the HA, suggesting that the enzyme is a glucosaminidase. Under certain conditions a disaccharide appears to be liberated, although the nature of the reaction products apparently varies with the source of the enzyme.[2] Since pure enzyme has not been studied, it is not clear whether the products so far isolated result from the action of one or of several enzymes. Meyer *et al.*[4] have suggested that the complete degradation of HA to its constituent monosaccharide units requires in addition to Hyase the participation of β-glucuronidase which brings about rupture of the glucuronide bond. It is not as yet clear whether any type of linkage other than the β-N-acetylglucosaminide or β-glucuronide bonds may play a role in the structure of high molecular weight mucopolysaccharides. Indeed there is evidence that treatment with alkali does cause some degradation, and in the case of CSA[1] Blix and Snellman[5] have shown that this degradation is accompanied by the release of carboxyl groups.

Certain problems regarding the specificity of Hyase are raised by the finding that, although bovine testicular Hyase has equal activity toward

[2] K. Meyer and M. M. Rapport, *Advances in Enzymol.* **13**, 199 (1952).

[3] H. Malmgren, *Biochim. et Biophys. Acta* **11**, 524 (1953).

[4] K. Meyer, A. Linker, and M. M. Rapport, *J. Biol. Chem.* **192**, 275 (1951).

[5] G. Blix and O. Snellman, *Arkiv Kemi, Minerol. Geol.* **19A**, No. 32 (1944).

HA and CSA,[6,7] Hyases prepared from filtrates of hemolytic streptococci and pneumococci are apparently without activity toward CSA.[2]

Although testicular Hyase apparently hydrolyzes CSA prepared from cartilage, other naturally occurring sulfated mucopolysaccharides act as inhibitors of this enzyme. Thus heparin has long been known to inhibit Hyase.[8] Recent studies have indicated that connective tissue ground substance is composed of a mixture of both sulfated and nonsulfated acid mucopolysaccharides. It is clear that some of the sulfated polysaccharides with properties which are similar to cartilage CSA are not acted on by the enzyme or are contaminated with inhibitory substances.[9]

The lack of definitive knowledge regarding the reaction catalyzed by Hyase poses obvious difficulties regarding an elucidation of the kinetics involved. A number of studies have shown that the reaction follows apparent first-order kinetics for only a short period of time.[2,10-13] This may be due to the impossibility of employing excess substrate, the presence of inhibitors in the substrate, inhibition by reaction products, or the participation of a variety of reactions in the production of the change being observed.

Although this discussion is primarily concerned with *in vitro* measurements of Hyase activity, brief mention should be made of the biological activity of Hyase. Since the original demonstration by Duran-Reynals of the existence of a spreading factor in testicular extracts and the subsequent identification of this activity with Hyase, it has been generally accepted that Hyase has the capacity of increasing the rate of spread of various dyes or suspensions of carbon particles when injected intradermally.[14] This biological potency forms the basis of the clinical use of Hyase to promote the absorption of fluids administered subcutaneously or the diffusion of drugs such as local anesthetics injected into the tissues. Although the spreading phenomenon is easily demonstrated, its quantitative measurement is on the whole unsatisfactory, despite numerous attempts at its improvement. In view of the importance of standardizing Hyase for clinical use, Humphrey and Jaques[15] have recently attempted to correlate the results of the method of Jaques[16] of testing spreading with

[6] K. Meyer and M. M. Rapport, *Arch. Biochem.* **27**, 287 (1950).

[7] M. B. Mathews, S. Roseman, and A. Dorfman, *J. Biol. Chem.* **188**, 327 (1951).

[8] D. McClean, *J. Pathol. Bacteriol.* **54**, 284 (1942).

[9] K. Meyer and M. M. Rapport, *Science* **113**, 596 (1951).

[10] A. Dorfman, *J. Biol. Chem.* **172**, 377 (1948).

[11] H. Gibian, *Z. physiol. Chem.* **289**, 1 (1951).

[12] J. G. Bachtold and L. P. Gebhardt, *J. Biol. Chem.* **194**, 635 (1952).

[13] R. H. Pearce, *Rev. can. biol.* **11**, 76 (1952).

[14] F. Duran-Reynals, *Bacteriol. Revs.* **6**, 197 (1942).

[15] J. H. Humphrey and R. Jaques, *Biochem. J.* **53**, 59 (1953).

[16] R. Jaques, *Biochem. J.* **53**, 56 (1953).

both the viscosity and turbidity methods of measuring enzyme activity. They found approximate correlations under certain conditions, although these are limited in their quantitative significance in view of the fact that the biological method employed showed an error of 70 to 120 %. Hyase has also been measured in terms of decapsulation of streptococci, but neither the specificity nor the accuracy of this method has been established.

Principles of *in Vitro* Measurements

Viscosity Reduction. Numerous investigators have demonstrated that the action of Hyase upon HA results in a decrease in viscosity, the rate of which is a function of enzyme concentration. This phenomenon has frequently been used as the basis of methods of assay of hyaluronidase derived from the original method of Madinaveitia and Quibell.[17] There is considerable evidence, however, that the absolute rate of viscosity decrease varies with different substrate preparations, depending on differences in initial viscosity and other unknown factors.[18] It is thus apparent that various viscosity units which have been defined cannot be used for absolute standardization of the enzyme. The viscosity method is also finding less application because of the tedious nature of the necessary repeated viscosity determinations to establish rates of viscosity decrease, although Swyer and Emmens[19] have suggested a simpler method which requires only one viscosity determination for each assay.

Mucin Clot Prevention. Robertson *et al.*[20] showed that the action of Hyase upon HA prevents the formation of a "mucin clot" when HA is mixed with albumin at an acid pH. This phenomenon has been studied in detail by McClean[21] and has been utilized for the quantitative determination of Hyase, particularly with respect to Hyase inhibitors. This method, although rapid and sensitive is of low accuracy.

"Spinnbarkeit." Gunter[22] has described a method for the assay of Hyase which depends on the fact that the action of the enzyme upon synovial fluid prevents the formation of fibers about a rotating rod (based upon "stringiness" of mucins). He demonstrated that the time required for reduction of "Spinnbarkeit" is inversely proportional to enzyme concentration. This method has not been sufficiently studied to determine its usefulness as a method of Hyase assay.

Turbidity Reducing Methods. Kass and Seastone[23] showed that, when HA is mixed with serum at an acid pH, a turbidity develops which can be

[17] J. Madinaveitia and T. H. H. Quibell, *Biochem. J.* **34**, 625 (1940).
[18] H. E. Alburn and R. W. Whitley, *J. Biol. Chem.* **192**, 379 (1951).
[19] G. I. M. Swyer and C. W. Emmens, *Biochem. J.* **41**, 29 (1947).
[20] W. van B. Robertson, M. W. Ropes, and W. Bauer, *J. Biol. Chem.* **133**, 261 (1940).
[21] D. McClean, *Biochem. J.* **37**, 169 (1943).
[22] G. S. Gunter, *Australian J. Exptl. Biol. Med. Sci.* **27**, 265 (1949).
[23] E. M. Kass and C. V. Seastone, *J. Exptl. Med.* **79**, 319 (1944).

prevented by the previous action of Hyase on HA. This method has been studied in considerable detail by a number of investigators who have introduced a variety of modifications.[2,12,18,24—28] The details of such a method are described below. The following considerations are of importance in the use of this method.

1. The method is not applicable to crude tissue extracts.

2. The development of turbidity is a function of concentration of HA, protein, pH, ionic strength, presence of specific ions, time, and temperature. If Hyase activity is to be successfully measured, it is important that all variables be rigidly controlled. Many of the modifications which have appeared in the literature have been concerned with redetermination of one or more of the above factors.

3. The action of the enzyme on the substrate is a function of the nature of the substrate preparation, pH, ionic strength, specific ionic composition, temperature, and enzyme concentration.

Detection of End Products. Earlier studies attempting to base Hyase activity on the release of acetylhexosamine or reducing sugars gave unsatisfactory results.[29] More recently Rapport *et al.*[30] have claimed a good correlation between turbidity reducing activity and measurement of reducing power by the method of Park and Johnson. The best enzyme preparations used by these authors, however, were relatively crude.

Precipitation by Acid Alcohol. Burnet[31] observed that the treatment of synovial fluid with Hyase prevented its precipitation by acid alcohol. Evans *et al.*[32] have attempted to apply this to the quantitative determination of Hyase. Their published results would indicate that, although this method is rapid and has qualitative validity, it does not give promise of a high degree of accuracy.

Standardization

In view of the difficulty of delineating the reaction catalyzed by the enzyme, it is at present impossible to define a unit of enzyme activity in kinetic terms with respect to a specific reaction. For this reason the enzyme is best standardized in terms of an arbitrary standard preparation.

[24] S. L. Leonard, P. L. Perlman, and R. Kurzrok, *Endocrinology* **39**, 261 (1946).

[25] A. Dorfman and M. L. Ott, *J. Biol. Chem.* **172**, 367 (1948).

[26] S. Tolksdorf, M. H. McCready, D. R. McCullagh, and E. Schwenk, *J. Lab. Clin. Med.* **34**, 74 (1949).

[27] K. Schmith and V. Faber, *Scand. J. Clin. Lab. Invest.* **2**, 292 (1950).

[28] R. L. Greif, *J. Biol. Chem.* **194**, 619 (1952).

[29] K. Meyer, *Physiol. Revs.* **27**, 335 (1947).

[30] M. M. Rapport, K. Meyer, and A. Linker, *J. Biol. Chem.* **186**, 615 (1950).

[31] F. M. Burnet, *Australian J. Exptl. Biol. Med. Sci.* **26**, 71 (1948).

[32] D. G. Evans, F. T. Perkins, and W. Gaisford, *Lancet* **260**, 1253 (1951).

All unknowns should then be compared to this standard, utilizing a fixed assay method and the same substrate preparations. Although highly purified substrate preparations are desirable for this purpose, the difficulty of preparing adequate amounts of such material makes necessary the use of cruder materials which can be utilized if the conditions outlined above are observed. Attempts are being made to agree on a standard preparation to become available through the U. S. Pharmacopeia.[33]

Standard enzyme preparations may be prepared according to the method outlined below and have been shown to be stable after lyophilization for at least three years.

Preparation of Enzyme.[34] All operations are carried out at 0°. 200 g. of bull testis is ground in a meat grinder and mixed with 190 ml. of 0.1 M acetic acid; 10.0 ml. of 2.0 N HCl is added slowly with vigorous mixing. This mixture is stirred for 4 hours, at the end of which time the pH should be approximately 3.6. After centrifugation the precipitate is discarded, and 21.2 g. of $(NH_4)_2SO_4$ is added to the supernatant. After standing for 1 hour, the solution is centrifuged. This precipitate contains a relatively small amount of enzyme of low activity. To the supernatant is added 28.2 g. of $(NH_4)_2SO_4$ per 100 ml. of the original supernatant, and the mixture is allowed to stand for approximately 12 hours before centrifugation. The precipitate is dissolved in H_2O and dialyzed free of $(NH_4)_2SO_4$. 65,000 units[35] with an activity of 1600 units/mg. of N is obtained from the original 200 g. of tissue. This preparation may be lyophilized and used for most purposes.

Further fractionation can be carried out by the addition of 300 g. of ammonium sulfate to a liter of solution containing 10 g. of crude enzyme. After 1 hour, the precipitate (about 10% of starting enzyme) is removed by filtration, and an additional 100 g. of ammonium sulfate is added to the supernatant. The second precipitate is dissolved in 100 ml. of water and dialyzed thoroughly against a phosphate-citrate (McIlvaine) buffer, pH

---

[33] At the time this article was prepared for publication the U.S.P. method of assay was not available. The following information was, however, obtained from Lloyd C. Miller, Director of Revision. (a) It has been agreed that a turbidity method will be used as the U.S.P. method. (b) A standard substrate will not be furnished. (c) A reference standard enzyme will be made available in a diluted stable form and can be obtained from U. S. Reference Standards, 46 Park Avenue, New York 16, New York. (d) The unit used by the author has been found to be equal to 1.56 units of the provisional U.S.P. standard when compared by the method of assay presented in this paper.

[34] This method was developed in the author's laboratory in collaboration with Dr. Martin B. Mathews.

[35] This unit is arbitrary, having been maintained constant by comparing to reference standard in the author's laboratory.

6.5, μ = 0.1, containing 0.28% sodium chloride. The yield is about 50% at a purity of 11,000 units/mg. of N.

Alcohol fractionation is carried out by the addition, with vigorous agitation, of 95% alcohol precooled to −6° (330 ml. measured at 25°) to 1 l. of a 1% enzyme solution which is maintained below 0.5° during alcohol addition. The enzyme-alcohol mixture is allowed to remain overnight at −6°. The precipitate is separated by centrifugation at −6°, dissolved in 300 ml. of the above-mentioned buffer, and dialyzed overnight at 1° against the same buffer. The yield is about 90% of starting enzyme with a purity of 32,000 units/mg. of N. Such preparations are still markedly heterogeneous.

Assay Method

Reagents

Buffer solution for enzyme: 0.02 M sodium phosphate buffer, pH 6.8 to 7.0; 0.45% NaCl; 0.01% bovine albumin (Armour Fraction V).[36]

Standard enzyme is dissolved in this solution to give concentration of 24 units/ml.

Hyaluronic Acid Solution. Crude HA is prepared from umbilical cords as previously described,[25] with the exception that the final product is dialyzed to remove excess salt. Methods of preparation of HA will be described in a subsequent volume. A stock solution of HA containing approximately 4 mg./ml. is prepared by dissolving the HA in a 0.30 M KH₂PO₄·Na₂HPO₄ buffer, pH 5.30 to 5.35. The HA is brought into solution with a Waring blendor or a mechanical shaker. Toluene is added to this solution as a preservative. A series of dilutions of this stock solution varying from 1:10 to 1:5 (thus containing approximately 0.4 to 2.0 mg. of HA per milliliter) are made. 1.0 ml. of each of these solutions is mixed with 1.0 ml. of the pH 7.0 phosphate buffer used for the solution of enzyme and incubated for 5 minutes, after which turbidity is developed and determined as described below. The dilution which gives a transmission of 54 to 56% is chosen, and the stock HA is diluted accordingly and is ready for use in the assay.

Acid Albumin Solution

32.6 g. of sodium acetate.
45.6 ml. of glacial acetic acid.
10.0 g. of albumin (Armour Fraction V).[36]
10 l. of H₂O.

[36] Available from the Armour Laboratories, Chicago, Ill.

The pH of the above solution is adjusted to 3.72 to 3.78 with 1:1 conc. HCl.

Procedure. Unknown enzyme preparations are dissolved in or diluted with phosphate buffer described above for standard enzyme solution. With each group of assays performed, a standard curve is included containing at least five points; 0, 2, 4, 6, and 8 units of enzyme per tube.[37] All determinations are performed in duplicate, and assays of unknown enzyme solutions are accepted only if two dilutions fall within the assay range. For best accuracy these should be between 2 and 8 units.

To perform assay, 1.0 ml. of enzyme solution is mixed with 1.0 ml. of HA solution (solutions are previously brought to 38°) and incubated at 38° for 45 minutes. At the end of this time 10 ml. of acid albumin solution (at room temperature) is added rapidly, and exactly 5 minutes later (by stopwatch) the optical density is determined in a photoelectric colorimeter. When a series of determinations is performed, reagents are added to different tubes at 30-second intervals. A Coleman Junior spectrophotometer set at 600 mμ with 15-mm. cuvettes has usually been employed, but the method is readily adapted to other comparable instruments.

Calculation of Results. A standard curve is constructed by plotting optical density against enzyme concentration per tube. Unknown samples are then determined from standard curve. The standard curve has been found to be constant from day to day, but it is redetermined daily to check on reagents.

Accuracy. Accuracy varies at different points on standard curve; at the 8-unit level the coefficient of variation has been found to be 5%.

[37] The activity is expressed in units per tube containing a reaction mixture of 2.0 ml. Thus the activity per milliliter as used in the previous publication[25] is one-half of that given above.

[20] Cellulase Preparation from *Helix pomatia* (Snails)

By GEORGE DE STEVENS[1]

Assay Method

Principle. The development of a method for recording the activity of cellulase preparations in absolute or reproducible terms has not been accomplished, owing to the heterogeneity of the reaction and also to the effect of the physical state of the cellulose substrate. Pringsheim[2] identi-

[1] The author wishes to express his appreciation to the authorities of Hall Laboratory of Chemistry, Wesleyan University, for the facilities placed at his disposal.
[2] H. Pringsheim, *Z. physiol. Chem.* **78**, 266 (1912).

fied the degradation product, glucose, as the osazone. Von Euler[3] and Karrer et al.[4] determined the extent of celluloysis through the Cu numbers. Karrer and Schubert[5] also followed the change by recording the viscosity of the solution at definite intervals. The procedure described herein has been used by Trager[6] and Ziese.[7] It is essentially the Willstätter-Schudel[8] method as modified by Kline and Acree.[9]

Reagents

    100 mg. of cellulose.
    2 ml. of phosphate buffer solution, pH 5.28.
    3 drops of toluene.
    0.4 ml. of enzyme solution.
    Add 5 ml. of water and store in the incubator at 36°.

Procedure. After the desired interval, titrate the reaction solution with NaOH or HCl solution until it is exactly neutral to phenolphthalein. Add 5 ml. of 0.1 N iodine from a buret, and then introduce dropwise 7.5 ml. of 0.1 N NaOH. Repeat this process until 22 ml. of I_2 solution and 35 ml. of NaOH have been added during 5 to 6 minutes. Let the reaction mixture stand for 5 minutes, make it acid with 0.1 N HCl, and titrate the liberated I_2 with $Na_2S_2O_3$. The amount of glucose present in solution is calculated from the amount of I_2 liberated.

Definition of Unit and Specific Activity. Karrer et al.[4] proposed that the cellulase unit be the amount of cellulase which in 50 ml. at a pH of 5.28 and a temperature of 36° will bring about 20% hydrolysis of 1 g. of fibrous cupraammonium rayon (Zellvag) in 96 hours.

Purification Procedure

Step 1 of the following procedure is based on the method described by Karrer,[10] and step 2 has been outlined by Grassmann et al.[11]

Step 1. The cellulose-splitting enzyme of *Helix pomatia* is found in the liver. The snails are suffocated by immersion in water in a closed vessel. The darkest-colored skin casing and the underlying light pale viscera

[3] H. von Euler, *Z. angew. Chem.* **25**, 250 (1912).
[4] P. Karrer, P. Schubert, and W. Wehrli, *Helv. Chim. Acta* **8**, 797 (1925).
[5] P. Karrer and P. Schubert, *Helv. Chim. Acta* **9**, 893 (1926).
[6] W. Trager, *Biochem. J.* **26**, 1762 (1932).
[7] W. Ziese, *Z. physiol. Chem.* **203**, 87 (1931).
[8] R. Willstätter and G. Schudel, *Ber.* **51**, 780 (1918).
[9] G. M. Kline and S. F. Acree, *Ind. Eng. Chem. Anal. Ed.* **2**, 413 (1930).
[10] P. Karrer, *Kolloid-Z.* **52**, 304 (1930).
[11] W. Grassmann, L. Zechmeister, G. Tóth, and R. Stadler, *Ann.* **503**, 167 (1933).

skin are then cut open. The gastrointestinal tract swells out of itself and can be easily detected. It is drawn out carefully, grasped on the back end with forceps, and separated in the region of the mouth. The posterior opening is then closed with pinch-clamps while the anterior opening closes of itself in the area of incision. The correct execution of this operation leads to a minimum loss of contents. The collected intestinal pieces are placed in a mortar with toluene, ground very fine with sand, covered with a small amount of water, and filtered through asbestos. The resulting brown solution consists of a mixture of various enzymes plus intestinal matter and sugar.

TABLE I

SPECIFICITY OF FRACTIONATED ENZYME PREPARATIONS[a]

| | | Cellulase | | Cellobiase | |
| | | Amount of hydrolysis in ml. of 0.02 N I_2 | | | |
Substrate	Mg.	8 hours	24 hours	8 hours	24 hours
Hydratcellulose	13.6	0.40	0.55	0.05	0.10
Cellodextrin	13.6	2.17	2.57	0.05	0.20
Cellotriose	21.0	0.00	0.05	1.85	2.70
Cellobiose	13.6	0.00	0.00	1.90	2.90

Note: Each solution contains 1.2 ml. of enzyme solution eluted to 4 ml. with 0.25 M phosphate buffer, pH 5.2. Temperature, 30°.

[a] W. Grassmann, L. Zechmeister, G. Tóth, and R. Stadler, *Ann.* **503**, 167 (1933).

The extraneous substituents are removed by dialyzing the solution for several days. Cellulase has the remarkable property of remaining quite stable in aqueous solution.

Step 2. 33 ml. of the dialyzed enzyme solution is combined with 20 ml. of 0.2 N acetate mixture at pH 3.5. This is then treated with a suspension of aluminum metahydroxide (corresponding to 370 mg. of Al_2O_3), and water is added to make a volume of 100 ml. After centrifugation, the residual solution is treated with 10 ml. of an alumina suspension (corresponding to 250 mg. of Al_2O_3) and centrifuged again. In this way a residual solution of high cellulase content with cellobiase being adsorbed on the Al_2O_3 is obtained.

In Table I are recorded the data demonstrating the degree of separation of enzymes attained.

Properties of Enzyme

Specificity. Since, to date, no method has been devised to isolate pure crystalline cellulase enzyme, very little can be said concerning its absolute specificity. However, the evidence does demonstrate that cellulase catalyzes the breakdown of polysaccharides (cellulose) to glucose. That this enzyme is also instrumental in the hydrolysis of cellobiose to glucose has not, as yet, been proved unequivocally, although Grassmann's[11] work does indicate that such is not the case.

Activators and Inhibitors. It has been shown[7] that 1% $CuSO_4$ and 3% HCN separately have practically no effect on the activity of snail cellulase, but the mixture is strongly inhibitory. Glutathione and cysteine, but not H_2S or $Na_2S_2O_3$, are inhibitory when phosphate buffer is used, but not when citrate buffer of the same pH is used. In contrast to the snail enzyme, malt cellulase is not inhibited by $CuSO_4$ plus HCN. Glutathione and cysteine, in phosphate or citrate buffer, also have no inhibitory effect on this enzyme preparations. Snail cellulase is also resistant to papain-HCN.

Walseth[12] demonstrated that *Aspergillus niger* cellulase preparation is resistant to pentachlorophenol but is inactivated by other antiseptics.

Effect of Temperature. The maximum temperature at which cellulase retains its activity has been found to differ with the source of the enzyme as well as with the method of preparation. Pringsheim[2] reports that bacterial cellulase has a temperature optimum at 46°, but its activity extends through a temperature range of 20 to 70°. The *Helix pomatia* cellulase[4] was found to exert its optimum activity at 36°, with complete deactivation at 60°. However, the snail cellulase prepared by Ziese[7] was found to be stable up to 100°. Walseth[12] maintained his *A. niger* cellulase preparation at a temperature of 47° for maximum enzyme activity. Kristiansson[13] found that barley malt cellulase loses two-thirds of its activity on heating at 60° and pH 5 for 5 minutes. The cellulase from the roach, *Cryptocercus punctulatus,*[6] was reported to give favorable results at a temperature of 26°, but inactivation occurred readily at 100° or heating for 1 hour at 60°.

Effect of pH. The data in Table II illustrate the optimal pH of cellulases from different sources. The apparent discrepancies outlined in this table are probably due to the fact that no pure enzyme preparation has been obtained. This becomes notable especially with regard to differences within the same species.

Kinetics of Enzyme Reaction. Although crude enzyme preparations have been used, several workers,[6,7,10,12] in the field have demonstrated that the reaction follows the unimolecular law. Karrer[10] followed the course of the reaction by determining the change in viscosity of the reaction mixture.

[12] C. Walseth, Thesis, Institute of Paper Chemistry, Appleton, Wisconsin, 1948.
[13] I. Kristiansson, *Svensk Kem. Tidskr.* **62**, 133 (1950).

TABLE II
Optimal pH of Various Cellulase Preparations

Source	Optimum pH	Reference
Aspergillus oryzae	4.5	Grassmann *et al.*[a]
Aspergillus oryzae	4.7	Freudenberg and Ploetz[b]
Aspergillus niger	4.5	Walseth[c]
Helix pomatia	5.2–5.5	Freudenberg and Ploetz[b]
Helix pomatia	5.28	Karrer *et al.*[d]
Barley malt cellulase	5.0	Kristiansson[e]
Helix pomatia	4.51	Ziese[f]
Malt cellulase	3.68	Ziese[f]
Merulius-lacrymans	4.7	Ploetz[g]
Cryptocerous punctulatus	5.3	Trager[h]
Earthworms	5.5	Tracey[i]

[a] W. Grassmann, L. Zechmeister, G. Tóth, and R. Stadler, *Ann.* **503**, 167 (1933).
[b] K. Freudenberg and T. Ploetz, *Z. physiol. Chem.* **259**, 19 (1939).
[c] C. Walseth, Thesis, Institute of Paper Chemistry, Appleton, Wisconsin, 1948.
[d] P. Karrer, P. Shubert, and W. Wehrli, *Helv. Chim. Acta* **8**, 797 (1925).
[e] I. Kristiansson, *Svensk Kem. Tidskr.* **62**, 133 (1950).
[f] W. Ziese, *Z. physiol. Chem.* **203**, 87 (1931).
[g] T. Ploetz, *Z. physiol. Chem.* **261**, 183 (1939).
[h] W. Trager, *Biochem. J.* **26**, 1762 (1932).
[i] M. V. Tracey, *Biochem. J.* **47**, 431 (1950); *Nature* **167**, 776 (1951).

Sources of Enzyme. The cellulase enzyme has also been reported to be present in potato sprouts,[14] in *Neurospora sitophila*,[15] and in numerous wood-destroying fungi of the brown rot type—e.g., *Polyporus sulfureous, Lenzites sepiaria, Daedalea quercina,* and *Lentinus lepideus.*[16,16a]

Cellobiase

The literature presents evidence for and against the separation of cellobiase from other enzymes. Pringsheim and his associates[17] found

[14] B. M. Singh, P. B. Mathar, and M. L. Mehta, *Current Sci. (India)* **7**, 281 (1938).
[15] E. Tamura and Y. Takai, *Japan. J. Nutrition* **8**, 129 (1950).
[16] G. de Stevens and F. F. Nord, *Fortschritte Chem. Forsch.* **3**, 70 (1954).
[16a] While this manuscript was in press, the work of D. R. Whitaker [*Arch. Biochem. and Biophys.* **43**, 253 (1953); **49**, 259 (1954)] on the purification of *Myrothecium verrucaria* cellulase was brought to the author's attention. Whitaker reports that his enzyme preparation is homogeneous electrophoretically and in the ultracentrifuge. His cellulase preparation is purified by a sequence of concentration, precipitation by ammonium sulfate, fractionation with ethanol, and precipitation with polymethacrylic acid.
[17] H. Pringsheim and J. Leibowitz, *Z. physiol Chem.* **131**, 267 (1923); H. Pringsheim and W. Kusenack, *ibid.* **137**, 265 (1924).

that aged solutions of malt retained considerable power to hydrolyze lichenin, whereas the cellobiase activity was lost. The work of Karrer and Lier[18] indicated that lichenase and cellobiase could be separated by fractional absorption on aluminum hydroxide. However, the products of hydrolysis were very complex and a simple product was not identified. Freudenberg and Ploetz[19] present similar results. Grassmann et al.[11] and later Zechmeister et al.[20] applied the adsorption technique for the solution of a cellobiase solution. The Grassmann procedure is a continuation of step 2 for the purification of cellulase.

Purification Procedure

The combined absorbed material (Al_2O_3 plus adsorbed enzyme fraction) was stirred with 20 ml. of 0.20 M sodium bicarbonate solution and 80 ml. of water and centrifuged. This neutralized supernatant solution was employed for the activity determination. The results of this determination are outlined in Table I.

Grassmann[11] further indicated that the cellobiase preparation from Aspergillus oryzae exhibited some lichenase activity ($2/3$ cellobiase potency to $1/3$ lichenase).

The pH optimum of cellobiase is between 4 and 5. Enders and Saji[21] found that the enzyme from malt and barley exhibited optimum activity at pH 4 to 4.5 and a temperature of 37°, whereas Grassmann et al.[11] and Zechmeister et al.[20] suggest working in the pH range of 4.5 to 5 and a temperature of 30°. Smith,[22] working in Nord's laboratory, found that the mold Merulius lachrymans hydrolyzed cellobiase initially to glucose at pH 4.28 and a temperature of 22°. Pringsheim[2] reports that cellobiase is deactivated at temperatures above 67°.

[18] P. Karrer and H. Lier, Helv. Chim. Acta 8, 248 (1925).
[19] K. Freudenberg and T. Ploetz, Z. physiol. Chem. 259, 19 (1939).
[20] L. Zechmeister, G. Tóth, P. Fürth, and J. Bársony, Enzymologia 9, 155 (1941).
[21] C. Enders and T. Saji, Biochem. Z. 306, 430 (1940).
[22] V. M. Smith, Arch. Biochem. 23, 446 (1949).

[21] Polysaccharide Synthesis from Disaccharides

By EDWARD J. HEHRE

Introductory Note

This section deals exclusively with enzymes known to catalyze, at least under certain conditions, the synthesis of high molecular weight polysaccharides from disaccharides. No attempt has been made to treat related biological systems that effect syntheses of series of oligosaccharides from disaccharides, or syntheses of polysaccharides from oligosaccharides other than disaccharides. For simplicity, all reactions have been considered principally from the forward or synthetic aspect. Moreover, all polymer products have been defined in generic terms, since none of the polysaccharide synthesizing enzymes has yet been prepared in a state of high purity, and it is not known to what extent accompanying enzymes operate to introduce branching or other structural irregularities in the polymers considered.

I. Dextransucrase

$$n\text{-}C_{12}H_{22}O_{11} + \quad HOR \quad \rightarrow H(C_6H_{10}O_5)_nOR + n\text{-}C_6H_{12}O_6$$

Sucrose   Acceptor   Dextran   Fructose

Assay Method

Principle. Measurement of fructose liberated under conditions providing a zero-order type reaction is the basis for estimating dextransucrase activity (Hehre[1]). The technique given below is that adopted by Koepsell[2] and by Tsuchiya *et al.*[3] for the assay of dextransucrase in culture fluids of *Leuconostoc mesenteroides.*

Reagents

0.04 N NaOH solution, containing phenolphthalein indicator (1 drop of indicator per 100 ml. of NaOH solution).

3 N acetic acid–NaOH buffer, pH 5.4.

Buffered sucrose solution (60%). 10 ml. of 3 M acetate buffer, pH 5.4, and 60 g. of sucrose are dissolved in distilled water and made up to 100 ml. The final pH should be 5.2.

Enzyme. Adjust the cell-free culture fluid to pH 5.0 to 5.2. Prepare an accurate dilution with water, if necessary, to obtain solution for assay containing less than 40 units of enzyme per milliliter. (See definition below.)

Procedure. Pipet 10 ml. of the enzyme solution, attempered to 30°, into a 1 × 8-inch test tube containing 2 ml. of buffered 60% sucrose solution also at 30°. Note time of beginning of the pipetting step. Mix thoroughly and incubate at 30°. At the end of 1 hour pipet 2 ml. of the mixture into a 100-ml. volumetric flask containing 10 ml. of 0.04 N NaOH solution.

[1] E. J. Hehre, *J. Biol. Chem.* **163**, 221 (1946).

[2] H. J. Koepsell, *in* U.S. Department of Agriculture Report of Working Conference on Dextran, Peoria, Illinois, p. 13, 1951; H. J. Koepsell and H. M. Tsuchiya, *J. Bacteriol.* **63**, 293 (1952).

[3] H. M. Tsuchiya, H. J. Koepsell, J. Corman, G. Bryant, M. O. Bogard, V. H. Feger, and R. W. Jackson, *J. Bacteriol.* **64**, 521 (1952); additional details were kindly furnished by Dr. Tsuchiya.

Add more NaOH solution if necessary to make the sample slightly alkaline to phenolphthalein. Make up to volume, mix, and determine the amount of fructose present in the sample by the Somogyi method for reducing sugars[4] (fructose factor = 1.108 × milligrams of glucose equivalent to 1 ml. of 0.005 N $Na_2S_2O_3$). Determine similarly the reducing power (as fructose) in a blank of enzyme made slightly alkaline to phenolphthalein with NaOH. Correct the amount of fructose present per milliliter of enzyme in the reaction mixture for the amount of reducing sugar (as fructose) per milliliter of enzyme in the blank. The difference, representing milligrams of fructose released per milliliter of enzyme, is multiplied by 1.9 or $\left(\dfrac{\text{mol. wt. sucrose}}{\text{mol. wt. fructose}}\right)$ to give the amount of sucrose converted to dextran per milliliter of enzyme.

Definition of Unit. One dextransucrase unit (DSU) is defined as the amount of enzyme which will convert 1 mg. of sucrose to dextran in 1 hour (releasing 0.52 mg. of fructose) under the above conditions.

Extent and Limitations of Application. Crude filtrates or fluids from cultures of various bacteria that form dextran from sucrose, as well as solutions of partially purified dextransucrase, may be assayed with a high degree of accuracy by the above method. Two sources of potential interference, however, are recognized.

1. The presence of any other enzyme capable of yielding reducing sugar from sucrose will, of course, render the assay inaccurate. Since certain strains of dextran-forming bacteria are able also to hydrolyze sucrose or convert it to levan plus glucose, it is important to determine that neither invertase nor levansucrase accompanies the dextransucrase under assay. Methods for differentiating various sucrose-derived products in crude bacterial culture fluids have been described.[5]

2. Accuracy of the dextransucrase assay also depends on freedom of the enzyme solution from high concentrations of certain sugars (sometimes employed to modify the synthesis so as to obtain oligosaccharides or low molecular weight dextrans as products) that affect the reaction velocity. The rate of fructose liberation is depressed, for example, by concentrations of sucrose higher than ca. 0.4 M.[1] Fructose and melibiose in high concentration likewise have a rate-depressing action, whereas isomaltose, maltose, α-methyl glucoside, and glucose have the opposite effect.[6] Hence, enzyme unitage should be determined either prior to the

---

[4] M. Somogyi, *J. Biol. Chem.* **160**, 61 (1945).

[5] E. J. Hehre and J. M. Neill, *J. Exptl. Med.* **83**, 147 (1946); W. G. C. Forsyth and D. M. Webley, *J. Gen. Microbiol.* **4**, 87 (1950).

[6] H. J. Koepsell, H. M. Tsuchiya, N. N. Hellman, A. Kazenko, C. A. Hoffman, E. S. Sharpe, and R. W. Jackson, *J. Biol. Chem.* **200**, 793 (1953).

addition of any sugar modifier or after separation of the enzyme from such sugar.

Supplementary Procedures. For purposes of orientation, dextransucrase activity may be crudely estimated by noting the speed of appearance (in mixtures with sucrose) of opalescence, serological activity, alcohol precipitable material, or other properties referable to dextran itself.[7] Use of the rate of dextran formation for assay of enzyme activity is proposed by Braswell *et al.*,[7a] who find linear relationships between reaction time and (1) the logarithm of the relative viscosity, and (2) the relative turbidity measured at an angle of 90° in a light scattering photometer.

Preparation

Biological Source. Although dextran is formed from sucrose by a number of different lactic acid bacteria, most dextransucrase preparations so far described have been made from *Leuconostoc mesenteroides.* Appreciable quantities of the enzyme are found extracellularly in broth cultures of these bacteria, provided that sucrose is present in the culture medium.[7] The higher the concentration of sucrose in the medium (at least up to 5%), the higher is the enzyme yield.[3]

High-Potency Culture Fluids. The following preparative procedure of Tsuchiya *et al.* (1952)[3] gives notably improved yields of enzyme over the earlier methods.[1,7] Slightly modified, it has proved suitable for the production of enzyme-rich fluids on an industrial scale.

Inoculate *L. mesenteroides* NURB B-512 at a 1 to 2% rate into flasks of culture medium, and incubate these at 25° for 24 hours on a reciprocating shaker. A suitable culture medium contains 2 to 3% sucrose, 2% corn steep solids or 0.5% yeast extract, 2 to 3% K_2HPO_4, 0.02% $MgSO_4 \cdot 7H_2O$, 0.001% NaCl, 0.001% $FeSO_4 \cdot 7H_2O$, and 0.001% $MnSO_4 \cdot H_2O$. Bacterial cells are removed from the grown cultures by filtration or centrifugation. The crude, essentially cell-free fluids assay 90 to 120 DSU/ml.

Separation of Enzyme from Preformed Dextran. Dextran and usually also sucrose and fructose are undesirable accompanying substances, present in the sucrose broth culture fluids. The following procedure has been used by the author to remove the greater part of these preformed reaction components in preparations from *L. mesenteroides*, strain B.[1]

To each liter of chilled culture fluid add 370 g. of ammonium sulfate, taking care to dissolve the salt completely. Centrifuge the mixture in the cold, and decant the supernatants containing most of the dextran and sugars. Drain the precipitates in the cold, then wash three times with

[7] E. J. Hehre and J. Y. Sugg, *J. Exptl. Med.* **75**, 339 (1942).

[7a] E. Braswell, G. Oster, and K. G. Stern, work reported at Meeting-in-Miniature, New York Section Am. Chem. Soc., Feb. 1954.

250-ml. portions of half-saturated ammonium sulfate in 0.1% acetic acid. For each washing, work the sticky precipitates from the bottoms and sides of the centrifuge tubes into suspension with the aid of a rubber policeman, centrifuge, and decant the wash fluid. Extract the final washed precipitate with 100 ml. of 0.025 M citrate buffer (pH 6.3), and clear by centrifugation in the cold.

Properties

Specificity. Sucrose is the only established "donor substrate" for polysaccharide formation by dextransucrase, with the Michaelis constant (K_s) recorded as 19 to 20 mM.[1] and 15.8 mM.[8] for enzymes of different source. None of a long list of other sugars and sugar derivatives has been found to serve in this capacity.[6,9] Isomaltose, dextran, and leucrose have, however, been reported to serve as donor substrates for oligosaccharide formation by the enzyme.[9a] The notion[10] that *Leuconostoc* systems form small amounts of dextran from glucose in the absence of sucrose stems from the use of a nonspecific (gravimetric) test method for dextran. The enzyme likewise has no detectable action upon inorganic phosphate or glucose-1-phosphate,[11] or upon glucose-6-phosphate, fructose-6-phosphate, fructose-1,6-diphosphate, or ATP.[8]

With respect to "acceptor substrate" specificity, the enzyme has, first of all, a strong preferential affinity for low molecular weight dextrans;[12–14] that is, the addition of fractions with average molecular weights of ca. 5000 to ca. 60,000 in small amount to enzyme-sucrose mixtures causes synthesis to be so modified that dextran molecules of low or intermediate size rather than of large size are formed as major reaction products.[15]

---

[8] W. W. Carlson, C. L. Rosano, and V. Whiteside-Carlson, *J. Bacteriol.* **65**, 136 (1953).

[9] E. J. Hehre, *Science* **93**, 237 (1941). Traces of material with the serological properties of dextran were formed in incubated enzyme-raffinose mixtures, suggesting that raffinose may be a donor substrate of low affinity. However, this effect may possibly have been due to enzymic action on traces of sucrose liberated in some way from the trisaccharide.

[9a] H. M. Tsuchiya and C. S. Stringer, *Bacteriol. Proc.*, p. 98 (1954).

[10] H. L. A. Tarr and H. Hibbert, *Can. J. Research* **5**, 414 (1931); Staff Report, *Chem. Eng. News* **29**, 650 (1951). The lack of formation of any dextran in incubated enzyme-glucose mixtures was shown by the use of serological tests of high specificity and sensitivity.[9]

[11] E. J. Hehre, *Proc. Soc. Exptl. Biol. Med.* **54**, 240 (1943).

[12] E. J. Hehre, *J. Am. Chem. Soc.* **75**, 4866 (1953).

[13] H. M. Tsuchiya, N. N. Hellman, and H. J. Koepsell, *J. Am. Chem. Soc.* **75**, 757 (1953).

[14] H. Nadel, C. I. Randles, and S. L. Stahly, *Appl. Microbiol.* **1**, 217 (1953).

[15] Reaction products of suitable molecular weight for use as a blood volume expander have been produced in good yield from mixtures (kept at pH 5.0 and 15°) containing

The enzyme also utilizes certain sugars (especially isomaltose and maltose, but also α-methyl glucoside and glucose) as acceptors for the glucosyl units from sucrose.[6] Enzymic action on equimolecular concentrations of sucrose and any one of these sugars differs from the action on sucrose alone in that the reaction rate (as measured by fructose liberation) is increased, oligosaccharides based on the sugar acceptor appear as major reaction products, and less high molecular weight dextran is synthesized. Fructose, leucrose, melibiose, and galactose also appear to act as acceptors, since in their presence (equimolecular with sucrose) some oligosaccharide formation occurs and the reaction rate is slightly depressed; the activity of both pyranose and furanose forms of fructose is indicated by the formation of two fructose-containing disaccharides, leucrose (5-D-glucopyranosyl-D-fructopyranose) and isomaltulose (6-D-glucopyranosyl-D-fructofuranose),[16] in mixtures containing added fructose.[6,17] Dextransucrase apparently does not utilize as acceptors xylose, D- and L-arabinose, ribose, rhamnose, mannose, sorbose, cellobiose, trehalose, lactose, melezitose, raffinose, inulin, inositol, mannitol, sorbitol, 2-ketogluconate, 5-ketogluconate, or glycerol.[6] High molecular weight (native) dextrans cannot be excluded as acceptors, but their addition does not affect the rate of dextran synthesis from sucrose.[1] The production of small amounts of glucose in enzyme-sucrose mixtures (Forsyth and Webley[5]) suggests that water may be an acceptor substrate of low affinity, unless traces of invertase contaminate the available enzyme preparations.

Equilibrium. The reaction in the synthetic direction proceeds beyond 99.5% completion;[1,2] reversal, although theoretically possible, has not been detected.

Activators and Inhibitors. Apart from the sugars noted above, no special activators or inhibitors of dextransucrase are known. The following common enzyme poisons are without detectable effect: 0.05 M fluoride

---

40 units/ml. dextransucrase, 100 mg./ml. sucrose, and from 2 to 20 mg./ml. (depending on the molecular weight) of low molecular weight dextran "acceptor." H. M. Tsuchiya, N. N. Hellman, H. J. Koepsell, J. Corman, C. S. Stringer, F. R. Senti, and R. W. Jackson, *Abstr.* 124th *Meeting Am. Chem. Soc.*, 35A (1953).

[16] F. H. Stodola, H. J. Koepsell, and E. S. Sharpe, *J. Am. Chem. Soc.* **74**, 3202 (1952); F. H. Stodola, E. S. Sharpe, and H. J. Koepsell, *Abstr. 126th Meeting Am. Chem. Soc.* (1954).

[17] Since fructose accumulates as a product of enzymic action upon sucrose, its effect may be observed even under "normal" circumstances. Leucrose in small amount appears on paper chromatograms in the terminal phases of action of dextransucrase from *L. mesenteroides* NURB B-512 upon 0.125 M sucrose; larger amounts appear when the enzyme operates on higher concentrations of sucrose (and larger amounts of fructose are released) or when fructose is added to enzyme-sucrose mixtures.[6]

or cyanide, 0.025 M azide or iodoacetate, 0.01 M pyridine or aniline, 0.001 M CuSO$_4$, ZnSO$_4$, or MnCl$_2$, or 0.00001 M AgNO$_3$.[1] Enzymic activity likewise is unaffected by dialysis or by the presence of sodium ethylenediamine tetraacetate (versene).[8] An early suggestion that dextransucrase may be activated by PABA has not been confirmed.

Effect of pH and Temperature. The enzyme is most active and stable between pH 4 and 6,[7] with the optimum reported as pH 5.0 to 5.2[2] or pH 5.6.[8] Activity is rapidly lost on exposure to pH 6.7 or above, even at 25 to 37°. Moreover, the enzyme catalyzes the formation of dextran over a wide temperature range (3 to 37°) but is, nevertheless, very heat labile.[7] For example, it shows high activity and stability at 30°, yet is destroyed in a few minutes at 40°, even at pH 5.0.[2] Lyophilized preparations, sealed *in vacuo* and stored at 5°, retain their potency for many years.

Reaction temperature has been observed to influence the nature of the dextran produced when the enzyme operates in the presence of added dextran acceptor. Lowering the reaction temperature of such systems from 30° to 15° greatly favors the formation of intermediate- as opposed to high-molecular weight dextran.[15]

II. Amylosucrase

$$n\text{-}C_{12}H_{22}O_{11} + \quad HOR \quad \rightarrow H(C_6H_{10}O_5)_n OR + n\text{-}C_6H_{12}O_6$$

Sucrose   Acceptor   Glycogen-type   Fructose
polysaccharide

Assay Method

A detailed procedure for estimating enzymic activity has not been reported, but a suitable basis is measurement of the fructose by-product released per unit of time under conditions yielding zero-order reaction kinetics. Such conditions are approached when enzyme solutions are incubated at 10° with an equal volume of 0.2 M sucrose in 0.025 M maleate buffer, pH 6.4.[18] Rough appraisals of activity may be obtained also from the speed of development of opalescence, color with iodine, alcohol-precipitable material, or other attribute of the glycogen-type polymer that is synthesized.

Preparation

The biological source is *Neisseria perflava*, a species of bacteria readily isolated from the throat.[19] Crude but cell-free enzyme preparations may be obtained by the author's method.[18] Cultivate *N. perflava* (strain 19–34)

[18] E. J. Hehre, *J. Biol. Chem.* **177**, 267 (1949).
[19] E. J. Hehre and D. M. Hamilton, *J. Biol. Chem.* **166**, 777 (1946); *J. Bacteriol.* **55**, 197 (1948).

for 5 days at 37° in unshaken flasks of broth (800 ml.) comprising 0.1% peptone, 0.15% sodium citrate, 0.02% yeast extract, 0.06% KH_2PO_4, 0.15% Na_2HPO_4, and 0.05% glucose. (Inoculate each flask with 10 ml. of a 1-day culture in the same medium.) Treat each 800 ml. of grown culture with 300 g. of ammonium sulfate, centrifuge the mixture, and decant the supernatant fluids. Suspend the sediments in 100 ml. of half-saturated ammonium sulfate previously adjusted to pH 6.4 with ammonia, centrifuge, and discard the wash fluid. Extract the washed and drained precipitates with 15 ml. of 0.025 M maleate buffer (pH 6.4), and clarify by centrifugation.

Properties

Specificity. The only known "donor substrate" for amylosucrase acting synthetically is sucrose. Neither glucose, maltose, lactose, trehalose, α-methyl glucoside, raffinose, melezitose, nor glucose-1-phosphate serves the enzyme in this capacity.[18,20]

Amylosucrase presumably utilizes a series of amylosaccharides as "acceptor substrates" in the course of building up polysaccharide molecules, but its affinity for various acceptors has not been systematically examined.

Equilibrium. The conversion of sucrose to polysaccharide and fructose proceeds to ca. 98% completion.[21] Reversal to the extent of about 1% can be demonstrated when the enzyme is incubated with fructose plus amylose, amylopectin, or glycogen.

Inhibitors. Synthesis of polysaccharide from sucrose is not impaired by the presence of inorganic phosphate in concentrations sufficiently high (8 moles P to 1 of substrate) to suppress all synthesis from glucose-1-phosphate by the phosphorylase in *N. perflava* extracts.[18,19] Traces of salivary amylase, however, inhibit the formation of polysaccharide and fructose from sucrose, possibly by destroying essential acceptor substrates.[21]

Effects of pH and Temperature. Satisfactory enzymic action occurs over the temperature range of 10 to 37° at pH 6.4. Amylosucrase (pH 6.4) appears stable at 5° for many weeks but is almost completely inactivated at 45° for 10 minutes (under which conditions the phosphorylase activity of the extracts remains unaffected[18]). Information is lacking on the activity and stability in different pH ranges.

---

[20] *N. perflava* preparations produce an amylaceous polysaccharide from glucose-1-phosphate, and also a glycogen-type polysaccharide from amylose, but these actions are attributable to an accompanying phosphorylase and to a branching enzyme of the Q type.

[21] E. J. Hehre, *Advances in Enzymol.* **11**, 297 (1951).

III. Levansucrase

$$n\text{-}C_{12}H_{22}O_{11} + \quad HOR \quad \rightarrow H(C_6H_{10}O_5)_n + n\text{-}C_6H_{12}O_6$$
$$\text{Sucrose} \quad \text{Acceptor} \quad \text{Levan} \quad \text{Glucose}$$

Assay Method

Principle. The circumstance that a variable but generally large amount of sucrose hydrolysis regularly accompanies levan formation in reactions catalyzed by available enzyme preparations has prevented development of an assay method based on the rate of liberation of glucose. The following procedure of Hestrin and Avineri-Shapiro[22] measures the rate of formation of highly polymerized levan itself under conditions where the synthesis follows an essentially linear course. (For rough appraisals of enzyme activity, the speed of release of reducing sugars and of development of features referable to levan, e.g., opalescence, viscosity, serological activity, or alcohol-precipitable polyfructoside, may be found useful.)

Procedure. The assay mixture comprises 1 ml. of enzyme solution, 1 ml. of 15% sucrose in distilled water, 1 ml. of Sørensen citrate buffer (pH 5.0), and 1 drop of chloroform containing thymol for maintaining sterility. Incubate at 37° until the levan concentration is in the range of 50 to 200 mg./100 ml. (as judged by comparison of the opalescence of the assay mixture with reference solutions containing known concentrations of levan), and then determine the levan content as follows.[23] Pipet 1.0 ml. of the assay solution into a centrifuge tube of 15-ml. capacity, and add 3.0 ml. of ethanol. If flocculation is slow, it can be hastened and rendered quantitative by the addition of a drop of 1% $CaCl_2$. The suspension is centrifuged, and the sedimented levan freed from reducing sugar and sucrose residues by twice-repeated solution in H_2O over a water-bath and precipitation each time with ethanol. Addition of a glass bead to the mixture facilitates these operations. The final sediment is taken up in 3.0 ml. of 0.5% oxalic acid and hydrolyzed in a boiling water bath for 1 hour. Evaporation is restricted during this treatment by placing a glass bulb at the tube aperture. The hydrolyzate is neutralized, cleared with $Zn(OH)_2$, and diluted to a suitable volume. The reducing power of the filtrate is estimated by the method of Somogyi. The amount of levan is calculated from the "glucose value" by multiplying by a factor which allows both for the small difference in reducing power between glucose and fructose and for the entry of H_2O during hydrolysis.

[22] S. Hestrin and S. Avineri-Shapiro, *Biochem. J.* **38**, 2 (1944); S. Avineri-Shapiro and S. Hestrin, *Biochem. J.* **39**, 167 (1945).

[23] S. Hestrin, S. Avineri-Shapiro, and M. Aschner, *Biochem. J.* **37**, 450 (1943); *Nature* **149**, 527 (1942).

Definition of Unit. One levansucrase unit (LU) is defined as the minimal amount of enzyme which induces production of 100 mg. of levan per 100 ml. in 2 hours under the above conditions. This definition is designed to circumvent the variation observed in the specific activity of levansucrase at different enzyme concentrations.

Limitations. The presence of levan-splitting enzymes (levanases) in certain enzyme preparations (e.g., from *Bacillus subtilis*) may lead to an erroneously low levansucrase assay figure. Likewise, the presence of pre-formed levan in crude sucrose broth culture fluids (unless corrected for) may give an error in the opposite direction. Extensive dextran production, which may occur under levansucrase assay conditions in the case of extracts from certain bacteria (especially certain streptococci), may also lead to unduly high assay values and should in any case be recognized.

Preparation

Bacteria of many kinds produce levan from sucrose or raffinose, and crude, cell-free levansucrase preparations have been obtained from varieties in such diverse genera as *Bacillus, Streptococcus,* and *Aerobacter.* Hestrin *et al.*[23] give the following method for preparing enzymes suitable for kinetic studies from *Aerobacter levanicum.* The bacterium (Jerusalem strain) is cultivated in 1-l. conical flasks at 30° in a two-phase medium consisting of a bottom solid layer of nutrient agar devoid of sucrose and a thin top layer of 2% aqueous sucrose. After 20 hours the cells are harvested, rinsed repeatedly in water to free them from medium constituents and levan, and finally left to autolyze at 37° under water containing a little thymol and chloroform. The volume ratio of culture medium to autolyzate fluid is 100:1. After 24 hours of autolysis, the suspension is centrifuged. The cell-free supernates may assay as high as 5 units/ml.

Various alternative methods have been used in preparing levansucrase from spore-forming bacilli or streptococci.[7,23-27] Since a procedure found applicable to one particular bacterial strain may not be suitable for another, the trial of several methods is advised.

Properties

Specificity. Levansucrase utilizes sucrose and raffinose (a galactosyl sucrose) as "donor substrates," with the Michaelis constant for sucrose in the range of 0.02 to 0.06 M. The enzyme does not utilize any other common sugar or other substance with a terminal fructofuranose group (e.g.,

[24] E. J. Hehre, *Proc. Soc. Exptl. Biol. Med.* **58,** 219 (1945).
[25] M. Doudoroff and R. O'Neal, *J. Biol. Chem.* **159,** 585 (1945).
[26] F. L. Horsfall, Jr., *J. Exptl. Med.* **93,** 229 (1951).
[27] R. Dedonder and Mme. Noblesse, *Ann. Inst. Pasteur* **85,** 356 (1953).

fructose-1,6-diphosphate, fructose-6-phosphate, methyl fructofuranoside, inulin), in this way. There is also no action upon "free" or "nascent" fructofuranose or upon inorganic phosphate.[22,28,29]

Little information is as yet available concerning the "acceptor substrate" specificity of levansucrase. A series of glucose-ended 2,6-linked polyfructosides is probably used in synthesis of the levan macromolecule.[29a] In addition, B. subtilis preparations seem to have some affinity for glucose (but not fructose) as an acceptor. The action on a mixture of 10% sucrose and 5% glucose, for example, differs from the action of 10% sucrose alone in that oligosaccharides containing glucose as well as fructose appear as prominent (although transient) reaction products, and high molecular weight levan is less rapidly synthesized.[27] (Cf. *Activators* and *Inhibitors*.)

Equilibrium. Because of the complication of concomitant hydrolytic breakdown of the substrate (sucrose) by the available enzyme preparations, the exact position of equilibrium in levan synthesis remains undefined. Sufficient data have been obtained,[22] however, to establish that the ratio of levan to sucrose at equilibrium is definitely greater than unity. Yields of levan as high as 62% of the theoretical maximum have been obtained under certain conditions. Since reversal of the reaction by the action of enzyme on levan and glucose has not been detected by direct means,[22] an equilibrium position far to the side of synthesis seems likely.

Activators and Inhibitors. Certain sugars, including D-glucose, L-arabinose, D-xylose, L-sorbose, lactose, maltose, α-methyl glucoside, and D-galactose (but not D-fructose, D-mannose, mannitol, sorbitol, glucosamine, or trehalose), retard the formation of highly polymerized levan from sucrose by A. levanicum preparations. The "inhibitory" effect increases, in the case of D-glucose at least, as its concentration relative to sucrose is increased; and levan formation is completely suppressed, for example, in mixtures containing 1% sucrose and 16% D-glucose.[22] It would appear most probable that this "inhibition" effect actually represents a modification of the synthetic reaction, and that D-glucose and the other sugars in question are "acceptor substrates." Some of them may actually stimulate the rate of fructosyl group transfer by the enzyme.

[28] It should be noted, however, that gentianose, verbascose, and several recently discovered sucrose-ended oligosaccharides passing terminal fructofuranosyl units, i.e., erlose, kestose, inulobiosyl glucose, have not been examined and may be able to serve as donor substrates for levansucrase.

[29] S. Hestrin, *Nature* **154**, 581 (1944).

[29a] Neither levanbiose, levantriose, nor levantetraose, however, has been found to serve as an acceptor substrate (or modifier of levan synthesis) in the case of *Aerobacter* levansucrase. See S. Hestrin, Instituto Superiore di Sanita' Symposium on Microbial Metabolism, Rome, p. 53, 1953.

Known poisons of respiration or of glycolysis, including 0.002 M cyanide, 0.02 M fluoride, 0.001 M monoiodoacetate, and 0.001 M phloridzin, fail to suppress or markedly inhibit levan formation. Dialysis similarly does not affect the activity of the enzyme, and $A.$ $levanicum$ extracts do not contain free adenylic acid.[22]

Effect of pH and Temperature. The pH-activity curve of levansucrase operating upon sucrose at 37° is regular and symmetrical, with its peak in the region of pH 5.0 to 5.8.[22] The relationship between activity and temperature has not been systematically studied, but the enzyme operates satisfactorily over the range of 5 to 37°. Enzyme stability (at pH 5.0) is very high at temperatures up to 30°. Storage for several days at this temperature, or for several months at 4°, has no deleterious effect. Gradual loss of activity at 37° is discernable over a period of days, and there is complete and almost immediate inactivation on exposure at 100°.[23] Preparations frozen and dried *in vacuo* are suitable for prolonged storage.

IV. Amylomaltase

$$n\text{-}C_{12}H_{22}O_{11} + \quad HOR \quad \rightarrow H(C_6H_{10}O_5)_nOR + n\text{-}C_6H_{12}O_6$$

Maltose   Acceptor   Amylose-type   Glucose
polysaccharide

Assay Method

Principle. The assay procedure (Monod and Torriani[30]) has as a basis determination of the glucose liberated under conditions where the reaction follows a kinetically zero-order course; the glucose is determined (in the presence of the maltose substrate) by measuring manometrically the uptake of O_2 during its specific oxidation by notatin.

Reagents

0.1 M phosphate buffer, pH 6.8

Notatin. Purified glucose oxidase of *Penicillium notatum.*[31]

Buffered maltose solution. The solution used as substrate for the assay contains 100 γ/ml. notatin, 0.033 M maltose, and 0.02 M sodium azide (for inactivation of catalase) in 0.1 M phosphate buffer of pH 6.8.

Enzyme. Prepare a known dilution, if necessary, to obtain an enzyme solution for assay that contains between 20 and 130 amylomaltase units/ml. (See definition below.)

[30] J. Monod and A. M. Torriani, *Ann. Inst. Pasteur* **78,** 65 (1950); *Compt. rend.* **227,** 240 (1948).

[31] C. E. Coulthard, R. Michaelis, W. F. Short, G. Sykes, G. E. H. Skrimshire, A. F. Standfast, J. H. Birkinshaw, and H. Raistrick, *Biochem. J.* **39,** 24 (1945); see Vol. I [45].

Procedure. Place 3.0 ml. of the buffered maltose solution containing notatin in the main chamber of a Warburg flask, and 0.1 ml. of enzyme in the side arm. After assembly of the manometric apparatus and equilibration at 28°, mix the contents of the flask. Measure the oxygen consumption at 10-minute intervals until a steady rate is achieved (usually 20 to 40 minutes). The O_2 uptake in microliters per hour divided by 2.26 gives the micromoles of O_2 consumed per hour per milliliter of enzyme and, hence, the micromoles of glucose formed per hour per milliliter of enzyme.

Definition of Unit. One amylomaltase unit is defined as the amount of enzyme which will liberate from maltose 1 μM. of glucose per hour under the above conditions. Specific activity is expressed as units per milligram of Kjeldahl N.[30]

Extent and Limitations of Application. The procedure has been used for assay of amylomaltase in toluene-treated suspensions of intact bacterial cells of the ML strain of *E. coli*.[32] For use with cells or with impure soluble preparations, however, the possibility of interference by the presence of enzymes yielding glucose from maltose or from amylosaccharides would, of course, have to be guarded against.

Preparation

The presence of amylomaltase has been described in two special mutant strains of *E. coli*, namely, the maltose- and lactose-positive form of strain ML,[30] and the maltose-positive lactose-negative form of strain K12.[33] The enzyme evidently is produced only when the bacteria are cultivated in the presence of maltose. Using strain ML, Monod and Torriani[30] prepare the cell-free enzyme as follows.

The bacteria are heavily inoculated into 250-ml. amounts of culture medium in 2-l. conical flasks. (The medium is made up of 2.72% KH_2PO_4, 0.4% $(NH_4)_2SO_4$, 0.04% $MgSO_4\cdot7H_2O$, 0.001% $CaCl_2$, 0.00005% $FeSO_4\cdot7H_2O$, and sufficient NaOH to attain pH 7.5; after sterilization, a filtered solution of maltose is added aseptically to give an 0.8% concentration.) The inoculated flasks are incubated, with agitation, at 34° for 14 hours. The bacterial cells are harvested by Sharples centrifugation, washed in the centrifuge with a volume of distilled water equal to that of the original culture medium (2 to 10 l.), then suspended in 0.05 M phosphate buffer, pH 6.8; 200 ml. of buffer is used for each 100 g. of the washed cells. Fine sand, 2 g. for each 1 ml. of suspension, is then added, and the mixture vigorously agitated in a shaking apparatus for 40 minutes.

The creamy mass is centrifuged at 12,000 r.p.m. for 15 minutes, and

[32] G. Cohen-Bazire and M. Jolet, *Ann. inst. Pasteur* **84**, 937 (1953).
[33] M. Doudoroff, W. Z. Hassid, E. W. Putman, A. L. Potter, and J. Lederberg, *J. Biol. Chem.* **179**, 921 (1949).

the supernatant fluid removed. The precipitate is resuspended in 100 ml. of buffer and again centrifuged. This operation is repeated a second time, and the supernatant fluids combined. Maltose is added to give a 0.05 M solution, followed by solid ammonium sulfate to 75% saturation (in the cold). After 2 hours, the precipitate is separated by centrifugation and redissolved in 100 ml. of buffer. The precipitation is repeated three times under similar conditions with sufficient solid ammonium sulfate to give 50% saturation. The final precipitate is taken up in about 20 ml. of buffer, and insoluble material eliminated by centrifugation. All operations are at 0°.

To remove accompanying amylase and phosphorylase, the solution is dialyzed in the cold against distilled water. A precipitate should form as a result of this treatment or be induced by careful acidification of the dialyzate with acetic acid to pH 5.2. The precipitate is collected by centrifugation and extracted with 5 ml. of 0.025 M veronal buffer, pH 6.8, containing 0.2 M sodium sulfate. After storage overnight at 0°, this final extract is cleared by centrifugation. The best preparations contain 5000 amylomaltase units/ml. (660 units/mg. N).[30]

Properties

Specificity. The enzyme has affinity for amylosaccharides as well as for maltose as "donor substrates." It is uncertain, however, whether any donor other than maltose is utilized in the forward or synthetic reaction. Lactose, sucrose, melibiose, cellobiose, α-methyl glucoside, β-methyl glucoside, trehalose, and glucose-1-phosphate definitely do not serve as donor substrates for the enzyme.[30]

Amylomaltase is presumed to utilize maltose and its higher homologs as "acceptor substrates" in the course of polysaccharide synthesis. Glucose definitely is utilized in this capacity, since enzymic action on maltose in the presence of glucose leads to the formation of oligosaccharides as major reaction products, whereas action in the absence of glucose (in systems containing notatin) leads to the formation of an amylose-type polysaccharide as the product.[30,33,34] The presence of glucose also permits the reverse reaction (polysaccharide degradation to maltose) to occur.[30] There is some indication that D-xylose and to some extent D-mannose (but not D-fructose, D-galactose, D-arabinose, or L-arabinose) can replace D-glucose as the acceptor substrate in this reverse reaction.[33]

Equilibrium. The polymerative action of amylomaltase upon maltose proceeds until about 60% of the substrate has been converted; a similar equilibrium position is reached in the reverse direction, i.e., after enzymic action on equimolecular amounts of polysaccharide and glucose.[30] (When

[34] S. A. Barker and E. J. Bourne, *J. Chem. Soc.* **1952**, 209.

the synthesis from maltose is conducted in the presence of notatin, an equilibrium is not observed and conversion to polysaccharide proceeds to completion.)

Activators and Inhibitors. Enzymic activity is unaffected by the presence or absence of orthophosphate, and it is retained after dialysis.[30] *E. coli* cells that contain amylomaltase are active in converting maltose to higher homologs in the presence of 0.002 M iodoacetate.[33,34]

Effect of pH and Temperature. Little has been recorded concerning the effect of pH and temperature upon enzyme activity and stability. Suitable activity is manifested in mixtures set at pH 6.8 and 28 to 32°.

[22] Phosphorylases from Plants

By W. J. WHELAN

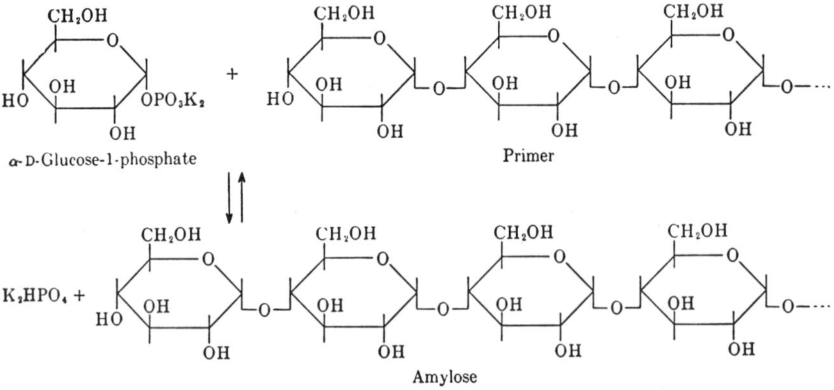

α-D-Glucose-1-phosphate   Primer

Amylose

Assay Method

Principle. Primer and G-1-P are mixed with enzyme, and the orthophosphate set free during amylose synthesis is measured colorimetrically. The following procedure is based on that of Green and Stumpf.[1]

Reagents

G-1-P solution (0.1 M) contains 37.22 g./l. of the dipotassium salt.[2]
5% soluble starch solution.
0.5 M citric acid—NaOH buffer, pH 6.0.

[1] D. E. Green and P. K. Stumpf, *J. Biol. Chem.* **142**, 355 (1942).
[2] Pure, primer-free G-1-P is conveniently prepared by the method of R. M. McCready and W. Z. Hassid, *J. Am. Chem. Soc.* **66**, 560 (1944).

Enzyme. Use between 5 and 10 units (for definition, see below) in a maximum volume of 1.8 ml.

5% TCA solution.

Procedure. Mix the starch solution (0.2 ml.), buffer (0.5 ml.), enzyme and water (if necessary) to 2.5 ml., equilibriate at 35°, and then add G-1-P solution (1 ml.). Stop the reaction at some convenient time (about 10 minutes) by adding TCA (5 ml.). Centrifuge and remove an aliquot for determination of inorganic phosphate[3] (Vol. III [147]). A control determination is used to measure orthophosphate in the reagents. The reactants, excluding the enzyme, are incubated together, TCA is added, and then phosphorylase. It is important that the G-1-P be treated with the acid for the same length of time in the determination as in the control. Weibull and Tiselius[4] have drawn attention to the slight coprecipitation of inorganic phosphate with protein by TCA and point out that, if little protein is present, the treatment with TCA can be omitted, the resulting turbidity of the molybdenum blue solution being allowed for in the control.

Definition of Unit of Activity. One unit of enzyme is defined as that amount which causes the liberation of 0.1 mg. of inorganic phosphorus in the above digest during 3 minutes.

Application of Assay Method to Crude Plant Extracts. Interference may be caused by the presence of glucose-1-phosphatase, but this interference is probably not serious because, during active synthesis of amylose, phosphorylase competes successfully for the G-1-P and suppresses phosphatase action.[5] The latter enzyme can be completely inhibited by addition of ammonium molybdate[6] or sodium fluoride.[5] β-Amylase is a powerful phosphorylase inhibitor, and, because of the presence of this enzyme, phosphorylase action cannot be demonstrated in crude extracts of cereals.[7,8] Traces of α-amylase or Q-enzyme accelerate the reaction by producing more primer chains, but high concentrations of the former enzyme cause inhibition because of destruction of the primer. Amylases and Q-enzyme can be selectively inhibited either by $HgCl_2$ [5,6,8] or by phenyl mercuric acetate.[8] The required concentrations of inhibitors for phosphatase and

---

[3] The author has used R. J. L. Allen's modification [*Biochem. J.* **34**, 858 (1940)] of the Fiske-SubbaRow method and by halving the specified amount of Amidol reagent the hydrolysis of unchanged G-1-P in the acid solution is largely prevented [W. J. Whelan and J. M. Bailey, *Biochem. J.* in press].

[4] C. Weibull and A. Tiselius, *Arkiv. Kemi, Mineral. Geol.* **A19**, No. 19 (1945).

[5] H. K. Porter, *J. Exptl. Botany* **4**, 44 (1953).

[6] J. M. Bailey, G. J. Thomas, and W. J. Whelan, *Biochem. J.* **49**, lvi (1951).

[7] H. K. Porter, *Biochem. J.* **47**, 476 (1950).

[8] M. Nakamura, K. Yamazaki, and B. Maruo, *J. Agr. Chem. Soc. Japan* **24**, 197, 299, 302 (1951).

amylase depend on the amount of foreign protein present, but amorphous preparations of potato phosphorylase will tolerate relatively large amounts (1×10^{-3} M and 3×10^{-5} M, respectively) of molybdate and mercuric ions.

Qualitative detection of phosphorylase can be made by using an achroic primer (e.g., partly acid-hydrolyzed starch) and testing for the formation of iodine-staining products. Many workers have used this method in the histochemical detection of plant phosphorylases.[9] Meeuse[10] has described a very sensitive test in which amylose synthesis is carried out on an agar plate.

Purification Procedure

Phosphorylase preparations have been obtained from the following plants: waxy maize,[11] sweet corn,[12] barley,[7,8] peas,[13] broad beans,[14] lima beans,[1] jack beans,[15] green gram,[16] potato,[1,4,17-19] *Ipomoea batatas* (sweet potato),[20] sugar beet,[21] squash,[22] and pumpkin.[20] Most methods involve fractional precipitation with $(NH_4)_2SO_4$. Two crystalline preparations have been isolated; both are from the potato. Fischer and Hilpert's material[18] represented a 300-fold purification as compared with the crude extract. The method of Baum and Gilbert[19] also results in a very high degree of purification. Final values for the absolute activity are not yet available. This latter method is as follows:[23]

"The preparation is most conveniently carried out from freeze-dried potato juice. Peeled potatoes are thinly sliced into $Na_2S_2O_4$ solution (7 g./l.) at room temperature and left to soak for 10 minutes. The slices

[9] H. C. Yin and C. N. Sun, *Science* **105**, 650 (1947); see K. Paech and E. Krech, *Planta* **41**, 391 (1953), for recent references to detection of phosphorylase in plastids.

[10] B. J. D. Meeuse, *Verslag. Gewone Vergader. Afdeel. Natuurk. Ned. Akad. Wetenschap.* **52**, 341 (1943).

[11] L. Bliss and N. M. Naylor, *Cereal Chem.* **23**, 177 (1946).

[12] S. Peat, W. J. Whelan, and J. R. Turvey, in press.

[13] C. S. Hanes, *Proc. Roy. Soc. (London)* **B128**, 421 (1940).

[14] P. N. Hobson, W. J. Whelan, and S. Peat, *J. Chem. Soc.* **1950**, 3566.

[15] J. B. Sumner, T. C. Chou, and A. T. Bever, *Arch. Biochem.* **26**, 1 (1950).

[16] J. S. Ram and K. V. Giri, *Arch. Biochem. and Biophys.* **38**, 231 (1952).

[17] S. A. Barker, E. J. Bourne, I. A. Wilkinson, and S. Peat, *J. Chem. Soc.* **1950**, 84.

[18] E. H. Fischer and H. M. Hilpert, *Experientia* **9**, 176 (1953).

[19] H. Baum and G. A. Gilbert, *Nature* **171**, 983 (1953).

[20] Y. Inoue and K. Onodera, *Repts. Inst. Chem. Research, Kyoto Univ.* **15**, 70 (1946); **16**, 28 (1947) [*C. A.* **45**, 4787 (1951)], M. Nakamura, *J. Agr. Chem. Soc. Japan* **25**, 413 (1952).

[21] A. Kursanov and O. Pavlovina, *Biokhimiya* **13**, 378 (1948) [*C. A.* **42**, 8842 (1948)].

[22] T. G. Phillips and W. Averill, *Plant Physiol.* **28**, 287 (1953).

[23] The author is indebted to Dr. Gilbert and Mr. Baum who have kindly amplified their published account[19] of the preparation.

are then washed well with tap water, cooled, and pulped at 0° for 1 minute in an Atomix blender. The juice is expressed through muslin, clarified by centrifugation at 0°, and freeze-dried. The product may be stored indefinitely at 0° over P_2O_5 without deterioration.

"The following reagents are prepared:

"*'50% ethanol':* 50% (w/v) ethanol solution, pH 6.0, containing 0.01 M citrate (1074 ml. of water, 1000 ml. of ethanol, 4.64 g. of sodium dihydrogen citrate, pH to 6.0 with solid NaOH. The pH is tested at 20° after fivefold dilution of a sample with 0.01 M KCl).

"*'11% ethanol':* 11% (w/v) ethanol solution, pH 6.0, containing 0.01 M citrate (0.282 ml. of '50% ethanol'/ml. of 0.01 M citrate, pH 6.0).

"*Amylose Solution.* Undried butanol-precipitated potato amylose,[24] (Blue Value[25] $\geq$ 1.1) is steam-distilled to remove butanol and the resulting solution (0.5%, w/v) made 0.01 M in citrate by the addition of 0.5 M citrate, pH 6.0.

"Freeze-dried potato juice is reconstituted with distilled water to give the original volume of juice and brought at 0° to 11% ethanol concentration with '50% ethanol,' (0.282 ml./ml. juice). After centrifugation at 0° the precipitate is discarded and the supernatant liquid is poured into the amylose solution (1 ml. supernatant/0.3 ml. amylose solution). The mixture is then brought to 11% ethanol concentration by adding '50% ethanol' (0.065 ml./ml. mixture). The precipitate, which contains nearly the whole of the Q-enzyme of the potato, is discarded. The supernatant liquid is poured into more amylose solution (1 ml. supernatant/2.5 ml. amylose solution), and the mixture again brought to 11% ethanol concentration with '50% ethanol' (0.201 ml./ml. mixture). After being stirred for 10 minutes the mixture is centrifuged and the precipitate washed three times with twenty times its volume of '11% ethanol,' all these operations and the following being performed at 0°. The precipitate is next extracted with gentle agitation for 15 minutes with 0.05 M citrate, pH 7.0 (0.5 ml./ml. of original juice), and the extract filtered carefully through a grade 3 sintered glass filter, avoiding foaming. The filtrate is brought to 5% (w/v) ammonium sulfate concentration by adding 50% (w/v) ammonium sulfate, pH 7.0, and the resulting precipitate of phosphorylase, which should barely be visible to the naked eye, is separated by

[24] A simple method of preparing this material is to boil a 1% dispersion of fresh potato starch, brought to pH 3 with HCl, for 30 minutes. Cool rapidly to 60°, stir in excess *n*-butanol, and cool. Centrifuge the precipitate, wash with butanol-saturated water in the centrifuge, and disperse the precipitate in hot water to give a 0.2% solution. Reprecipitate with *n*-butanol. (H. Baum and G. A. Gilbert, personal communication.)

[25] For definition see E. J. Bourne, W. N. Haworth, A. Macey, and S. Peat, *J. Chem. Soc.* **1948**, 924.

filtration (grade 4 glass filter). The precipitate is washed while on the filter with 5% (w/v) ammonium sulfate, pH 7.0, containing 0.01 M citrate, and is then washed through the filter with 0.05 M citrate, pH 7.0. Then 50% (w/v) ammonium sulfate, pH 7.0, is added through the filter to 5% (w/v) final concentration, and crystalline phosphorylase separates from the solution immediately. The preparation is conveniently carried out on 20 to 50 ml. of reconstituted juice.

"The phosphorylase exhibits no phosphatase activity when incubated with sodium glycerophosphate in the presence of Mg ions at pH 6.7.

"α-Amylase and Q-enzyme may be tested for by incubating the phosphorylase solution (5 units/ml.) for 16 hours at 20° in a phosphate-free 0.1% solution of amylose, 0.05 M in citrate, at pH 7.0. The intrinsic viscosity of the amylose, measured in 0.5 N NaOH at 20°, should not fall by more than corresponds to 0.01% hydrolysis of the amylose, and there should be no detectable change in the iodine stain of the solution or in its reducing power."

Properties

Unless otherwise stated, the following remarks apply only to the potato enzyme, which has received much more attention than any other plant phosphorylase. Such properties as have been described for phosphorylases from other sources are similar to those of the potato enzyme except that lima bean[1] and sweet potato[26] phosphorylases are said not to require a primer to initiate amylose synthesis (see, however, Nakamura[27]).

Specificity. In the synthesis of amylose the smallest molecule which will act as a primer is maltotriose,[4,28,29] but maltotetraose and higher members of the maltosaccharide series are much more efficient.[28,29] Apart from primers derived from starchlike polysaccharides, two dextrans[30-32] and the type III pneumococcus polysaccharide are reported to prime amylose synthesis. It is claimed that under certain conditions *Betacoccus arabinosaceous* dextran is as effective as starch;[31] the pneumococcus polysaccharide and dextran from *Leuconostoc mesenteroides* each has 40% of the activity of corn starch.[30] Other workers find that dextran from *Leuconostoc dextranicum*[33] and from an unspecified source[4] are without priming activity. Lichenin, from Iceland moss, is inactive.[4] Oxidation of the

[26] Y. Inoue, cited by M. Nakamura.[27]

[27] M. Nakamura, *Nature* **171**, 795 (1953).

[28] J. M. Bailey, W. J. Whelan, and S. Peat, *J. Chem. Soc.* **1950**, 3692.

[29] D. French and G. M. Wild, *J. Am. Chem. Soc.* **75**, 4490 (1953).

[30] E. C. Proehl and H. G. Day, *J. Biol. Chem.* **163**, 667 (1946).

[31] M. A. Swanson and C. F. Cori, *J. Biol. Chem.* **172**, 815 (1948).

[32] S. Hestrin, *J. Biol. Chem.* **179**, 943 (1949).

[33] E. J. Bourne, D. A. Sitch, and S. Peat, *J. Chem. Soc.* **1949**, 1448.

reducing group of a hexasaccharide primer (derived from Schardinger α-dextrin) only slightly diminished the activity.[32]

α-D-G-1-P cannot be replaced by β-D-G-1-P,[34] by α-L-G-1-P,[35] by maltose- or D-xylose-1-phosphates,[36] or by G-6-P, F-1-P, and FDP.[1]

In the breakdown of starch by phosphorylase, arsenate ion can replace phosphate.[37] The optimum pH of arsenolysis of amylopectin is the same as for phosphorolysis, but the relative rates of degradation are 1:8.[37] Arsenolysis of amylopectin is 1000 times as slow as β-amylolysis.[38] The product of arsenolysis is glucose; glucose-1-arsenate cannot be detected. Phosphorylase can transfer only α-1,4-linked glucopyranose units to phosphate.

Activators and Inhibitors. The activating and inhibiting actions of α- and β-amylase have already been mentioned. Green and Stumpf[1] report 80% inhibition by $1 \times 10^{-4} M$ AgNO$_3$, 24% by $3.7 \times 10^{-2} M$ NaF, and 18% by 0.3 saturated phloridzin (see also refs. 17 and 39), but no inhibition by glucose, maltose, sucrose, inulin, ZnSO$_4$, HgCl$_2$, lead acetate, iodoacetic acid, KCN, capryl alcohol, sulfanilamide, H$_2$O$_2$, or Na$_2$S$_2$O$_4$. The Schardinger α- and β-dextrins showed competitive inhibition with respect to the primer. Fischer and Hilpert[18] find that phloridzin inhibits the crude enzyme but not the purified enzyme, the reason for this difference being that the crude preparation contains a β-glucosidase which converts phloridzin into phloretin, a powerful inhibitor. The purified enzyme was destroyed when attempts were made to remove traces of α-amylase activity by addition of cupric, silver, zinc, mercuric, or molybdate ions. As mentioned above, the last two ions are without significant action on amorphous preparations.[6] Contrary to Green and Stumpf's finding, these authors report competitive inhibition by glucose toward G-1-P. The phosphorylases of jack bean[15] and green gram[16] are not inhibited by glucose; the latter enzyme is completely inhibited[16] by $1 \times 10^{-3} M$ HgCl$_2$ and 0.01 M AgNO$_3$. Adenylic acid, required for full activity of muscle phosphorylase (see next section), does not activate the potato[1] or jack bean[15] enzymes, and the latter, unlike the muscle enzyme, is not activated by cysteine. The jack bean preparation contains 4 γ of bound flavine adenine dinucleotide per milligram.[15] British anti-Lewisite ($2.5 \times 10^{-3} M$) caused 40% inhibition.[40] The phosphorylase activities of extracts of leaves and

[34] M. L. Wolfrom, C. S. Smith, and A. E. Brown, *J. Am. Chem. Soc.* **65**, 255 (1943).
[35] A. L. Potter, J. C. Sowden, W. Z. Hassid, and M. Doudoroff, *J. Am. Chem. Soc.* **70**, 1751 (1948).
[36] W. R. Meagher and W. Z. Hassid, *J. Am. Chem. Soc.* **68**, 2135 (1946).
[37] J. R. Katz and W. Z. Hassid, *Arch. Biochem.* **30**, 272 (1951).
[38] K. H. Meyer, R. M. Weil, and E. H. Fischer, *Helv. Chim. Acta* **35**, 247 (1952).
[39] J. M. Bailey and W. J. Whelan, *J. Chem. Soc.* **1950**, 3573.
[40] E. C. Webb and R. van Heyningen, *Biochem. J.* **41**, 74 (1947).

stems from red kidney beans were markedly lower from plants treated with 2,4-dichlorophenoxyacetic acid.[41] Sulfonamides with nitrogen-containing heterocyclic rings show marked inhibition, which is completely neutralized by addition of equimolar amounts of p-aminobenzoic acid or thiamine.[42] Sugar cane phosphorylase is inhibited by Brilliant Alizarin Blue and by Rosinduline GG.[43]

Effect of pH. Phosphorylase exhibits optimum activity between pH 5.9 and 6.1.[44] The pH value of the medium also influences the position of equilibrium in the reaction, since this is governed by either the ratio of any pair of corresponding inorganic orthophosphate and glucose-1-phosphate ions or the ratio of the un-ionized acids. The particular species with which phosphorylase reacts is not known. These ratios are independent of pH, but, since the concentrations of the various species vary with pH, the equilibrium ratio (inorganic orthophosphate):(G-1-P) must also vary. Values of this ratio show a progressive diminution with decreasing hydrogen ion concentration, being 10.8 at pH 5.0, 6.7 at pH 6.0, and 3.1 at pH 7.0.[44] The equilibrium ratios, which do not vary with pH, of (H_3PO_4):(G-1-P), $(H_2PO_4^-)$:(G-1-P$^-$), and (HPO_4^{--}):(G-1-P^{--}) are, respectively, 77, 10.7, and 2.2.[45] These calculations are based on values of k_1 and k_2 for glucose-1-phosphoric acid of 7.8×10^{-2} and 7.4×10^{-7} (ref. 46), and for orthophosphoric acid of 1.07×10^{-2} and 1.51×10^{-7}, respectively. Trevelyan *et al.*[47] have redetermined k_2 for glucose-1-phosphoric acid and find a value of 3.09×10^{-7}. The ionic equilibrium constant for the activity ratio of the monovalent ions is 11.35 ± 0.07 when extrapolated to zero ionic strength. The change in free energy during the reaction acid glucose-1-phosphate $\rightleftarrows$ acid orthophosphate + polysaccharide is -1460 cal./mole. The concentration of polysaccharide is without effect on the equilibrium unless, as has been shown for muscle phosphorylase,[32] the molar concentration of nonreducing primer end groups is greater than that of G-1-P.

Other Properties. Potato phosphorylase loses 61% of its activity when heated at 58° for 3 minutes, and 97% at 68° for 3 minutes.[1] Q_{10} for amylose synthesis is 2.3 between 20 and 50°. After a solution of the enzyme is stored at 4° for 53 days, the smallest loss of activity (approximately 25%)

[41] W. B. Neely, C. D. Ball, C. L. Hamner, and H. M. Sell, *Plant Physiol.* **25**, 525 (1950).

[42] M. Nakamura, *J. Agr. Chem. Soc. Japan* **24**, 5 (1950) [*C. A.* **47**, 1758 (1953)].

[43] C. E. Hartt, *Printed Repts. Hawaiian Sugar Planters' Assoc., Rept. Committee in Charge Expt. Sta.* **61**, 122 (1941) [*C. A.* **36**, 4858 (1942)].

[44] C. S. Hanes, *Proc. Roy. Soc. (London)* **B129**, 174 (1940).

[45] C. S. Hanes and E. J. Maskell, *Biochem. J.* **36**, 76 (1942).

[46] C. F. Cori, S. P. Colowick, and G. T. Cori, *J. Biol. Chem.* **121**, 465 (1937).

[47] W. E. Trevelyan, P. F. E. Mann, and J. S. Harrison, *Arch. Biochem. and Biophys.* **39**, 419 (1952).

occurs at pH 7. The activity decreases rapidly below pH 6.5, less than 10% remaining after this period of time at pH 6.[4] The enzyme is completely inactivated by dialysis against distilled water,[1,4,44] but reports of its behavior on dialysis against KCl are conflicting.[1,4] The destruction of the enzyme by freeze-drying[4] can be prevented by incorporation of citrate buffer (pH 6 to 7) in the enzyme solution.[17] Alcohol and acetone inactivate the enzyme at room temperature but not at temperatures around 0°.

The Michaelis constants for phosphorylase-orthophosphate, phosphorylase-G-1-P, phosphorylase-starch, and phosphorylase-hexasaccharide are 6.2×10^{-3} M, 2.6×10^{-3} M, 0.97 g./l., and 0.65 g./l., respectively,[4] and for phosphorylase-arsenate the value is 8×10^{-3} M.[37] The sedimentation constant $(S_{20} \times 10^{13})$ of phosphorylase is 9.2.[4]

Yeast Phosphorylase

Although it was among the first of the phosphorylases to be examined, this enzyme has received little attention in recent years. The few properties that are known are similar to those of the plant and animal phosphorylases. The usual source of the enzyme is baker's yeast (*Saccharomyces cerevisiae*). A solution of the crude enzyme has also been obtained from *Torulopsis rotundata* by grinding a water suspension of the cells with powdered glass.[48]

Purification Procedure

The following method is due to Meyer and Bernfeld.[49] Dried baker's yeast (20 g., "Feldschlossen") is treated with water (60 ml.) for 3 hours at 35°. The doughy mass is dialyzed through Cellophane against distilled water for 45 hours and after centrifugation gives 55 ml. of extract, pH 6.0.

All the following operations are carried out between 0 and +3°. The extract is adjusted to pH 7.8 with 2 N NaOH (0.8 ml.) when the turbidity disappears. N-Acetic acid (1.6 ml.) is added immediately, adjusting the pH to 6.7. The solution becomes turbid, and after a short time the precipitate flocculates and is removed on the centrifuge and rejected. The clear supernatant is acidified to pH 5.3 with N-acetic acid (1 ml.). The precipitate is separated on the centrifuge and triturated with 0.2 M sodium acetate (15 ml.); at the same time 0.5 N Na_2CO_3 is added dropwise until the pH is 6.9. After 10 minutes part of the precipitate has dissolved, and the undissolved material is removed on the centrifuge and rejected. The cloudy solution (about 20 ml.) contains the yeast phosphorylase. The preparation is contaminated with maltase and isomaltase but has little amylolytic activity and contains 0.02 mg. of phosphorus per milliliter.

[48] J. Mager, *Biochem. J.* **41**, 603 (1947).
[49] K. H. Meyer and P. Bernfeld, *Helv. Chim. Acta* **25**, 399 (1942).

The enzyme may be dialyzed against 0.3 saturated $(NH_4)_2SO_4$ without loss of activity.[50] It is not inhibited by iodoacetate or arsenate and is not activated by magnesium.[51]

[50] W. Kiessling, *Naturwissenschaften* **27**, 129 (1939).
[51] A. Schäffner and H. Specht, *Z. physiol. Chem.* **251**, 144 (1938).

[23] Muscle Phosphorylase

$$x \text{ Glucose-1-phosphate} + G_n \rightleftarrows G_{n+x} + x \text{ Inorganic Phosphate}$$

By GERTY T. CORI, BARBARA ILLINGWORTH, and PATRICIA J. KELLER

Assay Method

Principle. In the reversible bimolecular reaction catalyzed by phosphorylase,[1] one of the reactants in either direction is the terminal nonreducing glucose residue in the outer chains of glycogen (designated G_n, where n = average number of glucose residues in outer chains). The overall reaction for successive steps is given above. Within certain limits $[n > 3, (n + x) < 150]$, the equilibrium of the reaction is described by the ratio of concentrations, inorganic phosphate/glucose-1-phosphate. This ratio has a value of 3.55 at pH 6.8 and of 6.14 at pH 6.0, and the equilibrium is therefore in favor of synthesis in both instances. The rate of the reaction is at its maximum at pH 6.8 and is 30% lower at pH 6.0.

Phosphorylase activity is measured in the direction of synthesis by determination of the amount of inorganic phosphate formed from glucose-1-phosphate, and the conditions are so arranged that the reaction is kinetically of the first order. In crude enzyme solutions the reaction is carried out at pH 6 in order to minimize removal of glucose-1-phosphate by phosphoglucomutase action; after purification the enzyme is tested at pH 6.8.

Reagents

(1) 4% glycogen solution.
(2) 0.07 M sodium glycerophosphate (at pH 6.8 or 6.0).
(3) 0.06 M cysteine hydrochloride freshly neutralized to pH 6.8 or 6.0; (2) and (3) are mixed in equal volumes and used for dilution of the enzyme.
(4) 0.064 M crystalline dipotassium salt of glucose-1-phosphoric acid brought to pH 6.8 or 6.0.

[1] C. F. Cori, G. T. Cori, and A. A. Green, *J. Biol. Chem.* **151**, 39 (1943).

(5) Same as (4), but containing also 0.004 M adenosine-5-phosphate.

Procedure.[2] 0.1 ml. of enzyme (crude or crystal suspension) is diluted with the cysteine-glycerophosphate buffer. If the dilution is greater than 1:25, it is carried out in two or three steps. To 0.4 ml. of dilute enzyme (in duplicate) 0.2 ml. of (1) is added, and the solutions are incubated for 20 minutes at 30°. The reactions are started by adding to one of the samples 0.2 ml. of (4) and to the other 0.2 ml. of (5), prewarmed to 30°. After exactly 5, 10, and 15 minutes, 0.2-ml. aliquots are pipetted into 7 ml. of dilute H_2SO_4 (10.0 ml. 5 N H_2SO_4 + 690 ml. H_2O) in a Klett tube. The acid stops the enzymatic reaction but is too weak to cause appreciable hydrolysis of glucose-1-phosphate in 60 minutes at room temperature. In the determination of inorganic phosphate by the Fiske-SubbaRow method, only 0.9 ml. (instead of 1.0) of 5 N H_2SO_4 is added. The amount of inorganic phosphate present in the crude enzyme (after step 1) can be neglected; deproteinization is not necessary even at that stage. With the reagents used in this laboratory, the zero time blank of the reaction mixture reads 3 to 6 Klett units above the blank of the reagents used in the phosphate method.

After step 1 of the purification procedure as well as in the first and subsequent crystals, $\dfrac{\text{enzymatic activity without adenylate}}{\text{enzymatic activity with adenylate}} \times 100$ is between 55 and 70 at 30°. If this ratio is found to be below 40 after step 1, no or a very poor yield of crystals will be obtained by the method described for phosphorylase a.

A test for phosphoglucomutase is described below; were it present, it would interfere with the assay for phosphorylase activity. After incubation of the reaction mixture for 15 minutes, 0.1 ml. is pipetted into 5.0 ml. 1 N H_2SO_4 in a Klett tube which is placed for 5 minutes in boiling water. Under these conditions glucose-1-phosphate is completely hydrolyzed, whereas glucose-6-phosphate, the product of phosphoglucomutase action, is hardly hydrolyzed at all. The amount of inorganic phosphate after acid hydrolysis remains the same before and after incubation if no phosphoglucomutase action has taken place in the reaction mixture.

Calculation of Units. Under the conditions of the test here described, the reaction is kinetically first order and the rate is proportional to enzyme concentrations over a considerable range of enzyme dilutions.

$$K = \frac{1}{t} \log \frac{x_e}{x_e - x}$$

[2] B. Illingworth and G. T. Cori, *Biochem. Preparations* **3**, i (1953).

where x_e = % 1-ester converted to inorganic phosphate (or polysaccharide) at equilibrium.

   x = % converted at time t (in minutes).

The value of x_e varies with pH; hence, the pH of the reaction mixture
must be known and kept constant. At pH 6, x_e = 86; at pH 6.8, x_e = 78.

   Arbitrarily, $k \times 1000$ gives the total number of enzyme units present.
This is multiplied by the appropriate dilution factor to calculate the number of enzyme units present per milliliter of original enzyme. For example,
if the enzyme was first diluted 1:25 and then 1:1 in the actual test, the
multiplication factor would be 50. The number of units per milliliter of
original enzyme, divided by the milligrams of protein present per milliliter
of original enzyme, gives the specific enzyme activity per milligram of
protein.

Phosphorylases a and b

   Many tissues, particularly muscles, contain an enzyme (PR) which
converts phosphorylase a to phosphorylase b in a first-order reaction.
Phosphorylase b is enzymatically inactive unless adenylic acid is added
to the reaction mixture. In an extract of resting muscle prepared as described here, the a form prevails, whereas in the supernatant fluid of a
homogenate prepared in a blendor, the b form predominates. Rapid handling, and other precautions mentioned below, are necessary to prevent
PR action. The bulk of PR is separated from phosphorylase in the pH 6
precipitate described in step 1 of the purification procedure.

Purification and Crystallization of Phosphorylase a

   The method is based largely on the work of Green and Cori.[3] In the
procedure for crystallization and recrystallization, use is made of the high
temperature coefficient of solubility of phosphorylase a when dissolved in
a weak, buffered salt solution. The first crystallization is from a solution
in which the enzyme is only about 0.3 pure.

   A well-fed rabbit is deeply anesthetized by the intravenous injection
of about 5 ml. of sodium pentobarbital solution (60 mg./ml.) and bled.
There must be no convulsions or strong twitches. The animal is skinned
and the muscles of the leg and back are rapidly excised, weighed, and
taken to a cold room where all further steps are carried out, except as
stated.

   *Step 1. Preparation and Dialysis of Crude Extract, Followed by Removal
of a Precipitate at pH 6.0.* The muscles are ground twice in an ordinary
meat grinder, the hash transferred to a beaker, and 1 vol. of cold distilled
water immediately added. The suspension is stirred occasionally with a

[3] A. A. Green and G. T. Cori, *J. Biol. Chem.* **151**, 21 (1943).

glass rod; after 10 minutes it is poured on gauze which has been tied over the opening of a beaker in funnel shape, and the extract is allowed to collect in the beaker. The ground muscle is gently expressed with a wide spatula. After about 5 minutes little additional fluid continues to drip through, and the muscle hash is returned to the first beaker and extracted with another volume of water for about 10 minutes and then returned to the gauze funnel and treated as before. The gauze bag is not squeezed.

The filtrate is spun in a Servall angle centrifuge for 5 minutes at about 8000 r.p.m., and the supernatant fluid is filtered through a fluted filter. The pH at 10° of a small aliquot of the extract is measured with a glass electrode (temperature-compensating knob set at 10°). It is usually between 6.5 and 6.7 and is lowered to about 6.4 by the addition of 0.05 N HCl or, preferably, 1.0 M acetate, pH 4.6. The acid is added slowly with gentle stirring. The slightly turbid extract is dialyzed in Nojax casing (Visking Corporation, $1\frac{8}{32}$ inch) against running tap or distilled water (10° or below) for 3 to 4 hours. The casings are turned upside down several times. Dialysis should start not later than 2 hours after excision of the muscles.

At the termination of dialysis, the contents of the casings are collected in a beaker. Hydrochloric or acetic acid is again added with the proper precautions until the pH has dropped to 6.0. A flocculent precipitate forms within 5 to 10 minutes; it settles rapidly in a Servall centrifuge at about 8000 to 10,000 r.p.m. The clear supernatant fluid is filtered through a Whatman No. 1 fluted filter (18 cm. in diameter) placed in a wide-stemmed funnel. To the filtrate, solid $KHCO_3$ is added with stirring until the pH is 6.8.

Step. 2. Precipitation of Phosphorylase with Ammonium Sulfate. The salt solution, saturated at room temperature and the pH adjusted to 6.8 with strong ammonia water, is not precooled and is added in a slow stream with thorough but gentle mixing, adding 70 ml. of salt solution per 100 ml. of extract to bring the solution to 0.41 saturation. A flocculent precipitate settles slowly when left undisturbed overnight, when most of the supernatant fluid can be decanted. The precipitate is separated by centrifugation in a Servall (about 8000 to 10,000 r.p.m.) and eventually collected in two 45-ml. tubes; it packs well and is carefully drained.

Step 3. Crystallization. The precipitate, which contains phosphorylase at a purity level of about 0.3, is immediately dissolved in 10 to 15 ml. of distilled water. The slightly turbid solution is dialyzed in Nojax casing ($\frac{8}{32}$ inch) for 60 to 80 minutes against cold running tap or distilled water. At that time, the dialysis bags (usually two) are transferred to a cylinder (placed in an ice bath) which contains a solution of the following composition: 2.5 ml. of 0.3 M cysteine hydrochloride (or 0.015 M Versene),

100 ml. of 0.02 M Na$_2$ glycerophosphate; water is added to bring the volume to 250 ml., and the pH is adjusted to 6.6 to 6.8.[4] During 16 to 20 hours this solution is changed three to four times. A crystalline precipitate forms in the dialysis bag which gets progressively heavier.

Step 4. Recrystallizations. At the termination of dialysis the contents of the bags are collected and centrifuged in small plastic tubes in a high-speed head (International centrifuge) at about 8000 r.p.m. for 10 to 15 minutes. The supernatant fluid (first mother liquor) is almost clear, and to the well-packed and drained precipitate (first crystals) is added a total of 20 to 25 ml. of a buffered solution at 35° (2 ml. of 0.3 M cysteine hydrochloride or 0.0375 M Versene plus 48 ml. of 0.02 M Na glycerophosphate, pH adjusted to 6.6 to 6.8; for the first and second crystallizations NaF, 0.1 M, may be added to the solvent). The solvent should be added in small portions with constant gentle rubbing of the precipitate with a glass rod. After several minutes at 35° the first portion is centrifuged (about 10 to 15 ml.) at room temperature at about 8000 r.p.m. for about 5 minutes. The clear supernatant fluid is decanted into a glass vessel which is placed in an ice bath. The precipitate is re-extracted two to three times (each time with about 5 ml.) or until most of it has gone into solution. The final residue consists of cystine crystals (when cysteine was used) and denatured protein. The phosphorylase protein begins to crystallize as soon as the temperature approaches 0°. A silky sheen appears, which gets heavier rapidly. Under these conditions of rapid crystallization the crystals are very small.

After dissolving the third crystals the solution (preferably containing 0.1 M NaF) is added to washed Norit (100 mg. spun down from a suspension for 10 ml. of phosphorylase). After gentle agitation with a glass rod for 5 minutes the suspension is centrifuged and the supernatant fluid filtered. When the filtrate is chilled, crystals form rapidly. There is no detectable adsorption of phosphorylase on Norit, while traces of PR which still contaminate third crystals are removed, as well as traces of ribonucleic acid and nucleotides.

Phosphorylase a is stored as a crystal suspension at 0 to 4°. A small tube of toluene is suspended in the well-stoppered glass vessel.

Preparation of Phosphorylase b

A solution of crystalline phosphorylase a is incubated at 30° in 0.015 M cysteine solution, pH 7, with purified PR enzyme. At the highest purity

---

[4] Addition of NaF to give a concentration of 0.1 M in the dialysis fluid improves the yield of crystals. NaF inhibits the enzymatic conversion of phosphorylase a to b; see E. W. Sutherland, *in* "Phosphorus Metabolism" (McElroy and Glass, eds.), Vol. I, p. 53, Johns Hopkins Press, Baltimore, 1951.

yet obtained, 10 to 12 γ of PR converts 99% of phosphorylase a to b in 1 hour at 30° (see Vol. I [24]). Phosphorylase b has been crystallized from ammonium sulfate-cysteine solutions.[5] The b form is more soluble in dilute salt solution than the a form, both at 30° and at 0°. Although phosphorylase b is inactive without the addition of adenylic acid, it has the same specific activity at pH 6.6 as phosphorylase a when adenylic acid is present in 0.001 M concentration. In a solution containing both phosphorylases a and b, an enzymatic activity test in the presence and in the absence of adenylic acid is used to calculate the relative proportions of the two forms of the enzyme in the mixture.[6]

Properties of Muscle Phosphorylases a and b

Phosphorylase is absolutely specific for α-D-glucopyranosyl-1-phosphate and the α-1,4-maltosidic bond in the polysaccharide. Arsenate can be substituted for phosphate, in which case the reaction becomes irreversible and proceeds at a slower rate.

When glycogens or amylopectins are degraded, the reaction products are glucose-1-phosphate and a limit dextrin which has been characterized;[7] amylose is degraded at a much slower rate.[8] These polysaccharides are also added on to, whereas oligodextrins are practically not degraded or added on to. In this respect mammalian phosphorylase differs significantly from plant (potato) phosphorylase.

Glucose, phlorhizin, and phloretin inhibit muscle phosphorylase.[1] Phosphorylases a and b are homogeneous in the electrophoresis apparatus (pH 6 to 8) and the ultracentrifuge. For phosphorylase a, $S_{20,w}$ is 13.2; for b, it is 8.2. Calculated from sedimentation and diffusion constants, the molecular weight of a is 495,000; of b it is 242,000.[9]

SUMMARY OF PURIFICATION PROCEDURE
(Average of four preparations)

	Units/mg. protein	Protein, mg./ml.	Total units
After step 1	72	15.8	740,000
First crystals	2210	8.5	376,000
Second crystals	2300	6.6	303,000
Third crystals	2515	5.8	235,000

[5] C. F. Cori and G. T. Cori, *J. Biol. Chem.* **158**, 341 (1945).
[6] G. T. Cori and C. F. Cori, *J. Biol. Chem.* **158**, 321 (1945).
[7] G. T. Cori and J. Larner, *J. Biol. Chem.* **188**, 17 (1951).
[8] S. Hestrin, *J. Biol. Chem.* **179**, 943 (1949).
[9] P. J. Keller and G. T. Cori, *Biochem. et Biophys. Acta* **12**, 235 (1953).

[24] The PR Enzyme of Muscle

Phosphorylase $a \rightarrow$ Phosphorylase b

M.W. = 495,000   M.W. = 242,000

By Patricia J. Keller and Gerty T. Cori

Assay Method

The method of assay was described by Cori and Cori in 1945.[1]

Principle. PR catalyzes the conversion of phosphorylase a to phosphorylase b. The reaction is followed by assay for phosphorylase activity under conditions where only the a form is active, i.e., in the absence of added adenosine-5'-phosphate. That the disappearance of phosphorylase a is due to conversion to b rather than nonspecific losses in activity can be established by assaying also in the presence of adenylic acid. Under these conditions phosphorylase b is fully as active as a.

Reagents

(1) Phosphorylase a. A solution, free of NaF, of three (or more) times recrystallized, Norit-treated enzyme as described in the preceding article.[2] The Norit treatment is required for elimination of traces of PR which adhere to phosphorylase a through several recrystallizations.

(2) 0.03 M cysteine hydrochloride, freshly neutralized to pH 6.6. For assay of material from step 3 of the purification onward, the cysteine solution is made up to contain 100 mg.% bovine serum albumin.

(3) A 1:1 mixture of 0.064 M glucose-1-phosphate and 4% glycogen at pH 6.6. The crystalline dipotassium salt of glucose-1-phosphoric acid is used.

(4) 0.07 N sulfuric acid.

(5) Reagents for determining inorganic phosphate by the method of Fiske and SubbaRow.[3]

Procedure. The usual preparation of phosphorylase a contains about 5 mg. of protein per milliliter of crystal suspension. A fresh dilution (about 1:25) of such a preparation with 0.03 M cysteine, pH 6.6, is used

---

[1] G. T. Cori and C. F. Cori, *J. Biol. Chem.* **158**, 321 (1945).

[2] G. T. Cori, B. Illingworth, and P. J. Keller, see Vol. I [23].

[3] C. H. Fiske and Y. SubbaRow, *J. Biol. Chem.* **66**, 375 (1925).

for each PR assay. First, 1 ml. of the dilute phosphorylase solution is preincubated for 20 minutes in a 30° water bath. During this period three Klett tubes are prepared, each containing 0.1 ml. of the glucose-1-phosphate–glycogen mixture. Then, 1 ml. of PR, suitably diluted with 0.03 M cysteine, pH 6.6, is added to the phosphorylase solution. Immediately after mixing, 0.1 ml. is withdrawn from this reaction mixture and added to 0.1 ml. of prewarmed (30°) glycogen–glucose-1-phosphate mixture for phosphorylase assay. After 5 minutes at 30° the phosphorylase reaction is stopped by the addition of 7 ml. of 0.07 N sulfuric acid. Inorganic phosphate, released during phosphorylase action, is determined in the tube used for this reaction, by the method of Fiske and SubbaRow.[3] Only 0.9 ml. of 5 N sulfuric acid is added to bring the final acidity to 0.5 N.

It has been shown that phosphorylase a, in the presence of its substrates, is protected from PR action.[1] The initial, or zero time, level of a activity is thus measured in the first aliquot. Again at time intervals (10 and 20 minutes) after mixing the PR and phosphorylase solutions, 0.1-ml. aliquots are withdrawn and the residual phosphorylase a content is measured in the manner described above.

Application to Crude Extract. The crude supernatant fluid of a muscle homogenate can be assayed in the above manner, after dialysis for 3 hours against 0.08 M glycerophosphate–0.012 M cysteine at pH 9.0. Negligible amounts of phosphorylase a remain in such a dialyzed extract.

Definition of Unit and Specific Activity. The disappearance of phosphorylase a under the conditions described follows the rate equation for a first-order reaction.[1] The rate is proportional to enzyme concentration over a wide range. The unit of PR activity is expressed as a function of k, the first-order velocity constant.

The units of phosphorylase activity for the three time periods (0, 10, and 20 minutes after mixing with PR) are calculated.[2] These values are substituted into the equation:

$$k = \frac{1}{t} \log \frac{\text{units phosphorylase } a \text{ at } t = 0}{\text{units phosphorylase } a \text{ at time } t}$$

Arbitrarily, $k \times 1000$ is a PR unit. This value is multiplied by the dilution factor (the final dilution of PR in the PR-phosphorylase reaction mixture) to give the number of PR units per milliliter. Protein has been determined by the Lowry method.[4] The number of PR units per milliliter, divided by the protein concentration, in milligrams per milliliter, gives the specific activity (units per milligram).

[4] O. H. Lowry, N. J. Rosebrough, A. L. Farr, and R. J. Randall, *J. Biol. Chem.* **193**, 265 (1951).

The per cent of phosphorylase a converted to b at any time can be calculated from the initial level of a activity and the level at that time:

$$\% \text{ converted} = \frac{\text{Units of } a \text{ at } t = 0 - \text{units of } a \text{ at } t}{\text{Units of } a \text{ at } t = 0} \times 100$$

Alternatively, the ratio of activities without and with adenylic acid can be used to ascertain the relative proportions of a and b in a mixture. If a preparation of a, having initially 65% as much activity in the absence as in the presence of adenylic acid has, after incubation with PR, 39% as much:

$$\text{Per cent } a = \frac{0.39}{0.65} \times 100 = 60$$
$$\text{Per cent } b = 40$$

Purification Procedure

Steps 1 and 2 of the procedure entail extraction and concentration of the enzyme by means of an acid precipitation; steps 3 and 4 involve selective denaturation of accompanying proteins. The purification is conveniently carried out in 2 days. All operations after excision of the muscle are carried out in the cold room (5°) unless otherwise stated.

Step 1. Preparation of Crude Extract. Fresh rabbit muscle is passed once through a chilled meat grinder; the ground muscle is then homogenized for 3 minutes in a Waring blendor with 2 vol. of cold distilled water. The homogenate is centrifuged in a Servall angle centrifuge (8000 $\times$ g) for 5 to 10 minutes and the supernatant fluid filtered through coarse filter paper.

Step 2. Acid Precipitation. The pH of the extract is usually about 6.0. It is lowered to 5.65 [5] by the addition of 1.0 M acetate buffer, pH 4.6 (1 to 1.5 ml. of acetate buffer per 100 ml. of extract is usually required). After 5 to 10 minutes in the cold, the flocculent precipitate is centrifuged down at 8000 $\times$ g for about 10 minutes. The precipitate is collected into two tubes and washed once with about 80 ml. of cold distilled water, then dissolved in a cold solution of sodium glycerophosphate (0.08 M). The volume of the solution is adjusted to give a protein concentration of about 3% (ca. 30 to 50 ml.). The pH of the redissolved precipitate is usually near 7.0 and need not be further adjusted immediately. About an eightfold enrichment in PR activity is effected by the precipitation. The solution is quite turbid at this stage and can be frozen without loss of PR activity or taken directly through the next step in the purification.

Step 3. Alkaline Incubation. The pH of the glycerophosphate solution of PR is raised to 9.4 by the addition of 1.0 M sodium carbonate (about

---

[5] The pH meter is adjusted to the temperature of the extract

Summary of Purification Steps[a]

Fraction	Total volume, ml.	Units/ml.	Total units	Protein, mg./ml.	Total protein, mg.	Specific activity, units/mg.	Recovery, %	Purification
1. Crude extract	770	2,770	2,132,900	16.6	12,782	170		
2. Dissolved acid precipitate	30	40,800	1,224,000	28	840	1,456	57	×8.6
3. After alkaline incubation and removal of denatured protein	23	50,040	1,150,920	11	253	4,550	54	×27
4. Supernatant from acetone treatment	8	25,400	203,200	0.85	6.8	30,000	11	×177
5. 7.5 ml. of (4) lyophilized, taken up in 1.2 ml. H_2O	1.2	158,100	189,100	5.15	6.2	32,000	11	

[a] From 440 g. of muscle.

0.25 ml./10 ml.), and the solution is incubated in a 37° water bath for 60 minutes. Upon readjustment to neutrality a flocculent precipitate forms which can be discarded after centrifugation without loss of PR activity. About two-thirds of the total protein is removed in this way; hence a purification of about 3 over the acid precipitate is achieved. This material is stable to freezing at pH 9.4 and at neutrality.

Step 4. Acetone Treatment. The supernatant fluid from the preceding step is divided into portions of 10 ml. each in order to facilitate rapid handling. Each aliquot is cooled to 0°. An equal·volume of acetone (at −10°) is added at such a rate that the temperature never rises above 0° and is, at the end of the addition, about −5°. The heavy precipitate of protein is centrifuged immediately. About 2 minutes at 10,000 × g is sufficient to pack the precipitate well, while causing little rise in temperature. The well-drained precipitate is dispersed in 4 ml. of glycerophosphate (0.08 M)–Versene (0.01 M) at pH 9.0 and dialyzed against the same buffer for 1 hour. Much denatured protein remains insoluble during the dialysis and is centrifuged off. The supernatant fluid contains PR at a high level of purity.

The protein concentration is usually 0.1% or under at this stage. Losses are incurred on freezing and thawing. The solution can be lyophilized, however, and taken up in a more concentrated solution without loss in activity.

The final solution of PR prepared by this method contains nucleic acids. In the sample preparation cited, the optical density at 260 of the supernatant fluid from step 4 corresponded to 0.5 mg. of nucleic acids per milliliter. At these concentrations of protein and nucleic acids the contribution of nucleic acid to the protein test is small (under 5%) and has been ignored.

A Norit treatment can be included in the procedure if it is desired to reduce the nucleic acid content. The supernatant fluid from step 3 is stirred for 2 minutes at room temperature with Norit (0.14 ml. of a 10% suspension per milliliter of PR solution); the mixture is centrifuged immediately at room temperature (8000 r.p.m. for about 2 minutes), and the supernatant fluid is filtered through paper. No significant loss of PR activity occurs during this treatment,[6] whereas the nucleic acid content after step 4 is reduced to about one-half that found with untreated material. The same purification is achieved when step 4 is performed with a Norit-treated as with an untreated solution of PR.

---

[6] With larger amounts of charcoal, however, PR activity may be lost; see Vol. I [23].

[25] Amylo-1,6-glucosidase

Phosphorylase Limit Dextrin (Glycogen) → Glucose Plus Polysaccharide
in Which Penultimate
Branches Have Become
Outer Branches

By Gerty T. Cori

Assay Method

Principle. Glucosidase was described by Cori and Larner.[1] It splits the
α-1,6-glucosidic bonds in the branched polysaccharides glycogen and
amylopectin; it acts only after phosphorylase has degraded the outer
chains of these polysaccharides exhaustively, forming a limit dextrin
(LD). By hydrolytic action glucosidase then liberates glucose from this
LD. The hindrance to phosphorylase action is thus removed, and phosphorylase can act again and form a second LD. This process repeats itself
from tier to tier until the polysaccharide molecule is completely degraded.
Only glucose and glucose-1-phosphate are found as products of the combined activity of the two enzymes.

Glucosidase is allowed to act on LD in the presence of an excess of
muscle phosphorylase so that it becomes the limiting enzyme in degradation. Crude preparations contain phosphoglucomutase and phosphohexoisomerase, so that the reaction products are glucose, glucose-1-phos-
phate, and the equilibrium hexose-6-phosphate.

The method described can be used for assay of crude and purified
enzymes.

Reagents

0.2% solution of LD in H_2O.

1.0 M K-phosphate, pH 6.8.

Crystalline muscle phosphorylase (recrystallized six to eleven times,
and crystals suspended in 0.03 M Na-glycerophosphate–0.0012
M Versene, pH 6.8).

Schenk deproteinization mixture (2.5% $HgCl_2$ in 0.5 N HCl).

Alkaline copper reagent No. 60 of Shaffer and Somogyi.[2] Kl is
omitted, and $CuSO_4$ and alkali solutions are mixed just before use.

Nelson's[3] reagent for colorimetric determination of reducing sugars.

[1] G. T. Cori and J. Larner, *J. Biol. Chem.* **188,** 17 (1951).

[2] P. A. Shaffer and M. Somogyi, *J. Biol. Chem.* **100,** 695 (1933).

[3] N. Nelson, *J. Biol. Chem.* **153,** 375 (1944).

Reaction Mixture

   0.5 ml. of LD solution (1 mg.).

   0.05 ml. of 1.0 M phosphate, pH 6.8.

   0.35 ml. of phosphorylase (0.1 to 0.2 mg.) in 0.0012 M Versene solution, pH 6.8.

   0.1 ml. of glucosidase (dilution made with the same Versene).

Procedure. The phosphorylase is added just before the glucosidase; the latter addition starts the incubation period of 20 minutes at 30°. Each phosphorylase preparation has to be tested for contamination with glucosidase, using the test as described here. If there are still traces of glucosidase present, an appropriate correction has to be made. When crude glucosidase is used, copper reduction is also determined in a sample to which 0.05 ml. of 0.7 M NaCl is added instead of phosphate. This "amylase" blank is low even with crude extracts. After 20 minutes of incubation at 30°, 1.0 ml. of Schenk solution is added to the reaction mixture, the protein precipitate is removed by filtration, and H_2S is passed through the filtrate. HgS, which adsorbs the polysaccharide left in the reaction mixture, is removed by filtration; the filtrate is aerated and then heated to 100° for 7 minutes. HCl is present in about 0.3 N concentration, so that glucose-1-phosphate is completely hydrolyzed. The hexose-6-phosphates formed when crude glucosidase is used have the same molar reducing power as glucose for reagent No. 60, provided that the heating period is 30 minutes. After hydrolysis 0.5 ml. is pipetted into a Klett tube and neutralized with 0.3 N NaOH (against a trace of phenol red), and reduction is determined. The method becomes inaccurate when there is less than 30 γ of glucose present. Glucosidase concentration should be so adjusted that 20 to 70% of the LD (1000 γ) is digested in 20 minutes.

Definition of Unit and Specific Activity. Activity is proportional to the amount of glucosidase used, provided that phosphorylase is present in excess. 1 γ of glucose (sum of free and phosphorylated glucose) liberated in 10 minutes in 1 ml. of reaction mixture is designated as one unit. The units per milliliter divided by milligrams of protein per milliliter corresponds to specific activity of the glucosidase.

Purification Procedure

Preparation of crude glucosidase follows the procedure described for muscle phosphorylase[4] through step 1 (removal of PR). The neutralized extract is dialyzed for 20 hours against a 0.0012 M Versene–0.015 M glycerophosphate buffer, pH 7.2[5] at 0°. After dialysis, glucosidase activity

---

[4] G. T. Cori, B. Illingworth, and P. J. Keller, Vol. I [23].

[5] The same buffer-Versene solution is used in steps 4 and 5 of the purification procedure.

SAMPLE PROTOCOL OF PURIFICATION OF AMYLO-1,6-GLUCOSIDASE
(500 g. of rabbit muscle)

Fraction	Total ml.	Units/ml.	Total units	Protein, mg./ml.	Total protein, mg.	Units/mg.	Recovery of units, %	Purification
1. Crude dialyzed extract	700	1,720	1,200,000	14.3	10,000	120		
2. Mother liquor, first phosphorylase crystals	14	66,000	925,000	24.5	343	2,700	77	×22
3. Precipitate of 13 ml. of (2) at 0.33 saturation with (NH₄)₂SO₄ (dissolved, dialyzed)	6	108,000	650,000	25.85	155	4,160	76	×35
4. Precipitate of 3 ml. of (3) at 0.235 saturation with (NH₄)₂SO₄ (dissolved, dialyzed)	1.2	95,600	115,000	13.8	16.6	6,950	35	×58

213

was found no lower than in a frozen extract. Blank values (obtained in tests in which NaCl is substituted for phosphate) are low, indicating little amylase activity. The presence of inorganic P in the extract before the long dialysis makes it impossible to determine a "blank" value. In previous work,[1] α-amylase was removed by adsorption on starch; this procedure, particularly when repeated, leads to considerable losses of glucosidase. 0.1 ml. of undiluted dialyzed extract (about 1.4 mg. of protein) is added in the test.

Purification of Glucosidase. About 80% of the glucosidase present in the crude extract is recovered in the proteins which precipitate at 0.41 saturation with ammonium sulfate. The method for the preparation of crystalline phosphorylase[4] is followed through step 3; with the phosphorylase crystals one-fourth to one-third of the protein in this fraction is removed. In the mother liquor of the crystals the specific activity of glucosidase has been increased 20- to 30-fold over the starting material; 0.05 mg. of protein or less is now added in the test.

Step 4. The mother liquor is diluted with buffer so that the protein concentration is about 15 mg./ml. Neutral saturated ammonium sulfate solution is added to 0.33 saturation. The precipitate is centrifuged off and dissolved in about one-half of the original volume of buffer and dialyzed against the buffer at 0° for about 3 hours.

Step 5. The protein solution obtained in step 4 is diluted with buffer so that the concentration of protein is 5 to 6 mg./ml. Fractions which precipitate at 0.235 and 0.285 saturation with ammonium sulfate are considerably purer than the starting material. Only 0.02 to 0.03 mg. of protein is now required for the test.

Properties

Up to the level of 4000 units/mg., dialyzed buffered enzyme solutions can be frozen with little loss of activity. Versene preserves the enzyme activity when solutions are kept at 0° for several days, provided that the protein concentration is 1% or higher. Purified enzyme preparations (4000 to 8000 units/mg.) in the concentrations used in the test are free of phosphoglucomutase and α-amylase activity. They still contain phosphorylase.

Further steps of purification are being explored.

[26] Polysaccharide Phosphorylase, Liver[1]

Polysaccharide + P $\rightleftharpoons$ G-1-P

By EARL W. SUTHERLAND

Assay Method

Principle. Phosphorylase activity was determined by measurement of the rate of liberation of inorganic phosphate from G-1-P in the presence of glycogen;[2] i.e., phosphorylase activity was measured in the direction of synthesis of polysaccharide.

Reagents

Solution of potassium G-1-P (0.05 M, pH 6.1) containing 5.7 mg. of glycogen per milliliter and NaF (0.05 M).
HCl was used for pH adjustment.
Potassium salt of 5-AMP (0.02 M, pH 6.7).
NaF (0.1 M).
Enzyme. Dilute with cold NaF (0.1 M). When crude fractions were used, dilution was carried out immediately before testing.

Procedure. The reaction was started by the addition of 1.0 ml. of G-1-P–glycogen–NaF to 0.5 ml. of enzyme–5-AMP–NaF solution. The 0.5 ml. of enzyme mixture contained 0.1 ml. of 5-AMP (0.02 M), enzyme dilution, and 0.1 M NaF to the volume of 0.5 ml. A sample (0.5 ml.) was taken at zero time and after incubation for 10 minutes at 37° into TCA; the balance was used for an iodine starch test. Inorganic phosphate was determined by the method of Fiske and SubbaRow,[3] as adapted to the Klett-Summerson photometer. In crude preparations small corrections can be made for phosphatase activity by determination of glucose production due to the addition of G-1-P.

Units and Specific Activity. One unit of enzyme was defined as that amount which caused the liberation of 1.0 mg. of inorganic phosphorus in

---

[1] Purification procedures reported here were developed in the Department of Biochemistry, Washington University School of Medicine, and were outlined in a previous report [E. W. Sutherland, *Federation Proc.* **10**, 256 (1951)]. Current investigations by the author and Dr. Walter D. Wosilait at the present address indicate that the first two steps may be simplified greatly by extraction of a liver homogenate with a filter press, followed by precipitation of the enzyme with ammonium sulfate at 0.65 saturation.

[2] G. T. Cori and C. F. Cori, *J. Biol. Chem.* **135**, 733 (1940).

[3] C. H. Fiske and Y. SubbaRow, *J. Biol. Chem.* **66**, 375 (1925).

10 minutes when the per cent conversion of glucose-1-phosphate was in the range of 12 to 22%. Specific activity was expressed as units per milligram of protein. Protein was determined by a micromethod described previously[4] and by the Weichselbaum method.[5]

The purified enzyme may be assayed more accurately at pH 6.7 with a final glycogen concentration of 1.0% and G-1-P ($1.6 \times 10^{-2} M$). Under these conditions at 30°, first-order kinetics are followed, and 1 mg. of protein, which caused the liberation of 18 mg. of phosphorus in the usual test system, had an activity of about 2300 units, as defined by Cori et al.[6]

Purification Procedure

Step 1. Preparation of Acetone Powder. Medium-sized dogs of various mixed breeds were obtained from the pound and were well fed for one or two weeks. In most cases the dogs were used approximately 20 hours after the last feeding. 2 mg. of epinephrine in saline were given intraperitoneally 10 minutes before a lethal dose of Nembutal was injected intravenously. The carotid vessels were severed, and the thorax and abdomen were opened. A large cannula was inserted into the portal vein, and the inferior vena cava was cut just below the heart. Several liters of cold sodium fluoride (0.2 M) were perfused through the liver; intermittent pressure on the inferior vena cava just above the diaphragm increased the filling of the liver and the effectiveness of the perfusion. The chilled perfused liver was removed and, after separation of the gall bladder, was immersed in chipped ice and water. One-half of the liver (130 to 220 g.) was sliced rapidly into small pieces which were placed in a chilled Waring blendor containing 95 ml. of NaF (0.05 M) and 5.0 ml. of potassium phosphate (0.1 M, pH 7.4) and then homogenized for 2½ minutes. The homogenate (pH 7.0) was poured rapidly into 3 l. of acetone previously chilled to $-11°$ in an ice-salt bath. The acetone was swirled in a 4-l. Erlenmeyer as the homogenate was added, and, after the mixture was swirled for 1 to 2 minutes, the precipitate was collected on No. 1 Whatman paper in a large Büchner funnel with suction, in a cold room at 3°. The precipitate was washed with 1.5 l. of cold acetone before being transferred to a vacuum desiccator containing calcium chloride at room temperature. At this stage the second half of the liver was sliced, homogenized, and powdered in the same fashion. The acetone and water was removed from the precipitate by application of vacuum from a water pump for 3 to 4 hours, then vacuum from a high vacuum pump for 3 hours. The desiccators with the dried

[4] E. W. Sutherland, C. F. Cori, R. Haynes, and N. S. Olsen, *J. Biol. Chem.* **180**, 825 (1949).

[5] T. E. Weichselbaum, *Am. J. Clin. Pathol.* **16** (1946); *Tech. Suppl.* **10**, 40 (1946).

[6] C. F. Cori, G. T. Cori, and A. A. Green, *J. Biol. Chem.* **151**, 39 (1943).

powder were stored in a cold room overnight, or for periods of several weeks with only slight loss of activity. The yield of powder was often about 75 g. per liver. The fluoride was used in this and subsequent steps because it inhibited the enzymatic inactivation of liver phosphorylase.

Step 2. Extraction and Collection of Ammonium Sulfate Fraction (0.33 to 0.6). The dried powder was ground with mortar and pestle and divided into two weighed portions. Each portion was placed in a chilled Waring blendor with 9 vol. of cold NaF (0.1 M) and homogenized for 3 minutes. The homogenate was transferred to a beaker in an ice-water bath, and the pH of the homogenate was adjusted to 5.6 by the addition of 2% acetic acid. The homogenate was placed in 50-ml. plastic Servall tubes and centrifuged at 12,000 $\times$ g for 20 to 25 minutes in the cold. During all subsequent steps (except the heat step) the preparations were kept as near 3° as was possible. The supernatant was collected and neutralized to pH 7.0 to 7.1 by addition of 1.0 N KOH. The volume was measured, and to the extract was added 0.5 vol. of ammonium sulfate saturated at room temperature and neutralized to pH 6.8 with concentrated ammonium hydroxide. The 0.33 ammonium sulfate precipitate was removed by centrifugation at 12,000 $\times$ g for 15 to 20 minutes, and the Servall centrifuge was allowed to stop without braking, since the small precipitate obtained here did not always pack well. The supernatant was brought to 0.6 saturation by the addition of more saturated ammonium sulfate solution, and the 0.33 to 0.6 precipitate was collected by centrifugation at 10,000 $\times$ g for 20 minutes. The supernatant was discarded, and the tubes were drained for a few minutes. The precipitate was dissolved in cold NaF (0.1 M), 1 ml. for each gram of powder. The final volume of the solution was measured, and the increase in volume was taken as the measurement of the amount of ammonium sulfate present as a 0.6 saturated solution. The 0.33 to 0.6 fraction could be frozen at −20° and stored for several weeks with little or no loss of activity.

Step 3. Heat Denaturation of Inactivating Enzyme. Adenosine-5-phosphate protected liver phosphorylase against enzymatic destruction and also protected liver phosphorylase against heat denaturation. 0.1 vol. of 5-AMP (0.02 M) was added to the 0.33 to 0.6 fraction from one liver. The pH was adjusted to 7.0 to 7.1 by the addition of 0.2 N KOH. The preparation was transferred to a 2-l. Erlenmeyer flask and was washed in with 30 ml. of NaF, thus giving a final concentration of about 0.2 saturation with ammonium sulfate. The preparation was then heated with swirling in a 67° bath until the temperature rose to 55°; this temperature of 55° was reached within a few minutes. The temperature was maintained at 55° for 5 minutes; then the solution was chilled in an ice-water bath and more cold NaF (0.1 M) was added, in volume equal to 1.5 ml./g. of original powder. The preparation was centrifuged for 20 minutes at 12,000 $\times$ g in the cold, the supernatant was saved, and the precipitate, which varied considerably in size, was discarded. At times no precipitate was obtained at this stage.

Step 4. Collection of Ammonium Sulfate Fraction (0.42 to 0.8). The concentration of the ammonium sulfate present in the heated samples was calculated from the volume increase of the 0.33 to 0.6 fraction on solution and often was found to be from 0.10 to 0.17 saturated. Ordinarily heated extracts from several dogs were pooled at this stage. The saturated neutralized ammonium sulfate solution was added to bring the preparation to 0.42 saturation; the measured specific gravity at this calculated saturation was 1.13 at 2°. The precipitate obtained at 0.42 saturation contained an amount of phosphorylase which varied with the glycogen content of the preparation. Usually the 0.42 fractions obtained from extracts containing moderate or large amounts of glycogen were discarded or saved as crude side fractions, the bulk of the phosphorylase remaining in the supernatant. (In two preparations where the liver glycogen was very low, most of the phosphorylase appeared in the 0.42 precipitate.) Removal of the 0.42 fraction did not cause any substantial increase in specific activity but was carried out to eliminate protein which would otherwise precipitate with alcohol in the subsequent steps. The 0.42 to 0.8 fraction was obtained by adding solid ammonium sulfate to the supernatant of the 0.42 precipitate (28.5 g./100 ml. of supernatant) followed by centrifugation at 10,000 $\times$ g for 20 minutes. Precipitates from two centrifugations were collected in the same tubes; the 0.8 supernatant was discarded. It was found that the 0.42 to 0.8 fraction was stable in the cold overnight, so that centrifugation could be carried out the day after precipitation. The 0.42 to 0.8 precipitates were dissolved by the addition of 0.5 ml. of cold NaF (0.1 M) per gram of original powder. This solution could be frozen at −20° and kept for some days without serious loss of activity.

Step 5. Dialysis of 0.42 to 0.8 Ammonium Sulfate Fraction and Ethanol Fractionation. The ammonium sulfate concentration of the preparation was lowered before alcohol fractionation; otherwise inactivation of phosphorylase occurred. Since traces of inactivating enzyme sometimes remained, the dialysis was carried out in the cold with fluoride chilled in an ice bath so as to approach 0° as closely as possible. Dialysis was carried out in Visking casing (size $2\frac{7}{32}$) against 15 to 20 vol. of NaF (0.1 M) containing 3×10^{-4} N KOH, for 1 hour; then the dialysis was continued for another hour against a fresh NaF-KOH solution. After dialysis 0.1 vol. of 5-AMP (0.02 M, pH 7.0) was added, and the pH of the solution was brought to approximately 7.1 with KOH. The preparation was then ready for ethanol fractionation but could be kept frozen at −20°. Cold 95%

ethanol was added slowly with stirring to the preparation chilled in an ice bath until a 30% final concentration was reached. The preparation was chilled to 3° and centrifuged in the cold room in 220-ml. cups; the relatively small precipitate packed well at 400 × g for 18 minutes. (If the 0.42 precipitate was not removed satisfactorily, high speeds were required to pack the bulkier precipitate.) The hazy supernatant was discarded, and the precipitate was immediately taken into solution with 0.5 ml. of cold NaF (0.1 M) per gram of original powder plus one-twentieth this volume of 5-AMP (0.02 M, pH 7.0). The volume was recorded, the per cent ethanol concentration was estimated from the volume after solution, and cold 95% ethanol was added dropwise to the swirled solution to a final concentration of 20%. After being chilled to 3°, the precipitate was collected by centrifugation at 400 × g and dissolved in cold NaF (0.1 M, 0.4 ml. or less per gram of original powder) plus one-tenth this volume of 5-AMP (0.02 M, pH 7.0). (The supernatant of the 20% precipitate was taken to 35% ETOH with 95% ETOH, and the precipitate was saved. The 20 to 35 ETOH fraction rarely was found to contain considerable amounts of phosphorylase unless the 20% fraction had warmed during the centrifugation.) The temperature effect on the ETOH fractionation was very great, a large increase in solubility occurring with increasing temperature. Procedures so far discussed are summarized in Table I.

TABLE I

SUMMARY OF PURIFICATION PROCEDURE THROUGH EARLY STEPS

Fraction	Specific activity[a]	Original activity, %
1. Acetone powder extract	0.21	100
2. 0.33 to 0.6 ammonium sulfate	0.30	74
3. Heated at 55° for 5 minutes	0.35	50
4. 0.42 to 0.8 ammonium sulfate	0.38	43
5. Dialysis	0.36	39
6. 30% ETOH	3.1	32
7. 20% ETOH	4.2	29

[a] The specific activity of liver homogenates was approximately one-half that of the acetone powder extracts.

In most cases the specific activity of the 0 to 20% fraction was increased considerably by the removal of a relatively inactive fraction by high-speed centrifugation (Table II). At times it was necessary to increase the ethanol concentration to the formation of a strong haze at 3° (approximately to 14 to 16% ETOH), and then centrifugation on the Servall at 12,000 × g for 10 minutes caused separation of the relatively inactive precipitate. In either case a fraction could be obtained which had a specific activity from 50 to 70% of the activity of the best samples so far obtained. Such samples were stable and could be dialyzed, frozen, and kept for two years or more without much loss of activity. The solutions were opalescent because of the high glycogen content. Further ethanol fractionation at this stage was not very successful because the glycogen and protein tended to precipitate together.

TABLE II
FRACTIONATION OF THE 20% PRECIPITATE

(The 20% precipitate from 5 dogs contained 5200 units of liver phosphorylase. On solution the estimated ethanol concentration was 5.6%, and the relatively inactive precipitate was collected by centrifugation at 10,000 × g in a Servall. Increasing amounts of ethanol were added to the supernatant at 3° in order to collect the ethanol subfractions listed below.)

Fraction	Specific activity	Total activity, %
Pooled 20% precipitate	3.8	100
5.6 subfraction	1.0	12.1
5.6–15 subfraction	1.76	5.8
15–25 subfraction	11.5	66.0

Step 6.[7] *Further Ammonium Sulfate Fractionation.* The primary problem in this step was to free the protein from the large and variable amount of carbohydrate present after ethanol fractionation. The two most successful methods alone or in combination have been digestion of glycogen to the stage where precipitation of protein by ammonium sulfate can occur, and addition of crystalline bovine plasma albumin during the first precipitation with ammonium sulfate. Two examples of fractionation will be given.

The starting material was 100 ml. of ethanol fractions similar to the 15 to 25% fraction listed in Table II, containing 435 mg. of protein with specific activity of 9.6 and 5 g. of carbohydrate. This preparation was centrifuged in a Spinco preparative ultracentrifuge for 40 minutes at 100,000 × g. The supernatant containing 300 mg. of protein and 3.9 g. of carbohydrate was mixed with 4 ml. of saliva previously filtered, frozen, thawed, and centrifuged. The digestion mixture was incubated in NaCl (0.01 M, pH 7.0) at 30° for 1 hour; then it was washed into a dialysis sac with 3 ml. of potassium phosphate (0.1 M, pH 7.4) and dialyzed against 3 l. of buffer containing phosphate (0.025 M) and NaF (0.06 M) for $2\frac{1}{2}$

---

[7] This step is being investigated at the present time, since it has been variable and unsatisfactory.

hours. The preparation was centrifuged for 4 hours at 100,000 × g; the supernatant of 81 ml. contained 96 mg. of protein, with a specific activity of 8.7, and 1.9 g. of carbohydrate. An ammonium sulfate solution saturated at room temperature (pH 6.8) was added to bring the final concentration to 0.6 saturation. The precipitate was collected and brought to a volume of 5.0 ml. with NaF (0.1 M), and analysis showed 48 mg. of protein, with a specific activity of 14.1, and 8.0 mg. of carbohydrate. This was pooled with a similar sample, and, when centrifuged and reprecipitated at 0.5 saturation with ammonium sulfate, the specific activity rose to 17.2 and no carbohydrate was detected.

In another fractionation the procedure was almost identical through the digestion stage, except that dialysis was carried out against three changes of buffer containing more phosphate (0.05 M, pH 7.0). After incubation with amylase and inorganic phosphate, the preparation was taken to 0.7 saturation with solid ammonium sulfate. The precipitate collected here contained less than one-third of the total activity. To the supernatant (volume 120 ml.) was added 3.5 ml. of 10% crystalline bovine plasma albumin, at which time a heavy floc appeared. This was collected by centrifugation, and the precipitate was taken into solution with 16.5 ml. of 5-AMP (0.002 M, pH 7.2). The solution was precipitated by addition of ammonium sulfate solution to 0.5 saturation; then the fraction between 0.3 and 0.5 saturation was obtained. This fraction contained 68 mg. of protein, with a specific activity of 16.5, containing a trace of carbohydrate which could be removed by one reprecipitation between 0.3 to 0.5 saturation.

Properties

Liver phosphorylase with a specific activity of 17 to 18 was homogenous when examined by ultracentrifugal and electrophoretic techniques.[8] The molecular weight of the enzyme was 237,000 g., as determined by Miss Carmelita Lowrey. The enzyme was very soluble in water, and the solubility and activity was not noted to be influenced by cysteine. The pH optimum was relatively broad about 6.4, falling rapidly at pH 6.0 and 7.0. The activity was increased 20 to 40% by addition of 5-AMP or 5-IMP to the reaction mixture; aged preparations or occasional fresh preparations were activated to a greater extent. The purified enzyme catalyzed the formation of a polysaccharide which gave a blue color with iodine; this synthesis required the addition of glycogen to the reaction

[8] Recent experiments by the author and Dr. Walter D. Wosilait have shown that preparations with distinctly higher activities can be obtained. The purity of these early preparations therefore is questionable, even though these two techniques indicated homogeneity at this stage.

mixture. When freed of carbohydrate, the enzyme gradually lost activity on standing at 3°, and rapidly lost activity on freezing at −20° in the presence of several inorganic salts, such as NaCl. It was found, however, that such preparations were relatively stable when frozen at −20° in glycylglycine (0.2 M, pH 7.4).

[27] Branching Enzyme from Liver

Amylo-1,4 → 1,6-Transglucosidase

By Joseph Larner

Assay Method

Principle. The blue value determination of McCready and Hassid[1] as used by Bourne and Peat[2] serves as the basis of the assay method. Corn amylopectin is the substrate. The decrease in optical density of the iodine amylopectin complex (absorption maximum, 545 to 570 mμ) is taken as a measure of the branching action.

Reagents

    Corn amylopectin, 1.0%.[3]
    Tris buffer, 0.19 M, pH 7.9.
    I_2 + KI, 0.2% + 0.4%.
    Perchloric acid, 3.0%.

Procedure. 0.3 ml. of corn amylopectin and enzyme (about 150 units) are incubated at pH 7 in the absence of buffer at 30° in a total volume of 0.8 ml. At zero time and at subsequent time intervals (usually 5 to 10 minutes) 0.15-ml. aliquots are taken and deproteinized with equal volumes of perchloric acid. After shaking and standing for several minutes, the samples are centrifuged, and the supernatant fluids are decanted and neutralized to pH 7 with solid NaHCO₃.[4] 200 to 400 γ of polysaccharide is added to 1.0 ml. of Tris buffer, 0.15 ml. of I_2 + KI, and water to a final volume of 3.0 ml., mixed, and read in the Beckman Model DU spectrophotometer at 570 mμ within 10 minutes after mixing. The colors tend to

---

[1] R. M. McCready and W. Z. Hassid, *J. Am. Chem. Soc.* **65**, 1154 (1943).

[2] E. J. Bourne and S. Peat, *J. Chem. Soc.* **1945**, 877.

[3] We are indebted to Dr. R. W. Kerr, Corn Products Refining Company, Argo, Illinois, for the sample of corn amylopectin. *J. Am. Chem. Soc.* **73**, 111 (1951).

[4] KHCO₃ is not used because of the insolubility of potassium perchlorate.

fade after this time. A blank cuvette contains all components except polysaccharide. Under these conditions recovery of added polysaccharide was from 92.4 to 96.0%, and the reaction during its early stages usually followed a linear time course.

Definition of Unit and Specific Activity. One unit is defined as that amount of enzyme effecting a decrease in optical density of 0.001 per minute under the stated conditions. Specific activity is expressed as units per milligram of protein, as determined by the method of Robinson and Hogden.[5]

Application of Assay Method to Crude Tissue Preparations. The chief interfering enzyme is α-amylase, which is largely removed by starch adsorption.[6] Although the iodine color is extremely sensitive to experimental conditions,[7] the procedure as described has proved moderately satisfactory as an assay method.

Purification Procedure

Rat or rabbit livers have served as starting material. Rat livers (Anheuser-Busch strain used) yield the more active extracts. Liver extracts of the Sprague Dawley strain were very low in brancher activity after four starch treatments. Preliminary experiments indicate that the brancher of this strain is more readily adsorbed on starch than that of the Anheuser-Busch strain. The material at the end of step 3, after dialysis to remove inorganic phosphate, has been used in radioactive branching studies.[8] Step 4 has not been explored as completely and is not recommended as a routine procedure.

Animals are fasted to lower liver glycogen[9] and sacrificed by a blow on the head or by intravenous Nembutal (rabbits). The neck vessels are cut as rapidly as possible to allow free bleeding. Liver is perfused *in situ* with ice-cold 1% KCl until free of blood, which is rich in α-amylase.[10] All subsequent steps are carried out in the cold room at 3°.

Step 1. Preparation of Crude Extract. The livers are homogenized with 2 vol. of distilled water (w/v) in the Waring blendor for 3 minutes. The homogenate may be centrifuged at this step, but a clear supernatant fluid

---

[5] H. W. Robinson and C. G. Hogden, *J. Biol. Chem.* **135,** 727 (1940).

[6] Amylase activity may conveniently be followed by determination of reducing power in suitably deproteinized filtrates.

[7] D. L. Morris, *J. Biol. Chem.* **166,** 199 (1946).

[8] J. Larner, *J. Biol. Chem.* **202,** 491 (1953).

[9] Rats are fasted 24 hours at room temperature. Rabbits are fasted 24 hours at room temperature and 24 hours in the cold room at 3°.

[10] During the perfusion, the tissue at the edge of the lobes is freed of blood by gentle manual massage while back pressure is applied by clamping off the inferior vena cava above the diaphragm.

is not obtained. Usually, the homogenate is carried on through the acid precipitation (step 2) before centrifugation.

Step 2. Acid Precipitation. The pH is adjusted to 5.8 with 0.05 N HCl added dropwise with moderate rapidity (about 1 ml. required for 2 to 3 g. of liver). On acidification to pH 5.4 most of the brancher activity is lost.[11] The homogenate is centrifuged at top speed in the Servall angle head centrifuge, Model SS-1, for 20 minutes. The clear yellowish brown supernatant fluid is filtered through paper to remove any floating fatty debris and brought to pH 7 with solid $KHCO_3$. This acid step yields a clear supernatant fluid with almost a twofold increase in specific activity.

Step 3. Starch Adsorption.[12] USP corn starch (10% w/v) is added to the neutral extract at 0°, and the suspension is stirred manually for 10 minutes, allowing as little settling of the starch as possible. The starch is removed by centrifugation. After one or two such treatments the clear extract is filtered through paper to remove any remaining starch particles. Additional starch adsorption seems to result in no further lowering of α-amylase activity and in a decrease in brancher activity. Starch adsorption may be done at either an earlier or a later stage (see the table). The radioactive experiments have shown a separation of brancher from α-amylase by starch adsorption.[8]

FRACTIONATION OF RABBIT LIVER BRANCHING ENZYME

Fraction	Total volume, ml.	Units/ml.	Protein, mg./ml.	Specific activity, units/mg.	Recovery, %
1. Centrifuged homogenate	350	492	26.2	18.8	
3. Two starch adsorptions		403	25.8	15.6	
2. Acid precipitate removed	225	500	17.5	28.5	66
4. Ethanol fraction 0–40%, (48-ml. aliquot)	30	760	1.10	690.0	65

Step 4. Fractionation with Ethanol. To the extract is added neutral saturated ammonium sulfate[13] to 0.1 saturation. Cold absolute ethanol ($-10°$) is added slowly with constant stirring to 40% concentration with the temperature kept at -5 to $-8°$. The precipitate is collected in a

---

[11] The pH meter is set with standard phosphate buffer at room temperature. The pH of an aliquot of the cold extract is measured with the temperature-compensating knob set at 10°.

[12] G. T. Cori and J. Larner, J. Biol. Chem. **188**, 17 (1951).

[13] Saturated at room temperature and brought to pH 7 with concentrated NH_3 as determined with indicator after 1 to 5 dilution. Ammonium sulfate is added at this step because there is no precipitation with ethanol in the absence of salt.

refrigerated centrifuge and taken up in a small volume of 0.001 M cysteine at pH 7. The material is dialyzed for several hours against 1% KCl containing 0.001 M cysteine at pH 7, centrifuged to remove insoluble protein, and stored in the frozen state. Consistent results in this step have not as yet been achieved. Purifications vary from seven- to twentyfold, and recoveries up to 95% (average 60%).

[28] Disaccharide Phosphorylases

By M. DOUDOROFF

Sucrose Phosphorylase (α-Transglucosidase) of *Pseudomonas saccharophila* and *Leuconostoc mesenteroides*

$$\text{Sucrose} + P_i \longleftrightarrow \alpha\text{-}D\text{-Glucose-1-phosphate} + D\text{-Fructose}$$

Assay Method

Principle. The method is based on the measurement of the disappearance of inorganic phosphate in the presence of sucrose.

Reagents

0.2 M sucrose in 0.033 M pH 6.64 Sørensen phosphate buffer.

Enzyme. Dilute stock enzyme in 0.033 M Sørensen phosphate buffer at pH 6.64 to obtain between 5 and 35 units/ml. (final concentration of inorganic phosphate may vary from 0.025 M to 0.04 M, and pH may vary from 6.3 to 7.0 without altering the results substantially).

Procedure. Mix equal quantities of sucrose solution and enzyme solution at 30° and incubate for 20 minutes. Take out aliquots at 0, 10, and 20 minutes, stop the reaction with 5% trichloroacetic acid, and determine inorganic phosphate by the method of Fiske and SubbaRow.

Definition of Unit and Specific Activity. One unit of enzyme is defined as that amount which causes an initial uptake of 1 micromole of P_i in 20 minutes at 30° in 0.033 M phosphate and 0.1 M sucrose at pH 6.64. The computation of the activity of the enzyme in the determination described above is then made as follows: The sample showing an uptake of not less than 2.5 and not more than 10 micromoles of P_i per milliliter of final mixture is used. Specific activity is defined as units of enzyme per milligram protein. Protein is measured by the ratio of absorption at 280 and 260 mμ.

Modifications of Assay Procedure. (a) A curve for phosphate disappearance against enzyme concentration under the specified conditions can be obtained experimentally. Enzyme activities between 5 and 40 units/ml. of stock solution can then be determined with fair precision after a 20-minute period of incubation, and between 5 and 80 units if both 10- and 20-minute samples are taken.

(b) When sucrose phosphorylase preparations free of phosphoglucomutase, polysaccharide phosphorylase, and glucose-1-phosphatase are used, no initial P_i determination is necessary. The esterified phosphate in the sample can be measured by difference as inorganic phosphate liberated on 7-minute hydrolysis with N HCl at 100°.

Purification Procedure

A. Sucrose Phosphorylase from P. saccharophila. The organism is grown in a medium containing 2.5 g. of sucrose, 1.0 g. of N H$_4$Cl, 0.5 g. of MgSO$_4$·7H$_2$O, and 0.05 g. of ferric ammonium citrate per liter of 0.033 M Sørensen KH$_2$PO$_4$–Na$_2$HPO$_4$ buffer at pH 6.8, incubated with vigorous aeration at 30°. The cells are harvested by centrifugation immediately after the stationary phase is reached (Klett reading of 370 to 390 with 66 filter). Larger crops can be obtained if higher concentrations of sucrose (up to 0.5%) are used. However, in such cases aeration must be very vigorous and a careful control of pH maintained during the last hours of incubation. The preparation of enzyme extracts from dry cells has been described.[1] Another method, in which wet cell paste is used, is described below.

In this method, 10 g. of washed intact or frozen cells is ground for 10 minutes with 40 g. of alumina[2] and extracted with 100 ml. of 0.033 M pH 6.8 phosphate buffer in the cold. The extract is centrifuged at 20,000 $\times$ g for 20 minutes. The extract may be kept frozen for long periods of time. Next, 40 ml. of the extract is treated with 40 ml. of cold neutral ammonium sulfate solution saturated at 0°. The precipitate is removed by centrifugation, and 60 ml. of saturated ammonium sulfate is added. The precipitate is collected and redissolved in 40 ml. of 0.01 M pH 6.0 phosphate buffer, and 4 ml. of 2% protamine sulfate solution[3] at pH 5.0 is added slowly with stirring at 0 to 5°. The suspension is centrifuged, and the precipitate is discarded; 88 ml. of saturated ammonium sulfate is added, and the precipitate is collected and redissolved in 15 ml. of 0.033 M pH 6.8 phosphate buffer. Then 10 ml. of saturated ammonium sulfate is added. After centrifugation the precipitate is discarded and 5 ml. of

[1] M. Doudoroff, *J. Biol. Chem.* **151**, 351 (1943).
[2] Blue Label R. R. Alundum, electrically fused crystalline alumina, 60 mesh, Norton Co., Worcester, Mass.
[3] Protamine sulfate, Lilly and Co.

saturated ammonium sulfate is added to the supernatant. The precipitate is redissolved in tris(hydroxymethyl)aminomethane-maleic acid or phosphate buffer. The results of the above procedure are shown in the accompanying table. Actually, refractionation of the discarded precipitate and supernatant from the last ammonium sulfate fractionation can yield an additional 20% recovery of activity. Storing the enzyme preparations in 0.5 saturated ammonium sulfate for 2 or 3 days at 0° removes virtually all invertase and phosphatase activity. The enzyme may then be dialyzed and stored in the frozen state.

SUMMARY OF FRACTIONATION OF SUCROSE PHOSPHORYLASE

P. saccharophila	Ratio, $\frac{OD\,280}{OD\,260}$	Enzyme, units/ml.	Protein, mg./ml.	Enzyme, units/mg. protein	Recovery of enzyme, %
1. Crude extract (40 ml.)	0.5	12.9	5.35	2.42	100
2. After protamine, saturated ammonium sulfate (15 ml.)	1.37	31.2	2.34	13.3	90
3. After saturated ammonium sulfate fractionation (15 ml.)	1.42	19.0	0.62	30.6	55
L. mesenteroides					
1. Crude extract (690 ml.)	0.55	310	5.6	55.5	100
2. After protamine, ammonium sulfate precipitation (35 ml.)	1.57	3500	18.1	193	57
3. Ammonium sulfate fractionation, dialysis (30 ml.)	1.61	2750	11.0	250	39
4. Acid precipitation, dialysis (32 ml.)	1.62	1750	3.8	460	26
5. Calcium phosphate gel adsorption					
(a) 0.002 M phosphate eluate, dialyzed (6 ml.)	1.68	960	0.85	1130	2.7
(b) Pooled 0.033 M phosphate eluate, dialyzed (10 ml.)	1.81	2240	1.69	1320	10.2

B. *Sucrose Phosphorylase from L. mesenteroides. L. mesenteroides* (strain P-39) is grown without aeration in medium ACI-B[4] at 30° until the maximum stationary phase is just established (18 to 24 hours). The medium contains, per liter of distilled water, 10 g. of tryptone, 10 g. of yeast extract, 5 g. of K_2HPO_4, 10 g. of sucrose, and 10 ml. of vitamin-salt solution (0.1% thiamine HCl; 4.0% $MgSO_4 \cdot 7H_2O$; 0.1% $FeSO_4 \cdot 7H_2O$; 2.0% $MnSO_4 \cdot 4H_2O$; 0.5% ascorbic acid; sterilized by filtration and added

[4] R. D. De Moss, I. C. Gunsalus, and R. C. Bard, *J. Bacteriol.* **66**, 10 (1953).

aseptically to the rest of the medium which has been sterilized by autoclaving). The cells are harvested by centrifugation, washed, used immediately, or stored as a frozen paste.

Step 1. Extraction. 100 g. of fresh or frozen cell paste is ground with 400 g. of alumina[2] and extracted with 1 l. of 0.01 M pH 6.8 phosphate buffer. The extract is centrifuged at 20,000 $\times$ g for 20 minutes and may be stored in the frozen state.

Step 2. Preliminary Fractionation and Removal of Nucleic Acids. 100 ml. of extract is treated with 100 ml. of saturated ammonium sulfate (see above). After 15 minutes at 0° with constant stirring the precipitate is removed by centrifugation, and 24.8 g. of solid ammonium sulfate is added slowly at 0° with stirring to the supernatant. The precipitate is collected by centrifugation, redissolved in 50 ml. of 0.01 M phosphate buffer at pH 6.8, and 5 ml. of 2% protamine sulfate solution[3] is added slowly with vigorous stirring. The precipitate is removed by centrifugation, and the supernatant treated with 77 ml. of saturated ammonium sulfate. After 15 minutes at 0°, the precipitate is removed by centrifugation, and 20 g. of solid ammonium sulfate is added to the supernatant. The precipitate is collected and dissolved in 5 ml. of 0.033 M pH 6.8 phosphate buffer.

Step 3. Second Ammonium Sulfate Fractionation and Dialysis. 9.3 ml. of saturated ammonium sulfate is added. After centrifugation, the supernatant is treated with 1.51 g. of solid ammonium sulfate. The precipitate is redissolved in 5 ml. of phosphate buffer and refractionated twice in the same manner as the precipitate at the end of step 2.

The three precipitates obtained after the solid ammonium sulfate additions are pooled and redissolved in a total of 3 ml. of 0.033 M pH 6.8 phosphate buffer. The solution is dialyzed against demineralized water for 20 hours at 0° and centrifuged. The precipitate is discarded.

Step 4. Acid Precipitation. 0.5% acetic acid is added slowly with vigorous stirring at 0 to 5° to pH 4.5. The precipitate is removed, and the supernatant is dialyzed for 18 hours at 0° against 0.001 M phosphate at pH 5.8. If a precipitate is obtained, this is removed by centrifugation.

Step 5. Calcium Phosphate Gel Adsorption. The solution is treated with 4.4 mg. of $Ca_3(PO_4)_2$ gel (0.286 ml. of suspension containing 15.5 mg. of gel per milliliter). After 15 minutes at 0°, the gel is removed by centrifugation and 15.5 mg. (1 ml.) of gel is added to the supernatant.

After 15 minutes the suspension is centrifuged and the gel resuspended in 0.86 ml. of 0.002 M pH 6.8 buffer (see the table; 0.002 M phosphate eluate). The gel is again centrifuged and washed twice with 0.71-ml. portions of 0.033 M pH 6.8 phosphate buffer. The supernatants are combined, dialyzed for 20 hours at 0° against demineralized water, and centrifuged to remove inactive precipitate.

The preparations obtained in this manner may be stored in the frozen state for many months, although with some loss of activity.

Precautions and Variations. The method described above is only one of many variations which have been found suitable for purifying the enzyme. Better recoveries but lower specific activities have been obtained with other methods. The commercial source of protamine has been found to have profound effects on the purification. It has also been found that protamine sulfate treatment prior to ammonium sulfate precipitation may be used to advantage with the crude extract.

Acid precipitation may result in little or no purification if the previous steps are slightly altered.

Finally, adsorption and elution with $Ca_3(PO_4)_2$ gel varies greatly with the type of preparation used, and a preliminary test should be run before a large amount of enzyme is treated. Under some circumstances, it has been found that 0.1 M pH 6.0 phosphate buffer is a more suitable eluent of the enzyme than 0.033 M pH 6.8 buffer.

Properties

Both sucrose phosphorylases are transglucosidases, transferring glycosyl units between a variety of substrates.[5] Among glycosyl acceptors are D-fructose, L-sorbose, L-arabinose, and inorganic phosphate. The α-glucosidic linkage is preserved in the transfers. In the presence of arsenate, both sucrose and α-glucose-1-phosphate are rapidly hydrolyzed.[6] The *Leuconostoc* enzyme has a slight hydrolytic activity toward both sucrose and glucose-1-phosphate in the absence of arsenate. The rate of hydrolysis is approximately 1% of the rate of phosphorolysis. The same is probably true of the *P. saccharophila* enzyme.

P-Labeled glucose-1-phosphate or sucrose labeled in either the glucose or fructose moiety can be synthesized with the aid of either enzyme.[7]

The optimum pH for phosphorolysis is about 6.6 with both enzymes. No requirement for a dissociable coenzyme or metal cofactor has been detected. The purified enzymes are relatively stable in the frozen state.

Maltose Phosphorylase from *Neisseria meningitidis*

$$\text{D-Maltose} + P_i \longleftrightarrow \beta\text{-D-Glucose-1-phosphate} + \text{Glucose}$$

Assay Method

Principle. No assay method or unit of specific activity has been developed for this enzyme. In principle, a method similar to that estab-

---

[5] M. Doudoroff, H. A. Barker, and W. Z. Hassid, *J. Biol. Chem.* **168**, 725 (1947).

[6] M. Doudoroff, H. A. Barker, and W. Z. Hassid, *J. Biol. Chem.* **170**, 147 (1947).

[7] H. Wolochow, E. W. Putman, M. Doudoroff, W. Z. Hassid, and H. A. Barker, *J. Biol. Chem.* **180**, 1237 (1949).

lished for sucrose phosphorylase may be used. Since the equilibrium for the reaction is far to the left, ideally β-glucose-1-phosphate and glucose should be used as substrates and the evolution of inorganic phosphate measured. Owing to the relative unavailability of the ester, the disappearance of inorganic phosphate in the presence of a large excess of maltose is the more practicable measure of phosphorolysis. As a suggestion, different amounts of enzyme might be mixed with 0.1 M maltose and 0.033 M phosphate buffer at pH 6.5 to 6.6 and phosphate uptake determined after incubation for 60 minutes at 37°. The unit of activity can then be set arbitrarily and the activity of any given preparation determined from an experimentally derived curve.

Preparation of Crude Enzyme. A nonvirulent strain of *N. meningitidis* (strain 69, type I [8]) is grown on the surface of Trypticase Soy agar (Baltimore Biological Laboratory) for not more than 24 hours at 37°. The cells are washed off with distilled water, centrifuged, and washed several times by resuspension and centrifugation in distilled water. The cell paste is then ground in a mortar with three times its wet weight of alumina[2] and extracted with 10 vol. of tris(hydroxymethyl)aminomethane buffer at pH 7.0. The preparation is centrifuged at 4° at 5000 × g. The supernatant is fractionated several times between 0.4 and 0.7 saturation of ammonium sulfate in either phosphate or tris(hydroxymethyl)aminomethane buffer at pH 7.0.

Such preparations are virtually free of any interfering maltase and phosphatase activities. Other strains of *N. meningitidis* can undoubtedly be used, but the advantage of using strain 69 lies in its complete lack of pathogenicity. The enzyme is stable in pH 7.0 buffer in the frozen state. It can also be lyophilyzed and kept in the dry state.

Properties

The enzyme is a specific phosphorylase.[9] Maltose cannot be replaced by any of the better-known naturally occurring disaccharides or polysaccharides. In the reverse reaction, D-xylose can be substituted for glucose to produce a reducing disaccharide, glucosido-xylose. α-D-Glucose-1-phosphate cannot substitute for β-D-glucose-1-phosphate in the reverse reaction. In the presence of arsenate, maltose, but not β-glucose-1-phosphate, is hydrolyzed to glucose. The enzyme does not catalyze the exchange between inorganic phosphate and β-glucose-1-phosphate unless glucose is present, nor does it catalyze the exchange between free glucose and malt-

[8] H. W. Scherp and C. Fitting, *J. Bacteriol.* **58**, 1 (1949).
[9] C. Fitting and M. Doudoroff, *J. Biol. Chem.* **199**, 153 (1952).

ose unless phosphate is present. Maltose labeled with C^{14} in the glycosyl moiety can be synthesized from labeled β-glucose-1-phosphate and inactive glucose. Maltose labeled in the reducing glucose moiety can be synthesized from labeled glucose and inactive β-glucose-1-phosphate.[10]

The pH optimum for the phosphorolysis of maltose is about 6.5. There is no evidence for the necessity of an easily dissociable coenzyme or of a metal cofactor. Heating for 3 minutes at 98° destroys phosphorylase activity.

[10] C. Fitting and E. W. Putman, *J. Biol. Chem.* **199**, 573 (1952).

[29] Hexoside Hydrolases

By SHLOMO HESTRIN, DAVID S. FEINGOLD, and MICHAEL SCHRAMM

I. General Considerations

Review of Principal Assay Methods

The hydrolysis of any glycide may be followed in terms of the rates of the liberation of either the glycon or the aglycon determined separately or together. The enzymatic liberation of an aglycon from a glycide can be effected by phosphorolysis or by transglycosification as well as by hydrolysis. In general, therefore, a reliable assay of the hexoside hydrolase activity can be founded safely upon measurement of the rate of the glycon liberation. On the other hand, methods relying on the measurement of the rate of liberation of the aglycon alone or of the glycon and aglycon determined together are a permissible basis of hydrolase assay only if it can be shown that, in the conditions employed, hydrolysis of the glycosidic bond is the dominant mechanism of cleavage.[1] Since no analytical procedure can be singled out as possessing unique merit for universal use in hexosidase assays, the choice of the method best suited to any particular case is properly left at the present time to individual judgment.

Early workers showed a predilection for assaying hexosidase activity on the basis of the measurement of optical rotation. Some orienting data on optical rotation are assembled in Table I. Although the polarimetric

[1] The assay of the transglycosidasic activity of hexosidase is discussed by J. S. D. Bacon (Vol. I [30]). By determining the number and nature of the reaction products with the help of paper chromatography, it can be decided conveniently whether the cleavage observed is indeed essentially a hydrolysis (see the section on the analysis of carbohydrates by paper chromatography in Vol. III [11]).

method is simple, it suffers from the disadvantage that it is rather coarse and requires a liberal use of both substrate and enzyme.

TABLE I

OPTICAL ROTATION OF SOME IMPORTANT OLIGOSACCHARIDES AND OF THE PRODUCT OF THEIR TOTAL HYDROLYSIS[a]

| | Rotation[b] | |
Substrate	Nonhydrolyzed substrate	Total hydrolyzate
Lactose	18,950	24,000
Maltose	46,900	19,100
Melezitose	47,700	2,400
Melibiose	49,000	24,000
Raffinose	62,540	7,300
Sucrose	22,770	−7,200
Trehalose	67,460	19,100

[a] Values computed from F. J. Bates and associates, "Polarimetry, Saccharimetry, and the Sugars," National Bureau of Standards Circular C 440 (1942). These values are obtained at 20° with sodium D illumination after completion of mutarotation.

[b] Values for solutions containing 1 mole of oligosaccharide per milliliter.

Colorimetric and gasometric micromethods have won preference over the polarimetric method. If the substrate is a nonreducing glycide, the micromethods used to determine sugar in blood can be utilized readily as the basis of the hexosidase measurement.[2] If the substrate is a reducing glycide, hexose-selective copper reagents of the Barfoed type are advantageous.[3]

The aglycon may be the product of reaction most readily determined. If the aglycon is phenol, the method of Gottlieb and Marsh[4] based on color development with 4-amino-antipyrin is particularly useful. If the aglycon is nitrophenol, the estimation of the color density of a sample made alkaline with 0.1 M sodium carbonate affords an ideally simple and rapid form of analysis.[5]

Yeasts specific for zymohexose (glucose, mannose, fructose) enable

[2] Cf., for example, the method of N. Nelson, J. Biol. Chem. **153**, 375 (1944).

[3] H. Tauber and I. S. Kleiner, J. Biol. Chem. **99**, 249 (1932); S. Hestrin and C. C. Lindegren, Arch. Biochem. **29**, 315 (1950); R. Caputto, L. F. Leloir, and R. E. Trucco, Enzymologia **12**, 350 (1946–1948).

[4] S. Gottlieb and P. B. Marsh, Ind. Eng. Chem. (Anal. Ed.) **18**, 16 (1946); cf. D. J. Hockenhull, G. C. Ashton, K. H. Fantes, and B. K. Whitehead, Biochem. J. **57**, 93 (1954).

[5] J. Lederberg, J. Bacteriol. **60**, 381 (1950); S. A. Kuby and H. A. Lardy, J. Am. Chem. Soc. **75**, 890 (1953).

the *continuous* gasometric measurement of the course of hydrolysis of a zymohexoside.[6] If the released glycon or/and aglycon happens to be glucose, the very specific *Penicillium notatum* glucose oxidase (notatin) provides a satisfactory analytical tool for the continuous gasometric measurement of the hydrolysis.[7]

The above main types of assay are illustrated in the later sections: measurement of total reducing power (p. 234); measurement of hexose release (p. 249); utilization of a chromogenic substrate (β-*o*-nitrophenol galactoside) (p. 241); use of a specific glucose oxidase to estimate release of glucose (p. 242).

The Use of "Normalbedingungen" in Hexosidase Assay

Weidenhagen,[8] following in the steps of Willstätter, urged that the assay of all the hexosidases be conducted at a uniform substrate concentration (0.1388 M) and under "Normalbedingungen." The "Normalbedingungen" are fulfilled when the hexosidase is allowed to act at optimum pH on 2.500 g. of maltose monohydrate or on an equivalent quantity of another hexoside at 30° in a final total volume of 50 ml. In the acid pH range, acetate (0.160 M) serves as the buffer; in the neutral pH range, phosphate (0.04 M) serves as the buffer. For "Normalbedingungen," *time value* ("Zeitwert") is the time in minutes in which 1 g. of enzyme effects 50% hydrolysis of the substrate; *enzyme unit* ("Enzymeinheit") is the amount of the pure enzyme per gram of dry substance in an enzyme preparation of unit time value; and *enzyme value* ("Enzymwert") designates the number of the *enzyme units* per gram of the dry substance in the crude enzyme preparation assayed. In the special case in which the hydrolysis reaction is first order both as to substrate and enzyme,

$$\text{Enzyme value} = k/g(\log 2)$$

where g is the enzyme concentration expressed as grams (dry matter) of enzyme per 50 ml. of reaction mixture, and $k = \dfrac{1}{t} \log \dfrac{100}{100 - x}$, x being the per cent hydrolysis at time t in minutes. Since the hydrolysis reaction may not be truly monomolecular, it was recommended by Weidenhagen that the enzyme value be computed in general from the direct measurement of time value, i.e., on the basis: $k_{50\%}/g(\log 2)$.[9]

[6] S. Hestrin and C. C. Lindegren, *Arch. Biochem.* **29**, 315 (1950).

[7] D. Keilin and E. F. Hartree, *Biochem. J.* **42**, 230 (1948).

[8] R. Weidenhagen, *in* "Die Methoden der Fermentforschung" (E. Bamann and K. Myrbäck, eds.), Vol. 2, p. 1723, Georg Thieme Verlag, Leipzig, 1941.

[9] "Wertigkeit" or "enzyme efficiency" is computed in the same way except that the standard assay mixture contains a lower initial substrate concentration, generally 0.052 M (rather than 0.1388 M as in the "Normalbedingungen").

The reader should note that the selection of the "Normalbedingungen" for hexosidase assay was originally made on the basis of considerations of convenience particular to techniques of sugar analysis that were widely used in the early nineteen-twenties. The "Normalbedingungen" are objectionable in that they are wasteful of both the enzyme and the substrate; they are sometimes inapplicable in practice because the saccharide cannot be held in solution at the requisite high concentration; finally, the computation of activity from 50% hydrolysis time is a misleading one in the frequent cases in which substrate excess, product inhibition, thermal inactivation of enzyme, or other cause results in a serious divergence from first-order kinetics in the measured part of the hydrolysis. With the modern advent of rapid and sensitive micromethods for determining hexoside hydrolysis and of simple procedures for the evaluation of the constants of the Michaelis-Menten relationship, the utility of rigid observance of "Normalbedingungen" in hexosidase assay has become questionable.

II. β-Glucosidase from Sweet Almond Emulsin

Assay Method

Principle. Any one of a wide variety of β-glucosides may serve at convenient substrate concentration as the basis of the β-glucosidase assay.[10] Either glycose release or aglycon release may serve as the criterion of hydrolysis. The procedure given below utilizes β-phenyl glucoside as the test substrate and reductimetrically measures the glucose released.[11,12]

Reagents.

1. Somogyi's copper reagent:[13] This reagent is best prepared before use by the mixing of the following two stock solutions: (a) $CuSO_4 \cdot 5H_2O$ (analytical reagent), 10% in water; (b) phosphate tartrate solution made up as follows: 28 g. of anhydrous dibasic sodium phosphate (analytical reagent) is dissolved in 700 ml. of water, then 40 g. of crystalline sodium

[10] S. Veibel, *in* "The Enzymes" (J. B. Sumner and K. Myrbäck, eds.), Vol. 1, Part 1, p. 590, Academic Press, New York, 1950.

[11] R. L. Nath and H. N. Rydon, *Biochem. J.* **57**, 1 (1954).

[12] Even more conveniently, chromogenic β-nitrophenyl glucoside ("β-nipheglu"), e.g. β-*p*-nitrophenyl glucoside and the rapidly cleaved β-*o*-nitrophenyl glucoside, could be used as a test substrate for β-glucosidase activity [K. Aizawa, *J. Biochem. (Japan)* **30**, 89 (1939); K. Tapano and T. Miwa, *J. Biochem. (Japan)* **37**, 435 (1950); see also ref. 11]. For details of the colorimetric estimation of *o*-nitrophenol, see the section on β-galactosidase.

[13] M. Somogyi, *J. Biol. Chem.* **160**, 69 (1945).

potassium tartrate (tetrahydrate) is added. After solution is complete, 100 ml. of 1 N sodium hydroxide is added, followed by 120 g. of anhydrous sodium sulfate.[14] When solution is complete, the volume is made up to 900 ml. After 2 days the reagent is cleared of sediment by filtration through Whatman No. 4 paper.

To prepare the copper reagent 1 vol. of the copper sulfate solution is mixed with 9 vol. of the phosphate tartrate solution.

2. Nelson's arsenomolybdate color reagent:[15] 25 g. of ammonium molybdate $(NH_4)_6Mo_7O_{24}\cdot 4H_2O$ (analytical reagent) is dissolved in 450 ml. of water, and 21 ml. of concentrated H_2SO_4 (analytical reagent) is added and mixed. Three grams of crystalline disodium arsenate $(Na_2H\ AsO_4\cdot 7H_2O$, analytical reagent) is dissolved in 25 ml. of water and added. After 48 hours at 37°, the reagent is ready for use. It should be stored in brown glass-stoppered bottles.

3. Ba solution: 600 ml. of water is boiled in a Florence flask for 5 minutes, 45 g. of crystalline barium hydroxide $(Ba(OH)_2\cdot H_2O)$ is added and boiling continued until dissolved. The solution is stored in a container equipped with a siphon tube and soda lime tube to protect it from contact with carbon dioxide.

4. Zn solution: $ZnSO_4\cdot 7H_2O$, 5% in water. The concentration should be adjusted so that a mixture of 1.0 ml. each of Ba solution, Zn solution, and enzyme digest (see below) yields a neutral filtrate.

5. Substrate solution: β-phenyl-D-glucoside, 0.25 M in water.

6. Buffer solution: 0.05 M acetate (Na^+) buffer, pH 5.25.

7. Enzyme. An appropriate dilution is prepared in water.

Procedure. The assay system at 30° contains in final volume of 10 ml. the following: 2 ml. of substrate solution, 2 ml. of buffer solution, and the enzyme in an appropriate dilution. Withdraw at zero time and at appropriate intervals thereafter 1.0-ml. samples. To these add immediately 1.0 ml. each of the Zn and Ba solutions. Bring to 10.0 ml. and filter. Determine the reducing power of a suitable aliquot. The aliquot is added in a final volume of 2.0 ml. to a test tube containing 2.0 ml. of copper reagent. Cover the tube loosely with a glass bulb and heat it in a boiling water bath for 15 minutes. Cool to room temperature in tap water. Add 2.0 ml. of arsenomolybdate reagent. Mix by swirling vigorously. After 2 minutes add water to the 10-ml. mark and measure the blue color. In a Klett photoelectric colorimeter, filter 54 is suitable. Calibrate by reference to a standard curve prepared on known glucose solutions. A suitable range is 10 to 100 γ grams.

---

[14] The original directions specify 180 g. of Na_2SO_4, but this quantity is difficult both to dissolve completely and to maintain in solution.

[15] N. Nelson, *J. Biol. Chem.* **153**, 375 (1944).

Definition of Enzyme Activity. An over-all velocity constant, k_o, is obtained from the slope of the linear plot of the decadic logarithm of the concentration of unhydrolyzed substrate against time in minutes. The β-D-glucosidase activity is conveniently expressed in β-Ph. units defined as the first-order velocity constant (min.$^{-1}$) of the hydrolysis of β-phenyl-D-glucoside in 0.05 M concentration at 30° at pH 5.25 calculated for a standard enzyme concentration of 100 mg./l. (the reaction is first order with respect to the enzyme concentration). Units of β-glucosidase obtained by the use of different test substrates at any convenient concentration are interconvertible, provided that K_m and V_{max} values of the system used are known. The K_m value and V_{max} values for the sweet almond emulsin acting on β-phenyl-D-glucoside are, respectively, 0.027 M and 22 (for the basis of calculation of these values see footnote b of Table III).

Preparation and Purification

The accepted procedure for purification of sweet almond β-glucosidase is that of Helferich et al.,[16,17] which is quoted below as presented by Veibel.[18] The procedure consists essentially of (a) precipitation by tannin to obtain Helferich's "Rohferment" and (b) further purification by treatment with silver acetate. The steps of the purification are carried out rapidly in a cold room.

1. Preparation of "Rohferment" by Tannin Precipitation. One kilogram of finely powdered press cake of sweet almonds is dispersed in a solution of 50 g. of zinc sulfate (commercial hydrate, $7H_2O$) in 4.5 l. of water and left standing at 0° for 4 to 5 hours. The cold solution is then filtered through a cloth and well pressed on the filter. To the filtrate is added cautiously a solution of 1.4 g. of tannin in 500 ml. of water. A precipitate consisting mostly of impurities is removed by centrifugation and discarded. The bulk of the enzyme is then precipitated by slowly adding a solution of 15 g. of tannin in 500 ml. of water. The precipitate is isolated by centrifugation, freed from tannin by repeatedly dispersing it in acetone in the beakers, centrifuging it, and finally drying it in a desiccator over sulfuric acid. The crude enzyme ("Rohferment" of Helferich) obtained in this way has an activity of 4.44 β-Ph. units (Helferich "Wertigkeit" 0.30).[19]

[16] B. Helferich, S. Winkler, R. Gootz, O. Peters, and E. Günther, *Z. physiol. Chem.* **208**, 91 (1932).

[17] B. Helferich and S. Winkler, *Z. physiol. Chem.* **209**, 269 (1932).

[18] S. Veibel, *in* "The Enzymes" (J. B. Sumner and K. Myrbäck, eds.), Vol. 1, Part 1, p. 585, Academic Press, New York, 1950.

[19] Helferich "Wertigkeit" is approximately $k_o/(g \log 2)$, where g is milligrams of enzyme in 50 ml. of reaction mixture and k_o the first-order reaction velocity constant.

2. Purification by Treatment with Silver. Two grams of "Rohferment" is dissolved in 200 ml. of water, and the solution is cooled in ice water. An addition of 8.0 ml. of 0.02 N silver acetate immediately followed by the dropwise addition of 15.5 ml. of 0.01 N sodium hydroxide produces a precipitate practically free of β-glucosidase. It is removed by centrifugation and discarded. Without raising the temperature 14 ml. of 0.02 N silver acetate and then 27.5 ml. of 0.01 N sodium hydroxide are added. The precipitate is rapidly isolated by centrifugation, washed twice in ice-cold water without stirring, and dissolved in about 50 ml. of ice-cold 1/800 N ammonia, which is added in portions of about 2 ml. with efficient stirring. If some of the precipitate remains undissolved after 2 to 3 minutes it is removed by centrifugation. The somewhat turbid ice-cold liquid is precipitated by dropwise addition of 3.3 ml. of 0.2 N silver acetate. The precipitate contains the bulk of the enzyme and is isolated by centrifugation. The supernatant liquid is removed by decantation. The precipitate is suspended in 30 ml. of ice-cold water and hydrogen sulfide is introduced, just sufficient to cause a smell of hydrogen sulfide. The silver sulfide is removed by centrifugation. The enzyme is precipitated with an excess of cold acetone and washed with diethyl ether. The yield is 60 to 80 mg. with a β-Ph. value of 40. The recovery is 40 to 50% of the enzyme content of the crude enzyme ("Rohferment"), but only if the temperature is kept low and all operations are carried through rapidly.[20]

Properties

Specificity. This β-glucosidase preparation manifests hydrolytic activity toward α-D-galactosides, β-D-xylosides, and β-L-arabinosides, as well as α-D-mannosides, but is inert toward α-glucosides. Since the preparation is undoubtedly a mixture which probably still contains several hexosidases, a cleavage of any given substrate by this preparation cannot with certainty be referred to the β-glucosidase component of the preparation. On the other hand, failure of cleavage can be used with confidence as descriptive of the β-glucosidase entity.

The extensive researches of Helferich and his associates and of Pigman have thrown light on the limits of the absolute specificity of this enzyme.[21–23] Only a few of the salient results can be mentioned within the space of a short article. It is singularly unfortunate that much of the work

---

[20] Nath and Rydon[11] have obtained a much lower yield, using the same procedure.
[21] S. Veibel, *in* "The Enzymes" (J. B. Sumner and K. Myrbäck, eds.), Vol. 1, Part 1, p. 583, Academic Press, New York, 1950.
[22] W. W. Pigman, *Advances in Enzymol.* **4,** 41 (1944).
[23] B. Helferich, *Ergeb. Enzymforsch.* **7,** 83 (1938).

done has been devoted only to the comparison of rates of cleavage of series of substrates examined at one arbitrarily chosen substrate concentration and therefore does not lend itself to reliable generalization concerning the effect of the substrate structure on either K_m or V_{max}.

Substitution or configurational change either at C_2, C_3, or C_4 or a configurational change at C_1 of the β-D-glucopyranoside results in an absolute loss of cleavability by the β-glucosidase. On the other hand, glycon alteration at C_5 and C_6, like many changes in the nature of the aglycon, do not prevent cleavage of the substrate by the emulsin preparation. The values of K_m and V_{max} vary in magnitude with the structure of the aglycon (see Tables II and III).

TABLE II

INFLUENCE OF CARBON CHAIN LENGTH IN A NORMAL ALKYL RADICAL ON THE
HYDROLYSIS OF THE β-GLUCOSIDE BY SWEET ALMOND EMULSIN PREPARATION[a]

Aglucon radical	$K_m{}^b$	$V_{max}{}^b$
Methyl	0.62	1.78
Ethyl	0.25	1.54
n-Propyl	0.16	4.52
n-Butyl	0.03	1.71
n-Amyl	0.025	1.72

[a] Cited from S. Veibel, "Methoden der Fermentforschung," Vol. 2, p. 1779, Georg Thieme Verlag, Leipzig, 1941.

[b] $V_{max} = 10^2 \cdot k_{obs}(K_m + C)$, where k_{obs} is the observed first-order velocity constant (min.$^{-1}$), K_m the affinity constant in molarity, and C the molar substrate concentration.

The effect of substitution in the aromatic ring on the cleavage of β-phenylglucoside has been studied systematically in a series of twenty glucosides by Nath and Rydon,[11] who find that both the formation and breakdown of the enzyme-substrate complex are facilitated by electron-attracting substituents (Table III).

pH Relation. The pH activity curve on β-phenylglucoside shows an optimum near pH 5.[24] Significant shifts in the optimum pH value depending on the substrate structure have been recorded.[25] The pH stability curve shows an optimum at pH 7.1, inactivation being relatively rapid at pH > 8.2 and <4.1.[26]

Activation by Ions. The activity is a function of the ionic environment.[27] With several cations (added as nitrate 0.08 M) the rate of

[24] K. Josephson, *Z. physiol. Chem.* **147**, 1 (1925).
[25] S. Veibel and H. Lillelund, *Enzymologia* **9**, 161 (1941).
[26] B. Helferich, H. Brederech, and A. Schneidmüller, *Z. physiol. Chem.* **198**, 100 (1931).
[27] B. Helferich and E. Schmitz-Hillebrecht, *Z. physiol. Chem.* **234**, 54 (1935).

TABLE III

INFLUENCE OF A SUBSTITUENT IN THE PHENYL RADICAL ON THE HYDROLYSIS
OF SUBSTITUTED β-PHENYL-D-GLUCOSIDES BY SWEET ALMOND EMULSIN
PREPARATION[a]

Substituent	$K_m{}^b$	$V_{max}{}^b$
H	0.027	22
o-Methyl	0.022	904
m-Methyl	0.019	57
p-Methyl	0.013	17
o-Isopropyl	0.0029	56
p-Isopropyl	0.0042	9
o-Tertiary butyl	0.0076	6.7
p-Tertiary butyl	0.0064	3.6
2,4-Dimethyl	0.0058	25
2,6-Dimethyl	0.0347	13
o-Methoxyl	0.017	965
m-Methoxyl	0.026	136
p-Methoxyl	0.019	59
o-Chloro	0.0045	832
m-Chloro	0.0097	228
p-Chloro	0.0083	44
m-Cyano	0.0055	690
p-Cyano	0.0024	286
o-Nitro	0.0062	3550
m-Nitro	0.0044	460
p-Nitro	0.0020	3855

[a] Cited from R. L. Nath and N. Rydon, *Biochem. J.* **57**, 1 (1954).

[b] $V_{max} = 10^8 \cdot k_3'$ where $k_3' = k_0(K_m + S_0)/e_0'$, where k_0 is the observed over-all velocity constant of a first-order reaction (min.$^{-1}$), K_m is the affinity constant in molarity, S_0 is the initial substrate concentration in molarity, and e_0' is the enzyme concentration given in milligrams per liter.

hydrolysis is approximately doubled, e.g., with Na^+, K^+, Rb^+, Mg^{++}, Ca^{++}, Sr^{++}, Ba^{++}. Activation by the anions I^-, $(NO_3)^-$, $(ClO_4)^-$, and $(ClO_3)^-$ added as potassium or ammonium salt at 0.08 M is similarly high.

Temperature Relations; Equilibrium Constants. The equilibrium constant of the hydrolysis reaction has been studied in relation to dependence on aglucon structure at 4° and 30° with the results presented in Table IV.

The heat of activation of the enzymatic hydrolysis reaction is about

12,300 cal. in the case of glucosides of both primary and secondary alkyl alcohols and about 19,900 cal. in the case of glucosides of tertiary alkyl alcohols.[28] The heat of activation of the acid hydrolysis reaction, on the other hand, is near 31,000 for all these types of glucoside.

TABLE IV

INFLUENCE OF THE AGLUCON STRUCTURE ON THE EQUILIBRIUM CONSTANT AND HEAT OF REACTION OF β-GLUCOSIDE HYDROLYSIS IN THE PRESENCE OF β-GLUCOSIDASE[a]

Aglucon	$K_{30°}{}^b$	$K_{4°}$	$Q_{cal}{}^c$
Methanol	0.311	0.296	314
Ethanol	0.519	0.466	700
Propanol	0.657	0.587	722
Butanol	0.624	0.568	602
Isobutanol	0.785	0.738	397
Isopropanol	1.495	1.253	1131
sec-Butanol	1.391	1.221	835
tert-Butanol	3.761	2.967	1524

[a] S. Veibel, *Enzymologia* **1**, 124 (1936–1937).

[b] $K = \dfrac{[\text{glucose}][\text{alcohol}]}{[\text{glucoside}][H_2O]}$. The constant was determined in a reaction mixture containing initially glucose 0.15 M, alcohol 3.00 M, and water 15.00 M, acetone being used as the diluent.

[c] Heat of the reaction calculated from the equation $Q = \dfrac{RT_1T_2}{T_1 - T_2} \ln \left(\dfrac{K_{T_1}}{K_{T_2}} \right)$.

Inhibitors. Heavy metals (Ag, Cu, Hg) effect transiently reversible inactivation. Reversal can be effected at an early stage of inactivation by removal of the metal with hydrogen sulfide. Formaldehyde is a powerful inactivator, whose action may be reversed by prompt addition of sodium hydrogen sulfite.[29,30]

The enzyme is inactivated by oxidants, the inactivation being instantaneous with ozone[31,32] and relatively gradual with osmium tetroxide.[33] Toluene is a weak activator.[34]

[28] S. Veibel and E. Fredericksen, *Kgl. Danske Videnskat. Selskab. Mat.-fys. Medd.* **19**, No. 1 (1941), cited by S. Veibel, *in* "The Enzymes" (J. B. Sumner and K. Myrbäck, eds.), Vol. 1, Part 1, p. 610, Academic Press, New York, 1950.

[29] B. Helferich and S. Winkler, *Z. physiol. Chem.* **221**, 98 (1933).

[30] M. Mascré and A. Paris, *Bull. soc. chim. biol.* **15**, 918 (1933).

[31] B. Helferich and S. R. Petersen, *Z. physiol. Chem.* **233**, 75 (1935).

[32] B. Helferich, S. Winkler, E. Schmitz-Hillebrecht, and H. Bach, *Z. physiol. Chem.* **229**, 112 (1934).

[33] B. Helferich and F. Vorsatz, *Z. physiol. Chem.* **239**, 241 (1936).

[34] S. Veibel, *Enzymologia* **2**, 367 (1937–1938).

III. β-Galactosidase (Lactase) from *Escherichia coli*

Assay Method 1 (β-*o*-Nitrophenyl Galactoside as Substrate)

Principle. The chromogen β-*o*-nitrophenyl galactoside (β-niphegal) has been shown to be hydrolyzed by the same enzyme from *E. coli* as effects lactose hydrolysis[35] and thus can conveniently be used as the test substrate in *coli* lactase assay. The method described is that elaborated by Lederberg[36] as slightly modified by Kuby and Lardy.[37] The intact glycoside has a negligible optical density at visible wavelengths. On the other hand, the free *o*-nitrophenol is capable of a tautomeric change that gives it a yellow color in alkaline solution, with an absorption peak at 420 mμ. At a concentration of 0.001 *M* β-niphegal, the hydrolysis follows zero-order kinetics up to about 60% of total cleavage. The reaction rate is a first-order function of the enzyme concentration. At pH 7.25 in the presence of 0.14 *N* Na$^+$ the hydrolysis proceeds with maximum velocity.

Reagents

β-Niphegal solution, 0.01 *M*.[38]
Sodium carbonate solution, 1.0 *M*.
Sodium phosphate solution, 0.2 *N* Na$^+$ (phosphate, pH 7.25).
Enzyme solution.

Procedure. At 30° add 3.5 ml. of sodium phosphate solution, 0.5 ml. of β-niphegal solution, and 1.0 ml. of suitably diluted enzyme solution to the reaction tube. Remove an aliquot at zero time and at suitable times thereafter. Pipet samples directly into a colorimeter tube containing 1.0 ml. of sodium carbonate solution and add water to 10.0 ml. Read the density of the yellow color in a photoelectric colorimeter set at 420 mμ.[39] Over a wide range the color obeys Beer's law proximately. A reference curve is constructed on solutions of *o*-nitrophenol of known concentration.

Definition of Unit. The enzyme unit is defined as that amount of enzyme which will catalyze the hydrolysis of 0.001 *M* β-niphegal at pH 7.25 at 30° in the presence of 0.14 *N* Na$^+$ (phosphate) at the rate of 0.012 micromole per milliliter of reaction mixture per minute.

[35] M. Cohn and J. Monod, *Biochim. et Biophys. Acta* **7**, 153 (1951).
[36] J. Lederberg, *J. Bacteriol.* **60**, 381 (1950).
[37] S. A. Kuby and H. A. Lardy, *J. Am. Chem. Soc.* **75**, 890 (1953).
[38] For the synthesis of β-niphegal, see M. Seidman and K. P. Link, *J. Am. Chem. Soc.* **72**, 4324 (1950).
[39] In the conditions specified the color density determination is made at pH ≳ 10, i.e., a range in which the phenolic group of *o*-nitrophenol is completely dissociated (*pK* = 7.3) and the color density at maximum.

Assay Method 2 (Lactose as Substrate)

Principle. Hydrolysis of lactose has been followed often with reductimetric procedures. For this purpose hexose-selective copper reagents are of particular advantage. (See under section on α-galactosidase, p. 249.) In the gasometric method described below, glucose formed from hydrolysis of lactose is determined specifically with the help of *Penicillium notatum* glucose oxidase (notatin). This method, which has been worked out by Keilin and Hartree,[40] has also been used by Cohn and Monod.[35] Notatin oxidizes glucose even at low glucose concentration > 100 times as rapidly as it does any other hexose or oligosaccharide, even if the latter are at a relatively high concentration. In the absence of catalase, 1 mole of hydrogen peroxide and 1 mole of gluconate are formed from 1 mole of glucose with uptake of 1 mole of oxygen. In the presence of catalase, however, the peroxide is decomposed with evolution of 1 atom of oxygen per mole. Thus the net reaction in the presence of catalase involves an uptake of 1 atom of oxygen per mole of glucose oxidized. This net uptake is measured manometrically. In the assay system described below, a sufficient concentration of lactose $(0.1\ M)$ to permit approximately zero-order kinetics within the analyzed range of the hydrolysis reaction (up to 30% hydrolysis) is given.

Reagents

> Notatin. The pure enzyme is prepared according to Coulthard *et al.*[41] If necessary, crude preparations can be used.
> Catalase. This enzyme is prepared from horse liver.[42]
> Sodium phosphate, 0.35 N Na$^+$ (phosphate, pH 7.25).
> Lactose solution, 0.25 M.
> β-Galactosidase.
> Acetic acid, 0.2 N.

Procedure. At 30° add 4 ml. of sodium phosphate, 4 ml. of lactose solution, and 2 ml. of suitably diluted β-galactosidase to the reaction tube. Withdraw 1.0-ml. samples from the lactose-β-galactosidase mixture at zero time and at suitable intervals. Terminate the hydrolysis reaction by heating for 10 to 15 minutes in a boiling water bath. Adjust the samples to pH 5 to 6 by the addition of 0.4 ml. of acetic acid. Add water to a volume of 2 ml. and transfer a suitable aliquot in a volume of 1.0 ml. to a Warburg flask. For the determination of glucose, the aliquot should con-

[40] D. Keilin and E. F. Hartree, *Biochem. J.* **42**, 230 (1948).
[41] C. E. Coulthard, R. Michaelis, W. F. Short, G. Sykes, G. E. H. Skrimshire, A. F. Standfast, J. H. Birkinshaw, and H. Raistrick, *Biochem. J.* **39**, 24 (1945).
[42] D. Keilin and E. F. Hartree, *Biochem. J.* **39**, 148 (1945).

tain about 5 to 15 micromoles of glucose. After temperature equilibration, 300 γ of notatin and 0.1 ml. of catalase are introduced into the Warburg flask from the side arm. Follow the oxygen uptake until the rate of uptake falls to the steady level observed in the flask containing the zero-time sample. The difference of the total atom oxygen uptake of the two flasks represents in moles the glucose formed from lactose by hydrolysis.[43] A modified form of this procedure[35,40] permits of the use of notatin for the continuous measurement of the hydrolysis in which notatin in excess is incorporated directly in the lactose-lactase reaction mixture.

Purification of β-D-Galactosidase[37]

Because of the limited amount of source material available for isolation of the enzyme, and the very small amounts of protein dealt with, the fractionation scheme was so designed as to emphasize both yield and purification. The description below is quoted verbatim, from the original paper of Kuby and Lardy.[37]

Escherichia coli, strain K-12, was grown on a lactose-inorganic salts medium,[36] in an 80-gallon steel fermentor with strong aeration and stirring. After 18 to 20 hours the cells were harvested by means of the Sharples centrifuge. The *fresh* cell paste was washed with a minimum amount of distilled water required to form a homogeneous suspension and centrifuged hard in the Spinco preparative ultracentrifuge, Model L, head No. 20, at 20,000 r.p.m. for 0.5 hour to remove the liquid. The sedimented cells were then spread thin over the bottom of petri dishes and dried in evacuated desiccators over P_2O_5 at room temperature, with repeated changes of P_2O_5. After ca. 2 to 3 days, the drying was practically complete, and the cell residue was ground to a fine powder and allowed to dry an additional day. When stored in vacuum desiccators over P_2O_5 in the cold room, the dry cell powder showed only small loss in β-D-galactosidase activity after several months.

Fraction I. Twenty grams of dried cells and 50 ml. of 0.001 M sodium phosphate, pH 7.3, were ground in a large porcelain mortar to a homogeneous paste with a heavy porcelain pestle. The paste was then frozen (ca. $-10°$), in the mortar. After freezing, it was thawed somewhat (ca. 2°) and ground until the paste assumed the original consistency. The freezing and grinding procedure was repeated five times.

---

[43] In the presence of substances which can undergo coupled oxidation in an H_2O_2-catalase system, the evolution of oxygen from H_2O_2 is incomplete. To overcome this difficulty Keilin and Hartree suggest that 20 mg. of ethanol be added to the reaction mixture. In the presence of ethanol, H_2O_2 not utilized for other coupled oxidations will be utilized for oxidation of ethanol to aldehyde and the net uptake of oxygen will then be 1 mole per mole of glucose.

The mortar was then removed to room temperature, 75 ml. of the same buffer added to the mortar, and the paste ground to a homogeneous suspension which was diluted to 500 ml. with the buffer. The contents were placed in a 1-l. round-bottom flask, immersed in a 30° bath, and stirred (glass) mechanically for 3 hours.

The suspension was then centrifuged (Spinco Model L, head No. 20) for 0.5 hour at 15,000 r.p.m. and the supernatant (455 ml., fraction I) decanted. The residue was discarded. The pH of the supernatant should be around pH 6.0 (5.8 to 6.2); if not, it should be so adjusted with 0.5 M NaH$_2$PO$_4$.

Fraction II: Methanol Fractionation. Fraction I (455 ml.) was transferred to a 2-l., three-neck round-bottom flask equipped with a glass stirrer, 250-ml. graduated dropping funnel, and thermometer. With the contents at ca. 1°, about ⅕ vol. of absolute methanol was introduced at ca. 40 ml./min. with efficient mechanical stirring (foaming must be avoided). The flask was then immersed in a −10° bath and allowed to reach temperature equilibrium. Then absolute methanol was added at ca. 20 ml./min., with stirring. When a total of 372 ml. had been added, the contents of the flask were allowed to stand 0.5 hour at −10° and centrifuged at ca. −10° in the International refrigerated centrifuge (cat. No. 834 head) at maximum speed for 20 minutes. The slightly *turbid* supernatant was decanted and retained, and the small amount of precipitate was discarded. The supernatant was returned to the flask and allowed to reach −10° if any change in temperature had occurred. Then 386 ml. of methanol (abs.) was added as before. After allowing to stand 0.5 hour the suspension was centrifuged as before and the *clear* yellow supernatant discarded. The precipitate was resuspended in 100 ml. of 0.001 M sodium phosphate, pH 7.3, and dialyzed[44] vs. 4 l. of the same buffer at ca. 2° for 24 hours. The initial suspension which contained some "insoluble" material gradually became clear, with the protein redissolving, until finally at the end of the dialysis the solution (fraction II) was essentially clear and yellow in color. The fractionation was conducted between the levels of 2.45 to 5 ml. of absolute methanol for every 3 ml. of fraction I.

Fraction III: Precipitation of Nucleoproteins.[45] Fraction II (266 ml.) was made 0.0235 M with respect to MnSO$_4$ and allowed to stand at 2° for 38 hours. The precipitate which separated at 15,000 r.p.m. (Spinco, Model L, No. 20 head) for 0.5 hour was discarded; the clear supernatant solution (ca. 248 ml.) was retained as fraction III.

---

[44] Dialysis was conducted so that both the bag and the external fluid were stirred, and the external fluid changed about once every 8 hours.

[45] S. Kaufman, S. Korkes, and A. del Campillo, *J. Biol. Chem.* **192**, 301 (1951).

Fraction IV: Adsorption on $Ca_3(PO_4)_2$ *and Elution.* These steps were conducted in a cold room at 2°. To 248 ml. of fraction III, 9.93 ml. of $Ca_3(PO_4)_2$ gel[46] (gel concentration, 62.2 mg./ml; therefore 0.532 mg. gel/mg. protein in fraction III) was quickly added, mixed, and centrifuged immediately for ca. 4 minutes. Speed in this operation is essential; therefore this step was usually conducted in the 250-ml. centrifuge bottle. The supernatant was decanted (ppt. discarded), and 55.7 ml. of gel (2.91 mg. gel/mg. protein in fraction III) was added. After stirring for ca. 15 minutes the suspension was centrifuged for 10 minutes and the supernatant was discarded. The precipitate was washed with ca. 250 ml. of cold distilled H_2O and collected by centrifugation.

The enzyme was eluted from the gel with 62 ml. of a 25% saturated $(NH_4)_2SO_4$.[47] After stirring for 15 minutes the suspension was centrifuged. The eluate was saved and the precipitate re-extracted as above with another 62 ml. of 25% saturated $(NH_4)_2SO_4$. The eluates were combined (fraction IV).

Fractions V–VI: $(NH_4)_2SO_4$ *Fractionation. Step 1.* The 122 ml. of IV was brought to 45% saturation by adding 44.3 ml. of saturated $(NH_4)_2SO_4$, pH 6.3, slowly with stirring. After allowing to stand in the cold room for 48 hours, the small amount of precipitate was centrifuged off at 15,000 r.p.m. (Spinco, head No. 20) for 30 minutes, and the supernatant was retained for step 3.

Step 2. The precipitate was resuspended in 5 ml. of 25% saturated $(NH_4)_2SO_4$ and stirred for ca. 30 minutes. After centrifuging at 20,000 r.p.m. (Spinco, head No. 20) for 30 minutes, the supernatant was saved and the precipitate was re-extracted in the same manner. The supernatant from step 2 was combined (ca. 9.8 ml., fraction V), and the insoluble residue was discarded.

Step 3. The supernatant from step 1 (ca. 160 ml.) was brought to 50% saturation by addition of 15.9 ml. of saturated $(NH_4)_2SO_4$, pH 6.3. After 36 hours at 2° the material was centrifuged at 15,000 r.p.m. (Spinco No. 20 head) for 30 minutes and the supernatant discarded. The precipitate was eluted with two 10-ml. portions of 25% saturated $(NH_4)_2SO_4$ as described in step 2. The combined supernatants were designated V'.

Step 4. Fractions V and V' were combined (fraction V + V', 29.2 ml.) and brought to 45% saturation by adding 10.6 ml. of saturated $(NH_4)_2SO_4$, pH 6.3. After 28 hours at 2° the precipitate was collected at 20,000 r.p.m.

---

[46] The gel was prepared by a procedure based on that of S. Swingle and A. Tiselius, *Biochem. J.* **48**, 171 (1951).

[47] Obtained by dilution of saturated $(NH_4)_2SO_4$ (AR) solution which had been adjusted to pH 6.3 (Beckmann glass electrode) with NH_4OH.

(Spinco, No. 40 head) and was redissolved in 0.001 M sodium phosphate, pH 6.3 (fraction VI, ca. 4.5 ml.).

Fractions VII–X: Recycling. Fraction VI was brought to 0.0235 M MnSO$_4$, and after 21 hours at 2° the precipitate was collected by centrifugation at 20,000 r.p.m. (Spinco No. 40 head) for 30 minutes and discarded. The clear solution (4.4 ml., fraction VII) was then treated with Ca$_3$(PO$_4$)$_2$ exactly as above, using the same levels of gel per milligram of protein and eluted as before with two 2-ml. portions of 25% saturated (NH$_4$)$_2$SO$_4$. The combined eluates were designated fraction VIII (3.9 ml.). This solution was brought to 45% saturation by addition of 1.42 ml. of saturated (NH$_4$)$_2$SO$_4$ and allowed to stand at 2° for 24 hours. The precipitate which was collected at 20,000 r.p.m. (Spinco No. 40 head) for 30 minutes was resuspended in 1.5 ml. of 25% saturated (NH$_4$)$_2$SO$_4$. The very small amount of insoluble residue was removed by centrifugation as above. The supernatant was retained,[48] and the residue resuspended in 1.0 ml. of 25% saturated (NH$_4$)$_2$SO$_4$ and recentrifuged as above.

The combined supernatants (fraction IX, 2.4 ml.) were brought to 35% saturation by adding 0.37 ml. of saturated (NH$_4$)$_2$SO$_4$, pH 6.3. After 24 hours at 2° the precipitate was collected by centrifugation at 20,000 r.p.m. (Spinco head No. 40) for 30 minutes and redissolved in 0.082 M Na$^+$ phosphate buffer (pH 6.3) (fraction X, 1 ml.). It was then dialyzed against the same buffer at 2°. This preparation is stable for several months if stored just above 0°.

With some batches of cells, where the initial specific activity of fraction I was unusually low, the specific activity of all subsequent fractions was also low. In these cases X was brought to 33% saturation with (NH$_4$)$_2$SO$_4$. The precipitate contained about two-thirds of the total units, and the specific activity was raised to the normal range.

Table V summarizes a typical fractionation in terms of percentage recovery and purification for each individual step in the scheme. An overall purification (from I) of about 100-fold was achieved with about 8% recovery of the enzyme. The specific activity of fraction X varied from 7000 to 10,500 units/mg. in various experiments. Further fractionation by the use of solvents (methanol or acetone), salts (MgSO$_4$, Na$_2$SO$_4$, NaH$_2$PO$_4$), or adsorbents (alumina C$_\gamma$, starch, Dowex 2 anion exchanger) have not yielded higher specific activities.

Fraction X shows one component by microelectrophoresis at pH 6.3, but two by sedimentation. Because of the very small amounts of protein isolated, detailed and critical analyses of its homogeneity have not as yet been undertaken.

---

[48] For these very small volumes, the liquid phase was removed quantitatively by means of a capillary.

TABLE V
FRACTIONATION OF β-D-GALACTOSIDASE

Fraction no.	Total volume, ml.	Units/ ml.	Total units	Mg./ ml.	Units/ mg.	Recovery from preceding step, %	Purification over preceding step
I	455	1,635	744,000	16.2	101.0	—	—
II	266	2,285	608,000	8.17	279.5	82	3.61
III	248	2,195	545,000	4.68	469	89.8	1.68
IV	122	3,035	370,500	3.60	843	68.0	1.8
V	9.8	9,250	90,600	2.43	3810	24.5	4.52
V'	19.4	7,555	146,500	2.17	3488	39.6 (from IV)	4.14 (over IV)
V + V'	29.2	8,110	237,100	2.26	3590	64.0 (from IV)	4.26 (over IV)
VI	4.5	40,875	184,000	6.7	6100	77.6	1.70
VII	4.4	40,000	176,000	5.96	6715	95.7	1.10
VIII	3.9	30,000	117,000	3.75	8000	66.6	1.19
IX	2.4	35,350	84,250	3.82	9400	72.5	1.16
X	1.0	56,750	56,750	5.65	10050	67.0	1.09
						Over-all % recovery from I 7.63	Over-all purification over I 99.5

Properties

Specificity. This hydrolase acts on a wide range of β-galactosides.[37,49] The Michaelis-Menten relationship is obeyed. Affinity constants and V_{max} values for a selected group of substrates are given in Table VI, quoted from Kuby and Lardy.[37] Marked changes in the structure of the aglycon are compatible with enzyme activity. The removal of the C_6 atom of the galactose moiety does not render the derived substrate (the α-L-arabinoside) absolutely resistant to hydrolysis. Many other changes in the galactose moiety (substitution, reduction, oxidation, and changes in the steric configuration at the carbon positions) render the derived substrate inert. α-D-Galactosides, α- or β-D-glucosides, α-D-mannosides, and β-D-fructosides are not hydrolyzed.

Inhibitors and Activators. β-Phenylthiogalactoside is a powerful inhibitor which is not itself split by the enzyme.[50] Melibiose, galactose, glucose, and sucrose inhibit weakly. o-Nitrophenol and cysteine are inert.

The enzymatic activity is dependent strikingly on the ionic environment.[35-37] Cations have been arranged tentatively in an activity series of

---

[49] See Table VI; cf. J. Monod, G. Cohen-Bazire, and M. Cohn, *Biochim. et Biophys. Acta* **7**, 585 (1951).

[50] J. Monod and M. Cohn, *Advances in Enzymol.* **13**, 67 (1952).

the following form:

$$\text{Na}^+ > \begin{cases} \text{K}^+ & \text{Rb}^+ \\ & \geq \\ \text{Cs}^+ & \text{NH}_4^+ \end{cases} \geq \text{LI}^+ \geq \text{R—NH}_3^+$$

Thus the substituted ammonium ion (e.g., tris hydroxymethylamino-methane) is inhibitory in a system optimally activated by Na^+. The cation activation is in accord with the Michaelis-Menten relationship. Among divalent cations, Cu^{++}, Hg^{++}, and Zn^{++} at 10^{-3} M inhibit powerfully. A variety of common anions exert an inhibitory effect in a fully activated Na^+ system.[37]

TABLE VI

MICHAELIS CONSTANTS AND MAXIMAL VELOCITIES FOR SEVERAL SUBSTRATES

Compound	$K_s{}^b$	$V_{max}{}^a$
o-Nitrophenyl β-D-galactoside	1.8×10^{-4}	32×10^{-3}
p-Nitrophenyl β-D-galactoside	0.93×10^{-4}	5.8×10^{-3}
o-Nitrophenyl α-L-arabinoside	2.6×10^{-3}	2.2×10^{-3}
Phenyl β-D-galactoside	7.3×10^{-4}	5.0×10^{-3}
Methyl β-D-galactoside	6.9×10^{-3}	2.9×10^{-3}
n-Butyl β-D-galactoside	6.9×10^{-4}	2.3×10^{-3}
Lactose	$\sim 10^{-3}$	$\sim 3 \times 10^{-3}$
Lactositol	~ 0.03	$\sim 2.5 \times 10^{-4}$
Lactobionate	~ 0.04	$\sim 1.0 \times 10^{-4}$

^{a} V_{max} expressed in μm./ml./min./standard concentration of enzyme.
^{b} K_s in moles/l.

pH Relations. The pH activity curve has a peak at pH 7.25. Activity of 50% of optimum is observed on the acid side near pH 6 and on the alkaline side near pH 8. At 30° the purified enzyme is rapidly inactivated below pH 5.5.

Inducers. Genetic and physiologic bases governing the adaptive production of this enzyme have been studied extensively (for a review, see Monod and Cohn[50]). Among the galactosides acting as enzyme inducers, melibiose is of particular interest, since cells which fail to metabolize melibiose are nevertheless able to respond to this inducer with enzyme production.

Catalysis of Transglycosidation. The enzymatic hydrolysis of lactose is accompanied by reactions of galactosyl transfer which result in the synthesis of oligosaccharides (lactobiose, galactobiose, lactotriose).[51]

[51] K. Wallenfels and E. Bernt, *Ann.* **584,** 63 (1953).

IV. α-Galactosidase (Melibiase) from Sweet Almond Emulsin

Assay Method

Principle. Melibiose serves as a standard test substrate. The hydrolysis of melibiose is measured in terms of hexoses released, determined by means of a hexose-selective copper reagent.[52,53]

Reagents

1. Copper reagent: Dissolve 24 g. of copper acetate (c.p.) in 450 ml. of boiling water. If a precipitate forms, do not filter. Add immediately 25 ml. of 8.5% lactic acid. Shake. Cool. Dilute to 500 ml. Filter off impurities that settle out on standing.

2. Phosphomolybdate reagent: Place 150 g. of pure molybdic acid ("free of ammonium") in a conical flask. Add 75 g. of pure sodium carbonate (anhydrous). Add in small portions with cooling 500 ml. of water. Heat to boiling until molybdic acid is dissolved. Filter. Add 300 ml. of 85% phosphoric acid. Cool. Dilute to 1 l.

3. Zn reagent: Prepare a 5% solution of $ZnSO_4 \cdot 6H_2O$ in water.

4. Ba reagent: Prepare approximately 0.3 N solution of barium hydroxide. Adjust the concentration of this reagent so that a mixture containing 1.0 ml. of zinc reagent, 1.0 ml. of barium reagent, and 1.0 ml. of enzyme-substrate sample (see under Procedure) in 10.0 ml. of water yields a neutral filtrate.

5. Melibiose solution, 0.05 M.

6. Buffer, pH 4.5: 114 ml. of 0.1 M acetic acid is mixed with 86 ml. of 0.1 M sodium acetate solution.[54]

7. Enzyme.

Procedure. The assay mixture in a final volume of 10.0 ml. contains at 30° 2.0 ml. of substrate solution, 2.0 ml. of buffer, and 2.0 ml. of appropriately diluted enzyme. At zero time and thereafter at appropriate intervals 1.0-ml. samples are withdrawn and deproteinized forthwith by addition to a mixture of 1.0 ml. each of the Zn and Ba reagents. The suspensions are made up with water to 10.0 ml. and filtered. The reducing hexose is measured in samples containing not more than 1 mg. of meli-

---

[52] H. Tauber and I. S. Kleiner, *J. Biol. Chem.* **99**, 249 (1932).

[53] P. P. Gray and H. Rothchild, *Ind. Eng. Chem. (Anal. Ed.)* **13**, 902 (1941).

[54] Because phosphate as well as some other inorganic ions including chloride exert an inhibitory effect on the reduction of the cupric reagent, an acetate buffer solution has generally been specified for activity assays in which the analysis calls for the use of Tauber and Kleiner's copper reagent. However, in the procedure as given here, a phosphate buffer could be used equally well, since in the deproteinization by Zn-Ba the phosphate is effectively removed.

biose. A 2.0-ml. sample of the filtrate is heated in a test tube with 2.0 ml. of copper reagent in a boiling water bath for 8 minutes and cooled to room temperature. 2.0 ml. of phosphomolybdate reagent is then added with swirling, and the mixture is brought with water to 100 ml. Measure the blue color promptly. In the Klett-Summerson photoelectric colorimeter, filter 66 is suitable. The color density varies approximately as the hexose amount in the range 0.1 to 1.0 mg. of hexose. Calibrate by reference to a standard curve made with graded series of mixtures of hexose and disaccharide.

The above procedure is applicable to assay of melibiase in yeast as well as to melibiase in plant emulsin preparations. A modified form of the copper reagent in which anion interference is at least partly overcome has been described.[55]

Unit of Activity. An over-all velocity constant, k_o, is calculated from the slope of the linear plot of the decadic logarithm of the concentration of unhydrolyzed substrate against time in minutes. The melibiase activity of the preparation can then conveniently be expressed in k_o/g, where g is the enzyme concentration conventionally expressed as grams per 50 ml. reaction mixture. Alternatively, activity could be measured in "Wertigkeit."[19]

Procedure for Separation of the Sweet Almond Melibiase from Accompanying β-Glucosidase

A procedure recommended by Zechmeister *et al.*[56] depends on selective elution of melibiase from bauxite. Commercial (Hungarian) bauxite (18 g.) is mixed with 30% its weight of sieved sand and packed into a 2 × 10-cm. chromatograph tube to a height of 5 cm. The column is then washed with 10 ml. of acetate buffer, pH 4.7, which is drawn through under slight vacuum until the top surface is just covered. Emulsin (Merck, 1 g.) is added to 5 ml. of buffer, filtered, and the filtrate is made up to 25 ml. with water. 10 ml. of this solution is poured onto the column, and vacuum is applied until 9 to 10 ml. of an enzyme-free filtrate is collected and the top surface of the column is just covered with solvent. The remaining 15 ml. of enzyme solution is added and again driven through until 14 to 15 ml. has been collected. The solution contains melibiase and chitinase. Ninety-seven per cent of the original β-glucosidase activity is retained on the column, whereas 70% of the α-galactosidase activity is recovered in the filtrate.

Twenty-eight milliliters of filtrate is further chromatographed on a column of 10 g. of adsorbent 3.5 cm. deep. The column is first moistened

[55] R. Caputto, L. F. Leloir, and R. E. Trucco, *Enzymologia* **12**, 350 (1946–1948).
[56] L. Zechmeister, G. Tóth, and M. Balist, *Enzymologia* **5**, 302 (1938).

with 5 ml. of acetate buffer, then the filtrate is added and sucked through. The extruded and pulverized adsorbent is extracted with 13 ml. of 0.1 N ammonia solution, diluted with 13 ml. of water, centrifuged, decanted, and then washed twice with 25-ml. portions of water. The combined solution (75 ml.) is centrifuged to remove traces of adsorbent. The α-galactosidase thus obtained in a recovery of 78% is free from chitinase. For assay, the solution of enzyme is adjusted with acetic acid to the optimum pH 4.5.

Properties

The α-galactosidase of emulsin is more easily inactivated than the β-galactosidase occurring in the same source. Treatment of a mixture of the two enzymes with silver oxide completely destroys the α-galactosidase activity while leaving β activity unchanged. Heat treatment and tannin precipitation have likewise been shown to cause selective inactivation of α-galactopyranosidase.[57] It remains uncertain whether the enzymatic hydrolysis of melibiose and that of other α-galactosides is mediated by one enzyme entity or by several. Melibiase from yeast as well as from plant sources exhibits a broad optimum pH range of 3.5 to 5.3.

V. β-Fructofuranosidase (Invertase) from Yeast

Assay Method

Principle and Procedure. Hydrolysis of sucrose is followed reductimetrically. In the assay system proposed by Fischer and Kohtes[58] the enzyme acts on 2.5% sucrose in 0.01 M acetate (sodium) buffer, pH 4.77, at 20°. The amount of the enzyme is chosen such that at the time of the measurement of reducing power approximately 1 to 3% of the substrate is hydrolyzed. The method for the termination of the hydrolysis reaction and a microprocedure for the measurement of reducing power are given under the section on β-glucosidase.

Activity Unit (A). The A unit[58] is defined as an amount of enzyme which, in a reaction mixture of the above stated composition, liberates 1.0 mg. of reducing hexose (glucose plus fructose) in 3 minutes. Some other units of activity often used in the earlier literature have been tabulated by Pigman.[59,60]

---

[57] For a review of the literature, cf. S. Veibel, *in* "Die Methoden der Fermentforschung" (E. Bamann and K. Myrbäck, eds.), Vol. 2, p. 1791, Georg Thieme Verlag, Leipzig, 1941.
[58] E. Fischer and L. Kohtès, *Helv. Chim. Acta* **34**, 1123 (1951).
[59] W. W. Pigman, *J. Research Natl. Bur. Standards* **30**, 159 (1943).
[60] W. W. Pigman, *J. Research Natl. Bur. Standards* **30**, 257 (1943).

Preparation and Purification

A detailed procedure developed by Fischer and Kohtès[58] is presented.

1. *Aluminium hydroxide.* This reagent is freshly prepared and washed twice in distilled water by centrifugation. Centrifugation at a speed greater than 8000 r.p.m. is not recommended, since cakes which are excessively packed are difficult to resuspend and the product obtained has reduced adsorptive capacity. A motor-driven vibrator is useful in effecting the redispersion of the material. The washed suspension containing 15 to 30 mg. of $Al(OH)_3$ per milligram can be stored for about one month.

2. *Picrate.* Three grams of crystalline picric acid, 13 ml. of 1 N NaOH, and 2 ml. of 1 M acetate buffer, pH 3.7, are made up to 100 ml. with heating. This solution slowly crystallizes in the cold. Just before use it is warmed until free from turbidity and then cooled immediately to 0°. The solution must be perfectly clear before use.

3. *Acetone.* Redistilled over $KMnO_4 + Na_2CO_3$.

4. *Distilled water.* Distilled over barium hydroxide in a tin still.

All acetone precipitations are performed at 0° with efficient stirring, the acetone, previously cooled to −20°, being added in a fine stream. Unless otherwise stated, all centrifugations are made in the cold at 3000 r.p.m.

Extraction of Enzyme from the Cells.[61] Fresh baker's yeast serves as the enzyme source.

The extraction of invertase comprises the physical degradation of the yeast membrane structure by plasmolysis followed by a solubilization of invertase-binding components by means of autolysis. The procedure is as follows: 2.5 kg. of fresh baker's yeast is plasmolyzed at pH above 6 by mixing in a 5-l. vessel with 200 g. of sodium acetate. After the mass has liquefied (15 to 20 minutes), 375 ml. of toluene is added and the mixture is emulsified by a 10-minute treatment with a vibrator. A Vibro-Mischer (supplied by A. G. für Chemie Apparatenbau, Zürich) is recommended for this operation. The container is stoppered and held at 37° for 60 hours. At this time the pH is brought to 4.5 by the addition of about 400 ml. of

---

[61] High hexose concentration in the growth medium exerts a depressant effect on the hexosidase levels of yeast cells [S. Hestrin and C. Lindegren, *Arch. Biochim. et Biophys.* **38**, 317 (1952)]. It has been found that the invertase level can be raised severalfold by growing the yeast on a carbohydrate-poor medium or on one in which by use of raffinose as the carbon source the accumulation of an inhibiting concentration of fructose is prevented [R. Davies, *Biochem. J.* **55**, 484 (1953)]. Processes for the enrichment of yeast invertase content by slow feeding of sugar have been developed [R. Willstätter, C. D. Lowry, Jr., and K. Schneider, *Z. physiol. Chem.* **146**, 158 (1925)].

4 N acetic acid, and the mixture is centrifuged for $1\frac{1}{2}$ hours. Three layers are formed. The brownish middle layer contains the dissolved invertase. This layer is siphoned off. The cakes are taken up in water, again homogenized, and recentrifuged. The middle layer of this washing is removed and added to the first middle layer. The combined turbid brownish solutions have a volume of about 3.5 l. and a total activity of 1.3 to 1.7 $\times$ 10^6 A. The solution is perfectly stable when stored in the cold.

Precipitation of Inactive Proteins with Picrate Solution and Precipitation of the Enzyme by Acetone. To the enzyme solution at 0° is added, in one portion, 0.35 vol. of cold (0°) picrate solution. After at least 3 hours at 0° the mixture is centrifuged for 30 minutes (Sharples supercentrifuge). The sediment is discarded, and the solution is precipitated with 3 vol. of acetone. After an additional $\frac{1}{2}$ hour of agitation most of the solution is decanted and the remainder is centrifuged for 7 minutes.

The majority of the active material sticks to the stirrer and to the walls of the beaker. After the precipitate has been washed by twice triturating with $-20°$ acetone, the product (gummy) is dissolved in water. The aqueous solution is dialyzed in the cold for 48 to 72 hours with frequent changes of water (volume of outside water about 12 l.). When the outside external liquid is no longer yellow, the dialysis is terminated and the volume of the enzyme solution is made up to 5 l. Any turbidity is removed by rapid passage through the Sharples; a lightly colored solution is obtained. This fails to show some characteristics of typical protein solutions. Thus boiling fails to produce turbidity, and prolonged Sevag treatment does not result in inactivation.

Adsorption on Aluminium Hydroxide. The clear dialyzate is adjusted to pH 6.0 and diluted with enough water to bring the activity to 60 to 80 A/ml. In general 1 mg. of Al(OH)$_3$ suffices to adsorb 12 to 36 A. The minimum quantity of Al(OH)$_3$ at pH 6 necessary to completely adsorb all invertase in 3 minutes at 0° is determined with a small portion of the solution.[62] Then the solution in its entirety is so treated in small portions at 0°.

The suspension is well mixed, immediately distributed into centrifuge tubes, and centrifugation (not more than 3000 r.p.m.) is started exactly 3 minutes after the addition of the Al(OH)$_3$. The centrifuge cake is suspended in a minimum volume of water with the assistance of a vibrator. (The invertase on the Al(OH)$_3$ shows the same activity as an unadsorbed enzyme.)

[32] The conditions of adsorption are adjusted to permit of the highest recovery of purified invertase with minimum contamination by other components (protein, polysaccharide) which may be adsorbed by Al(OH)$_3$. Accordingly the adsorbent amount used is held to the permissible minimum, and other factors influencing the process (time, temperature, and enzyme concentration) are kept constant.

Elution from Al(OH)₃ by Lactate Solution. Since an overlong contact of the hydroxide suspension with an acid medium causes too much hydroxide to dissolve and gives a less pure product, the minimum time required to elute the enzyme must be determined experimentally. Three 1-ml. portions of the hydroxide suspension are diluted at 15° with 5 ml. of 4 N lactic acid-ammonium lactate buffer, pH 2.9, per gram of hydroxide. After 10, 15, and 20 minutes the suspensions are spun down at 12,000 r.p.m. The percentage of enzyme eluted is then determined in each of the supernatants. Usually no more than 15 minutes is required for complete elution. The main body of the suspension is then eluted for the period of time found to suffice for complete elution, and the aluminium hydroxide is removed by centrifugation (conveniently with the Sharples). The clear supernatant is brought to pH 4.8, and 3.5 vol. of acetone is added. The resulting suspension is centrifuged for 6 minutes, and the sediment is taken up in 300 ml. of water. A short centrifugation eliminates any turbidity caused by remaining hydroxide.

Finally the adsorption, elution and precipitation are repeated once, exactly as described above.

TABLE VII

SUMMARY OF PURIFICATION PROCEDURE[a]

| | Yield, % of recovery | | Purity[b] | |
| | On basis of preceding fraction | On basis of crude extract | | |
Purification step			A/mg. N	A/mg. carbohydrate[c]
1. Crude extract	100	100	20	10
2. Precipitation by picrate	80	80	150	20
3. Acetone precipitation and dialysis	90	72	150	20
4. Adsorption on Al(OH)₃	97	70	400	55
5. Elution	90	63	2800	97
6. Precipitation by acetone	95	60	2800	97
7. Repetition of steps 4–6	85	50	4000	250

[a] Cited from E. Fischer and L. Kohtès, *Helv. Chim. Acta* **34**, 1123 (1951).
[b] Nitrogen (Kjeldahl).
[c] Total carbohydrate (anthrone determination).

In Table VII the purification process is summarized step by step. The over-all yield of enzyme is about 50%. The enrichment factor in respect to nitrogen (Kjeldahl) is 200 and in respect to total polysaccharide is 25. Approximately 7×10^5 A units of invertase, representing 1.2 g. of protein and 3 g. of polysaccharide, are obtained from 2.5 kg. of baker's yeast. The product shows a characteristic single absorption peak at 265 mμ (pH of solution, 6 to 7). The solution may be dried by lyophilization or stored directly in the cold at pH 4 to 6.5 under toluene. The polysaccharide component of the purified preparation is a polymannan (molecular weight approximately 7770). The presence of the polysaccharide stabilizes the enzyme. Bentonite at pH 2.6 to 2.9 adsorbs the enzyme but not the polysaccharide. The enzyme-bentonite complex, however, rapidly becomes inactive.[63,64]

Properties

Specificity. β-Fructofuranosidase from yeast is the best known of the yeast invertases. It resembles the β-glucosidase from sweet almond emulsin and the β-galactosidase from *Escherichia coli* in respect to ability to tolerate considerable variation of aglycon structure while manifesting a much stricter specificity requirement in relation to the fructon moiety.

The highly purified invertase preparation described by Fischer and Kohtes mediates the transfer of the fructose residue from sucrose both to water and to acceptor carbohydrate, and this confirms the view that the same invertase functions both as hydrolase and as transfructosidase.[63,65,66] In the presence of the enzyme, alcohols (e.g., methanol, glycerol, benzyl alcohol, glucose, sucrose, and other oligosaccharides) function as acceptor for the fructofuranose moiety, which may be donated by different fructosides possessing an unchanged and unsubstituted β-fructofuranosidic terminal group, e.g., sucrose, raffinose, β-methylfructofuranoside. The variety of fructofuranosides which are thus formed by the transfer process are all also readily hydrolyzed by the enzyme. The cleavage catalyzed by invertase occurs at the bond between the fructose carbon and the bridge oxygen.[67] The early observation that certain free hexoses, glycides, and other alcohols inhibit the hydrolytic activity of the enzyme has to be reinterpreted in the light of the finding that these substances intervene stoichiometrically in the reaction as acceptors. A further discussion of this aspect is being presented in this volume by J. S. D. Bacon. The enzyme activity appears to be restricted absolutely to fructosides with an unsubstituted and unchanged β-fructofuranosidic terminal. Thus a partially purified preparation of yeast invertase has been found to split

---

[63] E. Fischer, L. Kohtès, and J. Fellig, *Helv. Chim. Acta* **34**, 1132 (1951).

[64] Highly active invertase preparations of high protein content and low carbohydrate content have been obtained by M. Adams and C. S. Hudson, *J. Am. Chem. Soc.* **65**, 1359 (1943). It is suggested that in such preparations a protein has substituted for polysaccharide in the role of enzyme stabilizer.

[65] J. Edelman, *Biochem. J.* **57**, 22 (1954).

[66] J. S. D. Bacon, *Biochem. J.* **57**, 320 (1954).

[67] D. E. Koshland, Jr., and S. S. Stein, *J. Biol. Chem.* **208**, 139 (1954).

raffinose and sucrose but not α-D-fructofuranoside (isosucrose), a substituted β-fructofuranoside (melezitose), β-D-galactosides (lactose, β-phenyl-D-galactoside) and α-D-glucosides (α-methyl-D-glucoside, α-phenyl-D-glucoside, trehalose, maltose, turanose), and it hydrolyzes some β-D-glucosides only at a negligible rate, presumably because of the presence of a trace of β-glucosidase as a contaminant.[68] The ability of the yeast invertase to hydrolyze inulins and levans probably varies, depending on the polymerization degree: whereas the short-chain members of these polymer homologous series are readily hydrolyzed by yeast invertase concentrates, the action on the corresponding native high polymers is extremely slow or negligible.[68,69]

K_m values for yeast invertase acting on sucrose and raffinose have been estimated to be approximately 0.016 M and 0.24 M, respectively.[70] Variations in the value depending on the yeast source have been reported. The decomposition of the invertase-sucrose complex proceeds about two times as fast as does that of the invertase-raffinose complex.[70,71,72]

pH and Temperature Relations. The pH activity curve on sucrose shows a broad optimum range of pH 4.5 to 5.5, activity tending to fall off rather rapidly on the alkaline side and more slowly on the acid.[68] The same curve is obtained when raffinose serves as the substrate. The pH optimum shows a slight shift toward the acid side when the substrate concentration is increased.[73] K_m increases with acidity on the acid side of the optimum pH point.

The invertase-mediated hydrolysis of sucrose follows the Arrhenius equation accurately from 0 to 35°, with an energy activation of 11,000 cal./mole.[74]

Molecular Weight. On the basis of diffusion data obtained on a crude preparation, the molecular weight has been estimated to be 60,000.[75] According to Dieu,[76] a component with a molecular weight of only 3500 is the active principle of a purified invertase system in which an enzymatically inactive substance with a molecular weight of 100,000 serves as a stabilizing principle. The dimensions of invertase, as determined by Pollard *et al.*[77] with the use of fast electrons and deuterons as a probe, are

[68] M. Adams, N. K. Richtmeyer, and C. S. Hudson, *J. Am. Chem. Soc.* **65**, 1369 (1943).
[69] S. Hestrin and J. Goldblum, *Nature* **172**, 1046 (1952).
[70] R. Kuhn, *Z. physiol. Chem.* **125**, 28 (1923).
[71] J. M. Nelson and M. P. Schubert, *J. Am. Chem. Soc.* **50**, 2188 (1928).
[72] K. Josephson, *Z. physiol. Chem.* **136**, 62 (1924).
[73] K. Myrbäck and E. Willstaedt, *Arkiv Kemi* **3**, 437 (1951).
[74] I. W. Sizer, *Enzymologia* **4**, 215 (1938).
[75] E. A. Moelwyn-Hughes, *Ergeb. Enzymforsch.* **2**, 1 (1933).
[76] H. A. Dieu, *Bull. soc. chim. belg.* **55**, 306, 327 (1946).
[77] E. Pollard, W. Powell, and S. Reaume, *Proc. Natl. Acad. Sci. U. S.* **38**, 173 (1952).

48 to 83 A. and the molecular weight computed on this basis is thought to be 123,000. Dieu[76] reports the isoelectric point of invertase as 5.

Precipitants and Adsorbents. The efficiency of any given precipitant will depend largely on the nature of the "carrier" material with which the invertase in a crude preparation happens to be associated. Depending on the degree of purification accomplished, the invertase may or may not be precipitated by ammonium sulfate, picric acid, or flavianic acid. From crude preparations invertase may be adsorbed by kaolin, lead phosphate, tricalcium phosphate, alumina, bentonite, zinc sulfide, strontium hydroxide, uranyl acetate, and zinc sulfate; these have served separately or jointly as the basis of purification procedures.[78] Alcohol and acetone precipitate the enzyme. The plant protein concanavalin A, a useful precipitant for polysaccharides, is a precipitant of a partially purified invertase.[79] A variety of substances, e.g., sugars, glycerol, tartrate, and others, have been used as stabilizing agents for invertase in commercal preparations of the enzyme.[80]

Inhibitors. Salts of heavy metals, notably Cu^{++}, Hg^{++}, and Ag^+, inactivate the enzyme reversibly, their effect being annulled by dialysis or addition of metal binding agents.[81] The combination of iodine with the enzyme results in a product still retaining considerable activity.[82] Fluoride, 0.3 M, does not significantly affect the activity. Fluoride, therefore, is useful as a means of suppressing fermentation in systems to be assayed for invertase. Nitrous acid causes irreversible inactivation, presumably by way of an attack on free amino groups.[83]

Dyestuffs (fuchsin, safranine, congo red) combine loosely with invertase and inactivate it but can be displaced from the enzyme site by sucrose.[84] Reagents which combine with carbonyl groups, e.g., aniline, *p*-toluidine, and phenylhydrazine, inhibit invertase noncompetitively, presumably by way of formation of a Schiff's base.[83] The enzyme is also subject to inactivation by an oxidizing enzyme tyrosinase.[85]

[78] For a review of preparative methods see R. Weidenhagen, *in* "Die Methoden der Fermentforschung" (E. Bamann and K. Myrbäck, eds.), Vol. 2, p. 1735, Georg Thieme Verlag, Leipzig, 1941.

[79] J. B. Sumner and D. J. O'Kane, *Enzymologia* **12,** 251 (1948).

[80] C. Neuberg and I. S. Roberts, Invertase, Scientific Report Series No. 4, Sugar Research Foundation, New York, 1946.

[81] M. Jacoby, *Biochem. Z.* **181,** 194 (1927).

[82] K. Myrbäck, *Z. physiol. Chem.* **159,** 1 (1926).

[83] J. B. Sumner and K. Myrbäck, *Z. physiol. Chem.* **189,** 2 (1930).

[84] J. H. Quastel and E. D. Yates, *Enzymologia* **1,** 60 (1936–1937).

[85] I. W. Sizer, *Science* **108,** 335 (1948).

[30] Methods for Measuring Transglycosylase Activity of Invertases

By J. S. D. BACON

The name "invertase" was given to the enzymes believed to hydrolyze sucrose to a mixture of glucose and fructose, thus producing an inversion of the sign of the optical rotation from positive to negative. However, since 1950[1] it has been known that the equation

$$\text{Sucrose} + H_2O \rightarrow \text{Glucose} + \text{fructose}$$

although it expresses the over-all reaction catalyzed by various invertase preparations, does not represent correctly the state of the reaction at intermediate stages. For example, in the case of some mold invertase preparations the early stages approximate to the reaction

$$\text{Sucrose} + \text{sucrose} \rightarrow \text{Trisaccharide} + \text{glucose}$$

the trisaccharide here consisting of one glucose and two fructose residues.[2] The distinctions between various invertases, first made on the basis of substrate specificity and inhibition by monosaccharides, have been supported by the discovery that yeast and mold invertases show two distinct types of transfructosylase activity,[3] whereas honey(bee) invertase has transglucosylase activity.[4] (For the present purpose it is assumed that both hydrolytic and transferring actions are properties of the same enzyme; this has not been proved.)

The ratio of hydrolytic to transferring action by invertase preparations varies with substrate concentration, and the composition of the reaction products changes with time toward the state of complete hydrolysis. However, if assays are carried out under standard conditions and the reaction is taken to about the same stage, measurements of optical rotation, or of reducing sugar formation, may give reliable indications of the amount of enzyme present. The measurement of the transferring activity of preparations presents some difficulties; it is most conveniently made by the use of paper partition chromatography.

[1] P. H. Blanchard and N. Albon, *Arch. Biochem.* **29,** 220 (1950); J. S. D. Bacon and J. Edelman, *Arch. Biochem.* **28,** 467 (1950).

[2] F. J. Bealing and J. S. D. Bacon, *Biochem. J.* **53,** 277 (1953).

[3] J. S. D. Bacon, *Biochem. J.* **57,** 320 (1954); J. Edelman, *Biochem. J.* **57,** 22 (1954).

[4] J. W. White, Jr., and J. Maher, *Arch. Biochem. and Biophys.* **42,** 360 (1953).

Choice of Conditions for the Assay

The choice of pH and substrate concentration at which to measure the enzymic activity requires a knowledge of the influence of these two factors on the enzyme in question. Except in the cases mentioned below, a preliminary investigation of this point will be necessary.

Choice of pH. The invertase of *Saccharomyces cerevisiae* has an optimum pH of 4.7 to 4.9,[5] and similar values have been given for several species of *Schizosaccharomyces.*[6] A pH of 5.0 is probably suitable for most invertase preparations from molds: varying estimates of pH optimum have been given,[6] but in many experiments the published data show very similar activities over a wide range of pH, from about 4.0 to 6.0. In the case of *Aspergillus oryzae* it has been shown that the optimum pH for fructose transfer is higher than that for fructose liberation.[2] This might be expected to lead to a dependence of the pH optimum for reducing sugar liberation on substrate concentration.

The pH optimum of bee invertase (from honey) is given as 5.5 to 6.2,[7] and a similar value, 5.5 to 6.0, has been found for the invertase of another insect, the cockroach, *Blattella germanica.*[8] Data on preparations from other sources were summarized by Hofmann.[6]

Choice of Sucrose Concentration. With increasing sucrose concentration the initial velocity of yeast invertase action reaches a maximum between 5 and 10% (w/v) sucrose and thereafter falls in a linear manner.[9] The effect of concentrations greater than 10% (w/v) has not been recorded for other invertases, except in the case of takadiastase and extracts of *Aspergillus oryzae* mycelium;[2] here there was little fall in velocity beyond a maximum at 10 to 20% (w/v), and half-maximal velocity was reached at 0.7 to 1.0% and 2 to 3% (w/v), respectively.

Assay Methods

1. By Polarimetry. The sugars liberated by hydrolysis of sucrose are α-D-glucopyranose and β-D-fructofuranose. These undergo mutarotation, and the final equilibrium mixture has a rotation depending on the temperature, such that

$$I = S(0.417 - 0.005t)$$

where S and I are, respectively, the numerical values of the rotations due

[5] J. M. Nelson and G. Bloomfield, *J. Am. Chem. Soc.* **46**, 1025 (1924).

[6] E. Hofmann, *Biochem. Z.* **275**, 320 (1935).

[7] J. M. Nelson and D. J. Cohn, *J. Biol. Chem.* **61**, 193 (1924).

[8] V. B. Wigglesworth, *Biochem. J.* **21**, 797 (1927).

[9] J. M. Nelson and M. P. Schubert, *J. Am. Chem. Soc.* **50**, 2188 (1928).

to the original sucrose, and to the mixture of glucose and fructose produced from it, and t is the temperature in degrees centigrade.[10]

If the rate of the enzymic reaction is rapid, the mutarotation of an appreciable proportion of the α-D-glucopyranose may continue after the enzyme has been inactivated. It is convenient to add alkali (e.g., Na_2CO_3· to a final concentration of 0.02 M [11]), which inactivates the enzyme and hastens the rate of mutarotation, but also leads eventually to epimerization and destruction of the reducing sugars. The rotation should thus be read when mutarotation is completed (after about 15 minutes) but not later than 2 hours after the addition of alkali.[11] From the change in optical rotation the apparent hydrolysis of the substrate may be calculated.

2. Measurement of Reducing Sugars. Any method of measurement of reducing sugar employing neutral or alkaline conditions may be used to measure the reducing substances produced, which are chiefly glucose and fructose. The enzymic action may be stopped by the addition of alkali or of mercuric chloride, or by raising the temperature quickly to the boiling point. The last method must not be used if the pH of the solution is less than 4.5; otherwise some hydrolysis of the sucrose will occur.

3. Measurement of Transferring Activity. The transferring activity of the enzymes is shown in the presence of sucrose by the formation of a number of oligosaccharides, including disaccharides other than sucrose.[3] An accurate measurement of the amounts of all these substances may not be possible because of the difficulty of separating them from sucrose, and from each other. The measurement of one or more selected components under standard conditions may, however, serve as a measure of the total transferring activity. The following procedure, using quantitative paper chromatography, is based on that developed by Bacon and Edelman[12] and is applicable to oligosaccharides containing fructose or having a free reducing group.

Reagents

Developing solvent (Partridge[13]). n-Butanol, glacial acetic acid, and water are mixed in the proportion (by volume) of $4:1:5$, a total quantity of 2 l. being shaken in a Winchester quart bottle and left with occasional shaking for 2 to 3 days before use.[14] The upper layer is used; most of this may be decanted as it is required, without recourse to a separating funnel.

Spraying reagent. Benzidine (0.5 g.) is dissolved in a mixture of 10 ml. of glacial acetic acid, 10 ml. of 40% (w/v) trichloroacetic acid,

[10] C. S. Hudson, *J. Ind. Eng. Chem.* **2**, 143 (1910).
[11] C. S. Hudson, *J. Am. Chem. Soc.* **30**, 1564 (1908).
[12] J. S. D. Bacon and J. Edelman, *Biochem. J.* **48**, 114 (1951).
[13] S. M. Partridge, *Biochem. J.* **42**, 238 (1948).
[14] E. C. Bate-Smith, *Biochem. Soc. Symposia, Cambridge, England* No. **3**, 62 (1949).

and 80 ml. of ethanol. The solution may be treated with a little decolorizing charcoal if it is at all strongly colored.

Procedure. Groups of four 5-µl. spots of the solution to be analyzed are applied with a micrometer syringe (e.g., Burroughs Wellcome Ltd. Agla syringe) to the starting line of a chromatogram, ruled in pencil on a sheet of Whatman No. 1 filter paper, 24 × 24 in.; each group should occupy a length of about 35 mm. Guide spots are placed in the middle of 80-mm. gaps left between the groups, and an 80-mm. gap is left at either edge. In this way four groups of spots may be accommodated, comprising 80 µl. of the original sample. When the spots are dry, a process that may be hastened by the use of a commercial hair-dryer, a further set of spots may be applied. It is not advisable to place in one position more solution than is equivalent to 5 µl. of a 20% (w/v) sucrose solution, or its inversion products, nor is it desirable to repeat the applications more than four times, since this results in distortion of the guide spots on the developed chromatogram.

The sheet of paper is then developed by the descending technique with the solvent described above. If the development is to be continued for more than 1 day at 15 to 20°, teeth should be cut along the bottom edge to help the solvent to drip off.

After development for a time sufficient to separate the components to be estimated, the paper is dried in an oven with a forced draught at a temperature not greater than 80°. The acetic acid remaining, which may sometimes affect estimations of reducing sugar, may be removed by allowing papers to stand for a day at room temperature.

Vertical strips, 65 mm. wide, carrying the groups of spots for analysis, are cut out, and the rest of the paper is sprayed with a suitable reagent. For most purposes the benzidine-trichloroacetic acid reagent is suitable; after spraying, the paper is heated at 100 to 110° for a few minutes. The whole sheet is then reassembled, and from the position of the spots on the guide strips the positions of the various substances on the unsprayed strips are deduced. The corresponding areas from the four strips are then cut out and placed together in a glass-stoppered tube. Sufficient water is added to give a final volume of filtrate convenient for the sugar estimation to be performed (10 sq. cm. of paper absorb about 1 ml. of water), the tubes are heated for 30 minutes in a water bath at 80°, cooled, the papers pressed down with a glass rod, and the extract decanted through a small filter paper.

For the estimation of ketose in the filtrate a modification of the Selivanoff reaction is used; that due to Cole[15] is very satisfactory and gives almost negligible paper blank values. For estimation of reducing

[15] See J. S. D. Bacon and D. J. Bell, *Biochem. J.* **42**, 397 (1948).

power the method of Nelson-Somogyi[16] is satisfactory, although a determination of the reducing value of blank areas of the developed chromatogram is necessary.

Expression of Activity. The activity of invertase preparations is still often expressed as the "time value," i.e., that time required to bring the rotation of a sucrose solution to zero under standard conditions;[17] zero rotation corresponds to 75.93% hydrolysis at 20°. A unit, the "inverton," based on polarimetric measurements, has been proposed by Johnston et al.,[18] being the amount of enzyme inverting 5 mg. of sucrose per minute, at zero time, from a 5% solution at pH 4.6 at 25°. This unit has not been accepted generally (cf. Sumner and O'Kane,[19] Fischer and Kohtès),[20] nor have the standard conditions suggested by Weidenhagen[21] been adopted at all widely.

[16] N. Nelson, *J. Biol. Chem.* **153**, 375 (1944); M. Somogyi, *J. Biol. Chem.* **117**, 771 (1937).

[17] Cf. K. Schneider, *in* "Die Methodik der Fermente" (Oppenheimer and Pincussen, eds.), p. 770, Thieme, Leipzig, 1929.

[18] W. R. Johnston, S. Redfern, and G. E. Miller, *Ind. Eng. Chem. Anal. Ed.* **7**, 82 (1935).

[19] J. B. Sumner and D. J. O'Kane, *Enzymologia* **12**, 251 (1946–48).

[20] E. H. Fischer and L. Kohtès, *Helv. Chim. Acta* **34**, 1123 (1951).

[21] R. Weidenhagen, *Ergeb. Enzymforsch.* **1**, 201 (1932).

[31] Glucuronidases

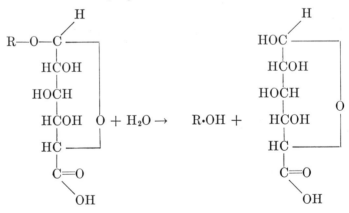

By WILLIAM H. FISHMAN and PETER BERNFELD

Assay Methods

Principle. The method described below is based on the colorimetric determination of phenolphthalein liberated from the substrate phenolphthalein mono-β-glucuronide by the action of β-glucuronidase employing the conditions of Fishman et al.,[1] which were developed from the original procedure of Talalay et al.[2] In alkaline solutions the glucuronide has only 0.18% of the light absorption of phenolphthalein at 552 mμ, causing only negligible interference.

Reagents

0.01 M phenolphthalein β-glucuronide solution (substrate), prepared according to directions given in Vol. III [9].

Acetate-acetic acid buffer, pH 5.0, ionic strength 0.07.

5% trichloroacetic acid solution.

Alkaline glycine solution. 1000 ml. of glycine-NaOH buffer [glycine 16.3 g., sodium chloride 12.65 g., 10.9 ml. of concentrated NaOH (100 g. of NaOH plus 100 ml. of H_2O) dissolved and made up to 1 l.] mixed with 250 ml. of 0.5 N NaOH solution.

Deoxyribonucleic acid solution. 3.10 g. of DNA (Krishell Laboratories, Inc.) is triturated with 20 ml. of 0.5 N NaOH until most of the product is dissolved. The pH of the solution is adjusted to 5.0 by the addition of 0.5 N acetic acid. The solution is made up to 100 ml. with distilled water, and the undissolved material is removed by centrifugation at 18,000 r.p.m. for 15 minutes. The solution is boiled under reflux for 5 minutes and stored in the cold.

Enzyme. The enzyme solution is diluted with acetate buffer so that 5 to 20 γ of phenolphthalein is liberated during the period of incubation. 0.1 ml. of DNA solution is added to 5.4 ml. of diluted enzyme, any resulting turbidity being removed by centrifugation. The addition of DNA is a requirement for the assay of purified β-glucuronidase. Assays of β-glucuronidase in crude homogenates of tissue, blood serum, vaginal fluid, etc., are done in the absence of added DNA at pH 4.5.

Procedure. 0.1 ml. of substrate and 0.4 ml. of acetate buffer are placed in a colorimeter tube. 0.5 ml. of diluted enzyme solution containing DNA is added, and the tube is stoppered and incubated in a water bath at 37° for 60 minutes. A control experiment, with 0.5 ml. of diluted enzyme to which no DNA has been added, is carried out simultaneously in a second tube. The reaction is stopped by the addition of 1.0 ml. of 5% trichloroacetic acid solution. Then 2.5 ml. of alkaline glycine solution and 1.5 ml. of H_2O are added, bringing the pH of the solution to 10.2 to 10.4, and the transmittance is determined in an Evelyn or a Klett-Summerson photoelectric colorimeter with a green filter No. 540 against a blank containing

[1] W. H. Fishman, B. Springer, and R. Brunetti, *J. Biol. Chem.* **173**, 449 (1948).
[2] P. Talalay, W. H. Fishman, and C. Huggins, *J. Biol. Chem.* **166**, 757 (1946).

no enzyme. From the observed transmittance values, the amounts of phenolphthalein liberated are read from a previously established calibration curve. The limit of error of this method is ±5%.

Definiton of Unit and Specific Activity. The unit is defined as being equivalent to 1 γ of phenolphthalein liberated in 1 hour at pH 5.0 and 37° from 0.001 M phenolphthalein mono-β-glucuronide solution in the presence of 0.028% DNA in a total volume of 1 ml. Specific activity is given by the quotient obtained by dividing enzyme activity by the protein content of the enzyme solution. Protein determinations were made with the Folin phenol reagent according to the micromethod of Lowry et al.,[3] with crystallized salt-free chymotrypsin (Worthington) as the reference protein.

Purification Procedure

The present procedure of Bernfeld et al.[4] for the purification of calf liver β-glucuronidase is based on the recent work of Bernfeld and Fishman[5] and earlier experiments of Fishman.[6] This procedure has been found to give satisfactory results on each of the many times it has been carried out in this laboratory.

Step 1. Preparation of Crude Extract. 5 lb. of fresh calf liver was transported, packed in ice, from the slaughterhouse to the laboratory. The tissue was dissected free of large blood vessels and fascia and was minced in a meat grinder. The liver pulp was then homogenized in portions of 300 ml. in a Waring blendor for 3 minutes with 200-ml. volumes of acetate buffer at pH 5.0 and ionic strength 0.07 and kept at 38° overnight. The following day the combined homogenates were centrifuged at room temperature in an International centrifuge No. 2 at 2800 to 3000 r.p.m. The supernatant solution contained the enzyme. Additional amounts of β-glucuronidase were recovered by twice washing the pulp, i.e., by suspending it in 800 ml. of acetate buffer, centrifuging the mixture, and combining all supernatant fluids. The total volume of the crude extract was 4400 ml.

Step 2. Ammonium Sulfate Precipitation. To this extract was added 1466 g. of solid ammonium sulfate, and the mixture was stirred for 30 minutes and kept overnight in the cold room (4°). The clear middle layer was then siphoned off, and the remainder centrifuged at 3000 r.p.m. for 20 minutes. The residue was suspended in 1000 ml. of 23.2% (w/v) ammonium sulfate solution, and the mixture was centrifuged again for 20

[3] O. H. Lowry, N. J. Rosebrough, A. L. Farr, and R. J. Randall, *J. Biol. Chem.* **193,** 265 (1951).

[4] P. Bernfeld, J. S. Nisselbaum, and W. H. Fishman, *J. Biol. Chem.* **202,** 763 (1953).

[5] P. Bernfeld and W. H. Fishman, *J. Biol. Chem.* **202,** 757 (1953).

[6] W. H. Fishman, *J. Biol. Chem.* **127,** 367 (1939).

minutes. The residue was washed once more by being suspended in 1000 ml. of 23.2% ammonium sulfate solution, and the mixture centrifuged for 30 minutes; all supernatant fluids were discarded. Then 1000 ml. of acetate buffer (pH 5.0, μ 0.07) was added to the combined residues to dissolve most of the solids, the mixture was incubated for 3 hours at 35°, and centrifuged for 15 minutes; the residue was triturated with 1000 ml. of acetate buffer, centrifuged, and the two supernatant solutions combined. The total volume was 1740 ml.

Step 3. First Alkaline Ammonium Sulfate Precipitation. This step and all subsequent ones were carried out at low temperatures, i.e., in an ice-water bath or in a refrigerated centrifuge at 0°. All precipitants were added dropwise from a funnel (not faster than 2 drops per second), with several dropping funnels being used simultaneously if necessary, and with slow mechanical stirring of the enzyme solution (not over 80 r.p.m.).

A saturated ammonium sulfate solution at pH 8.7 was prepared by addition of 360 ml. of concentrated ammonia water (density 0.9) to a warm solution (45°) of 9250 g. of ammonium sulfate in 12 l. of H_2O prepared at 60°. It was allowed to cool at 4°. In the following paragraphs, this solution is referred to as alkaline ammonium sulfate solution, AAS.

To 1.74 l. of the product of step 3, 100 ml. of alkaline ammonium sulfate solution was added, followed by sufficient 2 N aqueous ammonia (ca. 100 ml.) to adjust the solution to pH 8.73 $\pm$ 0.03. Then 1.632 l. of AAS was added to bring the ammonium sulfate concentration to 48.5% saturation. The precipitate was collected by centrifugation, dissolved in 500 ml. of acetate buffer, and saved for further purification.

Step 4. Alkaline Ammonium Sulfate Fractionation. To 740 ml. of the previous product, 50 ml. of AAS followed by 43 ml. of 2 N ammonium hydroxide was added to adjust the pH to 8.73. 211 ml. of AAS was then added, bringing the $(NH_4)_2SO_4$ concentration to 25% saturation. The precipitate was removed by centrifugation and discarded. To the supernatant, 260 ml. of AAS was added to adjust it to 40% saturation. The precipitate was collected by centrifugation and dissolved in 100 ml. of acetate buffer (pH 5.0). The supernatant was discarded.

Step 5. Ion Exchange. A glass column, 19 inches in length, $1\frac{1}{4}$ inches inside diameter, fitted with a ground-glass stopcock, was filled to a depth of $10\frac{1}{2}$ inches with citrate-charged Amberlite IRA-400. The resin was held in place by cotton plugs of about 1 inch in diameter. Before use the column was washed with 50 ml. of distilled H_2O. The effluent was collected at a rate of about 1 drop per second. When the level of the wash water had gone about half-way through the cotton plug at the top of the column, the passage of the enzyme solution was begun. The column was then washed with three portions of 3 ml. of distilled H_2O. The effluent was collected in 10-ml. portions, and the first and last colorless fractions were discarded. All fractions having a visible yellow color were combined and saved for further purification. If the enzyme solution gave a precipitate with barium ions, the ion exchange was repeated.

Step 6. First Methanol Fractionation. The product of the preceding ion-exchange step (184 ml.) was diluted with 500 ml. of 0.15 M secondary sodium citrate, pH 5.0, and 337 ml. of methanol was added, bringing the methanol concentration to 33 vol. %. The precipitate, containing most of the pigment, was removed by centrifugation and discarded. To the supernatant 223 ml. of methanol was added to adjust the concentration to 45%, and the precipitate was again centrifuged off and discarded. Although the first and second fractions were discarded, a precipitation of both these fractions in a single operation led to inferior potency of the third fraction. Finally, 605 ml. of methanol was added to bring the concentration to 63 vol. %. The mixture was centrifuged, the precipitate taken up in 100 ml. of secondary sodium citrate solution, and the supernatant discarded. A large amount of protein failed to dissolve in the citrate solution, but there was no necessity for removing it before proceeding with the purification, since it was removed with the first precipitate from the second methanol treatment.

Step 7. Second Methanol Fractionation. To 96 ml. of the product of step 6, 32 ml. of methanol was added to bring the methanol concentration to 25%. The precipitate was centrifuged off and discarded. 15.5 ml. of methanol was added to the supernatant to adjust the concentration to 33%, and the very light precipitate was discarded. Then 48.5 ml. of methanol was added to the supernatant, bringing the concentration to 50%. The mixture was centrifuged, and the precipitate was triturated with 6 ml. of 0.15 M secondary sodium citrate solution. A large amount of insoluble protein was removed by centrifugation and discarded. The supernatant solution contained about 12% of the β-glucuronidase activity present in the crude extract of liver and about 0.04% of the protein. Therefore, this final product represented a purification of some 250-fold over the initial extract.

As the extent of activation of a β-glucuronidase solution by DNA depends upon the enzyme concentration, the enzyme dilutions in the final digest during the activity determinations are indicated as follows: step 1, 1:200; step 2, 1:500; step 3, 1:2000; step 4, 1:5000; step 5, 1:2000; step 6, 1:2000; step 7, 1:15,000.

Purity of Calf Liver β-Glucuronidase. Electrophoresis of a preparation (purity, 107,000 γ of phenolphthalein per milligram of enzyme, step 7) at pH 3.6 in an electric field of 7.0 volts/cm. showed one main component which migrated toward the cathode and which amounted to 85 to

90% of the total protein present. Electrophoretic experiments at other pH values and with preparations of slightly lower purity all yielded the same value of enzyme activity divided by protein concentration for the pure enzyme, i.e., 110,000 to 120,000.

Properties[7]

Specificity.[7] β-Glucuronidase catalyzes the hydrolysis of a large variety of conjugated glucuronides such as the glucuronides of orcinol, phloroglucinol, menthol, phenol, borneol, β-naphthol, phenolphthalein, pregnanediol, 8-hydroxyquinoline, estriol, corticosteroids, and chlorophenol. It does not hydrolyze α- or β-glucosides.

Activators and Inhibitors. Dilute solutions of purified β-glucuronidase exhibit enhanced activity in the presence of deoxyribonucleic acid.[8] Albumin and, to a lesser extent, a number of low molecular weight diamines[9] behave as activators of β-glucuronidase.

Saccharo-1,4-lactone is an effective inhibitor of β-glucuronidase.[10] Glucuronic acid is a less potent inhibitor.[11] The existence of tissue and serum inhibitors of β-glucuronidase has been reported.[12,13]

Heavy metal ions such as Hg^{++} inhibit β-glucuronidase activity.

Effect of pH. The calf liver β-glucuronidase preparation described exhibits a maximal activity at pH 5.0. No peaks of activity other than this one have been observed in this laboratory.[14]

Isoelectric Point. From electrophoretic data, this is estimated to be between pH 5.0 and 6.0.[14]

Dissociation of β-Glucuronidase. The activity of both purified calf spleen and liver β-glucuronidase is not proportional to the enzyme concentration, especially at high enzyme dilutions. The ratio of enzyme activity to protein concentration when plotted against the logarithm of the protein concentration yields a typical S-shaped dissociation curve which coincides with a theoretical curve derived from the mass law.[8] The addition of a variety of substances to such dilute enzyme solutions restores the original activity seen before dilution. These include a filtered boiled

[7] The properties of β-glucuronidase prepared from other sources have been summarized previously; see W. H. Fishman, *in* "The Enzymes" (Sumner and Myrbäck, eds.), Vol. 1, Academic Press, New York, 1950.

[8] P. Bernfeld and W. H. Fishman, *Science* **112**, 653 (1950).

[9] P. Bernfeld, H. Bernfeld, and W. H. Fishman, *Federation Proc.* **12**, 177 (1953).

[10] G. A. Levvy, *Biochem. J.* **52**, 464 (1952).

[11] B. Spencer and R. T. Williams, *Biochem. J.* **48**, 537 (1951).

[12] W. H. Fishman, K. I. Altman, and B. Springer, *Federation Proc.* **7**, 154 (1948).

[13] P. G. Walker and G. A. Levvy, *Biochem. J.* **54**, 56 (1953).

[14] J. S. Nisselbaum, Ph. D. Thesis, The Purification and Properties of β-Glucuronidase from Calf Liver, Tufts College Medical School, Boston, 1953.

SUMMARY OF PURIFICATION OF CALF LIVER β-GLUCURONIDASE

Step of purification	Limit of fraction-ation[a]	Enzyme activity per ml. solution		Purity[b,c]		Total activity[b] × 10⁻³	
		Without addition of DNA	With DNA added	Main fraction	Dis-carded fraction	Main fraction	Dis-carded fraction
Liver homogenate		4,850	4,850	6.5[d]		15,400	
1		2,700	3,000	240		13,200	
2	0–60	3,200	4,850	2,500		8,450	
3	0–48.5	3,600	11,000	4,850		8,150	
4	0–25	4,400	9,000		2,550		1,650
	25–40	27,000	50,000	8,350		5,200	
5		16,000	28,000	10,000		5,200	
6	0–33	2,000	5,200		1,900		520
	33–45	3,500	10,000		16,000		980
	45–63	10,500	24,000	37,500		2,300	
7	0–25	55,000	95,000		29,500		760
	25–33	3,200	4,050		22,500		22
	33–50	140,000	245,000	60,000		1,600	
	50–67	25,000	59,000		45,500		515[e]

[a] Indicating lower and upper concentrations of precipitant at which each fraction is separated, expressed in per cent saturation of ammonium sulfate in steps 3 and 4, and in volume per cent of methanol in steps 6 and 7.

[b] Based on the enzyme activity with DNA added.

[c] Expressed as the quotient obtained by dividing enzyme activity by the protein concentration (mg./ml.) of the enzyme solution.

[d] Enzyme activity per milligram of liver.

[e] The over-all recovery of enzyme activity of 125% in step 7 might be explained by the removal of some inhibitor during this step.

purified calf liver β-glucuronidase solution, DNA, albumin, and diamino compounds.[9]

Energy of Activation. In the hydrolysis of phenolphthalein glucuronide by purified calf liver β-glucuronidase, the energy of activation was calculated to be 14,000 cal.[14]

Stability. The enzyme in crude extracts or in concentrated partially purified form is markedly stable in the cold.

Effect of Substrate Concentration. The initial velocity of hydrolysis bears a linear relation to substrate concentration at low substrate concentrations, then reaches a plateau, and at high substrate concentrations an inhibition is observed. The calf liver β-glucuronidase described has a K_m value of 1.43×10^{-4} M for the hydrolysis of phenolphthalein glucuronide.[14] The results obtained with other glucuronides have been summarized elsewhere.[7]

Turnover Number. A figure was calculated of 860 moles of substrate split per mole of enzyme (molecular weight assumed to be 100,000) per minute at 37 and pH 5.0.[13]

β-Glucuronidase Preparations Reported by Others. In 1946 Graham reported a procedure for purifying ox spleen β-glucuronidase using ammonium sulfate fractionation.[15] Smith and Mills[16] prepared ox liver β-glucuronidase using metal ions and acetone to fractionate the enzyme. Sarker and Sumner[17] used fractional precipitation with dioxane at low temperature, calcium phosphate adsorption, and ammonium sulfate fractionation to purify beef liver β-glucuronidase.

[15] A. F. Graham, *Biochem. J.* **40**, 603 (1946).
[16] E. E. B. Smith and G. T. Mills, *Biochem. J.* **54**, 164 (1953).
[17] N. K. Sarker and J. B. Sumner, *Arch. Biochem.* **27**, 453 (1950).

[32] Yeast Hexokinase

ATP + Hexose → Hexose-6-phosphate + ADP

By MARGARET R. McDONALD

Assay Method

Principle. The method is based on the fact that for each mole of phosphate transferred from ATP to glucose one acid equivalent is liberated.[1] Under the conditions described below, the rate of acid formation, as determined by direct titration,[2] is proportional to the hexokinase concentration for the first 75% of the ultimate amount of acid formed at the end of the reaction. The amount of acid formed can also be determined manometrically[1] or spectrophotometrically.[3]

Reagents

0.01 M NaOH containing 6.25 ml. of 0.1% phenol red per liter of solution.
0.01 M HCl.

[1] S. P. Colowick and H. M. Kalckar, *J. Biol. Chem.* **137**, 789 (1941); **148**, 117 (1943).
[2] M. Kunitz and M. R. McDonald, *J. Gen. Physiol.* **29**, 393 (1946).
[3] J. Wajzer, *Compt. rend.* **229**, 1270 (1949).

Solution A. 20 ml. of 0.045 M ATP[4] (prepared from dibarium salt) + 3 ml. of 0.05 M MgCl₂. Stored with thymol at 0 to 5°.

Solution B. 8 ml. of 5% glucose + 3 ml. of 0.5 M Sørensen's phosphate buffer (pH 7.5) + H₂O to 25 ml. Stored with thymol at 0 to 5°.

Solution C. 0.0025% phenol red.

Color standard. 2.5 ml. of 0.1 M phosphate buffer (pH 7.5) + 0.5 ml. of 0.0025% phenol red.

Reaction mixture. 0.5 ml. of solution A + 0.5 ml. of solution B + 0.5 ml. of solution C, mixed in 1.5 × 12-cm. test tubes.

Procedure. Cool the test tube containing the reaction mixture to 5°, then add 0.5 ml. of ice-cold aqueous hexokinase solution. Titrate the mixture immediately with 0.01 M NaOH or HCl from a microburet to the color of the standard. Leave the adjusted mixture at 5° for 30 minutes, then retitrate with 0.01 M NaOH to the color of the standard.

Definition of Unit and Specific Activity. One unit of hexokinase is defined as the amount of enzyme which catalyzes the formation of 1 × 10^{-8} acid equivalent per minute at 5° and pH 7.5 in the standard reaction mixture. Specific activity is expressed as hexokinase units per milligram of protein. Protein is determined by the method of Lowry *et al.*[5]

Application of Assay Method to Crude Preparations. The method is applicable to crude preparations of yeast hexokinase.

Purification Procedure

Three procedures have been described for the preparation of pure yeast hexokinase. The one given here is essentially that of Kunitz and McDonald.[2] Those of Berger *et al.*[6] and of Bailey and Webb[7] require repeated centrifugation of very large volumes of material, a tedious and time-consuming procedure for which facilities are often not available. The yields in all three methods are poor (about 1%); approximately 40 mg. of crystalline hexokinase is obtained from 25 lb. of yeast.

All operations, except where it is definitely stated otherwise, are performed between 0 and 5°. All filtrations, unless otherwise specified, are done with suction on Büchner funnels, with Eaton-Dikeman paper No. 612. Determinations of pH are made on a test plate by mixing 1 drop of an

[4] ATP is obtainable commercially from various sources. For methods of isolation and purification of this compound, see Vol. III [117].

[5] O. H. Lowry, N. J. Rosebrough, A. L. Farr, and R. J. Randall, *J. Biol. Chem.* **193**, 265 (1951); also see Vol. III [76].

[6] L. Berger, M. W. Slein, S. P. Colowick, and C. F. Cori, *J. Gen Physiol.* **29**, 379 (1946).

[7] K. Bailey and E. C. Webb, *Biochem. J.* **42**, 60 (1948).

appropriate indicator with 1 drop of the solution being tested and comparing the color with that of a standard prepared by mixing 1 drop of indicator with 1 drop of standard buffer. This method gives only apparent pH values but is adequate for reproducing the necessary conditions. The saturated ammonium sulfate solution (760 g./l. of H_2O) is prepared at 20 to 25°.

It is advantageous to carry out the purification process as far as step 4 with 25-lb. lots of yeast, storing fractions 0.63, 0.72, and 0.73 at 0 to 5° until an accumulated stock of approximately 100 g. of filter cake has been obtained for further treatment. The amount of hexokinase obtained varies considerably with individual lots of yeast; some even fail to yield any. The total yield can be increased by reworking the various inert crystals and Super-Cel residues of step 3, each according to its place in the general scheme.

Step 1. Plasmolysis and Extraction.[8] 25 lb. of fragmented fresh Fleischmann's baker's yeast[9] is macerated with 6 l. of toluene in a large aluminum pan placed in a water bath of about 45° until the suspension reaches 37°. At this point the yeast liquefies quickly and begins to "work," with rapid liberation of CO_2 and marked increase in volume. The mixture is left at room temperature for 2 to 3 hours, then cooled to 10°, diluted with 12 l. of cold (0 to 5°) H_2O, and left for 18 hours at 0 to 5°. A layer of an emulsion of toluene[10] and yeast stromata forms above the yeast-water suspension. The latter is siphoned off, and to it is added 100 g. of Hyflo Super-Cel[11] for each liter of suspension. The mixture is filtered with suction on four 32-cm. Büchner funnels. The residue on each funnel is washed once with 1 l. of cold H_2O.

Step 2. Fractionation with Ammonium Sulfate. The clear filtrate and washings are combined, and 313.6 g. of $(NH_4)_2SO_4$, 10 g. of standard Super-Cel, and 10 g. of Filter-Cel are added for each liter of solution (final concentration of $(NH_4)_2SO_4$, 0.5 saturation). The suspension is filtered, and the residue discarded. The clear filtrate is brought to 0.65 saturation by the addition of 99.3 g. of $(NH_4)_2SO_4$ per liter, and the resulting suspension is filtered. The filter cake, termed 0.6 fraction, is retained for the isolation of "yeast yellow protein" and hexokinase. The filtrate is brought to 0.7 saturation by the addition of 33.8 g. of $(NH_4)_2SO_4$ per liter, left at 0 to 5° for 18 hours, then filtered. The filter cake, termed the

[8] Based on procedure of O. Meyerhof, *Biochem. Z.* **183**, 176 (1927).
[9] Yeasts vary in their resistance to autolysis. If English yeast is used it may be necessary to allow longer times for autolysis.[7]
[10] The toluene can be partly recovered by filtering the toluene-stromata emulsion with the aid of 10% Hyflo Super-Cel.
[11] Supplied by Johns-Manville, 22 East 40th Street, New York, N. Y.

0.7 fraction, is retained for the isolation of "yeast proteins, Nos. 2 and 3" and hexokinase. The filtrate is discarded.

Step 3. Removal of "Inert" Crystalline Proteins. (a) "Yeast Protein, No. 2."[12] Each gram[13] of the 0.7 fraction is dissolved in 2 ml. of cold H_2O, the pH of the solution is adjusted to 7.4 with 1 M NaOH, and saturated $(NH_4)_2SO_4$ is added slowly with constant stirring until the solution becomes faintly turbid (approximately 2.5 ml./g. of filter cake). The solution is kept in an ice bath during this operation, after which it is left at 0 to 5° for 6 to 8 days. A gel of very fine crystals (needles) of "yeast protein, No. 2" gradually forms. This is filtered with the aid of 5 g. of standard Super-Cel per 100 ml. of suspension. The filtrate is brought to 0.85 saturation by the slow addition of 21.5 g. of $(NH_4)_2SO_4$ per 100 ml., and the resulting precipitate is filtered. The filtrate is discarded; the filter cake, termed the 0.71 fraction, is retained for the isolation of "yeast protein, No. 3" and hexokinase.

(b) "Yeast Protein, No. 3." Each gram of filter cake of fraction 0.71 is dissolved in 0.5 ml. of cold H_2O. Saturated $(NH_4)_2SO_4$ is added slowly with stirring until the solution becomes faintly turbid, then 1 M NaOH to pH 7.5. The solution is left at 0 to 5° for 5 to 6 days. Prismatic crystals of "yeast protein, No. 3" gradually form. The suspension is centrifuged. The supernatant is diluted with 2 vol. of 0.65 saturated $(NH_4)_2SO_4$ of pH 7.2 (containing 2 ml. of 5 M NaOH per liter); then 10 g. of standard Super-Cel and 43 ml. of saturated $(NH_4)_2SO_4$ are added for each 100 ml. of original supernatant, and the resulting suspension is filtered. The filter cake is resuspended twice in a volume of 0.65 saturated $(NH_4)_2SO_4$ of pH 7.2 equal to that of the first filtrate, and refiltered each time. The precipitate is discarded, and the combined filtrates are brought to 0.85 saturation by the addition of 143 g. of $(NH_4)_2SO_4$ per liter. The resulting precipitate is filtered, the filtrate discarded, and the filter cake, termed fraction 0.72,[14] used for the isolation of hexokinase.

(c) "Yeast Yellow Protein." Each gram of the 0.6 fraction is dissolved in 1 ml. of cold H_2O, and 1 ml. of saturated $(NH_4)_2SO_4$ is added; a slight precipitate usually forms. The turbid solution is left at 0 to 5° for 20 to 24 hours. It is then centrifuged, and the residue discarded. The supernatant is titrated with 0.5 M NaOH to pH 7.2 and stored at 0 to 5° for 5 to 7 days. A gelatinous precipitate forms. This is removed by filtration with the aid of 10% Hyflo Super-Cel. The filtrate is brought to 0.85

---

[12] This protein has been identified as D-glyceraldehyde 3-phosphate dehydrogenase; G. W. Rafter and E. G. Krebs, *Arch. Biochem.* **29**, 233 (1950).

[13] This expression is used to denote the relative volume of solvent in which the precipitate is dissolved. It does not mean that each gram is processed separately.

[14] If the specific activity of fraction 0.72 is less than 50, step 3b should be repeated before proceeding to step 4.

saturation by the slow addition of 233 ml. of saturated $(NH_4)_2SO_4$ per 100 ml. of filtrate. The resulting suspension is filtered, and the filtrate discarded. Each gram of filter cake is dissolved in 0.5 ml. of H_2O, and saturated $(NH_4)_2SO_4$ is added until the solution becomes faintly turbid. It is then titrated with 1 M NaOH to pH 7.5 and left at 0 to 5° for 1 to 2 weeks. Crystals of "yeast yellow protein" gradually form. The suspension is then centrifuged. The supernatant is diluted with 2 vol. of cold 0.65 saturated $(NH_4)_2SO_4$ of pH 7.2; 10 g. of standard Super-Cel and 43 ml. of saturated $(NH_4)_2SO_4$ are added for each 100 ml. of the original supernatant. The suspension is filtered with suction on Whatman No. 3 paper. The precipitate is resuspended twice in a volume of 0.65 saturated $(NH_4)_2$-SO_4 of pH 7.2 equal to that of the first filtrate, being refiltered each time as described above. The combined filtrates are brought to 0.85 saturation by the addition of 143 g. of $(NH_4)_2SO_4$ per liter and filtered. The filtrate is discarded. The filter cake, termed fraction 0.62, is treated as described in (b) to yield fraction 0.63. Some crystals of "yeast protein, No. 3" are generally formed during this step.

Step 4. Fractional Precipitation with Alcohol.[15] Each gram of filter cake of fraction 0.72, 0.73, or 0.63 is dissolved in 1 ml. of 1% glucose. The solution is then dialyzed[16] for 18 hours against slowly running 1% glucose. The dialyzed solution is diluted with cold 1% glucose to a volume in milliliters equal to 4.75 times the original weight in grams of the filter cake in step 4, and 1 ml. of 1 M acetate buffer, pH 5.4, is added per 19 ml. of solution. The mixture is cooled in a salt-ice bath to approximately $-2°$, and 35.7 ml. of cold 95% ethanol is added slowly with continuous stirring for each 100 ml. of solution (final concentration of ethanol, 25%). The resulting suspension is centrifuged, and the precipitate discarded. The supernatant is again cooled to $-2°$, and 55.5 ml. of 95% ethanol is added slowly with continuous stirring for each 100 ml. of supernatant (final concentration of ethanol, 50%). The suspension is centrifuged, and the supernatant discarded. The precipitate is resuspended in a volume of cold 1% glucose equal to twice the weight of the filter cake before dialysis, and recentrifuged. The residue is discarded. 5 ml. of 1 M acetate buffer, pH 5.4, is added for each 100 ml. of supernatant, and the ethanol fractionation repeated. (This repetition is unnecessary if the specific activity of the material before dialysis is above 100. In that case the residue left after centrifugation of the 50% alcohol fraction is suspended in a volume of cold H_2O equal to five times the weight of the filter cake before dialysis. The suspension is then centrifuged, and the supernatant is treated with $(NH_4)_2SO_4$ as described in step 5).

[15] Based on Berger *et al.*'s modification[6] of Meyerhof's procedure.[8]
[16] M. Kunitz and H. S. Simms, *J. Gen. Physiol.* **11**, 641 (1928).

Step 5. Crystallization. The final residue left after centrifugation of the 50% ethanol mixture is resuspended in a volume of cold H_2O equal to twice the weight of filter cake before dialysis in step 4. The suspension is recentrifuged, and the residue discarded. The supernatant is brought to 0.9 saturation by the addition of 66 g. of $(NH_4)_2SO_4$ per 100 ml., and the suspension is filtered. The filtration is slow, usually taking 1 to 2 days for completion. The filtrate is discarded. Each gram of filter cake is dissolved in 1 ml. of 0.1 M phosphate buffer, pH 7.0, at approximately 3°, and 1 ml. of saturated $(NH_4)_2SO_4$ is added slowly. If a heavy precipitate forms, a few drops of the phosphate buffer are added to incipient clearing. The solution is then centrifuged, and the clear supernatant left at approximately 5°. Crystals in the form of long prisms or fine needles gradually appear. (If crystals do not form within 7 to 10 days, the solution should be brought to 0.85 saturation with saturated $(NH_4)_2SO_4$ and filtered with suction. The filter cake is then reworked through steps 3b to 6.)

Step 6. Recrystallization. The suspension of crystals is centrifuged after 7 to 10 days. The residue is dissolved in a minimum amount of cold 0.1 M phosphate buffer, pH 7.0, and a volume of saturated $(NH_4)_2SO_4$ equal to 1.4 vol. of the buffer used is added. The solution is left at 5°. Crystallization is completed in 2 to 3 days. The crystals are then centrifuged or filtered with suction. They are relatively stable when stored at 0 to 5° in the form of crystalline filter cake.

Properties

Specificity. Crystalline yeast hexokinase catalyzes[2] the phosphorylation of D-glucose, D-fructose,[17] and D-mannose but not of L-arabinose, D-xylose, L-rhamnose, D-galactose, sucrose, D-lactose, maltose, trehalose, or raffinose. It also catalyzes the conversion of D-glucosamine to D-glucosamine-6-phosphate[18,19] but does not catalyze the phosphorylation of N-acetyl-D-glucosamine.[18] 2-Deoxy-D-glucose also serves as a substrate.[20] Experiments with sugar mixtures indicate that the sugars compete for a single enzyme.[6,21]

Kinetics. The following dissociation constants have been obtained for the hexokinase-hexose-ATP system:[21] mannose, 1.0×10^{-4}; glucose,[22] 1.5×10^{-4}; fructose,[22] 1.5×10^{-3}; ATP (glucose), 9.5×10^{-5}; and ATP

[17] Only the fructofuranose form is attacked: A. Gottschalk, *Biochem. J.* **41**, 478 (1947).
[18] D. H. Brown, *Biochim. et Biophys. Acta* **7**, 487 (1951).
[19] P. T. Grant and C. Long, *Biochem. J.* **50**, xx (1952).
[20] F. B. Cramer and G. E. Woodward, *J. Franklin Inst.* **253**, 354 (1952).
[21] M. W. Slein, G. T. Cori, and C. F. Cori, *J. Biol. Chem.* **186**, 763 (1950).
[22] R. van Heyningen[23] obtained 1×10^{-3} M as the dissociation constant for the glucose-enzyme complex and 2×10^{-3} M for the fructose-enzyme one; Wajzer[27] reports 5×10^{-4} M as the dissociation constant for both the glucose and fructose complexes.

(fructose), 4.2×10^{-5}. The turnover number[6] for hexokinase, when saturated with glucose, is 13,000 moles per 10^5 g. of protein per minute at pH 7.5 and 30° (about 13,000 molecules per molecule of enzyme per minute). This value is doubled when the enzyme is saturated with fructose[23] and halved with mannose as the substrate.[6] The turnover number with D-glucosamine[18] as substrate is approximately 12,000 moles per 10^5 g. of protein per minute at pH 7.8 and 30°. The temperature coefficient (Q_{10}) between 0 and 30° is approximately 2.[2,6]

Activators and Inhibitors. Crystalline hexokinase requires magnesium ions for its catalytic action, the dissociation constant for the magnesium-enzyme complex being 2.6×10^{-3} [6] Calcium ions do not affect the activity of the enzyme either in the presence or in the absence of magnesium ions.[7] Fluoride ions in concentrations as high as 0.125 M do not inhibit hexokinase with magnesium and phosphate concentrations of 6.5×10^{-3} and 1×10^{-3} M, respectively.[6] It does not require inorganic phosphate ions.[2]

Sorbose-1-phosphate,[24] suramin,[25] and sodium tripolyphosphate[26] inhibit hexokinase. Inhibition by sodium tripolyphosphate can be overcome by adding excess ATP or magnesium ions to the digestion mixture. Glucose-6-phosphate has been reported as both inhibitory[27] and noninhibitory.[28,29]

Hexokinase is irreversibly inactivated by treatment with mustard gas or the "nitrogen" mustards.[7,23] It is inhibited by —SH reagents and reactivated by thiols.[7,23] Berger *et al.*[6] found no evidence, however, for essential —SH groups in the enzyme.

The enzyme is inactivated by trypsin[2] (but not chymotrypsin) and by the proteolytic enzymes present in crude yeast extracts. Addition of glucose markedly prevents this type of inactivation. The pure enzyme undergoes instantaneous loss of a portion of its activity (up to 50%) in very dilute solutions. This type of inactivation, not progressive with time, can be prevented by dilution in the presence of various proteins, among which insulin is particularly effective.[6]

Physicochemical Properties.[2] Crystalline hexokinase is a protein of the albumin type with the following elementary analysis expressed in per cent dry weight: C, 52.16; H, 7.08; N, 15.62; P, 0.11; S, 0.91; ash, 0.36.

[23] Quoted by M. Dixon and D. M. Needham, *Nature* **158**, 432 (1946).

[24] H. A. Lardy, *in* "Respiratory Enzymes," rev. ed., p. 184, Burgess, Minneapolis, 1949.

[25] E. D. Wills and A. Wormall, *Biochem. J.* **47**, 158 (1950).

[26] W. Vishniac, *Arch. Biochem.* **26**, 167 (1950).

[27] J. Wajzer, *Compt. rend.* **236**, 2116 (1953).

[28] H. Weil-Malherbe and A. D. Bone, *Biochem. J.* **49**, 339 (1951).

[29] A. Sols and R. K. Crane, *Federation Proc.* **12**, 271 (1953).

Its isoelectric point is between pH 4.5 and 4.8. No evidence of a special prosthetic group is obtained from its ultraviolet absorption spectrum,[6] which shows a maximum molecular extinction coefficient of approximately 125,000 near 2780 A., and a minimum of 66,000 at 2500 A. Solubility, electrophoretic, and ultracentrifugation studies indicate that it is homogeneous. It is most stable at pH 5. Its diffusion constant (D_{20}°) at 1° in acetate buffer pH 5.5 is 2.9×10^{-7} sq. cm./sec.; its sedimentation constant (S_{20}°) under the same conditions is 3.1×10^{-13} cm./sec./dyne/g.; its molecular weight at pH 5.5, calculated from D_{20}° and S_{20}° (assuming a specific volume of 0.740), is 96,600.

The enzyme has a broad optimum for activity between pH 8 and 9.[23]

SUMMARY OF PURIFICATION PROCEDURE[a]

Fraction	Total units,[b] $\times 10^3$	Specific activity, units/mg. protein	Yield, %
1. Aqueous extract plus washings	6000	~20	100
2. Ammonium sulfate fractionation			
Fraction 0.6	2500	28	42
Fraction 0.7	1500	28	25
3. Removal of "inert" crystalline proteins			
Fraction 0.63	250	135	4
Fraction 0.72	450	85	8
4. Ethanol fractionation			
Fraction 0.63	100	250	1.7
Fraction 0.72	120	300	2.0
5. Crystallization			
Crystals, combined 0.63 and 0.72 fraction	150	900	2.5
Mother liquor	33	100	0.5
6. Recrystallization			
2 times crystallized	90	1100	1.5
3 times crystallized	54	1400	0.9
4 times crystallized	36	1350	0.6
5 times crystallized	24	1440	0.4

[a] M. Kunitz and M. R. McDonald, *J. Gen. Physiol.* **29**, 403 (1946).
[b] Based on 25 lb. of yeast.

[33] Animal Tissue Hexokinases

(Soluble and Particulate Forms)

Hexose + ATP → Hexose-6-P + ADP

By Robert K. Crane and Alberto Sols

Animal tissue hexokinases in soluble[1] and in particulate[2] forms have been described. The properties of these enzymes have been found to vary with the tissue of origin, and some tissues contain more than one hexokinase. None of the animal hexokinases has been highly purified, although brain hexokinase has been obtained in a state free of interfering enzymes. Procedures for making hexokinase preparations from several tissues are described.

Assay Methods

Principle. The hexokinase reaction, *at neutral pH*, may be described by the equation hexose + ATP → hexose-6-P + ADP + H^+. The methods[3] which have been used to follow this reaction are (1) hexose disappearance by analysis[4] after removal of hexose-6-P by precipitation with barium,[5] (2) acid-labile phosphorus[6] disappearance, (3) hexose-6-P formation by (a) specific enzymatic micromethods[1,7a,b] or (b) stable-P estimation,[6] and (4) acid production by (a) manometric,[8] (b) titrimetric,[9] or (c) photometric[10] procedures. Each of these methods possesses distinct advantages for specific purposes, as can best be determined by reference to the literature. Modifications of methods 3b and 4c which have been found useful are described below. Methods 1, 2, or 3 can be used with the following incubation procedure. The procedures in method 4 require special incubation media.

---

[1] M. W. Slein, G. T. Cori, and C. F. Cori, *J. Biol. Chem.* **186**, 763 (1950).

[2] R. K. Crane and A. Sols, *J. Biol. Chem.* **203**, 273 (1953).

[3] See also S. P. Colowick, *in* "The Enzymes" (Sumner and Myrbäck, eds.), Vol. 2, Part 1, p. 114, Academic Press, New York, 1951.

[4] M. Somogyi, *J. Biol. Chem.* **195**, 19 (1952); see Vol. III [12].

[5] M. Somogyi, *J. Biol. Chem.* **160**, 69 (1945).

[6] See Vol. III [147].

[7] (a) E. C. Slater, *Biochem. J.* **53**, 521 (1953); (b) E. Racker, *J. Biol. Chem.* **167**, 843 (1947).

[8] S. P. Colowick and H. M. Kalckar, *J. Biol. Chem.* **148**, 127 (1943).

[9] M. Kunitz and M. R. McDonald, *J. Gen. Physiol.* **29**, 393 (1946); see Vol. I [32].

[10] J. Wajzer, *Compt. rend.* **229**, 1270 (1949).

Reagents[11]

ATP-Mg mixture. Solutions of ATP and MgCl$_2$ of appropriate strength are mixed, neutralized to pH 7.0, and diluted to the final concentrations of 0.075 M ATP and 0.04 M MgCl$_2$. This solution is stable at $-15°$ for limited periods.

Buffer mixture. Solutions of histidine, Tris, EDTA,[12] and MgCl$_2$[13] of appropriate strength are mixed, neutralized to pH 7.0, and diluted to the respective concentrations of 0.1, 0.1, 0.01, and 0.01 M.

Glucose solution, 0.01 M.

Enzyme. The stock enzyme is diluted with 0.002 M EDTA, 0.002 M MgCl$_2$, pH 7.0, to obtain 5 to 25 units of enzyme per milliliter. (See definition below.)

Procedure. 0.2 ml. each of the ATP-Mg mixture, the buffer mixture, and glucose and 0.3 ml. of water are combined and warmed to 30°. 0.1 ml. of diluted enzyme is mixed in, and the mixture is incubated at 30° for 15 minutes. The reaction is stopped by the addition of the reagent appropriate to the analytical method chosen.[14] The analysis of the incubated mixture is compared to a similar analysis of a mixture stopped at zero time.

Definition of Unit and Specific Activity. One unit of enzyme is that amount which can catalyze the phosphorylation of 1 micromole of *glucose* in 15 minutes at 30° under otherwise optimal conditions. Specific activity is expressed as units per milligram of protein. Protein is determined by the method of Lowry *et al.*[15]

Modified Procedure for Method 3b. Incubation is carried out as described above. The reaction is terminated by the addition of 4 ml. of 5%

---

[11] The buffer pH and the amounts of enzyme, ATP-Mg, and glucose given are those specifically useful for the assay of purified preparations of brain hexokinase with which inhibition by G-6-P accumulation prevents complete utilization of the added substrate. As discussed below (under Application of Assay Methods) the conditions of test, as reflected in the concentration of the added components, as well as whether glucose may be profitably substituted by another sugar, depend upon which tissue is employed and what interfering enzymes are present. Likewise, attention must be given to the differences in pH optima of the enzymes. Consult the data given in a later part of this article.

[12] EDTA = ethylenediaminetetraacetic acid.

[13] MgCl$_2$ is added so that the later addition of EDTA to the incubation mixture will not reduce by virtue of complex formation the magnesium concentration calculated to be added with the ATP.

[14] If method 3a is chosen, it is convenient to add HCl to 1 N concentration, heat at 100° for 7 minutes, centrifuge, and analyze aliquots after neutralization.

[15] O. H. Lowry, N. J. Rosebrough, A. L. Farr, and R. J. Randall, *J. Biol. Chem.* **193**, 265 (1951); see Vol. III [73].

TCA. 0.3 to 0.5 g. of Norit-A charcoal[16] is stirred in, and the mixture is filtered. Analysis is then made for stable P.[6] Nucleotides are adsorbed quantitatively, whereas hexose phosphates are not at all adsorbed from TCA solutions.[17] If larger amounts of nucleotide are used, additional treatments with charcoal should be carried out.

Modified Procedure for Method 4c (Photometric Indicator Method). The method below provides a means for obtaining continuous rate measurement with low substrate utilization. It is particularly useful for estimation of K_m values. The method is as follows: 4.0 ml. of a mixture containing ATP, MgCl$_2$, and bromothymol blue, in amounts such as to give, when diluted to 5.0 ml., the respective concentrations of 0.0025 M, 0.005 M, and 0.003%, at pH 7.2, is mixed with 0.8 ml. of substrate solution in a Klett-Summerson photometer tube. 0.2 ml. of enzyme (about 10 units/ml.) is added and mixed in by inversion. The reaction is followed by the measurement of optical density with the 560 filter in the Klett-Summerson photometer, and incubating the tubes at 30° between readings. Initial readings of a series should agree within 10 to 20 Klett units. Owing to the changing buffer capacity of the mixture during reaction, the rates are not linear, and it is best to take the reciprocal of the time required for a given decrease in optical density, corrected for a blank without substrate, as a measure of the rate of reaction. The method may be calibrated by adding a measured amount (about 50 millimicromoles/ml.) of mannose (K_m for brain hexokinase $= 5 \times 10^{-6}$) and observing the total change in optical density for complete utilization. Under the above conditions, a change of 1 Klett unit is approximately equal to 1 millimicromole of phosphate transferred per millimeter. Cysteine or EDTA-MgCl$_2$ (at 0.0001 M concentration) may be added to stabilize the enzyme at some sacrifice in the sensitivity of the method. A different initial pH may be used by substituting an indicator having a suitable pK' value.

Application of Assay Methods to Various Tissue Preparations. Reference to the properties of the animal tissue hexokinases (see below) shows that they are, characteristically, inhibited by G-6-P and ADP. Thus, the principle interference in the assay of these enzymes is the production of G-6-P and ADP, either by the hexokinase itself or by contaminating enzymes. Since the animal hexokinases are quite stable in the presence of EDTA, interference from these sources is usually indicated by a nonlinear rate, although it should be emphasized that a linear rate does not assure the absence of significant G-6-P accumulation.[2] G-6-P accumulation with

---

[16] The charcoal is made essentially phosphate-free as follows: stir with 10 vol. of N HCl at 70 to 80° for 15 minutes. Wash with water until chloride-free. Dry at 110°.

[17] R. K. Crane and F. Lipmann, *J. Biol. Chem.* **201**, 235 (1953); J. O. Hutchens and B. Podolsky, personal communication.

crude enzymes is best eliminated by the addition of a large (100-fold) excess of phosphofructokinase[2,18] With the purified brain preparations, however, where G-6-P accumulation is quantitative owing to the complete absence of phosphofructokinase, the addition of the latter is not advisable. Instead the activity is corrected for G-6-P accumulation according to the formula $N_t = (1.33 N_o + 0.2)^2$, where N_t is the true activity and N_o is the observed activity, both in micromoles of glucose used per milliliter of solution.[2] Smaller amounts of enzyme than those given in the assay method are used when phosphofructokinase is added. The same equation *cannot* be used for the phosphofructokinase-free heart preparations, and the usefulness of a similar equation seems doubtful because of the greater inhibition by G-6-P in the case of heart. The problem of assay is greatly simplified by the use as substrates of mannose or 2-deoxyglucose, neither of which yields inhibitory esters.[19] With crude enzymes, however, phosphomannoseisomerase may cause interference when mannose is the substrate. Also, with 2-deoxyglucose, a more alkaline copper reagent than that usually employed for sugar analysis must be used.[20] When using enzymes from various tissues or different substrates with these enzymes, care should be taken to insure saturation of the enzyme throughout the incubation period. At least five times the K_m concentration of substrate should remain at the *termination* of incubation (some K_m values are given below).

Interference by ADP accumulation can be minimized by relatively large amounts of ATP in the assay mixture, since ADP inhibition is competitive.[20] Deoxycholate treatment of the brain enzyme (see below) completely destroys ATP-ase.[21] The ATP-ase activity of the enzyme preparations from other tissues (see below) is usually too low to interfere significantly. However, it is best that the ATP-ase activity be assayed to ensure the adequacy of the added ATP. Also, the presence of significant ATP-ase renders the photometric indicator method (see above) useless because of the high blank value. Attention should also be paid in the case of ATP to the apparent K_m value. For increased accuracy of the labile-P disappearance method, the ATP concentration given in the assay method may be reduced, care being taken to avoid the problems listed above and correction for ADP inhibition[20] being made when necessary.

[18] For the preparation of phosphofructokinase, see Vol. I [38].

[19] R. K. Crane and A. Sols, *J. Biol. Chem.* in press.

[20] A. Sols and R. K. Crane, *J. Biol. Chem.* **206**, 925 (1954).

[21] Acetone-powder extracts of brain have been employed [S. Ochoa, *J. Biol. Chem.* **141**, 245 (1941); S. P. Colowick, G. T. Cori, and M. W. Slein, *J. Biol. Chem.* **168**, 583 (1947); V. D. Wiebelhaus and H. A. Lardy, *Arch. Biochem.* **21**, 321 (1949)] because of the partial destruction of ATP-ase by the acetone treatment. These procedures have the disadvantage that they are attended by over 90% loss in total hexokinase activity and a reduction in specific activity.

The presence of G-6-P-ase in liver and kidney hinders accurate assay in crude preparations of these tissues. However, G-6-P-ase is associated with particulate matter,[22] as is much of the ATP-ase, and the removal of the greater part of these enzymes by centrifugation is relatively simple (see below). In crude homogenates of the various tissues, errors resulting from phosphatase activity, endogenous formation of glucose or other reducing sustances, and variations of activity with the duration of incubation may be expected. They usually can be eliminated by the use of appropriate inhibitors or controls.

Purification Procedures

Brain Hexokinase. The procedure described is essentially that of Crane and Sols[2] and is based on the fact that all the hexokinase of brain is in a particulate form. All operations are carried out at 0 to 4° unless stated otherwise.

Step 1. Preparation and Fractional Centrifugation of Homogenate (C). 2 to 3 lb. of fresh calf brain is chilled on ice, the adhering membranes are removed, and the cortex is scraped, as cleanly as possible, away from the white matter. The scrapings (250 g.) are homogenized, in portions, with 3 vol. of 0.1 M potassium phosphate, pH 6.8, for 3 minutes in the Waring blendor. The homogenate (1 l.) is immediately centrifuged for 20 minutes at 800 $\times$ g. The supernatant fluid is removed with suction and saved. The sediment is resuspended to 1 l. in phosphate and centrifuged, as before. The second supernatant fluid is removed, and the sediment is discarded. The supernatant fluids are combined and centrifuged at 3500 $\times$ g for 20 minutes. The supernatant fluid from this centrifugation is decanted and discarded. The sediment is suspended in 100 ml. of 0.05 M potassium phosphate, pH 6.2, collected by centrifugation at 5000 $\times$ g, and dispersed in 250 ml. of 0.05 M phosphate, pH 6.2.[23] (See step 2A, below.)

Step 2. Incubation with Lipase (CL). To the suspension is added an amount of lipase[24] containing 0.1 g. of protein. After 2 hours of incubation at 30° with occasional stirring, the mixture is centrifuged at 13,000 $\times$ g

[22] H. G. Hers, J. Berthet, L. Berthet, and C. de Duve, *Bull. soc. chim. biol.* **33**, 21 (1951).

[23] If it is convenient to stop at this point, the suspension may be kept in the cold room overnight without significant loss in activity.

[24] Prepared from pancreatic lipase powder (Nutritional Biochemicals Corp.) by the method of Willstätter and Waldschmidt-Leitz as described in "Die Methoden der Fermentforschung" (Bamann and Myrbäck, eds.), p. 1560, Leipzig, 1941. The procedure is followed through the first treatment with alumina gel, using alumina C. The ammonium phosphate eluate is dialyzed against 20% glycerol and clarified by centrifugation. The clear supernatant fluid is stored at −15° until used. For the preparation of alumina C, see Vol. I [11].

for 30 minutes. The sediment is washed by being dispersed in 100 ml. of 0.05 M phosphate, pH 7.2, and centrifuged again.

Step 3. Extraction with Deoxycholate (CLD). The washed sediment is thoroughly dispersed (Waring blendor) in 90 ml. of 0.05 M phosphate, pH 7.2. 10 ml. of 3.3% sodium deoxycholate, pH 7.5 to 8.0, is added, and, after stirring for 15 minutes, the mixture is centrifuged immediately at 13,000 × g for 30 minutes. The sediment is thoroughly dispersed in 100 ml. of 0.01 M histidine–0.01 M Tris–0.002 M EDTA, pH 7.0, and centrifuged, as before. The washing with histidine-Tris-EDTA buffer is repeated. The washed sediment is thoroughly dispersed in 25 ml. of histidine-Tris-EDTA buffer, and an equal volume of glycerol[25] is added. The preparation at this point is free of interfering enzymes.[26]

Step 2A. Direct Extraction with Deoxycholate (Lipase Omitted) (CD). Except for the presence of slightly more phosphohexoseisomerase, preparations made without lipase treatment, although less purified, are about as free of interfering enzymes as those treated with lipase. The procedure is as follows: The final sediment from step 1 is dispersed in buffer of pH 7.2 instead of 6.2. Step 3 is then followed, as described above.

Heart Muscle Hexokinase (Particulate). *Step 1. Preparation and Fractional Centrifugation of Homogenate.* Fresh calf heart is chilled on ice, trimmed of fat, and muscle from the left ventricle is excised and ground with a meat grinder. 100 g. of ground muscle is homogenized with 200 ml. of 0.115 M KCl–0.02 M potassium phosphate–0.01 M EDTA, pH 7.4,[27] for 3 minutes in the Waring blendor. The homogenate is diluted to 500 ml. and centrifuged at 1500 × g for 10 minutes. The supernatant fluid is removed with suction, and the sediment is discarded. The supernatant fluid is centrifuged at 18,000 × g for 30 minutes and then decanted.[28] The sediment is washed twice with 100-ml. aliquots of the above buffer and is finally suspended in 25 ml. of the same buffer.

[25] Suspensions in 50% glycerol are stable indefinitely at −15° and for several months at +4°. Glycerol at less than 10% concentration during incubation does not inhibit brain hexokinase.

[26] The following enzymatic activities are absent in CLD when tested for in the pH range 7 to 8 in the presence of the appropriate substrates and coenzymes: phosphomonoesterase, ATP-ase, adenylate kinase, phosphofructokinase, phosphoglucokinase, phosphomannoseisomerase, phosphoglucomutase, 5-adenylic acid deaminase, 5-adenylic nucleotidase, glycerol kinase, and inorganic pyrophosphatase. A trace (0.002 unit per unit hexokinase) of phosphohexoseisomerase remains.

[27] K. W. Cleland and E. C. Slater, *Biochem. J.* **53**, 547 (1953).

[28] The soluble hexokinase representing about 50% of the total can be concentrated and slightly purified as follows: The supernatant fluid is acidified to pH 5.5, and the precipitate which forms is removed by centrifugation and discarded. The pH 5.5 supernatant fluid is neutralized to pH 7.5 and brought to 66% saturation by the addition of saturated ammonium sulfate (pH 7.5). Fractionation of this precipitate

Step 2. Solubilization of the Hexokinase. The suspension of particulate enzyme (25 ml.) is mixed with 8 ml. of 0.1% Triton X-100,[29] to aid the removal of interfering enzymes, and centrifuged at 18,000 × *g* for 40 minutes. The sediment is resuspended to 25 ml. and mixed with 3 ml. of 10% Triton X-100 (1 mg. of Triton per milligram of protein). After standing for 1 hour, the mixture is centrifuged at 50,000 × *g* for 45 minutes. The fatty layer is removed, and the supernatant fluid containing the enzyme is decanted. The specific activity is about 4, and the recovery in this step is about 90%. The over-all recovery is about 15% of the particulate hexokinase. This solution may be kept at −15°, or the enzyme may be precipitated and the detergent removed by the addition of 4 vol. of acetone (−5°). The enzyme is recovered by filtration and drying *in vacuo*. The recovery in this step is about 20% at the same specific activity. The final preparation is essentially free of phosphofructokinase and contains very little ATP-ase.

Liver Preparation (Soluble). *Step 1. Preparation of Particle-Free Extract.* A fed (*ad lib.*) rat is killed by stunning, and the thorax and abdomen are opened. The liver is removed, weighed, and homogenized in 2 vol. of 0.15 *M* KCl–0.005 *M* EDTA–0.005 *M* MgCl₂, pH 7.0, in a Potter-Elvehjem homogenizer. The homogenate is centrifuged for 30 minutes at 100,000 × *g* in the Spinco Model L centrifuge. The clear pink supernatant fluid is removed by aspiration without disturbing the fatty layer, and the sediment is discarded. The preparation at this stage has 2 to 3 units of hexokinase activity per milliliter, an ATP-ase activity about one-fifth as much, and forms no reducing substance during incubation in the absence of ATP. G-6-P-ase activity is essentially absent.

Step 2. Concentration by Ammonium Sulfate Precipitation. The supernatant fluid from step 1 is brought to 50% saturation with ammonium sulfate by the addition, in the cold, of an equal volume of saturated (at room temperature) ammonium sulfate, pH 7.5. The precipitate is collected, dissolved in a small volume of 0.01 *M* Tris–0.005 *M* EDTA–0.005 *M* MgCl₂, pH 8, and dialyzed against a large volume of the same buffer. The dialyzed preparation may be stored at −15° for several weeks.

Skeletal Muscle Preparation. *Step 1. Preparation of Extract.* Skeletal muscle (back and legs) from the rat or dog is chilled on ice and ground in a Latapie mincer. The ground muscle is stirred for 15 minutes with 1.5 vol. of 0.01 *M* Tris–0.005 *M* EDTA, pH 8, and centrifuged at 7000 × *g* for

---

again with ammonium sulfate (at pH 6.2) yields between 45 and 55% saturation with ammonium sulfate, a preparation of specific activity about 1.5 and low in phosphofructokinase activity. The preparations are dialyzed against 0.002 *M* EDTA and stored at −15°.

[29] A nonionic detergent manufactured by Rohm and Haas Co., Philadelphia, Pa.

20 minutes. The supernatant fluid is decanted through glass wool. This extract contains about 3.5 units/ml. at a specific activity of about 0.15.

Step 2. Acetone Fractionation and Preparation of Acetone Powder. The extract is acidified to pH 6.5 with 1.0 M acetate, pH 4.0. 0.5 vol. of cold acetone is added, with stirring, to the extract, kept cold in an ice-salt bath. After 10 minutes' standing, the acetone mixture is centrifuged at 600 × g for 10 minutes. The supernatant fluid is discarded. The sediment is suspended in an equal volume of 33% acetone and slowly poured into 10 vol. of cold acetone, with stirring. The precipitate is recovered by suction filtration. The filter cake is spread over filter paper in a thin layer and dried *in vacuo.* The acetone powder may be stored at −15°. The recovery in this step is about 50% of the activity in the original extract. Extracts may be prepared by grinding the powder with 10 vol. of cold 0.01 M Tris–0.005 M EDTA buffer, pH 8, and centrifuging to remove the insoluble material. The clear supernatant fluid contains the hexokinase at a specific activity about 1.

Properties

Specificity. A total of sixteen sugars and sugar derivatives of the D series have been found to be phosphorylated by brain hexokinase.[30] Of these, the following are the more suitable for routine work (relative maximal rates and K_m values are given in parentheses): glucose (1.0, 8 × 10⁻⁶ M), mannose (0.5, 5 × 10⁻⁶ M), 2-deoxyglucose (1.0, 2.7 × 10⁻⁵ M), glucosamine at pH 7.5 (0.6, 8 × 10⁻⁵ M), fructose (1.4, 1.6 × 10⁻³ M) and 1,5-sorbitan (1.0, 3 × 10⁻² M). The specificity (K_m and maximal rate) of the heart hexokinases for the above listed substrates is about the same as that of brain hexokinase, but the specificity of the hexokinases from other tissues is relatively unknown. Skeletal muscle and liver preparations will phosphorylate glucose, mannose, fructose, glucosamine, and 2-deoxyglucose. Evidence that glucose and fructose are not phosphorylated by the same enzyme in liver preparations has been reported.[1,3] The K_m value of skeletal muscle hexokinase for glucose and mannose is about ten times as high as that of brain, and the K_m values of liver hexokinase are probably even greater. The K_m value of skeletal muscle hexokinase for fructose is only slightly higher than that of brain.

Activators and Inhibitors. The following sugars and sugar derivatives act as competitive inhibitors of brain hexokinase (K_i values in parentheses): N-acetyl-D-glucosamine (8 × 10⁻⁵ M), N-methyl-D-glucosamine (2 × 10⁻⁴ M), 6-deoxy-D-glucose (2 × 10⁻³ M), D-xylose (2 × 10⁻³ M), and D-lyxose (1.3 × 10⁻³ M).[30] The phosphorylation of fructose by the

[30] A. Sols and R. K. Crane, *J. Biol. Chem.* in press.

above preparations of dog skeletal or heart muscle is strongly inhibited by glucose and N-acetylglucosamine. These preparations do not phosphorylate L-sorbose.[31]

Half-maximal activity of brain hexokinase is obtained at 1.3×10^{-4} M ATP in the presence of 0.005 M MgCl$_2$, and at 8×10^{-4} M MgCl$_2$ in the presence of 0.01 M ATP. ADP inhibits competitively the activity of brain, heart, and skeletal muscle enzymes. For brain and heart, K_i ADP is equal to K_m ATP.

Glucose-6-phosphate inhibits noncompetitively the hexokinases of brain,[2,32] heart, skeletal muscle, kidney, red blood cell, and liver.[2] The K_i values of G-6-P and related compounds which have been found to inhibit brain hexokinase are: G-6-P, 4×10^{-4} M; 1,5-sorbitan-6-P, 6×10^{-4} M; L-sorbose-1-P, 6×10^{-4} M; α-glucose-1,6-diphosphate, 7×10^{-4} M; allose-6-P, 6×10^{-4} M; and 3-deoxyglucose-6-P, 1.7×10^{-2} M.[19,33] The K_i values of G-6-P, 1,5-sorbitan-6-P, and L-sorbose-1-P with the heart preparations are one-fourth or less than those with brain hexokinase.[34]

Purified brain hexokinase is inactivated by dilution. EDTA, cyanide, cysteine, and, to some extent, substrates, insulin, and serum albumin will prevent this inactivation.[20] EDTA also protects liver hexokinase from inactivation by dilution. p-Chloromercuribenzoate (10 moles/100,000 g. of protein) inhibits brain hexokinase completely.[20] Under certain conditions, cysteine will reverse the inhibition. o-Iodosobenzoate also inhibits brain hexokinase. Skeletal muscle hexokinase has been found to be inhibited by alloxan and reactivated by cysteine.[35]

Certain effects of ions[3,20] and tissue extracts[3] on the stabilization of the activity of animal hexokinases have been reported.

Effects of pH. Brain hexokinase has maximal activity over the pH range 6 to 8, whereas the liver and skeletal muscle enzymes appear to be most active in the vicinity of pH 8. The soluble heart hexokinases differ in their pH curves. That initially associated with particles resembles brain, whereas that in the first supernatant fluid appears to be more like liver and skeletal muscle.[34] The stability of brain hexokinase is maximal at pH values near 6. The variation in activity with pH of several other hexokinases has been studied by Long.[36]

---

[31] The purified fructokinase of liver and the similar enzyme from skeletal muscle, both of which phosphorylate L-sorbose, are not inhibited by glucose.[1] See also Vol. I [34].

[32] H. Weil-Malherbe and A. D. Bone, *Biochem. J.* **49**, 339 (1951).

[33] The value given for L-sorbose-1-P is about ten times as great as that indicated by the observations of H. A. Lardy, V. D. Wiebelhaus, and K. M. Mann, *J. Biol. Chem.* **187**, 325 (1950), on inhibition of hexokinase in extracts of acetone powders.

[34] A. Sols and R. K. Crane, *Federation Proc.* **13**, 301 (1954).

[35] M. Griffiths, *Arch. Biochem.* **20**, 21 (1949).

[36] C. Long, *Biochem. J.* **50**, 407 (1952).

Effect of Temperature. It has been observed that complete inactivation of skeletal muscle hexokinase occurs in 4 minutes at 55°,[7b] and of brain hexokinase in 15 minutes at the same temperature. The $Q_{10°}$ value for brain hexokinase in the range 30 to 40° is 2.2.

SUMMARY OF PURIFICATION PROCEDURE FOR BRAIN HEXOKINASE

Preparation	Total activity, units	Total protein, mg.	Specific activity, units/mg.	Recovery, %
Homogenate	25,000[a]	17,200	1.5	—
C	10,200[b]	1,125	9.2	41
CL	9,200[b]	410	22.5	37
CLD	8,700[b]	121	71	35

Activity assayed by hexose disappearance with glucose as substrate:
[a] In the presence of excess PFK.
[b] Corrected for G-6-P inhibition.

[34] Fructokinase (Ketohexokinase)

$$\text{Fructose} + \text{ATP} \rightarrow \text{Fructose-1-P} + \text{ADP}$$

By H. G. HERS

Liver

Assay Method

Principle. The method is based on the measurement of fructose disappearance under optimal conditions, as described by Hers.[1,2] Fructose is estimated by the colorimetric method developed by Roe *et al.*[3] for inulin and modified by Hers *et al.*[4]

Reagents

0.4% fructose solution.

0.05 M ATP solution. Dissolve 311 mg. of the sodium salt[5] in 10 ml. of water. This solution may be kept at −15° for many months without change.

[1] H. G. Hers, *Biochim. et Biophys. Acta* **8**, 416 (1952).
[2] H. G. Hers, *Biochim. et Biophys. Acta* **8**, 424 (1952).
[3] J. H. Roe, J. H. Epstein, and N. P. Goldstein, *J. Biol. Chem.* **178**, 839 (1949).
[4] H. G. Hers, H. Beaufays, and C. de Duve, *Biochim. et Biophys. Acta* **11**, 416 (1953).
[5] The sodium salt of ATP is now available commercially. It is easily prepared from rabbit muscle. The exact titer of the solution should be checked by suitable methods. For details, see Vol. III [117].

0.05 M MgAc$_2$.

4 M KAc.

0.5 M NaF.

0.15 M ZnSO$_4$.

0.3 N Ba(OH)$_2$.

Fructose standard solution. Fructose solutions do not keep well, even in the presence of benzoic acid. A stable standard is provided by a solution containing 19 mg. of sucrose and 250 mg. of benzoic acid per 100 ml. Just before use, 1 ml. of this solution is mixed with 1 ml. of 0.01 N HCl and heated during 20 minutes at 100°. In the analytical procedure described below it gives a color corresponding to that of 101 γ of fructose.[4]

Resorcinol-thiourea reagent. Dissolve 100 mg. of resorcinol and 250 mg. of thiourea in 100 ml. of very pure glacial acetic acid.[6]

30% HCl. To 4 vol. of HCl, sp. gr. 1.19, add 1 vol. of distilled water.

Enzyme. Dilute the enzyme with water to obtain 20 to 30 units/ml. (See definition below.)

Procedure. 0.1 ml. of the fructose solution is placed with great accuracy in the bottom of a test tube, and 0.1 ml. of ATP, 0.1 ml. of MgAc$_2$, and 0.3 ml. of KAc are added. After mixing, 0.1 ml. of NaF is added.[7] A control tube is set up similarly, except that 0.2 ml. of water is added instead of ATP and magnesium. The reaction is started by adding 0.5 ml. of enzyme to both tubes, which are incubated at 37° for 20 minutes. Deproteinization is effected by the successive addition of 3 ml. of the ZnSO$_4$ and Ba(OH)$_2$ solutions; the mixture is vigorously shaken and filtered or centrifuged. Phosphate esters are retained on the BaSO$_4$/Zn(OH)$_2$ precipitate, and free fructose can be estimated in the filtrates without interference. To 2 ml. of the filtrates and to the standard of hydrolyzed sucrose (see above), 1 ml. of the resorcinol-thiourea reagent and 7 ml. of 30% HCl are added, mixed, and placed for 15 minutes in a water bath at 75°. After cooling, extinction is read in the range of 500 to 520 mμ. The difference in fructose content of the two tubes incubated with and without ATP is proportional to the amount of enzyme added.

Definition of Unit and Specific Activity. One unit is defined as the amount of enzyme which phosphorylates 1 γ of fructose per minute. If

---

[6] Impurities present in some samples of acetic acid interfere with the determination procedure (Hers *et al.*[4]).

[7] NaF is added to inhibit phosphatases and may be omitted when purified enzyme preparations are used. It forms a flocculent precipitate in the presence of MgAc$_2$ and ATP, but the rate of the reaction is not influenced by this precipitate. When NaF is mixed with MgAc, before ATP has been added, however, an important inhibition obtains (Hers[1]).

sufficient amounts of substrates are present, the velocity of the reaction remains constant during the first 20 minutes. Specific activity is expressed as units per milligram of protein nitrogen (Kjeldahl).

Alternative Method. The manometric method described by Colowick and Kalckar[8] for hexokinase may be used with purified fructokinase.

Purification Procedure

A 25-fold purification from beef liver has been described by Hers.[1]

Step. 1. Preparation of Extract. 1 kg. of fresh beef liver is cut with scissors and homogenized in the Waring blendor with 3 l. of ice-cold water.

Step 2. Acid Denaturation of Inactive Material. The homogenate, maintained at 0°, is acidified to pH 4.5 (glass electrode) with 5 N HCl (10 to 20 ml.) and brought back immediately to pH 5.5 with 2.5 N NaOH. The bulky precipitate containing agglutinated particles and denatured proteins is centrifuged off. The clear supernatant is adjusted to pH 7.5.

Step 3. Fractionation with Ammonium Sulfate at pH 7.5. A solution of ammonium sulfate, saturated at 0° and adjusted to pH 7.5 with ammonia, is added under continuous stirring to give 0.4 saturation. After standing 1 hour, the precipitate is removed by centrifugation in the cold. The supernatant fluid is brought to 0.6 saturation, and, after centrifugation, the 0.4 to 0.6 fraction is dissolved in the least possible volume of water and dialyzed in the cold against 0.04% $NaHCO_3$ during 12 hours and then against distilled water for 24 hours, with several changes of water.

Step 4. Fractionation with Ammonium Sulfate at pH 4.6. The dialyzed solution is brought to pH 5.2 in a stepwise manner by the addition of HCl at 0°. When the pH reaches 6.5, 5.7, and 5.2, the solution is allowed to stand each time for 1 hour, and the isoelectric precipitate which is formed is removed by centrifugation. The last supernatant is then adjusted to pH 4.6 by the addition of acetate buffer (final concentration, 0.1 M) and quickly fractionated with a cold ammonium sulfate solution between 0.35 and 0.50 saturation. Centrifugation is performed immediately after the addition of salt, and the 0.35 to 0.50 fraction is rapidly dissolved in 30 ml. of 0.1 M cacodylate buffer, pH 6.5, to avoid excessive inactivation of the enzyme.

Step 5. Fractionation with Ammonium Sulfate at pH 6.5. Further purification is achieved by fractionating at pH 6.5 between 0.45 and 0.55 saturation of ammonium sulfate. The final precipitate is redissolved in water and dialyzed as in step 3. The resulting solution may be stored in the frozen state.

Alternative Method. An alternative method of purification from rat liver has been described by Staub and Vestling.[9]

[8] S. P. Colowick and H. M. Kalckar, *J. Biol. Chem.* **148**, 117 (1943).
[9] A. Staub and C. S. Vestling, *J. Biol. Chem.* **191**, 395 (1951).

Properties

Specificity. Liver fructokinase has been found to act on the three keto-hexoses: fructose, sorbose, and tagatose (Leuthardt and Testa,[10] Hers[1]). Fructose-6-phosphate and aldohexoses are not phosphorylated; neither do they exert any inhibition on the phosphorylation of fructose. Only the terminal phosphate of ATP is available as phosphate donor.

Activators. Fructokinase shows an absolute requirement for magnesium (or manganese) and is strongly activated by potassium ions (Hers[1,2]). Mg^{++} has been shown to act by forming with ATP a complex which appears to be the true substrate of the enzyme. When present in excess over ATP, it inhibits the reaction. The maximal rate is attained at low ionic strength with a Mg/ATP ratio of 0.5, at high ionic strength with a Mg/ATP ratio of 1. The concentration of K^+ ions necessary for optimal activity is extremely high, from 1 M to 2 M, depending on the amount of ATP and Mg^{++} present. This effect is mainly related to an increased affinity of the enzyme for the Mg-ATP complex in the presence of K^+.

In experiments by Cori *et al.*,[11] the activity of fructokinase in crude liver homogenates was increased several-fold over the anaerobic level, upon aerobic incubation in the presence of glutamate. ATP was not limiting in the anaerobic system.

Dissociation Constants of the Enzyme-Substrate Complexes. K_s values for ATP-Mg were found to be 10^{-3} M in the presence of K^+ and 5.4×10^{-3} M in the presence of Na^+. The affinity for fructose is very high, K_s being smaller than 5×10^{-4} M.

Effect of pH. Between 5.5 and 7.8, the pH is without effect on the velocity of the reaction.

Muscle

Partial purification of fructokinase from rabbit muscle has been described by Slein *et al.*[12] Ground rabbit muscle is extracted twice with 1 vol. of cold 0.03 M KOH and once with 0.5 vol. of distilled water. The filtrates obtained by straining the muscle suspension through gauze are combined and treated with saturated ammonium sulfate adjusted to pH 7.8. The fraction precipitating between 0.41 and 0.5 saturation contains fructokinase purified 12- to 40-fold. Glucokinase activity is practically absent, but both 1-phosphofructokinase and 6-phosphofructokinase are present in large quantities.

Some properties of the enzyme occurring in this fraction have been studied by Hers.[13] Like liver fructokinase, it forms fructose-1-phosphate

[10] F. Leuthardt and E. Testa, *Helv. Physiol. Acta* **8,** C 67 (1950).

[11] G. T. Cori, S. Ochoa, M. W. Slein, and C. F. Cori, *Biochim. et Biophys. Acta* **7,** 304 (1951).

[12] M. W. Slein, G. T. Cori, and C. F. Cori, *J. Biol. Chem.* **186,** 763 (1950).

[13] H. G. Hers, *Arch. intern. physiol.* **61,** 426 (1953).

and acts on sorbose. It appears, therefore, to be also a ketohexokinase. However, it is not activated by potassium ions. As already shown by Slein et al.,[12] it has a low affinity for fructose, the Michaelis constant for this substrate being approximately 0.2 M.

These facts complicate the assay (which should be made without added potassium and in the presence of a suitable buffer). At low fructose concentrations, the reaction obeys first-order kinetics. One can also use very high fructose concentrations (molar) and the manometric technique of Colowick and Kalckar,[8] but in that case 1-phosphofructokinase activity is included in the measurement.

SUMMARY OF PURIFICATION PROCEDURE

Fractions	Total volume, ml.	Units/ml.	Total units, thousands	Ni- trogen, mg./ml.	Specific activity, units/ mg. N	Recovery, %
1. Homogenate	4000	24.3	975	7.3	3.3	—
2. Supernatant after pH 4.5	2100	26	546	2.14	12.2	56
3. (NH₄)₂SO₄ fraction, pH 7.5	250	208	520	5.8	36	53.5
4. (NH₄)₂SO₄ fraction, pH 4.6	50	333	167	5.25	63.5	17
5. (NH₄)₂SO₄ fraction pH 6.5	40	338	135	4.55	74.5	13.9

[35] Galactokinase and Galactowaldenase

By LUIS F. LELOIR and RAÚL E. TRUCCO

Galactokinase

$$\text{Galactose} + \text{ATP} \rightarrow \text{Gal-1-P}^1 + \text{ADP}$$

Assay Method

Principle. The manometric[2] or titrimetric[3] methods described for hexokinase are applicable to galactokinase by simply substituting galactose for glucose. Another convenient procedure consists in measuring the

---

[1] The abbreviation Gal-1-P is used for galactose-1-phosphate.
[2] S. P. Colowick and H. M. Kalckar, *J. Biol. Chem.* **148**, 117 (1943).
[3] M. Kunitz and M. R. McDonald, *J. Gen. Physiol.* **29**, 393 (1946).

disappearance of galactose after precipitating the hexose phosphates with zinc sulfate and barium hydroxide. The method was originally used by Colowick et al.[4] for hexokinase, and by Trucco et al.[5] for galactokinase.

Reagents

> ATP solution, 40 micromoles/ml. Sodium salt.
> Galactose solution, 10 micromoles/ml.
> 0.1 M magnesium sulfate.
> Maleate buffer of pH 6.[6] Solution I, 1.160 g. of maleic acid dissolved in 100 ml. of 0.1 M NaOH; solution II, 1.160 g. of maleic acid dissolved in 100 ml. of 0.2 M NaOH. Mix 4 ml. of solution I and 6 ml. of solution II.
> 0.3 N barium hydroxide.
> 5% zinc sulfate.

Procedure. Mix 0.2 ml. of buffer, 0.2 ml. of magnesium sulfate, 0.2 ml. of galactose solution, 0.2 ml. of ATP, variable amounts of enzyme, and water. Total volume, 1 ml.

The reaction is started by adding the enzyme. After incubating 10 minutes at 30°, add 0.5 ml. of zinc sulfate, mix, and add 0.5 ml. of barium hydroxide; mix and centrifuge. Reducing sugar is estimated in an aliquot of supernatant by the Somogyi and Nelson procedure (see Vol. III [12]). The same mixture without galactose is incubated as control, and the galactose is added to it after the addition of the $ZnSO_4$ solution. The difference in reducing sugar between the experimental and control samples is taken as the amount of galactose phosphorylated.

Units. Results may be expressed in the same units as for hexokinase (see Vol. I [33]).

Application of Assay Method to Crude Preparations. The main source of trouble with crude extracts, especially those of liver, is the presence of glucose. For liver it has been found necessary to reduce the glycogen by fasting the animals and to carry out a partial purification with ammonium sulfate.[7] An alternative procedure would be to use galactosamine as substrate and estimate it by the Elson and Morgan procedure.[8] However, the identity of the enzyme-phosphorylating galactose and galactosamine has not been definitely proved.

[4] S. P. Colowick, G. T. Cori, and M. W. Slein, *J. Biol. Chem.* **168,** 583 (1947).

[5] R. E. Trucco, R. Caputto, L. F. Leloir, and N. Mittelman, *Arch. Biochem.* **18,** 137 (1948).

[6] G. Smits, *Biochim. et Biophys. Acta* **1,** 280 (1947).

[7] C. E. Cardini and L. F. Leloir, *Arch. Biochem. and Biophys.* **45,** 55 (1953).

[8] G. Blix, *Acta Chem. Scand.* **2,** 467 (1948).

Preparation of the Enzyme. The enzyme can be prepared from galactose-adapted yeasts.[9] A good source is *Saccharomyces fragilis*.[5] It can be grown on yeast extract-agar containing 2% galactose or lactose, or in a well-aerated liquid medium. Whey can also be used.[10] The cells are harvested after 48 hours at 30°, washed with water, spread in a 2-mm. layer and allowed to dry at room temperature for 3 to 4 days.

Extracts are prepared by suspending the dried yeast in three parts of 2.2% diammonium phosphate.[11] After standing at 5° for 20 to 24 hours with occasional stirring, the paste is centrifuged. 1 ml. of the supernatant catalyzes the esterification of about 50 micromoles of galactose per minute under the conditions described.

Methods for the purification of galactokinase are based on the adsorption of impurities with bentonite[5] or ammonium sulfate fractionation.[9]

Purification with Bentonite. Treatment of the crude extract with bentonite removes the hexokinase and part of the inactive protein. The procedure is as follows. 100 ml. of crude extract at pH 6.4 is treated with 3 to 5 g. of bentonite. The correct amount of bentonite is determined by small-scale trials in which galactokinase, hexokinase, and total proteins are estimated after treatment with different amounts of the adsorbent. The supernatant from the bentonite treatment is mixed with 1 vol. of saturated ammonium sulfate. The precipitate is discarded, and to the supernatant 0.25 vol. of saturated ammonium sulfate is added. The precipitate is then dissolved in $\frac{1}{20}$ of the original volume of water. The solution obtained by this procedure is usually five to six times as active as the crude extract. The ratio activity/total protein increases about fourfold, and the recovery is 30%.

Properties

Specificity. The crude yeast extracts catalyze the esterification of glucose at a rate four to five times as high as that of galactose and contain galactowaldenase and phosphoglucomutase.

Activators and Inhibitors. Magnesium or manganese ions are required for maximal activity, the optimum concentration being about 0.005 and 0.001 M, respectively. Fluoride does not inhibit appreciably at concentrations below 0.05 M.

The galactokinase of *S. fragilis* has been found to be inhibited by SH-blocking compounds such as *p*-chloromercuribenzoic acid, iodosobenzoic acid, and iodoacetamide,[12] and to be reactivated by glutathione.

[9] J. F. Wilkinson, *Biochem. J.* **44**, 460 (1949).
[10] R. Caputto, L. F. Leloir, and R. E. Trucco, *Enzymologia* **12**, 350 (1948).
[11] C. Neuberg and H. Lustig, *Arch. Biochem.* **1**, 191 (1942).
[12] M. Bacila, *Ciencia e cultura* **3**, 292 (1951).

Effect of pH. With maleate buffer the pH optimum has been found to be at pH 6, and the activity at pH 5 or 7 was about 80% of the maximum. With phosphate buffer the pH optimum is 6.8.

Galactowaldenase

$$\text{Gal-1-P} \rightleftarrows \text{G-1-P}$$

The enzymic reaction appears to take place in two steps (Leloir,[13] Kalckar *et al.*[14]):

$$\text{Gal-1-P} + \text{UDPG} \rightleftarrows \text{G-1-P} + \text{UDPGal}^{15}$$
$$\text{UDPGal} \rightleftarrows \text{UDPG}$$

The first enzyme has been called uridyl-transferase, and the second would be the waldenase proper, that is, the enzyme catalyzing the reaction in which the inversion of hydroxyl group 4 occurs.

Assay Method

Principle. The methods of estimation are based on the transformation of the G-1-P formed into G-6-P by an excess of phosphoglucomutase. The G-6-P can then be measured either by its reducing power or by adding TPN and Zwischenferment. The first procedure will be described here.

Reagents

  Gal-1-P, 20 micromoles/ml. Sodium salt.
  0.1 M magnesium chloride.
  Glucose-1,6-diphosphate, 1 micromole/ml. Sodium salt.
  UDPG, 10 micromoles/ml. Sodium salt.

Procedure. Mix the following components: 0.1 ml. of Gal-1-P, 0.02 ml. of magnesium chloride, 0.01 ml. of glucose diphosphate, 0.02 ml. of UDPG, enzyme, and water to complete 0.2 ml. With crude extracts of yeast it has not been found necessary to add phosphoglucomutase, but this would be necessary for studies of purification.

The reaction is started by adding the enzyme (about 0.01 ml. of *S. fragilis* extract), and, after 20 minutes at 37°, 1.5 ml. of Somogyi copper reagent (see Vol. III [12]) is added. A blank in which the reaction is stopped at $t = 0$ is run at the same time. The mixture is heated 10 minutes at 100°, cooled, and the 1.5 ml. of Nelson reagent and water to 7.5 ml. are added.

[13] L. F. Leloir, *Arch. Biochem. and Biophys.* **33**, 186 (1951).
[14] H. M. Kalckar, B. M. Braganca, and A. Munch-Petersen, *Nature* **172**, 1038 (1953).
[15] The abbreviation UDPGal is used for uridine diphosphate galactose.

Preparation of the Enzyme. A crude extract from *S. fragilis* (see "Galactokinase") is used as enzyme.

Properties

Equilibrium. The composition of the equilibrium mixture corresponds to about 25% of α-galactose-1-phosphate and 75% of α-glucose-1-phosphate.[16]

Inhibitors. Galactowaldenase is not affected by 0.028 M sodium arsenate. At this concentration phosphoglucomutase is completely inhibited.[16]

[16] L. F. Leloir, C. E. Cardini, and E. Cabib, *Anales asoc. quim. argentina* **40**, 228 (1952).

[36] Phosphoglucomutase from Muscle

By VICTOR A. NAJJAR

$$\text{Glucose-1-phosphate} \rightleftarrows \text{Glucose-6-phosphate}^1$$
$$(5.5\%) \qquad\qquad (94.5\%)$$

Glucose-1,6-diphosphate has been shown by Leloir *et al.*[2] to be the coenzyme for phosphoglucomutase and is needed in saturation quantities for maximum activity of the enzyme. In a strict sense the diphosphate is not a coenzyme, since it is formed *de novo* during enzyme activity.[3] The mechanism of action of phosphoglucomutase has recently been elucidated.[3,4]

Glucose-1-phosphate + phospho-enzyme I $\rightleftarrows$ Glucose-1,6-diphosphate + dephospho-enzyme   (1)

Glucose-1,6-diphosphate + dephospho-enzyme $\rightleftarrows$ Glucose-6-phosphate + phospho-enzyme II   (2)

Phospho-enzyme I = or $\rightleftarrows$ Phospho-enzyme II   (3)

Assay Method

Principle. The measure of enzyme activity is based on the different chemical properties of the two hexose phosphates. Glucose-1-phosphate is a nonreducing sugar with acid-labile phosphate. By contrast, glucose-6-phosphate has acid-stable phosphate and reduces copper reagents. For measuring enzyme activity glucose-1-phosphate is used as the substrate

[1] G. T. Cori, S. P. Colowick, and C. F. Cori, *J. Biol. Chem.* **123**, 375 (1938).

[2] L. F. Leloir, R. E. Trucco, C. E. Cardini, A. C. Paladini, and R. Caputto, *Arch. Biochem.* **19**, 339 (1948); see Vol. III [17].

[3] V. A. Najjar, *in* "Mechanism of Enzyme Action" (McElroy and Glass, eds.), p. 731, Johns Hopkins Press, Baltimore, 1954.

[4] V. A. Najjar and M. E. Pullman, *Science* **119**, 631 (1954).

because of the favorable equilibrium toward the forward reaction. The rate of the reaction, which is dependent on the amount of enzyme present, can then be followed by measuring the reduction in the hydrolyzable phosphate or the increase in the reducing sugar.

Reagents

Magnesium sulfate, $6 \times 10^{-3} M$.

Glucose-1-phosphate $2 \times 10^{-2} M$, pH 7.5.

Enzymatically prepared glucose-1-phosphate[5,6] usually contains over 2.5×10^{-4} micromole of glucose-1,6-diphosphate per micromole of the monoester. Under the conditions of the test this amount of contamination is sufficient to saturate the enzyme and consequently produce maximum activity. When the monoester is purified to the extent that it is free of the coenzyme, then glucose-1,6-diphosphate should be added to the glucose-1-phosphate stock solution in appropriate quantities to saturate the enzyme. $5 \times 10^{-7} M$ produces half maximal activity.[6]

Cysteine, 0.1 M, pH 7.5, freshly prepared.

Enzyme. Immediately before the reaction is started, the enzyme is diluted in water at 0° to contain about 0.1 to 0.3 unit/ml. In practice the amount of enzyme added should catalyze the formation of an amount of glucose-6-phosphate that falls within the range of 0.3 to 1.0 micromole under the conditions described below. In this range the amount of the 6-ester formed is directly proportional to the enzyme concentration.

Procedure. 0.1 ml. of each reagent (except the enzyme) is pipetted into a test tube and placed in a water bath at 30°. After temperature equilibration the reaction is started by adding 0.1 ml. of the enzyme dilution. Two or more dilutions are ordinarily used. After incubation for 5 minutes the reaction is stopped by the addition of 1 ml. of 5 N H_2SO_4. The volume is brought up to 5 ml. with water, and the reaction tubes are placed in a boiling water bath for 3 minutes to hydrolyze the remaining glucose-1-phosphate. To the sample 1 ml. of molybdate solution followed by 0.5 ml. of 1-amino-2-naphthol-4-sulfonic acid solution are added according to Fiske and SubbaRow.[7] The volume is made up to 10 ml., and the intensity of the color produced is measured in the Klett photocolorimeter with filter No. 66. The amount of stable P formed is the difference between the hydrolyzable P added and that remaining at the end of 5 minutes.

[5] C. F. Cori, S. P. Colowick, and G. T. Cori, *J. Biol. Chem.* **121**, 465 (1937); see also Vol. III [16].

[6] E. W. Sutherland, M. Cohn, T. Posternak, and C. F. Cori, *J. Biol. Chem.* **180**, 1285 (1949).

[7] C. H. Fiske and Y. SubbaRow, *J. Biol. Chem.* **66**, 375 (1925).

For the assay of enzyme activity by measuring copper reduction[8] by glucose-6-phosphate, the enzyme reaction is stopped by the addition of 1.5 ml. of Somogyi's sugar reagent.[9] The tubes are then heated for 10 minutes in a boiling water bath. After cooling, 1.5 ml. of Nelson's arseno-molybdate reagent[10] is added. The volume is made up to 7.5 ml. and the color intensity read against a standard using the Klett colorimeter with filter No. 54.

Definition of Unit and Specific Activity. One unit of enzyme is defined as that which, under the conditions described above, catalyzes the formation of 1 mg. of acid stable P or 32 micromoles of glucose-6-phosphate. Specific activity is expressed as units per milligram of protein. Protein is determined by the method of Lowry *et al.*[11]

Application of Assay Method to Crude Extracts. In crude extracts with low activity the quantity of protein in the reaction mixture may be sufficiently large as to result in turbid samples unsuitable for colorimetric measurements. In such instances the reaction is stopped by immersing the tubes in a boiling water bath. The coagulated protein is then centrifuged down, and the supernatant is used for the determination. The presence of phosphatases in the crude extracts may invalidate the assay.

Purification and Crystallization

The method of crystallization of the rabbit muscle enzyme, as described by Najjar in 1948,[12] has been successfully reproduced in other laboratories and on numerous occasions since by the author. Unless otherwise stated, all manipulations are carried out at 0 to 4° and at pH 5.0. A rabbit is killed by intravenous injection of Nembutal and bled thoroughly. The muscles of the back and hind legs are quickly excised and passed twice through a chilled meat grinder. After addition of 1 vol. of distilled water it is stirred thoroughly and allowed to extract for 20 minutes, then strained through gauze. The residue is then re-extracted with another volume of distilled water. The combined extracts, having an activity of about 0.5 unit/mg. of protein, are adjusted to pH 5.0 with 1 M acetic acid. The resulting precipitate is centrifuged down, and the clear supernatant is placed in a water bath at a temperature of 85 to 90°. When the temperature of the extract reaches 65°, it is immersed immediately in an ice bath until the temperature reaches 4°. Constant mechanical stirring is

[8] C. E. Cardini, A. C. Paladini, R. Caputto, L. F. Leloir, and R. E. Trucco, *Arch. Biochem.* **22**, 87 (1949).

[9] M. Somogyi, *J. Biol. Chem.* **160**, 61 (1945).

[10] N. Nelson, *J. Biol. Chem.* **153**, 375 (1944).

[11] O. H. Lowry, N. J. Rosebrough, A. L. Farr, and R. J. Randall, *J. Biol. Chem.* **193**, 265 (1951); see Vol. III [76].

[12] V. A. Najjar, *J. Biol. Chem.* **175**, 281 (1948).

maintained throughout the heating and cooling procedures. The extract is then filtered through coarse filter paper.

The filtrate, having an activity of about 2.3 units/mg., is brought to 0.65 saturation with solid ammonium sulfate, added gradually with gentle stirring. The precipitate is filtered off or centrifuged and taken up in an equal volume of 0.15 M acetate buffer, pH 5.0. The turbid solution containing ammonium sulfate at about 0.32 saturation is placed in a water bath at 65° and stirred constantly. The solution is allowed to reach 63° and to remain at that temperature for 3 minutes. The heavy precipitate which forms is centrifuged off. The supernatant at this stage has an activity of about 13 units/mg. of protein.

This solution is diluted with 0.32 saturated ammonium sulfate to obtain a protein concentration of about 3 to 5 mg./ml. The ammonium sulfate saturation is then raised to 0.50 by slow, dropwise addition of a saturated solution, and the mixture is allowed to stand in the cold for 30 minutes. The precipitate is centrifuged off and discarded. The clear supernatant fluid is then warmed to 30° and kept at this temperature for $\frac{1}{2}$ hour. A protein precipitate which may form is centrifuged off, care being taken to prevent cooling. The resulting supernatant fluid is placed in an ice bath, and the ammonium sulfate concentration is brought up gradually to 0.55 saturation by very slow addition of the saturated salt solution over a period of 1 hour. A faint turbidity appears, followed by a thin silky shimmer about half an hour later. The material is allowed to stand overnight.

The slow addition of ammonium sulfate is then resumed until the concentration gradually reaches 0.60. It is found that appreciable denaturation can be avoided by adding ammonium sulfate dropwise with a pipet, while the solution is stirred at the same time with a glass rod. The solution is left in the cold for 24 hours for maximum crystallization. The crystals are then centrifuged and dissolved in 0.15 M acetate buffer of pH 5 to give a protein concentration of about 2 to 3 mg./ml. The activity in the first crystals corresponds to 18 to 23 units/mg.

For the second crystallization the ammonium sulfate saturation is raised to 0.50 and the precipitate, if any, discarded. The supernatant fluid is again warmed to 30°, and any insoluble protein is separated by centrifugation at room temperature. The supernatant fluid is cooled to 0 to 4° and the ammonium sulfate concentration raised as previously described to 0.60, at which level a slight turbidity begins to form followed by the formation of crystals. After standing overnight, the crystals are separated. At this stage the activity is about 23 units/mg. of protein. The concentration of ammonium sulfate in the supernatant fluid is then raised gradually to 0.65, and, after standing 24 hours, further crystals form having an activity of 26 units/mg. of protein. After a third or fourth crystallization the activity remains at 26 units/mg., and the supernatant fluid has the same activity as the crystals.

Properties

The crystals are long, thin, very fragile plates. The crystalline phosphoglucomutase is free of phosphohexose isomerase, amylase, phosphorylase, aldolase, D-glyceraldehyde phosphate dehydrogenase, glycerophosphate dehydrogenase, triose mutase, enolase, and lactic dehydrogenase.

The absorption spectrum of the enzyme has a maximum at 278 mμ. The extinction of a 1% solution in a 1-cm. cell thickness ($E_{1\,cm.}^{1\,\%}$) at 278 mμ is 7.70. The molecular weight[4] is 77,700.

In acetate buffer, pH 5.0, of 0.1 ionic strength the enzyme had an electrophoretic mobility of 2.01 $\times$ 10^{-5} (sq. cm./volt/sec.). The protein is positively charged at that pH. Under optimal conditions the turnover number per 100,000 g. at 30° is calculated to be about 16,800 moles/min.

The enzyme is quite stable when dissolved in acetate buffer at pH 5.0. Its activity can be maintained for about 10 days at icebox temperature, then declines slowly over a period of 6 to 8 weeks. Below pH 4.5 and above pH 8.5 it is readily inactivated. In the crystalline state it is stable for months in 0.60 to 0.65 saturated ammonium sulfate containing 0.05 M acetate buffer, pH 5.0.

Activators and Inhibitors. The enzyme is practically inactive when diluted with water and tested in the absence of cysteine. Histidine, 8-hydroxyquinoline, and other metal-binding agents also stimulate the enzyme.[13] Maximum activity is obtained when the enzyme is diluted with ice-cold water and tested immediately after dilution in a reaction mixture containing 0.01 to 0.05 M cysteine.

The crystalline enzyme is active without added Mg^{++}. The optimum range of Mg^{++} concentration, between 0.0005 and 0.0025 M, produces a fourfold activation. Other ions activate the enzyme to a greater or lesser degree.[14] Some, however, are quite inhibitory, especially copper, zinc, lead, mercury, and silver.[13] The fluoride inhibition of phosphoglucomutase, however, is a special case and is dependent on the concentration of the three ions, magnesium, fluoride, and glucose-1-phosphate. Thus when the concentration of any one ion was varied a reasonably constant K was obtained with the following equation:

$$(\text{Mg})(\text{F})^2(\text{glucose-1-phosphate}) \times \frac{\%\ \text{residual activity}}{\%\ \text{inhibited activity}} = K$$

The value of the constant obtained under these conditions averaged

[13] E. W. Sutherland, *J. Biol. Chem.* **180**, 1279 (1949).
[14] G. T. Cori, S. P. Colowick, and C. F. Cori, *J. Biol. Chem.* **124**, 543 (1938).

1.7 × 10⁻¹². It is unlikely that glucose-1,6-diphosphate enters this complex, inasmuch as its concentration is less than 0.001 as much as the other components of the complex.

It appears that a magnesium-fluoro-phosphate complex is formed which is inhibitory. The phosphate in this complex is the substrate, glucose-1-phosphate. However, inorganic phosphate also forms an inhibitory complex with magnesium and fluoride,[15] inasmuch as the addition of inorganic phosphate increases the inhibition over and above that exerted by the magnesium-fluoro-1-ester complex. A similar type of inhibition with fluoride was also observed using *Cl. welchii* lecithinase with either calcium or magnesium as the metal activator.[16]

SAMPLE PROTOCOL FOR PREPARATION OF CRYSTALLINE PHOSPHOGLUCOMUTASE

	Total activity, units	Total protein, mg.	Specific activity, units/mg.
Water extract (900 g. muscle)	9850	19,700	0.5
First heat filtrate (65°)	7800	3,400	2.3
0.0 to 0.65 ammonium sulfate ppt.	7480	1,160	6.45
Second heat filtrate (63°)	4370	336	13.0
First crystallization, 0.50–0.60 ammonium sulfate	2030	112	18.0
Second crystallization, 0.50–0.60 ammonium sulfate	1080	47	23.0
Second crystallization, 0.60–0.65 ammonium sulfate	1040	40	26.0

[15] O. Warburg and W. Christian, *Biochem. Z.* **310**, 384 (1942).
[16] V. A. Najjar, *in* "Phosphorus Metabolism" (McElroy and Glass, eds.), Vol. I, p. 500, Johns Hopkins Press, Baltimore, 1951.

[37] Phosphohexoisomerases from Muscle

By MILTON W. SLEIN

Phosphomannose Isomerase[1]

$$\text{M-6-P} \rightleftarrows \text{F-6-P}$$
$$(40\%) \qquad (60\%)$$

Assay Methods

TPN Reduction. A spectrophotometric method based on the conversion of M-6-P to G-6-P which is dehydrogenated by TPN in the presence of G-6-P dehydrogenase was orginally used.[1] This test system depends on the presence of a large excess of G-6-P dehydrogenase and phosphoglucose isomerase[2] so that the rate of reduction of TPN, as measured in a spectrophotometer at 340 mμ, is proportional to the concentration of PMI. This assay method is advantageous for its rapidity and independence on varying amounts of PGI present as a contaminant in the PMI fractions. However, it has a disadvantage in that studies of the effects of pH, ions, etc., are complicated by the presence of two other enzymes in the chain of reactions.

The reaction mixture is made up in a cell having a 1-cm. light path:

> 1.0 ml. 0.1 M Tris, pH 8
> 1.2 ml. H$_2$O
> 0.2 ml. 0.001 M TPN[3]
> 0.2 ml. 0.1 M MgCl$_2$
> 0.1 ml. PGI[4]
> 0.1 ml. G-6-P dehydrogenase[5]
> 0.1 ml. 0.1 M M-6-P[6]
> 2.9 ml.

0.1 ml. (about 5 to 50 units) of PMI is added, and readings are made at 340 mμ at 30-second intervals. One unit of enzyme is defined as that amount which causes a rate of change in optical density (ΔE_{340}) of 0.001 per minute after maximal velocity is attained (in about 1 minute). Specific activity is expressed in units per milligram of protein. Protein is determined by the method of Weichselbaum[7] or Lowry *et al.*[8]

F-6-P Color. A colorimetric method is based on the conversion of M-6-P to F-6-P, which gives about 65% of the color of free fructose in the resorcinol method of Roe.[9] This assay method has the advantage of being

---

[1] A previous report on this enzyme, which will be referred to as PMI, has been published by M. W. Slein, *J. Biol. Chem.* **186,** 753 (1950).

[2] Phosphoglucose isomerase, described below, will be referred to as PGI.

[3] TPN is available commercially. For methods of isolation and purification, see Vol. III [124].

[4] This preparation should have no PMI activity. 0.25 mg. of a fraction having about 12,000 units/mg. is satisfactory.

[5] About 0.2 unit of the preparation described by A. Kornberg, *J. Biol. Chem.* **182,** 805 (1950); see Vol. I [42].

[6] For preparation and purification of this compound, see Volume III [20]. If traces of other hexosemonophosphates are present as impurities, it may be necessary to take preliminary readings for about 5 minutes until no further reduction of TPN occurs before adding the PMI.

[7] T. E. Weichselbaum, *Am. J. Clin. Pathol.* **10,** 40 (1946); see Vol. III [73].

[8] O. H. Lowry, N. J. Rosebrough, A. L. Farr, and R. J. Randall, *J. Biol. Chem.* **193,** 265 (1951); see Vol. III [73].

[9] J. H. Roe, *J. Biol. Chem.* **107,** 15 (1934); see Vol. III [12].

a direct measure of PMI activity and, therefore, is amenable to the study of pH and ion effects. Although it is less rapid than the TPN-reduction method, many fractions and concentrations of PMI may be assayed simultaneously. Only a simple colorimeter is needed, and M-6-P is the only uncommon substance required. Since several milligrams of protein are readily soluble in the strong HCl, there is no need to deproteinize the reaction mixture before color development. One disadvantage arises from the fact that, for several steps in the purification procedure, so much PGI activity is present as contaminant (see Table I) that it is difficult to assign accurate specific activities to the PMI. This effect is minimized by carrying out the assay at pH 5.5, the optimum for PMI. PGI has only about 10% of its full activity at this pH (see below). The specific activity roughly parallels that found with the TPN-reduction assay, so that it is possible to use the F-6-P color method as a guide for fractionation.

The reaction mixture consists of 1 micromole of M-6-P in 0.4 ml. of 0.06 M acetate buffer, pH 5.5. To this is added 0.1 ml. of PMI (2 to 18 units), and the mixture is incubated at 30° for 2 to 18 minutes. 3.5 ml. of approximately 8.3 M HCl is added to stop the reaction. 1 ml. of 0.1% resorcinol in 95% ethanol is added, and the mixture is heated for 10 minutes at 80°, then cooled in a water bath at room temperature. The color intensity is read in a colorimeter with a filter which absorbs at about 540 mμ.

One unit of enzyme for this assay is defined as the amount which causes a change of 1 Klett colorimeter scale unit per minute of incubation time. In order to obtain proportionality of color intensity to enzyme concentration, it is necessary to work within a colorimeter scale range of about 35 units, which corresponds to about 15% of the color change which occurs at equilibrium in the PMI reaction.

Purification Procedure

The fractionations with ammonium sulfate are very reproducible, but it has been found that better purification is obtained when the protein concentration is adjusted to about 5 to 10 mg./ml. before subfractionations than with the more concentrated solutions of 20 to 30 mg./ml. originally used.[1] The more dilute solutions probably result in less coprecipitation of proteins other than PMI.

It is possible to proceed with purification without dialyzing the fractions. One assumes that the precipitates contain the same concentration of ammonium sulfate as the solutions from which they came. However, measurement of activities during such fractionations is impaired by the presence of ammonium ions which inhibit PMI activity rather strongly (see below).

For the following procedure, a saturated solution of ammonium sulfate was prepared at 25° and adjusted to about pH 7.8 by the addition of about 0.01 vol. of 29% ammonia. The degree of saturation during fractionation was calculated without regard to the volume changes which occur when diluting the salt solution and, therefore, does not represent the true degree of saturation. Furthermore, temperature effects were also neglected; the 25°-saturated ammonium sulfate was added (while at room temperature) to extracts which were chilled in an ice bath.

Muscle was removed from the back and hind limbs of a rabbit and passed once through a meat grinder in the cold. 700 g. of muscle was extracted with 700 ml. of cold 0.02 M KOH for about 15 minutes with occasional stirring. The extract was filtered through gauze, and the residue was re-extracted with 700 ml. of KOH. The crude extract (1260 ml.) was treated with 1032 ml. of ammonium sulfate to give 0.45 saturation. After 15 minutes the precipitate was removed by centrifugation in the cold and discarded. The opalescent supernatant fluid (2105 ml.) was treated with 468 ml. of ammonium sulfate to give a fraction of 0.55 saturation which was centrifuged off after about 30 minutes in the cold. The 0.45 to 0.55 fraction was dissolved with 35 ml. of cold distilled water and dialyzed overnight against 20 l. of cold distilled water. A flocculent precipitate was discarded after centrifugation.

The supernatant solution was diluted to contain about 5 mg. of protein per milliliter and then refractionated at 0.45, 0.50, 0.525, and 0.575 saturation with ammonium sulfate. The fractions were dissolved with water, dialyzed overnight, and centrifuged to remove precipitates which were discarded.

The 0.50 to 0.525 fraction was diluted to about 10 mg. of protein per milliliter and refractionated at 0.475, 0.50, and 0.525 saturation. The fractions were dissolved with water, dialyzed, and centrifuged to remove small precipitates.

The 0.475 fraction was further purified by treatment with alumina C_γ [10] or calcium phosphate gel. [11] The protein was diluted to about 11 mg./ml. with water and treated in an ice bath with 0.4 vol. of adsorbent for about 5 minutes and centrifuged. The enzyme was eluted by stirring the precipitate with cold M NaHCO$_3$ (equal in volume to that of the adsorbent suspension used) for about 2 minutes. After centrifugation, the clear supernatant fluid was dialyzed for 3 hours against cold distilled water. Somewhat better results were obtained with alumina than with calcium

[10] For preparation, see Vol. I [11]. This preparation contained 20 mg. (dry weight) per milliliter and was about 4 years old when used here.

[11] For preparation, see Vol. I [11]. This gel contained about 13 mg. (dry weight) per milliliter and was 6 months old.

phosphate gel. When the elution was carried out in two stages, using one-half the volume of NaHCO₃ each time, higher specific activity was obtained in the first eluate, according to the F-6-P color assay. However, the more accurate TPN-reduction assay showed no difference between

TABLE I

SUMMARY OF PMI PURIFICATION PROCEDURE[a]

Fraction	Protein, mg.	TPN reduction assay (pH 8)		F-6-P color assay (pH 5.5)		PGI/PMI[b]
		Units/mg.	Recovery, %	Units/mg.	Recovery, %	
1. KOH extract	24,450	84	—	7.4	—	406
2. 0.45–0.55[c]	4,100	192	38.3	14.0	31.7	271
3. 0.55–0.65	4,000	40	7.8	2.2	4.9	3,520
Subfractions of 2:						
4. 0.0–0.45	49.5	600	1.4	73.3	2.0	27
5. 0.45–0.50	96	208	1.0	41.6	2.2	35
6. 0.50–0.525	1,270	224	13.8	24.0	16.8	63
7. 0.525–0.575	88	44	0.2	2.5	0.1	920
Subfractions of 6:						
8. 0.0–0.475	456	288	6.4	104.0	26.2[d]	5
9. 0.475–0.50	261	48	0.6	0.0	0.0	—
10. 0.50–0.525	194	8	<0.1	0.0	0.0	—
Subfractions of 8:						
11. Not adsorbed by alumina	329	0[e]	0.0	0.0	0.0	—
12. Alumina eluate 1	55.7	640[e]	1.7	460.0	14.1	0.9
13. Alumina eluate 2	35.1	640[e]	1.1	300.0	5.8	5.5

[a] The TPN and F-6-P assay units are not directly comparable (see Assay Methods).

[b] PMI and PGI measured by the F-6-P assay at the pH optimum of each, pH 5.5 and pH 9, respectively. These ratios are high, especially in the less pure fractions, since PGI interferes with the PMI assay to some extent.

[c] These figures refer to degree saturation with ammonium sulfate.

[d] Apparent recovery is excessive because of the relatively large amount of PGI in the less pure fractions.

[e] Measured at pH 7.5 instead of pH 8.

the two eluates. This discrepancy is explained by the presence of about six times as much PGI in the second eluate as in the first. The relatively large amount of PGI resulted in an apparently lower specific PMI activity in the F-6-P color method.

Properties

Equilibrium. Approximately 40% M-6-P and 60% F-6-P are present at equilibrium at 30°.

Stability. The PMI activity of all fractions, including the crude KOH extract, is fairly stable for a week or so when stored at 5°. Some fractions, especially eluates from adsorbents, are less stable when stored frozen.

Effect of Ions. There is no evidence for a metal requirement for PMI activity. The effects of ions on PMI activity were measured in Tris buffer at pH 7.5 in order to compare them with those on PGI at the same pH. The chlorides of cations were used. Potassium has no effect up to 0.04 M. Divalent cobalt, magnesium, manganese, and calcium produce an inhibition amounting to 25 to 35% at 0.04 M. Ammonium ions show a marked inhibitory effect, amounting to about 55% at 0.01 M and 90% at 0.04 M. This is in contrast to the lack of effect of 0.04 M ammonium ions on PGI.

Inorganic pyrophosphate and orthophosphate inhibit PMI about 25 and 30%, respectively, as compared with 0.1 M Tris at pH 7.5.

Effect of pH. PMI has a sharp optimum for activity at pH 5.5 in acetate or pyridine buffers. The activity is half-maximal at about pH 4.75 and pH 7.5, and becomes negligible at pH 4 and pH 9.5. Phosphate, Tris, and alanine buffers were used for the range pH 6 to 9.5.

It was first possible to demonstrate that F-6-P is the product of M-6-P isomerization by PMI when the F-6-P color assay method was used at low pH's where contaminating PGI has low activity. Previously, similar tests, carried out at pH 7.4, were not informative since the PGI activity was much greater than that of the PMI.[1]

Phosphoglucose Isomerase[12]

$$G\text{-}6\text{-}P \rightleftarrows F\text{-}6\text{-}P$$
$$(68\%) \qquad (32\%)$$

Assay Methods

F-6-P Color. This procedure is the same as that described above for PMI except that the sodium salt of G-6-P [13] is used instead of M-6-P, and Tris buffer, pH 9, replaces the acetate. Since there is much less PMI activity relative to that of PGI in muscle (see Table I) and PMI has only about 5% of its activity at pH 9, this assay method is quite adequate for the measurement of PGI. Even with the crude KOH extract of rabbit muscle, only 2 γ of protein need be incubated in the 0.5 ml. of reaction mixture for 4 minutes at 30° to give sufficient F-6-P for an accurate assay. The unit of PGI activity is the same as that for PMI.

---

[12] This is the phosphohexoisomerase originally described by K. Lohmann, *Biochem. Z.* **262**, 137 (1933). Since it has been found that F-6-P is the product of M-6-P isomerization catalyzed by PMI, Lohmann's isomerase may be more descriptively called phosphoglucose isomerase (PGI).

[13] The crystalline barium hydrate is available commercially. For preparation and purification of G-6-P, see Vol. III [19].

TPN Reduction. A rapid assay method similar to that described for PMI could be used for PGI measurement. G-6-P dehydrogenase prepared from brewer's yeast according to Kornberg[5] may contain some PGI activity which must be corrected for in rate determinations. However, this contaminating PGI could be removed by further purification of the dehydrogenase. Repeated use of the isoelectric precipitation step at pH 4.4 would probably be sufficient to remove PGI activity since, in muscle, considerable loss of both phosphohexoisomerase activities occurs by treatment at pH 4.5.

Purification Procedure

Although PGI has been known since 1933, is widespread in nature, and appears to be one of the more active enzymes in the glycolytic path, no purification studies have been published. The following procedure results in material which is six to seven times as active as the crude muscle extract.

TABLE II
SUMMARY OF PGI PURIFICATION PROCEDURE

Fraction	Protein, mg.	Units/mg., thousands	Recovery, %
1. KOH extract	24,450	3.0	—
2. 0.45–0.55[a]	4,100	3.8	21
3. 0.55–0.65	4,000	7.75	42
Subfractions of 3:			
4. 0.0–0.55	310	4.25	1.8
5. 0.55–0.60	705	7.5	7.2
6. 0.60–0.65	565	12.5	9.6
7. 0.65–0.70	895	8.85	10.8
Subfractions of 6:			
8. $Ca_3(PO_4)_2$ gel eluate	182	9.0	2.2
9. Not adsorbed by gel	286	17.0	6.6
10. Fraction 9, precipitated with $(NH_4)_2SO_4$	91	20.0	2.5

[a] These figures refer to degree saturation with ammonium sulfate.

The ammonium sulfate procedure described above for PMI is followed. The supernatant fluid, after removal of the 0.45 to 0.55 fraction containing PMI, was adjusted to 0.65 saturation with ammoniacal ammonium sulfate. The 0.55 to 0.65 fraction was dissolved with about 35 ml. of cold distilled water and dialyzed overnight against 20 l. of cold distilled water. A slight precipitate was removed by centrifugation, and the solution was diluted to about 10 mg. of protein per milliliter and refractionated at 0.55, 0.60, 0.65, and 0.70 saturation with ammonium sulfate. The fractions were dissolved with water, dialyzed, and centrifuged to remove precipitates.

The 0.60 to 0.65 fraction was treated with alumina C_γ or calcium phosphate gel and eluted with $NaHCO_3$ as described for PMI. The nonadsorbed protein solutions had the higher activities. About 30% of the protein in the dilute nonadsorbed fraction could be recovered by treatment with ammoniacal ammonium sulfate at 0.65 saturation.

Properties

Equilibrium. Approximately 68% G-6-P and 32% F-6-P are present at equilibrium at 30°.

Stability. All fractions of PGI appear to be fairly stable at 5° or when frozen. Even very dilute solutions of PGI containing only 10 γ of protein per milliliter of 0.025 M Tris, pH 7.5, lose no significant activity during storage for several days at 5°.

Effect of Ions. No metal or other cofactor for PGI activity is known. As with PMI, the effect of metal chlorides was tested at pH 7.5. Potassium, calcium, and ammonium ions have no significant effect on PGI activity in concentrations up to 0.04 M. The lack of effect of calcium and ammonium ions is in contrast to their effects on PMI. At 0.04 M, magnesium produced about 20% inhibition, and cobalt and manganese inhibited about 30%. Inorganic pyrophosphate and orthophosphate inhibit PGI about 25% as compared with 0.1 M Tris at pH 7.5.

Effect of pH. PGI has optimal activity at about pH 9. Half-maximal activity is obtained at pH 7. Only 10% remains at pH 5.5, the optimal pH for PMI, and, at pH 4.5, PGI is practically inactive.

[38] Phosphohexokinase

Fructose-6-phosphate + ATP → Fructose-1,6-diphosphate + ADP

By Kuo-Huang Ling, William L. Byrne, and Henry Lardy

Assay Method

Principle. The formation of fructose-1,6-diphosphate is followed by a spectrophotometric method involving the conversion of FDP to triose phosphates and the oxidation or reduction of the latter by pyridine nucleotide-requiring enzymes. Two different systems are used. In each, fructose-6-phosphate and ATP are the substrates; Mg^{++} is an essential cofactor. System a, which is used for most fractionation work, contains, in addition to the above constituents, aldolase, myogen a, and DPNH. The FDP formed is cleaved by aldolase, and the dihydroxyacetone-phosphate formed is reduced by DPNH to α-glycerophosphate. System b is suitable only when the phosphohexokinase has been purified sufficiently to free it of triosephosphate isomerase. In this system the glyceraldehyde-3-phosphate, formed by the action of aldolase on FDP, is oxidized by DPN in the presence of triosephosphate dehydrogenase and arsenate.

Reagents

0.2 M Tris-HCl buffer solution (pH 8.0).

0.02 M ATP(Na) (pH 7).

0.01 M F-6-P (pH 7).

0.01 M DPN from which 0.006 M DPNH is prepared by reduction with alcohol dehydrogenase.[1]

0.01 M MgCl$_2$.

0.2 M cysteine HCl (pH 7), freshly prepared.

0.05 M sodium arsenate.

Aldolase, prepared from rabbit muscle according to the method of Taylor et al.[2]

Myogen a, prepared according to the method of Baranowski and Niederland.[3]

Glyceraldehyde-3-phosphate dehydrogenase, prepared according to the method of Cori et al.[4]

All enzymes were stored in ammonium sulfate solution in crystalline form and diluted in cold 0.01 M Tris-HCl buffer (pH 8.0) before use.

Procedure. (a) To a 3-ml. Beckman cell, 0.5 ml. of Tris buffer, 0.30 ml. of ATP, 0.20 ml. of F-6-P, 0.3 ml. of MgCl$_2$, 0.1 ml. of cysteine HCl, 0.5 ml. of DPNH, 0.1 ml. of aldolase (about 100 γ), 0.1 ml. of myogen a (about 20 γ), and the enzyme sample to be assayed are added. The volume is made to 3 ml. with water. After mixing, the absorbance is measured at 340 mμ every minute for 5 minutes against a water blank. The disappearance of DPNH is also measured in a control cuvette containing all components except phosphohexokinase. The average decrease in absorbance

[1] E. C. Slater, *Biochem. J.* **53**, 159 (1953); E. Racker, *J. Biol. Chem.* **184**, 313 (1950); see also Vol. I [79].

[2] J. F. Taylor, A. A. Green, and G. T. Cori, *J. Biol. Chem.* **173**, 591 (1948); see also Vol. I [39].

[3] T. Baranowski and T. R. Niederland, *J. Biol. Chem.* **180**, 543 (1949); see also Vol. I [58].

[4] G. T. Cori, M. W. Slein, and C. F. Cori, *J. Biol. Chem.* **173**, 605 (1948); see Vol. I [60].

per minute occurring in the control is subtracted from that occurring in the experimental cuvette.

(b) To a 3-ml. Beckman cell, 0.5 ml. of Tris-HCl buffer, 0.3 ml. of ATP, 0.2 ml. of F-6-P, 0.3 ml. of $MgCl_2$, 0.1 ml. of DPN, 0.1 ml. of cysteine, 0.2 ml. of Na-arsenate, 0.03 ml. of glyceraldehyde-3-phosphate dehydrogenase (about 300 γ), 0.1 ml. of aldolase (about 100 γ), and the source of phosphohexokinase are added. The volume is made to 3 ml. with water. After mixing, the absorbance at 340 mμ is measured every minute for 5 minutes against a water blank. The average increase in absorbance per minute is corrected for the change occurring in an appropriate control as in procedure a.

Calculation of Activity. The activity of the enzyme is expressed in terms of units. One unit is that amount of enzyme which will phosphorylate 1 μM. of F-6-P per minute under the above conditions. The specific activity is expressed in terms of units per milligram of protein.

Procedure a. The myogen a used as a source of α-glycerophosphate dehydrogenase has considerable amounts of triosephosphate isomerase, and further addition of muscle triosephosphate isomerase prepared according to the method of Oesper and Meyerhof[5] did not increase the rate of DPNH oxidation in procedure a. Therefore triosephosphate isomerase was omitted. Studies of the stoichiometry have established that in this procedure each mole of FDP formed yields 2 moles of dihydroxyacetone phosphate and therefore oxidizes 2 moles of DPNH. If the extinction coefficient of DPNH reported by Horecker and Kornberg[6] is used, the factor for converting net absorbancy change to micromoles of FDP formed is 0.24.

Procedure b. The aldolase and glyceraldehyde-3-phosphate dehydrogenase used were recrystallized until free of detectable triosephosphate isomerase. When the phosphofructokinase has been purified until it also is free of isomerase, only one of the two trioses formed from FDP by aldolase should react. Quantitative studies confirm this expectation. Therefore, net absorbancy change per minute $\times$ 0.482 gives micromoles of FDP formed.

Purification Procedure

Step 1. Preparation of Crude Extract. A rabbit is anesthetized by intraperitoneal injection of 15 ml. of 25% magnesium sulfate and is decapitated. The back and leg muscles (590 g.) are removed and thrust into cracked ice. Further steps are conducted in the cold room. The muscle is passed through a meat grinder, extracted with 2 vol. (1180 ml.) of

[5] P. Oesper and O. Meyerhof, *Arch. Biochem.* **27**, 223 (1950).
[6] B. L. Horecker and A. Kornberg, *J. Biol. Chem.* **175**, 385 (1948).

cold 0.03 M KOH for 15 minutes, and squeezed through four layers of cheesecloth. The pH of the extract is about 9. It is then centrifuged to clear the solution of particulate material (fraction I).

Step 2. Purification with Ethanol. To the crude extract, solid magnesium acetate is added to a final concentration of 0.05 M, and the pH is adjusted to 7.5 with 0.2 N HCl (final volume 800 ml.). Alcohol fractionation is conducted in a cold bath ($-5°$), and the temperature of the solution is kept at $-2°$ to $-5°$. With stirring, 40 ml. of dry ice-cooled 95% ethanol is added very slowly. After addition of ethanol, the solution is stirred for 30 minutes in the cold bath. The precipitate is removed by centrifuging at $-2°$ for 30 minutes at 1000 × g. A second addition of 51.4 ml. of 95% ethanol is made as described above, and the precipitate (fraction II) collected by centrifuging.

Step 3. Purification with Ammonium Sulfate. Fraction II is suspended in a minimum amount of cold 0.1 M KHCO$_3$ (final volume of solution 71 ml.). To this solution 24.2 ml. of saturated ammoniacal ammonium sulfate (pH 8.2, saturated at 0°) is added and kept at 0° for 4 hours. The precipitate is removed by centrifuging for 15 minutes at 14,400 × g. To the supernatant a further 47 ml. of ammonium sulfate solution is added, and the mixture is allowed to stand at 0° for 4 hours. After centrifuging for 15 minutes at 14,400 × g, the precipitated protein is dissolved in a minimum amount of 0.1 M KHCO$_3$ (fraction III, 11 ml.). To fraction III, 4.7 ml. of ammonium sulfate is added, and the solution is kept at 0° for 4 hours. The precipitate is removed by centrifuging, and 2.7 ml. of ammonium sulfate is added to the supernatant. After 2 hours the precipitate is collected by centrifuging and is dissolved in 4 ml. of 0.1 M KHCO$_3$ (fraction IV). The enzyme is quite stable when kept frozen at the stage of fraction IV. Typical data of a purification are shown in the table.

PURIFICATION OF PHOSPHOFRUCTOKINASE

Fraction	Protein, mg.	Enzyme, units	Specific activity	Yield, %	Purification
I. KOH ext.	12,960	6350	0.49	100	
II. Mg-EtOH pptn.	292	5160	17.7	81	36×
III. (NH$_4$)$_2$SO$_4$, 0.25–0.5	144	3916	27.2	62	55×
IV. (NH$_4$)$_2$SO$_4$, 0.3–0.4	28	2870	104	45	210×

Properties

The preparation of specific activity of 104 is not pure. Activities up to 130 have been obtained but, as yet, not in high yield. At specific activity = **104,** 100,000 g. of enzyme phosphorylate 10,400 moles of substrate per minute at 26°. For this enzyme not only ATP but also ITP, UTP, and CTP can serve as phosphate donors. The K_s for each of these nucleotides is approximately 3×10^{-5} M.[7] In addition to fructose-6-phosphate, D-tagatose-6-phosphate and sedoheptulose-7-phosphate are phosphorylated by the purified enzyme.

[7] K.-H. Ling and H. A. Lardy, *J. Am. Chem. Soc.* **76**, 2842 (1954).

[39] Aldolase from Muscle

$$FDP \rightleftarrows GAP + DAP^1$$

By JOHN FULLER TAYLOR

Assay Method

Principle. Methods in use are based on the measurement of triose phosphate formed in the aldolase reaction. Triose phosphate may be determined either chemically or enzymatically.

1. Chemical Method. The method used to follow the isolation of crystalline aldolase from rabbit muscle[2] is based on the measurement of the alkali-labile phosphate of the triose phosphate formed in the reaction[3,4] which is carried out in the presence of cyanide to favor the formation of triose phosphate.

Reagents

> FDP solution (0.01 M). Prepare from the mono- or dibarium salt,[5] removing Ba with an equivalent amount of Na_2SO_4. Commercial samples may contain inorganic phosphate, causing a high blank. This can be removed by reprecipitation of the material as the monobarium salt (Neuberg *et al.*[6]). The 0.01 M solution is prepared from 0.1 M stock solution by dilution in 0.1 M glycine buffer, pH 9.0.
>
> 0.1 M KCN acidified to pH 9.0.
>
> 1.0 M NaOH freshly prepared each day.
>
> Reagents for phosphate determination.

[1] DAP = dihydroxy acetone phosphate.
[2] J. F. Taylor, A. A. Green, and G. T. Cori, *J. Biol. Chem.* **173**, 591 (1948).
[3] O. Meyerhof and K. Lohmann, *Biochem. Z.* **271**, 89 (1934).
[4] D. Herbert, A. H. Gordon, V. Subrahmanyan, and D. E. Green, *Biochem. J.* **34**, 1108 (1940).
[5] Fructose-1,6-diphosphate is available commercially. For methods of isolation and purification of this compound, see Vol. III [22].
[6] C. Neuberg, H. Lustig, and M. A. Rothenberg, *Arch. Biochem.* **3**, 33 (1943–44).

Enzyme solution, diluted with 0.1 KCN (pH 9) to contain approximately 0.1 unit in 1.0 ml. To preserve the full activity of the crystalline enzyme at greater dilution, it is advisable to add a protective protein, such as serum albumin, 0.2 mg./ml. of solution.

Procedure. To 0.1 ml. of FDP solution add 0.1 ml. of enzyme. Allow to incubate 5 minutes at 30°. Stop the reaction by the addition of 1.0 ml. of 1 M NaOH. Incubate for 15 minutes at room temperature. Exactly neutralize the NaOH, and proceed to determine phosphate by the method of Fiske and SubbaRow.[7] Appropriate blanks may be run to allow for phosphate in enzyme or reagents.

Other Chemical Methods. Triose phosphate may also be determined by means of colorimetric methods proposed by Sibley and Lehninger[8] and by Dounce and Beyer,[9] with modifications by Dounce *et al.*[10] and by Lowry *et al.*[11] These methods have been applied to crude extracts of various tissues as well as to crystalline aldolase. The different methods yield somewhat different activities under varying conditions.

2. Enzymatic Method. The method introduced by Warburg and Christian[12] is based on measurement of the reduction of DPN by the D-glyceraldehyde-3-phosphate formed in the aldolase reaction, in the presence of arsenate and excess D-glyceraldehyde-3-phosphate dehydrogenase.

Reagents

0.1 M FDP, adjusted to pH 7.6.
0.17 M sodium arsenate (5.4 g. of $Na_2HAsO_4 \cdot 7H_2O$ in 100 ml.).
0.005 M DPN.[13]
0.27 M glycine (2.0 g. in 100 ml.).
0.1 M cysteine or glutathione, pH 7.6.
D-glyceraldehyde-3-phosphate dehydrogenase, crystalline, from muscle or yeast.[14] The enzyme should be recrystallized until free of triose phosphate isomerase and α-glycerophosphate dehydrogenase.[15] Stock solution to contain 2.5 mg./ml.
Enzyme solution, diluted in water to contain less than 0.1 unit/ml.

[7] C. H. Fiske and Y. SubbaRow, *J. Biol. Chem.* **66**, 375 (1925); see Vol. III [147].
[8] J. A. Sibley and A. L. Lehninger, *J. Biol. Chem.* **177**, 859 (1949).
[9] A. L. Dounce and G. T. Beyer, *J. Biol. Chem.* **173**, 159 (1948).
[10] A. L. Dounce, S. R. Barnett, and G. T. Beyer, *J. Biol. Chem.* **185**, 769 (1950).
[11] O. H. Lowry, N. R. Roberts, M.-L. Wu, W. S. Hixon, and E. J. Crawford, *J. Biol. Chem.* **207**, 19 (1954).
[12] O. Warburg and W. Christian, *Biochem. Z.* **314**, 149 (1943).
[13] This compound is available commercially from various sources. For methods of isolation and purification, see Vol. III [124].
[14] For methods of isolation of these enzymes, see Vol. I [60, 61].
[15] T. Baranowski and T. R. Niederland, *J. Biol. Chem.* **180**, 543 (1949); see also Vol. I [58].

Procedure. Into a spectrophotometer cuvette place 0.5 ml. of FDP, 0.3 ml. of arsenate, 0.1 ml. of DPN, 0.3 ml. of glycine, 0.1 ml. of dehydrogenase, 0.6 ml. of cysteine, and 1.0 ml. of H_2O. Incubate for 5 to 10 minutes, start the reaction by the addition of 0.1 ml. of enzyme solution, and read the optical density (log I/I_o), at 340 mμ at 30-second intervals.

$(\log_{10} I_o/I$ in 1 minute$)/(6.22 \times 10^6)$ = Moles FDP transformed
per minute

Moles FDP per minute $\times 2 \times 31 \times 10^3$ = Mg. P transformed
per minute

Applicability of the Enzymatic Method. This method is subject to large errors from interfering enzymes which may be present in crude tissue extracts and also in insufficiently purified preparations. Baranowski and Niederland[15] have discussed the precautions necessary to obtain full activity by this method and have also described an alternative enzymatic method.

Definition of Unit and Specific Activity. One unit of enzyme was defined by Meyerhof and Beck[16] as that amount which transforms 1 mg. P in 1 minute. Specific activity is expressed as units per milligram of protein. Protein may be determined by the biuret method (Robinson and Hogden[17]) or by the method of Lowry *et al.*[18] The protein concentration in solutions of pure aldolase can also be measured optically at 280 mμ.

$(\log_{10} I_o/I)/0.91d$ = mg. protein per ml.[15]

The specific activity of crystalline rabbit muscle aldolase has been reported to be 1.74 units/mg., measured by the optical test at pH 7.3 and 30°. This corresponds to a turnover number of 4200 moles of FDP transformed per minute per 1.5×10^5 g. of proteins.[15] Specific activity measured by the chemical test at pH 9 was reported to be 30% lower.[15]

Purification Procedure

The following procedure is that developed and recommended by Taylor *et al.*[2] It has been used repeatedly in their laboratory and has been carried out successfully in many others, as described here or with minor modifications. Aldolase has also been crystallized from rat muscle by a more complicated procedure.[2,12]

[16] O. Meyerhof and L. V. Beck, *J. Biol. Chem.* **156**, 109 (1944).

[17] H. W. Robinson and C. G. Hogden, *J. Biol. Chem.* **135**, 727 (1940).

[18] O. H. Lowry, N. J. Rosebrough, A. L. Farr, and R. J. Randall, *J. Biol. Chem.* **193**, 265 (1951); see Vol. III [73].

Step 1. Extraction. Pass rabbit muscle through a chilled meat grinder, and extract twice by stirring with 1 vol. of cold distilled water[19] for 10 minutes and pressing gently through gauze.

Step 2. Fractionation with Ammoniacal Ammonium Sulfate. Adjust the pH of the cold extract to about 7.5 by cautious addition of cold 0.1 N alkali. Add an equal volume of ammonium sulfate solution, saturated at room temperature and adjusted to pH 7.5 by the addition of ammonia, stir gently, and cool to 0 to 5°. Remove the precipitate by filtration or centrifugation. Bring the supernatant to 0.52 saturation. Aldolase crystallizes in the course of some hours or several days. Initial crystallization may be hastened by seeding or by warming to room temperature and again cooling to 0 to 5°. Further standing in the cold for several days is recommended to obtain maximum yield.

Step 3. Recrystallization. Dissolve the crystalline precipitate, after centrifugation or filtration, in sufficient cold water to give 1 to 2% protein solution. Add saturated ammonium sulfate, pH 7.5, until precipitation just begins, and centrifuge to clarify. Continue to add saturated ammonium solution, pH 7.5, to bring about crystallization.

Other methods of recrystallization have been described.[2]

Properties

Specificity. In the condensation reactions catalyzed by aldolase, the muscle enzyme has been reported to be absolutely specific for dihydroxyacetone phosphate whereas D-glyceraldehyde phosphate can be replaced by other aldehydes.[20] Dounce *et al.*[10] have reported, however, that fructose-6-phosphate is split by the enzyme. Although only *trans*-linkage has been shown to be formed in the condensation, Lardy[21] has reported that tagatose-1,6-diphosphate, which has a *cis*-linkage, is split slowly.

Activators and Inhibitors. The activity of the muscle enzyme is inhibited by heavy metals, notably Ag and Cu, at concentrations of $2 \times 10^{-4} M$.[4] Iodine also inhibits the activity completely in low concentrations, but other oxidizing agents and reducing agents tested do not do so.[4] Iodoacetate does not inhibit.[4]

Glucose, fructose, and fructose-6-phosphate have been reported to inhibit the activity in a competitive fashion.[4]

[19] 0.03 M NaOH or KOH may be used instead of water. This reduces the amount of alkali which must be used in the subsequent neutralization and is an essential step if the extract is also to be used for the preparation of D-glyceraldehyde phosphate dehydrogenase or phosphofructokinase.

[20] O. Meyerhof, K. Lohmann, and P. Schuster, *Biochem. Z.* **286**, 301, 319 (1936); K. Lohmann, *Angew. Chem.* **49**, 327 (1936).

[21] H. A. Lardy *in* "Phosphorus Metabolism" (W. D. McElroy and B. Glass, eds.), Vol. I, p. 116. The Johns Hopkins Press, Baltimore, 1951.

The muscle enzyme is neither inhibited nor activated in the presence of metal-binding reagents.[12] The activity is apparently increased in the alkaline pH range in the presence of tris(hydroxymethyl)aminomethane buffer.[10]

Effect of pH. The enzyme was reported to be most active at pH 9 by Herbert *et al.*,[4] using the chemical method described above in which cyanide is used to trap the triose phosphate formed in the reaction. Sibley and Lehninger,[8] using hydrazine as a trapping agent, reported maximal activity at pH 8.6. With the optical enzymatic method, however, maximal activity was found at pH 7,[15] and with a chemical method in which no trapping agent was used the maximum was also close to pH 7.[9] It was pointed out[8,15] that this difference might arise because with the reagents used trapping is incomplete at pH 7. Dounce *et al.*[10] have now shown, however, that the use of tris(hydroxymethyl)aminomethane buffer in the hydrazine method produces a stimulatory effect near pH 9 and that with the use of collidine or other buffers, and hydrazine as a trapping agent, maximal activity is observed at pH 7.2.

Enzymatic Purity. Rabbit muscle aldolase has been shown to be free of phosphofructokinase, triose phosphate isomerase, D-glyceraldehyde-3-phosphate dehydrogenase, and α-glycerophosphate dehydrogenase activities, although eight or nine recrystallizations may be required.[2,15,22]

CRYSTALLIZATION OF ALDOLASE; SAMPLE PROTOCOL[a]

		Protein, g.	Enzyme activity, units	Units/mg. protein
Extract from 500 g. of rabbit skeletal muscle		24.0	2136	0.089
	Saturation			
Fractionation with	0 − 0.40	4.1	131	0.031
(NH$_4$)$_2$SO$_4$	0.40 − 0.50	0.6	91	0.152
	0.50 − 0.52[b]	2.43	1631	0.671
	Supernatant fluid	17.1	205	0.012
Total recovered		24.2	2058	
0.50–0.52 fraction, recrystallized		.		0.684

[a] J. F. Taylor, A. A. Green, and G. T. Cori, *J. Biol. Chem.* **173**, 591 (1948).
[b] Crystalline.

[22] E. Racker, *J. Biol. Chem.* **196**, 347 (1952).

Physical Properties. Crystalline rabbit muscle aldolase has been shown to move as a single component in the ultracentrifuge[23,24] and upon electrophoresis over the pH range 5 to 9.[2] It does not appear to be completely homogeneous when examined by the reversible boundary technique.[24] Physical constants[23,24] and amino acid composition[25] have been determined. The molecular weight is approximately 147,000, and the isoelectric point is about 6.1 in phosphate buffer, ionic strength 0.1. The isoelectric point varies greatly with buffer ion species and concentration,[2,26] extrapolating to an isoionic point at about pH 9.

[23] J. F. Taylor *in* "Phosphorus Metabolism" (W. D. McElroy and B. Glass, eds.), Vol. I, p. 104. The Johns Hopkins Press, Baltimore, 1951.
[24] J. F. Taylor, unpublished observations.
[25] S. F. Velick and E. Ronzoni, *J. Biol. Chem.* **173**, 627 (1948).
[26] S. F. Velick, *J. Phys. & Colloid Chem.* **53**, 135 (1949).

[40] Aldolase from Yeast

Fructose-1,6-diphosphate $\rightleftarrows$ D-Glyceraldehyde-3-phosphate
$$+ \text{Dihydroxyacetone phosphate}$$

By WALTER CHRISTIAN

Assay Method

Principle. The optical assay described by Warburg[1] is used. When aldolase is added to a solution containing FDP, DPN, arsenate, and excess GAP dehydrogenase (see Vol. I [60, 61]), the rate of increase in absorbance at 340 mμ is proportional to the rate of formation of glyceraldehyde-3-phosphate by aldolase.

Reagents. The composition of the assay mixture is given in Table I.

TABLE I
ASSAY MIXTURE FOR ALDOLASE

DPN	0.6 mg.	$3 \times 10^{-4} M$
Na HDP, pH 7.4	3.0 mg. p.	$1.67 \times 10^{-2} M$
Glycine	6.0 mg.	$2.66 \times 10^{-2} M$
$Na_2HAsO_4 \cdot 7H_2O$	16.2 mg.	$1.73 \times 10^{-2} M$
GAP dehydrogenase	0.25 mg.	
Distilled water	To 3.0 ml.	
pH	7.5	

[1] O. Warburg, "Wasserstoffübertragende Fermente," pp. 51 and 315, Dr. Werner Saenger, Berlin, 1948.

HDP serves as the buffer. Glycine is used to bind traces of copper which strongly inhibit GAP dehydrogenase. If the activity of yeast aldolase is to be tested at constant metal ion concentrations, or the inhibition and its reversal are to be investigated, the water is replaced by

$$\begin{pmatrix} 0.6 \text{ ml. } 0.1 \ M \text{ cysteine-HCl} \\ + \\ 0.6 \text{ ml. } 0.1 \ N \text{ NaOH} \end{pmatrix} \text{ or } \begin{pmatrix} 0.6 \text{ ml. } 0.1 \ M \text{ cysteine-HCl} \\ + \\ 0.6 \text{ ml. } 0.1 \ N \text{ NaOH} \\ + \\ 0.1 \text{ ml. } 1.74 \times 10^{-2} \ M \text{ ZnSO}_4 \end{pmatrix}$$

Procedure. 0.02 ml. of solution to be assayed is added to the cuvette, and the increase in absorbance measured. The same procedure is followed for assays in pyrophosphate buffer or in the presence of other complex-forming agents in the presence or absence of added metals. In determinations of the influence of oxygen, the solution is thoroughly flushed with oxygen or argon before aldolase is added. The concentration of metals must be kept so low that during the time of the experiment there will be no significant loss of cysteine through oxidation. In all cases GAP should be added to test for inhibition of GAP dehydrogenase or other reaction of GAP (e.g., with KCN).

TABLE II

ASSAY OF ALDOLASE

($d = 0.574$ cm., $\lambda = 340$ mμ, cysteine 2×10^{-2} M, ZnSO$_4$ 5.8×10^{-4} M)

Minutes	0.025 mg. protein $\Delta \ln \frac{I_0}{I}$	0.050 mg. $\Delta \ln \frac{I_0}{I}$	0.10 mg. $\Delta \ln \frac{I_0}{I}$
1	0.113	0.239	0.486
2	0.226	0.459	0.952

Purification Procedure

Aldolase was discovered by Meyerhof and Lohmann in 1934.[2] D. E. Green and co-workers[3] purified aldolase from muscle in 1940. Muscle aldolase was crystallized in 1942 by Warburg and Christian,[4] who found no requirement for coenzymes and no inhibition by complexing agents.

[2] O. Meyerhof and K. Lohmann, *Biochem. Z.* **271**, 102 (1934).

[3] D. Herbert, A. H. Gordon, V. Subrahmanyan, and D. E. Green, *Biochem. J.* **34**, 1108 (1940).

[4] O. Warburg and W. Christian, *Biochem. Z.* **314**, 149 (1943); for crystallization see O. Warburg and G. Krippahl, *Z. Naturforsch.* **9b**, 181 (1954).

Under the given assay conditions a different behavior was observed with yeast aldolase which was extensively purified by Warburg and co-workers[1,4] but has not yet been crystallized.

Step 1. Lebedew Juice. One part of washed and dried bottom yeast is ground with 3 parts of water according to the method of Lebedew,[5] then kept for 2 hours at 38° and centrifuged at high speed. The supernatant, pH 5.8, is decanted and cooled.

Step 2. Acetone Precipitation. To the solution is added 0.4 times its volume of cold acetone. The precipitate is discarded, and to the supernatant is added a further $\frac{1}{10}$ volume of cold acetone. This precipitate is dissolved in cold distilled water, made up to 0.4 the initial volume, and centrifuged.

Step 3. Alcohol Precipitation. To the supernatant solution from step 2 (pH 6.2) is added at 0° 0.45 its volume of cold alcohol. After centrifugation the residue is discarded. To the supernatant is added a further $\frac{1}{20}$ vol. of alcohol at 0°. This precipitate is dissolved in cold water and made up to $\frac{1}{5}$ the initial volume of Lebedew juice.

Step 4. Aluminum Hydroxide Adsorption and Elution. After the solution has been adjusted to pH 6.5, aldolase is adsorbed on a just sufficient volume of aluminum hydroxide (added as 30% suspension). The adsorbate is washed with a large volume of water and then completely eluted with a little 0.02 M pyrophosphate buffer, pH 7.1. The eluate is diluted with an equal volume of water and carefully acidified to pH 4.9. The inactive precipitate is centrifuged off.

Step 5. Alcohol Precipitation. To the supernatant is added carefully one-third its volume of cold alcohol, and the precipitate is centrifuged. The residue is suspended in a little water and brought into solution by addition of dilute NaOH.

Step 6. Drying. The frozen neutral solution is dried to a powder which keeps its activity well. This powder has about the same activity as muscle aldolase in the optical assay, but since the preparation is not crystalline no statement can be made about its purity.

Properties

Activators and Inhibitors. Yeast aldolase is inhibited by complexing agents, and this inhibition can be reversed by addition of metals. If these metals are capable of autoxidation, the activity of aldolase depends on the oxygen pressure. The inhibition by pyrophosphate at pH 7.4 is shown in Table III.

[5] A von Lebedev, *Z. physiol. Chem.* **73**, 447 (1911).

TABLE III
PYROPHOSPHATE INHIBITION OF ALDOLASE

Inhibition in the	Pyrophosphate concentration		
	$10^{-3} M$	$3.33 \times 10^{-3} M$	$10^{-2} M$
First minute	0%	21%	57%
Fifth minute	22.5%	66.5%	88%

α,α'-Dipyridyl complexes ferrous ions and shows the following inhibition;

TABLE IV
DIPYRIDYL INHIBITION OF ALDOLASE

Dipyridyl concentration, %	Inhibition, %	
	pH 6.4	pH 7.4
0.005	25	0
0.0167	77	0
0.050	94	38
0.167	—	80

0.02 M Cysteine extensively inhibits yeast aldolase at neutral pH and completely at slightly alkaline pH, whereas 0.02 M glutathione inhibits only at slightly alkaline pH. This difference is probably connected with the fact found by Kubowitz[6] that cysteine binds iron more firmly. Warburg and co-workers found that the cysteine inhibition is greater in oxygen than in argon. These results are summarized in Table V.

TABLE V
CYSTEINE AND GLUTATHIONE INHIBITION

Cysteine	pH 7.4				pH 8.5	
	0	0	0.02 M	0.02 M	Cysteine, 0.02 M	GSH, 0.02 M
Gas phase	Argon	O_2	Argon	O_2	Air	Air
Inhibition, %	0	0	75	90	97	31

According to a principle stated in 1923 by Warburg,[7] it may be concluded from those inhibitions that the enzyme activity is connected with a dissociable heavy metal complex. The heavy metal effect was also demonstrated by reactivating the inhibited aldolase by means of added metals. However, the following three precautions should be observed:

[6] F. Kubowitz, *Biochem. Z.* **282,** 278 (1935).
[7] O. Warburg and S. Sakuma, *Pflügers Arch. ges. Physiol.* **200,** 203 (1923).

1. The inhibition should not be due to irreversible inactivation.

2. The metal salt concentration should be so low that only a fraction of the complexing agent is bound and that during the experiment the concentration of complexing agent should not be significantly lowered through other reactions.

3. The metal itself should not inhibit the enzyme.

Warburg and co-workers noted that zinc, iron, cobalt, and copper reverse the cysteine inhibition, but magnesium, chromium, nickel, and vanadium are ineffective. Addition of zinc could prevent but not reverse inhibition by pyrophosphate. However, cysteine inhibition is reversible at every stage. Reversal by iron, cobalt, and copper was found only in argon, whereas in oxygen no or only slight reversal was observed.

TABLE VI
IRON REVERSAL OF CYSTEINE INHIBITION

Cysteine	0	0.02 M	0.02 M	0.02 M	0.02 M
Iron	0	0	$10^{-5} M$	0	$10^{-5} M$
Gas	O_2 or A	O_2	Argon	Argon	Argon
$\Delta \ln \frac{I_0}{I}$ in 5-min.	1.618	0.137	0.204	0.503	1.128

As zinc does not undergo a change of valence, the activation of yeast aldolase is independent of the oxygen pressure. At the same time zinc displaces other metals from aldolase and thus always produces the highest activity of aldolase under the assay conditions.

TABLE VII
ZINC REVERSAL OF CYSTEINE INHIBITION

Cysteine (M)	0	0.02	0.02	0.02	0.02	0.02
Zinc (M)	0	0	0	2.32×10^{-5}	2.32×10^{-4}	2.32×10^{-4}
Gas	O_2 or A	O_2	Argon	O_2	O_2	Argon
$\Delta \ln \frac{I_0}{I}$ in 5 min.	1.567	0.140	0.391	0.959	1.932	1.903

Under conditions of the aldolase assay ferrous and cobaltous ions reverse the inhibition, but ferric and cobaltic ions are ineffective. This behavior would suffice to explain the Pasteur effect of respiration on fermentation upon changing from aerobic to anaerobic conditions as being due to a change of valence of the metals in the yeast cell. However, Warburg and co-workers noted that *muscle* aldolase is not influenced by metals or complexing agents. So far the same enzyme activity in various cells has been found to require the same cofactors. This would be the first case in which the same activity is due to different active centers.

On the other hand, Lynen[8] and Holzer[9] have explained the Pasteur effect in living yeast on transition to anaerobiosis by a shift in the phosphate concentration. On the basis of their data they consider improbable an oxidative influence on yeast aldolase. Clarification of the mechanism of the various metal effects on yeast aldolase should extend our knowledge of enzymes and should remove some of the differences between yeast and muscle aldolase.

[8] F. Lynen and R. Koenigsberger, *Ann.* **573**, 60 (1951).
[9] H. Holzer and E. Holzer, *Z. physiol. Chem.* **292**, 232 (1953).

[41] Fructose-1-Phosphate Aldolase from Liver

$$\text{F-1-P} \rightleftarrows \text{PDA} + \text{GA}^1$$

By FRANZ LEUTHARDT and H. P. WOLF

Assay Methods

1. Optical Test. Principle. DPNH is dehydrogenated by PDA formed from F-1-P through the action of the aldolase. The rate of decrease of the extinction at 340 mμ may be taken as a measure of enzyme activity. The protein fractions used at present as a source of the aldolase always contain the enzyme transferring hydrogen from DPNH to PDA (α-glycerophosphate dehydrogenase).[2] In sufficiently purified enzyme solutions (not yet available), α-glycerophosphate dehydrogenase will have to be added. In protein fractions from liver there is another enzyme, transferring hydrogen from DPNH to the second split product of the aldolase reaction, glyceraldehyde (glycerol dehydrogenase).[3]

Reagents

DPNH solution about 5 μM/ml.
Veronal buffer, 0.1 M, pH 7.4.
F-1-P solution. (The barium salt of F-1-P is treated with a cation exchanger, e.g., Ionac C-240, filtered, and neutralized with NaOH.) The exact concentration of the solution may be checked by the method of Roe[4] or better by the determination of acid-labile phosphate (hydrolysis in 1 M HCl for 15 minutes at 100°).
Enzyme solution. See Purification Procedure.

[1] Abbreviations: PDA = phosphodihydroxyacetone, GA = glyceraldehyde.
[2] T. Baranowski, *J. Biol. Chem.* **180**, 535 (1949); see also Vol. I [58].
[3] H. P. Wolf and F. Leuthardt, *Helv. Chim. Acta* **36**, 1463 (1953).
[4] J. H. Roe, *J. Biol. Chem.* **107**, 15 (1934).

Procedure. To a quartz cell of the Beckman photometer (or another suitable instrument[5]) add 1.5 ml. of veronal buffer, 1.0 ml. of water, 0.4 ml. of the enzyme solution, and 0.1 ml. of DPNH (in the blank DPNH solution is replaced by water). The reaction is started at zero time by adding 0.4 ml. of the F-1-P solution (example, see Fig. 1).

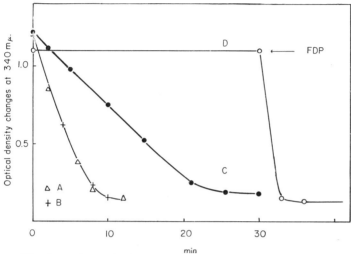

Fig. 1. The figure shows the difference between the F-1-P and FDP aldolase.
Curve A: action of FDP aldolase on FDP.
Curve B: action of F-1-P aldolase preparation on FDP (equal activity of both the enzymes on FDP).
Curve C: action of F-1-P aldolase preparation on F-1-P.
Curve D: action of FDP aldolase on F-1-P (no splitting). After 30 minutes FDP is added. The splitting reaction starts.
E represents optical density changes at 340 mμ.

2. The method proposed by Sibley and Lehninger[6] for measuring the activity of the FDP aldolase may also be used for testing F-1-P aldolase activity. F-1-P is incubated with the enzyme in the presence of hydrazine. After incubation the triosehydrazones are converted to the 2,4-dinitrophenylhydrazones which give a deep red color in alkaline solution. The absorption is measured at 540 mμ.

Purification Procedure

The F-1-P aldolase has not yet been prepared in the pure state. To obtain active solutions fresh rat liver is perfused with cold isotonic KCl

[5] A very suitable photometer for optical tests is the "Eppendorf" spectrophotometer of the Elektromedizinische Werkstätten GmbH, Hamburg, which uses as a source of monochromatic light the Hg line at 366 mμ.
[6] J. A. Sibley and A. L. Lehninger, *J. Biol. Chem.* **177**, 859 (1949).

solution and homogenized in the apparatus of Potter and Elvehjem.[7] The homogenate is centrifugated at about $25,000 \times g$ for 15 minutes, filtered through cotton, and fractionated at 0° with saturated ammonia sulfate solution, pH 7.8. The fraction between 45 and 55% saturation is dissolved in a small volume of 0.02 M NaHCO$_3$ and dialyzed for 3 hours against the same bicarbonate solution. After centrifugation a clear solution is obtained which may be stored for some 2 weeks at 0°.

Properties

Specificity. The protein fractions containing F-1-P aldolase always split FDP. As the enzyme has not yet been prepared in the pure state, it cannot be decided whether the same enzyme attacks both substrates or if the solutions also contain FDP aldolase besides F-1-P aldolase.

Pure FDP aldolase does not attack F-1-P at all, but it seems that several fractionated crystallizations of FDP aldolase are necessary to free the enzyme completely from F-1-P aldolase activity. (For preparation of pure FDP aldolase, see footnote 8.) From certain condensing reactions it may be concluded that F-1-P aldolase is not specific for F-1-P but may attack other 1-phosphoketoses (see below).

Equilibrium of the Reaction F-1-P $\rightleftharpoons PDA + GA$. The equilibrium strongly favors the fructose ester. Without removing the split products by other reactions (e.g., reduction in the optical test), no measurable splitting can be detected. If GA or some other unphosphorylated aldehydes are incubated with FDP and F-1-P aldolase solutions, the PDA formed from FDP condenses with the aldehyde to give 1-phosphoketoses:

$$\text{Glycolic aldehyde} + \text{PDA} \rightarrow \text{Ketopentose-1-phosphate}$$
$$\text{GA} + \text{PDA} \rightarrow \text{Ketohexose-1-phosphate}$$
$$\text{Erythrose} + \text{PDA} \rightarrow \text{Ketoheptose-1-phosphate}$$

This reaction is catalyzed only by the F-1-P aldolase and *not* by the FDP aldolase.[9]

Effect of pH. Between pH 6.7 and 7.8 there is no change in activity. At pH values lower than 6.4 activity decreases rapidly.

Other Sources for F-1-P Aldolase. The myogen A of Baranowski also contains F-1-P aldolase.[10] Cardini found splitting of F-1-P in extracts of jack bean seeds.[11]

[7] V. R. Potter and C. A. Elvehjem, *J. Biol. Chem.* **144**, 195 (1936).

[8] G. Beisenherz, H. J. Boltze, T. Bücher, R. Czök, K.-H. Garbade, E. Meyer-Arendt, and G. Pfleiderer, *Z. Naturforsch.* **8b**, 555 (1953).

[9] F. Leuthardt and H. P. Wolf, *Helv. Physiol. et Pharmacol. Acta* **11**, C 62 (1953).

[10] F. Leuthardt, E. Testa, and H. P. Wolf, *Helv. Chim. Acta* **36**, 227 (1953).

[11] C. E. Cardini, *Enzymologia* **15**, 303 (1952/53).

[42] Glucose-6-phosphate Dehydrogenase

Glucose-6-phosphate + TPN$^+$ $\rightleftarrows$ 6-Phosphogluconate + TPNH$^+$ + H$^+$

By A. Kornberg and B. L. Horecker

6-Phosphogluconic Dehydrogenase

6-Phosphogluconate + TPN $\rightleftarrows$ Ribulose-5-phosphate + CO$_2$
$$+ \text{TPNH} + \text{H}^+$$

By B. L. Horecker and P. Z. Smyrniotis

Assay Method

Principle. These enzymes were first described by Warburg and his co-workers,[1,2] who also demonstrated TPN to be the specific coenzyme. The reduction of the coenzyme is accompanied by the appearance of an absorption band with a maximum at 340 mμ which is absent in the oxidized form.[3] In the presence of excess substrate, the rate of reduction is proportional to the amount of enzyme present. Glucose-6-phosphate dehydrogenase is also known as Zwischenferment.[4]

Reagents

> Glucose-6-phosphate (0.02 M). Dissolve 522 mg. of the crystalline barium salt[5] (C$_6$H$_{11}$O$_9$PBa·7H$_2$O) in 25 ml. of 0.02 M acetic acid. Add 2.0 ml. of 0.57 M K$_2$SO$_4$. Remove the precipitated barium sulfate by centrifugation, neutralize the supernatant solution with 0.2 M KOH, and dilute to 42.5 ml. About 15% of the ester is adsorbed on the barium sulfate precipitate.

> 6-Phosphogluconate (0.02 M). Dissolve 200 mg. of barium phosphogluconate[6] in 2.0 ml. of 3.0 N acetic acid. Add 0.50 ml. of 2.0 N H$_2$SO$_4$. Remove the precipitated barium sulfate by centrifugation, and wash the precipitate with 0.6 ml. of water. Combine the supernatant solution and washing, neutralize with 5.0 N KOH, and dilute to 18 ml.

[1] O. Warburg and W. Christian, *Biochem. Z.* **242**, 206 (1931).

[2] O. Warburg, and W. Christian, *Biochem. Z.* **287**, 440 (1936).

[3] O. Warburg and W. Christian, *Biochem. Z.* **287**, 291 (1936).

[4] O. Warburg, W. Christian, and A. Griese, *Biochem. Z.* **282**, 157 (1935).

[5] W. A. Wood and B. L. Horecker, *Biochem. Preparations* **3**, 71 (1953).

[6] J. E. Seegmiller and B. L. Horecker, *J. Biol. Chem.* **192**, 175 (1951).

TPN $(1.5 \times 10^{-3}\ M)$. Dissolve 13 mg. of the free acid[7] (purity 85%) in 10 ml. of water.

0.1 M MgCl$_2$.

0.04 M glycylglycine buffer, pH 7.5.

Enzyme. Dilute the stock enzyme solution in cold water to obtain 0.2 to 0.8 unit/ml. (See definition below.)

Procedure. To 1.0 ml. of water in a quartz or Corex cell having a light path of 1 cm., add 0.1 ml. of TPN, 0.25 ml. of buffer, and 0.2 ml. of MgCl$_2$. Take readings at 340 mμ at 1-minute intervals immediately after the addition of 0.05 ml. of enzyme.

Definition of Unit and Specific Activity. A unit of enzyme is defined as that amount which causes an initial change in optical density of 1.000 per minute (room temperature 23 to 25°) under the above conditions. Specific activity is expressed as units per milligram of protein. Protein is determined by the turbidimetric method of Bücher,[8] standardized with a solution of crystalline serum albumin.

Purification Procedure for Glucose-6-phosphate Dehydrogenase

The procedure is that described by Kornberg[9] with several minor modifications. This preparation is free of 6-phosphogluconic dehydrogenase and is, therefore, suitable for such stoichiometric measurements as the determination of glucose-6-phosphate or (in conjunction with hexokinase) of ATP.

Dried brewer's yeast was obtained from Anheuser-Busch, Inc., St. Louis, or prepared from bottom yeast kindly furnished by the Chr. Heurich Brewing Co., Washington, D. C. In the latter case, wet yeast from the brewery is washed at 2° with 5 to 10 vol. of tap water and allowed to settle overnight. The thick slurry which remains when the wash water is poured off is pressed in muslin sacks in a lard press until cakes are formed which crumble readily. These are broken up by hand, spread on wire screens, and dried in a stream of air at room temperature.

Step 1. Yeast Autolyzate. 50 g. of dried brewer's yeast[10] is suspended in 150 ml. of 0.1 M sodium bicarbonate and kept at 40° for 5 hours. The clear autolyzate obtained by centrifugation can be stored overnight at 2° without loss. Subsequent operations are carried out at this temperature.

---

[7] A. Kornberg and B. L. Horecker, *Biochem. Preparations* **3**, 24 (1953).

[8] T. Bücher, *Biochim. et Biophys. Acta* **1**, 292 (1947).

[9] A. Kornberg, *J. Biol. Chem.* **182**, 805 (1950).

[10] The following procedure was developed for Anheuser-Busch yeast; with other brewer's yeasts, modification of the method may be necessary.

Step 2. Fractionation with Ammonium Sulfate. The autolyzate (91 ml.) is diluted with 265 ml. of 0.1 M sodium bicarbonate, and 125 g. of ammonium sulfate is added. The mixture is centrifuged, the precipitate discarded, and 26.7 g. of ammonium sulfate is added to the supernatant solution. The precipitate formed is collected by centrifugation and dissolved in water.

Step 3. Adsorption of Calcium Phosphate Gel.[11] Small trial runs are made to determine the minimum amount of gel required for adsorption of the enzyme. Best results are obtained with gel which is less than 1 month old. The ammonium sulfate fraction (114 ml.) is adsorbed with 85 ml. of calcium phosphate gel (1.6 g. of dry weight), kept for 5 minutes at 0°, centrifuged, and the supernatant solution discarded. The enzyme is eluted with 216 ml. of 0.1 M phosphate buffer, pH 7.5.

Step 4. Fractionation with Ammonium Sulfate. Ammonium sulfate (72 g.) is added to the eluate (215 ml.), and the precipitate formed is separated by centrifugation and discarded. Two more fractions are collected by the addition of 13 g. and 26 g. of ammonium sulfate, respectively. These are dissolved in water and analyzed separately. The third fraction usually contains the bulk of the activity.

Step 5. Precipitation at pH 4.5. The last ammonium sulfate fraction (11.0 ml.) is diluted with 34 ml. of water, and 220 ml. of 0.1 M acetate buffer, pH 4.5, is added. The mixture is kept at 0° for 25 minutes, and the precipitate is collected by centrifugation and dissolved in 5 ml. of 0.25 M glycylglycine buffer, pH 7.5. This step may be repeated to remove traces of phosphogluconic dehydrogenase activity. The final product is lyophilized and stored in a desiccator at 2°.

Properties of Glucose-6-phosphate Dehydrogenase

Specifity. Some 6-phosphogluconic dehydrogenase activity is removed by the calcium phosphate gel step, and most of the remainder is lost during pH 4.5 precipitation. The final product shows very weak activity with this substrate (see Table I). Other hexose or pentose esters are not attacked. DPN is not reduced.

Activators and Inhibitors. Phosphate has long been known to inhibit glucose-6-phosphate dehydrogenase, particularly at concentrations above 0.1 M.[12] Magnesium ions show a stimulating effect in glycylglycine buffer which is not evident in dilute phosphate buffer.[9]

Stability. The lyophilized preparations are stable indefinitely, provided that they are kept cold and dry. Aqueous solutions lose activity

[11] D. Keilin and E. F. Hartree, *Proc. Roy. Soc. (London)* **B124**, 397 (1938).
[12] E. Negelein and E. Haas, *Biochem. Z.* **282**, 206 (1935).

TABLE I
Summary of Purification Procedure (Glucose-6-phosphate Dehydrogenase)

Fraction	Total volume, ml.	Units/ ml.	Total units	Protein mg./ml.	Specific activity, units/ mg.	Recovery, %	R^a
1. Yeast autolyzate	91	16.4	1490	44.0	0.37	—	—
2. (NH₄)₂SO₄ fraction I	114	9.0	1025	11.0	0.82	69	0.80
3. Eluate from Ca₃(PO₄)₂	215	2.6	560	1.0	2.6	38	0.42
4. (NH₄)₂SO₄ fraction II	11.0	40.0	440	12.4	3.2	29	0.43
5. pH 4.5 precipitate I	5.2	64.6	336	12.0	5.4	23	0.015
6. pH 4.5 precipitate II	5.0	53.6	268	—	—	18	0.006

^{a} Ratio of activity with 6-phosphogluconate to that with glucose-6-phosphate.

when stored in the refrigerator but may be kept frozen at $-16°$ for weeks with only slight loss. Care must be taken to avoid shaking during freezing and thawing.

Purification Procedure for 6-Phosphogluconic Dehydrogenase

The procedure is based on the method described by Horecker and Smyrniotis.[13] Dried brewer's yeast is prepared as described for the preceding preparation.

Step 1. Yeast Autolyzate. Suspend 90 g. of dried yeast in 280 ml. of 0.1 M sodium carbonate, and autolyze for 4.5 hours at 34°. Centrifuge, and discard the residue.

Step 2. Acetone Precipitation. Dilute the yeast extract (170 ml.) with 940 ml. of 2 N acetic acid. Immediately centrifuge, discard the resulting precipitate, and to the supernatant solution add 514 ml. of 50% acetone, previously chilled to. $-10°$. During the acetone addition, which should require 6 to 10 minutes, cool the solution to $-8°$ in a freezing bath. After the addition is complete, keep the mixture at $-8°$ for 10 minutes, centrifuge, and dissolve the precipitate in 90 ml. of 0.01 M phosphate buffer, pH 6.4. Adjust the pH of the solution to 6.2 with 3.2 ml. of 0.1 N NaOH.

Step 3. Protamine Precipitation. To the acetone fraction add 10.0 ml. of protamine sulfate solution (Salmine, Lilly) containing 100 mg. Centrifuge the mixture, and discard the precipitate.

Step 4. Fractionation with Acid in Ammonium Sulfate. To the supernatant solution (102 ml.) add 37 g. of ammonium sulfate, and adjust the

[13] B. L. Horecker and P. Z. Smyrniotis, *J. Biol. Chem.* **193**, 371 (1951).

pH to 5.0 by the addition of 0.30 ml. of 1 N acetic acid. Remove the precipitate which forms by centrifugation, and adjust the supernatant solution to pH 4.5 with 1.1 ml. of 1 N acetic acid. Centrifuge, and dissolve the precipitate in 4.0 ml. of 0.25 M glycylglycine buffer, pH 7.5.

TABLE II

SUMMARY OF PURIFICATION PROCEDURE (6-PHOSPHOGLUCONIC DEHYDROGENASE)

Fraction	Total volume, ml.	Units/ml.	Total units	Protein, mg./ml.	Specific activity, units/mg.	Recovery, %
1. Yeast autolyzate	170	33.6	5,700	48.4	0.7	—
2. Acetone fraction	94	26.0	2,450	10.9	2.4	43
3. Protamine supernatant	102	22.4	2,290	4.7	4.8	40
4. Acid ammonium sulfate	4.2	386	1,630	37.5	10.3	29

Properties of 6-Phosphogluconic Dehydrogenase

Specificity. The preparation contains nearly as much glucose-6-phosphate dehydrogenase activity as 6-phosphogluconic dehydrogenase and can be employed for the assay of 6-phosphogluconic acid only when glucose-6-phosphate is not present in the sample. Significant amounts of isocitric dehydrogenase and pentose phosphate isomerase are also present.

Activators and Inhibitors. The enzyme is unstable to heavy metals, and most preparations can be activated by the addition of metal binders such as glycylglycine or cyanide. High concentrations of substrate have the same effect. In the presence of glycylglycine buffer some activation is obtained by the addition of magnesium ions. Fluoride, iodoacetate, malonate, and cyanide are without inhibitory effect.

Effect of pH. In glycylglycine buffer the enzyme is most active at pH 7.4, with about one-half maximal activity at pH 6.3 and 8.1.

Stability. Dried brewer's yeast may be stored for long periods without loss of 6-phosphogluconic dehydrogenase activity. However, during the purification procedure the enzyme becomes rather unstable, and the procedure must be carried through without interruption. The concentrated solutions obtained in the final step may be frozen and stored at $-16°$. Under these conditions only about 50% of the activity is lost in twelve months.

[43] Glucose-6-phosphate and 6-phosphogluconic Dehydrogenases from *Leuconostoc mesenteroides*

By R. D. DeMoss

Although this article is intended as a guide to the preparation and assay of two enzyme systems occurring in *Leuconostoc mesenteroides*, it should be emphasized that the primary consideration in preparing protein extracts of any bacterial species lies not in enzymological methods but in the environmental conditions of cell production. The extreme adaptability of most bacteria to various growth conditions is expressed as highly modified metabolic activities and, therefore, as markedly variable enzymatic composition. Metabolic activities and their rates also vary widely as a function of growth stage. A highly desirable, and in many instances necessary, preliminary to enzymatic investigations of bacteria is a careful physiological study of the organism involved.

The enzyme systems mentioned below have been partially purified from cell-free extracts of heterofermentative lactic acid coccus *Leuconostoc mesenteroides*.

Growth of Cells

The organism is grown at 30° in AC medium containing, per liter: 10 g. yeast extract (Difco), 10 g. tryptone (Difco), 10 g. glucose, 5 g. K_2HPO_4, 20 ml. salts solution (per 100 ml.: 0.8 g. $MgSO_4 \cdot 7H_2O$, 0.19 g. $MnCl_2 \cdot 4H_2O$, 0.04 g. NaCl, 0.04 g. $FeSO_4 \cdot 7H_2O$, 2.0 ml. 12 N HCl), and sterilized for 15 minutes at 120° (no pH adjustment is required).[1] Stock cultures as stabs in AC medium containing 2% agar are transferred monthly and stored at 2° after 24 hours of incubation. For large-scale cell production, a 10-ml. tube of AC medium is inoculated from the stock culture and transferred successively to similar tubes at three 24-hour intervals and one 12-hour interval, using 0.05-ml. inocula. From the last tube, after 12 hours of incubation, an 0.1-ml. inoculum is transferred to 100 ml. of AC medium in a 125-ml. Erlenmeyer flask, which, after 12 hours, serves as the inoculum source for large carboys of growth medium (1 ml. of inoculum per liter of growth medium). The carboys each contain 10 l. of slightly modified AC medium; during sterilization for 1 hour at 120°, the medium contains only 0.5 g. of glucose per liter. The remainder of the glucose is sterilized separately (1 hour at 120°) in aqueous solution (9.5 g./30 ml. of water) and added just prior to inoculation. After 12

[1] R. D. DeMoss, R. C. Bard, and I. C. Gunsalus, *J. Bacteriol.* **62**, 499 (1951).

hours of incubation of the culture, the cell crop is collected in a Sharples supercentrifuge and either used immediately or stored at $-20°$ without further treatment.

Preliminary Procedures

All manipulations are performed at $2°$ or in an ice bath, unless otherwise stated. These procedures constitute steps 1 and 2 in the purification of the enzymes described below.

Step 1. Cell-Free Extract. These bacterial cells are somewhat refractory to disintegration, apparently owing to their small size (0.9 to 1.2 μ diameter) and nearly spherical structure. Grinding techniques or sonic oscillation (for details, see Vol. I [7]) at 9 kc. for $1\frac{1}{2}$ to 3 hours results in disruption of only 30 to 60% of the cells. Sonic treatment for 3 hours also destroys a large portion of the enzymatic activity theoretically obtainable. However, treatment for 20 to 30 minutes in the 10-kc. oscillator disrupts about 80% of the cells with little concomitant loss of enzymatic activity. The maximum yields of both G-6-P dehydrogenase and 6-P-G dehydrogenase activities are obtained after 20 minutes of oscillation at a concentration of 1 g. of cells (wet weight) per 10 ml. of suspending fluid.

The concentration of cells for sonic treatment is usually about 1 g. of freshly harvested or frozen wet cells per 10 ml. of suspending fluid. For grinding techniques, about 2.5 g. of abrasive (Alumina A-301) is used per gram of wet cells,[2] and the ground mixture is extracted with 4 to 5 ml. of fluid per gram of cells.

Water or various buffers (bicarbonate, phosphate, gluconate, cysteine, Tris, Veronal, Versene, and others) have been used successfully as suspending or extracting fluids. In analogy to the hexokinase procedure (Vol. I [32]), the presence of substrate (or nonphosphorylated analog; i.e., gluconate in place of 6-phosphogluconate) has been found to protect against loss of activity in some cases.

Step 2. Nucleic Acid Precipitation. The crude cell-free extract is centrifuged (30 minutes at $15,000 \times g$), and the supernatant is dialyzed for 16 hours against 50 to 100 vol. of an appropriate buffer (see below) and carefully adjusted to pH 6.0 with N acetic acid. Manganous chloride ($\frac{1}{20}$ vol. of M MnCl$_2$) is added dropwise with stirring to remove nucleic acid. The precipitate is removed by centrifugation (10 to 15 minutes at $10,000 \times g$), and the clear supernatant solution (MgCl$_2$ supernatant, containing 4 to 7% nucleic acid) is used for subsequent fractionation procedures. Protein and nucleic acid concentrations are determined by the optical method of Warburg (see Vol. III [73]).

---

[2] H. McIlwain, *J. Gen. Microbiol.* **2**, 288 (1948).

Glucose-6-phosphate Dehydrogenase
Assay Method

Principle. This enzyme is active with either DPN or TPN as co-enzyme; hence the change in absorption at 340 mμ with time is a measure of the course of the reaction (Vol. III [128]). Previously used methods for this enzyme from other sources (see Vol. I [42]) have been based on (1) CO_2 evolution from bicarbonate buffer due to production of hydrogen ions (e.g., 6-phosphogluconic acid) and (2) reduction of dyes coupled with reoxidation of TPNH.

Reagents

> G-6-P solution (0.2 *M*). Dissolve 1.043 g. of the Ba salt of G-6-P ($C_6H_{11}O_9PBa \cdot 7H_2O$, mol. wt. 521.60) (see Vol. III [19]), available from Schwarz Laboratories, Inc., in approximately 5 ml. of 0.1 *N* HCl, remove Ba by addition of saturated Na_2SO_4 solution, wash $BaSO_4$ precipitate with 0.3 *M* $NaSO_4$, combine supernatants, adjust to pH 7.8, and dilute to 10.0 ml. with distilled water. This solution is stable for at least three years when stored at $-20°$.
> Tris buffer (0.1 *M*, pH 7.8). Mol. wt. 121.14.
> DPN (0.0027 *M*). 2 mg. DPN-90 (Sigma) per milliliter; mol. wt. 664.
> $MgCl_2$ (0.1 *M*).
> Enzyme. Dilute the stock enzyme with the Tris buffer or water to obtain 1 to 5 units of enzyme per milliliter. (See definition below.)

Procedure. To a spectrophotometric cell with a 1-cm. light path, add 1.5 ml. of Tris buffer, 0.2 ml. of DPN, 0.1 ml. of $MgCl_2$, 0.1 ml. of enzyme, and 1.05 ml. of water. Take readings at 340 mμ before and at 15-second intervals after mixing with 0.05 ml. of G-6-P solution (10 micromoles).

Definition of Unit and Specific Activity. One unit of enzyme effects a rate of change of optical density (ΔE_{340}) of 1.0 per minute during the 15- to 30-second interval under the above conditions. Specific activity is defined as enzyme units per milligram of protein.

Application of Assay Method to Crude Extracts. This method is a valid assay for crude extracts of *L. mesenteroides*, since aldolase (Vol. I [39]) is absent.[1] For other heterofermentative organisms, in which the presence or absence of aldolase is unknown, the method should be valid provided that ATP is excluded, since the equilibria of the hexose phosphate esters do not reduce the G-6-P concentration to a limiting value. Although 6-P-G dehydrogenase (see below; in *L. mesenteroides*, this enzyme is

DPN-linked) is also present, the 6-P-G concentration during the 15- to 30-second interval does not introduce an appreciable error.

Purification Procedure

The following procedure has been used many times in this laboratory with results similar to those of Table I. Use 0.1 M NaHCO$_3$ for the suspending fluid in step 1, and 0.02 M NaHCO$_3$ for dialyses. The procedure described herein is somewhat more convenient than the procedure employed previously.[3]

TABLE I

SUMMARY OF G-6-P DEHYDROGENASE PURIFICATION PROCEDURE[a]

Fraction	Specific activity	Total units
1. Crude extract (dialyzed)	1.6	576
2. MnCl$_2$ supernatant I	1.3	371
3. MnC$_2$ supernatant II	4.0	339
4. Ca$_3$(PO$_4$)$_2$ gel supernatant I	—	—
Ca$_3$(PO$_4$)$_2$ gel supernatant II	7.8	231
Ca$_3$(PO$_4$)$_2$ gel supernatant III	16.5	211

[a] DPN as coenzyme.

Step 3. Additional Nucleic Acid Precipitation. Step 2 (MnCl$_2$ treatment) is repeated after overnight dialysis, yielding the MnCl$_2$ supernatant II.

Step 4. Negative Calcium Phosphate Gel Adsorption. The MnCl$_2$ supernatant II is treated with three successive portions of Ca$_3$(PO$_4$)$_2$ gel (0.2 milligrams of gel per milligram of protein) yielding, respectively, Ca$_3$(PO$_4$)$_2$ gel supernatants I, II, and III. The last fraction contains about 40% of the original activity and represents an approximately 10-fold purification.

Properties

Specificity. The purified enzyme is active with either DPN or TPN (relative activity of DPN/TPN is 0.67 throughout the purification procedure). G-6-P appears to be the only naturally occurring substrate, since no pyridine nucleotide reduction is observed with F-6-P, FDP, G-1-P, or ribose-5-P.

The question of whether the activity is due to one or two enzymes is unresolved; it should be noted that transhydrogenase is not present.[4] The

[3] R. D. DeMoss, I. C. Gunsalus, and R. C. Bard, *J. Bacteriol.* **66**, 10 (1953).
[4] N. O. Kaplan, unpublished data.

constant activity ratio of DPN/TPN observed during purification, in addition to the observation that TPNH inhibits DPN reduction,[3] is taken as evidence for existence of a single enzyme, nonspecific in pyridine nucleotide requirement. However, Shuster and Kaplan[5] have adduced evidence against the single enzyme hypothesis, based on (1) the effects of adenylic acids (2, 3, and 5) on the enzymatic activity and (2) the ability of the enzyme to reduce 3-TPN. The single enzyme hypothesis is also contraindicated by the observation that crude cell-free extracts of *Leuconostoc*, grown with L(+)-arabinose as energy source, catalyze the reduction of DPN but not TPN with G-6-P as substrate.

Activators and Inhibitors. Orthophosphate at 5×10^{-3} M concentration inhibits the enzymatic activity about 50%. Hence, phosphate buffer should be avoided during the purification procedure. Magnesium ions $(3 \times 10^{-3}$ $M)$ stimulate activity, although an absolute requirement has not been demonstrated.

Effect of pH. The most rapid initial rate of pyridine nucleotide reduction occurs at pH 7.8 under the assay conditions described above.

Stability. The enzymatic activity of 20-fold purified preparations is stable for 3 days at 25° and for at least two years at −20°. Almost complete inactivation results from freezing enzyme solutions which contain concentrations of $MnCl_2$ or $MgCl_2$, at or above 0.05 M.

Dissociation Constants. When derived from Lineweaver-Burk[6] or Hofstee[7] plots, the dissociation constants are 3.1×10^{-5} M DPN, 9.9×10^{-6} M TPN, and 5.3×10^{-4} M G-6-P (with DPN as coenzyme).

6-Phosphogluconate Dehydrogenase

Assay Method

Principles. The enzyme specifically reacts with DPN as coenzyme, and the change in absorption at 340 mμ with time is a measure of the course of the reaction (see Vol. 1II [128]).

Reagents

6-P-G solution (0.1 M potassium salt). The barium salt of 6-P-G may be prepared from barium G-6-P (Vol. III [23]).
Glycylglycine buffer (0.4 M, pH 7.5).
DPN (0.0027 M).
$MgCl_2$ (0.02 M).
Enzyme. Dilute the stock enzyme with water to obtain 1 to 5 units of enzyme per milliliter. (See definition below.)

[5] L. Shuster and N. O. Kaplan, in press.
[5] H. Lineweaver and D. Burk, *J. Am. Chem. Soc.* **56**, 658 (1934)
[7] B. H. J. Hofstee, *Science* **116**, 329 (1952).

Procedure. To a spectrophotometric cell with a 1-cm. light path, add 1.5 ml. of glycylglycine buffer, 0.2 ml. of DPN, 0.1 ml. of MgCl₂, 0.1 ml. of enzyme, and 1.05 ml. of water. Take readings at 340 mμ before and at 15-second intervals after mixing with 0.05 ml. of 6-P-G solution (5 micromoles).

Definition of Unit and Specific Activity. One unit of enzyme effects a rate of change of optical density (ΔE_{340}) of 1.0 per minute during the 15- to 30-second interval under the above conditions. Specific activity is defined as enzyme units per milligram of protein.

Application of Assay Method to Crude Extracts. This method has been found to be a valid assay using crude extracts of *L. mesenteroides.* DPN is reduced stoichiometrically by 6-P-G; the rate of DPNH reoxidation in crude extracts does not introduce significant error.

Purification Procedure

The following procedure has been found to be reasonably reproducible in this laboratory (see Table II). Use 1% gluconate (pH 8.8) for the suspending fluid in step 1, and distilled water for dialyses.

TABLE II
SUMMARY OF 6-P-G DEHYDROGENASE PURIFICATION PROCEDURE

Fraction	Specific activity	Total units
1. Crude extract	0.12	210
2. MnCl₂ supernatant	0.11	159
3. First (NH₄)₂SO₄ ppt (0.45–0.75)	0.22	80
4. First Ca₃(PO₄)₂ gel (combined eluates)	0.30	59
5. Second (NH₄)₂SO₄ (0.0–0.6)	0.95	62
6. Second Ca₃(PO₄)₂ gel (eluate)	1.98	63

Step 3. Ammonium Sulfate Fractionation. The MnCl₂ supernatant (adjusted to about 10 mg. of protein per milliliter) is fractionally precipitated at 0.45, 0.60, and 0.75 saturation by addition of solid (NH₄)₂SO₄. The 0.45 to 0.75 fractions are dissolved in distilled water (10 mg. of protein per milliliter) and dialyzed for 16 hours against 40 vol. of distilled water at 2°.

Step 4. Calcium Phosphate Gel Adsorption and Elution. To the dialyzed enzyme solution, adjusted to pH 6.5, Ca₃(PO₄)₂ gel (0.1 mg./mg. protein) is added, and the mixture is stirred gently for 20 minutes. After centrifugation the supernatant is mixed with Ca₃(PO₄)₂ gel (1 mg./mg. of protein), and centrifuged. The supernatant is again mixed with Ca₃(PO₄)₂ gel (2.5 mg./mg. protein) and centrifuged. The sediments from the second and third adsorptions are combined and eluted successively with 15-ml. portions of 0.1 M phosphate buffer of pH 6.5 and pH 7.6.

Step 5. Ammonium Sulfate Fractionation. The combined $Ca_3(PO_4)_2$ gel eluates are fractionally precipitated by addition of solid $(NH_4)_2SO_4$ to 0.60 saturation. After centrifugation, the supernatant is discarded.

Step 6. Calcium Phosphate Gel Adsorption and Elution. The 0.0 to 0.6 fraction above is dissolved in distilled water (final pH 6.2) and mixed with $Ca_3(PO_4)_2$ gel (4 mg./mg. protein) and centrifuged. The sediment is eluted with 10 ml. of 0.1 M phosphate buffer, pH 7.0, centrifuged, and the sediment discarded. The supernatant contains 30% of the original activity and represents a 15-fold purification of the enzyme.

Analysis of one of these final purified preparations showed the presence of G-6-P dehydrogenase (see above), TPN-alcohol dehydrogenase (see Vol. I [80]), and D(−)lactic dehydrogenase[1] in addition to 6-P-G dehydrogenase.

Properties

Specificity. DPN reduction is catalyzed by the purified enzyme at twenty-five times the rate of TPN reduction with 6-P-G as substrate. The rate of DPN reduction with 6-P-G as substrate is about 2.5 that with ethylgluconate-6-P. The latter was prepared from ethylgluconate and ATP, using a gluconokinase from *Leuconostoc mesenteroides* (unpublished).

Activators and Inhibitors. After overnight dialysis at pH 7.5 against Versene (0.06 M), phosphate (0.05 M), Tris (0.05 M), or glycylglycine (0.04 M), Mg^{++} (6 × 10^{-4} M) stimulated the initial rate of DPN reduction 16, 22, 40, or 55% respectively. No absolute requirement for Mg^{++} was observed.

With the standard assay procedure, 20, 10, 36, or 63% inhibition was observed in the presence of 2 × 10^{-3} M iodoacetate, 5 × 10^{-4} M, 1 × 10^{-3} M, or 1.5 × 10^{-3} M p-chloromercuribenzoate, respectively. No inhibition was observed with 3 × 10^{-3} M NH_2OH.[8]

Effect of pH. The most rapid initial rate of DPN reduction occurs at pH 7.8 under the assay conditions described above.

Stability. After 2 months at −20°, no enzymatic activity had disappeared from the purified preparations.

Dissociation Constants. When derived from Lineweaver-Burk[6] or Hofstee[7] plots, the dissociation constants are 3.5 × 10^{-5} M DPN and 7.78 × 10^{-5} M 6-P-G.

[8] N. O. Kaplan and M. M. Ciotti, *J. Biol. Chem.* **201**, 785 (1953).

[44] Glucose Dehydrogenase from Liver

β-D-Glucose + DPN$^+$ $\rightleftarrows$ D-Gluconolactone + DPNH
                                      + H$^+$ (enzymatic)
D-Gluconolactone + H$_2$O $\rightleftarrows$ D-Gluconate$^-$ + H$^+$ (nonenzymatic)

β-D-Glucose + DPN$^+$ + H$_2$O $\rightleftarrows$ D-Gluconate$^-$ + DPNH + 2H$^+$

By HAROLD J. STRECKER

Assay Method

Principle. The formation of reduced DPN results in an increase of optical density at 340 mμ.

Reagents

  1 *M* glucose.
  DPN solution. 1.5 micromole/ml. in water.
  0.05 *M* potassium phosphate buffer, pH 7.6.
  Enzyme. The enzyme is diluted with distilled water so as to obtain
    a concentration of 200 to 500 units of enzyme per milliliter. (See
    definition below.)

Procedure. 2.6 ml. of buffer, 0.1 ml. of the DPN solution, and 0.1 ml. of the enzyme solution are mixed in a quartz cuvette with a 1.0-cm. light path. The same solution plus 0.2 ml. of H$_2$O in another cuvette serves as the blank. 0.2 ml. of the glucose solution is added to the experimental cuvette, and the solution is stirred with a glass rod. The optical density at 340 mμ is recorded at 30-second intervals.

Definition of Unit and Specific Activity. The increment in optical density (ΔE_{340}) between the reading at 30 seconds and that at 90 seconds after making the last addition is taken as the enzyme activity per minute. One enzyme unit is defined as that amount which causes a change in optical density of 0.001 per minute under the above conditions. Specific activity is expressed as units per milligram of protein. The protein concentration is determined spectrophotometrically by measurement of the light absorption at wavelenghts 280 and 260 mμ, a correction being made for the nucleic acid content from the data of Warburg and Christian.[1] The values reported in the table are for activities determined in potassium phosphate buffer at pH 7.6.

[1] O. Warburg and W. Christian, *Biochem. J.* **310**, 384 (1941–1942).

Other Methods of Assay. In crude systems methylene blue reduction or oxygen uptake can be used for determining activity.[2] As an alternative method, provided that a system for reoxidizing the DPNH is also present, the formation of CO_2 from $NaHCO_3$ buffer, which is proportional to the gluconic acid formed, can be determined.

Purification Procedure

Steps 1 and 2 are modifications of the procedures of Harrison.[2]

Step 1. Fresh beef liver obtained from the slaughterhouse is homogenized for 1 to 2 minutes in a Waring blendor with an approximately equal volume of acetone at −3° in 500-g. amounts. The acetone suspension is then poured into 10 vol. of acetone at 0 to −3°, stirred thoroughly, and filtered by suction on a large Büchner funnel. The cake of acetone powder thus formed is again homogenized in a Waring blendor and stirred into 10 vol. of cold acetone and filtered as before. The cake is sucked as dry as possible. These operations are performed in a cold room at 0 to 7°. The acetone powder cake is removed to room temperature, broken up, and dried rapidly by rubbing between the palms of the hands. The resulting powder may be kept in the cold for several hours. However, glucose dehydrogenase activity in the dry powder decreases fairly rapidly; approximately half the activity is lost in 24 hours. The dry liver powder, varying from 250 to 1000 g. in weight, is stirred for 30 minutes with 10 vol. of deionized water at room temperature. After allowing most of the insoluble material to settle (5 to 10 minutes), the supernatant fluid is decanted and filtered through four layers of cheesecloth. The residue is again stirred with 10 vol. of water for the same length of time, and the suspension filtered through cheesecloth as before. The solution at this stage contains fine suspended material.

All the following procedures are carried out at approximately +5°.

Step 2. First Ammonium Sulfate Fractionation. The solution, which is stable for at least 1 day at this stage, is chilled to +5° or lower, and the pH is adjusted to 6.0 by the addition of 10% acetic acid with stirring. The solution is then clarified by passage through a Sharples centrifuge, and the residue discarded. 30 g. of ammonium sulfate per 100 ml. of solution is added slowly with stirring until all the ammonium sulfate is dissolved. The solution is allowed to stand overnight, and the precipitate is collected by passage through a Sharples centrifuge. The supernatant fluid is discarded. The precipitate is dissolved in sufficient 0.05 M potassium phosphate buffer, pH 7.6, to yield a protein concentration of approximately 6%, and the insoluble residue discarded after centrifugation.

[2] D. C. Harrison, *Biochem. J.* **25**, 1016 (1931).

Step 3. Second Ammonium Sulfate Fractionation. The solution is adjusted to pH 5.5 by the addition of 10% acetic acid, and from 6 to 10 g. of ammonium sulfate per 100 ml. of solution is added. The exact amount required varies between these limits in different preparations and has to be determined by testing aliquots. In general, the amount of ammonium sulfate employed is that quantity which will precipitate no more than 5% of the total activity. After standing for 2 hours, the mixture is centrifuged, and the precipitate discarded. 18.0 g. of ammonium sulfate per 100 ml. of solution is then added, and the mixture is allowed to stand overnight. It is then centrifuged, and the supernatant fluid discarded. The precipitate is dissolved in 0.05 M phosphate buffer, pH 7.6, as before and adjusted to a protein concentration of approximately 5%.

Step 4. Acid Precipitation. 10% acetic acid is added slowly to the above solution to bring the pH to 4.7. The solution is allowed to stand for 4 hours. During this period, the pH slowly decreases, and it is advisable to readjust it hourly by addition of 1.0 N NaOH. The solution is then centrifuged, and the precipitate, consisting mainly of denatured protein is discarded. Some activity usually coprecipitates with nonenzymatic protein and can be recovered by washing the precipitate with a small quantity (about 10 ml.) of 0.05 M phosphate buffer, pH 7.6.

Step 5. Lead Acetate Precipitation. The solution is adjusted to pH 6.0 by the addition of 1.0 N NaOH and dialyzed for 4 to 5 hours against 10 vol. of deionized water with two changes of water. (Activity is lost when dialysis is carried out for longer than 5 hours.) Any precipitate which is formed is discarded. Saturated lead acetate is then added dropwise until no further precipitation occurs. This is most conveniently done by determining on a small aliquot the amount of lead acetate necessary for complete precipitation. The amount required is usually about one-tenth of the total volume. The solution is allowed to stand for 30 minutes and centrifuged; the precipitate is discarded.

Step 6. First Calcium Phosphate Gel Treatment. A few drops of 0.05 M phosphate buffer, pH 7.6, are added to precipitate any lead in solution, and the mixture is centrifuged. The pH is adjusted to 5.7, and calcium phosphate gel in a concentration of 23.5 mg./ml. (dry weight) is added to the solution with stirring. Sufficient gel is added to bring about maximum adsorption of inert protein with minimum adsorption of enzyme. The exact amount varies from 0.2 to 1 vol. of gel per volume of solution. After stirring for 15 minutes, the supension is centrifuged, and the gel discarded.

Step 7. Second Calcium Phosphate Gel Treatment. The solution is adjusted to pH 5.4, and sufficient calcium phosphate gel is added to adsorb the enzyme completely. From 0.2 to 0.6 vol. of gel for 1 vol. of enzyme solution is needed. The following procedure is used. The appropriate volume of gel is centrifuged, the supernatant fluid discarded, and the enzyme solution added directly to the sedimented gel and stirred for 15 minutes, centrifuged, and the supernatant discarded. The gel is stirred for 15 to 20 minutes with an amount of 0.05 M phosphate buffer, pH 7.0, equivalent to about one-fourth the volume of the solution after the first gel treatment. In one preparation this was not sufficient to elute the activity, and 0.1 M phosphate had to be used. The solution was centrifuged, and the gel discarded. The over-all yield at this stage was 30 to 35%; specific activity, 900 to 1100. These solutions have been kept frozen without loss of activity for more than a year. The activity of the solution at this stage is of the order 10,000 units/ml. When more concentrated solutions are necessary, the enzyme can be precipitated by adjusting the pH to 5.4 and adding ethanol at $-5°$ to a final concentration of 30%. The precipitate is then frozen, and portions are dissolved to the required concentration as desired. The pertinent data on the various steps in a typical procedure are summarized in the table.

PURIFICATION OF OX LIVER GLUCOSE DEHYDROGENASE
(250 g. of acetone powder)

Step	Volume of solution, ml.	Units	Protein, mg.	Specific activity, units/mg. protein	Yield, %
Aqueous extract	3600	1,500,000	90,000	10	100
First ammonium sulfate fractionation	700	960,000	36,750	26	65
Second ammonium sulfate fractionation	230	900,000	11,250	80	60
Acid precipitation	247	800,000	7,400	108	53
Lead acetate precipitation	270	700,000	4,070	172	47
First calcium phosphate gel treatment	378	600,000	2,320	258	40
Second calcium phosphate gel treatment	90	450,000	500	900	30

Properties[3-5]

Specificity. In the forward direction the purified enzyme is specific for β-glucose and β-xylose and inactive with the following sugars and deriva-

[3] H. J. Strecker and S. Korkes, *J. Biol. Chem.* **196**, 769 (1952).

[4] Beef liver glucose dehydrogenase has recently been purified by another method, resulting in a final product with a specific activity of 3400 units/mg. [N. G. Brink, *Acta Chem. Scand.* **7**, 1081 (1953)].

[5] N. G. Brink, *Acta Chem. Scand.* **7**, 1090 (1953).

tives thereof: raffinose, lactose, maltose, α-D-glucoheptose, β-D-glucoheptose, D-mannose, D-fructose, D-ribose, D-glyceraldehyde, glycolaldehyde, acetaldehyde, hexose diphosphate, glucose-6-phosphate, and ribose-5-phosphate. Galactose and arabinose were oxidized at about 3 to 4% of the rate with glucose, probably due to contamination with enzyme(s) specific for these sugars.

Both DPN and TPN are utilized, the rate being the same up to a pH of about 8.5.

In the reverse direction both the γ- and δ-gluconolactone are reduced.

Inhibitors. The activity of glucose dehydrogenase is competitively inhibited by fructose-1,6-diphosphate ($K_i = 6.2 \times 10^{-5}\ M$) and glucose-6-phosphate ($K_i = 2.5 \times 10^{-6}\ M$). Ribose-5-phosphate and fructose-6-phosphate are also inhibitory. In competition with DPN many compounds related to this coenzyme, including substituted pyridines and derivatives of adenine, are inhibitory, the most effective in each class being 4-pyridoxic acid ($K_i = 3 \times 10^{-4}$) and ATP ($K_i = 1.7 \times 10^{-3}$).

Effect of pH. The optimum pH in a buffer mixture of 0.025 M phosphate and 0.025 M Tris was 9.8 with DPN and 9.0 with TPN. At the optimum pH for each coenzyme, the rate with DPN was about twice that with TPN.

Enzyme-Substrate Dissociation Constants. The values for K_m were determined in phosphate buffer by application of the usual Lineweaver-Burk methods. These values change appreciably with pH.[6] Thus for glucose at pH 7.0, $K_m = 0.15\ M$ (0.031 M), and at pH 7.6, $K_m = 0.07\ M$; DPN, pH 7.0, $K_m = 1.5 \times 10^{-5}\ M$ (4.3 $\times 10^{-6}\ M$); pH 7.6, K_m below $10^{-5}\ M$ (TPN, pH 7.0, $K_m = 6.2 \times 10^{-6}\ M$); gluconolactone, pH 7.0, $K_m = 0.005\ M$; and DPNH, pH 7.0, $K_m = 8 \times 10^{-5}\ M$.

Equilibrium Constants.[6] For the reaction

$$\text{Glucose} + \text{DPN}^+ \rightleftarrows \text{Gluconolactone} + \text{DPNH}$$

at pH 6.7, $K = 15$ (1.5), and $\Delta F = -2410$ cal./mole.

For the reaction

$$\text{Gluconolactone} + \text{H}_2\text{O} \rightleftarrows \text{Gluconate}^- + \text{H}^+$$

ΔF at the above pH is -4700 cal./mole. Thus for the over-all reaction

$$\text{Glucose} + \text{DPN}^+ + \text{H}_2\text{O} \rightleftarrows \text{Gluconate}^- + \text{DPNH} + 2\text{H}^+$$

$\Delta F = -7110$ cal./mole, and for the reaction

$$\text{Glucose} + \text{H}_2\text{O} \rightleftarrows \text{Gluconate}^- + 2\text{H}^+ + 2e$$

$E_0' = 0.44$ volt.

[6] The values in brackets are those reported in ref. 4.

[45] Glucose Aerodehydrogenase (Glucose Oxidase)

$$C_6H_{12}O_6 \ + \ FAD \rightarrow C_6H_{10}O_6 + FAD\ H_2$$
$$C_6H_{10}O_6 \ + \ H_2O \rightarrow C_6H_{12}O_7$$
$$FAD\ H_2 \ + \ O_2 \rightarrow FAD \ + H_2O_2$$

$$C_6H_{12}O_6 + H_2O + O_2 \rightarrow C_6H_{12}O_7 + H_2O_2$$

By RONALD BENTLEY

Note on Nomenclature

This disconcertingly polyonymous enzyme obtained from mold cultures was originally named glucose oxidase.[1] The antibacterial activity of some *Penicillium notatum* culture filtrates was later shown to be due to H_2O_2 formed by action of this enzyme. It is virtually certain that the antibiotics notatin[2a,b] (originally penicillin A), penatin,[3] and penicillin B[4a,b] are all preparations of this enzyme.[5] The enzyme catalyzes the removal of two hydrogen atoms from β-D-glucopyranose,[6] forming as the primary product δ-D-gluconolactone.[7] The normal hydrogen acceptor is oxygen, and it would be correct to name the enzyme β-D-glucopyranose aerodehydrogenase. The short form, glucose aerodehydrogenase, will be used here.

Assay Method

Principle. A manometric method of assay, measuring oxygen uptake, was described in 1937.[8] Since the identification of the specific substrate

[1] D. Müller, *Biochem. Z.* **199**, 136 (1928).

[2a] C. E. Coulthard, R. Michaelis, W. F. Short, G. Sykes, G. E. H. Skrimshire, A. F. Standfast, J. H. Birkinshaw, and H. Raistrick, *Nature* **150**, 634 (1942).

[2b] C. E. Coulthard, R. Michaelis, W. F. Short, G. Sykes, G. E. H. Skrimshire, A. F. B. Standfast, J. H. Birkinshaw, and H. Raistrick, *Biochem. J.* **39**, 24 (1945).

[3] W. Kocholaty, *J. Bacteriol.* **44**, 143, 469 (1942); *Arch. Biochem.* **2**, 73 (1943).

[4a] E. C. Roberts, C. K. Cain, R. D. Muir, F. J. Reithel, W. L. Gaby, J. T. van Bruggen, D. M. Homan, P. A. Katzman, L. R. Jones, and E. A. Doisy, *J. Biol. Chem.* **147**, 47 (1943).

[4b] J. T. van Bruggen, F. J. Reithel, C. K. Cain, P. A. Katzman, E. A. Doisy, R. D. Muir, E. C. Roberts, W. L. Gaby, D. M. Homan, and L. R. Jones, *J. Biol. Chem.* **148**, 365 (1943).

[5] J. H. Birkinshaw and H. Raistrick, *J. Biol. Chem.* **148**, 459 (1943).

[6] D. Keilin and E. F. Hartree, *Biochem. J.* **50**, 331 (1952).

[7] R. Bentley and A. Neuberger, *Biochem. J.* **45**, 584 (1949).

[8] W. Franke and F. Lorenz, *Ann.* **532**, 1 (1937).

as β-glucose and since some preparations may contain both catalase and mutarotase,[9] it has become apparent that there may be possible difficulties in the manometric assay of preparations of varying activity. Much of the work leading to the preparation of highly purified samples was carried out in connection with the antibiotic properties of the enzyme; consequently, the assay method was to determine the limiting dilution at which inhibition of growth of *Staphylococcus aureus* was attained.[2-4] Paradoxically, the more specific and convenient manometric assay has not been reinvestigated, although the use of glucose aerodehydrogenase for glucose assay has been described.[10] The following is an attempt to indicate the practical considerations involved in assay of glucose aerodehydrogenase preparations of varying purity. Two important factors which would determine the experimental conditions are (1) the fate of hydrogen peroxide and (2) the mutarotation of glucose.

1. The fate of hydrogen peroxide will determine the ultimate oxygen uptake. Some decomposition of peroxide takes place even under the most favorable conditions and since glucose aerodehydrogenase preparations also frequently contain catalase,[2b] it would seem desirable to carry out all determinations in the presence of sufficient catalase to decompose rapidly all the hydrogen peroxide.

The following conditions have been used in studying glucose aerodehydrogenase:

(a) In the absence of catalase, 1 molecule of oxygen is required per molecule of glucose oxidized.

$$C_6H_{12}O_6 + H_2O + O_2 \rightarrow C_6H_{12}O_7 + H_2O_2$$

(b) In the presence of excess catalase, 1 atom of oxygen is consumed per molecule of glucose oxidized.

$$C_6H_{12}O_6 + \tfrac{1}{2}O_2 \rightarrow C_6H_{12}O_7$$

(c) In the presence of ethanol and catalase, hydrogen peroxide is used for the coupled oxidation of ethanol,[9] and 1 molecule of oxygen is consumed per molecule of glucose oxidized.

$$C_6H_{12}O_6 + C_2H_5OH + O_2 \rightarrow C_6H_{12}O_7 + CH_3CHO + H_2O$$

2. The manometric assay of glucose aerodehydrogenase is complicated by the mutarotation of the substrate which can lower the concentration of β-glucose to the point where the enzyme is not fully saturated and where the rate of oxygen consumption is not linear. Under the usual

[9] D. Keilin and E. F. Hartree, *Biochem. J.* **50**, 341 (1952).
[10] D. Keilin and E. F. Hartree, *Biochem. J.* **42**, 230 (1948).

experimental conditions (0.2 M phosphate buffer, pH 5.6) the mutarotation of glucose is four times as fast as in water alone. Further, glucose aerodehydrogenase is often accompanied by mutarotase,[9] which is the most powerful catalyst known for the mutarotation.

The following results exemplify conditions[6] where oxygen uptake has been found to be nonlinear. A glucose aerodehydrogenase preparation which contained 37% water-insoluble material ($Q_{O_2} = 40,000$ in air at 20° and pH 5.6) was used in the presence of 0.04 ml. of catalase (0.44 M with respect to hematin), 0.2 M phosphate buffer, pH 5.6, and 0.005 M β-glucose at 20°. This preparation did not contain mutarotase. With 10, 80, or 300 γ of this preparation, straight-line plots of $\log_{10}$ % glucose unoxidized against time were obtained; with 20 or 40 γ, however, the plots were not linear. This was a result of the $\beta \rightarrow \alpha$ change in the first part of the experiment so that in the later stages the change $\alpha \rightarrow \beta$ became limiting. With the higher enzyme concentrations all the β-glucose was rapidly oxidized.

It is necessary, therefore, to adjust the relative concentrations of β- (or equilibrium) glucose and glucose aerodehydrogenase so that the enzyme is saturated. The initial linear rates of oxygen consumption may then be used to calculate the Q_{O_2} value. Keilin and Hartree[6] report that the conditions are met by 0.1 M substrate and 10 γ of enzyme ($Q_{O_2} = 40,000$, determined in phosphate buffer, pH 5.6, in the presence of catalase at 20°). It can be calculated that for crude preparations of unknown activity, under the above conditions, the rate of oxygen uptake should be less than 400 to 500 μl./hr.

Almost all the kinetic studies on glucose aerodehydrogenase have been carried out at 20°. At 37° the rates of both the mutarotation and oxidation reactions will be increased. To work at this temperature it is necessary to make sure that the amounts of enzyme and substrate are so adjusted that during an initial period the rate of oxygen uptake is linear and that at several different enzyme concentrations the rates are proportional to the amount of enzyme.

Definition of Specific Activity. The description of enzyme activity was originally given as Q_{O_2} = microliters of oxygen absorbed per milligram dry weight per hour,[8] and an optimal substrate was defined.[11] In the light of the work by Keilin and Hartree[6] it would be better to standardize the conditions as follows:

0.1 M β-glucose.

0.2 M phosphate buffer, pH 5.6.

0.04 ml. of catalase (0.44 M with respect to hematin).

[11] W. Franke and M. Deffner, *Ann.* **541**, 117 (1939).

Low enzyme concentrations should be used as previously noted. It should be stated whether the determination was made in air or oxygen, since in the latter case the rate of oxygen uptake is two and one-half times that in air.

Purification Procedure

The following description is that of Coulthard et al.[2b] It has yielded preparations estimated to be 80 to 90% pure, containing only traces of two other components. Most of the recent studies on glucose aerodehydrogenase[6,7,9,12,13] have been carried out with material obtained in this way. Purification was followed by determination of antibacterial titer; this is the volume in which 1 ml. or 1 g. of material had to be dissolved to obtain the limiting active dilution. The final preparations were active at a dilution of 1 g. in 3.5×10^8 ml. and had $Q_{O_2} = 48,600$ at 15° in air (calculated $= 130,000$ at 30° from data of Franke and Lorenz[8]).

Step 1. Growth of Mold. *Penicillium notatum* Westling (an unspecified strain supplied by Sir Alexander Fleming) was grown as a surface culture at 20 to 22° on 500-ml. portions of sterilized culture medium, pH 6.0 to 6.5, of the following composition: $NaNO_3$, 2.0 g.; $MgSO_4 \cdot 7H_2O$, 0.5 g.; $FeSO_4 \cdot 7H_2O$, 0.01 g.; KH_2PO_4, 1.0 g.; KCl, 0.5 g.; glucose monohydrate, 40 g.; and water, 1 l. Maximum accumulation of enzyme occurred after 14 days; thereafter the medium became alkaline, and the enzyme disappeared from the solution.

Step 2. Concentration and Acetone Precipitation. 150 l. of culture fluid, having an antibacterial titer of 30,000 to 70,000, was filtered through cloth and concentrated to 30 l. *in vacuo* at an internal temperature not greater than 25°. The red-brown concentrate, pH 3.1 to 3.5, was treated with 2 vol. of acetone at 0° and allowed to stand overnight at 0°. The supernatant was decanted from the usually sticky precipitate; the latter was redissolved in 5 l. of water and reprecipitated by addition of an equal volume of acetone at 0°. The precipitate (antibacterial titer, 40 to 150 $\times 10^6$; recovery = 30 to 50%) dissolved in 2.5 l. water was similarly reprecipitated. This last precipitate was dissolved in 200 ml. of cold water and spun in a Sharples centrifuge to yield a clear orange-red concentrate, pH ca. 5.0.

Step 3. Precipitation with Tannic Acid. The 200 ml. of concentrate was diluted with 500 ml. of cold water and treated with a saturated aqueous solution of 8.0 g. of tannic acid at 0°. After standing for 0.5 hour the usually purple precipitate was spun down and decomposed by trituration

[12] D. Keilin and E. F. Hartree, *Biochem. J.* **42**, 221 (1948).
[13] R. Cecil and A. G. Ogston, *Biochem. J.* **42**, 229 (1948).

with two or three 50-ml. portions of acetone. The crude purple enzyme was centrifuged down and extracted at the centrifuge with 40-, 35-, and 20-ml. portions of water. The combined extracts were precipitated at 0° with an equal volume of acetone; the yellow precipitate was centrifuged, triturated three times with acetone, and dried quickly in air and then *in vacuo*. The yield was 3 to 4 g.; antibacterial activity, 3 to 4 × 10⁸ (20 to 40% recovery of activity calculated on the assay of the original culture filtrate).

$Step$ 4. *Ammonium Sulfate Precipitation*. A 2.5% solution of glucose aerodehydrogenase (antibacterial activity 2.9×10^8) in water or phosphate buffer, pH 7.0, was brought to 50% saturation with ammonium sulfate when a little inactive protein was removed. The supernatant was brought to 80 to 83% saturation, yielding a canary-yellow precipitate. This was dissolved in water and dialyzed for 56 hours against distilled water at 0°. The enzyme was recovered by precipitation with acetone at 0° (antibacterial activity, 3.5×10^8).

The following method used in the purification of penicillin B[4b] apparently offers a quicker route to highly active preparations and will therefore be described.

$Step$ 1. *Uranium Acetate Precipitation*. All operations were carried out in a cold room at 5°. Filtered culture medium from *P. notatum*, pH 3.5 to 3.9, was treated with 1% uranium acetate (15 to 20 ml. per liter of culture medium). The precipitate was allowed to settle, the supernatant siphoned off, and the residue centrifuged. The precipitate was washed twice with water and then extracted with a sufficient volume of 0.2 M phosphate buffer, pH 6.8, to give a mixture of creamy consistency. The material was centrifuged after being allowed to stand for several hours and the precipitate was reextracted similarly. The enzyme was salted out from the combined buffer extracts by addition of saturated ammonium sulfate, 3 vol. The precipitate was centrifuged, dissolved in water, and dialyzed until sulfate-free. The solid was recovered by lyophilization.

$Step$ 2. *Ammonium Sulfate Fractionation*. This step was carried out essentially as described under step 4 of the preceeding purification procedure. Material obtained in this way gave inhibition of bacterial growth in dilutions greater than 1 part in 6 billion.

Properties

Some of the properties of glucose aerodehydrogenase obtained prior to 1942 with material of low Q_{O_2} have been disputed. Unless stated specifically to the contrary, all the properties quoted here are those given by Coulthard *et al.*[2a,b] and Keilin and Hartree[6,9,10,12] for material of 80 to 90% purity.

Specificity. The enzyme is highly specific for β-D-glucopyranose; if the activity to this substrate is taken as 100, of approximately fifty other carbohydrates only the following had measurable activity.

4,6-Benzylidene glucose	1.90
6-Methyl glucose	1.85
Mannose	0.94
α-D-Glucopyranose	0.64
Altrose	0.16
Galactose	0.14

Activators and Inhibitors. The normal hydrogen acceptor is oxygen, and the activity in pure oxygen is two and one-half times as great as in air. 2,6-Dichlorophenolindophenol can act as hydrogen acceptor. The enzyme is not inhibited by HCN, H_2S, HN_3, NaF,[4b] or urethan.[4b] The oxidative activity is not affected at all or is inhibited by not more than 15% on addition of 10 equivalents of other sugars. α-Glucose does not inhibit the oxidation of β-glucose. 0.01 M 8-Hydroxyquinoline, sodium nitrate, and semicarbazide give, respectively, 11, 13, and 20% inhibition.

Effect of pH. Glucose aerodehydrogenase is most stable between pH 3.5 and 8.0. The optimum pH is 5.6, and activity is rapidly lost at pH > 8 and pH < 2.

Physical Properties. The purest enzyme preparations are pale-yellow powders which become colorless in the presence of glucose under anaerobic conditions. Samples of the dried enzyme are stable at 0° for up to 2 years. 0.1 to 0.2% aqueous solutions are stable for 1 week at 5°. The activity is lost on heating at temperatures greater than 39°.

The molecular weight is 149,000, and the enzyme contains 2 molecules of FAD per molecule of enzyme. Glucose aerodehydrogenase has absorption maxima at 377 and 455 mμ; it shows no fluorescence in ultraviolet light, but after heat or acid or alkali treatment the coenzyme shows the characteristic green fluorescence of FAD.

The Michaelis constant is 0.0096 M (determined with equilibrium glucose at 20° in the presence of catalase and 0.2 M phosphate buffer, pH 5.6).

Glucose aerodehydrogenase is not attacked by pepsin at pH 2.5 or by trypsin at pH 7.5.

[46] Hexose Phosphate and Hexose Reductase

A. D-Mannitol-1-phosphate Dehydrogenase from *E. coli*[1]

$$\text{F-6-P} + \text{DPNH} + \text{H}^+ \rightleftarrows \text{Mannitol-1-P} + \text{DPN}^+$$

By JOHN B. WOLFF and NATHAN O. KAPLAN

Assay Method

Principle. The method depends on the decrease in light absorbance at 340 mμ when reduced DPN is oxidized in the presence of fructose-6-phosphate.

Reagents

F-6-P stock solution (0.1 M). Dissolve 400 mg. of the barium salt in about 2 ml. of water. Add a volume of K_2SO_4 solution equivalent to the barium to be precipitated (2 ml. of 0.5 M K_2SO_4). Centrifuge, and wash the residue of $BaSO_4$ thoroughly with several 1-ml. portions of water to remove the adsorbed F-6-P. Combine all washings with supernatant liquid and make to a total volume of 10 ml. Assay for total phosphate by the method of Fiske and SubbaRow[2] and for fructose by the method of Roe[3] (F-6-P gives 61% of the amount of color of free fructose).

DPNH stock solution (0.003 M). Dissolve the barium salt (Vol. III [126]), in water, and remove the barium as above for F-6-P. Assay final concentration with acetaldehyde and yeast alcohol dehydrogenase (Vol. I [79]).

0.1 M phosphate *buffer*, pH 7.5, or 0.1 M $NaHCO_3$ buffer, pH 8.0.

Enzyme. Dilute concentrated solution with bicarbonate buffer to obtain 100 to 300 units of enzyme per milliliter.

Procedure. In a 3-ml. cell having a 1-cm. light path are placed 0.3 micromole of DPNH, 0.05 ml. of enzyme solution (5 to 15 units), and buffer to a total volume of 2.9 ml. Absorbance is read at 340 mμ before and at 60-second intervals after mixing with 0.1 ml. of F-6-P solution.

Definition of Unit and Specific Activity. One unit of enzyme is defined as that amount which causes an initial rate of change in absorbance (ΔA_{340}) of 0.010 per minute under the above conditions at 25°. Specific

---

[1] The procedure reported here is to be published by J. B. Wolff and N. O. Kaplan.

[2] C. H. Fiske and Y. SubbaRow, *J. Biol. Chem.* **66**, 375 (1925).

[3] J. H. Roe, *J. Biol. Chem.* **107**, 15 (1934); see also Vol. III [22].

activity is expressed as units per milligram of protein. Protein is determined by the method of Lowry et al.[4]

Application of Assay Method to Crude Tissue Preparations. In order to permit direct spectrophotometric measurements, it is usually necessary to dilute the extract to a point where its light absorption at 340 mμ is small enough to give an accurate measure of the rate of change of absorbance upon addition of substrate. The rate of endogenous oxidation of DPNH must be subtracted from the rate obtained in the presence of F-6-P.

Purification Procedure

Step 1. Preparation of Crude Bacterial Extract. 15 g. of freshly harvested or lyophilized bacteria is ground in a chilled mortar with 15 g. of powdered alumina (Alcoa A-301), then extracted by slow addition, during continued grinding, of 150 ml. of ice-cold 0.02 M NaHCO$_3$ buffer. The extract is centrifuged for 10 minutes at 0° at 2300 $\times$ g, and the residue is discarded. The supernatant liquid is dialyzed overnight against 4 l. of 0.02 M NaHCO$_3$ solution in the cold (4°).

Step 2. Acid Precipitation. The dialyzed solution from step 1 is placed in a pH meter and provided with a mechanical stirrer. Ice-cold 1 N acetic acid is added slowly until the pH has fallen to 4.7. The mixture is centrifuged at high speed, the residue discarded, and the supernatant fluid immediately neutralized by careful addition of cold 1 N NaOH solution.

Step 3. Fractionation with Ammonium Sulfate. To the solution from step 2 which is kept cold and stirred continuously is added very slowly finely powdered ammonium sulfate (35.1 g./100 ml.) to give 0.55 saturation. After high-speed centrifugation the supernatant is treated with additional ammonium sulfate (6.62 g./100 ml.) to 0.65 saturation. After centrifugation the residue, which contains the activity, is dissolved in a minimum amount of water. The preparation may be kept frozen for several months without loss of activity. It may be dialyzed against 0.02 M NaHCO$_3$ solution, but the activity subsequently decreases rapidly.

SUMMARY OF PURIFICATION PROCEDURE

Fraction	Total vol., ml.	Units/ml., thousands	Total units, thousands	Protein, mg./ml.	Specific activity, units/mg.	Recovery, %
1. NaHCO$_3$ extract	110	0.28	30.8	9.4	40	—
2. Acid precipitation, pH 4.7, supernatant	87	0.18	15.6	1.54	117	51
3. (NH$_4$)$_2$SO$_4$ fraction, 0.55–0.65	4	1.225	4.90	1.43	860	16

[4] O. H. Lowry, N. J. Rosebrough, A. Farr, and R. J. Randall, *J. Biol. Chem.* **193**, 265 (1951); see also Vol. III [73].

Properties

Specificity. The purified enzyme appears to be absolutely specific for F-6-P, having no action on F-1-P, FDP, mannose-6-P, ribulose-5-P, sedoheptulose-7-P, or free fructose. The slight activity with G-6-P is due to the presence of hexose phosphate isomerase which is difficult to remove. TPNH will not replace DPNH as the active coenzyme.

Activators and Inhibitors. The activity of the enzyme is independent of the type of buffer used. It appears to have no metal ion requirement. Fluoride, semicarbazide, and hydroxylamine (0.05 M) have no effect on the reaction, whereas 0.10 M NaHSO$_3$ or 0.015 M KCN inhibits the initial rate 50%. The reaction rate is also inhibited 80% by 5×10^{-4} M p-chloromercuribenzoate; this inhibition can be completely reversed by 10^{-3} M GSH.

Effect of pH. The activity exhibits no sharp optimum in the range of pH from 6 to 9, falling off gradually above pH 9. The stability of the enzyme to thermal denaturation increases from pH 5.2 to 8.6 in 0.1 M phosphate buffer. The activity is almost completely destroyed by heating the enzyme for 10 minutes at 60°. About 45% of activity is lost by a similar heating period at 55° and pH 7.0.

Reversibility. When DPN$^+$ and enzyme are added to a buffered solution of the product of the enzymatic reduction of F-6-P or synthetic D-mannitol-1-phosphate, DPNH and ketohexosephosphate are formed. The equilibrium appears to favor the reduction of F-6-P. The equilibrium constant for the reduction of F-6-P at pH 7 and 25° is approximately 2×10^2.

B. Sorbitol Dehydrogenase from Liver

$$\text{Sorbitol} + \text{DPN}^+ \rightleftarrows \text{D-Fructose} + \text{DPNH} + \text{H}^+$$

By JOHN B. WOLFF

Almost all the following material is taken from the article of Blakley.[5]

Assay Method

Principle. The method chiefly used by Blakley involves the manometric measurement of oxygen uptake, but the spectrophotometric determination of the increase in absorbance at 340 mμ, due to the reduction of DPN, is equally applicable.

Reagents

  Sorbitol stock solution (0.1 M). Dissolve 2.00 g. of D-sorbitol monohydrate in 100 ml. of water which is saturated with respect to

[5] R. L. Blakley, *Biochem. J.* **49**, 257 (1951).

benzoic acid (1 g.). The solution is stable at room temperature for many months.

DPN (0.005 M). Dissolve 36 mg. of 95% pure DPN in 10 ml. of water; assay exact concentration with yeast alcohol dehydrogenase and alcohol.

Phosphate buffer (0.1 M), pH 7.8 to 8.

Enzyme. Dilute the stock solution with phosphate buffer to obtain 500 to 1000 units of enzyme per milliliter.

Procedure. Place 0.1 ml. of sorbitol solution, 0.02 ml. of DPN solution, 1 ml. of buffer, and water to a total volume of 2.8 ml. in a cuvette having a 1-cm. light path. Take readings at 340 mμ before and at 1-minute intervals after mixing with 0.2 ml. of enzyme.

Definition of Unit. One unit of enzyme is defined as that amount which causes an initial rate of change in absorbance (ΔA_{340}) of 0.001 per minute under the above conditions.

Application of Assay Method to Crude Tissue Preparations. With liver homogenates it is usually necessary for maximum oxygen uptake to add 0.01 M (final concentration) nicotinamide, 0.64 $\times$ 10^{-4} M cytochrome c, and 10^{-3} M methylene blue. When the spectrophotometric method is used, the enzyme preparation should be diluted to permit accurate readings at 340 mμ.

Purification Procedure

Either a crude extract of mammalian liver or acetone-dried preparations extracted with 0.01 M bicarbonate buffer may be fractionated with ammonium sulfate between 50 and 60% saturation. Ammonium sulfate or ethanol fractionation does not increase the activity of purified preparations obtained as follows:

Rat livers are homogenized in 3 vol. of ice-cold 0.01 M phosphate buffer, pH 7.8, in a Waring blendor for 1.5 to 2 minutes. By slow addition of cold 2 N HCl the pH is lowered to 4.0, then raised again to 7.8, yielding an active supernatant after centrifugation. This supernatant is mixed with 0.25 vol. of cold ethanol and 0.25 vol. of cold chloroform, shaken in a stoppered vessel for 1 minute, and again centrifuged. The supernatant aqueous layer is pipetted off and thoroughly dialyzed against distilled water at 2°, giving a yellowish-brown solution representing about 8-fold purification and 45% recovery.

Properties

Specificity. In addition to oxidizing sorbitol to fructose, the enzyme will oxidize L-iditol to L-sorbose reversibly. The most highly purified enzyme preparations showed slight oxidation of D-glyceraldehyde but did not activate D-mannitol, dulcitol, inositol, D-glucose, glycerol, ethanol, lactate, dihydroxyacetone, or the corresponding phosphate esters.

Inhibitors. Sodium borate ($2 \times 10^{-2} M$, pH 7.8) inhibited the rate of oxygen uptake in the presence of sorbitol by 95%, whereas $10^{-2} M$ KCN produced a 53% inhibition. Mannitol and dulcitol ($2 \times 10^{-2} M$) have no effect on the rate of sorbitol oxidation.

Effect of pH and Temperature. The optimum activity of the enzyme occurs at pH 7.9 to 8.1 in phosphate or glycine buffer at 38 to 40°.

Equilibrium Constant. The constant determined by Blakley at 20° is 0.24 for the reversible oxidation of sorbitol at pH 8.0.

Distribution. The enzyme occurs in rat liver and kidney and in the livers of frog, mouse, guinea pig, cat, rabbit, and beef.[6]

[6] Williams-Ashman and Banks [H. G. Williams-Ashman and J. Banks, *Arch. Biochem. Biophys.* **50,** 513 (1954)] have recently purified the enzyme some 70 fold from rat liver. These authors also report the presence of the enzyme in the accessory male sex glands. It is also suggested, since the enzyme reduces a number of keto sugars, that the enzyme be called "ketose reductase."

[47] Gluconokinase

Gluconate + ATP → 6-Phosphogluconate + ADP

By SEYMOUR S. COHEN

Source

The enzyme catalyzing the above reaction was demonstrated by Cohen[1] in cell-free extracts of *E. coli*, strain B, adapted to growth in a synthetic medium containing potassium gluconate as sole carbon source. It was shown that bacteria grown on glucose were practically devoid of the enzyme. The mineral composition of the medium supporting growth of *E. coli* and production of the enzyme at 38° is given in per cent as follows: Na_2HPO_4, 1.64; KH_2PO_4, 0.15; $(NH_4)_2SO_4$, 0.2; $MgSO_4 \cdot 7H_2O$, 0.02; $CaCl_2$, 0.001; and $FeSO_4 \cdot 7H_2O$, 0.00005. Cells were grown overnight in this medium containing glucose (0.1%). The bacteria had stopped growth in the exponential phase at about 10^9 per milliliter and were inoculated into media containing K gluconate (0.13%) or mixtures of glucose (0.02%) and gluconate (0.11%). The purpose of the glucose is to start cell multiplication again without an extended lag phase. When begun on

[1] S. S. Cohen, *J. Biol. Chem.* **189,** 617 (1951).

glucose, the cells can grow on gluconate immediately after glucose is exhausted from the medium, adaptation to gluconate having taken place during the glucose phase of growth. Growth is followed turbidimetrically on a Klett-Summerson photoelectric colorimeter fitted with a 420 filter. The cells are harvested when in their exponential phase on gluconate, the mass doubling time being 55 to 60 minutes on that substrate. The bacteria are chilled, collected by centrifugation, and washed twice in 0.85% saline.

Gluconokinase has also been found in the maceration juice of brewer's and baker's yeast.[2,3]

Assay Methods

Principle. In the reaction catalyzed by gluconokinase 1 mole of acid is liberated per mole of gluconate phosphorylated. This has been followed by observing CO_2 evolution in a bicarbonate buffer in Warburg vessels in studies with *E. coli*, or by measuring the pH drop in studies with the yeast enzyme. The phosphorylation of gluconate also involves the decrease of acid-labile P from ATP, 6-phosphogluconate being relatively acid-stable. However, extracts of *E. coli* and yeast contain ATP-ase which also results in acid production and a decrease in acid-labile P. It is, therefore, necessary as a control to measure the ATP-ase activity of the enzyme preparation in the absence of gluconate or to purify the enzyme to remove ATP-ase.

Reagents for Manometry

K gluconate (0.05 M). Crystalline potassium salts of the onic acids can be prepared from the aldoses by hypoiodite oxidation.[4]

ATP (0.05 M). The disodium salt as purchased from a suitable commercial source is adjusted to pH 6.8 with KOH, and stored in the frozen state.

M $MgSO_4 \cdot 7H_2O$.

0.02 N $NaHCO_3$.

5% CO_2–95% N_2.

Manometric Procedure. The activities of the extracts are measured in Warburg respirometers at pH 7.4 at 38°. The main compartments of the vessels contain 10 micromoles of substrate, and small aliquots of enzyme (equivalent to 10^{10} bacteria) in 2.3 ml. of 0.02 N $NaHCO_3$. A side arm contains 5 micromoles of ATP and 0.01 M Mg^{++} in 0.2 ml. of 0.02 N

[2] H. Z. Sable and A. J. Guarino, *J. Biol. Chem.* **196**, 395 (1952).

[3] *A special active dried brewer's yeast* and a starch-free baker's yeast were obtained from Anheuser-Busch, Inc.

[4] S. Moore and K. P. Link, *J. Biol. Chem.* **133**, 293 (1940).

NaHCO₃. The atmosphere contains 5% CO_2 and 95% N_2. After equilibration in which gas evolution is observed to be essentially zero, the contents of the side arm are tipped into the main compartment. CO_2 production in the vessel containing gluconate in excess of that in the vessel missing substrate is a measure of gluconokinase activity.

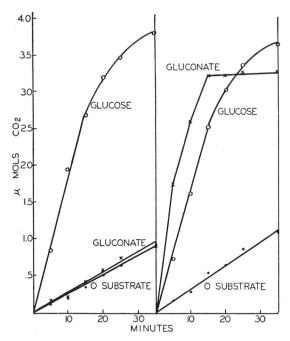

Fig. 1. Acid production in the enzymatic transphosphorylation from ATP to gluconate by extracts of adapted and nonadapted bacteria. Left, extract of unadapted cells, grown in glucose; right, extract of adapted cells, grown in gluconate.

As observed in Fig. 1, hexokinase and ATP-ase activity are present in extracts of adapted and unadapted cells, but gluconokinase is present only in adapted cells.

Formation of Acid-Stable Phosphate. The following instructions are given by Sable and Guarino[2] for measuring the formation of 6-phosphogluconate. 4 micromoles of gluconate and 6 micromoles of ATP were incubated at 30° with dialyzed yeast maceration juice, equivalent to 9.8 mg. of protein, 0.01 *M* Tris[5] buffer, pH 7.6, and 0.01 *M* Mg⁺⁺ in a final volume of 2 ml. After appropriate intervals, inorganic[6] and acid-labile

---

[5] Tris is tris(hydroxymethyl)aminomethane and may be obtained from the G. Frederick Smith Chemical Company.

[6] E. J. King, *Biochem. J.* **26**, 292 (1932).

phosphorus were determined in aliquots of protein-free filtrates. Acid-labile phosphate was determined by measuring the increase in inorganic orthophosphate after 10 minutes of hydrolysis in N H_2SO_4 at 100°.

Measurement of pH Change.[2] External electrodes of a Beckman pH meter were placed in incubation mixtures containing 50 mg. of protein, 0.02 M Mg^{++}, 0.007 M Tris buffer, pH 7.0, 15 micromoles of gluconate, and 20 micromoles of ATP in a final volume of 6 ml. According to Sable and Guarino this is the most convenient procedure for the estimation of the relative activity of enzyme preparations as a guide to their purification.

Enzymatic Identification of the End Product

Phosphogluconate may be determined by following the reduction of excess TPN at 340 mμ in the Beckman spectrophotometer in the presence of phosphogluconate dehydrogenase. The isolation of this enzyme from *E. coli* and conditions for its optimal reactivity are given by Scott and Cohen.[7] The isolation and properties of the enzyme from yeast are described by Horecker and Smyrniotis.[8]

Partial Purification of the Enzyme E. coli. All operations were carried out in a cold room. Wet pellets (of the order of 1 g.) of bacteria were mixed with 2.5 times their weight of alumina (A-301, Aluminum Corporation of America). The mixture was transferred to a chilled mortar and ground vigorously by hand for 2 to 3 minutes. The glutinous paste was suspended in 0.01 M glycylglycine or Tris buffer at pH 7.0 (10 ml./g. of wet bacteria), and sedimented for 30 minutes at 5000 r.p.m. Over 90% of the activity appeared in the supernatant fluid, the remainder being obtained in two further elutions. An equal volume of saturated $(NH_4)_2SO_4$ at pH 7.6 was added; after a half hour an inactive precipitate was removed by centrifugation. Gluconokinase was quantitatively precipitated by increasing the $(NH_4)_2SO_4$ concentration to 75% saturation. This fraction also contains some of the 6-phosphogluconate dehydrogenase.

Brewer's Yeast.[2] Maceration juice was prepared by incubating the dried yeast with 3 vol. of 0.067 M Na_2HPO_4 for 4 hours at 38°. To the juice (150 ml.) was added 183.5 ml. of a solution of $(NH_4)_2SO_4$, pH 7.6, saturated at 25°, to give a saturation of 0.55. After chilling for 20 to 30 minutes, the precipitate was removed by centrifugation. It was dissolved in 60 ml. of cold water and dialyzed for 18 hours at 2° against large volumes of distilled water. The solution was adjusted to pH 4.7 by addition of 0.1 N acetic acid, and, after standing, the precipitate was removed by

[7] D. B. M. Scott and S. S. Cohen, *Biochem. J.* **55**, 23 (1953).
[8] B. L. Horecker and P. Z. Smyrniotis, *J. Biol. Chem.* **193**, 371 (1951); see Vol. I [42].

centrifugation. The supernatant fluid was adjusted to 23% ethanol by addition of 27.5 ml. of cold 95% ethanol, and centrifuged immediately. The solution (108 ml.) was adjusted to 55% ethanol concentration by addition of 86 ml. of 95% ethanol. The mixture was centrifuged, and the precipitate suspended in 30 ml. of H_2O and adjusted to pH 6.8 with 0.1 N NaOH. The mixture was dialyzed with rotation for 4 hours against 2 l. of distilled water at 2° and frozen overnight. After thawing, the mixture was centrifuged, and the precipitate was discarded. At 6- to 8-fold purification was recorded. Hexokinase was present in only trace amounts.

Properties

Specificity. Gluconokinase from *E. coli* is inactive on D-mannonate, D-2-ketogluconate, D-altronate, D-galactonate, L-idonate, D-5-ketogluconate, D-xylonate, D-glucuronate, D-arabinose, D-isoascorbate, D-ribonate, D-arabonate, and D-ribose. It appeared absolutely specific for D-gluconate. The enzyme from yeast was inactive on glycerol, glycolaldehyde, D- and L-arabinose, and D-xylose.

Activators. The enzyme from *E. coli* was optimally active at 10^{-3} M Mg^{++}, that from yeast at 5 to 7 $\times$ 10^{-3} M Mg^{++}. Higher concentrations were inhibitory in both cases. The yeast enzyme was totally inactive in the absence of added magnesium. Mn^{++} also activated this enzyme, although to a lesser extent.

The Michaelis constants for the yeast enzyme are $K_m = 1.68 \times 10^{-3}$ M for magnesium, 1.74×10^{-3} M for gluconate, and 2.0×10^{-4} M for ATP.[2]

Effect of pH. A sharp optimum at pH 7.2 was obtained for the yeast enzyme.[2]

[48] Synthesis of Glucose-1,6-diphosphate

By Luis F. Leloir and Raúl E. Trucco

Glucose-1,6-diphosphate is synthesized in animal tissues and yeast according to the following reaction:[1]

$$G\text{-}1\text{-}P + ATP \rightarrow Glucose\text{-}1,6\text{-}diphosphate + ATP$$

This reaction is catalyzed by an enzyme different from hexokinase, which has been called glucose-1-phosphate kinase.

In some bacteria such as *Escherichia coli*, glucose-1;6-diphosphate can

[1] A. C. Paladini, R. Caputto, L. F. Leloir, R. E. Trucco, and C. E. Cardini, *Arch. Biochem.* **23**, 55 (1949).

be formed without the intervention of adenosine phosphate,[2] by a reaction which has been formulated as follows:

$$2\text{G-1-P} \rightarrow \text{Glucose-1,6-diphosphate} + \text{glucose}$$

It has not been decided if the enzyme catalyzing this reaction is identical to the bacterial phosphatase.

Glucose-1,6-diphosphate is also formed when G-1-P is incubated with alkaline phosphatase, but in a lower yield than with the bacterial enzyme.[3]

The glucose-1-phosphate kinase-catalyzed reaction, which gives a better yield, will be described here.

Assay Method

Principle. Enzymatic activity is estimated by measuring the glucose-1,6-diphosphate formed. The latter is determined enzymatically by the method of Cardini *et al.*[4] based on the acceleration of the phospho-glucomutase reaction.

Another procedure is the manometric method of Colowick and Kalckar[5] for hexokinase. However, with this procedure large amounts of substrates are required. The first procedure is as follows.

Reagents

G-1-P solution, 10 micromoles/ml.   Sodium salt.
ATP solution, 10 micromoles/ml.   Sodium salt.
0.1 M magnesium sulfate.

Procedure. Mix 0.05 ml. of G-1-P solution, 0.04 ml. of ATP, and 0.01 ml. of magnesium sulfate solution. Total volume, 0.1 ml.

The reaction is started by adding the enzyme. The mixture is incubated 10 minutes at 37° and heated at 100° for 1 minute in a boiling water bath. Glucose-1,6-diphosphate is then estimated by addition of G-1-P and phosphoglucomutase, with glucose-1,6-diphosphate as standard (see Vol. III [17]). With crude extracts it is necessary to centrifuge the heat-inactivated mixture and to take an aliquot of the supernatant. A control in which the kinase reaction is stopped at $t = 0$ gives the amount of pre-existing glucose-1,6-diphosphate. Another control to which no phos-

[2] L. F. Leloir, R. E. Trucco, C. E. Cardini, A. C. Paladini, and R. Caputto, *Arch. Biochem.* **24**, 65 (1949).

[3] C. E. Cardini, *Ciencia e invest. (Buenos Aires)* **7**, 235 (1951).

[4] C. E. Cardini, A. C. Paladini, R. Caputto, L. F. Leloir, and R. E. Trucco, *Arch. Biochem.* **22**, 87 (1949).

[5] S. P. Colowick and H. M. Kalckar, *J. Biol. Chem.* **148**, 117 (1943).

phoglucomutase is added gives the amount of reducing substances formed during the incubation.

Preparation of the Enzyme. Glucose-1-phosphate kinase can be prepared from brewer's yeast and rabbit muscle.[1] Purer preparations and higher yields are obtained from muscle. Rabbits are killed by a blow, and bled. The muscles from the limbs and back are chilled, passed through a meat grinder, and suspended in 2 vol. of cold water. After 10 minutes the suspension is strained through cheesecloth and extracted again with 1 vol. of water. The water extracts are discarded. A third extraction is made with 1 vol. of a solution of 0.1 M phosphate buffer of pH 7.4 in 0.5 M potassium chloride.

The saline extract is then fractionated with ammonium sulfate as follows. To 110 ml. of the extract, 80 ml. of saturated ammonium sulfate solution adjusted to pH 7.5 with ammonia is added. The myosin precipitate is separated by filtration through fluted paper in the cold. 150 ml. of clear filtrate is obtained, to which 25 ml. of saturated ammonium sulfate (pH 7.5) is added to give 0.50 saturation. The precipitate is collected by filtration and dissolved in 8 ml. of water. 0.01 ml. of this solution catalyzes the formation of glucose-1,6-diphosphate in amounts of the order of 1 to 4 $\times$ 10^{-3} micromole under the conditions specified for the assay method.

The progress of the purification may be followed by determining the ratio of activities glucose-1-phosphate kinase/phosphoglucomutase, or the maximum yield of glucose-1,6-diphosphate relative to the G-1-P added.

After this procedure a fourfold increase of the ratio activity/total protein is achieved, and the relation kinase/mutase increases forty times. With the best preparations, as much as 80% of G-1-P is transformed into glucose-1,6-diphosphate.

Properties

Specificity. The purified muscle enzyme forms glucose-1,6-diphosphate from G-1-P, but not from G-6-P, F-1-P, F-6-P, glucose, or fructose.

Activators and Inhibitors. Mg and Mn ions activate G-1-P kinase. The optimal concentrations for maximal activity are about 1.5 $\times$ 10^{-3} M and 2 $\times$ 10^{-3} M, respectively. ATP at concentrations higher than 4 $\times$ 10^{-3} M produces inhibition.

Effect of pH. With maleate buffer the pH optimum has been found to be at pH 6.8 for both the muscle and the yeast enzyme.

[49] Yeast Pentokinase

Ribose + ATP → R-5-P + ADP

By HENRY Z. SABLE

Assay Method

Principle. The assay depends on removal of protein and phosphorylated compounds with the $Ba(OH)_2$–$ZnSO_4$ reagent of Somogyi[1] and measurement of residual ribose in the filtrates by the alkaline copper reagent of Nelson.[2]

Reagents

ATP (0.1 M). Dissolve 655 mg. of $K_2ATP \cdot 4H_2O$[3] or 623 mg. of $Na_2ATP \cdot 4H_2O$[3] in 6 ml. of H_2O, adjust to pH 7.5 with 1 M KOH or NaOH, and dilute to 10 ml. This solution keeps well when stored in a deep-freeze.

Ribose 0.5%. Dissolve 250 mg. of ribose in water and dilute to 50 ml. This solution keeps well on storage at 3°.

0.4 M Tris-HCl buffer, pH 7.5.

0.1 M $MgCl_2$.

0.1 M NaF.

0.3 N $Ba(OH)_2$ and 5% $ZnSO_2 \cdot 7H_2O$.[4]

Enzyme. The undiluted solutions are used.

Procedure. A reaction mixture containing ribose (0.5 ml.), Tris (0.2 ml.), $MgCl_2$ (0.25 ml.), NaF (0.25 ml.), and H_2O (1.8 ml.) is incubated at 30°. The reaction is started by addition of 2.0 ml. of enzyme solution. At 1, 10, 25, and 60 minutes after addition of the enzyme, 1.0-ml. aliquots of the mixture are added to 2.0 ml. of $Ba(OH)_2$, mixed, 2.0 ml. of $ZnSO_4$ added, and the mixture shaken vigorously for 1 minute. After standing 10 to 30 minutes, the mixtures are filtered, and 1.0-ml. aliquots of the filtrates are analyzed for reducing sugar by the method of Nelson,[2] with ribose as a standard.[5] The rate of sugar utilization in the incubated mixture is determined graphically.

[1] M. Somogyi, *J. Biol. Chem.* **160**, 69 (1945).

[2] N. Nelson, *J. Biol. Chem.* **153**, 375 (1944).

[3] Various high-purity preparations of ATP are available commercially.

[4] Somogyi[1] points out that the accuracy of these concentrations is less important than the requirement that the alkali must neutralize the $ZnSO_4$ solution precisely, volume for volume. The author prefers to use a solution of $Ba(OH)_2$ which is 2 to 4% more concentrated than one which will just neutralize the $ZnSO_4$.

[5] 100 γ of ribose is an adequate working standard. There is some reduction of the copper reagent by the Tris, but a correction is easily made for any given colorimeter by carrying out the determination with a concentration of Tris equal to that present in the filtrate.

Definition of Unit and Specific Activity. A unit of ribokinase is the amount of enzyme which would catalyze the phosphorylation of 0.10 micromole of ribose in 2 hours at the initial rate observed. Specific activity is expressed as units per milligram of protein. Protein is determined by the method of Robinson and Hogden.[6]

Precautions in the Assay. The relatively low activity of the enzyme requires the use of large amounts of proteins. Consequently, the enzyme solutions should be dialyzed to remove interfering substances, e.g., sulfates and phosphates, which interfere with the deproteinization by precipitating with the $Ba(OH)_2$. In addition, autolyzates prepared from certain brands of yeast contain nondialyzable components which liberate reducing substances during the incubation and thus obscure the enzyme action. In such cases it is advisable to carry out the ammonium sulfate precipitation before assaying.

Purification Procedure

The procedure outlined is that reported previously by the author.[7] It has been carried out with yeast from various sources. The best results were obtained with a beer yeast obtained directly from the fermentation vat and washed and dried according to Lebedev.[8] Satisfactory results can be achieved with starch-free baker's yeast or with commercially available "special active" dried brewer's yeast.[9]

Step 1. Preparation of an Autolyzate. With commercial dried yeast, either of two methods of autolysis may be used. The yeast may be thoroughly suspended in 0.067 M Na_2HPO_4 and incubated for 4 hours at 37°. Alternatively, the yeast may be stirred for 5 days at 2° with 2.5 vol. of water and 0.15 vol. of toluene. At the end of the autolysis the solids are removed by centrifugation.

Step 2. Fractionation with Ammonium Sulfate. The autolyzate is chilled in ice, and a solution of ammonium sulfate (saturated at 25° and then adjusted to pH 7.6 with ammonia) is added slowly with constant stirring (0.82 ml. is added for each milliliter of the autolyzate). After chilling for an additional 30 minutes, the precipitate is removed by high-speed centrifugation and discarded. The solution, which is at 0.45 saturation, is now adjusted to 0.65 saturation by further addition of the ammonium sulfate solution (0.57 ml. of saturated ammonium sulfate for each milliliter of the 0.45 saturated solution). After chilling for 30 minutes the precipitate is collected by high-speed centrifugation and the supernatant solution

[6] H. W. Robinson and C. G. Hogden, *J. Biol. Chem.* **135**, 707, 737 (1940).

[7] H. Z. Sable, *Proc. Soc. Exptl. Biol. Med.* **75**, 215 (1950).

[8] A. von Lebedev, *Z. physiol. Chem.* **73**, 447 (1911).

[9] Product of Anheuser-Busch, Inc.

discarded. The precipitate is suspended in cold water (0.5 ml. for each milliliter of autolyzate) and stirred to dissolve. A large yield of crystals is frequently observed at this stage. These have been examined repeatedly and found to be $MgNH_4PO_4$ and may be discarded. With some brands of yeast the limits of fractionation may be narrowed without incurring any loss of enzyme; the first precipitation may be made at 0.50 saturation, and the second at 0.625 saturation.

SUMMARY OF PURIFICATION PROCEDURE[a]

Fraction	Total volume, ml.	Units/ml.	Total units, thousands	Protein, mg./ml.	Specific activity, units/mg.	Recovery, %
1. Autolyzate[b]	670	62.5	41.7	70.4	0.89	—
2. Ammonium sulfate fraction, 0.485–0.625	450	53.2	23.9	36.6	1.45	57
3. Acid precipitate, pH 4.6	239	51.3	12.2	10.8	4.75	29
4. Alcohol[c] fraction, 0–20%						

[a] Modified from the data reported previously.[7]

[b] From 500 g. dried yeast, autolyzed in the cold with toluene.

[c] Because of poor storage qualities of the alcohol fraction, this fraction is prepared as needed. In several experiments alcohol fractionation increased the purity by two- to threefold, with a recovery of 75% of the activity of the acid-precipitated fraction.

Step 3. Acid Precipitation. The solution obtained from step 2 is dialyzed with rotation for 12 to 14 hours against two changes of acetate buffer,[10] pH 6.0, ionic strength 0.05, and then for an additional 6 to 8 hours against acetate buffer, pH 4.5 to 4.6, ionic strength 0.02. The volume of buffer should be about two hundred times as great as the volume of solution being dialyzed. During the last dialysis a large amount of precipitate forms. This is collected by centrifugation and the solution discarded. The precipitate is hard and dissolves with difficulty. A small amount of cold water is added, and the solid is triturated, with gradual addition of 4 M sodium acetate until the pH reaches 6, and then 25% NaCl is added and trituration continued until most of the precipitate dissolves, or until all the precipitate is finely suspended. Frequently the precipitate contains irreversibly denatured protein which does not dissolve. The final solution, which is about 0.3 M with respect to salts, is then dialyzed against 100 vol. of acetate buffer, pH 6.0, ionic strength 0.30, the buffer being changed once in the course of an 8-hour dialysis. The insoluble material is removed by centrifugation.

[10] W. C. Boyd, *J. Am. Chem. Soc.* **67**, 1035 (1945).

Step 4. Alcohol Fractionation. A solution 80% by volume of ethanol in water, ionic strength 0.02 (NaCl), is used. The protein solution obtained in step 3 is maintained at -2 to $-5°$, and 0.33 vol. of the alcohol solution is added. After an additional 10 minutes' chilling at $-2°$, the precipitate is collected by centrifugation and dissolved by adding concentrated acetate buffer, pH 6.0, previously chilled to $-5°$ (0.2 to 0.5 ml./ml. of the step 3 solution). Excess salts and alcohol are removed by 4 to 5 hours' dialysis against dilute acetate buffer, pH 6.0; a precipitate which may form during this dialysis may be discarded. The final solution is usually bright yellow in color.

The ammonium sulfate fraction has the best storage properties, being fairly stable for several weeks at deep-freeze temperatures. The acid-precipitated fraction loses about one-third to one-half its activity in 2 weeks in a freezer.

Properties

Specificity. The enzyme is absolutely specific for ribose, having no action on D- or L-arabinose, D-xylose, or deoxyribose. K_s for ribose[11] is $1.2 \times 10^{-3} M$.

Contaminating Enzymes. ATP-ase is concentrated in the acid-precipitated fraction and is still present in the alcohol fraction. The latter fraction also contains phosphoribose isomerase. In step 3 when ribokinase is precipitated by acid, most of the hexokinase activity remains in the acid-soluble fraction.

Product of the Reaction. The product of the reaction has not been isolated, but there is considerable evidence[7,12] that the initial product is ribose-5-phosphate. There is no isomerization to a ketose prior to phosphorylation.[11]

Activators and Inhibitors.[11] The enzyme is inactive in the absence of Mg ions. There is a very narrow range of optimum Mg concentration at $5 \times 10^{-3} M$, less than 50% of the maximal activity being obtained at 10^{-3} or $10^{-2} M$. K_s for Mg ions is $3 \times 10^{-3} M$. Mn ions interfere with the copper reduction method, and their effect on the enzyme cannot be evaluated. Fluoride partially inhibits the ATP-ase but does not inhibit ribokinase.

Effect of pH.[11] There is very little difference in the activity of the enzyme at any pH between 5.8 and 8.4. The ATP-ase is more active at the higher pH values, but there appears to be no particular advantage in using any one pH in this range. The pH specified in the assay method is that at which most of the experiments were carried out.

[11] H. Z. Sable, unpublished experiments.
[12] H. Z. Sable, *Biochim. et Biophys. Acta* **8**, 687 (1952).

[50] Phosphoribomutase from Muscle

R-1-P $\rightleftarrows$ Ribose-5-phosphate

By HANS KLENOW

Assay Method

Principle. The assay is based on the fact that the ester bond of the substrate (R-1-P) is acid labile[1] in contrast to that of the product; i.e., the disappearance of acid-labile phosphate ester is followed as the enzyme reaction proceeds.

Reagents

The Ba salt of R-1-P is prepared by phosphorolytic cleavage of inosine.[1] It is purified by ion exchange chromatography with Dowex-1 formate.[2] The crystalline cyclohexylamine salt is prepared[3] by a method analogous to that used for deoxyribose-1-phosphate preparation.[4] Solution (0.03 M): Dissolve 13 mg. of cyclohexylamine salt per milliliter of water. This solution is stable when stored at low temperature.

0.25 M Tris-HCl buffer, pH 7.4.

0.01 M MgCl$_2$.

Enzyme. If necessary dilute the stock solution with water to give about 1 unit/ml.

Procedure. The procedure is essentially that described by Najjar[5] for the assay of phosphoglucomutase. To 30 μl. of Tris-HCl buffer is added 30 μl. of MgCl$_2$, 30 μl. of R-1-P, enzyme (about 0.01 unit), and water to a final volume of 330 μl. Aliquots of 0.1 ml. are taken out for analysis at 0, 5, and 10 minutes and are added to 1.5 ml. of 1 N H$_2$SO$_4$. After at least 10 minutes of hydrolysis at room temperature, the amount of inorganic phosphate is determined.

Definition of Unit and Specific Activity. One enzyme unit is defined as the amount which causes the disappearance of 1 mg. of acid-labile P in 5 minutes under the conditions used. The specific activity is number of

[1] H. M. Kalckar, *J. Biol. Chem.* **167,** 477 (1947); see Vol. III [25].

[2] J. W. Rowen and A. Kornberg, *J. Biol. Chem.* **193,** 497 (1951).

[3] H. Klenow, *Arch. Biochem. and Biophys.* **46,** 186 (1953).

[4] M. Friedkin, *J. Biol. Chem.* **184,** 449 (1950).

[5] V. A. Najjar, *J. Biol. Chem.* **175,** 281 (1948); see Vol. I [36].

units per milligram of protein. Protein is determined by the method of Bücher.[6]

Application of Assay Method to Crude Tissue Preparations. The specific activity of the enzyme in crude extracts is often so low that protein will precipitate in the acid solution for phosphate analysis. Therefore, it is necessary at appropriate time intervals to fix the aliquots to be analyzed by addition of perchloric acid to give a final concentration of 2%. The filtrate is analyzed directly.

Purification Procedure

The procedure described by Najjar[5] for the preparation of crystalline phosphoglucomutase is followed.

Properties

Specificity. Several properties of the enzyme suggest that it is identical with phosphoglucomutase.[3] The ratio of the activity toward R-1-P and G-1-P, respectively, is fairly constant during the fractionation[5] and is about 1:125 when the phosphoribomutase is assayed as described here and the phosphoglucomutase is assayed as described by Najjar.[5]

Activators and Inhibitors. Low concentrations (10^{-3} M) of 8-hydroxyquinoline inhibit the enzyme considerably. This inhibition, however, is counteracted by glucose-1,6-diphosphate, the addition of which restores the activity to a level slightly higher than that of the original enzyme solution. In the absence of 8-hydroxyquinoline, glucose-1,6-diphosphate has little or no effect on the enzyme activity. A phosphoribomutase from dialyzed yeast extract is almost inactive when assayed in the presence of only $MgCl_2$, R-1-P and buffer. After addition of glucose-1,6-diphosphate, however, the incubation mixture shows considerable enzyme activity. It is, therefore, presumed that glucose-1,6-diphosphate in both cases functions as coenzyme for the phosphoribomutase reaction in the sense that ribose-1,5-diphosphate is formed according to equation 1.

Glucose-1,6-diphosphate + ribose-1-phosphate

$$\rightleftarrows \text{Ribose-1,5-diphosphate} + \text{glucose-6-phosphate} \quad (1)$$

This equation is supported by the isolation[3] of an acid-labile ribose diphosphate from a reaction mixture consisting of R-1-P, glucose-1,6-diphosphate and phosphoglucomutase from muscle.

Magnesium ions in a final concentration of 10^{-3} M activate the enzyme by 20 to 30%.

[6] T. Bücher, *Biochim. et Biophys. Acta* **1**, 292 (1947).

The enzyme is apparently inhibited by salts in general. A salt concentration of 0.05 M reduces in most cases the enzyme activity to 30 to 60% of that of the uninhibited activity.

Effect of pH. The pH optimum of the enzyme reaction is about 7.4.

Equilibrium. The equilibrium of the reaction is much in favor of ribose-5-phosphate. More than 90% of the total phosphate is present as acid-stable phosphate at equilibrium.

[51] Pentose Phosphate Isomerase

Ribose-5-P $\rightleftarrows$ Ribulose-5-P

By BERNARD AXELROD

Principle

The method employed is based on the colorimetric determination of the ribulose-5-PO$_4$ which is formed by the action of the isomerase on R-5-P, and utilizes a modification of the carbazole procedure of Dische and Borenfreund.[1]

Reagents

R-5-P solution. Dissolve 0.5 mg. of Ba R-5-P·2H$_2$O in 0.5 ml. of buffer. Make up this solution only as needed, from the Ba R-5-P·2H$_2$O, which should be kept cold and dry. This ester undergoes a slow isomerization, even in the solid state.

Buffer solution. 0.1 M Tris, adjusted to pH 7.0 with conc. HCl.

H$_2$SO$_4$ solution. Mix 225 ml. of conc. H$_2$SO$_4$ with 95 ml. of H$_2$O. Cool before using.

Carbazole solution. 0.12% (w/v) carbazole in abs. EtOH.

Cysteine solution. 1.5% (w/v) cysteine·HCl in H$_2$O.

Enzyme. Dilute the stock enzyme solution with the buffer just prior to use so that 0.1 ml. contains about 0.2 to 0.6 unit.

Procedure.[2] Place 0.5 ml. of the substrate in a colorimeter tube, in the water bath at 37°. Add 0.1 ml. of the enzyme solution. Allow 10 minutes reaction time; then add 6 ml. of H$_2$SO$_4$ solution with a rapidly flowing pipet, followed by 0.2 ml. of carbazole and then 0.2 ml. of cysteine, mixing well between additions. Replace the tube in the bath. The time consumed

---

[1] Z. Dische and E. Borenfreund, *J. Biol. Chem.* **192**, 583 (1951).

[2] B. Axelrod and R. Jang, *Federation Proc.* **12**, 172 (1953); *J. Biol. Chem.* **209**, 847 (1954).

between the addition of the H_2SO_4 and the return of the tube to the bath should be 40 seconds. Read the color promptly at the end of 30 minutes, in a colorimeter fitted with a 520-mμ filter. Carry out a parallel determination using 0.1 ml. of heated enzyme for a zero-time blank.

Definition of Unit. One unit of enzyme is ten times that amount which catalyzes the formation of 0.1 micromole of ribulose-5-PO_4 under the standard assay conditions. A working curve is obtained by plotting the micromoles of ribulose-5-PO_4 formed and determined experimentally as a function of enzyme quantity. The relationship between the enzyme quantity and the standard unit is established from the initial portion of the curve. Although it would be desirable to use ribulose-5-PO_4 as a standard for the colorimetric determination, this compound is not easily obtained pure. However, one can readily relate color density to the concentration of this compound by measuring the color produced by a reaction mixture which has been allowed to reach equilibrium, since the equilibrium ratio of R-5-P to ribulose-5-PO_4 is known (75:25 at 37°).

Application of Assay Method to Crude Preparations. The presence of any other enzymes such as phosphatase or transketolase which can attack R-5-P or ribulose-5-PO_4 would interfere with this method. In the case of the several plant sources studied (alfalfa, orange flavedo, avocado, and spinch leaf) the isomerase activity is so high that the action of the interfering enzymes becomes negligible with the dilutions used.

Purification Procedure

The procedure is described in the summary of a purification performed with alfalfa press juice which had been stored at $-20°$ for four years. Fresh press juice may also be purified by the same procedure and is likely to be less laden with the gum-forming material which accompanies the enzyme in the earlier stages of purification. The heating indicated in step 1 should be done with caution. A stainless steel container equipped with a good stirrer is used, and the heating is confined to batches of 2 l. The temperature is raised rapidly to the point where coagulation just occurs and then maintained constant for 1 minute, after which the preparation is quickly cooled. Older preparations do not form as readily visible a coagulum as fresh preparations and may be overheated unless carefully watched. The elution, described in step 6 (see the table), results in full recovery of the enzyme; only the purest fraction is described here.

Properties

Specificity. The enzyme is active only on R-5-P (and ribulose-5-PO_4). No action occurs with the following compounds: R-3-P, G-6-P, DL-3-phosphoglyceraldehyde, dihydroxyacetone phosphate, and D-arabinose-5-phosphate.

Inhibitors. G-6-P, 5-AMP, 3-AMP, adenosine, adenine, and orthophosphate are appreciably inhibitory when present in concentrations of three to ten times that of the substrate. R-3-P, ribulose, and ethylenediaminetetraacetic acid are not inhibitory. 5-Phosphoribonic acid causes 50% inhibition in a concentration of 1.2×10^{-4} M.

pH Effect. The enzyme shows a sharp optimum at pH 7.0, the activity

PURIFICATION PROCEDURE

	Volume, ml.	Total units	Units/ mg. P.N.	Recovery, %
1. Alfalfa press juice	7100	47×10^5	380	100
Heat to 60–63°, 1 min., cool, filter.				
2. Heated filtrate	6000	31×10^5	3,860	67
Make 0.7 sat. with solid $(NH_4)_2SO_4$; centrifuge. Suspend ppt. in 300 ml. H_2O; dialyze 15 hr. vs. running tap H_2O.				
3. Dialyzate, 0.7 sat.	920	28×10^5	6,600	59
Make 0.35 sat. with solid $(NH_4)_2SO_4$; centrifuge. Adjust super. to 0.55 sat.; centrifuge; dissolve gummy ppt.				
4. Fraction, 0.35–0.55	150	26×10^5	12,000	57
Add 2 N H_2SO_4 to pH 4.5; centrifuge; discard ppt. Adjust pH of super. to pH 5.0 with 2 N NaOH. Make 0.35 sat. with $(NH_4)_2SO_4$; centrifuge. Dissolve ppt. in H_2O.				
5. Precipitate, 0.35.	40	13×10^5	16,900	29
Make 0.30 sat. with solid $(NH_4)_2SO_4$. Centrifuge; discard super. Dissolve ppt. in 150 ml. H_2O; adjust pH to 7.0. Fractionate between 0.40 and 0.50 sat., using sat. $(NH_4)_2SO_4$ solution, pH 7.0. Dissolve ppt. in about 15 ml. H_2O. Dialyze 4 hr. against running tap H_2O, then 15 hr. against 300 vol. dist. H_2O.				
6. Fraction, 0.40–0.50, dialyzed	26	4.7×10^5	34,500	10
Make 0.6 sat. with solid $(NH_4)_2SO_4$. Filter suspension through column (2.5-cm. dia. × 3.6 cm.) of Celite previously wetted with 0.6 sat. $(NH_4)_2SO_4$. Elute with 0.3 sat. $(NH_4)_2SO_4$; collect eluates in 1-ml. fractions.				
7. Eluate, fraction No. 16	1	1.3×10^5	145,000	2.8

falling away to 0 at pH 5.0. In the alkaline region the activity decreases more slowly, being 75% of the maximum value at pH 8.5.

Equilibrium. Higher reaction temperatures favor the formation of ribulose-5-phosphate, the equilibrium constants for R-5-P ⇆ ribulose-5-phosphate at 37°, 25°, and 0° being 0.323, 0.264, and 0.164, respectively. The $-\Delta F$ at 37° is approximately -700 calories.

Turnover Number. The number of moles of ribulose-5-phosphate formed per minute per 100,000 g. of protein calculated for the fraction obtained in step 7 is 240,000.

Sources of the Enzyme. Phosphopentose isomerase was first found in yeast by Horecker et al.[3] Pea leaves, spinach leaves, orange flavedo, avocado leaves, and avocado fruit also contain a similar enzyme.

[3] B. L. Horecker, P. Z. Smyrniotis, and J. E. Seegmiller, *J. Biol. Chem.* **193,** 383 (1953).

[52] Pentose Isomerases

By SEYMOUR S. COHEN

The following enzymatically catalyzed isomerizations of pentose to ketopentose have been described:

$$\text{D-Arabinose} \rightleftarrows \text{D-Ribulose}^1 \qquad (1)$$
$$\text{D-Xylose} \rightleftarrows \text{D-Xylulose}^{2,3} \qquad (2)$$

Sources

The isomerase for D-arabinose has been found in two strains of *E. coli.*[1] The synthesis of the enzyme is induced by exposure of the bacteria to the pentose. The induced biosynthesis of the enzyme has been found to occur in one strain of *E. coli* which is unable to grow on D-arabinose.[1]

The mineral medium and growth procedures described in the section on Gluconokinase (Vol. I [47]) were employed in studies of the arabinose isomerase of *E. coli.*[1] In addition, higher levels of bacterial growth were obtained by the procedure of rotation in media containing up to 1% of carbon source.[1] Cells were harvested during the exponential phase of growth, washed, disrupted by grinding with alumina, and extracted with buffer as described.

Xylose isomerase has been reported in *Pseudomonas hydrophila,*[2] *Lactobacillus pentosus,*[3] and *E. coli.*[4] The enzyme has been shown to be

[1] S. S. Cohen, *J. Biol. Chem.* **201,** 71 (1953).
[2] R. M. Hochster and R. W. Watson, *J. Am. Chem. Soc.* **75,** 3284 (1953).
[3] S. Mitsuhashi and J. O. Lampen, *J. Biol. Chem.* **204,** 999 (1953).
[4] S. S. Cohen and H. Barner, *Federation Proc.* **13,** 193 (1954).

adaptive in *L. pentosus* and in *E. coli;* it has not been found in cells grown on glucose.

In the preparation of the xylose isomerase of *Pseudomonas hydrophila,* the organism was grown on xylose as the sole carbon source in a medium containing 20 g. of D-xylose, 6 g. of $(NH_4)_2HPO_4$, 0.2 g. of KH_2PO_4, and 0.25 g. of $MgSO_4$ per liter.[5] The medium required for *L. pentosus* is more complex,[6] containing 4 g. of Difco yeast extract, 10 g. of Difco Nutrient Broth, 10 g. of sodium acetate, 10 g. of the sugar, 0.2 g. of $MgSO_4 \cdot 7H_2O$, 0.01 g. of NaCl, 0.01 g. of $FeSO_4 \cdot 7H_2O$ and 0.01 g. of $MnSO_4 \cdot 4H_2O$ per liter. Sugars were sterilized separately and added before use. In experiments with these organisms, the cultures were harvested after 24-hour intervals.

Cell-free extracts have been made by grinding with alumina[5,6] and by sonic disruption.[2,7] In the case of *P. hydrophila,* cell-free extracts prepared by sonic disruption were precipitated with 7 vol. of cold acetone. The precipitate was washed five times with cold acetone, dried quickly on a Büchner funnel, and *in vacuo* for an hour over $CaCl_2$. The isomerase was readily extractable from this material which was stored at 4°.

Assay Methods

Principle. The rate of formation of ketopentose from the readily available pentoses is estimated by the cysteine-carbazole reaction,[8] as adapted to ribulose[1] and xylulose.[7] Although the disappearance of pentose may be determined by the $FeCl_3$-orcinol reaction of Bial, this is less useful, since the ketopentoses also give significant amounts of color in that reaction, and impure enzyme preparations contain the Bial-reactive nucleic acids. Since the isomerization comes to equilibrium at about 13 to 17% of ketopentose, the maximal rate of reaction is established for the first quarter of the reaction or only about 3 to 4% of the pentose which disappears. The period of maximal reaction rate may be extended greatly by adding borate, which traps ketopentose. Borate may shift the reaction to form in excess of 80% yield of that product, but it does not increase the maximal reaction rate.

Reagents for Colorimetry

D-Arabinose or D-xylose (0.066 M). The sugars may be purchased commercially.

[5] R. M. Hochster and R. W. Watson, *Nature* **170,** 357 (1952).
[6] J. O. Lampen and H. R. Peterjohn, *J. Bacteriol.* **62,** 281 (1951).
[7] J. O. Lampen, *J. Biol. Chem.* **204,** 999 (1953).
[8] Z. Dische and E. Borenfreund, *J. Biol. Chem.* **192,** 583 (1951).

D-Ribulose (50 γ/ml.). The ketopentose can be prepared and isolated as the o-nitrophenylhydrazone by the method of Glatthaar
and Reichstein,[9] after isomerization of D-arabinose in refluxing
pyrimidine, or by the enzymatic isomerization of the aldopentose,
as described by Cohen.[1] The isolation of the free sugar has been
described.[9] It may be stored in solution at pH 5 in the frozen
state. It is relatively unstable in solution at or above pH 7.

1.5% cysteine hydrochloride.

Concentrated H_2SO_4 prepared by mixing 190 ml. of H_2O and 450
ml. of H_2SO_4.

0.12% carbazole in 95% ethanol.

Assay for Enzyme (D-*Arabinose Isomerase*). To 1.0 mg. of D-arabinose
in 0.9 ml. of 0.1 M borate buffer (pH 8.0) is added 0.1 ml. of the enzyme
preparation. Control tubes contain (1) buffer and enzyme without added
pentose and (2) buffer and pentose without enzyme. The tubes are incubated at 37°. Aliquots of 0.1 ml. are removed at 0-, 1-, 5-, and 10-minute
intervals and added to 0.9 ml. of 0.1 N HCl to stop the reaction. To the
mixture is added 0.2 ml. of 1.5% cysteine hydrochloride and 6.0 ml. of
H_2SO_4. These are mixed quickly, and 0.2 ml. of 0.12% carbazole is added
immediately afterward. The tubes are mixed and permitted to stand at
room temperature. Color formation with ribulose is complete in 15 minutes;[1] that with xylulose requires 1 hour.[10] The absorption maximum of
the colored product is at 540 mμ, and the optical density is directly proportional to concentration of ketopentose in the range of 0 to 40 γ.
D-Arabinose yields about $\frac{1}{60}$ the amount of color produced by an equivalent amount of ribulose. The color produced by the respective aliquots of
the control tubes is substracted from that produced by the aliquots of the
reaction mixture.

A reaction rate is determined for an isomerization in an interval in
which the rate does not fall; e.g., the rate from 1 to 5 minutes is found to
be equivalent to that in the interval of 5 to 10 minutes. The rate of
ribulose formation is proportional to enzyme concentration. By this assay,
it has been found that adapted *E. coli* may contain thirty to eighty times
as much enzyme as unadapted cells. A unit of D-arabinose isomerase has
been defined as the amount of enzyme producing 1 γ of ribulose per minute under the above conditions.

[9] C. Glatthaar and T. Reichstein, *Helv. Chim. Acta* **18**, 80 (1935).

[10] According to Lampen,[7] the color densities with 10-γ amounts of ribulose and xylose
   are equal after 1 hour of color development, suggesting that ribulose may be used
   as a standard in assay of xylose isomerase.

Purification Procedure

The procedure to be described was carried out at 4°. D-Arabinose isomerase is precipitable between 50 and 75% saturation with $(NH_4)_2SO_4$ at pH 7.6. In a typical experiment a moist pellet (2.6 g.) of *E. coli* was ground with alumina and extracted with 20 ml. of 0.04 *M* glycylglycine buffer, pH 7.4. The supernatant fluid (18 ml.) was removed after centrifugation at 5000 r.p.m. for 30 minutes. An equal volume of saturated $(NH_4)_2SO_4$ at pH 7.6 was added, and after 1 hour the mixture was sedimented at 5000 r.p.m. for 30 minutes. The precipitate was dissolved in 9 ml. of 0.04 glycylglycine buffer, pH 7.4. To the supernatant fluid was added 2 vol. of the saturated $(NH_4)_2SO_4$ at pH 7.6, and after 1 hour the precipitate was collected by centrifugation and dissolved in 10 ml. of buffer. To 68 ml. of the supernatant fluid was added solid $(NH_4)_2SO_4$ until saturation; the precipitate which formed after several hours was dissolved in 5 ml. The activities and protein contents of the fractions are presented in the table.

ISOMERASE ACTIVITY OF FRACTION OF AN EXTRACT OF *E. coli* ADAPTED TO D-ARABINOSE

Fraction % saturation with $(NH_4)_2SO_4$	Volume, ml.	Protein, mg./ml.	Activity, units/ml.
0–50	9	7.3	40
50–75	10	14.6	900
75–100	5	1.2	40

The purification of xylose isomerase has not been described. In the work of Lampen,[7] the nucleoprotein content of extracts was reduced by an initial precipitation with $MnCl_2$. In his studies of the pentose phosphates formed during xylose metabolism in extracts, ribonuclease had hydrolyzed cellular nucleoprotein, liberating acid-soluble pentose phosphate which tended to complicate the analysis.

Identification of Reaction Products. In studies with arabinose isomerase, it was found that, as D-arabinose disappeared, a ketopentose appeared which possessed the R_f of ribulose on paper chromatograms. Ribulose was isolated as the *o*-nitrophenylhydrazone after incubation of enzyme and D-arabinose in borate buffer.[1] Conversely, incubation of the enzyme with ribulose in glycylglycine buffer resulted in the production of a Bial-reactive compound with the R_f of arabinose. The D-arabinose diphenylhydrazone was isolated from the reaction mixture.[1]

In studies with xylose isomerase, incubation of enzyme with pentose resulted in the production of bromine-resistant ketopentose.[2,3] Residual xylose was converted to xylonic acid with Br_2, and xylonate was removed on an anion exchange resin.[2] Ketopentose was then readily demonstrable on paper chromatograms,[2] and was subsequently isolated as the pure syrup and as the *p*-bromophenylhydrazone.[2] Xylose and xylulose from a reaction mixture were also separated on a borate column,[3] after which appropriate properties were observed and characteristic derivatives were prepared.

Properties

Specificity. D-Arabinose isomerase was inactive on L-arabinose, D-lyxose, D-xylose, and D-ribose. However, L-fucose was converted by the enzyme to a substance which gives the cysteine-carbazole reaction.[1]

D-Xylose isomerase was inactive on D-lyxose, D-ribose, D-arabinose, L-arabinose, D-glucose, L-xylose, and xylose-5-phosphate.[3]

Activation. The pentose isomerases do not require phosphate. When xylose isomerase was dialyzed at 4°, 80% inactivation was obtained. Addition of 0.01 M Mn^{++} or Mg^{++} restored 92 and 70% of the lost activity, respectively.[2] When extracts containing D-arabinose isomerase were dialyzed against running distilled water at 4° for 20 hours, no loss of activity was observed.

Effect of pH. Arabinose isomerase had a broad optimum from pH 8.0 to 8.5. The activity fell rapidly on the alkaline side of pH 8.5, and more slowly on the acid side of pH 8.0.[1] The pH optimum of xylose isomerase is stated to be pH 7.5.

Reaction Equilibria and Their Displacement. In several determinations starting from D-arabinose or D-ribulose, the equilibrium mixture at pH 8.0 was found to be 83 and 87% D-arabinose and 17 to 13% D-ribulose.[1] A similar equilibrium of 84% D-xylose and 16% D-xylulose has been found for the xylose isomerase.[2]

As shown with D-arabinose isomerase, a borate/pentose ratio of 10:1 permits the equilibrium of the reaction to be shifted to 70 to 90% of ribulose.[1] In studies with xylose isomerase, the yield of ketose formed appeared to be a function of the xylose/borate ratio. Thus, a ratio of 4:1 permitted the accumulation of 45% xylulose.[3] However, in another study a larger amount of borate shifted the equilibrium to 81.5% of xylulose.[2]

[53] Transketolase from Liver and Spinach

Ribulose-5-P + Ribose-5-P $\rightleftarrows$ Sedoheptulose-7-P + Glyceraldehyde-3-P

By B. L. HORECKER and P. Z. SMYRNIOTIS

Assay Method

Principle. The method was originally introduced by Racker[1] for the assay of phosphofructokinase and is based on the oxidation of DPNH by dihydroxyacetone phosphate in the presence of α-glycerophosphate dehydrogenase. The rate of disappearance of DPNH is measured spectrophotometrically at 340 mμ.[2] A crude muscle fraction is a suitable source of α-glycerophosphate dehydrogenase as well as of triose phosphate isomerase, which is required to convert glyceraldehyde-3-P to dihydroxyacetone P.

Reagents

Ribulose-5-P (0.02 M). Dissolve 50 mg. of barium ribulose-5-P in 4.0 ml. of 0.02 M acetic acid, and add 0.30 ml. of 0.57 M K$_2$SO$_4$. Centrifuge, and discard the precipitated barium sulfate. Neutralize the supernatant solution with 2.0 M KOH. This solution contains 20 micromoles/ml. of pentose phosphate as measured in the orcinol reaction; about 75% of the total pentose is ribulose.

DPNH (0.003 M). Dissolve 6.0 mg. of DPNH (65% pure) in 2.0 ml. of 0.001 M NaOH.

Glycerophosphate dehydrogenase. Collect the precipitate which forms with ammonium sulfate between 52 and 72% saturation in the glyceraldehyde-3-P dehydrogenase preparation of Cori *et al.*[3] Dissolve 400 mg. of this paste in 2.0 ml. as needed. The dissolved enzyme preparation is stable for several weeks if stored at $-15°$.

0.01 M glycylglycine buffer, pH 7.5. Prepare this solution to contain 0.006 M cysteine.

Enzyme. Dilute the stock enzyme solution in cold buffer to obtain 0.5 to 2.0 units/ml. (See definition below.)

Procedure. To 0.96 ml. of buffer in a quartz or Corex cell with a light path of 1 cm. add 0.02 ml. of ribulose-5-P, 0.02 ml. of DPNH, and 0.01 ml. of glycerophosphate dehydrogenase. Take readings at 340 mμ at 1-minute intervals immediately after the addition of 0.02 ml. of enzyme.

[1] E. Racker, *J. Biol. Chem.* **167**, 843 (1947).

[2] O. Warburg and W Christian, *Biochem. Z.* **287**, 291 (1936).

[3] G. T. Cori, M. W. Slein, and C. F. Cori, *J. Biol. Chem.* **173**, 605 (1948).

Definition of Unit and Specific Activity. A unit of enzyme is defined as that amount which causes a change in optical density of 1.000 per minute at 25° under the above conditions, calculated from the 2- to 6-minute values. Specific activity is expressed as units per milligram of protein. Protein is determined by the turbidimetric method of Bücher,[4] standardized with a solution of crystalline serum albumin.

Purification from Liver

The enzyme is purified from acetone powder by the procedure of Horecker *et al.*[5] Homogenize 100 g. of rat liver in a Waring blendor for 2 minutes with 400 ml. of acetone chilled to −10°, and filter with suction through Whatman No. 1 filter paper with an 18-cm. Büchner funnel. Strip the pad from the paper, and rehomogenize as before. Scrape the residue from the paper, and dry in air. Store in a desiccator at 2°.

Step 1. Extraction. Extract 24 g. of acetone powder for 15 minutes at 0° with 265 ml. of 0.02 M K_2HPO_4, with occasional stirring. Centrifuge, and discard the residue.

Step 2. Ammonium Sulfate Fractionation. To the acetone powder extract (215 ml.) add 62.5 g. of ammonium sulfate, centrifuge, and discard the precipitate. Add 26.8 g. of ammonium sulfate to the supernatant solution, centrifuge, and dissolve the precipitate in 75 ml. of 0.01 M phosphate buffer, pH 7.6. Adjust to pH 7.7 with 2.5 ml. of 2 N ammonium hydroxide.

Step 3. Methanol Fractionation. Carry out the methanol fractionation with small batches to permit rapid handling. To 37 ml. of the ammonium sulfate fraction, diluted to 120 ml. with cold water, add 130 ml. of methanol, previously chilled to −12°. The additions should be complete in 30 seconds, during which time the solution is cooled in a −20° ice-ethanol freezing bath. Continue to cool the solution for 1 to 2 minutes, until its temperature is −13 to −16°. Centrifuge for 1 minute at −10° in a Servall SS-1 centrifuge, and treat the supernatant solution in the same manner with 95 ml. of cold methanol. Suspend the precipitate in 30 ml. of water, centrifuge, and discard the insoluble residue. Repeat the procedure until all the ammonium sulfate fraction is processed, and combine the solutions.

Step 4. Acetone Fractionation. Dilute the methanol fraction (58 ml.) with 12 ml. of water, and add 2 ml. of 4 M sodium acetate. Add 22 ml. of acetone, cooled to −10°, while cooling the solution in a freezing bath. The addition should require 2 minutes, during which time the solution is cooled to −8°. After 3 minutes longer at this temperature, centrifuge for

[4] T. Bücher, *Biochim. et Biophys. Acta* **1**, 292 (1947); see Vol. III [3].
[5] B. L. Horecker, P. Z. Smyrniotis, and H. Klenow, *J. Biol. Chem.* **205**, 661 (1953).

2 minutes (International, Size 2), and dissolve the precipitate in 8 ml. of cold water. Collect two more fractions in the same manner with the addition of 11 ml. of acetone each time. Assay the last two fractions separately; usually the last fraction will contain the bulk of the activity.

Step 5. Calcium Phosphate Gel Adsorption and Ammonium Sulfate Fractionation. Treat the acetone fraction (9.2 ml.) with 7.5 ml. of calcium phosphate gel[6] which has been aged three to six months (dry weight, 8 mg./ml.). Centrifuge, and discard the gel. Add 5.2 g. of ammonium sulfate to the supernatant solution (14.5 ml.), and centrifuge. Treat the supernatant

TABLE I
SUMMARY OF PURIFICATION PROCEDURE (LIVER)

Fraction	Total volume, ml.	Units/ml.	Total units	Protein, mg./ml.	Specific activity, units/mg.	Recovery, %
1. Extract	215	1.90	409	31.7	0.06	—
2. (NH₄)₂SO₄ fraction I	74	2.80	207	25.4	0.11	51
3. Methanol fraction	58	2.38	138	4.5	0.53	34
4. Acetone fraction	9.2	8.4	77	7.8	1.08	19
5. (NH₄)₂SO₄ fraction II	2.5	12.8	32	4.8	2.65	8

solution with 0.5 g. of ammonium sulfate, centrifuge, and dissolve the precipitate in 2.0 ml. of 0.02 M phosphate buffer, pH 8.0. Store in the frozen state at $-16°$.

Purification from Spinach

All operations are carried out at 2°.

Step 1. Extraction and Ammonium Sulfate Fractionation I. Homogenize 60 g. of spinach leaves, stems removed, for 3 minutes in a Waring blendor with 360 ml. of cold 50% saturated ammonium sulfate which has been adjusted to pH 7.8 with ammonium hydroxide. Filter with Schleicher and Schuell No. 588 fluted filter paper. Combine the filtrate from 60 batches (16.9 l.), add 3.82 kg. of ammonium sulfate, and filter overnight in the cold room. Dissolve the precipitate in 500 ml. of water, and adjust to pH 7.1 with 2.2 ml. of 2 N ammonium hydroxide.

Step 2. Ammonium Sulfate Fractionation II. Dilute the first ammonium sulfate fraction (600 ml.) with water (180 ml.) until the ammonium sulfate saturation is 0.10 as determined by conductivity measurement.[7] Add

---

[6] D. Keilin and E. F. Hartree, *Proc. Roy. Soc. (London)* **B124**, 397 (1938).

[7] For rapid conductivity determinations either the Barnstead Still and Sterilizer Co. Purity Meter Model PM-2 or the Industrial Instruments Solu Bridge Model RD-3 may be used.

176 g. of ammonium sulfate to the diluted solution, centrifuge, and discard the precipitate. To the supernatant solution add 78 g. of ammonium
sulfate. Collect the precipitate by centrifugation, dissolve it in 90 ml.
of water, and neutralize the solution with 0.3 ml. of 2 N ammonium
hydroxide.

Step 3. Calcium Phosphate Gel Adsorption and Ammonium Sulfate III.
Dilute the solution with 1188 ml. of water to bring the protein content to
1.2 mg./ml., and add 1130 ml. of calcium phosphate gel[6] (6.6 mg. dry
weight per milliliter). Centrifuge, discard the supernatant solution, and
elute the enzyme with 216 ml. of 0.01 M pyrophosphate buffer, pH 8.3. To
the eluate (262 ml.) add 81.0 g. of ammonium sulfate and centrifuge, and
to the supernatant solution add 30.0 g. of ammonium sulfate. Collect the
precipitate by centrifugation, dissolve in 25 ml. of water, and neutralize
with 0.65 ml. of 0.2 N ammonium hydroxide.

Step 4. Acetone Fractionation. Dialyze the last ammonium sulfate
fraction overnight against flowing 0.01 M sodium acetate, pH 7.2. Dilute
the dialyzed solution to 77.4 ml. with 0.1 M sodium acetate to bring the
protein content to 6.2 mg./ml. Slowly add 47.5 ml. of cold acetone
($-10°$), while cooling the solution in a freezing bath. Keep the mixture
at $-8°$ for 3 minutes, centrifuge for 2 to 3 minutes (International Size 2),
and dissolve the precipitate in 20 ml. of water. Collect three more fractions
in the same way by the addition of 14 ml., 15 ml., and 25 ml. of cold
acetone, respectively. Assay these fractions separately, and pool the
most active ones (usually fractions 3 and 4).

TABLE II
SUMMARY OF PURIFICATION PROCEDURE (SPINACH)

Fraction	Total volume, ml.	Units/ml.	Total units	Protein, mg./ml.	Specific activity, units/mg.	Recovery, %
1. $(NH_4)_2SO_4$ fraction I	600	28.5	17,100	17.8	1.6	—
2. $(NH_4)_2SO_4$ fraction II	108	129	14,000	14.0	9.2	82
3. $(NH_4)_2SO_4$ fraction III	29.0	325	9,440	17.8	18.2	55
4. Acetone fraction	42.0	173	7,270	4.7	37.0	42
5. $(NH_4)_2SO_4$ fraction IV	6.6	825	5,440	17.6	47.0	32

Step 5. Ammonium Sulfate Fractionation IV. Dilute the combined
acetone fractions (42 ml.) to 72 ml., and add 72 ml. of cold saturated
ammonium sulfate which has previously been adjusted to pH 7.8. Centrifuge, and to the supernatant solution add 8.6 g. of ammonium sulfate.
Collect the precipitate by centrifugation, and dissolve it in 3.0 ml. of
0.25 M glycylglycine buffer, pH 7.4. To the supernatant solution add

4.7 g. of ammonium sulfate, centrifuge, and dissolve the precipitate as before. Collect a fourth fraction in the same manner as the third. Fractions 3 and 4 are assayed separately. These usually contain the activity. To the combined fractions (6.6 ml.) add 3.0 g. of ammonium sulfate. Store at 2°. Centrifuge aliquots as needed, and dissolve the precipitates in water.

Properties

Specificity. Transketolase will react with a number of ketol substrates, including ribulose-5-P, sedoheptulose-7-P, L-erythrulose, hydroxypyruvate,[8] and fructose-6-P.[9] Acceptor aldehydes include ribose-5-P, glycolaldehyde,[8] D- or L-glyceraldehyde-3-P, and D-glyceraldehyde.

Coenzyme Requirement. The coenzyme of transketolase is thiamine pyrophosphate.[5,9] Enzyme preparations which have been stored for long periods may lose activity owing to dissociation of the coenzyme. When this occurs, the activity may be restored by the addition of $10^{-5} M$ coenzyme and $2 \times 10^{-3} M$ MgCl$_2$ to the test mixture.

[8] E. Racker, G. de la Haba, and I. G. Leder, *J. Am. Chem. Soc.* **75**, 1010 (1953).

[9] E. Racker, G. de la Haba, and I. G. Leder, *Arch. Biochem. and Biophys.* **48**, 238 (1954).

[54] Crystalline Transketolase from Baker's Yeast

By G. DE LA HABA and E. RACKER

$$CH_2OPO_3H_2(CHOH)_2COCH_2OH + CH_2OPO_3H_2(CHOH)_3CHO$$
$$\rightleftarrows CH_2OPO_3H_2(CHOH)_4COCH_2OH + CH_2OPO_3H_2CHOHCHO$$

Assay Method

Principle. Transketolase catalyzes the formation of glyceraldehyde-3-phosphate and heptulose phosphate from a mixture of ribulose-5-phosphate and ribose-5-phosphate. In the presence of triose phosphate isomerase and α-glycerophosphate dehydrogenase, triose phosphate is reduced to α-glycerophosphate by DPNH, the disappearance of which is measured spectrophotometrically at 340 mμ.[1]

Reagents

0.5 M glycylglycine buffer, pH 7.6.
0.002 M DPNH.

[1] E. Racker, *J. Biol. Chem.* **167**, 843 (1947).

52 to 72 ammonium sulfate fraction from rabbit muscle extract.[1]
Dissolve in water to a concentration of 10 mg. protein per milliliter (contains triose phosphate isomerase and α-glycerophosphate
dehydrogenase).

3% solution of equilibrium mixture of ribose and ribulose phosphates ("isomerase product," see below).

0.1% solution of thiamine pyrophosphate.

0.06 M solution of $MgCl_2$.

Enzyme. Dilute enzyme solution in water to obtain 200 to 2000
units/ml. (See definition of unit below.)

Procedure. Place into a microquartz cell having a 1-cm. light path
in a final volume of 1 ml., 0.65 ml. of distilled water, 0.05 ml. of buffer,
0.05 ml. of DPNH, 0.05 ml. of 52 to 72 ammonium sulfate rabbit muscle
fraction, 0.05 ml. of isomerase product, 0.05 ml. of thiamine pyrophosphate, and 0.05 ml. of magnesium chloride. Take density readings at 340
mμ at 30-second intervals. The course of the reaction is zero order for
several minutes. There is a slow DPNH oxidation (7 units/min.) without
addition of transketolase. This blank is subtracted from the activity obtained upon addition of transketolase.

Definition of Unit and Specific Activity. One unit of enzyme is defined
as that amount of enzyme which gives rise to a density change of 0.001
per minute under the above conditions. Specific activity is expressed as
units of enzyme activity per milligram of protein. Protein concentration
is determined spectrophotometrically.[2] In the presence of nucleic acid the
biuret test is used.[3]

Preparation of Substrate. 1. PREPARATION OF YEAST PENTOSE PHOS-
PHATE ISOMERASE. A crude baker's yeast extract is fractionated with
acetone as described below for the preparation of transketolase to obtain
fractions I (0 to 33% acetone) and II (33 to 50% acetone). Suspend
acetone II precipitate in 100 to 150 ml. of cold distilled water, and dialyze
for 12 hours in the cold room against 25 to 50 vol. of distilled water,
changing the water once during the process. Centrifuge the contents of
the bag at 18,000 $\times$ g in the cold. Dilute the supernatant solution 1:5
with distilled water at 60°. Bring the temperature of the mixture rapidly
to 60°, and maintain it there for 25 minutes. Cool to 5 to 10°, and spin in
the cold room. Add 29 g. of solid ammonium sulfate per 100 ml. of supernatant solutions at 5°, centrifuge, and discard the precipitate. To every
100 ml. of supernatant solution add 16.7 g. of ammonium sulfate, cen-

[2] O. Warburg and W. Christian, *Biochem. Z.* **310**, 384 (1941).

[3] H. W. Robinson and C. G. Hogden, *J. Biol. Chem.* **135**, 727 (1940).

trifuge, and dissolve the precipitate in 50 ml. of distilled water. This preparation of isomerase is now virtually free of transketolase. Store in a deep-freeze. Before use, dialyze against cold distilled water until free of sulfate.

2. PREPARATION OF ISOMERASE PRODUCT. Dissolve 1 g. of the barium salt of ribose-5-phosphate in 10 ml. of 0.1 N HCl. Heat in a boiling bath for 4 minutes. Cool, and remove barium with 1 M ammonium sulfate. Neutralize to pH 7.6. To this solution add 4 ml. of dialyzed isomerase and water to a final volume of 20 ml., and incubate at 38°. Follow isomerization every half hour by the carbazole reaction for keto sugars[4] until no further increase in ketopentose is noted. Add 3 ml. of 50% trichloroacetic acid; centrifuge, and discard the protein precipitate; adjust the pH of the supernatant solution to 8.0, add 5 ml. of 1 M barium acetate, and centrifuge. Wash the precipitate with water, combine washing and supernatant solution, and add 4 vol. of 95% ethanol. Place in an icebox for 2 to 3 hours. Collect the precipitate, wash it once with absolute ethanol, and dry it in a desiccator under vacuum. Store the barium salt in a desiccator at 2°.

Prepare a 3% solution of the barium salt of isomerase product by dissolving in 0.1 N HCl. Remove barium by addition of a slight excess of 1 M ammonium sulfate. Neutralize to pH 6.5. Store this solution at −20°.

Purification Procedure

Dried baker's yeast is used as starting material. This is prepared by crumbling 1-lb. cakes of yeast, spreading them between two sheets of paper to a thickness of about $\frac{1}{2}$ inch and allowing them to dry at room temperature for 4 days. The dried yeast is kept in stoppered bottles in a cold room.

Extract 300 g. of dried baker's yeast is with 900 ml. of 0.066 M disodium phosphate at 37° with occasional stirring for 2 hours, and then for 3 hours at room temperature. Centrifuge the extract at 3000 r.p.m. for 80 minutes (transketolase assays of crude extracts by the method outlined above are inaccurate and represent approximate values). The residue may be re-extracted (after storage at 0°) with 600 ml. of 0.066 M disodium phosphate at 37° for 2 hours and at room temperature for 1 hour. Adjust the extract obtained to pH 6.5 before proceeding to acetone fractionation. The yield and specific activity of transketolase in the second extract are as good or better than in the first extract.

Acetone Fractionation. Add 0.5 vol. of ice-cold acetone to the crude extract, maintaining the temperature of the extract at −2° in a dry

---

[4] Z. Dische and E. Borenfreund, *J. Biol. Chem.* **192**, 583 (1951).

ice-alcohol bath. Centrifuge for 15 to 20 minutes at this temperature, and discard the precipitate. To the clear supernatant solution add the same volume of acetone as above, and proceed in the same fashion. Dissolve the precipitate in 100 to 150 ml. of cold distilled water. Dialyze for 12 hours in a cold room against 25 to 50 vol. of distilled water, changing it once. Centrifuge the dialyzed preparation, and assay the supernatant solution.

Alcohol Fractionation. Determine the pH of the supernatant solution with a glass electrode at 10°. The pH should be 6.5 to 6.6—if necessary, adjust it with 1 N acetic acid. Add an equal volume of 95% ethanol, keeping the temperature of the mixture at $-6°$ in a dry ice-alcohol bath. Add alcohol slowly over a period of about 30 minutes. Centrifuge at $-6°$, discard the supernatant solution, and extract the precipitate with 60 ml. of water at room temperature for 10 minutes. Centrifuge in the cold, and discard the precipitate.

The supernatant solution is refractionated with alcohol exactly as above (pH should be 6.5 to 6.6) to a final concentration of 25%. Centrifuge at $-6°$. Dissolve the precipitate in 20 ml. of water (alcohol fraction II).

Nucleoprotein Fractionation. Adjust the pH of alcohol fraction II to 5.5 with 0.1 N acetic acid at 5 to 10°. Centrifuge at 0°, and discard the precipitate. To the supernatant solution at pH 5.5 add 0.39 ml. of 5% sodium nucleate (Schwarz Laboratories) for every 100 mg. of protein. Adjust the pH to 5.0 with 0.1 N acetic acid. Centrifuge at 0°, and discard the supernatant. Dissolve the nucleoprotein precipitate in 10 ml. of water by cautiously adjusting pH to 7.2 with 0.1 N NaOH at 5 to 10°. The protein concentration should be 20 to 25 mg./ml. Now adjust the pH to 5.3 to 5.4 with 0.1 N acetic acid. Centrifuge at 0°, and dissolve the precipitate in 5 ml. of water by adjusting the pH to 7.2. Adjust the pH of the supernatant solution to 5.0; centrifuge, and dissolve the precipitate in 5 ml. of water by adjusting the pH to 7.2. Assay these two fractions; transketolase should be present in the pH 5.3 precipitate. Transketolase is stable for weeks at this point if kept in an icebox.

Removal of Nucleic Acid with Protamine Sulfate. Adjust the nucleoprotein solution of transketolase to pH 6.7. For every 100 mg. of protein add 0.28 ml. of 6% protamine sulfate (Nutritional Biochemicals) which has been adjusted to pH 6.5 and warmed to 50°. Centrifuge, and discard the precipitate. Adjust the pH of supernatant solution to 7.2 with 0.1 N NaOH at 5 to 10°. Turbidity obtained due to excess protamine sulfate present is removed by centrifugation.

Alumina Gel C_γ Adsorption of Impurities. To the supernatant solution at room temperature add 0.1 vol. of C_γ (containing 16 mg. dry weight per milliliter). Stir occasionally for 5 minutes; centrifuge at room temperature, and assay the supernatant solution. Repeat the C_γ treatment once or twice as long as there is not more than 15 to 20% loss in transketolase activity. This step has been noted occasionally to give satisfactory results only if the solution is acidified to pH 5.8.

Ammonium Sulfate Fractionation and Crystallization. To the supernatant solution add saturated ammonium sulfate slowly at 5 to 10° until a final saturation of 50% is reached. After a few hours in the icebox remove the precipitate by centrifugation at high speed (13,000 to 16,000 $\times g$) in the cold. Discard the precipitate, which should not contain more than 10% of the activity. Add saturated ammonium sulfate to the 50% saturated supernatant solution in the cold with stirring until 70% saturation is reached. After several hours in the icebox or overnight, centrifuge the solution in the cold at 16,000 $\times g$, and dissolve the precipitate in cold water to give a final concentration of protein of 5 to 10 mg./ml. The solution of this precipitate should be assayed to ascertain recovery of transketolase. If recovery is not complete, it is necessary to raise the ammonium sulfate concentration of the supernatant solution to 80% saturation.

PURIFICATION OF TRANSKETOLASE FROM BAKER'S YEAST

Fraction	Units of enzyme	Specific activity, units/mg. protein
Crude extract	—	—
Acetone II, 50%	1,032,000	133
Alcohol II, 25%	800,000	643
Supernatant, pH 5.5 pptn.	893,000	1,240
Nucleoprotein pH 5.5–5.4	738,400	3,466
After protamine	576,000	8,000
After two treatments with C	490,000	25,925
AmSO₄, 65% satn. crystallization	280,000	50,000
Recrystallization, 58% satn.	273,000	54,400

Next, add saturated ammonium sulfate slowly with stirring to the solution of the precipitated enzyme until slight turbidity is visible (50 to 60% saturation). Allow crystallization of the enzyme to take place in the icebox overnight. The next day centrifuge the crystals, and recrystallize as above. It is necessary to recrystallize several times if removal of contaminating pentose phosphate isomerase is desired. A protocol of the purification procedure is presented in the table. Frequently the specific activity of acetone II and alcohol II is considerably higher than indicated in the table.

A modification of the above procedure of ammonium sulfate fractionation and crystallization has been used recently, resulting in better yields. Saturated ammonium sulfate at pH 7.6 (pH adjusted with concentrated ammonium hydroxide and pH determined with a glass electrode on a 1:50 dilution of the saturated solution) is used, and advantage is taken of the negative temperature coefficient of solubility exhibited by this enzyme in ammonium sulfate. The enzyme obtained by this procedure requires thiamine pyrophosphate, and specific activities as high as 150,000 have been observed.

Properties

As mentioned above, a requirement for thiamine pyrophosphate is readily obtained if the enzyme is fractionated or crystallized from slightly alkaline ammonium sulfate. A requirement for magnesium, however, is obtained only after prolonged dialysis against Versene-KCl solution.[5] Crystalline transketolase is stable under alkaline ammonium sulfate even at room temperature (crystalline suspensions have been kept at room temperature for 2 weeks without loss of activity). Water solutions of transketolase lose activity but are stable over a period of several weeks in the icebox in 0.008 M glycylglycine buffer of pH 7.4.[6]

Specificity.[5,7,8] Crystalline transketolase splits ribulose-5-phosphate only in the presence of an acceptor aldehyde such as ribose-5-phosphate, in which case the products of the reaction are glyceraldehyde-3-phosphate and a heptulose phosphate. Acceptor aldehydes and donors of "active glycolaldehyde," which have been found to react with yeast transketolase, are listed below.

Donors of "Active Glycolaldehyde." Ribulose-5-phosphate, sedoheptulose-7-phosphate, hydroxypyruvate, fructose-6-phosphate, 1-erythrulose.

Acceptor Aldehydes. Ribose-5-phosphate, D-glyceraldehyde-3-phosphate, glyceraldehyde, glycolaldehyde, deoxyribose-5-phosphate.

[5] E. Racker, G. de la Haba, and I. G. Leder, *J. Am. Chem. Soc.* **75**, 1010 (1953).
[6] H. L. Kornberg, unpublished experiments.
[7] G. de la Haba, I. G. Leder, and E. Racker, *Federation Proc.* **12**, 194 (1953).
[8] E. Racker, G. de la Haba, and I. G. Leder, *Arch. Biochem. and Biophys.* **48**, 238 (1954).

[55] Transaldolase

Sedoheptulose-7-P + Glyceraldehyde-3-P $\rightleftarrows$ Fructose-6-P + Tetrose-P

By B. L. HORECKER and P. Z. SMYRNIOTIS

Assay Method

Principle. Fructose-6-P produced in the reaction is converted to glucose-6-P by the addition of hexose phosphate isomerase. In the presence of glucose-6-P dehydrogenase this product reduces TPN, and the rate of reduction is measured spectrophotometrically at 340 mμ.

Reagents

Sedoheptulose-7-P (0.02 M). Dissolve 50 mg. of barium sedoheptulose-7-P in 3.5 ml. of 0.02 M acetic acid and 0.2 ml. of 0.57 M K_2SO_4. Remove the precipitated $BaSO_4$ by centrifugation, and neutralize the supernatant solution with 2 N KOH.

Fructose diphosphate (0.02 M). Dissolve 120 mg. of barium FDP in 8.0 ml. of 0.5 N acetic acid, and add 0.65 ml. of 0.57 M K_2SO_4. Centrifuge to remove the $BaSO_4$, and neutralize the supernatant solution with 2 N KOH.

Hexose phosphate isomerase. From the rabbit muscle aldolase preparation of Taylor *et al.*,[1] collect the precipitate which forms at 50% saturation, and store as a paste at 2°. Prepare solutions as needed by dissolving 400 mg. of paste in 2.0 ml. of water. Store these solutions at −16°.

Glucose-6-phosphate dehydrogenase. Prepare phosphogluconic dehydrogenase (Vol. I [42]). This preparation contains about 40 mg./ml. of protein and about 200 units/ml. of glucose-6-phosphate dehydrogenase activity.

Aldolase. Prepare by the method of Taylor *et al.*,[1] and recrystallize five times. Store as a suspension of crystals in $(NH_4)_2SO_4$. Centrifuge, and dissolve aliquots of the precipitate in water to obtain solutions containing about 5 mg. of protein per milliliter.

TPN (1.5 × 10⁻³ M). See Vol. I [42].

0.04 M triethanolamine buffer, pH 7.6. This solution is prepared in 0.01 M ethylenediaminetetraacetate.

Enzyme. Dilute the stock enzyme solution in cold buffer to obtain 0.2 to 0.8 unit/ml. (See definition below.)

[1] J. F. Taylor, A. A. Green, and G. T. Cori, *J. Biol. Chem.* **173**, 591 (1948); see also Vol. I [37].

Procedure. To 0.90 ml. of buffer in a quartz or Corex cell having a light path of 1 cm., add 0.05 ml. of TPN, 0.02 ml. of sedoheptulose-7-P, and 0.01 ml. each of FDP, aldolase, isomerase, and phosphogluconic dehydrogenase. Incubate at room temperature for 3 to 4 minutes, and add 0.02 ml. of diluted enzyme. Measure the density at 340 mμ at 1-minute intervals between 2 and 6 minutes. Correct for the rate obtained with a control cell without transaldolase.

Definition of Unit and Specific Activity. A unit of enzyme is defined as the amount which causes a change in optical density of 1.000 per minute at room temperature (23 to 25°). The rate falls off with increasing amounts of enzyme, and the true rate is estimated from a calibration curve obtained by plotting the rate of reaction against the amount of enzyme added. Specific activity is expressed as units per milligram of protein.[2]

Purification Procedure

Dried brewer's yeast is prepared as described for glucose-6-P dehydrogenase. (See Vol. I [42].)

Step 1. Yeast Autolyzate. Suspend 120 g. of dried yeast in 360 ml. of 0.1 M NaHCO$_3$, and autolyze for 7 hours at 25°. Add 2 l. of water, centrifuge (International, Size 2), and store the supernatant solution overnight at 2°. Perform the subsequent operations at this temperature unless otherwise specified.

Step 2. Acetone Fractionation. Adjust the yeast autolyzate (2150 ml.) to pH 4.85 with 36 ml. of 2 N acetic acid, and add 1130 ml. of acetone previously chilled to $-12°$. Add the acetone slowly (approximately 10 minutes), and during this time cool the mixture to $-5°$ in a freezing bath. Keep it as this temperature for 10 minutes longer, and centrifuge at 2° for 2 minutes (International, Size 2). Cool the supernatant solution to $-8°$, and treat with 460 ml. of cold acetone, added during 6 minutes. After 30 minutes at $-8°$ collect the precipitate by brief centrifugation, dissolve it in 180 ml. of water, and adjust the pH to 6.8 to 7.2 with 0.27 ml. of 5 N NH$_4$OH.

Step 3. Adsorption on Calcium Phosphate Gel. Dilute the acetone fraction (205 ml.) with an equal volume of water, and add 143 ml. of calcium phosphate gel (2.02 g. of dry weight, aged about 10 months). Collect the gel by centrifugation, and elute the enzyme with two 100-ml. portions of 0.15 M pyrophosphate buffer, pH 7.6.

Step 4. Ammonium Sulfate Fractionation. To the eluate (214 ml.) add 62 g. of ammonium sulfate, centrifuge the precipitate, and discard. Collect two more fractions by the addition of 30 g. of ammonium sulfate to

---

[2] T. Bücher, *Biochim. et Biophys. Acta* **1**, 292 (1947); see Vol. III [73].

the supernatant solutions. Dissolve these in 5.0 ml. of triethanolamine buffer, pH 7.6, and analyze separately. The last fraction usually contains most of the activity.

Step 5. pH Fractionation. Dilute the ammonium sulfate fraction (15.5 ml.) with an equal volume of water, and add 26 ml. of saturated ammonium sulfate solution, to bring the ammonium sulfate saturation to 55%. Adjust the pH to 4.4 with 5.9 ml. of 0.2 M H_2SO_4 which is 55% saturated with ammonium sulfate. Keep the solution for 5 minutes at 0°, centrifuge, and dissolve the precipitate in 2.0 ml. of glycylglycine buffer, pH 7.5. Adjust the supernatant solution to pH 3.14 with 2.8 ml. of the sulfuric acid-ammonium sulfate mixture. Collect the precipitate, and dissolve as before. Obtain another fraction by adding 7.7 ml. of sulfuric acid-ammonium sulfate solution, to pH 2.50. Analyze these fractions separately.

Step 6. Dialysis. Dialyze the pH fraction against running distilled water at 2°, and store the dialyzed solution in the frozen state at −16°.

SUMMARY OF PURIFICATION PROCEDURE

Fraction	Total volume, ml.	Units/ml.	Total units	Protein, mg./ml.	Specific activity, units/mg.	Recovery, %
1. Yeast autolyzate	2150	1.10	2370	9.8	0.11	—
2. Acetone fraction	205	8.10	1640	21.7	0.37	69
3. Gel eluate	214	9.6	2050	5.6	1.71	86
4. $(NH_4)_2SO_4$ fraction	8.0	210	1680	74.8	2.81	71
5. pH fraction	2.2	369	810	11.1	33.3	34
6. Dialyzed solution	4.5	173	780	4.6	37.6	33

Properties

Specificity. Transaldolase shows no activity with any other substrates tested. These include sedoheptulose-1,7-diphosphate and ribulose-5-phosphate, which are inactive as donors, and glycolaldehyde phosphate and D-glyceraldehyde, which do not act as acceptors. The purified preparation contains significant quantities of triose phosphate isomerase but is free of transketolase.

Stability. The dialyzed preparation loses activity slowly when stored in the frozen state at −15°.

[56] Deoxyribose Phosphate Aldolase (DR-Aldolase)

By E. RACKER

Assay Method

Principle.[1] When deoxyribose-5-phosphate is cleaved by DR-aldolase, glyceraldehyde-3-phosphate and acetaldehyde are formed. The products are reduced to the corresponding alcohols by the specific dehydrogenases (α-glycerophosphate dehydrogenase and alcohol dehydrogenase) in the presence of DPNH. The oxidation of the latter is measured spectrophotometrically at 340 mμ.

Reagents

1 M potassium phosphate buffer, pH 7.6.
α-Glycerophosphate dehydrogenase (cf. Vol. I [58]).
Triose phosphate isomerase (cf. Vol. I [57]) (instead of the pure enzyme a simple preparation of the partially purified enzyme may be used[1]).
Alcohol dehydrogenase (cf. Vol. I [79]).
0.1% solution of DPNH.
0.04 M solution of deoxyribose-5-phosphate.
Enzyme. Dilute stock solution of enzyme in 0.01 M phosphate buffer, pH 7.6, to obtain 200 to 1000 units of enzyme per milliliter.

Procedure. Place into a micro quartz cell having a 1-cm. light path, in a final volume of 1 ml., 0.05 ml. of buffer, 5 γ of alcohol dehydrogenase, 50 γ of the partially purified α-glycerophosphate dehydrogenase, 0.05 ml. of DPNH, 0.05 ml. of deoxyribose-5-phosphate, and the enzyme sample to be measured. Take density readings at 340 mμ at 30-second intervals.

Definition of Unit and Specific Activity. One unit of enzyme is defined as a change in density of 0.001 per minute under the above conditions. Since the total volume is only 1 ml., the units in this system are three times as large as those measured in standard Beckman cells in a 3-ml. volume. Specific activity is expressed as unit of enzyme activity per milligram of protein. Protein concentration is measured by the biuret test.[2]

Application of Assay Method to Crude Tissue Preparations. The spectrophotometric test is usually not applicable in crude tissue preparations because of the wide distribution of DPNH-oxidizing enzymes, which

[1] E. Racker, *J. Biol. Chem.* **196**, 347 (1952).
[2] H. W. Robinson and C. G. Hogden, *J. Biol. Chem.* **135**, 727 (1940).

interfere with the DR-aldolase assay. The less accurate colorimetric test, in which the formation of deoxyribose phosphate from acetaldehyde and triose phosphate is measured with diphenylamine, is suitable for assay in crude tissue extracts, but interfering side reactions have also been encountered.[1]

Purification Procedure

Lyophilized *E. coli* 4157 is ground in a mortar for 10 minutes with an equal weight of Alumina A-301 and mixed with 20 vol. of 0.9% KCl. The suspension is then exposed to sonic vibration[3] for 10 minutes. After centrifugation for 20 minutes at 13,000 r.p.m. in an International refrigerated

PURIFICATION OF DR-ALDOLASE FROM *E. coli* 4157

Fraction	Units	Specific activity, units/mg.	Yield, %
Crude extract (from 2 g. bacteria)	260,000	310	100
After MnCl₂ treatment	190,000	700	73
After pH fractionation	130,000	1,200	50
C𝛾 eluate	80,000	9,800	31
Ammonium sulfate ppt.	45,000	17,000	17

centrifuge, the supernatant solution is collected and dialyzed against 0.9% KCl overnight. To each 15 ml. of the dialyzed preparation, 1 ml. of 1 M MnCl₂ is added, and after 10 minutes at room temperature the precipitate is removed at high speed. The clear supernatant is dialyzed with constant mechanical stirring against distilled water to remove potassium chloride. The dialyzed preparation is diluted to about 5 mg. of protein per milliliter, and to the chilled preparation dilute acetic acid is carefully added until the pH drops to 4.5. The precipitate is rapidly collected at 13,000 r.p.m. in the refrigerated centrifuge, suspended in H₂O, and carefully alkalinized with dilute ammonia to pH 5.6. The mixture is centrifuged, and the precipitate discarded. To the supernatant solution for each gram of dried *E. coli* used, 20 mg. of protamine is added (the protamine solution is also adjusted to pH 5.6). The precipitate which forms is discarded, and the water-clear, slightly yellowish supernatant solution is treated with alumina gel C𝛾. For each 100 mg. of protein, 15 mg. of the gel is added, the mixture is centrifuged, and the precipitate discarded. Usually only a small amount of the enzyme is adsorbed on this first addition of gel. To the supernatant solution 30 mg. of gel is added, and nearly all the enzyme is adsorbed. The centrifuged gel is washed once with distilled water and then eluted three times with 0.01 M phosphate buffer at pH 7.6 with 0.25 of the volume of the supernatant solution for each elution. The combined eluates are then

[3] Raytheon sonic disintegrator.

fractionated with ammonium sulfate, and the precipitate between 40 and 65% saturation is collected and dialyzed against 0.01 M phosphate at pH 7.6. A typical protocol indicating yield and purification is presented in the table.

Properties

The enzyme is active at pH values ranging between 6.0 and 8.0 without exhibiting a sharp optimum. However, the activity drops on the acid and alkaline sides of this range. Tris(hydroxymethyl)aminomethane or phosphate buffer was used for the majority of the experiments. With crude enzyme preparations Tris buffer is distinctly superior, perhaps mainly because of side reactions which can occur in the presence of phosphate, since both glyceraldehyde-3-phosphate and acetaldehyde are diverted by phosphorylating oxidation in crude extracts.

DR-aldolase is heat labile and is rapidly inactivated at temperatures above 70°. It can, however, be heated for 3 minutes at 58° with small loss of activity. The enzyme loses activity on storage in the icebox but has been preserved for over 1 year when kept frozen at −20°.

There is no evidence for the participation of a coenzyme in the synthesis of deoxyribose-5-phosphate. The enzyme withstands extensive dialysis and can be treated with Dowex-1 Cl without appreciable loss in activity under conditions known to remove nucleotide coenzymes. With purified DR-aldolase, pyruvate, acetylphosphate, acetate plus adenosinetriphosphate (ATP), alcohol, and ethyl acetate cannot substitute for acetaldehyde.

The enzyme is quite resistant to a number of commonly used inhibitors. The following substances have no effect on the rate of deoxyribose-5-phosphate formation: sodium fluoride (10^{-2} M), iodoacetate (10^{-4} M), sodium azide (0.2%), indoleacetic acid (0.1%), sodium salicylate (0.1%), ethyl alcohol (2%), diphenylamine (0.1%). Inhibition was observed so far with only three compounds: octyl alcohol, chloral hydrate, and propionaldehyde. Substances known to interact with aldehydes were not tested extensively, since they inhibit the reaction by removal of substrate. Hydroxylamine (0.2 M), for instance, completely blocks the enzymatic formation of deoxyribose-5-phosphate. It cannot be stated whether it also has an inhibitory action on the enzyme itself. Among the active inhibitory substances chloral hydrate was studied in greater detail. This compound in concentrations which will markedly inhibit the condensation reaction (a final concentration of 0.5% chloral hydrate produces about 50% inhibition under the conditions of the test) has little effect on the breakdown of deoxyribose-5-phosphate by the same enzyme as measured spectrophotometrically.

[57] Triosephosphate Isomerase from Calf Muscle

D-3-Glyceraldehydephosphate $\rightleftarrows$ Dihydroxyacetonephosphate

By GERWIN BEISENHERZ

Assay Method

Principle. Chemical assay methods are based on the determination of alkali-labile phosphate before and after treatment with iodine in weakly alkaline solution[1] or on the polarimetric determination of phosphoglycerate after oxidation with iodine.[2] Optical assays have been made possible by coupling the triosephosphate isomerase with glyceraldehyde phosphate dehydrogenase before or after the main reaction.[3,4]

The following assay has been developed[5] in analogy with the optical assay given by Racker[6] for diphosphofructose aldolase. Glyceraldehyde phosphate is used as substrate, and α-glycerophosphate dehydrogenase as the coupling enzyme.

D-3-Glyceraldehyde phosphate
$\diagdown$ Triosephosphate isomerase
$\diagup$ α-Glycerophosphate dehydrogenase $\diagdown$
DPNH + H+ α-Glycerophosphate + DPN+

The reader is referred to the discussion of the fundamental principles of the optical assay given in connection with the isolation of α-glycerophosphate dehydrogenase (Vol. I [58]). Owing to the compound nature of the optical assay, α-glycerophosphate dehydrogenase must be added in excess before triosephosphate isomerase is added to the assay mixture.

Reagents. The composition of the assay mixture is shown in Table I.

TABLE I
ASSAY MIXTURE FOR TRIOSEPHOSPHATE ISOMERASE

DPNH	$8.5 \times 10^{-5} M$
Racemic glyceraldehyde-3-phosphate	$1.5 \times 10^{-4} M$ (D-isomer)
α-Glycerophosphate dehydrogenase (4000 units/mg.)	2 mg./l.
Triethanolamine-HCl buffer (pH 7.9)	$2 \times 10^{-2} M$
Cuvette: $d = 2$ cm., temperature 25°, wavelength 366 mμ	

[1] O. Meyerhof and W. Kiessling, *Biochem. Z.* **279**, 41 (1935).
[2] O. Meyerhof and R. Junowicz-Kocholaty, *J. Biol. Chem.* **149**, 76 (1943).
[3] O. Warburg and W. Christian, *Biochem. Z.* **314**, 159 (1942).
[4] P. Oesper and O. Meyerhof, *Arch. Biochem.* **27**, 224 (1950).
[5] T. Bücher and K.-H. Garbade, unpublished; E. Meyer-Arendt, G. Beisenherz, and T. Bücher, *Naturwissenschaften* **2**, 59 (1953).
[6] E. Racker, *J. Biol. Chem.* **167**, 843 (1947).

Triethanolamine-HCl buffer. This has been described in detail in this volume in connection with α-glycerophosphate dehydrogenase. Triosephosphate isomerase is about twice as active in this buffer as in bicarbonate-CO_2 buffer, pH 7.5.

Procedure. The reaction is started by pipetting into the mixture a preparation of triosephosphate isomerase, diluted in quartz-distilled water, corresponding to 3 to 4×10^{-3} γ of pure enzyme per milliliter. When the reaction has been allowed to proceed for a short while, the time is measured which is required for the absorbance to decrease by 0.1. Enzyme concentration is inversely proportional to the time measured.

The reaction rate in this assay is constant with time up to rather high rates. However, this property is shown only by very pure isomerase preparations. At lower stages of purification the rate *rises* within the first minute. That is one of the reasons for allowing a certain time to elapse after starting the reaction by addition of enzyme before starting the stop watch.

Definition of Unit and Specific Activity. A unit is defined in the same way as for α-glycerophosphate dehydrogenase (Vol. I [58]). The specific activity of purest enzyme (at 26°, light path 1 cm., wavelength 366 mμ) is 161,000 units per gram of protein per liter of assay mixture. (The rate of the enzyme reaction in this assay is only about 35% of that in the presence of very high substrate concentrations.)

Application of Assay Method to Crude Tissue Preparations. The above assay method has the advantage of being quite free from interference, even in crude extracts. Compound optical assays coupling with glyceraldehyde phosphate dehydrogenase as auxiliary enzyme are very sensitive to the presence of α-glycerophosphate dehydrogenase, among other things. The high activity of the isomerase makes the assay applicable to crude extracts. The enzyme occurs in plasma and serum.[5] Certain phenomena indicate that plasma contains an inhibitor of the enzyme.

Purification Procedure

Calf muscle is used as starting material. It is removed from the animal about 1 hour after sacrifice and skinning. All subsequent manipulations take place in the cold room at 3°. The muscle is freed of tendons and aponeuroses, diced, and put through a chilled grinder. The mass is then weighed.

Step 1. Preparation of Crude Aqueous Muscle Extract. 1000-g. portions of muscle are homogenized in a chilled Waring blendor with 1600 ml. of extracting solution (quartz-distilled water containing 0.5 g. of disodium ethylenediaminetetraacetate per liter, $t = 3°$). The resulting muscle homogenate is then extracted for 30 minutes with mechanical stirring and centrifuged in stainless steel cups for 30 minutes at 2200 × g. The supernatant is decanted, and the sediment is again homogenized with 1000 ml. of extracting solution (see above) in the blendor and extracted for 30 minutes with occasional manual stirring. The second extract is also centrifuged, and the supernatant is combined with that of the first extract. This yields about 2500 ml. of aqueous muscle extract which is tested for activity.

Step 2. Acetone Precipitation (to 30%). To the total extract in an ice bath provided with a mechanical stirrer reagent acetone (about 98%) is added from a separatory funnel in a fine stream to a concentration of 30% (3 parts of acetone for 7 parts of extract), over a 45-minute period. After addition of the acetone, stirring is continued for a further 15 minutes. After centrifugation for 30 minutes at 2500 × g in stainless steel cups, the supernatant is assayed for activity.

Step 3. Acetone Precipitation (to 50%). To the supernatant, acetone (⅖ Vol. 98%) is added as above over a 45-minute period, and the mixture is centrifuged as above. The supernatant is assayed for activity.

Step 4. Acetone Precipitation (to 60%). To the supernatant, acetone (¼ Vol. 98%) (as above) is added over a 45-minute period, followed by 15 minutes, centrifugation in stainless steel cups at 2500 × g. The *residue* is taken up in eight times the amount of quartz-distilled water containing 0.5 g. of disodium ethylenediaminetetraacetate per liter ($t = 3°$). A resulting turbidity is centrifuged in stainless steel cups for 10 minutes at high speed (15,000 r.p.m.) and discarded. The pH of the clear solution is 7.1.

Step 5. Ammonium Sulfate Precipitation. To 130 ml. of clear solution of the sediment from the last acetone precipitation is added 56 g. of ammonium sulfate (recrystallized from hot saturated aqueous solution containing 2 g. of disodium ethylenediaminetetraacetate per liter)[7] with vigorous stirring, during about 3 minutes (pH 7.1). The resulting precipitate is centrifuged for 30 minutes at 15,000 r.p.m. and discarded.

Step 6. Ammonium Sulfate Precipitation. To 140 ml. of clear supernatant which is stirred vigorously, 9.6 g. ammonium sulfate is added over 3 minutes. This is followed by 30 minutes' high-speed centrifugation (15,000 r.p.m.). The residue is dissolved by cautious, dropwise addition of the minimum amount of quartz-distilled water at 3° containing 0.5 g. of disodium ethylenediaminetetraacetate per liter. Any slight turbidity is removed by high-speed centrifugation.

Step 7. Crystallization. To the clear solution, which is stirred constantly, finely powdered, recrystallized ammonium sulfate is cautiously added until a delicate "silky luster" appears. The solution is kept in the

[7] See Vol. I [58].

cold room; the "silky luster" increases in the next hours. Saturation with ammonium sulfate is carefully brought to completion. The microscopic field shows fine, long crystals which under higher magnification are seen to be rectangular prisms.

Step 8. Washings. By washing with 3 *M* ammonium sulfate solution at 3° pH 7.0 and repeated crystallization it is possible to increase the degree of purity quite considerably and to free the preparation from contamination by other enzymes.

Step 9. Recrystallization. For recrystallization the crystals are centrifuged for 30 minutes at high speed and then dissolved in a minimum of quartz-distilled water (as above). Any possible turbidity is removed by high-speed centrifugation. Finely powdered ammonium sulfate is cautiously added until the "silky luster" reappears.

TABLE II
SUMMARY OF PURIFICATION PROCEDURE

Fraction	Total units, millions	Purity	Recovery, %
1. 2250 ml. crude extract	47	0.025	—
2–3. 2550 ml. acetone, supernatant	36	0.1	77
4. 1100 ml. acetone, residue	33	0.16	70
5. Acetone ppt. dissolved in 120 ml. H₂O, 56 g. ammonium sulfate, supernatant	30	0.33	64
6. 9.6 g. ammonium sulfate, residue	23	0.4	49
7. First crystallization from ammonium sulfate	23	0.4	49
8. Recrystallization	17	0.52	36
9. Washing with 3 *M* ammonium sulfate	17	0.52	36
Recrystallization	11	1.0	24

Properties

The Michaelis constant for glyceraldehyde phosphate is $3.9 \times 10^{-4}\,M$.

Inhibitors. The enzyme is inhibited by phosphate ions. 0.05 *M* phosphate reduces the activity to one-fourth.

Effect of pH. The enzyme activity varies little between pH 7 and 8, although it is about half at pH 6.3.

Equilibrium Constant. The equilibrium constant for substrate,

$$K = \frac{C_{\text{Dihydroxyacetone phosphate}}}{C_{\text{Glyceraldehyde phosphate}}} = 22$$

is independent of temperature and concentration.[2,4] The kinetics of the reaction are first-order in both directions. The quotient of velocity constants equals the above equilibrium constant.

The catalytic activity of the enzyme is extraordinarily great. When the enzyme is saturated with respect to substrate (limit in the Lineweaver-Burk representation), 10^5 g. of enzyme protein gives a turnover at 26° of 945,000 moles of substrate per minute. The turnover rate at 38° is about twice as great.

Distribution. Calf muscle contains 47×10^6 units of enzyme per kilogram. The relative content (muscle = 100) of various tissues is taken from a paper by P. Oesper and O. Meyerhof:[4]

Yeast	260	Rat sarcoma	23
Brain	36	Mouse tumor	12

[58] α-Glycerophosphate Dehydrogenase from Rabbit Muscle

Dihydroxyacetone Phosphate + DPNH + H^+
$$\rightleftharpoons \text{L-}\alpha\text{-Glycerophosphate} + DPN^+$$

By GERWIN BEISENHERZ, THEODOR BÜCHER, and KARL-HEINZ GARBADE

Assay Method

Principle. H. von Euler, E. Adler, and co-workers[1] recognized the enzyme as a DPN enzyme. The assay given in principle by Baranowski[2] corresponds to the optical assay of O. Warburg.[3] The specific absorption band of reduced DPN at 340 mμ is measured spectrophotometrically. The following assay procedure has been worked out[4] for measurements at the easily accessible wavelength of 366 mμ (mercury vapor line). Measurements at other wavelengths can be readily interconverted by use of the following table of molar extinction coefficients (difference between reduced and oxidized DPN).

TABLE I
$\epsilon^{DPNH} - \epsilon^{DPN}$ AT VARIOUS WAVELENGTHS

Wavelength	ϵ ($\log_{10}$, light path 1 cm.)
340 mμ	$6.3 \times 10^3 \ M^{-1}$
334 mμ (Hg line)	$6.0 \times 10^3 \ M^{-1}$
366 mμ (Hg line)	$3.4 \times 10^3 \ M^{-1}$

[1] H. von Euler, G. Günther, and H. Hellström, *Z. physiol. Chem.* **245**, 217 (1937); H. von Euler, E. Adler, and G. Günther, *Z. physiol. Chem.* **249**, 1 (1937).

[2] T. Baranowski, *J. Biol. Chem.* **180**, 535 (1949).

[3] O. Warburg, "Wasserstoffübertragende Fermente," Editio Cantor, Freiburg, 1949.

[4] G. Beisenherz, H. J. Boltze, T. Bücher, R. Czók, K.-H. Garbade, E. Meyer-Arendt, and G. Pfleiderer, *Z. Naturforsch.* **8b**, 555 (1953).

TABLE II
ASSAY MIXTURE FOR GLYCEROPHOSPHATE DEHYDROGENASE

DPNH	$1.35 \times 10^{-4} M$
Dihydroxyacetone-P	$1.0 \times 10^{-4} M$
Triethanolamine-HCl buffer (pH 7.5)	$5.0 \times 10^{-2} M$

Reagents. The composition of the assay mixture is shown in Table II.
   Triethanolamine buffer.[4] 7.6 g. of redistilled triethanolamine (boiling point 195° at 5 mm. Hg), 15 ml. of 2 N HCl, and 2.0 g. of disodium ethylenediaminetetraacetate are made to 1 l. with heavy metal-free water. The pH of the buffer is 7.5. In this buffer the enzyme is about twice as active as in bicarbonate-CO_2 buffer of the same pH (16.8 g. of sodium bicarbonate and 1.0 g. of disodium ethylenediaminetetraacetate per liter, saturated with 20% CO_2 in nitrogen).
   DPNH solution.[4] 100 mg. of DPN (about 70%) and 80 mg. of sodium bicarbonate are dissolved in 7 ml. of water, to which 40 mg. of sodium dithionite is added immediately before the mixture is immersed in a boiling water bath. After vigorous shaking for 90 seconds in the water bath, the mixture is cooled in an ice bath. 2.5 ml. of a solution containing 1 g. of sodium bicarbonate and 1 g. of sodium carbonate in 100 ml. of water is added, and the mixture is thoroughly aerated for 5 minutes at the water aspirator pump.
   Dihydroxyacetone phosphate solution. This is obtained by splitting FDP with aldolase as follows: 10 ml. of 0.1 M FDP (428 mg. of sodium salt made to 10 ml. with water), 10 ml. of 1 M hydrazine (1.3 g. of hydrazine sulfate neutralized with 2 N NaOH), and 10 ml. of bicarbonate-CO_2 buffer (see above) are mixed, and at 40° 25 mg. of crystalline aldolase[3] is added. After 30 minutes a further 10 ml. of 1 M hydrazine solution and the same amount of aldolase are added. Thirty minutes later the solution is deproteinized with 2 ml. of 70% perchloric acid and, after the protein is centrifuged, neutralized with 2 N KOH. $KClO_4$ is centrifuged, and the residue washed. The combined supernatants are shaken four times with 20 ml. of benzaldehyde, extracted three times with ether, and then the ether is removed in a desiccator.

The principle of these directions was given by Meyerhof.[5] The above modifications yield a solution containing no fructose diphosphate and equal amounts of dihydroxyacetone phosphate and glyceraldehyde phosphate.

[5] O. Meyerhof, *Bull. soc. chim. biol.* **20**, 1033, 1045 (1938); **21**, 965 (1938).

Procedure. By a preliminary experiment it is determined how much of the DPNH stock solution is required to give an absorbance of 0.75 in the assay mixture. The assay is started with this absorbance, and enzyme preparation corresponding to about 1 γ of pure enzyme (diluted in heavy metal-free water) is added. The specific activity of purest crystalline enzyme (25°, light path 1 cm.) is 5700 units per gram of protein per liter of assay solution.

Under the conditions of the assay the enzyme is not saturated with respect to substrate; the turnover number of pure enzyme is about one-third of that in the presence of very high substrate concentrations. However, the assay is sufficiently reproducible even under these substrate-saving conditions.

Definition of Unit. In all optical tests the time is determined which the reaction requires for a change in absorbance of 0.1. The reaction is allowed to proceed for ½ minute before the stop watch is started.

EXAMPLE: Absorbance on adding enzyme: 0.75. Measured the time between absorbances 0.7 and 0.6. A unit is defined as follows:

$$U = \frac{100}{\text{Seconds for change of absorbance of 0.1}}$$

With a 1-cm. light path at 366 mμ, this unit corresponds to a turnover of 1.76×10^{-5} M per minute.

Purification Procedure

General Considerations. The use of quartz-distilled water to dilute enzyme preparations in the following purification procedure does not usually lead to inactivation of the enzyme upon dilution, but this point should be checked occasionally.

The following directions are based on more comprehensive ones[4] for the preparation of several enzymes from rabbit muscle in one procedure. There have been a few improvements in detail over the published method. All steps (including centrifugations), except sacrificing the rabbit, are carried out in the cold room at 2°. Water for extraction and dilution of enzymes is quartz-distilled and contains 0.5 g. of disodium ethylenediaminetetraacetate per liter.

For the first fractionation of the muscle extract, Merck reagent ammonium sulfate is used; but for further fractionation, for the crude and fine crystallizations, this ammonium sulfate is recrystallized (hot saturated aqueous solution of ammonium sulfate containing 2 g. of ethylenediamine tetraacetate per liter, made weakly basic with ammonia).

Protein is determined by the biuret method [after TCA precipitation:[4] absorbance ($d = 2$ cm.) at 546 mμ $\times$ 1.7 $\times$ final volume of colored solution = milligrams of protein in the determination]. Ammonium sulfate concentrations are nesslerized[4] [absorbance (d = 1 cm.) at 436 mμ $\times$ dilution of sample in colored solution $\times$ 2.4 $\times$ 10^{-4} = molarity of ammonium sulfate].

Step 1. Preparation of Muscle Extract. Three heavy, muscular rabbits are killed by a blow on the neck, bled by opening the carotid artery, skinned, and the muscle removed at room temperature. The muscle is minced in the cold room and put through a grinder, then weighed. Portions are homogenized with twice the weight of ice-cold extracting solution (see above) in a Waring blendor that has been previously chilled. The resultant thick homogenate is stirred mechanically and vigorously for 30 minutes in the cold room, then centrifuged in stainless steel cups at 2200 $\times$ g for 60 minutes. The supernatant is filtered through glass wool to remove fat. The residue is again homogenized with half the weight of extracting solution, stirred by hand for 15 minutes, and centrifuged.

Step 2. Ammonium Sulfate Precipitation. To the combined extracts (7 to 8 l.) is added over a 40-minute period 1 kg. of ammonium sulfate; then within the following 3 hours ammonium sulfate is gradually added until a concentration of 1.75 M has been attained (total, 260 g./l. of extract). The ammonium sulfate concentration can be checked by ammonia determination with Nessler's reagent. For the addition of ammonium sulfate the authors use a mechanical salt shaker.[4] Addition of salt is made in the cold room in a temperature range of 4 to 6°. The resulting precipitate is centrifuged (15 minutes at 2200 $\times$ g) and discarded. To the measured volume of supernatant, ammonium sulfate is added over a 2-hour period to bring the molarity to 2.4 (105 g./l. of extract). The resultant precipitate is centrifuged down. It contains 60% of the enzyme activity and is used for further purification. The supernatant may be used for the preparation of the enzymes pyruvate kinase and glyceraldehyde phosphate dehydrogenase.[4]

Step 3. Separation of Aldolase. The sediment (150 ml.) is dissolved in the above extracting solution which has been made $M/500$ in acetate buffer, pH 4.62, and made up to a total volume of 550 ml. Finely powdered ammonium sulfate (see above) is added, and the pH is adjusted to 6.1 by adding 1 N ammonia. 78 g. of ammonium sulfate leads to a molarity of 1.4 (determined by nesslerization). The protein content of the solution (biuret method) is 55 to 60 g./l. The resulting rather small turbidity is removed by centrifugation (60 minutes at 1200 $\times$ g) and discarded.

To the clear supernatant (solution I, Table III) 28 g. of finely powdered ammonium sulfate are added with rapid agitation over a 2-hour period. The pH are maintained at 6.1 by addition of ammonia. The ammonium sulfate molarity is now 1.8. The solution is covered and allowed to stand at 2 to 4° for 2 days, whereupon a heavy sediment of aldolase crystals (bipyramids) forms. By very slow addition of about 2.5 g. of ammonium sulfate the molarity of the latter is raised to 1.9, and the solution is again allowed to stand in the cold room overnight. The crude aldolase crystals are centrifuged. The supernatant (solution II, Table III) contains 70 to 80% of glycerophosphate dehydrogenase activity relative to the first ammonium sulfate precipitate.

TABLE III
SUMMARY OF PURIFICATION PROCEDURE
Yields and Purity in the Course of Isolating α-Glycerophosphate Dehydrogenase

	Aldolase		Glycerophosphate dehydrogenase		Lactic dehydrogenase	
	Total, units	Specific, units/mg. protein	Total	Specific	Total	Specific
Solution I	10.2×10^6	309	3.5×10^6	107	28×10^6	855
Solution II	1.6×10^6	79	2.73×10^6	134	29×10^6	1450
Crude crystals	10^4	4	1.8×10^6	730	2.6×10^6	105
Pure crystals I	—		9×10^5	3400	—	
Pure crystals II	—		8.6×10^5	4500	—	

If the activity of the supernatant is considerably lower, the enzyme has crystallized earlier together with aldolase. In that case the very much smaller crystals of glycerophosphate dehydrogenase may be separated to a great extent by fractional centrifugation of the aldolase crystals suspended in 1.9 M ammonium sulfate solution.[6]

Step 4. Crude Crystallization of Glycerophosphate Dehydrogenase. Crystallization of the enzyme is initiated by slow addition of 8 g. of finely powdered ammonium sulfate over a 2-hour period to 2.1 M and adjusting the pH to 6.2.[7] This produces an intense "silky luster." Microscopic examination reveals rather thick needles and long or elongated plates, as well as a few bipyramids (aldolase) and a fine noncrystalline background. The resultant crude crystals are centrifuged. They contain

---

[6] Aldolase crystals can be removed also in this way: suspension of crude crystals in 1 M ammonium sulfate solution and heating for 15 min. to 30 degrees C. The aldolase crystals dissolve. Glycerophosphate-dehydrogenase crystals are centrifuged.

[7] In our experience we have obtained better crystallization of the enzyme, and higher yields, if at this point, and also in the recrystallization, 20 mg. of a DPN preparation (from yeast, 70% pure) in neutral solution is added. However, the crystalline enzyme does not contain DPN but another nucleotide which may be present in our DPN preparation as an impurity.

50 to 60% the activity of the first ammonium sulfate precipitate (crude crystals, Table III).

Step 5. Purification of Crude Crystals. In addition to a few aldolase bipyramids,[6] the crude crystals contain very fine crystals of lactic dehydrogenase. The latter crystals, responsible for the intensive "silky luster," may be removed by fractional centrifugation. The crude crystals are suspended in 1.9 M ammonium sulfate solution and finely divided. Centrifugation for 5 minutes at 2000 r.p.m. firmly sediments most of the glycerophosphate dehydrogenase crystals, leaving a turbid supernatant. The process is repeated twice with 15 ml. of 1.9 M ammonium sulfate solution. Besides fractional centrifugation this process represents a washing; therefore the molarity of the ammonium sulfate must be carefully maintained.

For recrystallization the crystals are dissolved in 15 ml. of water and any turbidity is centrifuged off. Finely powdered ammonium sulfate is added slowly with constant stirring until the ammonium sulfate concentration is reached which just produces a visible turbidity. Any possible amorphous precipitate should always be redissolved by addition of a few drops of water. The pH is adjusted to neutrality (indicator paper) by addition of ammonia. After a short time the "silky luster" becomes apparent. Crystallization is completed overnight in the cold room (pure crystals I, Table III). The purity of the preparation may be slightly increased further by washing in 1.9 M ammonium sulfate solution and recrystallization under the conditions just described (pure crystals II, Table III).

Properties

Specificity. Pure crystals II contain 0.1% lactic dehydrogenase activity, 0.03% aldolase activity, and less than 0.01% triose isomerase activity. Other enzymes of the glycolytic scheme could not be detected. FDP, F-6-P, G-6-P, pyruvate, α-ketoglutarate, and glyceraldehyde phosphate, in the presence of DPNH, if acted upon at all, are attacked with an activity less then $\frac{1}{1000}$ rate. To our knowledge F-1-P and dihydroxyacetone have not yet been tested.

Turnover Number. Baranowski,[2] under conditions where enzyme is saturated with substrate, gives at 20° 26,500 moles $\times$ min.$^{-1}$/10^5 g. of enzyme protein. At 25° and under conditions where competition by FDP was not entirely excluded, we have found a turnover of 30,000.

Other Properties. The pure enzyme dissolved in heavy metal-free water will maintain its activity for weeks in the refrigerator even with low protein concentrations. The content of Kjeldahl nitrogen relative to protein dried at 110° is 13.5%.

The absorption spectrum of the pure crystals is not a pure protein spectrum but indicates an additional prosthetic group absorbing at 260 mμ that cannot be removed by recrystallization but only by perchlorate precipitation of the protein. The nature of the component is at present unknown. It is not DPN, although the absorption spectrum of the enzyme is very similar to that of glyceraldehyde phosphate dehydrogenase. The extinction coefficients of a solution of 1 g. protein per liter and light path 1 cm. are as follows (relative to Kjeldahl-N $\times$ 6.25):

280 mμ	0.75
270 mμ	0.81
260 mμ	0.72
250 mμ	0.53

[59] Glycerol Dehydrogenase from *Aerobactor aerogenes*[1,2]

$$
\begin{array}{c}
\text{H}_2\text{C}-\text{OH} \\
| \\
\text{H}-\text{C}-\text{OH} + \text{DPN} \rightleftarrows \\
| \\
\text{H}_2\text{C}-\text{OH}
\end{array}
\qquad
\begin{array}{c}
\text{H}_2\text{C}-\text{OH} \\
| \\
\text{C}=\text{O} + \text{DPNH} + \text{H}^+ \\
| \\
\text{H}_2\text{C}-\text{OH}
\end{array}
$$

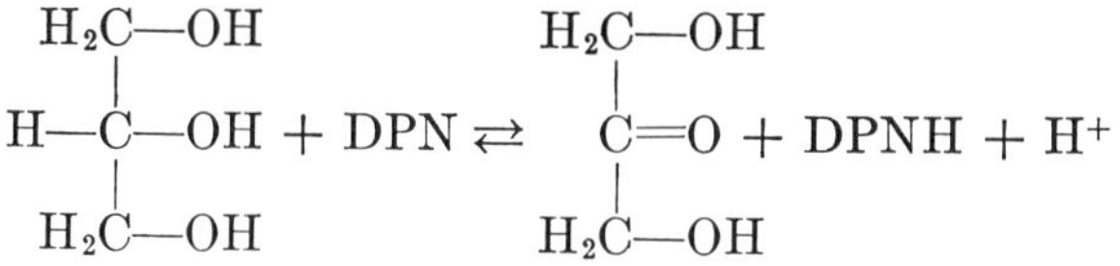

By ROBERT MAIN BURTON

Assay Method

Principle. The increase in the optical density at 340 mμ, due to the formation of DPNH,[3] is utilized to determine the rate of the reaction under standardized conditions.

Reagents

Glycerol (1.0 M aqueous solution). This is best prepared by weighing the glycerol and diluting to an appropriate volume. This should be refrigerated to prevent contamination by microorganisms.

0.1 M sodium pyrophosphate.

0.005 M DPN.

---

[1] R. M. Burton and N. O. Kaplan, *J. Am. Chem. Soc.* **75**, 1005 (1953); details are reported in R. M. Burton, N. O. Kaplan, and F. Stolzenbach, *J. Biol. Chem.* in press.

[2] Other glycerol dehydrogenases have been reported by R. E. Asnis and A. F. Brodie, *J. Biol. Chem.* **203**, 153 (1953), in *Escherichia coli*, and by H. P. Wolf and F. Leuthardt, *Helv. Chim. Acta* **36**, 1463 (1953), in rat and swine liver.

[3] See Vol. III [128].

Enzyme. The enzyme preparation should be diluted, prior to assaying, so that 0.1 ml. will contain approximately 0.1 unit as defined below.

Procedure. 1.0 ml. of sodium pyrophosphate, 0.5 ml. of glycerol, and 0.1 ml. of the enzyme preparation are added to water to give a total volume of 2.9 ml. in a spectrophotometer cuvette having a light path of 1 cm. After an initial determination of optical density of this solution, 0.1 ml. of the DPN solution is added, and the optical density is recorded at 30-second intervals for several minutes.

Definition of Unit and Specific Activity. One unit of enzyme is defined as that amount which causes an initial rate of change in optical density at 340 mμ of 1.000 per minute under the conditions specified above. The specific activity is expressed as units of enzyme per milligram of protein. Protein is determined by the optical density of the enzyme preparation at 280 mμ after correcting for nucleic acid absorption at 260 mμ.[4]

Application of Assay Method to Cell Suspensions or Crude Extracts. This assay is based on the rate of production of DPNH; a considerable source of error in assaying may be introduced, since many cell suspension and crude extracts have pyridine nucleotide oxidase activity.[5] This would result in either detecting no glycerol dehydrogenase activity or in finding low activity in crude preparations. The accompanying table indicates a total recovery of 180% of the enzyme activity in the second step of the purification procedure. This may be explained as above, since *A. aerogenes* does possess high DPNH oxidase activity in crude extracts. The direct spectrophotometric method for protein determination is not always applicable to crude extracts, owing to their high nucleic acid content. The turbidometric method, with sulfosalicylic acid as the precipitating agent, is fast and reliable enough for the purification procedure.[6]

Purification Procedure

Medium. *A. aerogenes* (American Type Culture Collection No. 8724) was grown on a medium of Dagley *et al.*[7] consisting of 0.54% KH_2PO_4, 0.12% $(NH_4)_2SO_4$, and 0.04% $MgSO_4 \cdot 7H_2O$, made to volume with tap water. The carbon source, normally glucose, was 1.5% glycerol; the final pH, 7.1. The glycerol may be autoclaved in the medium without carbonization. A 25-ml. actively growing innoculum was used per 20 l. of carboy. After 18 to 24 hours of growth with vigorous aeration, the cells were

---

[4] O. Warburg and W. Christian, *Biochem. Z.* **310**, 384 (1941–2); see Vol. III [73].

[5] See Vol. II [125].

[6] Klett-Summerson Photoelectric Colorimeter Chemical Manual, Klett Manufacturing Company, New York.

[7] S. Dagley, E. A. Dawes, and G. A. Morrison, *J. Bacteriol.* **60**, 369 (1950).

collected with a refrigerated Sharples steam-driven centrifuge. They were washed twice with 2 vol. of cold water and immediately extracted as described below.

Step 1. 21 g. of wet cells was suspended in 60 ml. of cold water and placed in a Raytheon sonic oscillator (9 kc.) for 40 minutes with continuous ice-water cooling. The cell debris was removed by centrifuging for 20 minutes in a refrigerated Servall centrifuge[8] at maximum speed. The residue was discarded, and the supernatant protein solution (76 ml.) was frozen until used. For purification, the solution was diluted to 130 ml. with water to give a protein concentration of about 10 mg./ml.

Step 2. Cold saturated ammonium sulfate (adjusted to pH 7.0 to 7.2 with NH_4OH) was added to the diluted cell free extract to obtain a final concentration of 40% in respect to ammonium sulfate. The precipitate was collected by centrifuging after allowing 5 to 10 minutes at 0° for equilibration. The precipitate was washed twice with 10 to 15 ml. of cold potassium phosphate buffer (0.05 M, pH 7.6) and recentrifuged. The washings contain the enzyme activity; after assaying, the precipitate may be discarded.

Step 3. Calcium phosphate gel[9] (25 mg./ml., pH 7.0) was added to the combined washings to give a gel/protein ratio of 1.5 (dry weight basis). After 5 to 10 minutes at 0° the gel was removed by centrifuging and discarded.

Step 4. The supernatant liquid from the gel centrifugation was placed in a water bath at 60° for 5 minutes and then rapidly cooled in ice. The precipitated protein was removed by centrifugation. The supernatant protein solution contains the partially purified glycerol dehydrogenase. The enzyme is stored frozen, since it loses activity rapidly at 0 to 5°.

SUMMARY OF PURIFICATION PROCEDURE

	Volume, ml.	Protein, mg./ml.	Specific activity, units/mg.	Total units	Recovery of activity, %	Purification
1. Crude extract	130	9.9	1.130	1660		
2. Protein eluates						
(1)	11.6	12.0	12.2	1655⎱	180	12×
(2)	15.0	5.1	17.2	1320⎰		
3. Gel supernatant fluid	31.5	4.2	16.5	2140	130	15×
4. Heating step (60°)	29.0	0.93	82.0	2210	133	73×

[8] Unless otherwise specified, all centrifugations were performed with a refrigerated Servall high-speed centrifuge.

[9] D. Keilin and E. F. Hartree, *Proc. Roy. Soc.* (*London*) **B124**, 397 (1938); see Vol. I [11].

Properties

Stability. The purified enzyme retains about 42% of its initial activity after being frozen for 2 months. However, only 12% activity is retained if the enzyme preparation has been frozen and thawed daily over the 2-month interval. Maximum stability to 10 minutes at 25° is shown at pH 7.0 to 7.5 with retention of 90% of its activity; 40% of activity is lost at pH 10, and 30% at pH 4.

Specificity. The glycerol dehydrogenase is free of alcohol dehydrogenase and DPNH oxidase activities. The enzyme is specific for DPN; TPN or deamino-DPN cannot replace DPN. Glycerol, 1,2-propandiol, and 2,3-butandiol are oxidized at the maximum rate by the glycerol dehydrogenase. The relative rates for other alcohols are as follows: 1,3-propandiol, 37%; ethylene glycol, 20%; 1,4-butanediol, 17%; isopropyl alcohol, 17%; i-inositol, 18%; sorbitol, 3%; glycerol-α-chlorohydrin, 26%; diglycerol, 21%; α-glycerol phosphate, 11%; β-glycerol phosphate, 2%; ethanol, 1%; n-propyl alcohol, D-ribose, glucose, <1%. Glycolic acid, ascorbic acid, mannitol, and i-erythritol are not oxidized. The maximum rate of reduction with DPNH is obtained with dihydroxyacetone. Glyceraldehyde is reduced at the relative rate of 14%, hydroxy-2-propanone acetate at 27%, and methylglyoxal at 56%. Acetone, chloroacetone, and s-dichloroacetone were not reduced.

Effect of pH and Activators. The oxidation of glycerol by the enzyme shows a constant increase in the initial rates of reaction upon increasing the pH from 6.8 to 11.0. Past pH 11 the initial rate falls rapidly; at pH 11 the initial rate is very rapid, but after several minutes the rate drops below the comparable rate at pH 10. This is probably due to denaturation of the enzyme protein. Ammonium ions show a slight stimulatory effect from pH 6.8 to 8.8, but the reaction is inhibited at pH > 8.9. The reduction of dihydroxyacetone by the enzyme shows no response to changes in pH from 6.7 to 8.3; at pH > 8.3 the initial rate of reaction decreases. Ammonium ions show a twofold stimulatory effect on the reduction at pH 8 to 9 which increases to threefold at pH 6.7 to 7. Sodium sulfate, glycine, and urea cannot replace ammonium ions.

Physical Constants. The equilibrium for the reaction as written at the beginning of the paper is about 5.1×10^{-12}. This is increased slightly in the presence of ammonium sulfate but not in the presence of sodium sulfate.

The K_m for some of the substrates are as follows: DPN, $1.5 \times 10^{-4} M$; DPNH, $1.4 \times 10^{-5} M$; glycerol, $3.9 \times 10^{-2} M$; 1,2-propanediol, $5.1 \times 10^{-3} M$; 2,3-butandiol, $5.9 \times 10^{-3} M$; and dihydroxyacetone, $1.3 \times 10^{-3} M$.

[60] Glyceraldehyde-3-phosphate Dehydrogenase from Muscle

$$GAP + DPN + P \longleftrightarrow PGA\text{-}P + DPNH + H^+$$

$$GAP + DPN + H_2O \xrightarrow{\quad HAsO_4^{--} \quad} PGA + DPNH + H^+$$

By SIDNEY F. VELICK

Assay Method

Principle. The reaction with purified enzyme is most conveniently followed spectrophotometrically by measuring the rate of change of optical density at 340 mμ, the absorption maximum of DPNH. The oxidative phosphorylation is an equilibrium reaction which, because of the liberation of a proton, is shifted to the right as the pH is increased. In the oxidation involving arsenate the reaction will approach completion if the aldehyde or DPN is in excess. There are many possible variations of the test system which may be selected for convenience according to the circumstances or to emphasize particular aspects of the reaction sequence. Of the two methods described below, the first was used by Cori *et al.*[1] and is essentially the same as the method originally described by Warburg and Christian[2] for the dehydrogenase from yeast. The second method involves a relatively high concentration of enzyme which approximates that found in skeletal muscle. Since the natural substrate reacts too rapidly for convenient rate measurements, under the latter conditions the nonphosphorylated glyceraldehyde is used as substrate in activity tests at high enzyme concentration.

Reagents

GAP stock solution. Dissolve 21.6 mg. of DL-glyceraldehyde 1-bromide 3-phosphoric acid dioxane addition compound[3] in 1.5 ml. of water. Adjust the pH to 7.0 with 2.0 *N* NaOH, and dilute to 2.0 ml. Hydrolysis of the bromide is rapid, and the bromide and dioxane do not interfere with subsequent tests. The stock solution, approximately 0.01 *M* as the D-form, may be kept for several days in the refrigerator or longer in the deep-freeze, with only slight decomposition. The concentration of the D-aldehyde may be determined in the enzyme test system described be-

[1] G. T. Cori, M. W. Slein, and C. F. Cori, *J. Biol. Chem.* **173**, 605 (1948).

[2] O. Warburg and W. Christian, *Biochem. Z.* **303**, 40 (1939).

[3] E. Baer, *Biochem. Preparations* **1**, 50 (1949).

low, with arsenate, an excess of DPN, and limiting amounts of aldehyde. Sufficient enzyme should be used to insure rapid completion of the reaction. A dilute stock solution of the aldehyde, 0.0075 M, in the appropriate buffer is made up shortly before use. In order to avoid formation of addition compounds, cysteine should not be added to stock aldehyde solutions.

DPN stock solution. Coenzyme of high purity is now available from several commercial sources. A 0.003 M solution is made up by weight, and its concentration checked in the enzyme test system with arsenate and excess GAP.

Sodium pyrophosphate, 0.03 M, adjusted to pH 8.4 with HCl.

Cysteine. 0.004 M in 0.03 M sodium pyrophosphate, pH 8.4, made up shortly before use. Cysteine is unstable at this pH.

Enzyme, stock solution. Crystals are centrifuged from the mother liquor at 3° in a high-speed centrifuge, drained from the supernatant solution, and dissolved in cold 0.03 M pyrophosphate. After the protein concentration has been determined spectrophotometrically at 276 mμ in a diluted aliquot, a dilute stock solution is made containing 120 γg. of protein per milliliter of cysteine pyrophosphate. The dilute stock solution is kept in an ice bath and used promptly.

Disodium arsenate, 0.4 M.

Procedure. To 2 6 ml. of the cysteine pyrophosphate in a silica cuvette of 1-cm. light path is added 0.1 ml. each of DPN, arsenate, and enzyme. After 7 minutes at room temperature the optical density is measured at 340 mμ, and reaction is initiated by rapidly stirring in 0.1 ml. of the dilute GAP solution. Readings are taken at 20-second intervals for 2 minutes.

The second test system contains 2 to 4 mg. of enzyme per milliliter (0.17 to 0.34) $\times$ 10^{-7} M, 0.05 M sodium arsenate, pH 7.6, and 0.003 M cysteine. The 2 equivalents of DPN bound to the enzyme (mol. wt. 120,000) suffice for the optical test, and no extra DPN need be added. After incubation for 1 hour at 0°, 2.9 ml. of the test solution is brought to room temperature in a silica cuvette, and the reaction is initiated by the addition of 0.1 ml. of 0.12 M DL-glyceraldehyde. The aldehyde should be of high purity because trace impurities may be inhibitory at high substrate concentration. The optical density at 340 mμ prior to addition of the aldehyde is relatively high under the test conditions because of the characteristic absorption of the enzyme-DPN complex.[4]

Specific Activity. Under the conditions described for the first assay method the reaction follows a second-order rate law with respect to GAP

[4] S. F. Velick, *J. Biol. Chem.* **203**, 563 (1953).

and DPN. The specific activity has been defined as $k/[E]$, where k is the second-order rate constant and $[E]$ is the enzyme concentration in milligrams per milliliter. In an average case, with 0.0031 mg. of enzyme per milliliter, k, calculated from the first- and second-minute intervals, was 2.59×10^3 (liter mole^{-1} mm.$^{-1}$); $k = 2.3/[t(a - b)] \times \log [b(a - x)]/[a(b - x)]$ where a is the initial concentration of aldehyde, b that of DPN, and x the concentration of DPNH produced at time t (minutes), all expressed in moles per liter. The specific activity is therefore $(2.59 \times 10^3)/0.0031 = 8.36 \times 10^5$ at 24°. The turnover number of the enzyme calculated from initial rates at GAP and DPN concentrations of $4.8 \times 10^{-4} M$ is about 10,300 per minute per molecule of enzyme (mol. wt. 120,000) at pH 8.6 and 27°.

The reaction rate with the more concentrated enzyme is controlled by the glyceraldehyde concentration and in the range described in the second test system is proportional to the aldehyde concentration. The rate of the reaction calculated from the optical density change at 340 mμ between the 15- and 30-second interval is 1.3 moles of DPNH formed per mole of enzyme per minute at 27°.[5]

Assay of Enzyme in Crude Preparations. Quantitative application of the above methods to extracts and crude enzyme fractions is unsatisfactory because of the ubiquity of enzymes which may involve substrate, coenzyme, and products in other reactions. Each situation therefore requires special control from this point of view, and some preliminary fractionation is usually necessary. The high but difficultly controlled specific rotation of molybdate complexes of PGA have been used to follow the reaction in crude systems.[6]

Purification Procedure

Step 1. The method described is that of Cori *et al.*[1] Rapidly excised hind leg and back muscle (300 to 500 g.) from an adult rabbit, killed by intravenous injection of 5 ml. of 5% Nembutal, is weighed and put through a chilled meat grinder. The tissue is then *immediately* extracted with occasional stirring for 10 minutes with 1 vol. of cold 0.03 M KOH, and the mixture is poured onto two layers of cheesecloth tied over an 800-ml. beaker. The cloth is gathered and squeezed. Most of the solvent is retained by the tissue pulp in the first extraction. The process is repeated with 1 vol. of KOH, based upon the original tissue weight, and then with 0.5 vol. of cold distilled water. These and subsequent operations are carried out in the cold room. The pH of the combined extracts should be between 6.8 and 7.2.

[5] S. F. Velick, J. E. Hayes, Jr., and J. Harting, *J. Biol. Chem.* **203**, 527 (1953).
[6] O. Meyerhof, and W. Schulz, *Biochem. Z.* **297**, 60 (1938).

Step 2. Intermediate ammonium sulfate fractions may be removed for the isolation of other enzymes (Aldolase, Vol. I [**39**]). If aldolase is not desired, add, for each 100 ml. of extract, 108 ml. of ammonium sulfate solution saturated at 26° and adjusted to pH 7.5 with concentrated ammonium hydroxide. The precipitate is removed by gravity filtration on large fluted filters (Whatman No. 1). To each 100 ml. of the clear filtrate is added 13 g. of solid ammonium sulfate. When the salt has dissolved, the precipitated protein is removed by gravity filtration. The pH of the filtrate is adjusted to 8.4 with ammonium hydroxide (1:1 dilution of the concentrated reagent with distilled water). Ammonium sulfate solutions are diluted with 4 vol. of distilled water before a pH measurement is made with the glass electrode. Crystals appear overnight and increase on standing in the cold for 3 to 5 days. Crystals are separated from the mother liquor by centrifugation in the cold in an angle centrifuge at about 12,000 × *g*.

Step 3. The crystal cakes in the centrifuge tubes are dissolved in small portions of water and combined (total volume, 40 to 80 ml.). After clarification by centrifugation 2 vol. of saturated ammonium sulfate, pH 8.4, is added. A heavy yield of crystals is obtained when the solution is allowed to stand overnight in the cold. Three recrystallizations are usually necessary to remove traces of heme protein as indicated by diminution of a Soret band. The specific activity does not increase appreciably after the first recrystallization. Somewhat greater yields are obtained if the extraction and subsequent recrystallizations are done in the presence of 0.001 *M* ethylenediaminetetraacetic acid (EDTA). Enzyme so protected exhibits high activity without activation by cysteine.

Properties

Specificity. The enzyme acts most effectively upon glyceraldehyde-3-phosphate and is relatively specific for the D-isomer. At higher concentrations of both enzyme and aldehyde, glyceraldehyde, acetaldehyde, propionaldehyde, butyraldehyde,[7] and other nonphosphorylated aldehydes serve as substrates. TPN cannot be substituted for DPN, but deamino-DPN reacts slowly.[8] With the natural substrate at low enzyme concentration no reaction can be detected unless phosphate or arsenate is added. At enzyme concentrations comparable with those of GAP and DPN an oxidation equilibrium is rapidly established in the absence of phosphate or arsenate, with the enzyme acting as the acyl acceptor. Subsequent addition of phosphate permits reversible acyl transfer from enzyme to phosphate. The equilibria of the consecutive steps may be meas-

---

[7] J. Harting and S. F. Velick, *J. Biol. Chem.* **207**, 857 (1954).
[8] M. E. Pullman, S. P. Colowick, and N. O. Kaplan, *J. Biol. Chem.* **194**, 593 (1952).

ured separately.[9] Acyl enzyme intermediates in the oxidation of the nonphosphorylated aldehydes are less stable in the absence of suitable acceptors and appear to break down rapidly by hydrolysis to form the free acids. In such cases aldehyde oxidation in the absence of arsenate or phosphate occurs with enzyme "turnover" and can easily be detected at enzyme concentrations of the order of 1 mg./ml. Various thiols may act as acetyl acceptors in the oxidation of acetyldehyde or with acetyl phosphate as substrate, but the rates are relatively slow and require high enzyme and acceptor concentrations. In the oxidation of GAP at low or high enzyme concentrations the acceptor activity of thiols has not been detected, and if it occurs it is insignificant compared to the activity of orthophosphate. Since the phosphorylation step is a reversible acyl transfer reaction between enzyme and orthophosphate ion, appropriate acyl phosphates undergo phosphate exchange with labeled orthophosphate ion in the presence of the enzyme.[10] This activity requires the presence of the bound coenzyme but appears to be a true exchange in which the coenzyme does not function as a hydrogen donor or acceptor in the usual sense.

Activators and Inhibitors. Heavy metal ions, depending on the metal, its concentration, and environmental conditions, may lead to either reversible or irreversible inactivation of the enzyme. Loss of activity of the enzyme or recrystallization and reactivation by thiols is frequently attributed to conversion of the enzyme to an inactive "oxidized" state and reconversion to an active "reduced" state. The effect may actually be due in part to competitive binding of inhibitory metals, since EDTA not only stabilizes the activity but also partially reactivates an inactivated preparation. Disulfides such as oxidized glutathione are inhibitory under certain conditions, and the inhibition is reversed by thiols. The inhibition in this case appears to result from disulfide formation between the essential thiol groups of the enzyme, either with each other or by exchange reactions with the inhibiting disulfide. Inhibition by oxidizing agents such as o-iodosobenzoate or peroxides is irreversible, as is the well-known inhibition by iodoacetate. Three equivalents of p-chlormercuribenzoate per mole of enzyme reduce the activity to a low level and promote the release of bound DPN.[4] Activity is completely restored by the prompt addition of cysteine, but denaturation is promoted, particularly at room temperature, if the cysteine addition is delayed. Some types of inactivation, presumably those involving disulfide formation, are not reversed instantaneously by cysteine but require incubations for various lengths of time. The enzyme may be stored and tested in a variety of

[9] S. F. Velick and J. E. Hayes, Jr., *J. Biol. Chem.* **203**, 545 (1953).
[10] J. Harting and S. F. Velick, *J. Biol. Chem.* **207**, 867 (1954).

buffers. Veronal, trishydroxymethylaminomethane, and acetate are somewhat inhibitory.

pH Optimum. The relative initial rates of GAP oxidation at pH 9.0, 8.6, 8.1, 7.7, and 7.1 are, respectively, 100, 100, 81, 57, and 15 when tested with arsenate in cysteine pyrophosphate.[1]

Physical Characteristics. The molecular weight of the enzyme from sedimentation and diffusion measurements is 120,000.[11] The isoelectric point of the protein is a strong function of the type and concentration of the buffer. In phosphate buffers of ionic strength 0.05, 0.1, and 0.2, the isoelectric points are, respectively, 7.2, 6.6, and 5.9. The shift in the acid direction as the buffer concentration is increased is the result of phosphate binding.[9] The isoelectric point extrapolated to zero ionic strength is above pH 8 in the region expected from the amino acid composition. The protein crystallizes as a complex with 2 moles of bound DPN which may be removed with activated charcoal but not by recrystallization. The molar absorption coefficient at the 276-mμ absorption maximum is 12.7×10^4. The enzyme DPN complex in the "active" state exhibits a broad absorption band with a maximum around 365 mμ. This absorption is decreased about 60% by the addition of p-chloromercuribenzoate or by the removal of the bound DPN with charcoal and is restored to the original value in the two cases, respectively, by the addition of cysteine and DPN. The dissociation constant of the enzyme DPN complex from binding studies is of the order of 10^{-7} M or less. There is some evidence for cooperative interaction in the binding. The Michaelis constant for DPN is influenced by steady-state conditions, and in test system number 1 described above, with arsenate and GAP concentrations that give nearly maximal rates, the Michaelis constant for DPN is 3.9×10^{-5} M. Under analogous conditions the Michaelis constant for GAP is 5.1×10^{-5} M.

Crystal suspensions of the enzyme in ammonium sulfate mother liquor may be kept for months in the cold with losses in activity that are partially restored by cysteine. The stability of the protein in solution is favored by neutral salts, metal binding agents, and high protein concentration. Loss of activity at room temperature is retarded by cysteine and EDTA. Solutions of the enzyme at 0.1% concentration or higher are stable for hours at 0° in neutral phosphate or arsenate buffers and may show spontaneous increases in activity. The stability of the protein rapidly decreases below pH 6.

[11] J. F. Taylor, *in* "Phosphorus Metabolism" (McElroy and Glass, eds.), Vol. I, p. 104. Johns Hopkins Press, Baltimore, 1951.

[61] Glyceraldehyde-3-phosphate Dehydrogenase from Yeast

GAP + DPN + P $\leftrightarrows$ 1,3-Diphosphoglycerate + DPNH + H$^+$

By EDWIN G. KREBS

Assay Method

Principle. The rate of reduction of DPN resulting from GAP oxidation is followed, using the spectrophotometer as described by Warburg and Christian.[1] If GAP is not available, it may be generated with FDP and crystalline muscle aldolase. A definition of units in a system of the latter type has been described.[2] The method given below is identical to that of Warburg and Christian,[1] except that cysteine has been included in the reaction mixture, and the initiation of the reaction is brought about by addition of GAP-arsenate mixture instead of enzyme as the last component. This assay system serves equally as well to measure *muscle* triosephosphate dehydrogenase activity.

Reagents

GAP solution (7.6 × 10^{-3} *M*). Prepare a solution of the dioxane addition compound of DL-glyceraldehyde-1-bromide-3-phosphoric acid (dimer)[3] approximately three times as concentrated in the reactive D-isomer as the indicated molarity. Assay an appropriate aliquot in a test system[4] with excess DPN, and dilute to 7.6 10^{-3} *M* D-GAP. This solution may be stored at −10° for at least 3 weeks without change.

DPN solution (7.6 × 10^{-3} *M*). Assayed by standard methods.[5]

0.03 *M* pyrophosphate–0.006 *M* cysteine buffer, pH 8.5. Prepare immediately before use by diluting 0.3 *M* cysteine hydrochloride solution (stable for 1 to 2 weeks at +3°) 1:50 with 0.03 *M* sodium pyrophosphate solution and adjusting the pH to 8.5 with HCl.

0.17 *M* sodium arsenate.

Enzyme. A solution of the enzyme is diluted with pyrophosphate, cysteine buffer to obtain 0.010 to 0.060 mg. of crystalline enzyme

---

[1] O. Warburg and W. Christian, *Biochem. Z.* **303**, 40 (1939).

[2] E. G. Krebs, G. W. Rafter, and J. McBroom Junge, *J. Biol. Chem.* **200**, 479 (1953).

[3] See Vol. III [31].

[4] The test system described for activity measurement is satisfactory for this purpose except that the cysteine should be omitted and at least ten times as much enzyme used.

[5] See Vol. III [128].

per milliliter (or the equivalent in enzyme activity). This solution is stable for several hours at 0°.

Procedure. In a cuvette having a 1-cm. light path prepare an incomplete reaction mixture as follows: 2.6 ml. of pyrophosphate-cysteine buffer, 0.1 ml. of DPN, 0.1 ml. of enzyme. Incubate at room temperature for 7 minutes to insure complete activation of the enzyme by cysteine. A reference cell containing 2.8 ml. of pyrophosphate-cysteine buffer is used. The initial absorption of the reaction mixture is observed, and the final component, 0.2 ml. of a 1:1 mixture of GAP and arsenate,[6] is added to both cells. Readings at 340 mμ are taken at 30-second intervals after initiation of the reaction; they are corrected for any initial absorption.

Definition of Specific Activity. Under the conditions of the test, the rate is not linear, but the reaction is approximately second order. The velocity constant calculated by means of equation 1 is proportional to enzyme concentration.

$$k = \frac{1}{t} \cdot \frac{C_0 - C}{C_0 C} \tag{1}$$

where C_0 is the initial concentration of DPN, and C is the concentration of DPN at time t. Specific activity is obtained by dividing k by the concentration of enzyme in the reaction mixture (mg./ml.). Warburg and Christian[1] found a value of 8.3×10^8 for the crystalline yeast enzyme as tested at 20°. Enzyme isolated by the present procedure gave a value of 12×10^8 when assayed at 25° under these conditions.

Application of the Assay Method to Crude Tissue Preparations. In crude yeast extracts the concentration of this enzyme is sufficiently high so that no difficulty is encountered with the spectrophotometric assay system. The interpretation of the assay as applied to crude extracts is difficult, since other reactions (triose isomerase and glycerophosphate dehydrogenase) may interfere. Independent assays by immunological techniques[2] indicate that errors of the order of 50 to 100% may result in the determination of enzyme concentration in crude extracts by activity measurements.

Purification Procedure

Two different procedures are available for the isolation of yeast triose-phosphate dehydrogenase in the crystalline form. Warburg and Christian[1]

---

[6] Arsenate is used instead of phosphate in the activity test. This makes the reaction irreversible, since the product, 1-arseno-3-phosphoglycerate, is unstable.[1] The particular method of adding the arsenate is convenient to use, since the dioxane addition compound of the bromide derivative of GAP is decomposed to yield free GAP in the alkaline solution. The GAP-arsenate mixture is stable for several hours.

originally isolated the pure enzyme by acetone fractionation of yeast juice[7] followed by precipitation of proteins with nucleic acid at varying pH. Kunitz and McDonald[8] isolated several unidentified "inert" crystalline yeast proteins. One of these, which they designated as yeast protein 2, was later[2] found to possess triosephosphate dehydrogenase activity equal to or slightly higher than the crystalline enzyme of Warburg and Christian. Various physical and chemical characteristics of protein 2 and the original crystalline yeast triosephosphate dehydrogenase have been found to be identical.[2,9] The procedure to be described is based on the simpler method of Kunitz and McDonald, and the product will be referred to simply as triosephosphate dehydrogenase.

Step 1. Plasmolysis and Extraction of Yeast. To 10 lb. of fresh, starch-free baker's yeast is added 2400 ml. of toluene that has been warmed to 40°. The mixture is incubated in a water bath at 45° with occasional stirring until liquefaction has occurred. This usually requires 60 to 90 minutes. The mixture is allowed to stand for 3 hours at room temperature (25°) and is then cooled in an ice bath to about 10°. 4800 ml. of cold distilled water is added, and the mixture is stirred occasionally during the next hour and then allowed to stand for 16 hours at 3°. The turbid extract (lower layer) is collected with a siphon. For each liter of extract, 100 g. of Hyflo Super-Cel[10] is added, and the mixture is filtered[11] with suction at 3°.

Step 2. Fractionation with Ammonium Sulfate. To the clear yellow filtrate is added 314 g. of solid ammonium sulfate per liter of filtrate, bringing the solution to 0.5 saturation.[12] The small amount of precipitate formed is removed by filtration at 3° with suction with the aid of 10 g. of standard Super-Cel,[10] plus 10 g. of Filter-Cel[10] per liter of suspension. The precipitate is discarded. The filtrate is brought to 0.7 saturation by addition of 135 g. of solid ammonium sulfate per liter of filtrate. This mixture is allowed to stand 10 to 15 hours at 3° and is then filtered with suction in the cold. No filter aids are used in this step, and it is necessary to pour back the first portion filtered to obtain good retention. Suction is continued until cracks appear in the filter cake.

Step 3. Crystallization. The filter cake, which weighs about 300 g., is dissolved in an equal weight of water, and the resulting yellow-brown

---

[7] Yeast juice was made by incubating dry yeast with Na_2HPO_4 solution.

[8] M. Kunitz and M. R. McDonald, *J. Gen. Physiol.* **29**, 393 (1946).

[9] S. F. Velick and S. Udenfriend, *J. Biol. Chem.* **203**, 575 (1953).

[10] Johns-Manville, Sales Corporation, Celite Division, San Francisco, California.

[11] Whatman No. 2, 24.0-cm. paper, or its equivalent, may be used. Several filters are set up in order to speed the process.

[12] Ammonium sulfate saturation is calculated using the convention adopted by Kunitz. See M. Kunitz, *J. Gen. Physiol.* **35**, 423 (1952).

solution is brought to very faint turbidity by the careful addition of saturated ammonium sulfate previously adjusted to pH 8.2.[13] A volume of ammonium sulfate solution approximately two-thirds of the volume of water used in dissolving the cake is necessary. The pH of the mixture is then raised to 8.2 by addition of 4% NH_4OH. Crystallization starts within 10 to 15 minutes and is complete after 2 to 3 days at 3°. The first crystals may be collected by filtration employing filter aids[6] but are more conveniently separated in the Spinco preparative ultracentrifuge at 20,000 r.p.m. with the No. 21 rotor. For recrystallization, the protein[14] is dissolved in 150 ml. of H_2O, and the solution is brought to turbidity with ammonium sulfate; pH adjustments are carried out as described for the first crystallization. Two or three recrystallizations result in maximum activity. The enzyme is stored at 3° as the suspension of crystals in the ammonium sulfate solution and retains almost full activity for at least 3 months as determined in a test system with cysteine present. The activity, as determined in the absence of cysteine, decreases with storage. For use, a sample of the enzyme is centrifuged and dissolved in water. A 2% protein solution made in this manner is stable at 3° for at least 2 days.

Properties

Physical and Chemical Properties. Enzyme recrystallized three times gives a 280 mμ/260 mμ absorption ratio of 1.9. It contains no detectable bound DPN. The molar extinction coefficient at 280 mμ is 10.9×10^4 referred to a molecular weight of 120,000.[15] Electrophoretically, the protein migrates as a single peak in phosphate buffer of 0.1 ionic strength, pH 7.25, with a mobility of -4.0×10^{-5} cm.^{2} volt^{-1} sec.$^{-1}$. There is evidence that the mobility of the enzyme is markedly influenced by phosphate binding. In barbiturate buffer at pH 8.5, two peaks are apparent. Immunochemically the protein appears homogeneous. Velick and Udenfriend[9] have determined that there are two valine amino end groups in yeast and muscle triosephosphate dehydrogenase.

Specificity. The purified yeast enzyme has been shown[1] to be active with glyceraldehyde as well as GAP. With the former substrate, approximately 1000 times as much enzyme is required for comparable rates.

Activators and Inhibitors. Maximal activity of the enzyme has been observed only in the presence of cysteine. Freshly isolated enzyme is

[13] The pH of solutions with high salt concentrations is determined on diluted aliquots by means of indicators.
[14] Approximately 14 g. of crystalline protein, as measured by the method of Robinson and Hogden [*J. Biol. Chem.* **135**, 727 (1940)], are obtained in the first crystallization.
[15] J. F. Taylor, private communication.

usually 60 to 80% active without cysteine, but the requirement increases as the protein is stored. Two equivalents of *p*-chloromercuribenzoate completely inhibit the enzyme, but the activity is restored in the presence of cysteine. The mercuribenzoate-enzyme complex can be crystallized.[16] Iodoacetate inactivates the enzyme irreversibly.

Effect of pH. The pH optimum for the yeast enzyme is 8.3 to 8.5. Activity falls off sharply above this range but decreases more gradually on the low pH side.

SUMMARY OF PURIFICATION PROCEDURE

Fraction	Total volume, ml.	Total protein, g.	Specific activity,[b] % of third crystals
1. Original extract[a]	6000	140	13
2. 0.5–0.7 sat. $(NH_4)_2SO_4$ fraction dissolved in H_2O	540	76	17
3. First crystals	—	14	44
4. Second crystals	—	8.6	59
5. Third crystals	—	4.0	100
6. Fourth crystals	—	3.5	96

[a] Filtrate obtained at the end of step 1.

[b] The activity tests in this preparation were carried out using the more complicated assay system in which GAP is generated using aldolase and HDP.[2]

[16] S. F. Velick, *J. Biol. Chem.* **203**, 563 (1953).

[62] TPN Triosephosphate Dehydrogenase from Plant Tissue

D-Glyceraldehyde-3-phosphate + PO_4 + TPN
$$\rightleftarrows 1,3\text{-Diphosphoglyceric acid} + TPNH$$

By MARTIN GIBBS

Plant tissue contains two triosephosphate dehydrogenases, one using TPN as cofactor,[1,2] the other DPN.[3] The leaves possess both enzymes, whereas roots and seeds possess only the DPN enzyme. Only the purification of the TPN enzyme will be described.

[1] M. Gibbs, *Nature* **170**, 164 (1952).
[2] D. I. Arnon, *Science* **116**, 635 (1952).
[3] P. K. Stumpf, *J. Biol. Chem.* **182**, 261 (1950).

Assay Method

Principle. The method described below was developed by Warburg[4] and is based on the fact that TPNH and DPNH absorb light of wavelength 340 mμ, whereas TPN and DPN do not. The DPN enzyme has also been assayed[3] by a dismutation reaction between the enzyme and yeast alcohol dehydrogenase in the presence of bicarbonate ion. One equivalent of CO_2 is released for each equivalent of G-3-P oxidized to yield PGA.

Reagents

> Tris buffer, pH 8.5, 0.2 *M*.
> Sodium arsenate, $Na_2HAsO_4 \cdot 7H_2O$, 0.17 *M*.
> Cysteine, 0.04 *M*. Dissolve 19 mg. of cysteine·HCl in 3 ml. of H_2O, and neutralize to pH 8 with KOH. Prepare each day. TPN.
> Sodium fluoride, 0.1 *M*.
> G-3-P, 0.03 *M*. Since dioxane and bromide ion do not interfere with the assay procedure, it is best to keep the G-3-P as the dioxane compound of glyceraldehyde 1-bromide 3-phosphoric acid.[5] The dioxane compound (49 mg.) is dissolved in 5 ml. of cold water and adjusted carefully to pH 7 with alkali. The compound is labile in alkaline solution. If G-3-P is not available, FDP and aldolase may be substituted (see precautions for crude extracts).
> Enzyme. Dilute the stock enzyme with Tris buffer to obtain 4 to 20 units of enzyme per milliliter.

Procedure. Reactions are run in Beckmann cuvettes ($d = 1$ cm.) with a total volume of 3 ml., and readings are made against a blank containing all reactants except G-3-P or FDP. The increase in optical density at 340 mμ is read at 30-second intervals for 2 minutes. The reaction mixture contains 100 micromoles of Tris buffer, 51 micromoles of sodium arsenate, 12 micromoles of cysteine, 60 micromoles of sodium fluoride, 0.12 micromole of TPN, enzyme (0.4 to 2.0 unit), and aldolase (if FDP is used as source of G-3-P). The reaction is started by adding 1.5 micromoles of G-3-P or FDP.

Definition of Unit and Specific Activity. One unit of enzyme is defined as the amount of enzyme causing an increase in optical density at 340 mμ of 0.01 per minute. Specific activity is expressed as units per milligram

---

[4] O. Warburg, *Ergeb. Enzymforsch.* **7**, 210 (1938).
[5] E. Baer, *Biochem. Preparations* **1**, 50 (1949).

of protein. Protein concentration is determined spectrophotometrically by measurement of light absorption at wavelengths 280 and 260 mμ. A correction for the nucleic acid content is made according to Warburg and Christian.[6]

Application of Assay Method to Crude Tissue Preparations. Since G-3-P is scarce, it is common to employ FDP and aldolase as sources of G-3-P. This procedure can be used if the DPN enzyme is to be assayed; however, most crude plant extracts readily convert FDP to G-6-P, and a TPN reduction occurs catalyzed by glucose-6-phosphate dehydrogenase. Before the TPN enzyme is assayed and if G-3-P is unavailable, the crude extract is incubated with 0.01 M IAA for 5 minutes to poison triosephosphate dehydrogenase, FDP is added, and if reduction of TPN does not occur, then the TPN enzyme can be assayed with HDP and aldolase.

With extracts of most green tissue, the amount of material required is such that light absorption by the enzyme preparations itself is too high to permit direct spectrophotometric measurements. Buffered extracts (Tris or PO$_4$) of acetone powders are recommended.

Purification Procedure

This procedure is based upon young pea leaves as starting material. The enzyme can also be isolated from spinach leaves. The enzyme has been prepared routinely through step 4. Steps 5 and 6 have not been investigated thoroughly. The last two steps have yielded fractions of varying specific activity.

Step 1. Preparation of Crude Extract. 1 kg. of pea leaves (10 to 15 days old) separated from stems are ground in a cold mortar with sharp sand, 100 ml. of 0.5 M KHCO$_3$, and 100 ml. of sodium versenate (30 mg./ml., pH 8.0). The brei is squeezed through a cheesecloth and then centrifuged in the cold at 15,000 r.p.m. The crude extract contains approximately twice as much TPN as DPN enzyme.

Step 2. Acid Fractionation. The pH of the supernatant solution is lowered to 5.0 with cold 2 N acetic acid. The heavy precipitate is removed by high-speed centrifugation. The pH of the supernatant solution is raised to 8.0 with 2 N KOH. If a precipitate forms, it is removed by centrifugation. The solution is now a light tan.

Step 3. Fractionation with Solid Ammonium Sulfate. The solution from step 2 is brought to 0.31 saturation. After 10 minutes, the precipitate is removed by centrifugation in the cold at 15,000 r.p.m. and discarded. The supernatant fluid is brought to 0.42 saturation, and, after centrifugation, the 0.31 to 0.41 fraction is dissolved in 0.1 M Tris buffer, pH 8.0.

[6] O. Warburg and W. Christian, *Biochem. Z.* **310**, 384 (1941–42).

Step 4. Fractionation with Acetone. To the solution from step 3 an equal volume of acetone, previously chilled to $-18°$, is added slowly. During the addition, which requires about 10 minutes, the mixture is cooled to $-5°$. The solution is centrifuged, and the precipitate is extracted three times with 0.033 M PO_4, pH 7.6.

Step 5. Fractionation with Alumina—Alkaline pH. 3 ml. of alumina C_γ (dry weight 25 mg./ml.) is added slowly to the fluid from step 4. The suspension is kept at $0°$ for 15 minutes with occasional stirring. It is then centrifuged in the cold at 15,000 r.p.m., and the precipitate is discarded.

Step 6. Fractionation with Alumina—Acid pH. After the pH of the supernatant is lowered to 5.6 with 2 N acetic acid, it is treated successively with three 15-ml. portions of alumina. Generally the first two batches of gel contain most of the enzyme. In one run, a third treatment with gel was necessary. The enzyme is eluted with three 10-ml. portions of 0.05 M phosphate buffer, pH 7.6.

Properties

Specificity. The purified enzyme (step 6) is specific for TPN. The DPN enzyme is removed completely during alumina C_γ treatment. The activity of the enzyme when tested in Tris buffer or glycylglycine buffer is highest between pH 8.5 to 8.8. At pH 7 activity is about 0.4 that of the optimum, whereas at pH 9.1 activity is about 0.9.

Activators and Inhibitors. For full activity the enzyme has to be preincubated with cysteine or glutathione. The concentration of iodoacetamide to produce 50% inhibition is 2×10^{-3} M; a concentration of 3.3×10^{-3} M inhibits the enzyme completely. The DPN-pea seed enzyme is inhibited 81% by 1×10^{-3} iodoacetamide.[3] Inhibition by iodoacetamide is not instantaneous. Iodoacetamide inactivates the system irreversibly. However, enzyme inactivated owing to storage may be reactivated by cysteine or glutathione.

Certain heavy metals (Cu) inhibit the enzyme. This inhibition is reversable by Versene.

Fluoride, azide, and indoleacetic acid, each at 0.01 M, cause no detectable effect.

In crude extracts, the DPN enzyme is greatly accelerated by arsenate; however, arsenate has no effect in these same extracts when TPN is substituted for DPN. Fractions (step 6) show an activity 0.3 to 0.5 that of the control when arsenate is omitted from the assay procedure.

Reversibility. TPNH can be oxidized in the assay procedure[1] by generating 1,3-diphosphoglyceric acid using PGA, crystalline Bücher enzyme (yeast), and ATP. Without ATP, no reaction occurs, indicating that PGA must be phosphorylated to a diphosphorylated acid. Using fractions of step 6, triose phosphate (HDP and aldolase) yields PGA, the reaction being coupled to lactic acid dehydrogenase.

SUMMARY OF PURIFICATION PROCEDURE

Step	Fraction	Total units	Specific activity, units/mg. protein
1	Leaf extract	45,000	0.7
2	Acid fractionation	41,860	1.4
3	$(NH_4)_2SO_4$ fraction (0.31–0.42)	25,000	9.0
4	Acetone fraction	21,000	25.0
5	Alumina purification—alkaline pH	15,000	34.1
6	Eluate from alumina	500	67.2–115.0

[63] Phosphoglycerate Kinase from Brewer's Yeast

D-1,3-Diphosphoglycerate + ADP ⇌ D-3-Phosphoglycerate + ATP

By THEODOR BÜCHER

Assay Method

Principle. According to the problem at hand, the forward or back reaction may be used in the assay. Both can be followed by means of the compound optical assay coupling with glyceraldehyde phosphate dehydrogenase as coupling enzyme.

ADP   PGA + ATP
  ↖   PGA kinase   ↗
    ↘   Mg⁺⁺   ↗
      ↗ GAP dehydrogenase ↘
GAP + P_i + DPN⁺   DPNH + H⁺

In the back reaction the equilibrium of substrates is less favorable for the assay; cysteine is added to trap the GAP formed. The assay using the forward reaction was used almost exclusively in the course of isolating the enzyme.[1] It has the advantage of being more sensitive and reliable with rather pure enzyme solutions. Its disadvantages are (1) sensitivity

[1] T. Bücher, *Biochim. et Biophys. Acta* 1, 292 (1947).

to interference from certain enzymes (isomerase and glycerophosphate dehydrogenase) and (2) the fact that the equilibrium of the coupling reaction precedes the main reaction (see below).

The assay using the back reaction has the disadvantage of low sensitivity and some interference between requisite magnesium ions and substrate concentrations, making it essential to observe the assay medium exactly in order to obtain a high degree of reproducibility. It has the advantage of being little affected by other enzymes. The reaction velocity in this assay is about four times lower than in the assay with the forward reaction. The reader is referred to a discussion of the fundamentals of procedure and interpretation of the optical assay in connection with glycerophosphate dehydrogenase (also pyruvate kinase and triosephosphate isomerase) (Vol. I [58]).

A. Forward Reaction

Reagents. The composition of the assay mixture is given in Table I.

TABLE I
ASSAY MIXTURE FOR PHOSPHOGLYCERATE KINASE

K-Na phosphate buffer (pH 6.9)	$5 \times 10^{-2}\ M$
DL-3-GAP	$8.3 \times 10^{-4}\ M$
DPN	$4.15 \times 10^{-4}\ M$
ADP	$2.5 \times 10^{-4}\ M$
MgSO₄	$5 \times 10^{-3}\ M$*
GAP dehydrogenase	100 γ/ml. (see Vol. I [60, 61])
Glycine	10 g./1.

Temperature 25°, pH 6.9

* The concentration of MgSO₄ given in the original paper is incorrect as a result of a misprint.

Procedure. The course of an assay is illustrated in Fig. 1. The increase in absorbance upon adding the auxiliary enzyme, due to adjustment of the equilibrium of the oxidative reaction of glycolysis, is about 0.15 (366 mμ, 1-cm. light path). The enzyme to be tested is considerably diluted in heavy-metal-free water. The special peculiarity of the compound optical assay coupling with a preceding auxiliary reaction causes the increase in absorbance after addition of the enzyme to be less than the real reaction rate catalyzed by the enzyme. Details[1] may be seen from the legend of Fig. 1. Under conditions of the assay the enzyme is not completely saturated with substrate (in the sense of the Michaelis-Menten theory). The maximum turnover number is about twice as great.

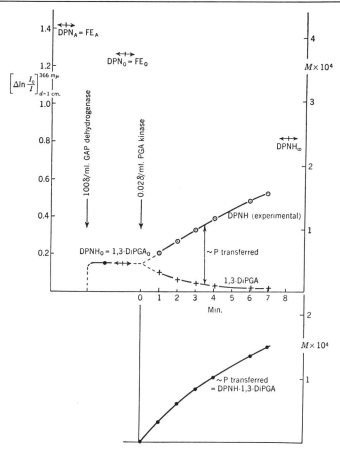

Fig. 1. Assay of phosphate transfer coupled with preceding auxiliary reaction (oxidative phosphorylation of GAP). The concentrations of P_i and H^+ in this experimental arrangement remain approximately constant throughout the assay.

The upper curve of the upper figure (circles) indicates experimentally measured points. The lower curve of the same figure (crosses) indicates the calculated concentration of the intermediate product (1,3-diphosphoglycerate):

$$C_{1,3\text{-DiPGA}} = \frac{C_{\text{DPNH}0}^2}{C_{\text{DPNH}}} \times \frac{C_{\text{DPN}}^2}{C_{\text{DPN}0}^2}$$

The curve of the lower figure indicates the amount of phosphate transferred by the enzyme, calculated from the difference between the curves in the upper figure:

$$C_{\text{phosphate transferred}} = C_{\text{DPNH}} - C_{1,3\text{DiPGA}}$$

B. Back Reaction

Reagents. The composition of the assay mixture is given in Table II.

TABLE II
ASSAY MIXTURE FOR BACK REACTION

D-3-PGA	$5 \times 10^{-3} M$
ATP	$3 \times 10^{-4} M$
MgSO$_4$	$5 \times 10^{-3} M$
DPNH	$2.5 \times 10^{-4} M$
NaHCO$_3$	$1.75 \times 10^{-2} M$
Cysteine (freshly neutralized)	$2 \times 10^{-2} M$
Glycine	8 mg./ml.
GAP dehydrogenase	0.1 mg./ml.

The solution is saturated with a gas mixture containing 10% CO_2.
Temperature 25°, pH 6.9

Definition of Unit and Specific Activity. A unit is defined in the same manner as in the case of α-glycerophosphate dehydrogenase (Vol. I [58]) (and other optical assays described by the author in this volume):

$$U = \frac{100}{\text{Seconds for change in absorbance by } 0.1}$$

It should, however, be remembered that, in contrast with the other assays, a closer relationship exists between the reaction rate and the unit, owing to the peculiarities of the assays arising from coupling with a preceding coupling reaction. One unit in the back reaction assay corresponds to four units of the forward reaction assay. The activity of crystalline enzyme is 55,000 units per gram of protein per liter assay mixture.

Application of Assay Method to Crude Tissue Preparations. The back reaction assay has been used to test tissue extracts and serum.

Purification Procedure

The enzyme has so far been shown to occur in muscle, tumor cells, yeast, and serum. The purification procedure has been worked out for Lebedew juice from washed and air-dried brewer's yeast, strain K, from the Schultheis-Patzenhofer Brewery, Tempelhof. For other yeasts the following may be stated: (1) brewer's yeast from the Holsten Brewery, Hamburg, yields about the same activity in the Lebedew juice as the yeast used for isolation of the enzyme. (2) The following purification procedure was repeated with Lipmann juice[2] from baker's yeast obtained from the North German Yeast Industry, Hamburg. The initial activity in the juice is decisive for the purification, since the activity varies very

---

[2] F. Lipmann, *Compt. rend. trav. lab. Carlsberg* (Sörensen Festschrift) **22**, 317 (1937).

widely for unknown reasons. In some purifications the enzyme was crystallized; in others a noncrystallizable product of almost the same degree of purity was obtained. Beisenherz et al.[3] make some statements about the fate of the activity during ammonium sulfate fractionation of muscle juice.

Step 1. Washing. Dried yeast[4] is washed for 48 hours with running tap water in countercurrent flow (the water rises so slowly that the yeast cells settle at the same rate). When the effluent is clear and colorless, the yeast is placed in a cloth, allowed to drain, and the water pressed out at 50 atmospheres. The mass is divided into small pieces, air-dried for 4 to 6 days at room temperature, then pulverized for 24 hours in a ball mill. The powder is stored at 0°.

Step 2. Acid Fractionation. 1.15 l. of Lebedew juice prepared from 850 g. of dried yeast and 2.55 l. of water (3 hours at 36°) are chilled in an ice-salt bath to $-2°$. With constant stirring, 230 ml. of alcohol and 138 ml. of nucleic acid solution (5 g. of nucleic acid, Merck, and 7.5 ml. of N NaOH in 100 ml.) are added in a fine stream, and the pH is lowered to 5.2 by adding about 40 ml. of 2 N acetic acid. The precipitate is centrifuged in the cold for 20 minutes at 3000 r.p.m., and the supernatant is acidified at $-2°$ with about 42 ml. of 2 N acetic acid to pH 4.8. This precipitates most of the activity. After cold centrifugation for 20 minutes at 3000 r.p.m. the residue is washed once with cold 9% alcohol.

Step 3. Preparation of Dry Powder. The precipitate is dissolved in a cold mixture of 10 ml. of 0.5 M Na_2HPO_4, 2.0 ml. of 2 N NH_4OH, and 3.0 ml. of 0.2 M $Na_4P_2O_7$. The clear solution is frozen in vacuo over silica gel and dried overnight at room temperature. Yield, 11.2 g. Activity, see Table III. The dry powder remains active when stored in a desiccator at room temperature.

Step 4. Heating to 50°. 8 g. of dry powder is dissolved with stirring in a cold mixture of 8 ml. of 0.2 M $Na_4P_2O_7$ and 68 ml. of water. The solution is warmed in a water bath for 20 minutes at 50°, whereupon it coagulates. The voluminous, inactive coagulate is centrifuged off for 10 minutes at 3000 r.p.m., digested for 10 minutes with 73 ml. of cold water, and again centrifuged.

Step 5. Ammonium Sulfate Fractionation. The supernatant and washing are combined (126 ml.). To the stirred solution are added at the same time 1.2 ml. of 2 N ammonia and 38 g. of powdered ammonium sulfate. After the latter has completely dissolved, 35 ml. of 0°-saturated ammonium sulfate solution is gradually added. Centrifugation for 60 minutes

[3] G. Beisenherz, H. J. Boltze, T. Bücher, R. Czök, K.-H. Garbade, E. Meyer-Arendt, and G. Pfleiderer, Z. Naturforsch. 8b, 555 (1953).
[4] O. Warburg and W. Christian, Biochem. Z. 303, 55 (1939).

at 3000 r.p.m. yields 173 ml. of clear solution which is heated for 10 minutes to 63°, cooled, and stirred with 57 ml. of saturated ammonium sulfate solution. After 60 minutes' centrifugation at 3000 r.p.m. the residue is discarded.

Step 6. Filtration. To the clear supernatant 38 g. of powdered ammonium sulfate is added and the precipitate filtered with suction on a Büchner funnel (Schleicher and Schüll paper 602h, diameter 7 cm.). The first turbid 70 ml. of filtrate is passed through the filter again when the latter has become covered. As soon as the precipitate is dry, but before cracks begin to form, the funnel is rinsed with 20 ml. of 0.875 saturated ammonium sulfate solution, and the precipitate is washed with the same volume. The whole step takes 2 hours. The filter cake is made to stick together by folding the filter paper once. The paper is removed with a stainless steel spatula, yielding 3.8 g. of a firm, white mass which can be stored in a wide-necked Erlenmeyer flask overnight in the refrigerator.

Step 7. Heating with Alcohol. 3.8 g. of the filter cake is dissolved in a cold mixture of 10 ml. of water, 5 ml. of 0.2 M $Na_4P_2O_7$, and a drop of 2 N ammonia. To the cold solution 7.5 ml. of 45% alcohol is added gradually; then the mixture is heated for 16 minutes at 32° and centrifuged at high speed.

Step 8. Acid Precipitation at pH 3.5. At $-4°$ the enzyme is precipitated from the clear solution by acidification to pH 3.5 with 1 ml. of 2 N acetic acid and 0.65 ml. of 2 N sulfuric acid, centrifuged in the cold for 7 minutes at 3000 r.p.m., and washed twice with cold 15% alcohol. The residue is dissolved in 4 ml. of water and neutralized to litmus with 0.35 ml. of 2 N ammonia, yielding a protein solution containing little salt, as required for removing nucleic acid.

Step 9. Protamine Precipitation. To 4.6 ml. of enzyme solution is added in portions about 7 ml. of protamine solution (2 g. of protamine sulfate and 3.2 ml. of 0.1 N $NaOH$ in 100 ml.) until the supernatant after a short, cold centrifugation is fairly clear. Addition of protamine is continued until spot tests for both protamine and nucleic acid are negative. 1.2 ml. of 30 vol. % aluminum hydroxide is added to the clear supernatant and centrifuged after a few minutes at 15,000 r.p.m. for 10 minutes.

Step 10. Precipitation and Filtration. 16 ml. of cold protein solution is stirred with slow addition of 4.0 ml. of 0.2 M $Na_4P_2O_7$ and a mixture of 30 ml. of saturated ammonium sulfate solution containing 8.5 g. of solid ammonium sulfate (freed of gas under reduced pressure). The precipitate is filtered (Schleicher and Schüll paper 602h, diameter 5 cm.), yielding 1.5 g. of solid filter cake.

Step 11. Crystallization. 1.5 g. of filter cake is suspended in 4 ml. of the following mixture and placed in the refrigerator overnight: 6 ml. of saturated ammonium sulfate solution, 2 ml. of 0.2 M $Na_4P_2O_7$, 0.17 ml. of 2 N ammonia made to 10 ml. with water. The next morning some amorphous and large amounts of unspecific "whetstone-shaped" crystals are found. On further standing both disappear and instead rectangular plates and rods form. They are centrifuged after 4 days and washed with 4 ml. of the above mixture. The crystals when suspended in that mixture will retain their activity in the refrigerator. To recrystallize, they are dissolved in water and ammonium sulfate is added until turbidity begins to appear. Crystallization begins at once and is completed on standing in the refrigerator. Repeated recrystallization raises neither the activity nor the ratio of extinction coefficients at 280 and 260 mμ.

TABLE III
ASSAY VALUES DURING PURIFICATION

	Material, g.	Units/g. enzyme/ l. assay mixture
Lebedew juice	160	940
Dry powder	11.2	9000
Crystals	0.37	55000

Properties[1]

Specificity. The enzyme will not react (or only to an insignificant extent) with AMP, phosphoenolpyruvic acid, or 2,3-diphosphoglycerate (Greenwald ester).

Activators. The activity of phosphoglycerate kinase depends on the presence of divalent metal ions such as magnesium. Magnesium may be replaced by manganese which is bound more tightly but produces less stimulation at higher concentrations. The Michaelis constant for magnesium, measured by the back reaction assay (5×10^{-3} M PGA, 6×10^{-5} M ATP, bicarbonate-CO_2 buffer, pH 6.9) is 2.5×10^{-4} M. This value was obtained in the absence of phosphate. Whether it is influenced by substrate concentration, and to what extent, has not yet been determined.

Equilibrium Constant. The constant for the substrate reaction at 25°, pH 6.9 (without correcting for binding of substrate by magnesium or for activities), under the conditions of assay mixture shown in Table I is

$$\frac{C_{ADP} \times C_{D-1,3-DiPGA}}{C_{ATP} \times C_{D-3-PGA}} = 3.1 \times 10^{-4}$$

Making the appropriate corrections, Burton and Krebs[5] find 3.2×10^{-4}.

[5] K. Burton and H. A. Krebs, *Biochem. J.* **54**, 94 (1953).

Michaelis Constants for Substrates. The following constants were determined under the assay conditions given above:

TABLE IV
MICHAELIS CONSTANTS FOR PGA KINASE FROM YEAST

ADP	$2 \times 10^{-4}\,M$
ATP	$1.1 \times 10^{-4}\,M$
D-1,3-DiPGA	$1.8 \times 10^{-6}\,M$
D-3-PGA	$2 \times 10^{-4}\,M$

The extent of interdependence of Michaelis constants and concentration of other substrates has been investigated only for ADP and ATP. On the basis of the Michaelis-Menten theory, the results lead to the postulate that the two nucleotides do not compete for the same active site on the apoenzyme. Bücher has based a working hypothesis on this unexpected finding concerning structural relationships at the active site on the enzyme, according to which the active site is composed of at least "three active loci." However, it has recently been emphasized that conclusions from kinetic data to substrate affinities in the sense of the Michaelis-Menten theory may not be made directly.

Turnover Number. With optimal substrate concentrations (25°, pH 6.9) the turnover number for the forward reaction is

$$320,000\ \frac{\text{moles}}{\text{minute} \times 10^5\ \text{g. enzyme protein}}$$

for the back reaction, $36,000\ \dfrac{\text{moles}}{\text{minute} \times 10^5\ \text{g. enzyme protein}}$.

Analytical Data. The enzyme contains 0.312% S (10 g. atoms in 10^5 g. of protein). The absorption spectrum is shown in Table V. The qualitative and quantitative agreement with the spectrum of pyruvate kinase from muscle of various sources (see Vol. I [**66**]) is to be noted.

TABLE V
ABSORPTION SPECTRUM OF PGA KINASE

λ, mμ	ϵ
290	0.30
280	0.50
270	0.44
260	0.31
250	0.24
240	0.80

[64] Phosphoglyceric Acid Mutase

By PETER OESPER

A. 3-Phosphoglyceric Acid Mutase

3-PGA → 2-PGA

Assay Method

Principle. The method of Meyerhof and Schulz,[1] as modified by Sutherland *et al.*,[2] is based on the change in optical rotation when 2-PGA is converted into 3-PGA. In neutral molybdate solution the $[\alpha]_D$ for D-3-PGA is $-745°$, whereas that of D-2-PGA is $-68°$, so that the fraction of 3-PGA in a mixture of 2- and 3-acids is $\dfrac{-[\alpha] - 68}{677}$.

Reagents

D-2-PGA, 0.05 *M*, neutral.
2,3-DiPGA, 2×10^{-4} *M*.
NaF, 1 *M*.
NaHCO$_3$, 2.6%, saturated with N$_2$–5% CO$_2$. The pH is approximately 7.3. Other buffers may be used, except phosphate, which reacts with the molybdate.
TCA, 20%.
Ammonium molybdate, 25%.

Procedure. Place 5 ml. of 2-PGA, 1 ml. of 2,3-DiPGA, 1 ml. of NaF, and 1 ml. of NaHCO$_3$ in a tube at any convenient temperature up to 60°. Add 2 ml. of suitably diluted enzyme. At 3-minute intervals remove 2-ml. aliquots, and precipitate the protein with 1 ml. of 20% TCA. Centrifuge, and neutralize a 1.5-ml. portion with NaOH. Add 0.5 vol. of ammonium molybdate, and determine the rotation immediately in a 2-dm. tube. Determine the total PGA in another aliquot (acid-stable P, 60 minutes, 100° in *N* HCl), and calculate the specific rotation. The fraction, and hence the amount, of 3-PGA formed is calculated from the formula given above.

No units of activity have been defined for this enzyme.

Application of the Assay Method to Various Preparations. Extensively purified enzyme preparations require 2,3-DiPGA as coenzyme[2,3] but do

---

[1] O. Meyerhof and W. Schulz, *Biochem. Z.* **297**, 60 (1938).
[2] E. W. Sutherland, T. Posternak, and C. F. Cori, *J. Biol. Chem.* **181**, 153 (1949).
[3] E. W. Sutherland, T. Posternak, and C. F. Cori, *J. Biol. Chem.* **179**, 501 (1949).

not require NaF, which is used to inhibit partially the enolase accompanying crude fractions. The method is applicable to dialyzed muscle extracts, yeast maceration juice, etc.

Modifications of the Method. (1) Racemic 2-PGA may be used if a correction is made for the rotation of the biologically inactive L-form which remains after the enzyme reaction. $[\alpha] = +68°$.

(2) The rotation may be measured in 1 N HCl instead of in neutral molybdate. The rotation in molybdate is very sensitive to neutral salts; moreover, small amounts of P gradually form molybdenum blue, which disturbs the measurement. On the other hand, the rotations in molybdate are much larger. In N HCl, $[\alpha]_D = -14.5°$ for D-3-PGA, and $+24.3°$ for D-2-PGA. The fraction of 3-PGA is then $\dfrac{-[\alpha] + 24.3}{38.8}$.

(3) The reaction rate can be measured from the side of 3-PGA, which is more readily available than 2-PGA. This entails the disadvantage that the equilibrium point is reached after the formation of only 15% 2-PGA.

Alternative Procedures. The mutase reaction can also be followed spectrophotometrically. In one procedure[2] phosphopyruvate is treated with an excess of enolase; after equilibrium has been established, the mutase is added and the decrease in phosphopyruvate caused by the reaction sequence ph-pyruvate → 2-PGA → 3-PGA is followed by the decrease in absorption at 240 mμ.

Another procedure[2] starts with 3-PGA and an excess of the enzymes leading from 2-PGA to lactate (crude muscle preparations contain mutase plus such an excess of other enzymes). The DPNH oxidized during formation of lactate is measured spectrophotometrically at 340 mμ.

Purification Procedure

The mutase occurs in all glycolyzing tissues. A partial purification is due to Sutherland *et al.*[2]

Step 1. Ground rabbit muscle is extracted with 1.5 vol. of water and again with 1 vol. The extracts are combined and dialyzed 24 hours against cold running tap water.

Step 2. The dialyzed extract is brought to 0.4 saturation with ammonium sulfate, and the precipitate is discarded. The supernatant is brought to 0.5 saturation, and the precipitate taken up in water and dialyzed against cold tap water.

The resultant preparation is nearly free of enolase and 2,3-DiPGA phosphatase. The authors do not say what degree of purification is obtained, but it is certainly minimal.

Properties

Activators and Inhibitors. 2,3-DiPGA acts as coenzyme.[2] The maximal effect is obtained at 2.4×10^{-5} M. Crude preparations which have not been extensively dialyzed usually contain sufficient DiPGA for activation.

No other activators or inhibitors are known. IAA does not inhibit the mutase (7×10^{-4} M).

Equilibrium Constant. The equilibrium constant of the reaction 2-PGA → 3-PGA has been found[4] to be about 6, but this value may be slightly in error; indirect calculations[5] give values as low as 2.6.

Other properties of the enzyme have not been investigated.

B. Diphosphoglyceric Acid Mutase

$$1,3\text{-DiPGA} \rightarrow 2,3\text{-DiPGA}$$

Assay Method

Principle. The spectrophotometric method of Rapoport and Luebering[6] is based on the fact that the mutase disturbs the equilibrium

$$\text{GAP} + \text{DPN}^+ + \text{H}_3\text{PO}_4 \rightleftarrows 1,3\text{-DiPGA} + \text{DPNH} + \text{H}^+$$

by removing 1,3-DiPGA. The resulting change in DPNH concentration is followed by measuring the absorption at 340 mμ. The authors generated GAP from FDP by means of aldolase.

Reagents

Phosphate-cysteine buffer, 0.024 M in phosphate, 0.0036 M in cysteine, pH 7.2 (contains 6.5 g. of Na$_2$HPO$_4$·7H$_2$O and 0.57 g. of cysteine hydrochloride per liter; the pH must be adjusted with HCl).

FDP, 0.15 M, neutral to litmus (about 75 mg./ml. of hexosediphosphate barium).

DPN, 4 mg./ml.

3-PGA, 0.003 M.

Aldolase. Crystalline aldolase,[7] dissolved to a concentration of 2.4 mg./ml.

Glyceraldehyde phosphate dehydrogenase. Crystalline enzyme,[8] dissolved to a concentration of 3 mg./ml.

Procedure. To 2.5 ml. of buffer in a Beckmann spectrophotometer cuvette, add 0.2 ml. of FDP, 0.2 ml. of DPN, and 0.05 ml. of 3-PGA. The

[4] O. Meyerhof and P. Oesper, *J. Biol. Chem.* **179**, 1371 (1949).

[5] O. Warburg and W. Christian, *Biochem. Z.* **310**, 384 (1941).

[6] S. Rapoport and J. Luebering, *J. Biol. Chem.* **196**, 583 (1952).

[7] J. F. Taylor, A. A. Green, and G. T. Cori, *J. Biol. Chem.* **173**, 591 (1948).

[8] G. T. Cori, M. W. Slein, and C. F. Cori, *J. Biol. Chem.* **173**, 605 (1948).

density should not change on addition of 0.05 ml. of aldolase, but the subsequent addition of 0.05 ml. of GAP-dehydrogenase causes the density to increase as the GAP is oxidized. When the equilibrium is established (order of 1 minute), the mutase is added and the rate of further reduction of DPN is observed. When the rate of change of density lies within the range 0.03 per minute to 0.10 per minute, it is approximately proportional to the concentration of the mutase.

No units of activity have been defined for this enzyme.

Application to Crude Tissue Preparations. The assay method described above is not applicable to most crude tissue preparations, which contain, among many other interfering enzymes, triosephosphate isomerase, which would upset the oxidation equilibrium by converting GAP to dihydroxy-acetone phosphate.

Alternative Procedures. The conversion of 1,3-DiPGA to 2,3-DiPGA can be measured directly by stopping the reaction with TCA and determining the decrease in inorganic P, since the Lohmann-Jendrassik procedure[9] measures the 1-P of 1,3-DiPGA as inorganic phosphate, but not the 2-P of 2,3-DiPGA. This method is applicable to crude preparations which have been dialyzed free of coenzymes; appreciable amounts of ADP could interfere by forming ATP + 3-PGA. The absence of 2,3-DiPGA phosphatase[10] should be established by a suitable control experiment.

If 1,3-DiPGA is not available, it may be produced[11] *in situ* from ATP + 3-PGA. Crude mutase preparations contain sufficient Bücher enzyme[12] to produce 1,3-DiPGA faster than it is consumed by the mutase; purified preparations require the addition of one-tenth their volume of 10% suspension of acetone powder of muscle. Since the net reaction in this case is conversion of ATP to 2,3-DiPGA, the decrease in easily hydrolyzable P is measured. Conversion of 3-PGA to phosphopyruvate is prevented by the addition of 0.025 M fluoride.

Purification Procedure

The enzyme has thus far been found to occur only in erythrocytes containing 2,3-DiPGA. It does not occur in beef cells. The following procedure of Rapoport and Luebering[11] effects a limited purification. (The procedure has apparently not been checked by other authors.)

Step 1. Wash rabbit erythrocytes twice with cold 0.9% NaCl. Hemolyze by freezing and adding 1 vol. of H_2O.

[9] K. Lohmann and L. Jendrassik, *Biochem. Z.* **178**, 419 (1926).
[10] S. Rapoport and J. Luebering, *J. Biol. Chem.* **189**, 683 (1951).
[11] S. Rapoport and J. Luebering, *J. Biol. Chem.* **183**, 507 (1950).
[12] T. Bücher, *Biochim. et Biophys. Acta* **1**, 292 (1947).

Step 2. Treat the hemolyzate (pH 6.9) with ⅓ vol. of saturated ammonium sulfate. Centrifuge, and discard the precipitate.

Step 3. Treat the supernatant with sufficient saturated ammonium sulfate to make it ½ saturated. Centrifuge down the precipitate, and collect another fraction precipitating at ⅔ saturation. Dissolve the precipitates in small quantities of water, and dialyze overnight against running distilled water. Discard the small precipitates which form. These two fractions contain two-thirds of the activity of the original hemolyzate. They are combined and submitted to step 4.

Step 4. Bring to pH 5.8 by the addition of 0.1 vol. of 0.2 M acetate buffer. Repeat steps 2 and 3, discarding the fraction precipitated in ⅓ saturated ammonium sulfate, and collecting fractions precipitating at ½ and ⅔ saturation. The authors do not state what degree of purification is obtained, but a great increase in activity per milligram of nitrogen can be inferred from their report that the material obtained at step 3 retains only a little of the original hemoglobin of the cells. The preparation obtained at step 4 was found to contain no 3-phosphoglyceric acid kinase, and much less phosphatase than the initial hemolyzate.

Properties

Effect of pH. The activity of the mutase has a sharp maximum at pH 7.2 and decreases about 50% at pH 6.5 or 8.0. There is almost no activity below pH 6.0.

Activators and Inhibitors. 3-PGA activates the enzyme,[7] probably by acting as a coenzyme. The maximum rate is observed at a concentration of 5×10^{-5} M but the rate is half-maximal even without added 3-PGA.

2,3-DiPGA acts as an inhibitor.[7] The inhibition amounts to 90% at a concentration of 2×10^{-3} M.

[65] Enolase from Brewer's Yeast

$$\text{D}(-)\text{-2-PGA} \rightleftharpoons \text{PEP} + H_2O$$

By THEODOR BÜCHER

Assay Method

The method for the determination of phosphoenolpyruvic acid (PEP) originally developed by Lohmann and Meyerhof[1] makes use of the lability of the phosphate group to acid hydrolysis, oxidation by hypoiodite, or the presence of Hg ions.

[1] K. Lohmann and O. Meyerhof, *Biochem. Z.* **273**, 60 (1934).

The optical test described here was developed by O. Warburg and W. Christian[2] during the isolation and crystallization of the enzyme. It is based on the absorption of light by PEP at 240 mμ, a region in which the extinction of 2-PGA is negligible. The extinction coefficient of PEP in this range depends greatly on the wavelength and, to a lesser degree, on the pH and the ionic medium (magnesium ions). At pH 7.4 and a magnesium ion concentration of 2.7×10^{-3} M, it is ($\log_{10}$, length of light path in centimeters)

$$\epsilon_{PEP}^{240\,m\mu} = 1.7 \times 10^3 \ M$$

Reagents

Buffer. The bicarbonate-carbonate buffer consists of 5×10^{-2} M bicarbonate saturated with a gas mixture of 10 vol.% carbon dioxide in nitrogen. The remaining components are made up as follows:

D,L-2-PGA[2] approximately	3×10^{-3} M
Glycine	2.7×10^{-2} M
Magnesium	2.7×10^{-3} M
pH	7.3–7.4

Enzyme. With pure preparations 5 to 10 γ/ml. is used in the test; with impure preparations, correspondingly larger amounts are needed. The absorption due to the added enzyme protein is negligible when pure enzyme is used, but it must be determined in the assay mixture (in the absence of substrate) and subtracted in the case of impure preparations.

Procedure. The following data refer to a temperature of 20°. A quartz cuvette ($d = 0.5$ cm.) is filled with the assay mixture, and the initial absorption is read (to be subtracted from all subsequent readings). To a small glass spatula is applied exactly 0.02 ml. of enzyme solution which is then stirred into the mixture, and readings are taken at 60-second intervals.

Results and Specific Activity. The reaction occurring after addition of enzyme does not lead to the complete disappearance of added PGA but to an equilibrium corresponding to the equation given at the beginning. The equilibrium constant[2] at 20° and pH 7.34 is

$$K_I = \frac{C_{PEP}}{C_{PGA}} = 1.4$$

The kinetics of the reaction under assay conditions indicate attainment of the equilibrium by two opposing first-order reactions (the forward and

[2] O. Warburg and W. Christian, *Biochem. Z.* **310**, 384 (1941), see also Vol. III [35].

back reactions). Therefore the following time concentration law, derived by Warburg and Christian,[2] applies to the course of the reaction:

$$k = \frac{1}{\Delta t} \ln \frac{C_g - C_{t_1}}{C_g - C_{t_2}} \text{ min.}^{-1}$$

Here k is proportional to the enzyme activity used in the assay (in the derivation by Warburg and Christian the term $(\omega + \rho)$ replaces k). Δt is the time elapsed between two readings in which the concentrations of PEP (C_{t_1}, C_{t_2}) are determined. C_g is the equilibrium concentration of PEP at the end of the reaction. Since the absorptions read in the optical test are proportional to the concentrations of PEP, the extinctions may be substituted for concentrations in evaluating the assay results:

$$k = \frac{1}{\Delta t} \ln \frac{E_g - E_{t_1}}{E_g - E_{t_2}} \text{ min.}^{-1}$$

In practice one can determine the final extinction, E_g, for a given solution of PGA under assay conditions once for several assays. In the enzyme assay, extinctions in two to three 1-minute intervals (E_{t_1}, E_{t_2}) are subtracted from the final reading (E_g). When these differences are plotted on semilogarithmic paper (extinctions logarithmic, time linear) the readings should lie on a straight line intersecting the ordinate at E_g (Fig. 1). The half-time value $t_{1/2}$, obtained graphically for the extinction $E_g/2$, is proportional to the constant k:

$$k = \frac{\ln 2}{t_{1/2}} = \frac{0.69}{t_{1/2}} \text{ min.}^{-1}$$

As mentioned above, k is proportional to the added enzyme activity and may be used as a unit during purification.

The initial velocity in the assay is calculated from k:

$$\left(\frac{dc}{dt}\right)_{t=0} = kC_g = k \frac{E_g}{\epsilon_{PEP}^{240 \, m\mu} \times d}$$

Under assay conditions the enzyme is saturated with substrate (concentration of naturally occurring $(-)$-PGA in mixture used $= 1.5 \times 10^{-3} M$; $K_M^{PGA} = 1.5 \times 10^{-4} M$), and therefore one can obtain the turnover number of the enzyme by dividing the initial velocity by the concentration of enzyme protein. According to the data of Warburg and Christian[2] this yields for the purest enzyme from brewer's yeast:

$$W = \left(\frac{dc}{dt}\right)_{t=0} \frac{\text{l. assay mixture}}{\text{g. enzyme protein}} = 0.99 \text{ mole/min./g. enzyme}$$

Application of Assay Method to Crude Tissue Preparations. With crude tissue extracts the assay may be initiated by the absorption of the extract

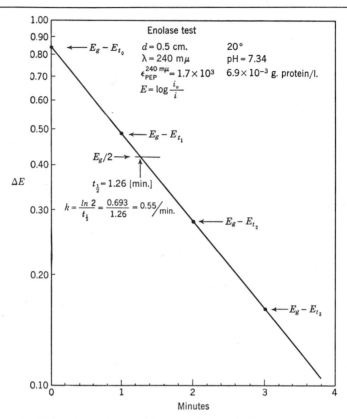

FIG. 1. Semilogarithmic plot of enolase test demonstrating monomolecular kinetics of reaction. See text for derivations.

$$k = \frac{\ln 2}{t_{1/2}} = \frac{0.693}{1.26} = 0.55 \ (\text{min.}^{-1})$$

$$\left(\frac{dc}{dt}\right)_{t=0} = k \frac{E_g}{\epsilon_{PEP}^{240m\mu} \cdot d} = 0.55 \frac{0.84}{1.7 \times 10^3 \times 0.5} = 0.54 \times 10^{-3} \left(\frac{M}{\text{min.}}\right)$$

$$W = \frac{0.54 \times 10^{-3}}{6.9 \times 10^{-3}} = 0.79 \left(\frac{\text{mole}}{\text{g. protein/min.}}\right)$$

itself; especially nucleotides and nucleic acids interfere. Caution is advised in evaluating the assay of impure preparations because of the interference of other enzymes attacking the substrate, such as phosphoglyceromutase and phosphatases.

Purification Procedure

The following procedure conforms exactly with the data of Warburg and Christian. In the hands of the author it has repeatedly proved reliable. Our experience is limited to brewer's yeast.

Step 1. Lebedew Juice. 3.3 kg. of washed and air-dried yeast (compare Vol. I [**63**]) is stirred with 10 l. of water for 2½ hours at 38° and then centrifuged for 3 to 6 hours at 3000 r.p.m. The volume of Lebedew juice is about 5 l. Before further processing it is kept 24 hours at about 0°.

General prefatory note: During the following fractionations with organic solvents, the solution must not become warm. Solutions are kept at 0° and centrifuged cold. All steps up to "Warming to 53°" should follow each other without interruption. Precipitates settled rapidly and cleanly at 3000 r.p.m.

Step 2. Fractionation with Acetone. 5.1 l. of Lebedew juice is stirred rapidly at 0° while 2550 ml. of ice-cold acetone is added. The mixture is centrifuged, the precipitate discarded, and the supernatant again treated with 2550 ml. of acetone at 0°. After centrifugation, the precipitate containing the enzyme is dissolved in cold water. Insoluble material is again centrifuged off, and the supernatant (2400 ml.) contains the activity.

Step 3. First Fractionation with Alcohol. The solution at 0° is brought to pH 4.76 with about 130 ml. of molar acetic acid and then treated with 2000 ml. of ice-cold ethanol. The inactive precipitate is centrifuged, and to the supernatant a further 910 ml. of ice-cold ethanol is added. The precipitate after centrifugation contains most of the activity and is stirred with 1000 ml. of ice-cold water. The inactive insoluble material is centrifuged, and the supernatant (1100 ml.) subjected to a second alcohol fractionation.

Step 4. Second Fractionation with Alcohol. Solution is stirred with 635 ml. of ethanol at 0°. After centrifugation the small precipitate is discarded. A further 310 ml. of alcohol is added, and the enzyme-containing precipitate *centrifuged* and dissolved in 900 ml. of cold water. Suspended material is centrifuged and discarded.

Step 5. Nucleic Acid Precipitation. The solution (900 ml.) at 0° is brought to pH 4.65 with 22 ml. of molar acetate buffer, pH 4.5, and 94 ml. of 0.1 N acetic acid. From the acidified solution the enzyme is precipitated with 120 ml. of the following nucleic acid solution: 1 g. of yeast nucleic acid (Merck), 39 ml. of water, and 1.2 ml. of normal sodium hydroxide to dissolve and to adjust the pH to 4.5. The precipitate of nucleoprotein containing the activity is centrifuged and washed twice (with centrifugation) with 400 ml. of ice-cold 0.2 M acetate buffer, pH 4.5.

Step 6. Removal of Nucleic Acid with Protamine. The washed nucleoprotein precipitate is finely divided in 300 ml. of water and just dissolved by careful addition of about 24 ml. of 0.1 N NaOH. After solution, the reaction should be slightly acid to litmus; otherwise inactivation results and protein is carried down in the precipitation of nucleic acid. To this solution is added about 25 ml. of 5% solution of protamine sulfate that has been neutralized with sodium hydroxide until slightly acid to litmus. During precipitation the pH should be about 4.6. The amount of protamine is determined by spot testing with solutions of nucleic acid and protamine so as to keep the excess of protamine to the minimum. The precipitate is centrifuged (spot tests with trial centrifugations), and the supernatant adjusted to pH 6.2 with 0.1 N sodium hydroxide. It is again centrifuged, and the precipitate is discarded.

Step 7. *Warming to 53°*. To 400 ml. of solution 20 g. of magnesium sulfate ($MgSO_4 \cdot 7H_2O$) is added, and the mixture is kept for 17 hours at 38°. Without centrifuging, the solution is heated for 15 minutes at 53°. The resultant massive precipitate is centrifuged and discarded.

Step 8. *Dry Powder I*. 400 ml. of enzyme solution from the previous step is dialyzed 24 hours at 0° against 20 l. of 10^{-3} N ammonia. The turbid solution is centrifuged, and the precipitate discarded. The supernatant is lyophilized (frozen-dried). The dry material, which contains some salt, weighs about 9 g. and is about 50% pure. This preparation may be used when it is desired to crystallize the enzyme.

Step 9. *Amorphous Dry Preparations*. 9 g. of dry powder I is dissolved in 90 ml. of water. After centrifugation, the supernatant is dialyzed at 0° for 7 days against an aqueous solution of 18 vol.% of ethanol. After this the solution is frozen and dried *in vacuo*. The yield is about 6 g. of 75% pure material containing 10% moisture (removed at 60° under high vacuum). This preparation still contains small amounts of nucleic acid. It is stable for a long time in a desiccator at 0° and is free of mutase.

Step 10. *Crystallization of Enolase as Mercury Complex*. 2 g. of dry powder I is dissolved in 20 ml. of water. The insoluble matter is discarded. The enzyme is precipitated by addition of 60 ml. of saturated ammonium sulfate solution, then centrifuged at high speed. The precipitate containing the activity is stirred with 20 ml. of half-saturated ammonium sulfate solution containing 0.1 N free ammonia, and just dissolved by dropwise addition of a small volume of water. The solution (40 ml.) is stirred 5 minutes at room temperature with 1.2 g. of fibrous alumina (Faserton, Merck). To the supernatant from the alumina is added 0.1 vol. of a mercuric sulfate solution (1.5 g. of $HgSO_4$ in 100 ml. of half-saturated ammonium sulfate solution made weakly alkaline by addition of ammonia). To the clear solution 0.05 vol. of normal ammonia is added, and the whole is stirred a few hours at room temperature. The solution remains clear. A saturated solution of ammonium sulfate is added very slowly with stirring until a slight turbidity appears. This turbidity is made to disappear by very careful and very patient addition of water. Pure needles soon begin to separate, and after a few hours a copious precipitate appears. This can be increased by placing the solution in a closed desiccator with slow evaporation (without vacuum) in the course of 24 hours. In any case the loss of too much ammonia is to be avoided. The crystals are centrifuged and washed twice with 65% saturated (at 0°) ammonium sulfate solution. This removes the last traces of impurities (nucleic acid). The yield from 2 g. of dry powder is about 0.8 g.; the recovery of activity, 80%. The crystalline preparation is not active in the assay in this state, because the mercury complex denatures when it is dissolved in the absence of a high concentration of ammonium sulfate. In the presence of the latter the crystals can be safely dissolved (suspend crystals in half-saturated ammonium sulfate solution containing 0.1 N ammonia, and add water carefully).

Step 11. Recrystallization. Although recrystallization is easy to carry out, it does not increase the purity of the enzyme. The crystals are suspended in half-saturated ammonium sulfate solution containing per liter 0.1 mole of free ammonia and 2 g. of Hg. By careful addition of water the crystals are slowly but completely dissolved and then allowed to crystallize as above.

Step 12. Recovery of Free Enolase from the Mercury Complex. The best procedure for liberating the enzyme from the mercury complex is dialysis against potassium cyanide solution.[3] 30 ml. of a solution of the crystals (as described above) containing 800 mg. of protein is dialyzed first overnight at 0° in a Cellophane bag against a fixed volume of 2000 ml. of half-saturated ammonium sulfate solution containing 0.1 N free ammonia and 0.01 N potassium cyanide.[2] During the day the dialyzate is replaced by a fresh solution of the same composition. This dialysis may be followed by dialyses against 10^{-3} N ammonia in glass-distilled (free of heavy metals) water (48 hours), 18 vol.% ethanol (24 hours), and pure water (as long as desired). Crystalline enzyme preparations suffer no loss of activity after dialysis at +2° for a long duration. After dialysis against 18% ethanol the enzyme can be frozen and dried *in vacuo*.

Properties

Specificity. As far as we know, it has not been tested whether the enolase enzyme can catalyze reactions other than the one cited at the beginning. For general reasons, the reactions may be considered to have considerable specificity.

Activators. The activity of enolase[1,2] depends on the presence of certain *divalent metal ions*. The crystalline protein is inactive in the absence of these ions which may be present in small amounts in the substrate or in the other components of the assay mixture. In low concentrations *zinc and manganese ions* are more effective than magnesium ions,

[3] F, Kubowitz, *Biochem. Z,* **299**, 32 (1938),

Apparently their affinity for the apoenzyme is greater. At higher concentrations, *magnesium* becomes more effective, and it is assumed that magnesium is the physiological activator.

According to data of Warburg and Christian, the K_M for magnesium in bicarbonate buffer at pH 7.34 is 0.61×10^{-3} M; in phosphate at pH 6.74 it is 2.8×10^{-3} M. The variation of K_M under varied conditions is primarily an effect of hydrogen ion concentration.

Inhibitors. Calcium and strontium ions inhibit competitively.[4] The inhibition of fermentation by *fluoride*[5] is primarily due to inhibition of enolase activity.[1] Warburg and Christian[2] have shown that this inhibitor requires the presence of phosphate.[6] In this way it may be possible to offer a physiological explanation for the competitive relation between the presence of substrate and fluoride inhibition in living cells.[7,8] The competitive inhibitor is a magnesium-fluorophosphate complex.[2] Arsenate shows an effect similar to that of phosphate. In the range of fluoride concentration of 10^{-2} to 10^{-4} M the inhibition follows the relation:

$$C_{Mg} \times C_{PO_4} \times C_F^2 \times \frac{\text{Residual activity}}{\text{Inhibited activity}} = 3.2 \times 10^{-12} \ (M^4)$$

Absorption Spectrum. The absorption spectrum of enolase shows no abnormalities. The absorbance ($\log_{10}$, concentration in grams of protein per liter) at 260 mμ is 0.51; at 280 mμ, 0.90.

Molecular Weight. The molecular weight of the enzyme as determined by different methods is summarized in the following table:[9]

Method	Molecular weight
Light scattering	66,000
Hg determination (1 atom Hg)	64,000–68,000
Elementary analysis (8 atoms of S)	67,300
Sedimentation and diffusion[10]	63,700
Enrichment in Mg during preparative ultracentrifugation[2]	52,000

From the agreement of the values it may be concluded that the enzyme contains 8 atoms of sulfur and that in the mercury complex 1 atom of mercury is bound. Presumably the active protein has 1 atom of magnesium attached.

[4] P. Ohlmeyer and R. Duſait, *Naturwissenschaften* **29**, 672 (1941).

[5] J. Effront, *Bull. soc. chim. France* (3) **4**, 337 (1890).

[6] O. Meyerhof and W. Schulz, *Biochem. Z.* **297**, 60 (1938).

[7] J. Runnström and T. Hemberg, *Naturwissenschaften* **25**, 74 (1937).

[8] J. Runnström and E. Sperber, *Biochem. Z.* **298**, 340 (1938).

[9] T. Bücher, *Biochim. et Biophys. Acta* **1**, 467 (1947).

[10] G. Bergold, *Z. Naturforsch.* **1**, 100 (1946).

The interpretation of light-scattering measurements on enolase solutions containing little electrolyte as a splitting of the protein into smaller fragments (Bücher[9]) can no longer be considered valid in view of the newer knowledge on light-scattering effects of proteins in solutions poor in electrolytes.[11]

Density and Refractive Index of enolase in aqueous solution:

$$d = 1.358 \text{ g./ml.}$$
$$n^{546 \, m\mu} = 1.612$$

[11] John T. Edsall, personal communication; cf. P. Doty and R. F. Steiner, *J. Chem. Phys.* **17**, 743 (1949); J. T. Edsall, H. Edelhoch, R. Lontie, and P. R. Morrison, *J. Am. Chem. Soc.* **72**, 4641 (1950).

[66] Pyruvate Kinase from Muscle

(Pyruvate Phosphokinase, Pyruvic Phosphoferase, Phosphopyruvate Transphosphorylase, Phosphate—Transferring Enzyme II, etc.)

Phosphoenolpyruvate + ADP $\rightleftharpoons$ Pyruvate + ATP

By THEODOR BÜCHER *and* G. PFLEIDERER

Assay Method

Principle. The compound optical assay in which the pyruvate kinase reaction is coupled with that of lactic dehydrogenase was first developed by Negelein:[1]

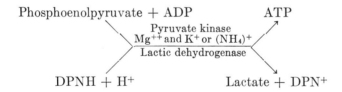

The reader is referred to a discussion of the fundamentals of the optical assay in connection with the isolation of glycerophosphate dehydrogenase (Vol. I [58]). As the pyruvate kinase assay is a compound assay, the coupling enzyme lactic dehydrogenase must be added in excess to the assay mixture before addition of pyruvate kinase.

[1] E. Negelein, unpublished observations; cf. F. Kubowitz and P. Ott, *Biochem. Z.* **317**, 193 (1944).

TABLE I
ASSAY MIXTURE FOR PYRUVATE KINASE

DPNH	$1.5 \times 10^{-4} M$
ADP	$2.3 \times 10^{-4} M$
Phosphoenolpyruvic acid	$7.8 \times 10^{-4} M$
MgSO$_4$	$8.0 \times 10^{-3} M$
KCl	$7.5 \times 10^{-2} M$
Lactic dehydrogenase[2] (18,000 units/mg.)	35×10^{-3} g./l.
$M/20$ Triethanolamine-HCl buffer (pH 7.5)	$5.0 \times 10^{-2} M$

Reagents. The composition of the assay mixture is shown in Table I.

Triethanolamine buffer. The preparation of this buffer solution is described in connection with the enzyme α-glycerophosphate dehydrogenase (Vol. I [58]).

Procedure. The reaction is started by pipetting into the assay mixture a preparation of enzyme, diluted in quartz-distilled water, corresponding to 0.2 γ/ml. of pure enzyme. When the reaction has proceeded for a short while, the time is measured which is required for the absorbance to decrease by 0.1. Enzyme concentration is inversely proportional to the measured time.

Definition of Unit and Specific Activity. A unit is defined in the same way as for glycerophosphate dehydrogenase. The specific activity of purest enzyme (at 25°, 1-cm. light path, wavelength 366 mμ) is 6900 units per gram of protein per liter of assay mixture.[2]

Application of Assay Method to Crude Tissue Preparations. Under conditions of the assay the enzyme is saturated with substrate. If a sufficiently high excess of substrate is used, the assay in crude tissue extracts should be rather insensitive to interference. The assay has yielded good results with crude muscle extract.

Purification Procedure

A. From Rabbit Muscle. Pyruvate kinase was first obtained in crystalline form from rat muscle by Negelein.[1] His protocols and manuscript related to this work were lost in the confusion at the end of World War II. Methods for the purification of the enzyme have been given by Kornberg and Pricer[3] as well as by Kachmar and Boyer.[4] The following directions were worked out as part of a general procedure for obtaining several enzymes from rabbit muscle at the same time.[2]

[2] G. Beisenherz, H. J. Boltze, T. Bücher, R. Czök, K.-H. Garbade, E. Meyer-Arendt, and G. Pfleiderer, *Z. Naturforsch.* **8b**, 555 (1953); see also Vol. I [67].

[3] A. Kornberg and W. E. Pricer, Jr., *J. Biol. Chem.* **193**, 481 (1951).

[4] J. F. Kachmar and P. D. Boyer, *J. Biol. Chem.* **200**, 669 (1953).

For general preliminary remarks and *preparation of crude extract* (*step 1*) the reader is referred to our paper on α-glycerophosphate dehydrogenase (Vol. I [58]).

Step 2. Ammonium Sulfate Precipitation. The supernatant of the ammonium sulfate precipitation (2.4 M, pH 5.8) used for the preparation of α-glycerophosphate dehydrogenase is brought to 2.6 M by addition of 39 g. of ammonium sulfate per liter. Centrifugation at 2200 $\times$ g for 45 minutes yields a residue containing one-third the pyruvate kinase activity of muscle extract (10.1 g. of protein by biuret reaction).

Step 3. "Crystallization." The residue (56 ml.) is dissolved in 100 ml. of 0.02 M acetate buffer, pH 4.62 (made with quartz-distilled water), to give a clear orange solution. To this solution (pH 5.4, 0.82 M in ammonium sulfate, 10.0 g. of protein) are added 21 g. of ammonium sulfate and 2 ml. of 0.5 M acetate buffer, pH 4.62 (pH 5.15, 1.7 M in ammonium sulfate). Overnight 1.65 g. of protein "crystallizes," containing 52% of the initial pyruvate kinase and 12% lactic dehydrogenase activity. The suspension when agitated appears to have an intense "silky luster." No crystals could be detected under the microscope. After centrifugation for 30 minutes at high speed (16,000 $\times$ g), the resulting pellet is suspended in 60 ml. of (unbuffered) 1.95 M ammonium sulfate solution. The supernatant (144 ml.) is brought to 1.95 M by addition of 5 g. of ammonium sulfate. On standing, an additional 1.34 g. of protein "crystallizes," which contains 28% of the initial activity. After centrifuging this is suspended in 53 ml. of ammonium sulfate solution (as above) and combined with the first crystals.

Step 4. "Recrystallization." The combined crystals are dissolved in 70 ml. of quartz-distilled water, and any turbidity is removed by high-speed centrifugation (2.78 g. of protein, pH 5.2). To the solution 20 g. of ammonium sulfate is added (1.65 M, pH 5.1). Immediately after the addition the solution is centrifuged for 5 minutes at high speed, and the residue discarded. The colorless supernatant begins to show the "silky luster" which increases considerably when the solution is allowed to stand for 2 hours at room temperature. The protein is separated from the solution and suspended in 25 ml. of 1.95 M ammonium sulfate solution. A second recrystallization is made under exactly the same conditions. The yield in this last step is 0.44 g. of enzyme protein. Even in these "crystallizations" no crystals are to be seen under the microscope, although the "silky luster" as well as the behavior of the enzyme on washing indicate a nonamorphous state.

B. From Human Muscle. Kubowitz and Ott[5] give a procedure for the isolation of the enzyme from thigh muscle. As the authors originally

[5] F. Kubowitz and P. Ott, *Biochem. Z.* **317**, 193 (1944).

sought to isolate lactic dehydrogenase by the method, the crystallization of pyruvate kinase is due to chance. However, the yield is remarkably favorable (about 10% of initial activity in the extract).

Step 1. Preparation of Crude Extract. Extraction is made with three times the weight of distilled water, pH 6, 0°, for 30 minutes.

Step 2. Acetone Precipitation. 67% of the volume of the extract of ice-cold acetone is added, and the precipitate is centrifuged in the cold and then dissolved in 0.1 M phosphate buffer, pH 7.2. Acetone is removed under reduced pressure. Insoluble material is centrifuged off, and the volume is made up with water to 500 ml.

Step 3. Fractional Precipitation with Nucleic Acid. The solution is brought to pH 5 with acetic acid, and the precipitate is removed by centrifugation. To the supernatant is added 10 ml. of nucleic acid solution (5 g. of nucleic acid, Merck, dissolved in about 7.5 ml. of 1 M NaOH solution and made up to 100 ml. with distilled water; pH about 5), and the precipitate is discarded. 15 ml. of nucleic acid solution is added, the pH is adjusted to 4.7 with 1 M acetate buffer, pH 4, and the precipitate is again discarded. Finally the pH is lowered to 4.6 with 1 M sodium acetate buffer, pH 4. The resulting precipitate (between pH 4.7 and 4.6) contains the activity and is centrifuged for 30 minutes at 0°, then dissolved in water and 8 ml. of 0.5 M phosphate buffer, pH 7.2, and made to 50 ml. with water (pH 6.0).

Step 4. Heating. The solution (pH 6.0), which contains about 60% the activity of the extract and does not keep well, is heated for 3 minutes in a 60° bath (solution, 1 minute at 60°) and cooled immediately. After high-speed centrifugation the clear solution will keep at 0° for a limited time.

Step 5. Removal of Nucleic Acid. Protamine solution (2 g. of protamine sulfate dissolved at room temperature in 3.2 ml. of 0.1 M NaOH solution and water to 100 ml., pH 6.8) is added carefully, using spot tests, to a very slight excess.

Step 6. Acetone Precipitation. The solution is adjusted to pH 6.6 with secondary phosphate solution, and 56.5% of the resulting volume of ice-cold acetone is added. The precipitate is dissolved in 0.1 M phosphate buffer, pH 7.2, to a total volume of 25 ml. Acetone is removed under reduced pressure. Yield about 30% of extract activity.

Step 7. Crystallization. The solution is diluted to 50 ml. and simultaneously acidified to pH 5.1 with M sodium acetate buffer (pH 4.0). Protein is 20 mg./ml.; activity 30% of initial extract. At 0° 55 ml., then very slowly 23 ml., of saturated (0°) ammonium sulfate solution is added. The enzyme crystallizes in cubes. The crystals are washed with a small volume of 0.225 saturated ice-cold ammonium sulfate solution.

Step 8. Recrystallization. The crystals are dissolved in $M/30$ phosphate buffer, the pH adjusted to 5.5, and ammonium sulfate added very slowly to 0.6 saturation.

Properties

Specificity. The preparation made from rabbit muscle according to the above directions contains the following amounts of impurities[2] (referred to activity in optical assay); aldolase 0.05%, lactic dehydrogenase 0.2%, glycerophosphate dehydrogenase 0.09%. The ability of the enzyme preparation to catalyze the reaction of phosphoglycerate kinase[5] or to react with AMP instead of ADP is below the level of the above impurities. Preparations showing the latter property contain contaminations by myokinase.

Activators and Inhibitors. The activity of pyruvate kinase is linked to the presence of divalent metals such as magnesium.[6] Calcium ions inhibit.[7,8] The extent of the inhibition depends on the potassium concentration;[4] it has not been investigated whether it also depends on the magnesium concentration or that of single substrates (for Mg^{++}–Ca^{++} antagonism, see Greville and Lehmann[9]). At a K^+ concentration of $5 \times K_M$, 2×10^3 M Ca^{++} produces a 50% inhibition.[4] In addition to divalent ions, the muscle enzyme requires monovalent ions like K^+, NH_4^+, and Rb^+,[7,8,10] although the yeast enzyme is reported not to have such requirements.[11] The Michaelis constants for the three different alkali ions are almost identical. Concentrations above 0.15 M lower the enzyme activity; therefore an activation maximum lies within this range. Sodium or lithium ions inhibit. In the absence of potassium, sodium is reported to have a small but significant stimulatory effect.[4] At a potassium concentration of $5 \times K_M$, 10^{-1} M sodium ions inhibit about 20%, the same concentration of lithium ions about 40%.[4] The Michaelis constants of phosphoenolpyruvate and potassium have little effect on each other;[4] it is assumed that the two substances are attached to the protein at different loci.

Equilibrium Constant. The constant for the substrate reaction[1] at 30° is[12]

$$K = \frac{C_{\text{pyruvate}} \times C_{\text{ATP}}}{C_{\text{phosphoenolpyruvate}} \times C_{\text{ADP}}} = 2 \times 10^3$$

[6] K. Lohmann and O. Meyerhof, *Biochem. Z.* **273**, 60 (1934).

[7] P. D. Boyer, H. A. Lardy, and P. H. Phillips, *J. Biol. Chem.* **146**, 673 (1942).

[8] P. D. Boyer, H. A. Lardy, and P. H. Phillips, *J. Biol. Chem.* **149**, 529 (1943).

[9] G. D. Greville and H. Lehmann, *Nature* **152**, 81 (1943).

[10] H. A. Lardy and J. Ziegler, *J. Biol. Chem.* **159**, 343 (1945).

[11] J. Muntz and J. Hegler, *J. Biol. Chem.* **171**, 653 (1947).

[12] O. Meyerhof and P. Oesper, *J. Biol. Chem.* **179**, 1371 (1949).

With the aid of the heat of reaction (-3.5 kcal.) a value of K at 20° can be calculated from that at 30°: $K = 1.65 \times 10^3$, corresponding to a change in free energy $-\Delta F° = 4.3$ kcal. (With respect to pH for these values, the only indication is that bicarbonate buffer was used.)

Michaelis Constants. For phosphoenolpyruvate, $K_M = 8.6 \times 10^{-5}\ M$; for potassium ($NH_4^+$, Rb^+) ions, $K_M = 1.1 \times 10^{-2}\ M$.

Colowick[13] assumed that phosphoenolpyruvate is bound very firmly; the quotient for the initial velocities of the forward and back reactions is 500, so that it is possible that the situation with respect to pyruvate kinase is similar to that with phosphoglycerate kinase. As in the case of the latter enzyme,[14] kinetic data indicate no competition between ATP and ADP for one and the same site on the enzyme.[1]

TABLE II

ABSORPTION SPECTRUM OF PYRUVATE KINASE FROM VARIOUS SOURCES

$$\epsilon \left[\frac{cm.^2}{mg.\ protein\ (dry\ weight)} \right]$$

λ, mμ	Rabbit[2]	Man[5]	Rat[1]
300	0.10	0.05	
290	0.37	0.32	
280	0.54	0.54	0.55
270	0.46	0.47	
260	0.31	0.32	
250	0.19	0.22	
240	0.48	0.71	

Absorption Spectrum. The absorption spectrum of pyruvate kinase is given in Table II. The extinction coefficients are lower than with the average of proteins. Besides the agreement in the low values for pyruvate kinases from rather different sources, it is to be noted that the pyruvate kinases show an absorption spectrum (as does phosphoglycerate kinase[2,14]) which is thoroughly representative of the "tyrosine type."[2]

Molecular Weight. The following unpublished measurements were made by Bücher in 1942, using the method of light scattering,[15] on Negelein's enzyme from rat muscle: $M = 166,000$ g./mole; density of dissolved protein, $d = 1.341$ g./ml.; refractive index (H_2O solution) $n = 1.606$ ($\lambda = 546$ mμ).

[13] S. P. Colowick *in* "The Enzymes" (J. B. Sumner and K. Myrbäck, eds.), Vol. II, Part 1, Academic Press, New York, 1951.

[14] T. Bücher, *Biochim. et Biophys. Acta* **1**, 292 (1947).

[15] T. Bücher, *Angew. Chem.* **56**, 328 (1943); *Biochim. et Biophys. Acta* **1**, 467 (1947).

[67] Lactic Dehydrogenase of Muscle

Pyruvate + DPNH $\rightleftarrows$ Lactate + DPN

By ARTHUR KORNBERG

Assay Method

Principle. The oxidation of DPNH is observed at 340 mμ and in the absence of interfering reactions is a direct measure of the reduction of pyruvate to lactate. This assay method was introduced by Kubowitz and Ott.[1] Since the equilibrium of the reaction greatly favors pyruvate reduction, the reaction is most conveniently measured in this way. From the known extinction coefficients of the pyridine nucleotides,[2] the extent of reaction can be calculated.

Reagents

Na pyruvate (0.01 M).
DPNH (0.002 M, kept at slightly alkaline pH).
KH_2PO_4–K_2HPO_4 buffer (0.1 M, pH 7.4).
NaCl (0.01 M).
Enzyme. Make dilutions in cold NaCl solution.

Procedure. In a cuvette of 1-cm. light path and 3-ml. capacity, place 0.1 ml. of pyruvate, 0.1 ml. of DPNH, 1.0 ml. of phosphate buffer, enzyme solution, and water to make a volume of 3.0 ml. Determine the optical density values at 30-second intervals for 3 minutes. Density changes of 0.020 to 0.050 per minute led to the most reliable values.

Definition of Unit and Specific Activity. One unit of enzyme is defined as that amount which causes an initial rate of oxidation of 1 micromole of DPNH per minute. Specific activity is expressed as units per milligram of protein. Protein is determined by the method of Lowry *et al.*[3] or by a nephelometric method.[4]

Purification Procedure

This method[5] is recommended for the relative simplicity of the procedure (especially when the starting material is made available as a

[1] F. Kubowitz and P. Ott, *Biochem. Z.* **314**, 94 (1943).

[2] B. L. Horecker and A. Kornberg, *J. Biol. Chem.* **175**, 385 (1948).

[3] O. H. Lowry, N. J. Rosebrough, A. L. Farr, and R. J. Randall, *J. Biol. Chem.* **193**, 265 (1951).

[4] T. Bücher, *Biochim. et Biophys. Acta* **1**, 292 (1947).

[5] A. Kornberg and W. E. Pricer, Jr., *J. Biol. Chem.* **193**, 481 (1951).

by-product in the preparation of crystalline aldolase[6] and triosephosphate dehydrogenase[7] from rabbit muscle) and the high yield of enzyme. The over-all yield was approximately 50% of the enzyme determined in the starting material, and the purity, based on the turnover number cited by Kubowitz and Ott[1] for the crystalline enzyme from rat muscle, was estimated to be about 80%. Lactic dehydrogenase (once recrystallized) prepared from beef heart, according to Straub's procedure,[8] had a specific activity estimated at only about 50% as great as that of the Kubowitz and Ott preparation, and the yield referred to fresh weight of muscle was only one-tenth as great as that obtained from rabbit muscle by our procedure. A modification of Straub's procedure has been reported by Meister.[9] Racker has reported the crystallization of lactic dehydrogenase from rabbit muscle.[10] It should be emphasized that the product obtained by our procedure is not pure.[11] Therefore, in those instances where the presence of an impurity constitutes a nuisance, the use of crystallization or some other purification procedure should be undertaken.

Step 1. Preparation of 0.52 to 0.72 Paste. Rabbit muscle extract was prepared and fractionated with ammonium sulfate between saturations of 0.52 and 0.72, according to Cori *et al.*,[7] with or without prior crystallization of aldolase. The precipitate may be stored as a paste at 3° for many months. 57.5 g. of the 0.52 to 0.72 fraction, derived from 500 g. of muscle, was dissolved in cold water, and the small amount of insoluble material was removed by filtration. The volume of the filtrate was 92 ml.

Step 2. Ammonium Sulfate Refractionations. Several (usually three) successive ammonium sulfate fractions were carried out. The protocol that follows should serve merely as a guide.

This procedure depends on the fact that lactic dehydrogenase is among the least soluble proteins in this fraction, and the success of the procedure relies on collecting the very earliest precipitate with care to avoid excessive loss of enzyme. Precipitation is often induced by a brief warming of the cold solution, lactic dehydrogenase being more insoluble at 12° than at 3°.

(a) To 92 ml. of the dissolved 52 to 72% paste was added 39 ml. of ammonium sulfate (saturated at 2° and kept cold). The addition was made slowly with mechanical stirring, and the temperature was maintained at 0 to 2°. After exposure to room temperature for 15 minutes, the

---

[6] J. F. Taylor, A. A. Green, and G. T. Cori, *J. Biol. Chem.* **173**, 591 (1948).

[7] G. T. Cori, M. W. Slein, and C. F. Cori, *J. Biol. Chem.* **173**, 605 (1948).

[8] F. B. Straub, *Biochem. J.* **34**, 483 (1940).

[9] A. Meister, *Biochem. Preparations* **2**, 18 (1952).

[10] E. Racker, *J. Biol. Chem.* **196**, 347 (1951).

[11] Pyruvate phosphokinase contaminates many of the preparations.

temperature of the solution rose to 12°, and the resulting precipitate was collected by centrifugation and dissolved in water to a volume of 16.5 ml.

(b) To 15 ml. of this fraction, 5.0 ml. of saturated ammonium sulfate was added dropwise, and the slightly turbid solution was exposed to room temperature for 30 minutes (during which time the temperature of the solution rose to 18°). The resulting precipitate was centrifuged and dissolved in water to a volume of 15 ml.

(c) The preceding fractionation was then repeated, yielding the final fraction in a volume of 7.5 ml. This fraction is stable for months at 3° or −15° but is unstable in high dilutions.[12]

SUMMARY OF PURIFICATION PROCEDURE

Fraction	Total volume, ml.	Units/ml.	Total	Protein, mg./ml.	Specific activity, units/mg.	Recovery, %
1. 52–72 % paste	92	725	6660	60.3	12	—
2. Ammonium sulfate refractionation						
(a)	16.5	3430	5660	63.2	54	85
(b)	15	2950	4430	25.9	114	73
(c)	7.5	4180	3130	19.0	220	52

Properties

TPNH can be substituted for DPNH but reacts at less than 1% of the rate.[13] Meister[9] reports that the crystalline enzyme reduced α-keto-butyrate, α-ketovalerate, α-ketocaproate, and phenylpyruvate, but acts on oxalacetate, α-ketoglutarate, and β-keto acids at negligible rates or not at all.

---

[12] A new method for the crystallization of the enzyme has been reported by Bücher et al. [G. Beisenherz, H. J. Boltze, T. Bücher, R. Czók, K.-H. Garbade, E. Meyer-Arendt, and G. Pfleiderer, Z. Naturforsch. **8b**, 555 (1953)].

[13] A. H. Mehler, A. Kornberg, S. Grisolia, and S. Ochoa, J. Biol. Chem. **174**, 961 (1948).

[68] Lactic Dehydrogenase from Yeast

$$\text{L-Lactate} + A = \text{Pyruvate} + AH_2$$
(where A is a suitable hydrogen acceptor)

By MALCOLM DIXON

Assay Method

Principle. This enzyme differs from the lactic dehydrogenase of muscle in being independent of nucleotide coenzymes and also in requiring no fixative (e.g., HCN) for the pyruvate formed, as this does not inhibit the reaction. This greatly simplifies the test. Methylene blue is the most suitable acceptor. The procedure is that of Bach *et al.*[1]

Reagents

0.0005 M methylene blue stock solution.

10% sodium DL-lactate.

0.1 M acetate buffer, pH 5.2.

Enzyme. Dilute to give a suitable activity; about 1000 units/ml. is convenient, but wide variations are permissible.

Procedure. Make up a stock test solution by mixing 50 ml. of the methylene blue, 50 ml. of the lactate, and 200 ml. of the acetate buffer. This will keep for some time. Pipet 3.0 ml. of test solution into a Thunberg tube, and measure accurately a suitable quantity (0.1 or 0.5 ml.) of enzyme solution into the hollow stopper. Evacuate the tube, fill with O_2-free N_2, and re-evacuate.[2] Immerse the whole tube in the water bath at 37°, and after 2 minutes mix the contents. Measure the decoloration time with a stop watch from the moment of mixing.

For best results the amount of enzyme taken should give a reduction time of between 10 seconds and 1 minute. Duplicates should agree within 3%, and the reciprocal of the reduction time should be exactly proportional to the amount of enzyme. Except with crude extracts, no reduction occurs in the absence of lactate. When it occurs, it must be allowed for.

Definition of Unit and Specific Activity. The specific activity is measured by the Q_{MB}, which denotes the amount of MB (in microliters) reduced per hour per milligram of protein. One micromole of MB is taken as being equivalent to 22.4 μl. Since 5.6 μl. of MB is used in the test,

$$Q_{MB} = \frac{60 \times 5.6}{t_R \times w}$$

[1] S. J. Bach, M. Dixon, and L. G. Zerfas, *Biochem. J.* **40**, 229 (1946).

[2] D. E. Green and M. Dixon, *Biochem. J.* **28**, 237 (1934).

where t_R is the reduction time in *minutes* given by w mg. of protein. Protein was determined by Kjeldahl estimations, after precipitating with trichloroacetic acid and washing free from ammonium salts with dilute trichloroacetic acid.

The unit of enzyme is defined as the amount of enzyme present in 1 mg. of a preparation with a Q_{MB} of 1. Then Q_{MB} = units per milligram.

Alternative Methods. Instead of methylene blue, either cytochrome c or ferricyanide may be used and their reduction followed in the Beckman spectrophotometer.[3] The reaction with cytochrome c is carried out at pH 7 and with ferricyanide at pH 8.

Purification Procedure

The only published method is that of Bach *et al.*[1] Very recently the enzyme has been obtained pure and crystalline by Appleby and Morton,[3] but the method is not available at the time of writing, although it is known to include a butanol treatment at an early stage.

The method of Bach *et al.* is based on the following facts. (1) The enzyme is liberated from the cells during the early stages of autolysis of dried baker's yeast. (2) The substrate, i.e., lactate, has a specific stabilizing action on the enzyme, so that in its presence the solution can be heated to a temperature high enough to coagulate much of the other protein. (3) The enzyme is adsorbed by calcium phosphate but is not eluted by slightly alkaline phosphate solutions. Large quantities of other material are thus eluted, however, and the enzyme can afterwards be eluted by ammonium sulfate. (4) In the later stages, when the enzyme is concentrated, an absorption band at 556.5 mμ becomes visible on the addition of lactate. The strength of this band thereafter runs parallel with the enzyme activity, and this enables the later fractionations to be carried out rapidly, using a spectroscope to follow the enzyme.

The success of the method depends on the use of a suitable yeast. The enzyme could not be demonstrated in a number of brewer's yeasts, and some baker's yeasts do not autolyze satisfactorily, although they contain the enzyme. It is advisable to carry out preliminary small-scale tests on different yeast samples to obtain the best results. Delft baker's yeast was found to be very satisfactory.

As in all enzyme preparations, the fractionations must be controlled by activity tests, and it is possible that the quantities given below may be found to require some modification with a different yeast.

Step 1. Drying the Yeast. The fresh yeast cake is passed through a "potato-masher" or mincer with holes about 2 mm. in diameter, to reduce it to a vermicelli-like form which will dry quickly and uniformly.

[3] C. A. Appleby and R. K. Morton, *Nature* **173,** 749 (1954).

It is then spread out and air-dried at room temperature. The dried yeast retains its activity for many weeks, but not indefinitely.

Step 2. Autolysis. 3 kg. of dried yeast is dispersed in 6 l. of warm tap water and kept at 37° for $4\frac{1}{2}$ hours (or until the Q_{MB} of the extract reaches a maximum). 4.5 l. of cold tap water is then added, and the mixture is stirred and allowed to stand 15 minutes at room temperature. It is centrifuged for 20 minutes at 3000 r.p.m., and the supernatant Lebedew juice is sucked off (6250 ml.). The Q_{MB} in different preparations varied between 30 and 60. The juice is colored reddish brown with cytochrome c, flavoproteins, and other yellow pigments.

Step 3. Treatment with Alumina. To 6250 ml. of Lebedew juice about 1 l. of alumina Cγ suspension (containing 15 mg. of dry weight per milliliter) is added in portions, until about 15% of the enzyme is adsorbed, keeping the pH at 5 to 6. The alumina (deep red by transmitted light with cytochrome c) is centrifuged off and discarded. The solution is pure yellow and shows no absorption band of cytochrome c.

Step 4. Heating. To 7 l. of solution 140 ml. of 50% sodium lactate is added. The mixture is heated rapidly (in 500-ml. portions) to 53° and held at this temperature for 7 minutes. The bulky precipitate of coagulated protein is centrifuged off. (Without lactate the enzyme is detroyed by heating to 40°, and only a little protein is coagulated at this temperature.)

Step 5. Adsorption on Calcium Phosphate. To 7 l. of solution 3.5 l. of calcium phosphate gel suspension (40 mg. dry weight per milliliter, prepared according to Keilin and Hartree[4]) is added. After centrifuging, the supernatant contains very little enzyme but much protein. The calcium phosphate is eluted five times in succession, each time with 2.5 to 3 l. of $M/15$ phosphate buffer, pH 7.2. The eluates contain no enzyme but much protein and practically all the yellow pigments. The enzyme is then eluted from the calcium phosphate by three successive extractions with pH 7.2 phosphate buffer containing 10% of ammonium sulfate, 2.8 l. being used in all. The eluate is water-clear and can be kept cold overnight without loss of activity.

Step 6. Concentration. The 2.8 l. of eluate is then concentrated to 130 ml. on a battery of six Bechhold ultrafilters fitted with an automatic feed and with soft brushes lightly sweeping the membranes in order to increase the rate of filtration. With "3%" Schleicher collodion ultrafilter membranes, 100 ml. can be filtered per hour by each filter. The process requires about 5 hours, and some loss of activity occurs. Instability first becomes noticeable at this point and thereafter increases in proportion with the Q_{MB}.

[4] D. Keilin and E. F. Hartree, *Proc. Roy. Soc.* (*London*) **B124**, 397 (1938); see also Vol. I [11].

Step 7. Ammonium Sulfate Fractionation at pH 7.2. To the solution (130 ml., now colored pink, mainly with cytochrome c) 31 g. of solid AR ammonium sulfate is added (0.39 saturation), and after a few minutes the resulting precipitate is centrifuged off and discarded. To the supernatant a further 18.5 g. of ammonium sulfate (0.6 saturation) is added, and the precipitate containing the enzyme is centrifuged down and dissolved in water (volume now 57 ml.). The clear red solution shows the spectrum of cytochrome c, but on addition of a small amount of lactate the band at 556.5 mμ also appears. This band is due to the cytochrome b_2 of Bach *et al.;*[1] they suggested that this cytochrome might be identical with the lactic dehydrogenase, and this has now been shown by Morton.[3]

Step 8. Ammonium Sulfate Fractionation at pH 4.8. The cytochrome c is completely eliminated, and the enzyme is further purified as follows. 53 ml. of the solution is mixed with 26 ml. of $M/5$ acetate buffer, pH 4.8, and 53 ml. of water. 36.8 g. of ammonium sulfate are added (0.45 saturation), and the precipitate is centrifuged off and discarded. A further 14.0 g. of ammonium sulfate is added (0.6 saturation), and the precipitate is centrifuged off and dissolved in 24 ml. of water. The solution is perfectly clear, deep red, free from cytochrome c, and very active. It shows the pure spectrum of cytochrome b_2, the α-band giving an optical density of about 0.6 in a 1-cm. cell.

Having reached a Q_{MB} of 2500, the enzyme has become so unstable that further purification is difficult by this method, although in one case a small amount of Q_{MB} 3000 was obtained. The recent results of Morton[3] on the pure enzyme show that at this point the purity was about 7%.

SUMMARY OF PURIFICATION PROCEDURE

Fraction	Total volume, ml.	T_R,[a] sec.	Units/ml., thousands	Total units, thousands	Protein, mg./ml.	Specific activity, units/mg., = Q_{MB}	Recovery, %
1. Lebedew juice	6250	14	1.44	9000	36	40	100
2. After alumina	7000	17	1.18	8260	—	—	92
3. After heating	7000	19	1.06	7420	—	—	82.5
4. Eluate from Ca phosphate	2800	16	1.26	3530	—	—	39
5. Ultrafilter concentrate	130	1.25	16.13	2100	—	—	23
6. (NH$_4$)$_2$SO$_4$ fraction pH 7.2	57	0.9	22.40	1280	14.45	1550	14
7. (NH$_4$)$_2$SO$_4$ fraction pH 4.8	24	0.8	25.20	610	10.1	2500	6.8

[a] T_R denotes the reduction time in *seconds* corresponding to *1 ml.* of enzyme solution.

Properties

pH Optimum. With methylene blue there is a well-marked optimum at pH 5.2.[1] The curve is unsymmetrical, owing to destruction of the enzyme below pH 4. With ferricyanide, on the other hand, the optimum is at pH 8.0.[3]

Specific Activity. At 38° the Q_{MB} of the pure enzyme is 47,000, which corresponds to a reduction of 2100 micromoles of methylene blue per hour per milligram dry weight,[3] at the optimum pH (5.2). At 20° the figures are, for methylene blue at pH 5.2, 740 micromoles; for cytochrome c at pH 7.0, 1800 micromoles, and for ferricyanide at pH 8.0, 7000 micromoles.[3] In comparing these figures it must be remembered that 1 mole of methylene blue corresponds to 2 moles of cytochrome c or ferricyanide.

Coenzymes. The yeast lactic dehydrogenase, unlike that of muscle, does not involve any dissociable coenzyme. The usual coenzymes are completely removed in the early stages of the purification, and their addition does not increase the activity.

Prosthetic Groups. Bach *et al.*[1] found that preparations of the enzyme always contained a new soluble cytochrome, which they named cytochrome b_2. Its prosthetic group was identified as protohem. It shows, when reduced by lactate, a typical cytochrome spectrum, with α-, β-, and γ-bands at 556.5, 528.0 and 424.0 mμ, respectively. Bach *et al.* found that the intensity of the spectrum was proportional to the enzyme activity in different preparations and concluded that cytochrome b_2 was an essential component of the system. Morton[3] has now shown that the dehydrogenase is identical with cytochrome b_2, and the red crystals of the pure enzyme show a strong b_2 spectrum.

The best fractions of Bach *et al.* still contained flavoprotein, but they did not regard this as forming part of the system. Morton, however, has shown that in addition to the b_2 hem the enzyme contains a second prosthetic group, namely flavin mononucleotide. This is liberated on denaturation and was identified chromatographically. The enzyme contains one flavin group for each hem group.

The enzyme is therefore both a flavoprotein and a cytochrome and may be described as a "flavocytochrome." The ratio of flavin to hem remains unaltered by recrystallization, and all attempts to resolve the crystalline enzyme (which is electrophoretically homogeneous) into two different proteins have failed. Both the flavin and the hem groups are instantly reduced on the addition of lactate.

Instability. The instability of the purified enzyme noted by Bach *et al.* has been shown[3] to be due to a progressive loss of flavin mononucleotide from the enzyme. Freshly prepared solutions of the crystals are non-fluorescent, showing that the flavin is firmly bound to the protein. On standing, a fluorescence due to the liberation of free flavin develops, and this is parallelled by the decline in activity.

When the activity has fallen to zero the b_2 hem group is no longer reduced by lactate, but in other respects it appears to be unchanged. It is still reduced by dithionite, giving its normal spectrum. This suggests that the reduction of the b_2 group by lactate takes place through the flavin group. The reaction would then take place as follows,

$$(\text{Lactate} \rightarrow \text{flavin} \rightarrow \text{cytochrome } b_2) \rightarrow \text{Cytochrome c}$$

where the arrows represent reduction reactions and the components within the parentheses are combined with the enzyme protein.

Turnover Number. Observations on the partially purified enzyme[1] gave a turnover number for cytochrome b_2 of 2900 per minute at 37°.

[69] Lactic Dehydrogenase of Heart Muscle

$$\text{L}(+)\text{-Lactate} + \text{DPN} \leftrightarrow \text{Pyruvate} + \text{DPNH} + \text{H}^+$$

By J. B. NEILANDS

Assay Method

Principle. The most convenient assay for lactic dehydrogenase is a spectrophotometric measurement of the rate of appearance of the band at 340 mμ in DPNH. The assay is carried out at pH 10, since the reaction produces 1 equivalent of acid. The amount of enzyme needed for the test is of the order of 0.1 γ, and hence the method may be applied directly to crude extracts without danger of nonspecific reduction of DPN by contaminating substrates in the enzyme solution.

Reagents

0.5 M sodium DL-lactate. Dilute commercial 85% lactic acid with an equal volume of water. To 10 ml. of the diluted solution add 5 N NaOH in 2.0-ml. portions until the solution tests alkaline to an outside indicator (phenolphthalein). Heat the solution to 80° in order to hydrolyze inner ester. The cautious addition of NaOH followed by heating is continued until the solution remains neutral. Dilute to 94 ml.

2×10^{-2} M DPN. Dissolve 133 mg. of DPN[1] in 5 ml. of water, and
add N NaOH carefully from a 0.1-ml. pipet until the pH is about
6.0. Dilute the solution to 10 ml. and store at 5°.

0.1 M glycine–NaOH buffer, pH 10.0.

0.1 M phosphate buffer, pH 7.0.

Enzyme. Dilute the stock enzyme solution with 0.1 M phosphate
buffer, pH 7.0, to obtain about 0.05 millimicromole/ml.

Procedure. The activity measurements are carried out with a modified
Beckman DUR quartz spectrophotometer equipped with a Minneapolis-
Honeywell Model 153X17V-X-9 strip chart recorder with an optical
density range of 0.0000 to 0.0458.[2] A thermospacer arrangement is used
to maintain the temperature of the cell compartment at 25°.

Exactly 1.8 ml. of glycine buffer, 0.1 ml. of lactate solution, and 0.1 ml.
of DPN solution are placed in a 1-cm. quartz cell. The mixture is stirred
with a smooth glass rod, and the cell is placed in the holder in the instrument. The cell is positioned so that the light beam passes through the
solution.

The enzyme solution, 0.02 ml., is pipetted into an excavation on the
end of a glass rod and rapidly stirred into the reaction mixture. The
shutter is opened, and the optical density change at 340 mμ, ΔE_{340}, is
recorded as a function of time.

Definition of Turnover Number. The turnover number is defined as the
moles of DPNH formed per 135,000 g. of protein per minute at 25°. The
activity measurements are derived from the elapsed time for a ΔE_{340} of
0.0300, i.e., over the optical density range 0.0100 to 0.0400. Assuming a
molecular extinction coefficient of 6.2×10^3 for DPNH,[3] the ΔE_{340} of
0.0300 corresponds to the formation of 9.7 millimicromoles of DPNH in
the 2-ml. test volume. Protein is determined by Kjeldahl nitrogen $\times$ 6.25.

Isolation Procedure. The original method of Straub[4] is simple, and,
in our hands, it has never failed to yield a crystalline preparation. The
following procedure for eight beef hearts is adopted from that of Straub,
which is for four hearts. All steps, unless otherwise stated, were performed
at 5° using 40-l. stainless steel stock pots as containers.

Step 1. Extraction. Eight fresh beef hearts are freed from fat and
passed through a commercial meat grinder. The mince, 7.7 kg., is gently
stirred for 15 minutes with 30 l. of ice-cold distilled water. The muscle is
removed by straining through a double layer of cheesecloth.

[1] Pabst DPN may be assumed to be 100% pure as a first approximation.

[2] The manually operated Beckman DU spectrophotometer may also be used, although less accurate results will be obtained.

[3] B. L. Horecker and A. Kornberg. *J. Biol. Chem.* **175**, 385 (1948).

[4] F. B. Straub, *Biochem. J.* **34**, 483 (1940).

Step 2. Adsorption and Elution. The extract is stirred for 5 minutes with 6 l. of calcium phosphate gel[5] and then allowed to settle for 30 minutes. About 20 l. of clear supernatant fluid can now be siphoned off and discarded. The remaining fluid, containing the gel, is centrifuged for 10 minutes at 2000 × *g* in a refrigerated 13-l. International serum centrifuge. The gel is thoroughly stirred with 4 l. of 0.2 *M* phosphate buffer, pH 7.2, and centrifuged. The supernatant is saved, and the gel again extracted with 3 l. of phosphate buffer. The gel is centrifuged and discarded. The combined eluates have a volume of 6.4 l.

Step 3. First Ammonium Sulfate Precipitation. To 6.4 l. of the above eluate is added 2.5 kg. of solid ammonium sulfate (about 0.6 saturation). The precipitate which forms is filtered off on fluted papers overnight. The next morning the precipitate is scraped off the papers and dissolved in 800 ml. of 0.1 *M* phosphate buffer, pH 7.2. A brownish solution, 960 ml., is obtained.

Step 4. First Acetone Precipitation. To the 960 ml. of solution is added 580 ml. of −15° acetone, and the temperature is allowed to rise to 13°. After 10 minutes the suspension is centrifuged at 3200 × *g* in 250-ml. polyethylene cups. The major fluid layer and a minor fluid layer under the precipitate are carefully discarded. The precipitate is suspended in 400 ml. of 0.3 saturated ammonium sulfate solution (125 g./640 ml.). After the precipitate is thoroughly stirred, the suspension is centrifuged for 10 minutes at 7500 × *g*, and the voluminous precipitate discarded.

Step 5. Second Ammonium Sulfate Precipitation. The supernatant solution, 400 ml., is brought up to 0.5 saturation by the addition of 50 g. of solid ammonium sulfate. The solution is allowed to stand for 6 hours at 5° and then centrifuged for 20 minutes at 7500 × *g*. The supernatant is discarded, and the precipitate dissolved in 110 ml. of distilled water.

Step 6. Second Acetone Precipitation. To the 110 ml. of solution obtained above is added 66 ml. of −15° acetone. The temperature is held at 18° for 10 minutes, and the suspension is then centrifuged for 10 minutes at 3200 × *g*, again using the polyethylene cups. The precipitate is suspended in 50 ml. of 0.3 saturated ammonium sulfate and dialyzed against 1 l. of this same solvent for 12 hours. After clarification by centrifugation, 44 ml. of slightly opalescent solution is obtained.

Step 7. Crystallization. Exactly 2.6 g. of ammonium sulfate is added to the 44 ml., and the solution is immediately centrifuged at 7500 × *g* for 5 minutes. A further 2.6 g. of ammonium sulfate is added to the clear supernatant solution in 200-mg. portions over a period of several hours.

[5] The calcium phosphate gel is prepared from 3 l. of 0.44 *M* Na_2HPO_4 and 3 l. of 0.66 *M* $CaCl_2$, according to the directions of A. Meister, *Biochem. Preparations* **2**, 18 (1952); see also Vol. I [11].

After addition of the first 200 mg. of salt the enzyme begins to crystallize at once. The crystals are centrifuged, washed with 0.5 saturated ammonium sulfate, and dialyzed against 0.1 M phosphate buffer, pH 7.0. The resulting clear solution is dispensed in test tubes and stored by freezing.

SUMMARY OF ISOLATION PROCEDURE

Fraction	Total enzyme, μM.	Total protein, g.	Turnover number, moles substrate/ 135,000 g. protein/ min. at pH 10 and 25°	Recovery, %
1. Water extract	38	132	470	—
2. Ca$_3$(PO$_4$)$_2$ eluate	20	38	850	53
3. First (NH$_4$)$_2$SO$_4$ ppt.	18	31	940	47
4. First acetone ppt.	11	4.1	4300	29
5. Second (NH$_4$)$_2$SO$_4$ ppt.	10	3.6	4700	26
6. Second acetone ppt.	8	1.5	8700	21
7. Crystals	6	0.8	12300	16

Properties

Purity. The crystalline enzyme shows a single boundary in the ultracentrifuge, but both salting-out and electrophoresis experiments demonstrate the presence of two catalytically active components.[6-8] These fractions have been separated by electrophoresis and designated as A and C, corresponding to the anode and cathode limbs of the separation cell. Fraction A, the major component, makes up about 75% of the total protein. The two fractions cannot be separated by recrystallization. Lactic dehydrogenase has been prepared without the use of calcium phosphate gel, but two components were still found in the crystals. Fraction A, after recrystallization from ammonium sulfate, migrates in the electrophoresis cell as a single component.

Physical and Chemical Properties. Meister[7] has reported the sedimentation constant, S_{20}, to be 6.48 and 6.36 $\times$ 10^{-13} second. We found a value of 7.0 $\times$ 10^{-13} second. The free diffusion constant, D_{20}, was 5.3 $\times$ 10^{-7} sq. cm. sec.$^{-1}$ for fraction A. From these data the molecular weight has been estimated to be 135,000 $\pm$ 15,000.[8]

Both fractions precipitate from solution below pH 4.5. By extrapolation of the mobility at higher pH values, isoelectric points of 4.5 and 4.8 have been estimated for fractions A and C, respectively. The extinction

[6] J. B. Neilands, *Science* **115**, 143 (1952).
[7] A. Meister, *J. Biol. Chem.* **184**, 117 (1950).
[8] J. B. Neilands, *J. Biol. Chem.* **199**, 373 (1952).

coefficient at 280 mμ has been found to be 1.49 for 1 mg./ml. in a 1-cm. cell.[8] Hakala and Schwert[9] report a value of 1.3 for a preparation which they believe to contain one-third active enzyme. Careful ammonium sulfate fractionation by the technique of Falconer and Taylor[10] shows that one fraction begins to precipitate at 0.40 saturation and the other begins to precipitate at 0.43 saturation.[8]

Specificity. The enzyme oxidizes L(+)-lactate; there is no detectable activity with D(−)-lactate.[8]

Pyruvic acid is the substrate which is most rapidly reduced. According to Meister,[7] the enzyme also reduces several α-keto and α,γ-diketo acids. However, α-ketoisocaproate, oxalacetate, α-ketoglutarate, β-keto and δ-keto acids were reduced at negligible rates or not at all. Other inactive carbonyl compounds are acetaldehyde, acetone, methyl ethyl ketone, and 3,6-dimethyl-2,5-p-dioxanedione.[11]

Activators and Inhibitors. Dilute solutions of the crystalline enzyme retain full activity for several days in 0.1 M phosphate buffer, pH 7.0, at room temperature.

Since protons participate in the catalyzed reaction, the activity of the enzyme is enormously dependent on pH. Thus the reduction of DPN is routinely carried out at pH 10, whereas the reverse reaction, the oxidation of DPNH, is conveniently studied at pH 7.0.

The enzyme exhibits great stability toward the common inhibitors. Recently[11] it has been possible to show that a 30-minute preincubation with p-chloromercuribenzoate leads to almost complete inactivation. The activity may be restored with cysteine. Under comparable conditions, other thiol reagents such as o-iodosobenzoate, iodoacetate, and N-ethylmaleimide are without effect. Iodine in dilute solution is a powerful inhibitor, as are also oxalic and oxamic acids.[11,12]

Von Euler[13] found the enzyme to be competitively inhibited by pyridine-3-sulfonate. Apparently the D(−)-lactate does not interfere with the oxidation of the L(+)-isomer.

Activity. The turnover numbers in the assay system used previously have been reported as 12,800 and 8100 moles of substrate per minute per 135,000 g. of protein at 25° for fractions A and C, respectively.[8]

The turnover number at pH 7.0 with DPNH and pyruvate as substrates is considerably larger. A figure of 37,000 has been reported for 135,000 g. of the unresolved enzyme.[11] Hakala *et al.*[12] give 2.9 × 10⁴ moles

[9] M. T. Hakala and G. W. Schwert, *Arch. Biochem. and Biophys.* **38**, 55 (1952).

[10] J. S. Falconer and D. B. Taylor, *Nature* **155**, 303 (1945).

[11] J. B. Neilands, *J. Biol. Chem.* **208**, 225 (1954).

[12] M. T. Hakala, A. J. Glaid, and G. W. Schwert, *Federation Proc.* **12**, 213 (1953).

[13] H. von Euler, *Ber.* **75**, 1876 (1942).

per 10^5 g. of enzyme at pH 6.8 and 25°. Meister[7] found 3.0×10^4 moles per 10^5 g. of protein at 26° and pH 7.2.

The Michaelis constant, K_m, for pyruvate has been reported as $1.7 \times 10^{-5} M$[12] and $5.2 \times 10^{-5} M$.[7] The values for DPNH are $2.6 \times 10^{-6} M$[12] and an estimated figure of $10^{-5} M$.[7]

Mechanism of Action. Lactic dehydrogenase forms a compound with DPNH in the same manner and with the same molecular extinction coefficient as in horse liver alcohol dehydrogenase.[14,15] The apparent dissociation constant of the complex is $7 \times 10^{-6} M$ at 5°. The lactic dehydrogenase DPNH compound is, however, much more highly dissociated than the corresponding alcohol dehydrogenase complex. This fact probably accounts for the higher turnover number of lactic dehydrogenase.

The Chicago group[16] has shown the stereochemical specificity requirements of yeast alcohol dehydrogenase and heart lactic dehydrogenase to be strictly comparable.

[14] B. Chance and J. B. Neilands, *J. Biol. Chem.* **199**, 383 (1952).
[15] H. Theorell and B. Chance, *Acta Chem. Scand.* **5**, 1127 (1951).
[16] F. A. Loewus, P. Ofner, H. F. Fisher, F. H. Westheimer, and B. Vennesland, *J. Biol. Chem.* **202**, 699 (1953).

[70] Glyoxalases

By E. RACKER

Glyoxalase I from Baker's Yeast

Assay Method

Principle. The method is based on measurement of ultraviolet light absorption due to the thiol ester formed by condensation between the keto aldehydes and GSH.[1]

Reagents

0.1 M potassium phosphate buffer, pH 6.6.

2% solution of glutathione neutralized to pH 6.6.

2.5% solution of methylglyoxal (redistilled).

0.01% solution of bovine serum albumin in 0.01 M phosphate buffer, pH 7.4.

Enzyme. Dilute stock enzyme solution in the bovine serum albumin solution to obtain 200 to 2000 units of enzyme per milliliter. (See definition of unit below.)

[1] E. Racker, *J. Biol. Chem.* **190**, 685 (1951).

Procedure. Place into a quartz cell having a 1-cm. light path 2.7 ml. of distilled water, 0.1 ml. of buffer, 0.05 ml. of GSH, and 0.05 ml. of methylglyoxal, and after mixing and stabilization of density readings add 0.1 ml. of enzyme. Take readings of density increment at 240 mμ at 30-second intervals.

Definition of Unit and Specific Activity. One unit of enzyme is defined as a change in log I_0/I of 0.001 per minute under the above conditions. Specific activity is expressed as units of enzyme activity per milligram of protein. Protein concentration is determined spectrophotometrically.[2] With cruder preparations which contain considerable amounts of nucleic acid, the biuret test[3] is used.

Application of Assay Method to Crude Tissue Preparations. Glyoxalase I can be determined spectrophotometrically only if the crude preparations contain enough enzyme activity to permit dilution and readings at 240 mμ. The test is suitable for yeast extracts but unsuitable for liver extracts containing active glyoxalase II which interferes with the quantitative determination of glyoxalase I activity. As an alternative method for crude preparation a manometric assay similar to that used for GSH[1] can be used. To make glyoxalase I the limiting factor, methylglyoxal, GSH, and purified glyoxalase II (free of glyoxalase I) are added in excess.

Purification Procedure[1]

Fleischmann's baker's yeast is crumbled and dried between two sheets of paper in a thin layer for 4 to 5 days at room temperature. The dry yeast can be stored in the cold room in well-stoppered containers.

Step 1. Initial Yeast Extracts. 300 g. of the dry yeast is extracted with 900 ml. of 0.066 M disodium phosphate for 2 hours at 37° with occasional stirring, followed by extraction at room temperature for an additional 3 hours. The mixture is centrifuged for 80 minutes at 3000 r.p.m., and the supernatant decanted and stored in the deep-freeze. The residue is re-extracted the next day with 300 ml. of 0.066 M disodium phosphate for 2 hours at 37° and then centrifuged.

Step 2. Acetone Fractionation. To each 100 ml. of the combined yeast extracts, 50 ml. of ice-cold acetone is slowly added while the mixture is maintained at about −2° in a dry ice-alcohol bath. The resulting precipitate is separated by centrifugation at 0° and discarded. To the supernatant solution 50 ml. of cold acetone is added for each 100 ml. of the yeast extract, the temperature being kept at −2°. The mixture is centrifuged at 0°, and the supernatant solution discarded. The precipitate is

[2] O. Warburg and W. Christian, *Biochem. Z.* **310**, 384 (1941).
[3] H. W. Robinson and C. G. Hogden, *J. Biol. Chem.* **135**, 727 (1940).

suspended in about 150 ml. of cold water and dialyzed overnight against running tap water. The precipitate is removed by centrifugation, and the clear supernatant solution is stored in the deep-freeze at −20° overnight.

Step 3. Alcohol Fractionation. After thawing, the enzyme is precipitated by the slow addition of an equal volume of 95% ethyl alcohol at −6° and centrifuged at 3000 r.p.m. for 45 minutes. The precipitate is taken up in about 60 ml. of cold water and extracted for 15 minutes at room temperature. After centrifugation the precipitate is re-extracted with about 20 ml. of water.

Step 4. Heating. The combined extracts are brought rapidly to 45° in a water bath and maintained at this temperature for 5 minutes. The mixture is cooled and centrifuged. The bulky precipitate is washed once with 20 ml. of water, and the supernatants are combined.

Steps 5 and 6 are repetitions of steps 3 and 4, except that 80 ml. of alcohol is used for each 100 ml. of solution, the precipitate is taken up in 30 ml. of water, and the heating step is carried out at 50°.

Steps 7 and 8 are repetitions of steps 3 and 4, except that 50 ml. of alcohol is used for each 100 ml. of solution, the precipitate is taken up with 15 ml. of water and washed with 5 ml. of water, and the extracts are heated at 54° for 8 minutes. At this stage the enzyme can be kept for several months in the deep-freeze at −20° with little loss of activity.

Step 9. Adsorption and Elution with Alumina C_γ. The enzyme solution is diluted to contain 15 mg. of protein per milliliter, and 35 ml. of alumina C_γ (20 mg. of dry weight per milliliter) is added for each 100 ml. of solution. After 5 minutes the mixture is centrifuged in the cold room, and the precipitate washed once with 60 ml. (per 100 ml. of solution) of 0.01 M potassium phosphate buffer, pH 7.4. This washing is discarded. The enzyme is then eluted by the addition of 0.1 M potassium phosphate, pH 7.4, at 37° for 5 minutes. Two elutions with 30 ml. of this buffer and three elutions with 20 ml. of buffer resulted in nearly quantitative recovery of the adsorbed enzyme.

Step 10. Ammonium Sulfate Fractionation. To each 100 ml. of the combined eluates 40 g. of solid ammonium sulfate is added, and, after standing for 15 minutes, the mixture is centrifuged at 12,000 r.p.m. The precipitate (which contains crystalline alcohol dehydrogenase of about 95% purity) is discarded. To the supernatant 30 g. of ammonium sulfate is added to bring the solution to saturation. The mixture is kept at −20° overnight and is centrifuged the next day for 25 minutes at 15,000 r.p.m. in the refrigerated centrifuge. The precipitate is taken up in a small volume of water, and ammoniacal ammonium sulfate (pH 8.0) is added until a fairly heavy precipitate appears. The mixture is left in the icebox overnight, and crystals usually appear on the next day.

A typical protocol, showing data on the specific activity of the various fractions and the yield obtained, is given in Table I.

TABLE I
PURIFICATION OF GLYOXALASE I

Fraction	Total units	Specific activity	Yield, %
1. Combined crude extracts	65,000,000	1,500	100
2. Second acetone ppt.	48,000,000	4,000	74
4. First alcohol ppt.	35,000,000	8,000	54
6. Second alcohol ppt.	26,000,000	27,000	40
8. Third alcohol ppt. after heating	21,000,000	43,000	32
9. Combined eluates	15,000,000	85,000	23
10. Second ammonium sulfate ppt.	11,000,000	250,000	17
Crystalline ppt.	6,500,000	350,000	10

Properties

The purified enzyme reacts with a number of keto aldehydes. The following compounds are reported to react with the purified enzyme:[4] methylglyoxal, phenylglyoxal, hydroxypyruvic aldehyde, and glyoxal. Since phenylglyoxal exhibits a pronounced ultraviolet absorption, which changes in the course of the reaction in a manner different from that of other keto aldehydes, the glyoxalase I reaction with phenylglyoxal as substrate is more suitably studied colorimetrically with hydroxylamine,[5] which reacts in characteristic manner with the product of the reaction.[4]

Glyoxalase I shows a fairly broad optimum of activity between pH 6 and 8. The enzyme is quite stable if kept frozen at $-20°$. Such preparations have been kept frozen over a period of 1 year without appreciable loss in activity. However, dilute solutions of the enzyme are quite unstable and must be protected by the addition of a protein such as bovine serum albumin. For example, a dilute enzyme solution containing 920 units/ml. dropped to 420 units/ml. in the course of 3 hours, although kept in an ice bath during this period. The same enzyme preparation showed no loss of activity in a solution containing 100 γ of serum albumin per milliliter. Temperatures above 65° lead to rapid inactivation of the enzyme, which is, however, quite stable below 60°.

There is no evidence that a coenzyme is involved in glyoxalase I activity. The enzyme withstands extensive dialysis without loss of activity. Partially purified preparations of glyoxalase I frequently show activity in the absence of added glutathione,[1] indicating the possibility that

[4] E. Racker, *Biochim. et Biophys. Acta* **9**, 577 (1952).
[5] F. Lipmann and L. C. Tuttle, *J. Biol. Chem.* **159**, 21 (1945).

glutathione is associated with the enzyme. However, on further purification, these blank values disappear. The turnover number of the purest fractions obtained is 35,000 (assuming a molecular weight of 100,000).

Glyoxalase II from Beef Liver

Assay Method

Principle. Thiol esters exhibit a characteristic ultraviolet absorption spectrum. The enzyme assay is based on the disappearance of ultraviolet absorption after addition of glyoxalase II.[1]

Reagents

> 0.012 M solution of substrate. The substrates are prepared either enzymatically with glyoxalase I [1,4] or chemically.[6]
> 1 M potassium phosphate buffer, pH 6.6.
> Enzyme. Dilute the enzyme in distilled water to obtain 200 to 1000 units of enzyme per milliliter. (Definition of units and specific activity same as for glyoxalase I.)

Procedure. Place into a quartz cell having a 1-cm. light path 2.7 ml. of water, 0.1 ml. of buffer, 0.1 ml. of substrate (free of barium), and 0.1 ml. of enzyme. Activity measurements are carried out as described for glyoxalase I except that the decrease rather than the increase in absorption is measured. In the case of glyoxalase II the density decrease at 340 mμ obtained between 30 and 60 seconds after enzyme addition is multiplied by 2 to obtain the activity of the enzyme sample.

Application of Assay Method to Crude Tissue Preparations. In most tissues glyoxalase II activity is too low to permit spectrophotometric measurement at 240 mμ because of the high absorption of crude preparations in this region. A convenient alternative method applicable to crude tissue preparations is based on the hydroxylamine reaction.[5] Disappearance of the thiol ester can be thus measured colorimetrically and is proportional to enzyme concentration when measured in phosphate buffer at pH 6.6.

Purification Procedure[1]

Frozen beef liver, stored in a deep-freeze, is used as starting material. An acetone-dried powder is prepared as follows: The thawed liver is mixed with 2 to 3 vol. of ice-cold acetone in a Waring blendor, and the mixture poured into 8 vol. of acetone and rapidly centrifuged in the cold room. The precipitate is washed once with 8 vol. of acetone, then pressed out between

---

[6] T. Wieland and H. Koeppe, *Ann.* **581**, 1 (1953).

paper towels, and crumbled into a fine powder which dries rapidly when distributed on large filter papers.

Step 1. Initial Liver Extract. 40 g. of this powder is extracted with 8 vol. of 0.02 *M* phosphate buffer, pH 7.4, at room temperature for 45 minutes with intermittent shaking.

Step 2. Ammonium Sulfate Fractionation. After centrifugation 40 g. of ammonium sulfate is added to each 100 ml. of the supernatant solution, and, after 15 minutes at room temperature, the mixture is centrifuged 30 minutes at 13,000 r.p.m. and the supernatant discarded. The precipitate is extracted with 100 ml. of 30% saturated ammonium sulfate of pH 8.0. After centrifugation at 13,000 r.p.m. for 25 minutes, the clear supernatant is brought to a specific gravity of 1.150 by the addition of saturated ammonium sulfate. The precipitate is collected with 50 ml. of water and dialyzed 3 hours against running tap water.

TABLE II

PURIFICATION OF GLYOXALASE II

Fraction	Total units	Specific activity	Yield, %
1. Crude extract	9,000,000	540	100
2. After ammonium sulfate	5,000,000	1,200	55
3. After heating	3,000,000	2,100	33
4. After alcohol precipitation	1,400,000	4,000	15
5. C_γ eluates	800,000	11,000	9

Step 3. Heating. The dialyzed solution is heated for 5 minutes at 54°, the mixture rapidly cooled, and the precipitate removed by centrifugation.

Step 4. Alcohol Fractionation. To the supernatant of step 3, 0.5 vol. of ethyl alcohol is slowly added; the mixture is kept cold in a dry ice-alcohol bath and then centrifuged in a refrigerated centrifuge at −10°. The precipitate is taken up in 0.01 *M* phosphate buffer, pH 7.2, and diluted with buffer to contain 6 mg. of protein per milliliter.

Step 5a. The diluted enzyme is then adsorbed by the addition of 0.3 vol. of alumina C_γ (20 mg. of dry weight per milliliter) and eluted three times with 0.5 vol. of 0.1 *M* phosphate buffer, pH 7.4. This preparation is dialyzed for 3 hours with stirring against distilled water and then stored in the deep-freeze.

Step 5b. As an alternative method to the use of alumina C_γ, which does not give very reproducible results at this step, calcium phosphate gel adsorption has been utilized. Calcium phosphate gel (19 mg. of dry weight per milliliter) is added to the dilute enzyme solution (15 mg. of protein per milliliter) to give a gel-protein ratio of 0.5. With this excess of gel most of the enzyme is adsorbed and is readily eluted with 0.1 M potassium phosphate buffer at pH 6.7.

A typical protocol, showing data on the specific activity of the various fractions and the yield obtained, is given in Table II.

Properties

Glyoxalase II prepared from beef liver appears to be less stable than glyoxalase I. Some preparations lose activity quite rapidly, even when kept frozen. However, several highly purified preparations have been kept for more than 3 months at $-20°$ with little loss in activity. The stability is not entirely predictable, since some preparations of apparently equal purity lost 40 to 50% of their activity during an equal period of storage. Glyoxalase II is also less stable to dialysis than is glyoxalase I, and considerable loss in activity results from this procedure. No evidence, however, could be obtained for the participation of a coenzyme in the reaction. Preparations of glyoxalase II can be heated for 3 minutes at 56° without loss of activity, but at higher temperatures the enzyme is rapidly inactivated.

[71] Plant Carboxylases

$$RCOCOOH \rightarrow RCHO + CO_2 \qquad (1)$$
$$RCOCOOH + R'CHO \rightarrow RCHOHCOR' + CO_2 \qquad (2)$$
$$RCHO + R'CHO \rightarrow RCHOHCOR' \qquad (3)$$

By Thomas P. Singer

α-Carboxylase from Yeast

Assay Method

Principle. The enzyme catalyzes reactions 1 and 2, but not 3.[1-3] Reaction 4, which may also be observed with the yeast enzyme, as with other plant carboxylases, is a summation of reactions 1 and 2.

$$2RCOCOOH \rightarrow RCHOHCOCOOH + 2CO_2 \qquad (4)$$

Commonly used assay methods, such as that of Green *et al.*,[4] outlined below, are based upon the manometric measurement of CO_2 liberation from pyruvate as substrate. The relative contribution of reactions 1 and 4,

[1] C. Neuberg and L. Karczag, *Biochem. Z.* **36**, 68 (1911).
[2] E. Juni, *J. Biol. Chem.* **195**, 727 (1952).
[3] W. Dirscherl, *Z. physiol. Chem.* **188**, 225 (1930).
[4] D. E. Green, D. Herbert, and V. Subrahmanyan, *J. Biol. Chem.* **138**, 327 (1941).

respectively, to the CO_2 formation depends upon the experimental conditions, but reaction 1 always predominates.

Reagents

1 M pyruvate. Prepared from commercial crystalline Na pyruvate or from redistilled pyruvic acid, neutralized with NaOH or KOH. This solution may be stored for several weeks at $-15°$.

0.5 M citric acid–NaOH buffer, pH 6.0.

Enzyme. Dilute the stock enzyme so as to contain 0.25 to 10 units/ml. (See definition below.)

Procedure. The test is carried out in conventional Warburg vessels of 15- to 30-ml. capacity. The main compartment contains 0.3 ml. of 0.5 M citrate buffer, pH 6.0, and 0.5 to 2 units of enzyme in a total volume of 2.8 ml. The side arms contain 0.5 ml. of M pyruvate. CO_2 evolution is followed for 3 minutes at 30° after tipping in the pyruvate. Under these conditions the CO_2 evolution is strictly proportional to the enzyme concentration. In experiments of longer duration usually no proportionality is observed.

Definition of Unit and Activity Ratio. One unit of carboxylase is defined as that amount which catalyzes the production of 100 μl. of CO_2 in 3 minutes at 30°. Protein concentration is expressed as optical density at 280 mμ in a cell of 1-cm. light path. Activity ratio is optical density at 280 mμ divided by units per milliliter.

Application to Preparations of the Apoenzyme. The enzyme, as isolated,[4] contains sufficient divalent cation and TPP for full activity. In the manometric assay of the resolved apoenzyme, 0.03 mg. of Mn^{++} and 0.05 mg. of TPP are added to the apoenzyme, in that order, and 10 minutes of incubation at room temperature is allowed to permit maximal recombination of the components.

Purification Procedure

Moderate degrees of purification of the yeast enzyme have been attained by numerous workers.[2,5-9] Extensive purification has been reported by Kubowitz and Lüttgens[10] and by Green *et al.*,[4] although the degree of

[5] W. Langenbeck, R. Jütterman, O. Schäfer, and H. Wrede, *Z. physiol. Chem.* **221**, 1 (1933).

[6] W. Langenbeck, H. Wrede, and W. Schlockermann, *Z. physiol. Chem.* **227**, 263 (1934).

[7] F. Axmacher and H. Bergstermann, *Biochem. Z.* **272**, 259 (1934).

[8] H. Westerkamp, *Biochem. Z.* **263**, 239 (1933).

[9] J. L. Melnick and K. G. Stern, *Enzymologia* **8**, 129 (1940).

[10] F. Kubowitz and W. Lüttgens, *Biochem. Z.* **307**, 170 (1941).

purity achieved is not known. Since the detailed procedure of the former group has never been published, only the latter procedure will be included here.

Step 1. 100 g. of thoroughly washed, low-temperature-dried ale yeast[11] is slowly added to 300 ml. of 0.066 M phosphate buffer, pH 7.2, with mechanical stirring, and the suspension is incubated for 1 hour at 37°. After dilution with 400 ml. of water, the insoluble material is removed by centrifugation. Subsequent steps are carried out at 0 to 5°.

Step 2. To the supernatant solution 80 ml. of 0.5 M phosphate buffer, pH 7.2, and 40 ml. of M calcium acetate are added under mechanical stirring. The precipitate is centrifuged off and washed with 1.5 vol. of water. The supernatant solution and wash are combined (700 ml.).

Step 3. For each 100 ml. of solution 38 g. of ammonium sulfate is added. After 15 minutes of stirring, the precipitated enzyme is centrifuged off. Since at slower speeds the precipitate is packed tightly only by prolonged centrifugation, a high-speed centrifuge (10,000 $\times$ g) is recommended in this step. The precipitate is dissolved in 300 ml. of 0.04 M citrate buffer, pH 6.0.

Step 4. For each 100 ml. of solution 38 g. of ammonium sulfate is added, and the precipitate is packed tightly by centrifugation, as above. The precipitate is dissolved in 126 ml. of 0.04 M citrate buffer, pH 6.0. The resulting solution (162 ml.) is 0.11 saturated with respect to ammonium sulfate.

Step 5. The solution obtained in step 4 is fractionated with saturated ammonium sulfate as indicated in Table I. Fractions I and II are discarded, and fraction III is redissolved in 200 ml. of 0.025 M citrate buffer, pH 6.0, 0.25 saturated in ammonium sulfate. Saturated ammonium sulfate is added gradually (Table I, fractions IIIa, IIIb, and IIIc). Fraction IIIb (the precipitate collected between 0.47 to 0.52 saturated ammonium sulfate) has the lowest activity ratio. It is redissolved in 200 ml. of 0.025 M citrate buffer, pH 6.0, 0.25 saturated in ammonium sulfate, and the solution is refractionated as shown in Table I, fractions IIIb$_1$, IIIb$_2$, IIIb$_3$. The precipitate collected between 0.47 and 0.53 saturated ammonium sulfate is stated to contain the yeast enzyme 27-fold purified compared with the original extract. Since the progress of fractionation in step 5 may vary with the kind of yeast, it is advisable to check the activity ratio after each cycle of ammonium sulfate precipitations and to select the one having the lowest activity ratio as starting material for the next

[11] Available from Anheuser-Busch, Inc. If it is desired to use locally available ale yeasts, the yeast should be thoroughly washed on a basket centrifuge, spread out in a thin layer on glass plates in a cool place, and dried by blowing a stream of air over it with the aid of a fan.

fractionation. It is essential to carry the fractionation through step 3 with a minimum of delay, since crude preparations of carboxylase lose activity rapidly.

TABLE I
SUMMARY OF PURIFICATION OF YEAST CARBOXYLASE[a]

Step	Fraction	Saturated ammonium sulfate added, vol.	Degree of saturation	Total volume, ml.	Total units	Activity ratio	Recovery, %
1				530	9650	10.9	
2				700	8170	8.8	84
3				344	7250	3.6	75
4				162	6300	1.5	65
5	I	0.35	0.35		168	6.5	1.7
	II	0.52	0.42		960	0.90	10
	III	0.88	0.52		4450	0.61	46
	IIIa	0.44	0.47		940	0.57	9.7
	IIIb	0.55	0.52		2060	0.45	21
	IIIc	0.77	0.58		1270	0.67	13
	IIIb$_1$	0.41	0.47		440	0.44	4.6
	IIIb$_2$	0.58	0.53		587	0.39	6.1
	IIIb$_3$	1.0	0.63		116	0.62	1.3

[a] D. E. Green, D. Herbert, and V. Subrahmanyan, *J. Biol. Chem.* **138**, 327 (1941).

Properties

Stability. After step 3 the enzyme is rather stable and may be stored for several weeks at 0° in 0.5 saturated ammonium sulfate.[4]

Prosthetic Group. The most purified fraction of Green et al.[4] (Q_{CO_2} = 12,100) contains 1 mole of TPP and about 5.3 moles of Mg per 100,000 g. of protein. The best preparation obtained by Kubowitz and Lüttgens[10] is reported to contain 1 mole of TPP and 1 mole of Mg per 75,000 g. of protein. This apparent discrepancy may be ascribed to the probability that neither preparation is homogeneous. The metal requirements of the resolved apoenzyme are equally well satisfied by Mg^{++} and Mn^{++}; Co^{++}, Cd^{++}, Zn^{++}, Ca^{++}, Fe^{++}, Al^{+++}, and Fe^{+++} are also active, in decreasing order of effectiveness.[4,12] Thiamine triphosphate replaces TPP in activation of the apoenzyme, but the saturation requirements for thiamine triphosphate are higher than for TPP, and the maximal activity considerably lower.[13] Thiamine monophosphate is inactive in this test.

[12] A. J. Kossel, *Z. physiol. Chem.* **276**, 251 (1942).
[13] L. Velluz and M. Herbain, *J. Biol. Chem.* **190**, 241 (1951).

Specificity. A number of higher homologs of pyruvic acid (up to C_6) are decarboxylated by the enzyme, and the rate decreases with increasing chain length.[4,14] β-Keto acids and phenyl-substituted keto acids are not acted upon. In the acyloin condensation, as catalyzed by yeast carboxylase (reaction 2), the aldehyde which condenses with the product of the enzymatic decarboxylation of the α-keto acid (i.e., an enzyme–TPP–aldehyde complex[15]) may be acetaldehyde, propionaldehyde, or substituted aromatic aldehydes but not butyraldehyde.[16–19] The reasons for the inactivity of certain aldehydes and the mechanism of acyloin synthesis (carboligase action) have been discussed elsewhere.[15]

Inhibitors. Yeast carboyxlase is a typical sulfhydryl enzyme, being reversibly inhibited by low concentrations of p-chloromercuribenzoate, trivalent organic arsenicals, and mercaptide-forming cations (Ag^+, Hg^{++}, Cu^{++}).[4,20] At a concentration of 1.5×10^{-2} M, acetaldehyde inhibits the initial rate of decarboxylation of 1.5×10^{-1} M pyruvate 21%.[4] At lower concentrations of pyruvate and over longer periods the inhibition becomes more pronounced. Oxythiamine triphosphate is an effective inhibitor of yeast carboxylase. Using the apoenzyme prepared from unfractionated yeast, Velluz and Herbain[13] have shown that, in the presence of 0.005 mg. of TPP per milliliter, 0.05 mg. of oxythiamine triphosphosphate per milliliter produces 50% inhibition.

Effect of pH. The enzyme exhibits optimum activity in citrate buffer at approximately pH 6. The activity in phosphate buffer at the same pH is considerably lower.[4]

Saturation Requirements. Under the conditions outlined, the enzyme is saturated at about 1.6×10^{-1} M concentration of pyruvate, and half-saturated at 3×10^{-2} M.[4] The apoenzyme (cf. below) requires 0.03 mg. of Mn^{++} and 0.05 mg. of TPP for saturation in the assay outlined. These concentrations are much higher than the TPP and divalent cation content of the unresolved holoenzyme. This may suggest that the linkages of TPP and cation to the protein are different in the intact and reconstructed holoenzymes.[4,10]

Preparation of Apoenzyme. The holoenzyme has been reversibly split into apoenzyme and prosthetic group by treatment with phosphate buffer at pH 8.1[10] or with ammoniacal ammonium sulfate.[4]

[14] S. Kobayasi, *J. Biochem. Japan* **33**, 301 (1941).
[15] T. P. Singer and J. Pensky, *Biochim. et Biophys. Acta* **9**, 316 (1952).
[16] C. Neuberg and E. Reinfurth, *Biochem. Z.* **143**, 553 (1923).
[17] C. Neuberg and L. Liebermann, *Biochem. Z.* **121**, 311 (1921).
[18] M. Behrens and N. N. Iwanoff, *Biochem. Z.* **169**, 478 (1926).
[19] E. Hofmann, *Biochem. Z.* **243**, 429 (1931).
[20] E. S. G. Barron and T. P. Singer, *J. Biol. Chem.* **157**, 221 (1945).

α-Carboxylase from Wheat Germ
Assay Method

Principle. The enzyme catalyzes reactions 1 to 4. It may be assayed manometrically[21] by CO_2 evolution in the presence of pyruvate as substrate (reactions 1 and 4) or colorimetrically,[15] with pyruvate plus acetaldehyde (reaction 2) or acetaldehyde alone (reaction 3) as substrate. The choice of method is a matter of personal preference. The author prefers the manometric estimation in view of its speed and the small amount of enzyme required. The details of the manometric assay only are given here. The experimental details of the colorometric assay of carboxylase by acetoin estimation are described elsewhere.[15]

In the assay outlined below, over 90% of the CO_2 formation originates in reaction 1.

Reagents

> 0.5 M sodium pyruvate.[22]
> 0.2 M succinate buffer, pH 6.0.
> 0.05 M dimedon. Dissolve 701 mg. of dimedon in 95 ml. of water, adjust to pH 6, and dilute to 100 ml.
> 5.8×10^{-4} M TPP. Dissolve 3 mg. of diphosphothiamine chloride[23] in 10 ml. of water. This solution may be kept frozen for several months without change.
> 0.01 M $MgSO_4$.
> 1% bovine serum albumin. Crystalline or "fraction V" serum albumin[24] is dissolved in water, adjusted to pH 6, and diluted to a concentration of 1%. The solution may be kept frozen for several months.
> Enzyme. The stock enzyme is diluted to contain 50 to 500 units/ml. (See definition below.)

Procedure. CO_2 evolution is followed in a standard Warburg apparatus for 5 minutes at 30°. The main compartment of the vessel contains, in a total volume of 2.8 ml., 1 ml. of succinate buffer, 0.6 ml. of dimedon, 0.1 ml. of TPP, 0.2 ml. of $MgSO_4$, 0.1 ml. of serum albumin solution, and 30 to 80 units of enzyme. The side arm contains 0.2 ml. of pyruvate, which is dumped at zero time.

[21] T. P. Singer and J. Pensky, *J. Biol. Chem.* **196**, 375 (1952).
[22] Cf. section on yeast enzyme.
[23] TPP is commercially available from several sources.
[24] Armour Laboratories, Chicago.

The dimedon and the serum albumin serve to trap most of the acetaldehyde liberated in the reaction, the accumulation of which strongly inhibits further decarboxylation of pyruvate. When the enzyme activity exceeds 50 μl. of CO_2 per 5 minutes, the observed CO_2 evolution is no longer proportional to the enzyme concentration, even in the presence of aldehyde fixatives, and a correction factor has to be introduced, as illustrated in Fig. 1. This correction curve, obtained with a purified enzyme

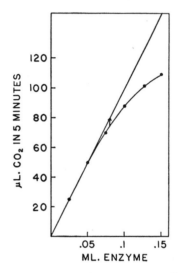

FIG. 1. Relation of enzyme concentration to apparent CO_2 evolution. The circles are experimental points; the straight line is the extrapolated activity to which observed activity is corrected. The deviation from linearity is caused by the accumulation of acetaldehyde.

preparation, holds true throughout the purification procedure and is accurately reproducible at every stage thereof. When experimental values in excess of 50 μl. of CO_2 per 5 minutes are obtained, the value is corrected by taking the corresponding figure from the extrapolated line. Thus, in Fig. 1, as indicated by the arrow, 74 μl. of apparent activity corresponds to 80 μl. of corrected activity.

Definition of Unit and Activity Ratio. One unit of enzyme is defined as that amount which causes the liberation of 1 μl. of CO_2 in 5 minutes under the conditions described. Activity ratio denotes units of activity per milligram of protein. Protein is determined by dry weight in the initial extract and by light absorption at 280 mμ thereafter. The absorption of a 1% solution of the enzyme at various stages of purification, in a 1-cm. cell, $D_{280}^{1\%}$, is recorded in the procedure.

Application of Assay Method to Crude Preparations. The method outlined is equally applicable to a homogenate or extract of wheat germ and to the homogeneous enzyme.

Purification Procedure

α-Carboxylases have been extracted from various higher plants,[25-29] but significant purification has been achieved only with the enzyme from wheat germ. The procedure[21] described below has been carried out numerous times in the author's laboratory with uniform results. The main precaution to be observed is exact pH adjustment; the electrodes should be standardized, and the pH values measured at the temperature of the enzyme solution (2 to 4°).

Step 1. Acetone Powder. 1 kg. of wheat germ is defatted with 10 vol. of reagent-grade acetone at −10°. A portion of the acetone is used to homogenize the germ in a Waring blendor, and the resulting suspension is poured into the rest of the acetone. After 10 minutes of stirring, the suspension is filtered rapidly, and the washing and filtration are repeated. While still moist, the defatted germ is placed in desiccators and dried *in vacuo* over H_2SO_4 in the cold.

Step 2. Extraction. The following procedure is workable on a 250- to 2500-g. scale. Unless otherwise indicated, it is carried out at 5 to 7°. All pH values are measured at the temperature at which the step is carried out. 1 kg. of acetone powder is extracted with 5 l. of H_2O by vigorous stirring for 15 minutes, followed by centrifugation for 30 to 45 minutes at 2600 r.p.m. The cloudy supernatant (2 l.) contains all the enzyme (750,000 to 780,000 units; activity ratio = 2.1, based on dry weight).

Step 3. Isoelectric Precipitation. To the water extract 1 N acetic acid is added under vigorous stirring to exactly pH 5.20 at 0 to 4°. The suspension is centrifuged immediately (2600 r.p.m. for 30 minutes), yielding an almost clear, yellow supernatant. The solution is chilled to about 3°, and the enzyme is then precipitated from the supernatant by further addition of 1 N acetic acid to exactly pH 4.90. The suspension is kept overnight at 0° to complete the precipitation. The enzyme is quite stable at this stage, since it is insoluble at pH 4.90, whereas at pH 5.20 it is unstable. The suspension at pH 4.90 is centrifuged for 30 minutes at 2600 r.p.m. The precipitate is resuspended in 900 ml. of 0.1 M succinate

[25] B. Anderson, *Z. physiol. Chem.* **210**, 15 (1932).
[26] B. Tankó and L. Munk, *Z. physiol. Chem.* **262**, 144 (1939).
[27] A. H. Bunting and W. O. James, *New Phytologist* **40**, 262 (1941).
[28] P. P. Cohen, *J. Biol. Chem.* **164**, 685 (1946).
[29] B. Vennesland and R. Z. Felsher, *Arch. Biochem.* **11**, 279 (1946).

buffer, pH 6.0, with the aid of glass homogenizers, and stirred for 20 minutes. Centrifugation for 30 minutes in an angle head centrifuge at 5000 r.p.m. yields a clear, yellow supernatant, containing 530,000 to 612,000 units ($D_{280}^{1\%} = 11.2$; activity ratio = 77 to 99).

Step 4. Alcohol Fractionation. The enzyme solution is chilled to 0° and adjusted to pH 5.50 with 1 N acetic acid. Ethanol (95%, −15°) is added dropwise, with stirring, to a final concentration of 15% by volume, while the enzyme solution is maintained at 0 to −1°. The stirring is continued for 15 minutes after the addition of alcohol, and the precipitated enzyme is separated by centrifugation for 45 minutes at 2500 r.p.m. at 0 to −1°. The moist precipitate is immediately dried in a desiccator at 0° over H_2SO_4 (high vacuum). The resulting powder is stable for months in the cold. About 530 mg. of powder, containing 490,000 to 560,000 units, are obtained per kilogram of acetone powder. Activity ratio = 1365 to 1510 on the basis of $D_{280}^{1\%} = 11.0$.

Step 5. $(NH_4)_2SO_4$ Precipitation and Dialysis. The dry enzyme is dissolved in 0.05 M Tris buffer, pH 7.7, at 6°, 20 ml. of buffer being used for each 100 mg. of powder. The insoluble part is removed by centrifugation. The solution is adjusted to pH 6.0 with 1 N acetic acid at 0°, and saturated $(NH_4)_2SO_4$ is added dropwise to give 0.33 saturation. After 15 minutes of stirring, the precipitated enzyme is centrifuged (45 minutes at 5000 r.p.m.). The precipitate is resuspended in about 10 ml. of H_2O for each kilogram of acetone powder used and dialyzed for 24 to 48 hours against glass-distilled water. The enzyme precipitates quantitatively during dialysis. After a brief centrifugation at 5000 r.p.m., the precipitated enzyme is dissolved in 0.1 M imidazole buffer, pH 6.8, at 5° (10 ml. of buffer for each 100 mg. of alcohol-precipitated enzyme). Insoluble protein particles are removed by centrifugation at 18,000 r.p.m. In the best preparations, the clear, colorless imidazole extract contains 560,000 units of enzyme for each kilogram of acetone powder used (activity ratio = 5600, based on $D_{280}^{1\%} = 10.7$). The over-all yield is, therefore, 75%, and the losses are due mainly to occlusion of the soluble enzyme in the voluminous precipitates. The purification is about 2700-fold compared with the first extract (activity ratio = 2.1) and 11,000-fold compared with the acetone powder (activity ratio = 0.5).

Occasionally, the imidazole extract has a slightly lower activity ratio (4000 to 4300) and a total yield of about 400,000 units. The purity of such preparations can be raised by treatment with $Ca_3(PO_4)_2$ gel.[30] In a typical experiment 144 mg. of protein in 10 ml. of imidazole buffer (based on $D_{280}^{1\%} = 10.7$), at an activity ratio of 4000, was treated dropwise with

---

[30] For the preparation of $Ca_3(PO_4)_2$ gel, see Vol. I [11].

70 mg. of $Ca_3(PO_4)_2$ gel, until 28% of the initial activity and 41% of the protein were adsorbed; the purity in the unadsorbed portion was raised to an activity ratio of 4850.

TABLE II

SUMMARY OF PURIFICATION OF WHEAT GERM CARBOXYLASE[a]

Fraction	Total units, thousands	Activity ratio, units/mg.	Recovery, %
Water extract	750–780	2.1	—
Succinate extract of isoelectric precipitate	530–612	77–99	69–80
Alcohol precipitate	560	1365–1510	73
Final imidazole extract	560	5600	73

[a] T. P. Singer and J. Pensky, *J. Biol. Chem.* **196**, 375 (1952).

Properties

Physicochemical Properties. At the activity ratio of 5600 the enzyme is electrophoretically homogeneous throughout the pH range of its stability. In the ultracentrifuge a 7% impurity may be detected. This may be removed in a preparative ultracentrifuge, and the resulting preparation is homogeneous in both electrophoretic and ultracentrifugal analyses.[21] The high sedimentation velocity ($S_{20} = 29.0 \times 10^{-13}$ cm. sec.$^{-1}$) indicates a molecular weight in excess of 1 million.

Stability. Concentrated solutions of the enzyme in imidazole buffer, pH 6.8, are stable for about a week at 5°. If precipitated with solid $(NH_4)_2SO_4$ (18.6 g. per 100 ml. of solution) or by exhaustive dialysis against H_2O, the enzyme retains its activity for several weeks in the cold. Freezing or lyophilizing of enzyme solutions causes considerable inactivation. The best way to store the purified preparations for prolonged periods is in form of the alcohol-precipitated dry powder.

Coenzyme Requirements. The α-carboxylase of wheat germ requires TPP for enzymatic activity. The TPP moiety is lost early in purification; thus, at the end of step 4 there is very little or no activity without added TPP, and any remaining TPP is completely removed by dialysis in step 5. Half-saturation is reached at 1.35×10^{-6} M TPP in the standard assay.

Besides TPP, the enzyme requires a divalent metal ion for full activity. Maximal activity is reached with Mg^{++}, Zn^{++}, and Co^{++} (1×10^{-3} M); Fe^{++}, Mn^{++}, Ni^{++}, and Cd^{++} yield lower maximal activities, in decreasing order. Half-saturation is reached at the following concentrations: Mg^{++}, 1.0×10^{-4} M; Co^{++}, 1.9×10^{-5} M; Zn^{++}, 1.3×10^{-5} M; Fe^{++}, 4.5×10^{-5} M; and Mn^{++}, 7.9×10^{-6} M. Ba^{++}, Ca^{++}, and Sr^{++} are inactive. In the absence of added cations and after exhaustive dialysis and passage through cation exchange resins, some 20% of the maximal activity remains. Such preparations have been shown spectrographically to be free from traces of cations which could act as activators. It has been suggested that in the metal-free enzyme TPP may be bound to the protein to a limited extent, resulting in submaximal activity, whereas in the complete holoenzyme TPP is bound to the protein via the metal.[21]

Specificity and Substrate Requirements. The enzyme decarboxylates pyruvate and α-ketobutyrate at identical rates; α-ketoglutarate is acted upon very slowly, and pyruvamide is not attacked. In the decarboxylation reaction $3 \times 10^{-2} M$ pyruvate saturates and $3.6 \times 10^{-3} M$ pyruvate half-saturates the enzyme. In acetoin synthesis from pyruvate plus acetaldehyde maximal velocity is reached at $5 \times 10^{-3} M$ pyruvate and $5 \times 10^{-2} M$ acetaldehyde, and half-saturation is reached at $1.3 \times 10^{-3} M$ pyruvate and $8.6 \times 10^{-3} M$ acetaldehyde. With acetaldehyde alone as substrate for acetoin synthesis, saturation is reached at a concentration of $2 \times 10^{-1} M$, and half-saturation at $2 \times 10^{-2} M$.[15]

Effect of pH and Temperature. The apparent pH optimum at 30° is pH 6.3 to 6.4. The activity of the enzyme is the same in imidazole, succinate, and histidine buffers. In phosphate and citrate buffers the activity is slightly lower, but full activity may be regained by raising the effective Mg^{++} concentration.[21] The temperature optimum is above 45°.

At 0° the enzyme is completely stable for at least 1 hour between pH 5.7 and 7.9. Above pH 8.3 extensive inactivation occurs. At 30° the pH of optimum stability is 6.0.

Inhibitors. The enzyme is strongly inhibited by p-chloromercuribenzoate, and the sensitivity of the enzyme to this reagent increases greatly in the course of purification. At an activity ratio of 1400 in the presence of 0.058 mg. of protein per 3 ml., $1 \times 10^{-5} M$ p-chloromercuribenzoate inhibits the enzyme 82%. The decarboxylation of pyruvate is strongly inhibited by acetaldehyde. The extent of inhibition obtained by a given concentration of acetaldehyde is the same whether it is added at the beginning of the experiment or accumulated by decarboxylation of pyruvate. Propionaldehyde is somewhat less inhibitory. In the standard assay, but in the absence of dimedon and albumin, $1 \times 10^{-2} M$ acetaldehyde inhibits 70% and the same concentration of propionaldehyde 41%. In general, the greater the inhibitory action of a given aldehyde, the more active it is as substrate for acyloin synthesis. A full explanation of this phenomenon has been given elsewhere.[15]

Fe^{+++} and Al^{+++} at $1 \times 10^{-3} M$ concentration inhibit the "blank" activity of thoroughly dialyzed preparations (in the absence of added divalent cations) 40 and 60%, respectively.

Other Properties. Acetoin synthesis by the enzyme is irreversible. Furthermore, the acetoin synthesized from either acetaldehyde alone or acetaldehyde plus pyruvate is *partially* asymmetric: it consists of 72% of the (+) and 28% of the (−) isomer.[31]

[31] T. P. Singer, *Biochim. et Biophys. Acta* **8**, 108 (1952).

[72] Acetoin Formation in Bacteria

$$2CH_3COCOOH \rightarrow CH_3COH(COCH_3)COOH + CO_2$$
$$CH_3COH(COCH_3)COOH \rightarrow CH_3CHOHCOCH_3 + CO_2$$

$$2CH_3COCOOH \rightarrow CH_3CHOHCOCH_3 + 2CO_2$$

By ELLIOT JUNI

Assay Method

Principle. The reactions are followed manometrically by the measurement of CO_2 evolution. Acetoin formation may also be followed by its colorimetric determination according to the method of Westerfeld.[1] α-Acetolactic acid is assayed manometrically by decarboxylation with H_2SO_4[2] or with α-acetolactic acid decarboxylase.[2]

Reagents

Sodium pyruvate (1.0 *M*).[3]
Sodium α-acetolactate (0.5 *M*).[4]
TPP (200 γ/ml.).[5]
0.13 *M* phosphoric acid–KOH buffer, pH 5.7.
Enzymes. The stock enzymes should be diluted with phosphate buffer to obtain solutions each milliliter of which is capable of decarboxylating approximately 30 micromoles of substrate in 5 minutes.

A Warburg vessel with a fluid volume of 2 ml. should contain 1.5 ml. of phosphate buffer, 0.2 ml. of sodium pyruvate or 0.2 ml. of sodium

[1] W. W. Westerfeld, *J. Biol. Chem.* **161**, 495 (1945).
[2] E. Juni, *J. Biol. Chem.* **195**, 715 (1952).
[3] For the preparation of sodium pyruvate, see Vol. III [65]. Crystalline sodium pyruvate is obtainable commercially.
[4] Krampitz, L. O., *Arch. Biochem.* **17**, 81 (1948); see Vol. III [50].
[5] TPP is obtainable commercially; see Vol. III [139].

α-acetolactate, and 0.2 ml. of TPP (required only when activity on sodium pyruvate is being measured and may be replaced with 0.2 ml. of H_2O when sodium α-acetolactate is the substrate) in the main compartment. The side arm should contain 0.1 ml. of enzyme. Incubate at 30° with an air atmosphere. Take readings at 5-minute intervals after adding the enzyme from the side arm of the vessel.

Definition of Specific Activity. Specific activity is expressed as microliters of CO_2 evolved per hour per milligram dry weight of enzyme (Q_{CO_2}). Dry weight is determined after placing a given volume of enzyme at 110° for 24 hours and correcting for the weight of salt present.

The formation of acetoin from pyruvic acid requires a mixture of the α-acetolactic acid-forming enzyme system and the α-acetolactic acid decarboxylase. These two enzymes normally occur together in crude extracts. The formation of α-acetolactic acid from pyruvic acid involves the liberation of $\frac{1}{2}$ mole of CO_2 per mole of pyruvic acid used. For the total reaction 1 mole of CO_2 is released per mole of pyruvic acid used.

Application of Assay Method to Intact Cells or Crude Tissue Preparations. Washed intact bacterial suspensions, crude acetone powder suspensions, and similar preparations may be assayed as described. Certain crude animal tissue preparations such as the pig heart system of Green et al.[6] and the filarial homogenates of Berl and Bueding[7] appear to decarboxylate sodium α-acetolactic acid nonenzymatically. Such extracts exhibit catalytic activity even when boiled.

Purification Procedure

Steps 1 and 2 are based on the method of Silverman and Werkman.[8] The entire procedure has been repeated successfully more than a dozen times. The various fractions obtained are stable for several months when kept in the frozen state and may be preserved by lyophilization.

Step 1. Growth of Bacterial Culture. A laboratory strain of *Aerobacter aerogenes* is grown on a medium consisting of 1.0% glucose, 0.3% proteose peptone, and 0.8% K_2HPO_4. The proteose peptone and K_2HPO_4 are dissolved in 9 l. of distilled water contained in a 12-l. Florence flask and autoclaved for 45 minutes at 120°. The glucose is dissolved in 1 l. of distilled water and sterilized separately by autoclaving for 20 minutes at 120°. The solutions are then mixed by asceptic addition of the glucose solution to the 12-l. flask. The medium[9] is inoculated with the 24-hour

---

[6] D. E. Green, W. W. Westerfeld, B. Vennesland, and W. E. Knox, *J. Biol. Chem.* **145**, 69 (1942).

[7] S. Berl and E. Bueding, *J. Biol. Chem.* **191**, 401 (1951).

[8] M. Silverman and C. H. Werkman, *J. Biol. Chem.* **138**, 35 (1941).

[9] The medium should be incubated at 30° for 24 hours before inoculation to achieve adequate temperature equilibration.

growth of organisms scraped from a nutrient agar slant and incubated at
30° without shaking or aeration for 18 to 24 hours. The cells are removed
by centrifugation in a continuous Sharples centrifuge at about 35,000
r.p.m. The yield of wet paste of cells is approximately 25 g.

Step 2. Preparation of Crude Extract. The unwashed cell paste (25 g.)
is mixed with 50 g. of a 60 U.S. mesh powdered glass[10] in a beaker immersed in an ice bath, and 1 ml. of H_2O[11] is added to give a consistency
such that a path made with a spatula on the surface of the paste will
gradually disappear in 2 to 4 minutes. The cells are crushed by passing
the cell-glass mixture once through power-driven glass cones as described
by Kalnitsky *et al.*[12,13] The crushed cell mass is extracted twice by mixing
with 25-ml. portions of cold (0 to 5°) 0.13 M potassium phosphate buffer,
pH 5.5, and centrifuging for 5 minutes at 20,000 × g in the cold. The
extracts are combined and recentrifuged for 15 to 30 minutes to remove
any intact cells or large cell fragments.

Step 3. Adsorption of α-Acetolactate Forming System by Alumina. To
40 ml. of crude enzyme preparation is added 6 g. of alumina[14] with
stirring. The suspension is kept in an ice bath, and the stirring is continued for 15 minutes. The alumina is then centrifuged in the cold, and the
supernatant fluid is saved. The alumina is washed three times by suspending with 25-ml. portions of ice-cold 0.13 M potassium phosphate buffer,
pH 5.5, centrifuging, and discarding the washes. The alumina is eluted
twice by suspending with 10-ml. portions of 0.13 M K_2HPO_4, centrifuging,
and adjusting the combined eluates to pH 7.0 with ice-cold 1 M H_3PO_4.

Step 4. Ethanol Precipitation. To 20 ml. of eluate is added dropwise
with stirring 16 ml. of cold (−18°) 95% ethyl alcohol.[15] The enzyme-
alcohol solution is maintained at −3° by use of an ice-salt bath. Precipita-

---

[10] Corning Glass Co.

[11] The exact amount of water required to give the proper consistency will depend upon
the amount of water in the centrifuged cell paste.

[12] G. Kalnitsky, M. F. Utter, and C. H. Werkman, *J. Bacteriol.* **49**, 595 (1945).

[13] Cells may also be disrupted by sonic vibration or crushing with alumina. These
methods are not recommended, however, since they usually yield crude extracts
that decarboxylate α-acetolactic acid at a greater rate than pyruvic acid. The ultimate resolution of the α-acetolactic acid-forming system from the α-acetolactic
decarboxylase is dependent upon an initial favorable ratio of the rates of pyruvic
acid to α-acetolactic acid decarboxylation.

[14] Alumina A-303, Aluminum Company of America. Some samples were found to give
an alkaline reaction when suspended in water. These were washed several times
before use.

[15] In some cases it was necessary to add as much as 40 ml. of ethyl alcohol before
precipitation took place. Sufficient alcohol must be added to cause precipitation of
nearly all the protein.

tion is allowed to take place for 12 hours at $-18°$. The mixture is then centrifuged at 20,000 $\times$ g for 15 minutes at $-5°$. The supernatant fluid is discarded, and the precipitate is drained and triturated with 10 ml. of 0.07 M potassium phosphate buffer, pH 6.0. Only a small amount of the precipitate redissolves. The insoluble material is removed by centrifugation and discarded.

Step 5. Heat Inactivation of Crude Enzyme. A preparation active for the decarboxylation of α-acetolactic acid (α-acetolactic acid decarboxylase) but completely inactive on pyruvic acid is obtained by heating 10 ml. of the crude enzyme preparation in a test tube at 70° for 2 to 3 minutes in a water bath and removing the heat-denatured protein by centrifugation.

Properties

Specificity. The α-acetolactic acid-forming system is specific for pyruvic acid and will not decarboxylate α-ketobutyric acid or α-ketoglutaric acid. These α-keto acids do not act as competitive inhibitors for pyruvic acid. α-Acetolactic acid decarboxylase is specific for dextrorotatory α-acetolactic acid and attacks the levorotatory isomer at one-twentieth the rate of decarboxylation of the natural isomer. Oxalacetic acid and acetoacetic acid are not decarboxylated.

Activators and Inhibitors. The activity of the α-acetolactic acid-forming system can be doubled by the addition of TPP. These preparations are not sufficiently resolved to show a cationic requirement. Dolin and Gunsalus,[16] working with an acetoin-forming system from *Streptococcus faecalis*, have demonstrated a requirement for TPP ($K_m = 1 \times 10^{-6} M$) and for manganous ions ($K_m = 2 \times 10^{-5} M$). Magnesium and cobaltous ions showed 65% and 25% of the manganese activity, respectively, other common divalent ions showing virtually no activity. Using such preparations resolved for TPP and manganese (but not for activities on pyruvic acid and α-acetolactic acid), no activity was obtained with pyruvate as substrate, but full activity was obtained with α-acetolactic acid as substrate. Treatment of the α-acetolactic acid with 8-hydroxyquinoline followed by chloroform extraction to remove metals did not influence the rate of decarboxylation of α-acetolactic acid. Watt and Krampitz[17] have reported a manganous ion stimulation of decarboxylation of α-acetolactic acid by extracts from *Staphylococcus aureus*. Milhaud *et al.*[18] obtained an inhibition of 65% in the rate of pyruvate decarboxylation with 1×10^{-4}

[16] M. I. Dolin and I. C. Gunsalus, *J. Bacteriol.* **62**, 199 (1951).
[17] D. Watt and L. O. Krampitz, *Federation Proc.* **6**, 301 (1947).
[18] G. Milhaud, J. P. Aubert, and R. Gavard, *Compt. rend.* **234**, 2026 (1952).

M manganous ion, using an acetoin-forming cell-free extract obtained from *Bacillus subtilis*. Other divalent ions were without effect.

The following inhibitors have been tested for activity in limiting the rates of CO_2 evolution from pyruvic acid by a crude extract from *Acetobacter aerogenes:*[8] 0.05 M sodium fluoride, 36% inhibition; 0.05 M ethyl carbamate, 61%; 0.05% potassium cyanide, 45%; 0.125 M α-bromo-propionic acid, 39%; 0.033 M sodium arsenite, 83%; 0.044 M sodium azide, 88%; and 0.007 M iodoacetate, 85%. The α-acetolactic acid-forming component is completely inhibited by 0.001 M mercuric chloride. The *Streptococcus faecalis* system is inhibited 75% by 0.01 M phenylpyruvate.[16]

Effect of pH. Both enzyme components have optima at pH 6.0, in phosphate buffer, the activities falling to one-half maximum at approximately pH 5.0 and pH 7.0.

Substrate Concentrations. The K_m values for pyruvic acid and α-acetolactic acid are 0.037 M and 0.002 M, respectively. The K_m values obtained from the *Streptococcus faecalis* system are 0.09 M and 0.003 M, respectively.

SUMMARY OF PURIFICATION PROCEDURE[a]

Fraction	Total volume, ml.	CO_2, μl. in 15 minutes[b]		Q_{CO^2} (pyruvate)	Activity of crude enzyme in fraction, %	
		Pyruvate	α-Aceto-lactate		Pyruvate	α-Aceto-lactate
1. Crude extract	40	445	277	395	100	100
2. Alumina super-natant[c]	40	59	243	—	13.3	87.7
3. Alumina eluate	20	537	220	—	60.3	39.8[d]
4. Soluble part of ethanol precipitate	10	244	19	7000	27.4[e]	1.7
5. Heat inactivation of crude enzyme	—	0	220	—	0	85

[a] E. Juni, *J. Biol. Chem.* **195**, 715 (1952).

[b] Gas values are from 0.1 ml. of the respective enzyme fraction.

[c] The alumina eluate will contain virtually no activity on pyruvic acid if 7 to 8 g. of alumina is used (see step 3). Approximately 50 to 70% of the α-acetolactic acid decarboxylase activity will remain in the alumina supernatant fluid.

[d] It has been found that subjecting the crude enzyme preparation to adsorption on alumina results in a net increase in the sum of the α-acetolactic acid decarboxylase activities of all the fractions as compared with that of the crude extract.

[e] Calculated by assuming that twice the amount of CO_2 indicated would have been observed if sufficient α-acetolactic acid decarboxylase had been present, as was the case in the crude extract.

[73] Phosphoroclastic Split of Pyruvate, Yielding Formate (*E. coli*)

$$CH_3COCOOH + H_3PO_4 \rightleftarrows CH_3COOPO_3H_2 + HCOOH$$
$$CH_3COOPO_3H_2 + H_2O \rightleftarrows CH_3COOH + H_3PO_4$$

$$CH_3COCOOH + H_2O \rightleftarrows CH_3COOH + HCOOH$$

By HAROLD J. STRECKER

Assay Method

The system of enzymes which catalyze the above reaction has not as yet been purified, nor have any of the component enzymes been isolated. As a result, the reaction is poorly understood, and the methods used for assay are applicable only in terms of our limited comprehension of the mechanism. Two general assay procedures have been employed. The first method depends on the evolution of CO_2 from $NaHCO_3$ buffer, generated by the extra equivalent of acid produced by the over-all reaction.[1] Since there are enzymes present in the mixture which oxidize pyruvate and formate to CO_2, correction for this must be applied by subtracting the quantity of CO_2 evolved in phosphate buffer. The correction is not practical, however, when the rate of CO_2 evolution from these side reactions approaches the rate of an evolution from the main reaction. The second method of asay depends on the incorporation of labeled formate into the carboxyl group of pyruvate.[2,3] The limitations of this method arise from the possibility that the rate of incorporation is only partially dependent on the extent of splitting of pyruvate.

Method I

Reagents

0.25 M $NaHCO_3$.
0.5 M sodium pyruvate, 55.5 mg./ml. in H_2O adjusted to pH 6.8. (This solution should be kept cold and made up fresh every few days.)
0.5 M potassium phosphate buffer, pH 6.8.
Enzyme in buffer solution (see method of preparation).

[1] G. Kalnitsky and C. H. Werkman, *Arch. Biochem.* **2**, 113 (1943).
[2] M. F. Utter, F. Lipmann, and C. H. Werkman, *J. Biol. Chem.* **158**, 521 (1945).
[3] H, J, Strecker, H. G. Wood, and L O. Krampitz, *J. Biol. Chem.* **182**, 525 (1950).

Procedure. From 0.2 to 0.8 ml. of enzyme solution is incubated together with 0.4 ml. of the $NaHCO_3$ solution, 0.2 ml. of the potassium phosphate solution, and water to a volume of 1.8 ml. in the main compartment of a respirometer vessel. 0.2 ml. of the pyruvate solution is placed in the side arm. The vessel is then gassed for 10 minutes with a mixture of 50% CO_2 and 50% N_2 while shaking. The incubation is carried out at 30°. The pyruvate solution is tipped in after equilibration, and manometer readings are taken at 10-minute intervals.

Method II

Reagents

Same as method I except for omission of $NaHCO_3$.
0.5 M $HC^{14}OONa$ containing 10,000 to 50,000 counts per minute per mole.
2.5 N NaOH.
Saturated ceric sulfate in 6 N H_2SO_4.[4]

Procedure. 0.2 to 0.8 ml. of the enzyme solution is incubated together with 0.2 ml. of the potassium phosphate solution, 0.2 ml. of the sodium formate solution, and water to a total volume of 1.8 ml. in the main compartment of a respirometer vessel having two side arms. 0.2 ml. of the pyruvate solution is placed in one side arm, and 0.3 ml. of 3 N H_2SO_4 in the other. The vessel is gassed for 7 to 10 minutes with N_2 and incubated at 30°, the pyruvate solution being tipped in at the conclusion of gassing. After 5 to 15 minutes of incubation, the H_2SO_4 is tipped to stop the reaction. The solution is centrifuged free of denatured protein, and a 1.0- to 1.5-ml. aliquot is transferred to the main compartment of another respirometer vessel with 0.3 ml. of 2.5 N NaOH in the center well and a total of 0.6 to 0.8 ml. of ceric sulfate in the side arms. The ceric sulfate solution is tipped in cautiously in order to decarboxylate the residual pyruvate. The CO_2 thus formed is trapped in the NaOH over a period of 1 hour while shaking. The NaOH solution is then quantitatively removed with washing and reacted with $Ba(OH)_2$ solution to precipitate $BaC^{14}O_3$ which is plated and counted by standard methods. The usual techniques in working with $C^{14}O_2$ are adhered to throughout.

Preparation of Enzyme[1]

Escherichia coli (E-26) are grown at 37° under continuous aeration in a liquid medium containing 0.4% beef extract, 0.4% peptone, 0.2% yeast extract, 0.2% NaCl, and 10% tap water. After 16 to 18 hours of growth,

[4] H. A. Krebs and W. A. Johnson, *Biochem. J.* **31,** 645 (1937).

the cells (ca. 2.5 g./l., wet weight) are harvested with the aid of a Sharples centrifuge and ground with glass.[5] The suspension thus obtained is stirred for 15 to 20 minutes with ice-cold 0.001 M potassium phosphate buffer, pH 6.8, using 1.5 to 2.0 ml. of buffer per gram of wet cells. The extract is centrifuged free of cell debris at 13,000 r.p.m. for 30 to 45 minutes. The supernatant is a yellow opalescent solution which may be kept frozen for several weeks with slight loss of activity.

Properties[1,6]

Activators. Dialysis of the enzyme preparation against distilled water for periods of time shorter than 3 hours results in considerable decrease of activity which can be partially restored by addition of the following components: inorganic phosphate, 0.02 M; coenzyme A, 2 units/ml.; cocarboxylase, 20 γ/ml.; Mn^{++}, 5.0×10^{-4} M, and a boiled yeast extract. The addition of all of these components cannot fully restore activity. Dialysis for longer than 3 to 4 hours results in complete inactivation not restorable by the above components. The extent of inactivation by dialysis of short duration and restoration of activity by cofactors is variable and appears to depend partially on the age of the enzyme preparation.

Addition of DPN, adenylic acid, ATP, Mg^{++}, riboflavin, fumarate, and biotin does not stimulate the activity of a dialyzed preparation.

Inhibitors. Activity with regard to fixation of C^{14}-labeled formate is inhibited by bicarbonate, arsenate, imidazole, 2,4,6-collidine, and high salt concentrations. The inhibition by arsenate is not reversed by phosphate.

pH optimum. The optimum pH for both CO_2 evolution and $HCOO^{14}H$ fixation is 6.8.

[5] M. F. Utter and C. H. Werkman, *Biochem. J.* **386**, 485 (1942).
[6] H. J. Strecker, *J. Biol. Chem.* **189**, 815 (1951).

[74] Phosphoroclastic Split of Pyruvate, Yielding Hydrogen
(*Clostridium butylicum*)

$$\underset{\substack{\text{O}\\\|}}{\text{CH}_3\text{C}}\text{COOH} + \text{H}_3\text{PO}_4 \leftrightarrows \underset{\substack{\text{O}\\\|}}{\text{CH}_3\text{C}}\text{—OPO}_3\text{H} + \text{H}_2 + \text{CO}_2$$

By H. J. KOEPSELL

Assay Method

Principle. Enzyme activity is measured as Q_{H_2}, and as Q_{CO_2} if desired, in splitting pyruvate in phosphate buffer, with heated rat liver extract as activator, under a nitrogen atmosphere.

Reagents

0.05 M sodium pyruvate solution.

0.125 M sodium phosphate buffer, pH 6.5.

Rat liver extract. Liver from a freshly sacrificed adult animal is homogenized in an equal weight of water and heated in a boiling water bath for 5 minutes. The centrifuged extract is adjusted to pH 6.5 and stored frozen.

Enzyme solution. A solution containing 30 to 50 mg. of lyophilized enzyme powder per 0.3 ml. is prepared with freshly boiled ice-cold 0.05 M phosphate buffer and used immediately. Cell suspensions or other preparations are prepared in the same manner.

Procedure. Place 1 ml. each of pyruvate and buffer solutions, 0.3 ml. of liver extract, and 0.4 ml. of water in the main compartment, and 0.3 ml. of enzyme solution in the side arm of a standard 15-ml. Warburg vessel. For measurement of H_2 evolution only, place NaOH and filter paper in the center well. For concomitant measurement of CO_2 evolution, omit NaOH from a second vessel. Control vessels receive no pyruvate. Attach the vessels to manometers in a 37° bath, and replace the atmosphere with N_2. Equilibrate to temperature, mix the compartments, and determine the rate of gas evolution by standard manometric practice.

Wolfe and O'Kane[1] report that rat liver extract may be replaced by CoA, but an assay procedure with this modification has not been described.

Alternative Assay Methods. A method based on reduction of neotetrazolium is given by Wolfe and O'Kane.[1] Acetyl phosphate formation

[1] R. S. Wolfe and D. J. O'Kane, *J. Biol. Chem.* **205**, 755 (1953).

may be determined colorimetrically by the method of Lipmann and Tuttle.[2] Pyruvate utilization may also be determined.

Preparation Procedure

The enzyme system involved is obtained as a stable, dry, lyophilized powder from species of saccharolytic *Clostridia*. *Clostridium butylicum*, strain 21 of the University of Wisconsin collection,[3] was employed by Koepsell and Johnson.[4] Another strain of this species, and a different mode of preparation, was used by Wilson *et al.*[5] Use of *Clostridium butyricum*, strain 6014 of the American Type Culture Collection, is described by Wolfe and O'Kane.[1] The procedure given below is for laboratory preparation of the powder and is based on the work of Koepsell and Johnson.[4] A larger scale preparation has been described by Koepsell *et al.*[6]

Step 1. Culturing the Microorganisms. Spores of *Clostridium butylicum* strain 21, carried on soil, are germinated at 37° in sterile medium containing 2% glucose, 0.5% tryptone, and 0.1% reduced iron. The iron aids germination. Upon vigorous growth as indicated by gassing, the culture is inoculated at a rate of 2 to 5% into sterile propagation medium. This medium (16 l. in 5-gallon bottles is convenient) contains 1.5% glucose, 0.1% blackstrap molasses, 0.125% solubilized liver powder, 0.075% Difco yeast extract, 0.05% dibasic ammonium phosphate, 0.15% ammonium sulfate, and 0.0025% manganese sulfate. The ammonium salts of the medium are sterilized separately to minimize formation of inert precipitate on autoclaving. The separate solutions should be adjusted to pH 6.8 to 7.0 before sterilization. The inoculated medium is incubated at 37° with occasional shaking.

This strain tends to form a heavy slime at about 18 hours. If harvested after that time, the cells are centrifuged with difficulty and have a high endogenous gas evolution. The cells are therefore harvested, on a Sharples supercentrifuge with minimal washing, while the culture is gassing actively but before slime formation is excessive. The tightly packed cell cake is placed in stoppered vessels in a freeze chest and held frozen.

Step 2. Preparation of Active Dry Extract. While frozen, the cells are slowly ruptured by growth of endocellular ice crystals, and cell contents are released. Progress of this phenomenon is determined by periodically

[2] F. Lipmann and L. C. Tuttle, *J. Biol. Chem.* **158**, 505 (1945).
[3] This strain may be obtained from the Department of Bacteriology, University of Wisconsin, Madison 6, Wisconsin.
[4] H. J. Koepsell and M. J. Johnson, *J. Biol. Chem.* **145**, 379 (1942).
[5] J. Wilson, L. O. Krampitz, and C. H. Werkman, *Biochem. J.* **42**, 598 (1948).
[6] H. J. Koepsell, M. J. Johnson, and J. S. Meek, *J. Biol. Chem.* **154**, 535 (1944).

thawing a small portion of the cake, centrifuging to remove cell debris, and determining the enzyme activity of the supernatant liquid as compared to the activity of the suspended uncentrifuged cake. After 10 to 14 days the cake will have lost about 20% of its original activity, and half or more of the activity will be found in the supernatant. The cake is now suspended in freshly boiled ice-cold water to make a thin cream, allowed to stand about 10 minutes, and centrifuged in the cold to remove cell debris. Wolfe and O'Kane[1] report that addition of potassium ascorbate (1 mg./ml.) to the cream enhances the activity of the extract. The slightly cloudy, amber supernatant liquid is poured off and lyophilized. The grayish-brown, hygroscopic powder obtained is stored in tightly stoppered vessels in the refrigerator. It represents about one-third of the activity of the original frozen cells and has a Q_{H_2} of about 65.

Properties

Stability. Lyophilized powder is stable for months if kept dry. Refrigeration is helpful, although a portion retained ability to transphosphorylate butyrate from acetyl phosphate after 7 years' storage, during much of which time it was kept stoppered at room temperature.[7]

Enzyme solutions, however, are extremely labile and may lose 75 to 95% activity if kept in an ice bath and exposed to air.[1] Covering with hydrogen or nitrogen tends to reduce loss.[1] Up to 80% activity can be maintained for 3 hours by addition of potassium ascorbate or ferrous sulfate (1 mg./ml.).[1]

Further Purification. Reported attempts to purify the enzyme preparations further have been frustrated by lability of enzyme solutions. Partial purifications to free preparations of CoA, TPP, and Fe are described by Wolfe and O'Kane.[1]

Specificity and Types of Enzyme Activity. Enzyme preparation will not act on glucose, 3-PGA, lactate, or succinate.[4] Formate is not attacked,[4] and labeled formate does not exchange with pyruvate.[5] Transphosphorylation from acetyl phosphate to butyrate,[6] and from ATP to acetate and butyrate,[8] by the preparations has been described, with an explanation by Stadtman[9] of the mechanism involved. Glucose, if present, is phosphorylated with increase in over-all reaction rate.[6]

pH. Optimum is at pH 6.5, with about half activity at pH 5.8 and 7.5.[4]

Cofactors. CoA and TPP participate in the phosphoclastic reaction but are present in excess in the enzyme preparations.[1] Ferrous ions

---

[7] E. R. Stadtman and H. A. Barker, *J. Biol. Chem.* **184**, 769 (1950).

[8] F. Lipmann, *J. Biol. Chem.* **155**, 55 (1944).

[9] E. R. Stadtman, *J. Biol. Chem.* **203**, 501 (1953).

activate, cobaltous ions may be substituted but are less effective.[1] Pyruvate oxidation factor is apparently not involved.[1] ATP,[1] and possibly CoA,[9] mediate in transphosphorylation of fatty acids.

Inhibitors. Hydrogen gas inhibits the forward reaction.[2] Arsenite shows slight inhibition, and streptomycin sulfate is without effect.[1]

Alternate Hydrogen Acceptors. Instead of combining to form H_2, hydrogen atoms may be deposited on some alternate acceptors. Riboflavin, FMN, FAD, certain nitrofuraldehyde semicarbazones, triphenyltetrazolium chloride, and neotetrazolium can act as such acceptors.[1] MB is slightly active.[1] The flavins are autoxidizable and permit use of O_2 as ultimate hydrogen acceptor; with FMN the product was water rather than hydrogen peroxide.[1]

Reversibility. The reaction has been shown to be reversible by Wilson et al.[5] Hydrogen gas enhances the reverse reaction.

[75] Phosphate-Linked Pyruvic Acid Oxidase from *Lactobacillus delbrückii*

(*Bacillus acidificans longissimus*, Lafar)

$$\text{Pyruvate} + P + O_2 \rightarrow Ac{\sim}P + CO_2 + H_2O_2$$

By L. P. HAGER and FRITZ LIPMANN

Introduction

The enzyme is derived from a special strain of *Lactobacillus delbrückii* obtained originally from the Institut für Garungsgewerbe in Berlin, Germany.[1] After having been lost here during the recent war, the original strain was relocated in 1953 again in Germany with the kind help of Dr. Feodor Lynen. A large variety of other strains have been tried without success.

The enzyme is extracted from dry preparations or from fresh paste by various methods. It can be separated by acid ammonium sulfate precipitation into an apoenzyme fraction and several cofactors.[1] The cofactors include: (1) thiamine pyrophosphate, (2) flavin adenine dinucleotide, (3) a divalent cation such as maganese, magnesium, or cobalt, and (4) inorganic phosphate.

During considerable purification in a manner described later on, no resolution of the system into separate enzyme units pertaining to TPP

---

[1] F. Lipmann, *Cold Spring Harbor Symposia Quant. Biol.* **7**, 248 (1939).

and FAD, respectively, has been obtained, and we feel inclined to assume that a single enzyme protein may carry the two coenzymes as two prosthetic groups.

The oxidation of pyruvate is phosphate-dependent; the products are acetyl phosphate and carbon dioxide.[2] In a secondary nonenzymatic reaction the hydrogen peroxide formed in the flavin oxidation reacts with a second mole of pyruvate to form free acetate and CO_2.

Equations

$$CH_3\!-\!\overset{O}{\overset{\|}{C}}\!-\!COOH + P + O_2 \;\rightarrow\; CH_3\!-\!\overset{O}{\overset{\|}{C}}\!\sim\!P + CO_2 + H_2O_2 \qquad (1)$$

$$CH_3\!-\!\overset{O}{\overset{\|}{C}}\!-\!COOH + H_2O_2 \;\;\rightarrow\; CH_3\!-\!\overset{O}{\overset{\|}{C}}\!-\!OH + CO_2 + H_2O \qquad (2)$$

Sum:

$$2CH_3\!-\!\overset{O}{\overset{\|}{C}}\!-\!COOH + P + O_2 \rightarrow CH_3\!-\!\overset{O}{\overset{\|}{C}}\!\sim\!P + CH_3\!-\!\overset{O}{\overset{\|}{C}}\!-\!OH + 2CO_2$$
$$+ H_2O \quad (3)$$

Assay Method

Principle. The assay is based on the measurement of oxygen uptake during the oxidative decarboxylation of pyruvate. The oxygen uptake values can be confirmed by the measurement of Ac~P formation using the hydroxamic acid method.[3] In addition the enzyme may be assayed by using ferricyanide, methylene blue, and tetrazolium as electron acceptors.[4]

Reagents

1.0 M K orthophosphate buffer, pH 6.0.
0.1 M $MgCl_2$.
0.01 M TPP.
0.5 M K pyruvate.
5 N KOH.
Enzyme. Use an amount of enzyme to give a rate of 40 to 200 μl. of O_2 uptake per 30 minutes.

Procedure. Each Warburg cup contains 200 micromoles of phosphate buffer, 10 micromoles of $MgCl_2$, 0.2 micromole of TPP, 50 micromoles of pyruvate in the side arm, and 750 micromoles of KOH in the center well

[2] F. Lipmann, *J. Biol. Chem.* **155**, 55 (1944).
[3] F. Lipmann and L. C. Tuttle, *J. Biol. Chem.* **159**, 21 (1945).
[4] L. P. Hager, D. M. Geller, and F. Lipmann, *Federation Proc.* **13**, 11 (1954).

with a filter paper wick. After temperature equilibration, the pyruvate is tipped in from the side arm to start the reaction.

Definition of Unit and Specific Activity. One unit of enzyme is defined as that amount which catalyzes one microliter of oxygen uptake per 30 minutes under the above conditions. Specific activity is expressed as units per milligram of protein. Before protamine treatment (see Purification) protein is determined by the TCA method of Stadtman *et al.*;[5] after protamine treatment, protein is determined by the optical method of Warburg and Christian.[6]

Application of Assay Method to Crude Preparation. Before extraction, dried cells give low results in the assay. However, the enzyme content of crude sonic extracts and the purified preparations can be readily estimated by this assay. Crude extracts are usually capable of hydrolyzing Ac~P, and therefore Ac~P formation is rarely equivalent to oxygen uptake in the crude extracts. NaF[1] depresses the hydrolysis of Ac~P and may be used in the assay of crude extracts. During purification, the enzymes responsible for the hydrolysis Ac~P are lost, and the theoretical 1:1 ratio between Ac~P and oxygen uptake is achieved.

Preparation of Enzyme

Growth Medium. The growth medium contains per liter: 15 g. of Difco malt extract broth (dissolved first and adjusted to pH 6.5), 5 g. of Difco yeast extract, 10 g. of Bacto tryptose, 3 g. of glucose, 10 g. of $CaCO_3$, 5 ml. of salts B (salts B contains per 250 ml.: 10 g. of $MgSO_4·7H_2O$, 0.5 g. of NaCl, 0.5 g. of $FeSO_4·7H_2O$, 0.5 g. of $MnSO_4·4 H_2O$, and of 0.5 g. ascorbic acid) and 100 ml. of tomato juice (prepared by steaming canned tomatoes and removing the pulp by filtration on a Buchner funnel using filter aid).

Inoculum. Lactobacillus delbrückii (Bacillus acidificans longissimus, Lafar) is grown at 37° in the above medium. Actively fermenting cells (log phase) are inoculated at a 10% level into 5-gallon Pyrex fermentor jugs containing 15 l. of growth medium.

Growth. The 5-gallon jugs are incubated with stirring and under slight nitrogen pressure for 20 to 22 hours. The cells are harvested by centrifugation and washed once with distilled water. The resulting cell paste can be frozen directly or dried *in vacuo.*

Purification Procedure

Step 1. Preparation of Cell Extract. 10 g. of cell paste (or the equivalent in dried cells, which is 3 g.) plus 10 g. of alumina is suspended in 40 ml. of

[5] E. R. Stadtman, G. D. Novelli, and F. Lipmann, *J. Biol. Chem.* **191**, 365 (1951).
[6] O. Warburg and W. Christian, *Biochem. Z.* **310**, 384 (1941).

0.02 M K phosphate buffer, pH 7.0, and oscillated for 20 minutes in a Raytheon 10-kc sonic oscillator. The cell debris is removed by centrifugation for 30 minutes at top speed in the Servall SS-1 centrifuge at 0°. The supernatant is frozen, thawed, and recentrifuged.

Step 2. First Ammonium Sulfate Precipitation. The cell-free extract is adjusted to 0.75 ammonium sulfate saturation. The precipitate is collected and suspended in a volume of buffer equal to the original volume.

Step 3. Removal of Nucleic Acid. Protamine sulfate is added to the 0 to 0.75 ammonium sulfate fraction until a 280/260-mμ absorption ratio of 0.9 to 1.0 is obtained on centrifugation. The protamine nucleate precipitate is removed by centrifugation and discarded.

SUMMARY OF PURIFICATION PROCEDURE

Fraction	Total volume, ml.	Units/ml., thousands	Total units, thousands	Protein, mg./ml.	Specific activity, units/mg.	Recovery, %
1. Sonic extract	600	0.532	319	26.2	20.0	—
2. (NH$_4$)$_2$SO$_4$ fraction, 0–0.75	590	0.537	317	12.3	43.6	99
3. After protamine	600	0.505	301	9.7	52.0	95
4. (NH$_4$)$_2$SO$_4$ fraction, 0–0.45	176	0.995	182	15.9	62.5	57
5. Supernatant after acid	176	0.955	180	4.6	208	56
6. Eluate from gel	35	2.51	88	4.3	580	28

Step 4. Second Ammonium Sulfate Fractionation. The supernatant from step 3 is adjusted to 0.45 ammonium sulfate saturation to precipitate the enzyme.

Step 5. Acid Precipitation. The resuspended precipitate from step 4 is adjusted with acetic acid to pH 4.4 and then to pH 4.1. The precipitates are removed by centrifugation and discarded. The supernatant is adjusted to pH 6.6 with NaOH.

Step 6. Calcium Phosphate Gel. Inactive protein is removed from the supernatant from step 5 by adsorption on calcium phosphate gel[7] (1.5 mg./mg. protein). The supernatant is then adjusted to pH 5, and the pyruvic oxidase is adsorbed on the gel. The gel is first washed with 0.02 M K phosphate buffer, pH 7.0, and then the enzyme is eluted from the gel with 0.1 M K phosphate, pH 7.0.

[7] D. Keilin and E. F. Hartree, *Proc. Roy. Soc. (London)* **B124,** 397 (1938); see also Vol. I [11].

Properties

Cofactors. The apoenzyme may be prepared by acid ammonium sulfate precipitation in the following manner: the extract is brought to 0.15 ammonium sulfate saturation with saturated ammonium sulfate and then to 0.3 saturation with saturated ammonium sulfate containing 3 ml. of concentrated H_2SO_4 per liter. The precipitate is discarded, and the supernatant is brought to 0.5 saturation with the acid ammonium sulfate. The precipitate is suspended and neutralized. This procedure may be repeated if the 0.3 to 0.5 acid ammonium sulfate fraction is not completely dependent upon TPP and FAD.

Specificity. The enzyme will oxidatively decarboxylate α-ketobutyrate to propionyl phosphate at one-third the rate of pyruvate oxidation. It is inactive toward glyoxyllic acid.

Comments

In addition to the mentioned cofactors, no involvement of CoA or lipoic acid is indicated, nor does the reaction show the characteristics of an SH-linked reaction such as sensitivity to iodoacetate or mercury compounds. A phosphotransacetylase function found with cruder preparation was shown to be due to a CoA-dependent transacetylase which can be separated from the pyruvic oxidase and appears not to be connected with the oxidation reaction proper.[4] In the absence of oxygen or other hydrogen acceptor, the system was considered nonreactive until recently, when a slow acetoin formation was found to occur (unpublished). The rate of this reaction is approximately one hundred times slower than the oxidative breakdown.

[76] Coenzyme A-Linked Pyruvic Oxidase (Animal)

Pyruvate + DPN$^+$ + CoA $\rightleftarrows$ Ac$\sim$SCoA + DPNH + H$^+$ + CO$_2$

By S. KORKES

Assay Method

Principle. Pyruvate is converted to Ac$\sim$SCoA and CO_2 in the presence of DPN and CoA. The reaction may be followed by measurement of pyruvate disappearance, acetyl phosphate formation, or CO_2 evolution in the presence of transacetylase and orthophosphate, and lactic dehydrogenase.[1,2]

[1] S. Korkes, A. del Campillo, and S. Ochoa, *J. Biol. Chem.* **195**, 541 (1952).
[2] See Vol. I [77].

Procedure. The reaction mixture contains (in addition to the enzyme to be assayed) the following reactants (in micromoles): potassium phosphate, pH 7.4 (100); $MnCl_2$ (2.5); L-cysteine (15); DPN (0.15); potassium pyruvate (50); CoA (0.10); and TPP (0.2). 2000 units of lactic dehydrogenase[3] and 15 units of transacetylase[3] are added. Total volume, 2 ml. The simplest index of reaction is the measurement of acetyl phosphate formation by conversion to acetohydroxamic acid.[4,5] For this purpose it is convenient to run the reaction in stoppered tubes gassed with N_2 at 30° for 1 hour. A unit of enzyme may be defined as the amount required to form 1 micromole of acetyl phosphate per hour. CO_2 liberation may be measured manometrically with the same reaction mixture. Activity measured manometrically may be greater than by estimation of acetyl phosphate formation, since, if prepared from pig heart, the enzyme may contain some acyl phosphatase, whereas that prepared from pigeon breast muscle may form acetoin and carbon dioxide as a side reaction. If pyruvate disappearance is to be measured,[5] it should be borne in mind that 2 moles of pyruvate disappear per mole of acetyl phosphate formed, since the assay is measuring the reaction

$$2 \text{ Pyruvate} + P \rightarrow \text{Acetyl phosphate} + \text{lactate} + CO_2$$

Preparation of Enzyme. A crude preparation from pig heart may be prepared[1] by mincing fresh chilled hearts, washing the mince five times each with 6 vol. of cold water, and draining through cheesecloth. The mince is extracted in a Waring blendor with 3.5 vol. of ice-cold 0.2 M K_2HPO_4 for 5 to 6 minutes and then allowed to stand in the cold for 1 hour, with occasional stirring. After straining through cheesecloth, the extract is centrifuged at 3000 × g at 0° for 10 minutes. The residue is re-extracted with one-third the volume of fluid used originally, and the supernatant is combined with the first extract. Solid ammonium sulfate (35 g./100 ml.) is added, and the mixture is centrifuged at 15,000 × g in a Servall SS-1 centrifuge in the cold. The precipitate is dissolved in 0.02 M potassium phosphate buffer, pH 7, and dialyzed overnight against the same buffer at 0°. The dialyzed solutions are clarified by high-speed centrifugation. These preparations are 5- to 10-fold purified relative to the initial extract and have a specific activity of 0.6 to 1.0 units/mg. of protein. The activity decreases on storage, but more slowly at −18° than at 0°.

---

[3] See Vol. I [67, 98].
[4] F. Lipmann and L. C. Tuttle, *J. Biol. Chem.* **159,** 21 (1945).
[5] See Vol. III [39, 66].

An enzyme preparation which is about fifteen times as active as the one described above, when assayed with ferricyanide as oxidant, may be prepared from pigeon breast muscle by the following procedure.[6]

Step 1. Homogenization and Extraction. Frozen breast muscle from twelve pigeons (700 g.) is minced, suspended in 3 to 4 l. of ice-cold redistilled water, and washed several times by decantation and finally by gentle squeezing through cheesecloth. The washed muscle is suspended in 3 l. of 0.01 M phosphate buffer, pH 7.5, and aliquots are homogenized in a Waring blendor for 2 minutes. The blendor should be filled to minimize foaming. Homogenization and subsequent operations are performed at 0 to 5°. The homogenate is then centrifuged for 1 hour at 4000 $\times$ g. The supernatant fluid, which should be only slightly turbid, is poured through cheesecloth to remove large particles of fat. The precipitate is discarded. Extract, 1900 ml.; specific activity = 0.1 to 0.3; yield = 10,000 to 12,000 units.

Step 2. Precipitation at pH 5.4. The extract (pH 6.5) is adjusted to pH 5.4 by the addition of 10% acetic acid, dropwise with stirring. Owing to the lability of the enzyme under acid conditions, the solution is kept at 0°. The extract is then centrifuged for 30 minutes at 4000 $\times$ g. The supernatant fluid is discarded, and the precipitate is taken up in 50 ml. of 0.01 M phosphate buffer, pH 7.5, and adjusted to pH 6.5 with 1.0 N sodium hydroxide. The thick suspension is homogenized by hand in a Potter-Elvehjem homogenizer to break up lumps. Fraction, 70 ml.; specific activity = 0.5 to 1.5; yield 8000 to 10,000 units.

Step 3. Freezing and Thawing. Fraction 1 is frozen in a cold box at −15°. The enzyme is frozen and thawed several times over a period of 24 to 48 hours. This treatment results in protein coagulation; hence, the original thick cream will show a definite precipitate and a clear supernatant fluid on standing. The suspension is then centrifuged for 30 minutes at 20,000 $\times$ g. The precipitate is discarded, and the supernatant solution filtered to remove small fat particles. Fraction 2, 50 ml.; specific activity = 5.0 to 7.0; yield = 7000 to 9000 units.

Step 4. Fractionation at pH 6.2. Fraction 2 is adjusted to pH 5.4 with 1% acetic acid and centrifuged for 20 minutes at 4000 $\times$ g. The supernatant fluid is discarded, and the precipitate is taken up in 0.1 M phosphate buffer, pH 6.3, and water so that the final phosphate concentration is 0.05 M and the enzyme concentration is 500 to 550 units/ml. The pH

---

[6] V. Jagannathan and R. S. Schweet, *J. Biol. Chem.* **196**, 551 (1952). The authors employed an assay in which activity is expressed as micromoles of CO_2 liberated per hour in the presence of bicarbonate and ferricyanide. Three micromoles of CO_2 are produced per micromole of pyruvate oxidized under these conditions. In this assay, the product is free acetate; DPN and CoA do not augment activity.

is 6.2 to 6.3. The suspension is assayed before the final adjustment to the required volume. The thick creamy material is then frozen overnight. After thawing, the suspension is centrifuged for 30 minutes at 20,000 × g. The precipitate is discarded, and the viscous, golden supernatant fluid is stored. Fraction 3, 15 ml.; specific activity = 8.0 to 10.0; yield = 6000 to 8000 units.

Step 5. Fractionation at pH 5.9. Fraction 3 is diluted to 50 ml. (enzyme concentration 150 units/ml.) and dialyzed against 0.01 M phosphate buffer, pH 5.9, for 24 to 36 hours. Four 500-ml. portions of buffer are used, and the enzyme solution is stirred slowly, but continuously, during this period to promote more efficient dialysis and coagulation of metastable protein. The enzyme solution is then at pH 5.9. The precipitated protein is removed by centrifugation for 30 minutes at 4000 × g. Fraction 4, 50 ml.; specific activity = 17 to 20; yield = 4000 to 5000 units.

Step 6. Fractionation at pH 5.7. Fraction 4 is adjusted to a phosphate concentration of 0.1 M by the addition of 1.0 M phosphate, pH 5.9. Phosphoric acid (0.1 M) is then added dropwise with rapid stirring until the first traces of precipitate begin to form (pH 5.65 to 5.7). The solution is then centrifuged for 15 minutes at 5000 × g. The precipitate is taken up in 10 ml. of 0.01 M phosphate buffer, pH 7.5. The supernatant solution is then adjusted to pH 5.4, and the precipitate is removed and taken up in the same way. This latter fraction contains the most highly purified enzyme. Fraction 5, 10 ml.; specific activity = 35 to 50; yield = 1500 to 2500 units.

Properties

Pyruvic oxidase prepared from pigeon breast muscle, as described, is reported to be homogeneous both electrophoretically and by sedimentation, with a molecular weight of four million.[7] The enzyme contains lipoic acid at the level of 1 protogen unit per microgram of protein. In addition to its ability to oxidize pyruvate, it also catalyzes the formation of 1 mole of acetoin from 2 moles of pyruvate.[6]

The pig heart enzyme makes no acetoin under equivalent conditions. According to Schweet et al.,[8] it does not react with artificial acceptors, such as ferricyanide and dyes.

The animal tissue preparations are contaminated with varying amounts of lactic dehydrogenase. In the pigeon breast muscle prepara-

---

[7] R. S. Schweet, B. Katchman, R. M. Bock, and V. Jagannathan, *J. Biol. Chem.* **196**, 563 (1952).

[8] R. S. Schweet, M. Fuld, K. Cheslock, and M. H. Paul, *in* "Phosphorus Metabolism" (McElroy and Glass, eds.), Vol. I, p. 246, The Johns Hopkins Press, Baltimore, 1951.

tion, DPN reduction can be demonstrated spectrophotometrically, as in the case of the purified bacterial preparations,[2] but the presence of lactic dehydrogenase makes the accurate quantitative assay of pyruvic oxidase by this procedure unfeasible.

Neither the pig heart nor the pigeon breast muscle preparations has thus far been resolved into two or more protein components, in contrast to the bacterial pyruvic oxidase.[2,9] However, the demonstration in pig heart of a protein fraction which activates fraction A of the bacterial system[1] strongly suggests that the animal pyruvic oxidase complex may be resolved into separate, single-step catalysts.[2]

[9] S. Korkes, A. del Campillo, I. C. Gunsalus, and S. Ochoa, J. Biol. Chem. **193**, 721 (1951).

[77] Coenzyme A-Linked Pyruvic Oxidase (Bacterial)

$$\text{Pyruvate}^+ + \text{DPN}^+ + \text{CoA} \rightleftharpoons \text{Ac}{\sim}\text{SCoA} + \text{CO}_2 + \text{DPNH} + \text{H}^+$$

By S. KORKES

Assay Method

Principle. The dismutation assay[1] may be employed at all levels of purity of the enzyme. It is based on measurement of CO_2 liberation or estimation of acetyl phosphate in the presence of transacetylase. The spectrophotometric assay is applicable to preparations free of DPNH oxidase and lactic dehydrogenase.

Procedure. DISMUTATION ASSAY. The reaction mixture[2] contains the following (in micromoles): potassium phosphate, pH 6.0 (150); $MnSO_4$ (1.0); TPP (0.2); DPN (0.4); CoA (0.1); cysteine (10); and potassium pyruvate (50). Lactic dehydrogenase[3] (2000 units), transacetylase[3] (10 units), and appropriate amounts of enzyme fractions are added. The reaction is carried out in the Warburg apparatus at 30° under nitrogen. The total fluid volume is 1.4 ml. The evolved CO_2 is measured. Independent assay of fraction A (to be described below) is carried out in the presence of excess fraction B, and vice versa.

SPECTROPHOTOMETRIC ASSAY.[2] The reaction mixture contains the following (in micromoles): potassium phosphate buffer, pH 7.0 (200);

[1] S. Korkes, A. del Campillo, I. C. Gunsalus, and S. Ochoa, J. Biol. Chem. **193**, 721 (1951).
[2] L. P. Hager and I. C. Gunsalus, in press. The description given here is largely derived from the Ph.D. thesis of L. P. Hager.
[3] See Vol. I [67, 98].

MgCl₂ (0.3); CoA (0.1); DPN (0.4); GSH (10); TPP (0.2); and potassium pyruvate (5). Appropriate amounts of fractions A and B are added to give measurable rates in the Beckman spectrophotometer. The rate of reduction of DPN at 340 mμ is measured in the initial phase of the reaction and calculated as micromoles per hour, employing a value of 6.6×10^3 as the molar extinction coefficient of DPNH.[2] In this assay the reaction measured is that described by the equation above; in the dismutation, the CoA and DPN act catalytically and are regenerated by the actions of transacetylase and lactic dehydrogenase, respectively.

Definition of Unit and Specific Activity. One unit of either fraction A or B is that amount which catalyzes the conversion of 1 micromole of pyruvate to oxidation products per hour, in the presence of an excess of the other component. Specific activity is expressed in units per milligram of protein, the latter determined by light absorption[4] when the 280/260-mμ absorption ratio is greater than 0.6. For preparations where this condition is not met, protein may be determined turbidimetrically, or by the method of Lowry *et al.*[5]

Purification Procedure

The method to be described is that of Hager and Gunsalus.[2] *E. coli*, Crookes strain, is grown at 30° in a medium consisting of 1% tryptone, 1% yeast extract, 0.5% K₂HPO₄, and 0.3% glucose, and 0.5 ml. of an 8- to 15-hour culture is used as an inoculum into a 2-l. Erlenmeyer flask containing 500 ml. of the following medium: 0.2% NH₄Cl; 0.4% cerelose; 0.25% sodium glutamate; 0.0005% yeast extract (Fleischmann No. 3); 0.15% KH₂PO₄; 1.35% Na₂HPO₄; 0.02% MgSO₄·7H₂O, 0.001% CaCl₂; and 0.00005% FeSO₄·7H₂O. The phosphate, glutamate, and glucose are sterilized in separate containers and added to the salt solution after the latter is autoclaved. After overnight incubation with shaking at 30°, the contents of two such flasks are used as an inoculum for 15 l. of the same medium contained in 5-gallon Pyrex jugs. These are incubated at 30° with vigorous aeration (1 vol. of air per volume of medium per minute). Maximal growth is obtained in 6 to 7 hours, and the cells are harvested by means of a Sharples centrifuge. The yield is 10 g. of cell paste per liter of medium.

Step 1. Preparation of Extracts. 30 g. of cell paste is suspended in 150 ml. of 0.02 *M* phosphate buffer, pH 7.0, containing 0.5 mg. of GSH per milliliter. The suspension is subjected to the action of a Raytheon

[4] O. Warburg and W. Christian, *Biochem. Z.* **310**, 384 (1941).
[5] O. H. Lowry, N. J. Rosebrough, A. L. Farr, and R. J. Randall, *J. Biol. Chem.* **193**, 265 (1951).

10-kc. oscillator for 20 minutes, followed by centrifugation in a Servall SS-1 centrifuge at maximum speed for 1 hour at 0°. The supernatant contains 40 mg. protein per milliliter. All subsequent steps are carried out in the cold, and fractions are stored at −15°.

Step 2. First Ammonium Sulfate Fractionation. The sonic extract is diluted to contain 20 mg. protein per milliliter (300 ml.), and solid ammonium sulfate is added to 0.2 saturation. The precipitate is removed by centrifugation and discarded. The supernatant is brought to 0.75 ammonium sulfate saturation, and the precipitate collected by centrifugation and suspended in a volume of phosphate-GSH buffer (0.02 M potassium phosphate, pH 7.0, 0.5 mg. of GSH per milliliter)[6] equivalent to five-sixths the starting volume of extract (250 ml.).

Step 3. Removal of Nucleic Acid with Protamine. Fraction 1 is adjusted to pH 6.0 with 1 N acetic acid. 2% protamine sulfate, pH 5.0, is added with vigorous stirring until the 280/260-mμ absorption ratio of 1.0 to 1.1 is obtained in a centrifuged aliquot. After centrifugation, the supernatant is dialyzed against 0.02 M phosphate, pH 7.0; 0.005 M cysteine. The precipitate appearing on dialysis is centrifuged off and discarded. Volume, 295 ml.

Step 4. Second Ammonium Sulfate Fractionation. Saturated ammonium sulfate adjusted to pH 7.0 with ammonia is added to fraction 2, and three successive fractions are collected by centrifugation. The first (0 to 0.36 saturation) contains the enzyme for ketoglutarate oxidation analogous to fraction A. Fraction A itself is concentrated in the 0.36 to 0.48 fraction. The fraction between 0.48 and 0.60 is discarded. The supernatant is saturated with solid ammonium sulfate, and the fraction so obtained (0.6 to 1.0) is designated fraction B. The fractions are dissolved in a volume of buffer one-fourth the volume of initial extract (60 to 70 ml.).

Step 5. Removal of Transacetylase by Heating. Fraction A (0.36 to 0.48) and fraction B (0.6 to 1.0) are treated in an identical manner. The pH is adjusted to 6.0 with 1 N acetic acid, and the fraction heated to 50° and held at that temperature for 10 minutes. The denatured protein is removed by centrifugation. Transacetylase is effectively removed from each of the fractions by the treatment.

Step 6A. Refractionation of A with Ammonium Sulfate. Saturated ammonium sulfate is added to heat-treated fraction A, and fractions are collected as follows: 0 to 0.36; 0.36 to 0.40; 0.40 to 0.44; and 0.44 to 0.50. Usually the fraction 0.36 to 0.40 contains about 90% of the activity. A

---

[6] All subsequent ammonium sulfate fractions are dissolved in this buffer unless otherwise specified. For assay of such fractions, it is necessary to dialyze as in step 3 to remove ammonium sulfate, which is inhibitory.

pilot run is advisable to determine how an individual batch may fractionate. The fraction is taken up in sufficient buffer to give a protein concentration of approximately 5 to 6 mg./ml. (9 to 10 ml.).

Step 7A. Adsorption of A on Calcium Phosphate Gel. The fraction from step 6A is adjusted to pH 6.3 with 1 N acetic acid and 0.1 to 0.2 vol. of calcium phosphate gel (25 mg. dry weight per milliliter) is added with stirring. After 15 minutes, the gel is removed by centrifugation and treated in succession with 0.1 M potassium phosphate buffer, pH 6.5, 7.0, and 7.5. Each eluant is stirred with the gel for 15 minutes and is employed in a volume equal to that of the starting fraction for this step. The three eluates are discarded, and the gel is eluted with a volume of 0.25 saturated ammonium sulfate equivalent to one-half the starting fraction (5 ml.). This eluate contains the fraction A activity.

Step 6B. Refractionation of B with Ammonium Sulfate. The heated 0.6 to 1.0 fraction of step 6 is subjected to saturated ammonium sulfate, and fractions corresponding to 0 to 0.5, 0.5 to 0.6, and 0.6 to 0.7 saturation are collected. The activity is found almost entirely in the 0.5 to 0.6[7] saturation fraction. Volume, 17 ml.

Step 7B. Adsorption of B on Calcium Phosphate Gel. The fraction from step 6B is adsorbed under conditions identical with those described for fraction A. The elution is performed in the same manner, and the activity is found in the pH 7.0 and pH 7.5 phosphate eluates, which are combined and constitute purified fraction B (32 ml.).

Properties

Fraction A. The low ammonium sulfate fraction appears to be the site of decarboxylation of pyruvic acid, as manifested by the exchange of CO_2 into the carboxyl of pyruvate.[8] This requires the presence of TPP, but none of the other components (DPN, CoA, fraction B) possesses any ability to augment the rate of exchange. When ferricyanide is employed as artifical oxidant, again the only requirements are for fractions A and TPP, as is true of the reaction with 2,6-dichlorophenolindophenol. The product of these oxidations is acetate. The inference that fraction A may catalyze decarboxylation without oxidation is supported by the observation that only this fraction and TPP are required to form acetoin from pyruvate and acetaldehyde. On the other hand, this fraction gives rise to no demonstrable acetaldehyde and CO_2 formation when incubated with pyruvate alone, in contrast to yeast carboxylase.

[7] In one run the activity was found in the 0.6 to 0.7 fraction.
[8] S. Korkes, *Brookhaven Symposia Biol.* No. **5**, 192 (1952).

Lipoic acid is found to be concentrated in the A fraction. The reversible transfer of acetyl from $Ac\sim SCoA$ to lipoic acid is also catalyzed by the A fraction. This latter activity has been referred to as lipoic transacetylase.[9]

Fraction B. The high ammonium sulfate fraction functions as a lipoic dehydrogenase,[10] catalyzing the reduction of DPN by reduced lipoic acid. This reaction appears to be reversible. This reaction proceeds with both the D- and L-isomers of lipoic acid, in contrast to the lipoic transacetylase, which shows optical specificity, presumably for the D-isomer. Unlike the mutant enzymes of Reed and DeBusk,[11] LTPP cannot substitute for the B fraction in the dismutation, nor can it substitute for TPP when the latter has been removed by dialysis against Versene or alkaline pyrophosphate.[2]

The behavior of these fractions is consistent with the following stepwise pathway:

$$\text{Pyruvate} + \text{TPP} \rightleftarrows \text{``Aldehyde-TPP''} + CO_2 \qquad (1)$$

$$\text{Aldehyde-TPP} + \begin{matrix} S \\ | \\ S \end{matrix} R \rightleftarrows \begin{matrix} Ac\sim S \\ \\ HS \end{matrix} R + \text{TPP} \qquad (2)$$

$$\begin{matrix} Ac\sim S \\ \\ HS \end{matrix} R + \text{CoA} \rightleftarrows Ac\sim SCoA + \begin{matrix} HS \\ \\ HS \end{matrix} R \qquad (3)$$

$$\begin{matrix} HS \\ \\ HS \end{matrix} R + DPN^+ \rightleftarrows \begin{matrix} S \\ | \\ S \end{matrix} R + DPNH + H^+ \qquad (4)$$

Reactions 1 and 3 are catalyzed by the A fraction, and step 4 by the B fraction. $\begin{matrix} S \\ / | \\ R \\ \backslash | \\ S \end{matrix}$ would appear to be LTPP in the case of the *E. coli* mutant of Reed and DeBusk.[11] However, the inactivity of LTPP in the system described above[2] leaves open the possibility of other coenzyme forms of lipoic acid.

[9] I. C. Gunsalus, *in* "The Mechanism of Enzyme Action" (McElroy and Glass, eds.), p. 545, The Johns Hopkins Press, Baltimore, 1954.

[10] L. P. Hager and I. C. Gunsalus, *J. Am. Chem. Soc.* **75**, 5767 (1953).

[11] L. J. Reed and B. G. DeBusk, *J. Am. Chem. Soc.* **74**, 4727 (1952).

SUMMARY OF PURIFICATION PROCEDURE

	Volume, ml.	Total protein, mg.	Total units, thousands	Specific activity, units/mg.	Yield, %
Fraction A					
1. Cell extract	300	6300	11.6	1.8	100
2. 0.20–0.75 AS[a]	250	5950	9.0	1.5	77
3. Protamine supernatant	295	2380	13.6	5.7	118
4. 0.36–0.48 AS	60	387	6.90	17.8	60
5. Heat-treated supernatant	60	143	3.19	22.0	28
6. 0.40–0.44 AS	9	54	2.58	48	22
7. Gel eluate	5	57	1.95	340	17
Fraction B					
1. Cell extract	300	6300	18.0	2.9	100
2. 0.20–0.75 AS	250	5950	10.5	1.7	59
3. Protamine supernatant	295	2380	8.8	3.7	49
4. 0.60–1.0 AS	70	520	10.4	20.0	58
5. Heat-treated supernatant	67	308	9.5	31	53
6. 0.50–0.60 AS	17	144	8.2	56	45
7. Gel eluate	32	9.4	3.84	410	21

[a] Ammonium sulfate.

[78] Liver Alcohol Dehydrogenase

$$\text{Alcohol} + \text{DPN} \underset{\text{ADH}}{\rightleftharpoons} \text{Acetaldehyde} + \text{DPNH} + \text{H}^+$$

By ROGER K. BONNICHSEN and NORMAN G. BRINK

Assay Method

Principle. Routine activity determination as described by Theorell and Bonnichsen[1] is based on spectrophotometric measurement of the amount of DPN being reduced in 3 minutes at pH 9.6 in the presence of excess of alcohol.

Reagents

0.1 M glycine-sodium hydroxide buffer, pH 9.6.
DPN stock solution containing about 10 mg. of pure DPN per milliliter, or its equivalent in less pure material.

[1] H. Theorell and R. K. Bonnichsen, *Acta Chem. Scand.* **5**, 1105 (1951); H. Theorell and B. Chance, *ibid.* **5**, 1127 (1951).

Ethanol, 96%.

Enzyme. The enzyme should be diluted so that it contains 10 to 50 γ of enzyme per milliliter.

Procedure. 3 ml. of the buffer, 0.1 ml. of the DPN solution, and 0.1 ml. of ethanol are mixed in a glass or plastic cell. 1 to 5 γ of enzyme is added, and the optical density at 340 mμ is promptly read; $t = 0$. After 3 minutes the change of the optical density is recorded.

Definition of Units. 1 γ of enzyme per milliliter gives in 1 cm. an increase in density of 0.045 in 3 minutes. The protein is measured at 280 mμ; 1 mg. of enzyme per milliliter gives in a 1-cm. cell an optical density of 0.455.

Application of Assay Method to Crude Tissue Preparations. The above method is used, but in this case a blank determination is also carried out. The change in density is measured under the same conditions but without addition of ethanol. This value is subtracted from the increase in density found in the presence of the substrate.

Purification Procedure

The following procedure is a modified form of the method described by Bonnichsen.[2] It has been carried out many times by different people and found to be reliable. The starting material is horse liver. Fresh material is the best, but preparations of frozen liver have been successful.

Step 1. The liver is ground in a meat grinder. A Turmix should not be used. The liver is then extracted for 3 to 4 hours with double the amount of water. The extract can be left overnight at 4°. The extract is centrifuged or filtered conveniently through cheesecloth.

Step 2. The red, turbid solution is heated to 52° and kept at this temperature for 15 minutes. It is important for further purification that this temperature be kept for fully 15 minutes. The solution is cooled to 20 to 25° and centrifuged or filtered.

Step 3. This is a fractionation with ammonium sulfate at room temperature, the enzyme being collected between a saturation of 0.55 and 0.80. First, 349 g. of ammonium sulfate is added per liter of solution. After about 30 minutes the solution is centrifuged. It is important that the solution should be completely clear after the centrifugation. The precipitate is discarded, 208 g. of ammonium sulfate is added, and after 30 minutes the solution is centrifuged. The supernatant is discarded. The precipitate is dissolved in a small amount of phosphate buffer, pH 7, 0.01 M. The enzyme solution is then dialyzed against a phosphate buffer, pH 7, 0.01 M. The buffer should be changed several times, and

[2] R. K. Bonnichsen, *Acta Chem. Scand.* **4**, 715 (1950).

the dialysis continued until there is less than 2% ammonium sulfate left in the enzyme solution.

When the ammonium sulfate is added, small amounts of ammonia should be added to maintain the pH at about 6, as the enzyme is not stable below pH 5.40. At this point the only colored material the solution may contain is hemoglobin and very small amounts of catalase.

Step 4. This step is intended to remove the hemoglobin. The method of Tsuchihashi[3] is used. To the dialyzed solution which contains 0.01 M phosphate, pH 7, is added 240 ml. of a alcohol-chloroform mixture (160 ml. of alcohol and 80 ml. of chloroform) to each liter of solution, and the mixture is shaken vigorously for 10 minutes. The solution is centrifuged, and the denatured protein is discarded. The solution is immediately concentrated *in vacuo*. Most of the alcohol and the chloroform should be evaporated. It is convenient to concentrate to a small volume. The enzyme solution is dialyzed overnight against a phosphate buffer as above. After the dialysis the solution is centrifuged. The purity at this step should at least be 10%. It is possible at this point to get a rather pure enzyme by repeated ammonium sulfate fractionations.

Step 5. The procedure involves a fractionation with ethanol at low temperature at the isoelectric point of the enzyme, pH 7. The ethanol used should be diluted to 90% v/v before use and chilled to −10°. To the enzyme solution is added phosphate buffer, pH 7, to make the final concentration 0.02 M. It is then cooled to 0°. The ethanol is added slowly, and the solution is cooled to −5 to −10° as soon as possible. The temperature should never be allowed to rise above 0°. At 20 to 25% of ethanol the enzyme begins to precipitate. The solution always contains some hemoglobin, and, as the enzyme is precipitated together with it, the hemoglobin color can be used as an indicator for the precipitation of the enzyme. The solution is allowed to stand for about an hour and is then centrifuged at −10°. The precipitate that contains the enzyme is suspended in ice-cold phosphate buffer, 0.02 M, pH 7. While the solution is being stirred, additional drops of buffer are added until all the hemoglobin is dissolved. The solution still remains cloudy. After some 30 to 60 minutes the enzyme begins to crystallize. It is left overnight in the cold room, and then the crystals are centrifuged at 0°. The crystals are washed several times with cold phosphate buffer, 0.02 M, pH 7, containing about 6% ethanol. If the enzyme does not crystallize as described above, small amounts of ethanol can be added to the solution with continued stirring. If there is still difficulty in crystallizing the enzyme, it is probably because too much buffer has been used in dissolving the precipitate. In this case all the protein should be precipitated at −10° with 30% ethanol, the

[3] M. Tsuchihashi, *Biochem. Z.* **140**, 62 (1923).

precipitate this time taken up in a smaller amount of buffer, and the procedure described above repeated. The yield is about 800 mg. of crystalline enzyme from 5 kg. of liver.

Properties

Specificity. In addition to ethanol, the enzyme reacts with higher aliphatic alcohols. Bliss[4] has shown that it also oxidizes vitamin A.

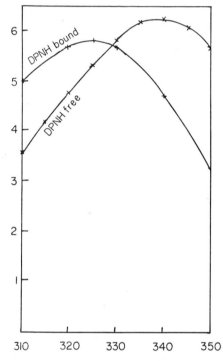

FIG. 1. Millimolar extinction of free and ADH-bound DPNH [H. Theorell and R. Bonnichsen, *Acta Chem. Scand.* **5**, 1105 (1951)].

Protein. The pure enzyme is colorless. The molecular weight is 73,000.[1] The content of sulfur is 1.7%, of nitrogen 14.9%. The number of free SH groups titrable with silver is seven per mole.[5]

Inhibition. The enzyme depends on free SH groups for its activity and is inhibited by SH-inhibiting reagents. Contrary to the yeast, ADH is not inhibited by 0.01 *M* iodoacetate. The enzyme is strongly inhibited by hydroxylamine.[6]

[4] A. F. Bliss, *Arch. Biochem. and Biophys.* **31**, 197 (1951).
[5] R. K. Bonnichsen, in 4 Mosbacher Colloquiums, Springer Verlag, in press.
[6] N. O. Kaplan and M. M. Ciotti, *J. Biol. Chem.* **201**, 785 (1953).

pH Optimum. The Michaelis constant, K_m, for alcohol varies with pH, reaching a minimum at pH 8, 5×10^{-4} M. With small substrate concentrations the pH optimum will be at the same pH.

Mode of Action. One ADH molecule combines with 2 DPNH molecules at pH 7 and with 1 molecule at pH 10 to form a ADH-DPNH complex.[1] The spectrum of DPNH, as shown by Bonnichsen and Theorell,[7]

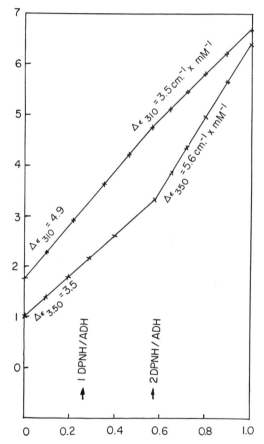

Fig. 2. Spectrophotometric titration of ADH with increasing amounts of DPNH, pH 7 [H. Theorell and R. Bonnichsen, *Acta Chem. Scand.* **5**, 1105 (1951)].

is changed by the complex formation, as seen in Fig. 1. This spectrum change can be used for an accurate determination of the enzyme. Figure 2 shows a titration of the ADH with DPNH. If the enzyme solution to be

[7] R. K. Bonnichsen and H. Theorell, see H. Theorell, *8e Conseil Chim. Inst. Intern. Solvay Bruxelles* 395 (1950).

tested is titrated with DPNH, the break of the curve as seen in the figure indicates when the solution is saturated with DPNH and when the amount of active enzyme present can be calculated.

[79] Alcohol Dehydrogenase from Baker's Yeast

$$\text{RCH}_2\text{OH} + \text{DPN}^+ \rightleftarrows \text{RCHO} + \text{DPNH} + \text{H}^+$$

By E. RACKER

Assay Method[1]

Principle. The method is based on the absorption of DPNH at 340 mμ. With ethanol in excess the rate of DPN reduction is proportional to enzyme concentration.

Reagents

> 3 M ethanol.[2]
> 0.06 M sodium pyrophosphate, pH 8.5.
> 0.0015 M DPN[3] (enzymatically assayed).
> Enzyme. Dilute stock solution of enzyme in 0.1% bovine serum albumin in 0.01 M potassium phosphate, pH 7.5, to obtain about 200 to 1000 units of enzyme per milliliter.

Procedure. Place 2.2 ml. of distilled water, 0.5 ml. of pyrophosphate buffer, 0.1 ml. of ethanol,[2] and 0.1 ml. of DPN[3] into a quartz cell having a 1-cm. light path, and start the reaction by addition of 0.1 ml. of enzyme. The control cell contains all reagents except the substrate. The first density reading is taken 15 seconds after addition of the substrate, and further readings are recorded at 15-second intervals. The increment in density between the 15- and 45-second readings times 2 is taken as enzyme activity per minute.

Definition of Unit and Specific Activity. One unit of enzyme is defined as that amount which causes a change in optical density of 0.001 per minute under the above conditions. Specific activity is expressed as units

---

[1] E. Racker, *J. Biol. Chem.* **184**, 313 (1950).

[2] In the original paper[1] the final alcohol concentration was erroneously given as 0.01 M instead of 0.1 M.

[3] The amount of DPN used in this assay is not saturating but gives reproducible assay values.

per milligram of protein. Protein is determined spectrophotometrically[4] and with crude preparations by the biuret reaction.[5]

Application of Assay Method to Crude Tissue Preparations. Except in liver, yeasts, and some plant tissues the concentrations of alcohol dehydrogenase is too small to permit accurate spectrophotometric determination. In general, for crude preparation the measurement of the reaction in the reverse direction, with DPNH and acetaldehyde as substrate, is more suitable and permits more accurate corrections for oxidation of DPNH due to side reactions. It should be noted, however, that the activity ratio of alcohol oxidation/acetaldehyde reduction varies with different alcohol dehydrogenases.

Purification Procedure

Fleischmann's baker's yeast is obtained in 1-lb. cakes, crumbled, and dried between two sheets of paper in a thin layer for 4 to 5 days at room temperature. The dry yeast is finely ground in a ball mill at 0° for 16 hours and stored in the cold room in a well-stoppered container. 200 g. of the dry yeast powder is extracted with 600 ml. of 0.066 M disodium phosphate for 2 hours at 37° with continuous stirring, followed by extraction at room temperature for an additional 3 hours. The yeast residue is then removed by centrifugation at 13,000 r.p.m. for 15 minutes. The supernatant solution is quickly brought to 55° and maintained at this temperature in a water bath for 15 minutes; after cooling, the mixture is centrifuged. At this stage the clear supernatant fluid can be stored at 0° overnight.

To each 100 ml. of the yeast extract, 50 ml. of ice-cold acetone is slowly added, while the temperature of the mixture is maintained at about −2° in a dry ice-alcohol bath. The resulting precipitate is separated by centrifugation at 0° and discarded. To the supernatant solution, 55 ml. of cold acetone is added for each 100 ml. of the yeast extract, the temperature being kept at −2°. The mixture is centrifuged at 0°, and the supernatant solution discarded. The precipitate is suspended in about 50 ml. of cold water and dialyzed for 3 hours against running tap water.[6] The precipitate is removed by centrifugation. To the clear supernatant solution 36 g. of solid ammonium sulfate is added per 100 ml. of solution. After standing at 0° for 30 minutes, the mixture is centrifuged at 15,000 r.p.m.

---

[4] O. Warburg and W. Christian, *Biochem. Z.* **310**, 384 (1941).

[5] H. W. Robinson and C. G. Hogden, *J. Biol. Chem.* **135**, 727 (1940).

[6] In some locations tap water cannot be used because of a high chlorine content. Instead, dialysis against large volumes of 0.001 M potassium phosphate buffer, pH 7.5, in the cold room with stirring of the dialysis bag and several changes of the buffer has been substituted.

for 20 minutes at 0°. The precipitate is dissolved in 20 ml. of distilled water, and 4 g. of ammonium sulfate is added. The precipitate is centrifuged off and discarded. From the supernatant solution, the alcohol dehydrogenase enzyme starts to crystallize a few minutes after the further addition of ammonium sulfate, which is added in small portions during the course of several hours until about 60 per cent of ammonium sulfate saturation is reached. The crystals are collected by centrifugation and resuspended in a small volume of distilled water, from which they start to recrystallize[7] almost immediately on addition of small volumes of saturated ammonium sulfate solution. A typical protocol showing data on the specific activity of the various fractions and the yields obtained is presented in the table.

The enzyme is most conveniently stored in 50% saturated ammonium sulfate at −20°.

PURIFICATION OF ALCOHOL DEHYDROGENASE FROM BAKER'S YEAST

	Total units	Specific activity, units/mg. protein	Yield, %
First extract[a]	55,000,000	3,700	
After heating	62,000,000		100
Acetone ppt.	56,000,000	55,000	90
First crystalline preparation	38,000,000	120,000	61
Recrystallized preparation	26,000,000	158,000	42

[a] Prepared from 200 g. of dried yeast. The activity measurements of the first extract are usually low, probably owing to side reactions occurring during the test.

Properties

Specificity. The enzyme has been shown by spectrophotometric measurements to react with the following alcohols: ethyl alcohol = allyl alcohol < n-propyl alcohol < n-butyl alcohol < n-amyl alcohol < isopropyl alcohol at low protein concentration (4 γ). With twenty to fifty times as much enzyme, oxidation of ethylene glycol, isobutyl alcohol, methyl alcohol, sec-butyl alcohol, octyl alcohol, and glycerol was observed. Substitution on the methyl group resulted in inactivation of substrate activity, e.g., $CH_2NH_2CH_2OH$. Substitution with halogen resulted in the formation of an analog, e.g., fluoroethanol with considerable inhibitory activity for alcohol dehydrogenase.[8]

[7] A preparation which was recrystallized at 40% ammonium sulfate saturation and dialyzed against veronal buffer at pH 7.8 was found to consist of one major electrophoretic component (85% of total).

[8] E. S. G. Barron and S. Levine, *Arch. Biochem. and Biophys.* **41**, 175 (1952).

Recrystallized preparations of alcohol dehydrogenase show negligible activity when TPN is used instead of DPN as coenzyme. Since there is another TPN-linked alcohol dehydrogenase in yeast,[9] small residual activity in the crystalline preparation may be due to contamination with the TPN-linked alcohol dehydrogenase.

Activators and Inhibitors. Alcohol dehydrogenase from yeast is an SH enzyme which is quite sensitive to iodoacetate. The enzyme can be oxidized by shaking in air and reactivated by GSH.[10] Both substrate and DPN [8,9] protect the enzyme against IAA (in contrast to triosephosphate dehydrogenase which is protected only by substrate). With hydroxylamine an inhibition of alcohol dehydrogenase was described, which was overcome by addition of excess alcohol.[11]

Physical-Chemical Properties. In a classical paper Negelein and Wulff[12] reported the first crystallization of alcohol dehydrogenase from brewer's yeast and described the kinetic properties of the enzyme. More recent data, which include the determination of molecular weight, have been obtained by Hayes and Velick[13] with the above described preparation of alcohol dehydrogenase from baker's yeast. Four molecules of DPN were found to be bound by the enzyme. The turnover number for ethanol oxidation was estimated to be 27,000 [14] at pH 7.9 at 26°. The K_m for DPN is 1.7×10^{-4}; for DPNH, 2.3×10^{-5}. The dissociation constant (determined by direct binding measurements at 5°) for DPN is 2.6×10^{-4}; for DPNH, 1.3×10^{-5}. The sedimentation constant $S_{20,w}$ is 6.72×10^{-13}, the diffusion constant is 4.7×10^{-7}, and the molecular weight is 150,000.

[9] E. Racker, unpublished observation.
[10] T. Wagner-Jauregg and E. F. Möller, *Z. physiol. Chem.* **236**, 222 (1935).
[11] N. O. Kaplan and M. M. Ciotti, *J. Biol. Chem.* **201**, 785 (1953).
[12] E. Negelein and H. J. Wulff, *Biochem. Z.* **293**, 351 (1937).
[13] J. E. Hayes, Jr., and S. F. Velick, *J. Biol. Chem.* in press.
[14] Considerable variations in the turnover number (at saturating concentrations of DPN and alcohol) have been observed with different preparations of the crystalline enzyme, and values higher than the above reported were occasionally observed. Inactivation of the enzyme due to contact with rubber stoppers has been observed (K. Wallenfels, personal communication).

[80] TPN-Alcohol Dehydrogenase from *Leuconostic mesenteroides*

By R. D. DeMoss

Alcohol dehydrogenases which catalyze direct reduction of TPN have recently been reported to occur in higher plants.[1] Observations with cell-free extracts indicate the presence in baker's yeast of a TPN-alcohol dehydrogenase[2] in addition to the DPN-linked enzyme.[3] A TPN-specific alcohol dehydrogenase from *Leuconostoc mesenteroides* has been studied[4] and constitutes the subject of the present report. It should be noted that the role of the enzyme in the metabolic activities of the organism is in doubt, in view of the presence of both DPN and TPN linked alcohol dehydrogenases.

Assay Method

Principle. The oxidation of ethanol catalyzed by this enzyme is accompanied by TPN reduction and may be followed as the change in optical density (ΔE_{340}) in the spectrophotometer.

Reagents

Ethanol (5 M). Add 1.0 ml. of 95% ethanol to 2.33 ml. of water.
Tris buffer (0.2 M, pH 7.5). Mol. wt. 121.14.
TPN (0.0024 M). 2 mg. of TPN-80 (Sigma) per milliliter, mol. wt. 744.
Enzyme. Dilute stock enzyme with water or cysteine (0.001 M, pH 6.5) to obtain 1 to 5 units/ml.

Procedure. To a spectrophotometric cell with a 1-cm. light path, add 1.5 ml. of Tris buffer, 0.2 ml. of TPN, 0.1 ml. of enzyme, and 0.9 ml. of water. Take readings at 340 mμ before and at 15-second intervals after mixing with 0.3 ml. of ethanol solution. This assay procedure cannot be used either in the presence of added Mn^{++} ion or with MnCl$_2$ supernatants (step 2), owing to the high absorption apparently produced by reaction of Tris and Mn^{++} ions. Under such circumstances, phosphate buffer (0.2 M) or pyrophosphate buffer (0.02 M) may be substituted for Tris.

[1] H. A. Stafford and B. Vennesland, *Arch. Biochem. and Biophys.* **44**, 404 (1953).
[2] H. Z. Sable, *Biochem. et Biophys. Acta* **8**, 687 (1952).
[3] E. Racker, *J. Biol. Chem.* **184**, 313 (1950).
[4] R. D. DeMoss, *J. Bacteriol.* **68**, 252 (1954).

Application of Assay Method to Crude Extracts. The above procedure may be used for assay of crude extracts, since no TPNH oxidizing systems interfere under the experimental conditions described. No TPN-linked aldehyde dehydrogenase has been observed in extracts of *L. mesenteroides*. If endogenous substrates are present, the extract should be dialyzed before the assay.

Growth of Cells. See Vol. I [43].

Purification Procedure

All manipulations are performed at 2° or in an ice bath, unless otherwise stated. The enzymatic activity is extremely labile after step 1 but is protected by the presence of cysteine. Other similar reagents (not tested) may possibly afford protection, but neither cyanide nor ethanol is effective. Reactivation of the enzyme after loss of activity is not effected by cysteine at any step of purification. Table I summarizes the results of the purification procedure.

TABLE I

SUMMARY OF TPN-ALCOHOL DEHYDROGENASE PURIFICATION PROCEDURE

Fraction	Specific activity		Total units	
	TPN	DPN	TPN	DPN
1. Crude extract	0.234	0.234	357	357
2. Protamine supernatant[a]	0.014	0.036	0.5	12
3. Ammonium sulfate fraction				
0.0–0.475	0.176	0.222	8.5	10.8
0.475–0.750	0.472	0.296	133	83
4. Ca$_3$(PO$_4$)$_2$ gel eluate	1.89	1.07	158	90

[a] These fractions were assayed after overnight storage at −20° and clearly demonstrate the marked instability of the enzyme in the absence of cysteine.

Protein concentration is determined by the optical method of Warburg (see Vol. III [73]).

Step 1. Cell-Free Extract. These bacterial cells are somewhat refractory to disintegration, apparently because of their small size (0.9 to 1.2 μ in diameter) and nearly spherical structure. Grinding techniques or sonic oscillation for 1.5 to 3 hours (for details, see Vol. I [7]) at 9 kc. results in disruption of only 30 to 60% of the cells. Three-hour sonic treatment also destroys a large portion of the enzymatic activity theoretically obtainable. However, a 20-minute treatment in the 10-kc. oscillator disrupts about 80% of the cells with little concomitant loss of enzymatic activity.

The concentration of cells for sonic treatment is usually about 1 g. of freshly harvested or frozen wet cells per 10 ml. of potassium gluconate (1%, pH 9.0, containing 0.01 M cysteine). For grinding techniques, about 2.5 g. of abrasive is used per gram of wet cells,[5] and the ground mixture is extracted with 4 to 5 ml. of fluid per gram of cells.

Step 2. Nucleic Acid Precipitation. The crude cell-free extract is centrifuged (30 minutes at 15,000 × g), and the supernatant dialyzed for 16 hours against 100 vol. of phosphate buffer (0.02 M, pH 6.5, containing 0.001 M cysteine) and carefully adjusted to pH 6.0 with N acetic acid. Protamine sulfate (17 mg./ml., pH 5.0; use 0.01 ml./mg. extract protein) is added dropwise with stirring to remove nucleic acid. The precipitate is removed by centrifugation (10 to 15 minutes at 10,000 × g), and the clear supernatant solution (protamine supernatant, containing 4 to 7% nucleic acid) is used for subsequent steps.

Step 3. Ammonium Sulfate Precipitation. The protamine supernatant is fractionally precipitated by addition of ammonium sulfate solution (saturated at 2°); the 0.475 to 0.75 fraction is retained and dissolved in 2 ml. of phosphate buffer (0.02 M, pH 6.5, containing cysteine, 0.001 M) per gram of original cell weight.

Step 4. Calcium Phosphate Gel; Adsorption and Elution. To the 0.475 to 0.75 ammonium sulfate fraction, calcium phosphate gel (1 mg./mg. protein) is added, stirred for 20 minutes, and centrifuged. The sediment is eluted with 0.05 M phosphate buffer-0.001 M cysteine (pH 7.3, use 2 ml./g. original cell weight) by strong stirring (or homogenization in a Potter-Elvehjem grinder) for 20 minutes and centrifuged. The supernate contains most of the enzymatic activity, although a small additional portion of activity may be obtained by elution of the sediment with the same buffer at pH 8.5.

Properties

Specificity. Under the assay conditions described, either DPN or TPN serves as coenzyme for ethanol oxidation. However, the change in relative activity of the two nucleotides observed during the purification procedure indicates that two enzymes are present, each specific for one pyridine nucleotide (see Table I).

Table II shows the relative activities of several alcohols which serve as substrates for TPN reduction; the following hydroxy compounds are inactive: malate, glycollate, glycerol, isopropanol, *sec*-butanol, *tert*-butanol, propanediol-1,2, and ethanolamine. TPNH oxidation is catalyzed by the enzyme in the presence of the aldehydes listed in Table III; the

[5] H. McIlwain, *J. Gen. Microbiol.* **2**, 288 (1948).

TABLE II
TPN REDUCTION WITH ALCOHOL DEHYDROGENASE

Alcohol, 0.5 M	ΔE_{340} per minute	Relative activity
Ethanol	0.410	100
n-Propanol	0.093	23
n-Butanol	0.040	10
n-Pentanol	0.010	3
Methanol	0.008	2
Ethanediol-1.2	0.005	1

following carbonyl compounds are ineffective as substrates: acetone, pyruvate, glyceraldehyde, salicylaldehyde, hydroxypyruvate, and dihydroxyacetone.

TABLE III
TPNH OXIDATION WITH ALCOHOL DEHYDROGENASE

Aldehyde, 0.007 M	ΔE_{340} per minute	Relative activity
Acetaldehyde	0.088	100
n-Butyraldehyde	0.028	32
Glyoxal	0.014	16
Glycolaldehyde	0.012	14
Formaldehyde	0.006	8
Acetaldehyde + formaldehyde	0.028	32

Activators and Inhibitors. The rate of TPN reduction under the assay conditions described is inhibited completely by 10^{-2} M AgNO$_3$, HgCl$_2$, or CaCl$_2$ and is stimulated 50 to 100% by 10^{-3} M CaCl$_2$, or 10^{-2} M CuCl$_2$, MgCl$_2$, KI, or NaCl. Complete inhibition of TPN reduction is also effected by preincubation for 7 minutes with 10^{-3} M iodoacetate, by 10^{-4} M p-chloromercuribenzoate, or by 10^{-2} M hydroxylamine (see Vol. I [78, 79]).

Effect of pH. Under the assay conditions described, the initial rate of TPN reduction increases nearly linearly from pH 7.2 to at least pH 9.2. However, when pyrophosphate buffer (0.01 M final concentration) replaces Tris, the most rapid initial rate is observed at pH 7.5. The rate of TPN reduction in pyrophosphate buffer decreases sharply below pH 7.1 and gradually above pH 7.5 to a value of zero at pH 11.4.

Stability. The stability of the enzyme has been discussed under purification procedures.

Dissociation Constant. When derived from Lineweaver-Burk[6] or Hofstee[7] plots, the dissociation constant is 1.6×10^{-4} M TPN.

[6] H. Lineweaver and D. Burk, *J. Am. Chem. Soc.* **56**, 658 (1934).
[7] B. H. J. Hofstee, *Science* **116**, 329 (1952).

[81] Potassium-Activated Yeast Aldehyde Dehydrogenase

$$RCHO + \begin{array}{c} DPN^+ \\ or \\ TPN^+ \end{array} . + H_2O \rightarrow RCOOH + \begin{array}{c} DPNH \\ or \\ TPNH \end{array} + H^+$$

By SIMON BLACK

Assay Method

Principle. Spectrophotometric measurement of DPNH (or TPNH) formation is used to determine enzyme activity.

Reagents

> 1.0 M Tris-HCl buffer, pH 8.0.
> 3.0 M KCl.
> 0.01 M DPN, brought to pH 7.5 with Tris base.
> 0.10 M 2-mercaptoethanol (Eastman).
> 0.01 M acetaldehyde. This is prepared about once a week from a 2.0 M stock solution. The latter is made with acetaldehyde redistilled in the cold room and can be kept indefinitely at $-20°$.
> Enzyme. Dilute in 0.1 M potassium phosphate–0.001 M cysteine, pH 6.0, containing 5% sodium nucleate (Nutritional Biochemicals), to 500 to 2000 units/ml. Keep in an ice bath.

Procedure. The solutions are added to a spectrophotometer cell to give a final volume of 3.0 ml. and the following final molar concentrations: Tris-HCl buffer, 0.10; KCl, 0.10; DPN (or TPN), 0.0005; mercaptoethanol, 0.001; and acetaldehyde, 0.00017. The final pH should be 8.0. The reaction is started by addition of acetaldehyde; optical density readings at 340 mμ are made every 30 seconds for several minutes.

One unit is defined as the amount of enzyme required to produce an initial increase in optical density of 0.001 per minute.

Application to Crude Extracts. With DPN, the assay procedure is not applicable to crude yeast extracts, since these contain alcohol dehydrogenase, which, in the presence of acetaldehyde, reoxidizes DPNH. When

TPN is used, the activity is measured as a difference between two determinations, KCl being omitted in one. The activity observed in the absence of potassium is due to another enzyme.[1] The potassium-dependent enzyme is about one-tenth as active with TPN as with DPN.

Purification Procedure[2]

Step 1. Preparation of Crude Extract. One pound of fresh Red Star baker's yeast[3] is crumbled by hand and frozen by dropping into 2 l. of liquid nitrogen. The excess nitrogen is decanted or allowed to evaporate, and the yeast is suspended in 450 ml. of 0.3 M K_2HPO_4. After warming to about 3°, the pH is brought to 8.6 by slow addition, with stirring, of concentrated NH_4OH. Slow stirring is continued at 2 to 3° for 5 days, after which the suspension is centrifuged in the cold, and the sediment discarded.

Step 2. Heat Coagulation of Inactive Protein. The cold extract is brought to pH 6.6 by slow addition, with stirring, of 1 M citric acid. With vigorous stirring on a hot-water bath it is rapidly warmed to 55°, maintained at this temperature for 15 minutes, then rapidly cooled to 2 to 4°. The precipitate is removed by centrifugation at this temperature.

Step 3. Acid Precipitation. The pH of the solution obtained in step 2 is brought to 5.0 by gradual addition of 1 M citric acid, and the precipitate containing the enzyme is centrifuged in the cold. This precipitate is suspended evenly in a solution containing 0.025 M K_2HPO_4 and 0.001 M cysteine hydrochloride. The volume is brought to 80 ml., and the pH is adjusted to 6.3 with 3 M NH_4OH. The suspension is warmed rapidly to 50°, held at this temperature for 15 minutes, and cooled to less than 5°. Centrifugation at 15,000 $\times$ g in the cold yields a clear solution of the enzyme.

This solution has been prepared successfully many times. In the frozen state it has retained its activity for more than two years. Because of its greater stability, it is recommended in preference to the product of step 4. It contains, however, a large amount of TPN-specific aldehyde dehydrogenase.

Step 4. Sodium Chloride Precipitation. To the cold solution obtained in step 3 are added 0.15 vol. of 0.01 M acetaldehyde, 0.06 vol. of 0.006 M DPN (neutralized with Tris base), and 0.35 g. of NaCl per milliliter of final solution. The mixture is stirred at 0 to 3° until all the NaCl dissolves, and the precipitated enzyme is centrifuged in the cold. It is dis-

---

[1] See Vol. I [82].

[2] S. Black, *Arch. Biochem. and Biophys.* **34**, 86 (1951).

[3] Anheuser-Busch yeast can also be extracted by this procedure. Fleischmann's is not suitable.

solved for use in cold buffered sodium nucleate (see Reagents). This preparation has not been successfully stored. The relative activity of TPN-specific enzyme is 5% of that found in the step 3 product.

SUMMARY OF PURIFICATION PROCEDURE

Fraction	Total volume, ml.	Units/ml., thousands	Total units, thousands	Protein, mg./ml.	Specific activity, units/mg.	Recovery, %
1. Extract	620	9.1	5640	61.0	149	—
2. 55° super-natant	490	8.5	4160	9.0	945	74
3. 50° super-natant of acid pre-cipitate	70	43.0	3010	6.0	7,170	53
4. NaCl pre-cipitate	12	230.0	2760	6.5	35,300	49

Properties

Specificity. The potassium-activated enzyme reacts with DPN ten times as rapidly as with TPN.[4] Its TPN activity is distinguished from Seegmiller's[1] TPN-specific enzyme by its (1) activation by thiols, (2) requirement for potassium ions, and (3) activity with benzaldehyde as substrate. Acetaldehyde is the most rapidly oxidized substrate, but the following aldehydes, in order of decreasing activity, are utilized: propion-aldehyde, n-butyraldehyde, benzaldehyde, succinic semialdehyde, croton-aldehyde, formaldehyde, and DL-glyceraldehyde. Glyoxylic acid and salicylaldehyde are inactive.

Activators and Inhibitors. Potassium ions are essential for the enzyme activity. Rubidium or ammonium ions can be substituted for potassium, but not lithium, cesium, or sodium. The latter three ions are moderately active inhibitors.

Although some activity may be observed in deionized water in the absence of thiols (cysteine, glutathione, 2-mercaptoethanol), 0.001 M concentrations of the latter invariably produce marked activation.

p-Chloromercuribenzoate (10^{-4} M) reduces enzyme activity 50%, and this inhibition is reversed by 10^{-3} M glutathione.

The enzyme is extremely sensitive to small traces of heavy metals, particularly copper. Reagents such as ethylenediaminetetraacetic acid (Versene) give ineffective protection; thiols are much superior.

[4] The observation of nearly equivalent rates with DPN and TPN,[2] made with the product of step 3, must be attributed to the presence of TPN-specific enzyme.

Solutions of the enzyme deteriorate rapidly (in a few hours) in the absence of nucleic acid salts which have a marked stabilizing effect. A large amount of nucleic acid is present in the crude extract and is carried through step 3 in the fractionation procedure.

Effect of pH. Maximum initial rates are obtained in several buffers at pH 8.75. However, the initial rate is sustained for longer periods at pH 8.0.

Substrate and Potassium Concentrations. A sharp maximum in rate is observed with an acetaldehyde concentration of 1.7×10^{-4} M. Half maximum rates are obtained with DPN concentrations of 1.3×10^{-4} and 0.3×10^{-4} M at pH values of 8.7 and 7.7, respectively.

Fifty per cent activity results with a KCl concentration of 0.01 M; saturation with potassium requires about 0.2 M KCl.

[82] TPN-Linked Aldehyde Dehydrogenase from Yeast

$$CH_3CHO + TPN^+ + H_2O \rightarrow CH_3COOH + TPNH + H^+$$

By J. EDWIN SEEGMILLER

Assay Method

Principle. The method is based on the spectrophotometric determination of TPNH formation as shown by the increase in optical density at its absorption maximum of 340 mμ.

Reagents

Acetaldehyde stock solution (0.2 M). Commercial acetaldehyde is distilled and 1.13 ml. diluted to 100 ml. with water, giving a 0.2 M solution which may be stored at 2° for several weeks without significant change. Dilutions freshly prepared from acetaldehyde contain an enzyme inhibitor which disappears after storage at 2° for a day.

1.3 mM TPN (purity 80%).[1]

0.25 M glycylglycine–NaOH buffer, pH 7.7.

0.1 M MgCl₂.

Enzyme. Dilute enzyme with 0.01 M phosphate buffer, pH 7.5, to obtain 0.3 to 1.0 units of enzyme per milliliter. (See definition below.)

Procedure. Prepare water dilutions of stock acetaldehyde solution 1:20 fresh daily. To a Corex cell with a 1-cm. light path and volume of

[1] A. Kornberg, *Biochem. Preparations* **3**, 24 (1953); see also Vol. III [124].

1.05 ml. add 0.57 ml. of water, 0.15 ml. of 0.1 M $MgCl_2$, 0.2 ml. of 0.25 M glycylglycine buffer, pH 7.7, and 0.05 ml. of 0.01 M acetaldehyde and 0.05 ml. of 1.3 mM TPN immediately before assay. Take readings at 340 mμ before and at 30-second intervals for 3 minutes after mixing with 0.03 ml. of diluted enzyme.

Definition of Unit and Specific Activity. One unit of enzyme activity is defined as that amount of enzyme which would give an initial rate of change of 1.0 optical density unit per minute under the above conditions. Specific activity is expressed as units per milligram of protein. Protein is determined by the turbidimetric method of Bücher.[2]

Application of Assay Method to Crude Tissue Preparations. Crude extracts of fresh yeast obtained by shaking with sand in the cold[3] are readily assayed by the above method by correcting for a slight endogenous reduction of TPN in the absence of acetaldehyde. The enzyme does not survive well the disruption of yeast cells by toluene autolysis, NaCl plasmolysis, or grinding with alumina.

Purification Procedure

The substitution of citrate buffers for the acetate buffers in the following purification procedure has resulted in but little change in the yield or purity of the final product. The final extraction of step 5 can be made with 0.25 M glycylglycine buffer, pH 7.7, with much improved recovery but slightly less purification.

Step 1. Extraction. 40 g. of fresh Fleischmann's bakers' yeast, 200 g. of 40- to 80-mesh washed sea sand, and 200 ml. of chilled 0.1 M $NaHCO_3$ are added to each of four 540-ml. screw-cap bottles. Cells are broken by shaking at maximal speed in an International shaking machine for 1½ hours in a cold room. The extract obtained after centrifugation for 10 minutes at 11,000 r.p.m. in a Servall angle centrifuge in the cold loses a third of its activity in 2 hours at 0°.

Step 2. Ammonium Sulfate, pH Fractionation. 194 g. of ammonium sulfate is added to 596 ml. of extract. The precipitate is centrifuged and discarded, and the supernatant solution brought to pH 3.0 by the slow addition of 145 ml. of cold 0.5 N H_2SO_4 with stirring in an ice bath over a period of 5 minutes. The precipitate is collected by centrifugation and washed with 36 ml. of 0.1 M acetate buffer, pH 4.0, containing 5 g. of ammonium sulfate per 100 ml. The active enzyme is extracted from the residue with 0.1 M phosphate buffer, pH 7.0, in successive 20- and 10-ml. quantities, which are then combined. This preparation can be frozen and stored overnight at $-17°$ without significant loss of activity.

[2] T. Bücher, *Biochim. et Biophys. Acta* **1**, 292 (1947).

[3] A. Kornberg and W. E. Pricer, Jr., *J. Biol. Chem.* **189**, 123 (1951); see also Vol. I [118].

Step 3. First Acetone Precipitation. The pH fraction is treated with 3 vol. (96 ml.) of 0.1 M acetate buffer, pH 5.3, and 96 ml. of $-10°$ acetone is added. The addition requires 3 minutes, during which time the solution is cooled to $-3°$. The precipitate is immediately centrifuged for 5 minutes and extracted with 6 ml. of 0.01 M acetate buffer, pH 5.3.

Step 4. Aluminum Hydroxide Gel Adsorption. The enzyme is adsorbed by the addition of 7 ml. (71 mg., dry weight) of aluminum hydroxide gel.[4] The gel is centrifuged, washed with 10 ml. of 0.01 M phosphate buffer, pH 6.6, and the enzyme eluted three times with 20-ml. quantities of 0.01 M phosphate buffer, pH 7.1.

Step 5. Second Acetone Precipitation. This step serves mainly to concentrate the enzyme. 20 ml. of 0.1 M acetate buffer, pH 5.3, is added to the eluate, and 60 ml. of $-10°$ acetone is added as before. The precipitate is extracted with two 1.0-ml. quantities of 0.01 M acetate buffer, pH 5.3. The extracts are combined and neutralized by the addition of 0.21 ml. of 0.25 M glycylglycine buffer, pH 7.7. 20 mg. of purified bovine albumin is added to increase the stability, and the enzyme is frozen and stored at $-16°$. Under these conditions it loses about 10% of its activity in 5 days.

SUMMARY OF PURIFICATION PROCEDURE

Fraction	Total volume, ml.	Units/ml.	Total units	Protein, mg./ml.	Specific activity, units/mg.	Recovery, %
1. Extract	596	2.7	1610	6.8	0.40	
2. pH fraction	32	21.6	691	9.0	2.4	43
3. Acetone I	6	77	462	10.5	7.3	29
4. Aluminum hydroxide eluate	64	6.8	435	0.33	21	27
5. Acetone II	2.2	80	176	2.6	31	11

Properties[5]

Specificity. The purified enzyme shows greatest activity with acetaldehyde as substrate, although glycolaldehyde,[6] propionaldehyde, and formaldehyde are oxidized at 70, 50, and 30%, respectively, of the rate shown with acetaldehyde. The ratio of activity with glycolaldehyde and acetaldehyde does not change during purification, and their rates of oxidation are not additive. The enzyme did not act on *n*-butyraldehyde, chloral hydrate, benzaldehyde, phosphoglycolaldehyde,[7] D-glucose, D-ribose, D-arabinose, xanthine, and crotonaldehyde.

[4] R. Willstätter and H. Kraut, *Ber.* **56**, 1117 (1923).
[5] J. E. Seegmiller, *J. Biol. Chem.* **201**, 629 (1953).
[6] H. O. L. Fischer and C. Taube, *Ber.* **60B**, 1704 (1927).
[7] P. Fleury, J. Courtois, and A. Desjobert, *Bull. soc. chim. France* p. 694 (1948).

Coenzyme Requirement. The enzyme has a specific requirement for TPN. Although the enzyme after step 2 shows reduction of DPN at a rate 4% of that shown with TPN, the subsequent purification steps reduce this to a value in the final preparations 0.1 to 0.4% of the activity shown with TPN. There was no evidence for the participation of coenzyme A or Ac~SCoA in this oxidation, as shown by the failure to form citrate in the presence of condensing enzyme of Ochoa[8] and oxalacetate.

Activators and Inhibitors. In the presence of the divalent cations Ca^{++} or Mg^{++} the activity of the enzyme was increased four- to sixfold; Mn^{++} or Ba^{++} gave a threefold activation. Na^+, K^+, and NH_4^+ gave no significant activation. Crotonaldehyde inhibited markedly the oxidation of acetaldehyde.

Effect of pH. The purified enzyme showed a relatively broad pH optimum between pH 7.5 and 8.0 with 50% of maximal activity at pH 6.8 and 8.7.

Michaelis Constants. The Michaelis constants are 1.4×10^{-5} for TPN, 3.5×10^{-5} for acetaldehyde, and 2.2×10^{-3} for glycolaldehyde. The high affinity for TPN and for acetaldehyde permits the use of this enzyme for the spectrophotometric assay of acetaldehyde and of TPN.

[8] S. Ochoa, J. R. Stern, and M. C. Schneider, *J. Biol. Chem.* **193**, 691 (1951); see also Vol. I [114].

[83] Liver Aldehyde Dehydrogenase

$$RCHO + DPN^+ \rightarrow RCOOH + DPNH + H^+$$

By E. RACKER

Assay Method

Principle. In the presence of enzyme, aldehyde, and DPN, the latter is reduced to DPNH which is measured spectrophotometrically at 340 mμ.[1]

Reagents

0.1 M sodium pyrophosphate buffer, pH 9.3.

0.2% solution of neutralized DPN.

0.5% solution of acetaldehyde.

Enzyme. Dilute stock solution of enzyme in 0.01 M potassium phosphate buffer, pH 7.9, containing 0.6 mg. of Versene per milliliter to obtain 100 to 1000 units of enzyme per milliliter. (See definition of unit below.)

[1] E. Racker, *J. Biol. Chem.* **177**, 883 (1949).

Procedure. Place into a quartz cell having a 1-cm. light path 2.4 ml. of distilled water, 0.3 ml. of buffer, 0.1 ml. of DPN, 0.1 ml. of acetaldehyde, and 0.1 ml. of enzyme solution. Take density readings at 340 mμ at 30-second intervals. The course of the reaction is zero order for several minutes.

Definition of Unit and Specific Activity. One unit of enzyme is defined as a density increment of 0.001 per minute under the above conditions. Specific activity is expressed as units of enzyme activity per milligram of protein. Protein concentration is determined spectrophotometrically.[2] In the presence of nucleic acid the biuret test is used.[3]

Application of Assay Method to Crude Tissue Preparations. The enzyme assay described above is not suitable for crude tissue preparations which contain alcohol dehydrogenase or DPNH oxidase activity. A manometric assay can be used instead. This test is based on the liberation of CO_2 from bicarbonate buffer when acid is produced. For the determination of aldehyde dehydrogenase activity the enzyme preparation, together with an excess of crystalline alcohol dehydrogenase (about 300 γ), is placed in the main compartment of a Warburg vessel. To this is added 1 mg. of DPN, bicarbonate buffer to a final concentration of 0.05 M, and distilled water to make the volume 2.5 ml. Into the side arm is placed 0.5 ml. of a 1% solution of acetaldehyde. Gas with nitrogen containing 5% CO_2 for 7 minutes, and determine the CO_2 evolution in a conventional Warburg apparatus after the content of the side arm has been tipped into the main compartment.

This method has been used only for crude liver preparations but should be applicable to other tissues as well.

Purification Procedure

The procedure described below is essentially as described previously[1] except that all steps are carried out in the presence of 600 γ of neutralized Versene per milliliter, and the pH of the dialyzed enzyme is adjusted to pH 6.0 prior to centrifugation. These modifications were introduced by Dr. I. Leder. Some variability in the effectiveness of purification will be encountered at the nucleic acid step, depending on the source and purity of the nucleic acid used. With most commercial preparations the specific activity of the aldehyde enzyme after nucleic acid fractionation is 400 to 500 rather than 700 (see table).

Frozen beef liver is used as starting material. An acetone-dried powder is prepared as follows: The thawed liver is mixed with 2 to 3 vol. of ice-cold acetone in a Waring blendor, and the mixture poured into 8 vol. of

[2] O. Warburg and W. Christian, *Biochem. Z.* **310**, 384 (1941).
[3] H. W. Robinson and C. G. Hogden, *J. Biol. Chem.* **135**, 727 (1940).

acetone and then rapidly centrifuged in the cold room. The precipitate is washed once with 8 vol. of acetone, then pressed out between paper towels and crumbled into a fine powder which dries rapidly when distributed on large filter papers.

This powder is extracted with 8 vol. of distilled water containing 600 γ of neutralized Versene per milliliter at room temperature for 45 minutes under continuous mechanical stirring. From this point on, all operations are carried out in the cold room. The mixture is centrifuged, and to each 100 ml. of the supernatant 70 ml. of cold 95% ethyl alcohol is added in the course of 10 minutes. The temperature is allowed to rise to 12 to 14° during this procedure. After standing for an additional 20 minutes, the preparation is centrifuged, and to the clear supernatant 40 ml. of cold ethyl alcohol for each 100 ml. of the original extract is added slowly. After standing for 10 minutes at 0°, the precipitate is centrifuged off and dissolved in dilute Versene. The solution is dialyzed for 2 hours with mechanical stirring against large volumes of dilute Versene. After the pH has been adjusted to 6.0 the resulting precipitate is centrifuged off and discarded. For each gram of protein present in the supernatant, 2 ml. of a neutralized 5% solution of nucleic acid (Merck) is added, and the mixture is carefully adjusted to pH 5.2 with 0.1 M acetic acid. The precipitate formed is discarded even though considerable quantities of the aldehyde enzyme are present in this fraction. To the clear supernatant, half the amount of nucleic acid used in the above step is added, the pH readjusted to 5.2, and the mixture centrifuged. The precipitate is dissolved in dilute alkali, and protamine sulfate (Squibb) is added in small fractions at pH 6.5 until the absorption ratio of 280 mμ:260 mμ indicates the removal of nucleic acid.

The degree of purification obtained by this method is usually from 20- to 30-fold, with a yield of 20 to 40%. A typical protocol of purification and yields achieved is given in the table. The enzyme thus purified is free of alcohol dehydrogenase and xanthine oxidase activity.

PURIFICATION OF ALDEHYDE DEHYDROGENASE FROM BEEF LIVER ACETONE POWDER

Fraction	Volume of solution, ml.	Units	Protein, mg.	Specific activity, units/mg. protein	Yield, %
Aqueous extract	300	360,000	12,000	30	100
Second alcohol ppt.	100	300,000	2,000	150	83
Second nucleic acid ppt. (after nucleic acid removal)	10	140,000	200	700	38

Properties

Specificity. Although the enzyme can only reduce DPN and is inactive with TPN, a number of aldehydes are oxidized by the liver enzyme. These are formaldehyde, acetaldehyde, glycolaldehyde, propionaldehyde, butyl-aldehyde, isovaleraldehyde, salicylaldehyde, and benzaldehyde. The latter compound is oxidized at a comparatively slow rate[4] and inhibits markedly the oxidation of acetaldehyde.[1] The enzyme has not been explored in its action on numerous other aliphatic and aromatic aldehydes of physiological interest. Betaine aldehyde was found to be oxidized by the enzyme.[5]

Inhibitors. Aldehyde dehydrogenase is a sulfhydryl enzyme. After prolonged dialysis the enzyme is inactivated, and activity can be partially restored by addition of cysteine. The addition of either cysteine or Versene to the water used for dialysis affords full protection against inactivation. Tetraethylthiuram disulfide was shown to inhibit the enzyme at very low concentrations[5,6] noncompetitively; chloralhydrate inhibits by competing for the substrate.[5]

Stability, pH Optimum and Other Properties. Partially purified preparations (after alcohol precipitation) have been kept at $-20°$ for several years and have retained over 25% of their original activity. On further purification the enzyme appears to become more labile, although considerable stabilization is achieved by addition of Versene. The enzyme exhibits a sharp pH optimum at 9.3 when tested in 0.01 M pyrophosphate buffer. Both DPN as well as the substrates are utilized at very low concentrations, which permits a convenient assay method for their determination. In view of the variety of substrates attacked, the method lacks specificity. Besides the high substrate affinity the liver enzyme shows other properties which clearly differentiate it from glyceraldehyde-3-phosphate dehydrogenase. Whereas the latter enzyme catalyzes the reduction of acylphosphates by DPNH, the liver enzyme does not. Moreover no stimulation by either phosphate or arsenate can be found. Finally, highly purified liver preparations have little or no activity with glyceraldehyde-3-phosphate as substrate. In view of these findings it is difficult to support the view[7] that these two enzymes may be identical.

[4] H. Tabor, personal communication.
[5] E. Racker, *Ann. Repts. Long Island Biol. Assoc.* Cold Spring Harbor, **1949**, 38.
[6] W. D. Graham, *J. Pharm. Pharmacol.* **3**, 160 (1951).
[7] A. P. Nygaard and J. B. Sumner, *Arch. Biochem. and Biophys.* **39**, 119 (1952).

[84] Aldehyde Dehydrogenase from *Clostridium kluyveri*

$$CH_3CHO + DPN^+ + CoA \leftrightarrows Ac{\sim}SCoA + DPNH + H^+$$

By E. R. STADTMAN and ROBERT MAIN BURTON

Assay Method

Principle. The method was previously described[1] and is based on the fact that the dismutation of acetaldehyde to $Ac{\sim}P$ and ethanol in *Cl. kluyveri* occurs by the following coupled reactions:

$$CH_3CHO + CoA + DPN^+ \xleftrightarrow{\text{aldehyde dehydrogenase}} Ac{\sim}SCoA + DPNH + H^+ \tag{1}$$

$$DPNH + H^+ + CH_3CHO \xleftrightarrow{\text{alcohol dehydrogenase}} C_2H_5OH + DPN^+ \tag{2}$$

$$Ac{\sim}SCoA + Pi \xleftrightarrow{\text{phosphotransacetylase}} Ac{\sim}P + CoA \tag{3}$$

$$2CH_3CHO + Pi \leftrightarrows Ac{\sim}P + C_2H_5OH \tag{4}$$

In the presence of an excess of alcohol dehydrogenase and phosphotransacetylase the rate of the over-all dismutation (reaction 4) is determined by the concentration of aldehyde dehydrogenase. Reaction 4 is followed by measuring the formation of $Ac{\sim}P$.

Reagents

1.0 *M* acetaldehyde + 1.0 *M* potassium phosphate (pH 8.1)
0.2 *M* Tris buffer, pH 8.1.
0.05 *M* GSH.
0.01 *M* DPN.
0.0005 *M* CoA (150 units/ml.).
Enzymes. Yeast alcohol dehydrogenase,[2] 13,000 units/ml.; phosphotransacetylase,[3] 100 units/ml.; aldehyde dehydrogenase—dilute the stock solution or crude bacterial extract with water to obtain 0.5 to 1.5 units/ml. (See definition below.)

Procedure. Mix 0.1-ml. aliquots of all the above reagents and 0.2 ml. of water (final volume, 1.0 ml.), and incubate at 30° for 30 minutes. After

---

[1] R. M. Burton and E. R. Stadtman, *J. Biol. Chem.* **202**, 873 (1953).
[2] E. Racker, *J. Biol. Chem.* **184**, 313 (1950); see Vol. I [79].
[3] E. R. Stadtman, *J. Biol. Chem.* **196**, 527 (1952); see Vol. I [98].

incubation determine the amount of Ac~P formed, using the hydroxamic acid method of Lipmann and Tuttle, as modified for a final volume of 3.0 ml.[4]

Definition of Unit and Specific Activity. One unit of enzyme is defined as that amount which causes the formation of 1 micromole of Ac~P in 30 minutes under the above conditions. Specific activity is expressed as units per milligram of protein.

Purification Procedure[1]

Step 1. Preparation of Cell-Free Extract.[5] The vacuum-dried cells (see below for preparation) are suspended (10 g. in 100 ml.) in 0.01 M potassium phosphate buffer (pH 8.0). The suspension is incubated at 28°, and at hourly intervals a 1-ml. aliquot is removed and centrifuged. 0.1 ml. of the clear supernatant solution is used for protein determination. The incubation is continued until there is no further increase in the protein content of the cell-free extract. The time required for extraction varies (2 to 8 hours) with different cell preparations. When extraction is complete, the cell suspension is centrifuged for 15 minutes at 12,000 r.p.m. in a Servall centrifuge. The residue is washed with water (50 ml./10 g. of bacteria), and the cell-free supernatant solutions are combined. If kept at −15° these extracts are stable for months.

Step 2. First Fractionation with Ammonium Sulfate. The extract from step 1 is diluted with cold distilled water to obtain a solution containing 11 mg. of protein per milliliter. This step and all subsequent steps are performed in a cold room at 2° with reagents chilled to this temperature. The solution is adjusted to pH 7.0 with 0.5 M K_3PO_4, and the slight precipitate which forms is removed by centrifugation in a Servall SS-1 centrifuge at maximal speed. A solution of ammonium sulfate, saturated at 2° and adjusted to pH 7.0 with ammonium hydroxide, is added to the supernatant to give 0.35 ammonium sulfate saturation. The suspension is stirred for 10 minutes and then centrifuged. The supernatant solution is discarded, and the precipitate is dissolved in 0.002 M potassium phosphate buffer (pH 8.0) to obtain a solution containing 20 mg. of protein per milliliter.

Step 3. Adsorption of Aldehyde Dehydrogenase by Calcium Phosphate Gel. Calcium phosphate gel (25 mg./ml., pH 6.7) is added to the dissolved precipitate from step 2 to give a gel-protein ratio of 0.25 (dry weight basis). The suspension is stirred for 10 minutes and then centrifuged, and the precipitate is discarded. The supernatant solution is adjusted to a gel-

---

[4] F. Lipmann and L. C. Tuttle, *J. Biol. Chem.* **159**, 21 (1945); see Vol. III [39].

[5] This method was previously described by E. R. Stadtman and H. A. Barker, *J. Biol. Chem.* **180**, 1085 (1949).

protein ratio of 1.0, and after 11 minutes' equilibration the gel is recovered by centrifugation at 2500 r.p.m. The protein is eluted from the gel with three portions (1.2 ml./100 mg. of gel) of 0.1 M potassium phosphate (pH 8.0). The combined eluates are dialyzed overnight against 2 l. of solution containing 0.002 M Tris, 0.001 M cysteine, and 0.12 M potassium chloride (pH 7.3).

Step 4. Second Ammonium Sulfate Fractionation. The dialyzed solution from step 3 is diluted to contain 5 mg. of protein per milliliter and is adjusted to 0.17 ammonium sulfate saturation by the addition of ammonium sulfate solution saturated at 2° and adjusted to pH 7.0 with ammonium hydroxide. After stirring for 30 minutes, the precipitate is recovered by centrifugation and is finally dissolved in 0.002 M imidazole·HCl buffer at pH 8.0.

SUMMARY OF PURIFICATION PROCEDURE[a]

Fraction	Total volume, ml.	Protein, mg./ml.	Aldehyde dehydrogenase activity		Recovery of activity, %
			Units/mg.	Total units	
1. Cell-free extract	1680	11.0	0.71	12,600	—
2. First ammonium sulfate fraction (0–0.35 sat.)	72	19.6	3.65	5,120	41
3. Dialyzed eluate from gel	49	5.8	15.2	4,300	34
4. Second ammonium sulfate fraction (0–0.17 sat.)	1.4	3.6	261.0	1,330	11

[a] R. M. Burton and E. R. Stadtman, *J. Biol. Chem.* **202**, 873 (1953).

Properties[1]

Spectrophotometric Measurement of Activity. With the purified enzyme, reaction 1 may be followed spectrophotometrically by measuring the change in optical density at 340 mμ that is associated with the reduction of DPN. The reaction mixture contains reduced CoA, 0.1 to 0.2 micromole; potassium veronal buffer (pH 8.8), 100 micromoles; GSH, 2 micromoles; acetaldehyde, 100 micromoles; DPN, 0.2 micromole; and purified aldehyde dehydrogenase, 10 units. Total volume, 3.0 ml. The same reaction mixture without acetaldehyde serves as the reference solution in the spectrophotometer. The spectrophotometric method cannot be used with unpurified enzyme preparations, since extracts of *Cl. kluyveri* contain a very active DPNH oxidase system.

Specificity. The purified enzyme will oxidize acetaldehyde ($K_m = 1.5 \times 10^{-3}$), propionaldehyde ($K_m = 4.5 \times 10^{-3}$), butyraldehyde ($K_m = 11.0 \times 10^{-3}$), and glycoaldehyde ($K_m = 20 \times 10^{-3}$). With concentrations of substrates that saturate the enzyme the maximal velocity of acetaldehyde oxidation is about three times as great as that obtained with the other aldehydes. Benzaldehyde, glyceraldehyde, formaldehyde, and chloral are not oxidized at a significant rate under comparable conditions.

Activators and Inhibitors. No metal ion requirements have been observed. Sulfhydryl reducing agents frequently show a stimulating effect, owing presumably to the fact that they prevent oxidation of reduced CoA.

Effect of pH. The reaction is strongly pH dependent, owing to the fact that H^+ is produced in the reaction. Maximum rates of oxidation are observed in the pH range of 7.5 to 8.8.

Equilibrium Constant. Reaction 1 is freely reversible, and the equilibrium constant is:

$$K = \frac{(Ac{\sim}S\ CoA)(DPNH)(H^+)}{(CH_3CHO)(DPN^+)(CoA)} = 1.2 \times 10^{-4}\ (25°)$$

Thus at pH 7.0 (25°), $K = 1.2 \times 10^3$.

Growth of Cl. kluyveri.[5,6] Under the highly anaerobic conditions required for the growth of *Cl. kluyveri* the completely synthetic medium used to support growth is relatively selective so that unsterilized media may be employed if an active 5 to 10% inoculum is used. Building up an active inoculum for the large-scale culture may be a slow process, and sterilized media are required.

95% ethanol	20 ml.
50% KAc or an equivalent amount of NaAc·3H₂O	10 ml.
Glacial acetic acid	2.5 ml.
10% NH₄Cl	2.5 ml.
M/1 phosphate buffer, pH 7.0	3.5 ml.
1% CaCl₂	1.0 ml.
20% MgSO₄·7H₂O	1.0 ml.
0.1% Na₂MoO₄·2H₂O	2.0 ml.
0.1% MmSO₄·4H₂O	2.0 ml.
1.0% FeSO₄·7H₂O	0.2 ml.
Yeast extract (Difco)[a]	0.7 g.
60% K₂CO₃·1.5H₂O	8.0–10 ml.
Sodium thioglycollate[a]	0.75 g.
Na₂S₂O₄ (dry powder)[b]	35 mg.
3 mg.% biotin[b]	0.3 ml.
50 mg.% p-aminobenzoic acid[b]	0.4 ml.
Tap water	855 ml.

[a] Used in preparation of sterilized medium only.
[b] Used in preparation of unsterilized medium only.

Procedure for Growth in Sterilized Medium (Preparation of Inoculum).
All the constituents except the thioglycollate and the K₂CO₃ are mixed.

[6] B. T. Bornstein and H. A. Barker, *J. Bacteriol.* **55**, 223 (1948).

The thioglycollate is added immediately prior to autoclaving. The solution is autoclaved for 20 minutes at 15 lb. pressure and is then cooled rapidly in a water bath, to about 35 to 45°. The pH is adjusted to 7.0 to 7.1 by the addition of the K_2CO_3 solution which has been sterilized separately.

For test-tube cultures, 9- to 10-ml. aliquots of the medium are transferred rapidly to sterile test tubes. After inoculation with a transfer of *Cl. kluyveri*, a small wad of adsorbent cotton is pushed down over the sterile test-tube plug and 0.8 to 1.0 ml. of Oxsorbent (Burrell Corporation) is pipetted deep into the cotton. The tube is sealed immediately with a rubber stopper that should fit tightly in order to prevent its being blown out during growth of the bacterium.

If the inoculum is from an old culture of *Cl. kluyveri*, 10 days to several weeks may be required for the organism to get started. When an inoculum from an actively growing culture is used, 24 to 48 hours is usually sufficient to provide heavy growth.

A test-tube culture is used to inoculate 500 ml. of the same medium sterilized in 500-ml. flasks with long slender necks (volumetric flasks, Pyrex). The cultures are incubated at 35 to 37°.

Large-Scale Culture

Unsterilized Medium. The composition of this medium for large-scale cultures is the same as that given above with the following modifications. For the large-scale cultures, the yeast extract and the thioglycollate are omitted. In place of yeast extract, biotin (0.3 ml. of a 3-mg. % solution) and *p*-aminobenzoic acid (0.4 ml. of a 50-mg. % solution) are added per liter, and in place of thioglycollate 35 mg. of $Na_2S_2O_4$ per liter is used to obtain anaerobic conditions.

Procedure. All reagents except the $Na_2S_2O_4$ and K_2CO_3 are added to warm (35 to 37°) tap water. The solution is mixed and deaerated by passing through a vigorous stream of nitrogen for several minutes. After deaeration, the pH is adjusted to 7.2 to 7.4 by stepwise addition of the K_2CO_3 solution, and finally the solid $Na_2S_2O_4$ is added and the solution is agitated briefly with a stream of nitrogen to insure good mixing.

Ten liters of the resulting medium is inoculated immediately with 500 ml. of the inoculum grown in the sterilized medium previously described. If necessary the volume of the medium should be increased by the addition of more water until the level is within about 2 inches of the top of the vessel. The vessel is finally stoppered with a rubber stopper fitted with a piece of glass tubing which is connected to a water seal so as to allow the H_2 produced in the fermentation to escape. When good growth is obtained (36 to 70 hours), the 10-l. culture may be used to inoculate 200 l.

of the same medium prepared in the same manner. For these larger fermentations, a 55-gallon drum (No. 3-16 stainless steel preferred, but iron satisfactory) is used as the container.

Caution! Since large volumes of H_2 are evolved during the fermentation, care must be taken to direct the evolved gas outside the building.

Growth Measurement. At 4- to 12-hour intervals, 5- to 10-ml. aliquots of the culture solution are removed and the optical density is measured in a photoelectric colorimeter. When maximum growth is obtained (i.e., when the change in optical density with time becomes relatively small), the culture is harvested. When measured in a Klett-Summerson photoelectric colorimeter, maximum growth is usually obtained when the optical density (with a filter No. 540) is 170 to 200.

Preparation of Dried Cells. When maximum growth is reached, the culture solution is siphoned into a Sharples centrifuge operating at about 30,000 r.p.m. The rate of flow is adjusted to about 1 to 1.5 l./min.

Not more than 100 l. of culture should be centrifuged before the bacteria are removed from the centrifuge bowl. The cells are suspended in 10 vol. of 0.03 % sodium sulfide solution, the pH is adjusted to 7.0, and the bacteria are collected again by centrifugation for 15 minutes in a Servall centrifuge. The cell paste is spread out thinly (1 to 5 mm.) in a large evaporating dish and is dried *in vacuo* over $CaCl_2$ at room temperature. The dried cells are relatively stable at room temperature and retain most of their enzymatic activity for years at $-10°$. The yield of dry cells is about 275 mg./l. of culture.

[85] Flavin-Linked Aldehyde Oxidase

$$RCHO + 3OH^- \rightarrow RCO_2^- + 2H_2O + 2e^-$$

By HENRY R. MAHLER

Assay Method

Principle. The method originally developed by Gordon *et al.*[1] was based on methylene blue reduction carried out anaerobically in Thunberg tubes. The present method was developed in this laboratory[2] and is contingent on the fact that 2,6-dichlorophenolindophenol and cytochrome c are reducible by acetaldehyde in the presence of the enzyme, even under

[1] A. H. Gordon, D. E. Green, and V. Subrahmanyan, *Biochem. J.* **34**, 764 (1940).

[2] H. R. Mahler, B. Mackler, and D. E. Green, *J. Biol. Chem.* **210**, 465 (1954).

aerobic conditions. The reduction of these oxidizing agents is followed spectrophotometrically at 600 mμ and 550 mμ, respectively. Catalase and ethanol are added to destroy any H_2O_2 formed by reaction with oxygen. Two different assay systems are used, since the enzyme shows a difference in properties depending on the nature of the electron acceptor (see below).

Reagents

1 M acetaldehyde (redistilled).
0.2 M phosphate buffer, pH 7.1.
1% aqueous cytochrome c (Fe^{+++}) solution.
0.1% catalase[3] solution in 0.02 M $KHCO_3$.
95% ethyl alcohol.
0.001 M MoO_3.
0.01% aqueous 2,6-dichlorophenolindophenol.
Enzyme. Dilute the enzyme with phosphate buffer to obtain 1 to 5 cytochrome units of enzyme per milliliter (see definition below).

Procedure. CYTOCHROME c ASSAY. Place 0.40 ml. of phosphate buffer, 0.1 ml. of cytochrome c, 0.02 ml. of enzyme, 0.01 ml. of catalase, 0.02 ml. of ethyl alcohol, and 0.45 ml. of MoO_3 solution in a cuvette and determine the reading at 550 mμ against a blank cell containing 0.45 ml. of buffer, 0.1 ml. of cytochrome c, and 0.45 ml. of MoO_3 solution. Add 0.02 ml. of acetaldehyde to both cells, and determine the rise in optical density (ΔE_{550}) every 15 seconds for 2 minutes.

INDOPHENOL ASSAY. Place 0.40 ml. of phosphate buffer, 0.1 ml. of indophenol, 0.02 ml. of enzyme, 0.01 ml. of catalase, 0.02 ml. of ethyl alcohol, and 0.45 ml. of water in a cuvette. Determine the initial reading, and take readings every 15 seconds after the addition of acetaldehyde to determine the drop in optical density ($-\Delta E_{600}$).

Definition of Units and Specific Activity. One cytochrome unit of enzyme is defined as the amount causing an initial ΔE_{550} of 1.00 per minute. In practice, the period from 15 to 75 seconds is used. Specific activity equals units per milligram of protein, as determined by the biuret reaction.[4] Similarly, one indophenol unit is defined as that amount which causes a $-\Delta E_{600}$ of 1.00. Under the conditions described above, with an essentially homogenous enzyme, one cytochrome c unit equals 0.4 indophenol unit.[5]

---

[3] Cytochrome c and catalase are obtainable commercially from various sources. For their preparation and purification see Vol. II [133, 137].

[4] A. G. Gornall, C. J. Bardawill, and M. M. David, *J. Biol. Chem.* **177**, 751 (1949).

[5] One cytochrome c unit also equals 1.7 units as defined by Gordon *et al.*[1]

Purification Procedure

Steps 1, 2, and 3 of the following procedure are those of Gordon et al.[1] The subsequent modifications have been developed recently,[2] and have been successfully employed in numerous preparations of highly purified enzyme.

Step 1. Extraction and Heat Treatment. 1.8 kg. of minced pig liver is homogenized for 20 seconds in Waring blendors with a mixture of 4 l. of water and 1880 ml. of 95% ethyl alcohol. The suspension is maintained at 48° for 5 minutes and then rapidly cooled to 20° by the addition of crushed ice. The heating must be so regulated that the temperature reaches 48° within 4 minutes of application of heat. Heating is best carried out in a round-bottom flask immersed in a boiling water bath. Stirring must be vigorous to prevent local overheating. After chilling, the mixture is centrifuged, and the denatured protein is discarded.

Step 2. Lead Acetate Precipitation. The clear, pale-red supernatant (5.6 l.) is treated with 40 ml. of 25% basic lead acetate. The precipitate after centrifugation is decomposed by thorough shaking with 400 ml. of saturated Na_2HPO_4. The precipitated lead phosphate is removed by centrifugation, and the supernatant fluid (570 ml.) is made 40% saturated with respect to neutral $(NH_4)_2SO_4$. The precipitate is dissolved in 100 ml. of water. The subsequent steps are carried out at 0°.

Step 3. Alkaline Ammonium Sulfate Fractionation. The enzyme solution is made 27% saturated with respect to a solution prepared by adding 6 ml. of concentrated NH_4OH (sp. gr., 0.880) to 94 ml. of saturated $(NH_4)_2$-SO_4. The precipitate is centrifuged off and discarded. The saturation with respect to $(NH_4)_2SO_4$ is raised to 40%, and the precipitate collected. It is dissolved in sufficient water to give a protein concentration of 30 to 40 mg./ml.

Step 4. Aging in Alkaline Ammonium Sulfate. The solution from step 3 is made 30% saturated with respect to the ammoniacal $(NH_4)_2SO_4$ described above and allowed to stand at 0° for an extended period. As precipitate forms, it is removed by centrifugation. When no further precipitation occurs during a 24-hour period, the saturation of the solution is raised to 40% with respect to $(NH_4)_2SO_4$, and the precipitate formed is dissolved in a small volume of water. If the enzyme still contains impurities, the aging procedure is repeated. The aging period necessary will vary from preparation to preparation. Occasionally as much as a week will be necessary to remove all the contaminating materials. Two optical density ratios—the E_{410}/E_{450} and the E_{280}/E_{450}—are useful guides throughout the purification and especially during the last step. The former should be ≤ 1.5, whereas the latter equals 10.8 for an enzyme approaching homogeneity.[6]

SUMMARY OF PURIFICATION PROCEDURE[a,b]

Fraction	Total volume, ml.	Total protein, mg.	Total units	Specific activity, cytochrome units/mg.	Recovery, %
1. Heat-treated extract	5600				—
2. Lead precipitation and (NH₄)₂SO₄ fractionation	100	6–14,000	2750	0.2–0.5	100
3. Alkaline (NH₄)₂SO₄ fractionation	20	1330	1600	1.2	58
4. Aging and final fractionation	10	200	800	4.0	29

[a] A. H. Gordon, D. E. Green, and V. Subrahmanyan, *Biochem. J.* **34**, 764 (1940).
[b] H. R. Mahler, B. Mackler, and D. E. Green, *J. Biol. Chem.* **210**, 465 (1954).

Properties

Homogeneity and Physical Properties. The enzyme as described above is better than 90% homogenous on electrophoresis in phosphate buffer at pH 6.8 and 7.9.[7] It is, however, inhomogeneous on ultracentrifugation, although the various fractions have identical absorption spectra and specific activities.

Absorption Spectrum. The enzyme shows an atypical flavoprotein spectrum with maxima at 278, 350 (diffuse), and 405 mμ, with shoulders at 450, 530, and 630 mμ. On reduction with aldehydes or hydrosulfite, the peak at 350 mμ disappears, the peak at 405 mμ is lowered, the shoulder at 450 mμ disappears (the E_{450} is lowered to 30 to 50% of the oxidized value), and peaks at 560 and 620 mμ appear.

Prosthetic Groups. The flavin component of the enzyme has been identified as FAD, which can be irreversibly removed by a variety of treatments.[1] In addition, the enzyme contains varying amounts of molybdenum, depending on its method of preparation.[2] Our purest preparation contained 0.5 g. atoms of molybdenum per mole of flavin, but this is not to be taken as the actual metal content of the native enzyme, since molybdenum is split off under conditions approximating those actually employed during isolation. The molybdenum can be removed by dialysis first against 0.001 M NH₄OH, followed by dialysis against Tris, pH 8.1.

[6] If these ratios are higher than the limits shown here, refractionation with alkaline ammonium sulfate between the limits of 30 and 35% of saturation, frequently yields a satisfactory enzyme.
[7] R. M. Bock, to be published.

The role of the metal will be discussed below. In addition the enzyme has associated with it an iron-porphyrin component,[2] not identical with catalase, in definite and constant proportion.

Specificity of Substrate. The enzyme is active with a large variety of aldehydes. The following values relative to acetaldehyde as 100 have been determined by Gordon et al.:[1] propionaldehyde, 17; butyraldehyde, 13; crotonaldehyde, 66; benzaldehyde, 25; glycollic aldehyde, 8; and salicylaldehyde 2. The enzyme is inactive with ketones, keto acids, alcohols, hypoxanthine, xanthine, and pteridines. It shows slight activity (approximately 0.01 that of acetaldehyde) with DPNH and cinchonidine.[2] The K_m for crotonaldehyde equals 0.007 M.[1]

Specificity of Electron Acceptor. In the presence of the enzyme aldehydes can be oxidized by molecular oxygen and oxido-reduction dyes, such as MB, or indophenols at approximately equal rate.[1] The reaction with cytochrome c is slow with the enzyme as isolated but can be raised to the values for the other acceptors by the addition of MoO_3.[2] Nitrate can function as an acceptor anaerobically, but it is relatively inefficient.

Activators and Inhibitors.[2] No activators are necessary if indophenol, MB, or O_2 is used as hydrogen acceptor. If cytochrome c is the oxidizing agent, the presence of molybdenum and phosphate ions is obligatory. With an enzyme split with respect to molybdenum, the dissociation constant for the metal component as MoO_3 has been determined as 2.8 × 10^{-4} M. Tungstate will partially replace molybdate, but no other metal tried can be substituted. The phosphate requirement is completely satisfied at a concentration of 0.05 M. Arsenate can partially substitute for phosphate at higher concentrations. The reaction with dyes (or oxygen) is inhibited quantitatively and instantaneously by 1 × 10^{-4} M p-chloromercuribenzoate and 5 × 10^{-4} M arsenite; 2 × 10^{-4} M cyanide and 10^{-3} M azide do not inhibit significantly, unless the enzyme is preincubated for 10 minutes with the inhibitor. 10^{-5} M quinacrine inhibits 60%; this inhibition can be completely overcome by FAD (2 γ/ml.) and partially by RMP at the same concentration. None of the other inhibitors to be mentioned below is effective in this system.

The reaction with cytochrome c is completely and instantaneously abolished by 1 × 10^{-4} M p-chloromercuribenzoate, 5 × 10^{-4} M arsenite, 10^{-3} M iodoacetate, and 10^{-5} M azide. 2 × 10^{-4} M cyanide inhibits 80%, 10^{-3} M citrate or ethylenediaminetetraacetate shows 67% inhibition, and 8-oxyquinoline and o-phenanthroline at the same concentration inhibit 60%. Quinacrine at 10^{-4} M inhibits 100%, and 58% at 10^{-5} M. Preincubation with acetaldehyde, the substrate, affords complete protection against iodoacetate or arsenite inhibition. The inhibition by metal-binding reagents can be reversed by the addition of MoO_3, even for enzymes which do not ordinarily require added molybdenum, whereas the quinacrine inhibition is overcome by FAD. Cyanide and azide inhibition appears to be irreversible.

Effect of pH.[1] The enzyme functions optimally in the pH range 7.0 to 8.0. Rates at 6.0 and 9.0 are approximately 50% those at the optimum; at pH 5.0 and 10.0 they are 25% and 40% of optimum, respectively.

[86] Glycolic Acid Oxidase and Glyoxylic Acid Reductase

By ISRAEL ZELITCH

Glycolic Acid Oxidase from Spinach Leaves

$$CH_2OH—COO' + O_2 \rightarrow CHO—COO' + H_2O_2 \rightarrow H—COO' + CO_2 + H_2O$$
Glycolate   Glyoxylate   Formate

Assay Method

Principle. The original method used by Clagett *et al.*[1] was based upon measurement of the rate of oxygen uptake in Warburg vessels. The rapid assay described below depends on measurement of the rate of reduction of 2,6-dichlorophenolindophenol at 620 mμ in the Beckman spectrophotometer.[2]

Reagents

Potassium glycolate (0.04 M). 304 mg. of glycolic acid is neutralized with potassium hydroxide and brought to a final volume of 100 ml.

0.1 M potassium phosphate buffer, pH 8.0.

0.01% 2,6-dichlorophenolindophenol.

0.1 M potassium cyanide in 0.01 M ammonium hydroxide.

Enzyme. The solution containing the enzyme is diluted with phosphate buffer if necessary.

Procedure. To a cell of 1-cm. light path are added 1.0 ml. of phosphate buffer, 0.3 ml. of 2,6-dichlorophenolindophenol, 0.01 ml. of potassium cyanide, 0.5 ml. of potassium glycolate, and water to bring the final volume to 3.0 ml. Readings are taken at 620 mμ, against a cell containing all components except dye, at 15-second intervals for 1 minute after adding the enzyme; from 1 to 15 units can be used in the assay.

[1] C. O. Clagett, N. E. Tolbert, and R. H. Burris, *J. Biol. Chem.* **178**, 977 (1949); N. E. Tolbert, C. O. Clagett, and R. H. Burris, *J. Biol. Chem.* **181**, 905 (1949).

[2] I. Zelitch and S. Ochoa, *J. Biol. Chem.* **201**, 707 (1953).

Definition of Unit and Specific Activity. One enzyme unit is defined as the amount which causes a decrease in optical density of 0.01 per minute. Specific activity is expressed in units per milligram of protein. With purified enzyme preparations, 1 unit in the dye assay is equivalent to about 1.5 μl. of O_2 uptake in 10 minutes in the manometric assay when measured without adding FMN. Protein is determined from measurement of the light absorption at wavelengths 260 mμ and 280 mμ by the method of Warburg and Christian.[3]

Application of Assay Method to Crude Tissue Extracts. Spinach leaf extracts give rise to a considerable reduction of dye in the absence of added glycolate. This prohibits the use of the assay before concentration of the enzyme by salt fractionation. The addition of cyanide is necessary to inhibit the indophenol oxidase activity that is present during the early steps of purification.

The purified enzyme causes 1 mole of oxygen to be consumed per mole of glycolate oxidized. In crude preparations the oxygen uptake may be as little as one-half mole or as high as several moles per mole of glycolate added, depending on the amount of catalase present and on the endogenous substrates being oxidized through the glycolate-glyoxylate hydrogen transport system.[2,4]

The optical method of assay can be used only with the undissociated enzyme. If the prosthetic group (FMN) is removed, activation by FMN is observed only if the test is conducted under anaerobic conditions. If the enzyme appears to lose activity during storage, it should be assayed manometrically in the presence of excess FMN.[2]

Purification Procedure

Glycolic oxidase from tobacco leaves has been partially purified by salt fractionation,[1] and an enzyme of similar function from mammalian liver has been purified by a procedure like the one below.[5]

The following steps have been carried out successfully many times, starting with amounts of 4.5 to 70 kg. of spinach leaves.

Step 1. Preparation of Crude Extract. Spinach leaves are washed with tap water and ground in a Waring blendor, a juice extractor, or a Nixtamal mill. The juice is expressed with a hydraulic press and is cooled to 0 to 5°. The temperature is maintained between these limits throughout the procedure unless otherwise stated. The pH of the extract (about 6.3) is lowered to 5.4 by the addition of 10% acetic acid. The small precipitate

---

[3] O. Warburg and W. Christian, *Biochem. Z.* **310**, 384 (1941).

[4] I. Zelitch, *J. Biol. Chem.* **201**, 719 (1953).

[5] E. Kun, *Federation Proc.* **11**, 364 (1952); **12**, 338 (1953).

formed is removed by centrifugation at high speed or by filtration through a pad of filter paper on a Büchner funnel.[6] The protein concentration should be adjusted by dilution with water, if necessary, so as not to exceed 50 mg./ml.

Step 2. Fractionation with Ammonium Sulfate. To each liter of crude extract is added 140 g. of ammonium sulfate to give 0.2 saturation. Stirring is continued for at least 30 minutes, and the precipitate is then removed by high-speed centrifugation or by filtration through fluted filter papers. The supernatant liquid is brought to 0.3 saturation by the addition of 70 g. of ammonium sulfate per liter. The precipitate, which contains the enzyme, is collected by centrifugation or filtration. It is dissolved in 0.02 M potassium phosphate buffer of pH 8.0, using about one-tenth of the volume of the crude extract.

Step 3. First Precipitation of Inactive Protein at pH 4.9. The enzyme solution is dialyzed with stirring for 4 hours against 25 vol. of 0.02 M potassium phosphate buffer, pH 8.0. The pH of the dialyzate is slowly lowered to exactly 4.9 by the addition of 10% acetic acid. The large precipitate is removed by centrifugation and discarded.

Step 4. Fractionation with Ethanol. Absolute ethanol, cooled to $-50°$, is slowly added to the supernatant liquid at $0°$ to a final concentration of 8% by volume. The final temperature is $-1°$. The precipitate is removed by centrifugation at the same temperature. Ethanol is than added to the supernatant fluid to a final concentration of 20% by volume, and the temperature is lowered gradually to $-5°$. The precipitate is collected by centrifugation and dissolved in 0.02 M potassium phosphate buffer of pH 8.0, using about one-fifth of the volume at the end of step 3.

Step 5. Second Precipitation of Inactive Protein at pH 4.9. The pH of the enzyme solution is slowly lowered to pH 4.9 by the dropwise addition of 10% acetic acid. The precipitate is removed by centrifuging at high speed.

Step 6. Adsorption and Elution from Calcium Phosphate Gel. For the fractionation with calcium phosphate gel, the reagent[7] (25 to 30 mg. of calcium phosphate per milliliter) is added stepwise, with centrifugation and assay of the supernatant fluid after each step. About 1 ml. of gel is added for each 50 mg. of protein. The suspension is stirred at $0°$ for about 15 minutes and is centrifuged. The first four adsorptions usually remove inactive material, and the enzyme is adsorbed in the course of the succeeding four treatments, at which point little activity remains in the supernatant fluid. The gel with adsorbed enzyme is eluted twice, first with

[6] T. B. Osborne, *in* "Handbuch der Biochemischen Arbeitsmethoden" (Abderhalden, ed.), Vol. II, p. 278, Urban and Schwarzenberg, Berlin, 1910.

[7] D. Keilin and E. F. Hartree, *Proc. Roy. Soc. (London)* **B124**, 397 (1938).

0.033 M potassium phosphate buffer of pH 8.0, using about five times the original volume of the gel used to adsorb the enzyme. Glycolic acid oxidase with a specific activity of at least 120 is liberated. The gel is next eluted with 0.1 M potassium phosphate buffer, pH 8.0, in the same manner. The second elution removes enzyme with a specific activity of 289 or higher.

The activity of these preparations is stimulated markedly by the addition of excess FMN in tests of rate of oxygen uptake. This indicates that the enzyme is at least partly dissociated. If all the steps are conducted within a period of 36 hours, a preparation of specific activity as high as 530 by the optical assay can be obtained. The preparations can be preserved by the addition of ammonium sulfate added to 0.5 of saturation. Inactivation appears to take place when the enzyme is stored at 0° under these conditions. The apoenzyme, however, is stable for at least 1 year and can be reactivated by the addition of excess of FMN.

Properties

Purity. The glycolic acid oxidase so prepared is free from catalase. If excess catalase is added, the glyoxylate formed in the reaction is not oxidized further. The purified enzyme is free from glyoxylic acid reductase[4] activity. In the ultracentrifuge, the glycolic oxidase showed a major component (about 80% of the total protein) and a minor component. Upon electrophoresis in 0.05 M potassium phosphate buffer, pH 7.0, two components were also observed. The larger one (comprising 74% of the protein) was the glycolic acid oxidase.

Prosthetic Group. The absorption spectra of the enzyme and of its prosthetic group, the equal effectiveness of the enzyme flavin nucleotide and of FMN in activating the apoenzyme, and the enzymatic conversion of the prosthetic group by ATP to form FAD all indicate that the prosthetic group is FMN. For maximum activity in the manometric test, the apoenzyme requires a concentration of FMN of 20 μM. FAD is less active than FMN in activating the apoenzyme.

Specificity. The purified enzyme is specific in its action on glycolate and L-lactate. The rate of oxidation of L-lactate is about half of the rate with glycolate. No activity is observed with glyoxylate, D-lactate, DL-alanine, glycine, L-leucine, glucose, DPNH, DL-glycerate, 3-phospho-D-glycerate, DL-mandelate, ethylene glycol, or glycoladeldehyde. Clagett et al.,[1] using a tenfold purified enzyme from tobacco leaves, found no activity with D-lactate, L-malate, D- or L- or *meso*-tartrate, DL-glycerate, or α-hydroxyisobutyrate. They found the oxidation of DL-α-hydroxy-*n*-butyrate to be very slow.

Activators and Inhibitors. Clagett et al.,[1] using a partially purified glycolic acid oxidase from tobacco leaves, reported a stimulation of oxygen uptake in the presence of cyanide. We have also observed increases in activity from 50 to 100% in the presence of 1×10^{-2} M cyanide at all stages of purification. At a concentration of 1×10^{-2} M, malonate gave 30%, iodoacetate 48%, and hydroxylamine 30% inhibition of glycolate oxidation.[1] Kun[5] has reported that the reactivation of the glycolic oxidase apoenzyme from mammalian liver is inhibited by iodoacetate which does not inhibit the holenzyme and that the oxidation of glycolate is inhibited 80 to 100% by hydroxylamine at 5×10^{-3} M.

TABLE I

SUMMARY OF PROCEDURE FOR PURIFYING GLYCOLIC ACID OXIDASE FROM SPINACH LEAVES[a]

Fraction	Total volume, ml.	Units/ml.	Total units, thousands	Protein, mg./ml.	Specific activity, units/mg.	Recovery, %
1. Extract, pH 5.4	25000	25[b]	625	42.0	0.6[b]	—
2. (NH₄)₂SO₄ fraction 0.2–0.3 satn.	2600	180	468	97.0	1.9	75
3. First supernatant fluid, pH 4.9	2020	220	444	20.0	11	71
4. Ethanol fraction 8–20% by volume	830	417	346	10.8	39	56
5. Second supernatant fluid, pH 4.9	800	365	292	5.2	70	47
6. Second Ca₃(PO₄)₂ gel eluate	200	173	34.6	0.60	289[c]	6

[a] Data from a representative experiment, I. Zelitch and S. Ochoa, *J. Biol. Chem.* **210**, 707 (1953).

[b] Approximate figures.

[c] The activity of this preparation in the manometric test was stimulated 2.7-fold by the addition of FMN.

Other Properties. Glycolic acid oxidase has a rather sharp pH optimum at 8.3 with Tris buffer, using glycolate as substrate. The oxidation of L-lactate shows a broader optimum between pH 8.0 and 8.5. The Michaelis constants are 3.8×10^{-4} M for glycolate and 2.0×10^{-3} M for L-lactate.

Glyoxylic Acid Reductase from Tobacco Leaves

$$CHO—COO' + DPNH + H^+ \rightarrow CH_2OH—COO' + DPN$$
Glyoxylate   Glycolate

Assay Method

Principle. The assay is based upon one originally used for the tenfold purification of an enzyme of similar function from spinach leaves.[1] The rate of oxidation of DPNH is measured by following the decrease in optical density at 340 mμ in the Beckman spectrophotometer.

Reagents

Sodium glyoxylate (0.17 *M*). 400 mg. of three-times-crystallized glyoxylate bisulfite addition product[8] is converted to sodium glyoxylate by being boiled for 2 minutes in excess 2 *N* HCl. The solution is cooled and neutralized with NaOH and brought to a final volume of 10 ml.

DPNH (5 mM.). DPN at a concentration of 4 mg./ml. is reduced with dithionite.[9]

0.1 *M* potassium phosphate buffer, pH 6.4.

Enzyme. If necessary, the enzyme solution is diluted with 0.005 *M* ethylenediaminetetraacetate at pH 7.0 containing 0.1% gelatin.

Procedure. To a cell of 1-cm. light path are added 1.0 ml. of phosphate buffer, 0.3 ml. of sodium glyoxylate, enzyme solution, and water to bring the final volume to 3.0 ml. The DPNH, 0.03 to 0.05 ml., is added at zero time, and readings are taken at 15-second intervals for 1 minute at 340 mμ. From 1 to 25 units of enzyme can be used in the assay.

Definition of Unit and Specific Activity. One unit is taken as the amount of enzyme that causes a decrease in optical density of 0.01 per minute at 340 mμ. Protein is determined by measuring the light absorption at wavelengths 260 mμ and 280 mμ.[3]

Application of Assay Method to Crude Extracts. Leaf extracts usually oxidize DPNH slowly without added substrates. This activity is generally not great compared with the rate of DPNH oxidation in the presence of glyoxylate, and a suitable correction for endogenous activity can be applied.

Purification Procedure

The steps below effect a partial purification of glyoxylic reductase by procedures that are reproducible with tobacco leaves from the field or the greenhouse. Purer enzyme preparations than those described here, with specific activites of the order of 3600, have been obtained. Unfortunately the procedures followed to achieve this higher level of purity are not fully reproducible, and the recoveries are poor. These further steps are being investigated.

Step 1. Preparation of Crude Extract. Large tobacco leaves from mature plants are washed with tap water and ground in a Nixtamal mill. The

[8] S. Weinhouse and B. Friedmann, *J. Biol. Chem.* **191**, 707 (1951).
[9] See Vol. III [126].

pulp is collected in a chilled container, mixed with filter paper clippings, and the juice expressed in a hydraulic press. The press juice is centrifuged at high speed or filtered through a pad of paper pulp on a Büchner funnel[6] in the cold. The clear liquid, containing the enzyme, is usually about equal in volume (in milliliters) to the original fresh weight (in grams) of the leaves. The temperature is maintained between 0 and 5° throughout the rest of the procedure.

Step 2. Fractionation with Ammonium Sulfate. 168 g. of ammonium sulfate is added to each liter of extract to give 0.24 saturation. Stirring is continued for at least one-half hour, and the precipitate is then removed by centrifugation or by filtration through fluted filter papers and discarded. 112 g. of ammonium sulfate is added for each liter of supernatant liquid to give 0.4 saturation. The precipitate, containing the enzyme, is collected by centrifugation or filtration. It is dissolved in 0.01 M potassium phosphate buffer containing 0.005 M ethylenediaminetetraacetate, pH 7.0, using one-seventh of the volume of the crude extract.

Step 3. Removal of Inactive Protein with Protamine Sulfate. Protamine sulfate solution (20 mg./ml., pH 5) is added to the amber-colored enzyme solution with stirring. About 1mg. of protamine sulfate is added for every 15 mg. of protein. Stirring is continued for one-half hour, and the precipitate is then removed by centrifugation or filtration and discarded.

Step 4. Second Fractionation with Ammonium Sulfate. 168 g. of ammonium sulfate is added to each liter as before to give 0.24 saturation, and the precipitate is discarded. 91 g. of ammonium sulfate is added to each liter of supernatant fluid to give 0.37 saturation. The precipitate is collected and dissolved in 0.001 M potassium phosphate buffer, pH 7.0, using one-fourth of the volume at the end of step 3. Calculations based on the increase in volume of the enzyme solution usually indicate that about 0.03 saturation of ammonium sulfate is present. If necessary, ammonium sulfate is added to bring the salt concentration to this level.

Step 5. Adsorption of Inactive Protein by Calcium Phosphate Gel. 1 ml. of calcium phosphate gel[7] (25 to 30 mg. of solids per milliliter) is added for each 30 mg. of protein. Stirring is continued for 15 minutes, and the gel is then removed by centrifugation and discarded. 1 ml. of gel is added for each 15 mg. of protein in the supernatant liquid, and the gel removed as before.

Step 6. Concentration with Ammonium Sulfate. 28 g. of ammonium sulfate is added for each 100 ml. of enzyme solution to give 0.4 saturation. The precipitate is dissolved in 0.005 M ethylenediaminetetraacetate, pH 7.5, with one-seventieth of the volume at the end of step 5. This preparation is stable for months if stored under refrigeration.

Properties

Specificity. Crude preparations act with TPNH at about one-fifth of the rate with DPNH. The partially purified glyoxylic acid reductase is specific in its action with DPNH. Acetaldehyde or pyruvate have no activity as a substrate, indicating that alcohol dehydrogenase and lactic dehydrogenase are not present in the enzyme preparation. It has been possible to demonstrate the reduction of DPN with glycolate as substrate. The equilibrium of the reaction is very far in the direction of formation of glycolate.

Inhibitors. Carbonyl-binding reagents such as semicarbazide and phenylhydrazine at 1×10^{-2} M inhibit about 65% under the conditions of the assay. Iodoacetate at 1×10^{-2} M is without effect, but p-chloromercuribenzoate at 1×10^{-4} M inhibits 68%.

Other Properties. The Michaelis constant for glyoxylate is approximately 5.6×10^{-3} M. Glyoxylic acid reductase exhibits a fairly broad pH optimum with a maximum at 6.4 under the conditions of the assay.

TABLE II

SUMMARY OF PROCEDURE FOR PARTIAL PURIFICATION OF GLYOXYLIC REDUCTASE FROM TOBACCO LEAVES

Fraction	Total volume, ml.	Units/ml.	Total units, thousands	Protein, mg./ml.	Specific activity, units/mg.	Recovery, %
1. Clear extract	20500	120	2460	24.6	4.9	—
2. (NH₄)₂SO₄ fraction, 0.24–0.4 satn.	3935	440	1720	25.2	17.4	70
3. Protamine sulfate, supernatant fluid	3900	450	1750	19.6	23.0	71
4. (NH₄)₂SO₄ fraction 0.24–0.37 satn.	1065	1380	1470	26.2	53.0	60
5. Calcium phosphate gel, supernatant liquid	2150	475	1025	1.29	369	42
6. (NH₄)₂SO₄ fraction, 0–0.4 satn.	37	27800	1025	37.9	730	42

[87] Formic Dehydrogenase from Peas[1-3]

$$HCOOH + DPN^+ \rightarrow DPNH + CO_2 + H^+$$

By ALVIN NASON and HENRY N. LITTLE

Assay Method

Principle. The enzyme is determined by measuring the rate of DPN reduction spectrophotometrically as observed by the increase in optical density at 340 mμ.

Reagents

DPN approximately 2 μM./ml. Dissolve 10 mg. (70 to 90% purity) in 5 ml. of H_2O.
0.5 M sodium formate. Dissolve 3.4 g. in 100 ml. of H_2O and adjust to pH 6.5 with HCl.
0.5 M phosphate-NaOH buffer, pH 6.5.
Enzyme. At least 40 enzyme units (see definition below) should be used in the assay procedure.

Procedure. The reaction is started by the addition of 0.3 ml. of 0.5 M sodium formate to a mixture containing 1.0 ml. of phosphate buffer, 0.1 ml. of DPN (2 μM./ml.), formic dehydrogenase, and water to give a final volume of 3.0 ml. in a cell having a 1-cm. light path. The increase in optical density is measured at 340 mμ at 15-second intervals for 2 minutes.

Definition of Unit and Specific Activity. One unit of formic dehydrogenase is defined as that amount of enzyme which results in an increase in log I_0/I of 0.001 per minute calculated from the change between the 15- and 45-second readings. Specific activity is expressed as units per milligram of protein. Protein is determined by the method of Lowry *et al.*[4]

Purification Procedure

Step 1. Preparation of Crude Extract. Cell-free extracts are prepared from dried pea seeds, *Pisum sativum* var. "Laxton's Progress," which have been soaked in tap water at room temperature overnight and then blended with an approximately equal volume of 0.1 M Na_2HPO_4 in a

[1] D. C. Davison, *Proc. Linnean Soc. N. S. Wales* **74,** 26 (1949).
[2] M. B. Mathews and B. Vennesland, *J. Biol. Chem.* **186,** 667 (1950).
[3] D. C. Davison, *Biochem. J.* **49,** 520 (1951).
[4] O. H. Lowry, N. J. Rosebrough, A. L. Farr, and R. J. Randall, *J. Biol. Chem.* **193,** 265 (1951).

Waring blendor for 1 to 2 minutes. As an alternate procedure the dried seeds may first be ground or milled to a powder and extracted with three to five times its weight of 0.1 M Na_2HPO_4 by stirring at room temperature as indicated below. The blend is kept at room temperature for 2 to 3 hours with occasional stirring and is then pressed by hand through four layers of cheesecloth in order to remove the coarse material. The mixture is centrifuged at room temperature for 30 minutes at 3000 $\times$ g, resulting in a turbid, light-green supernatant solution which represents the crude extract (fraction 1).

Step 2. Fractionation with Ammonium Sulfate. To 5600 ml. of crude extract (fraction 1) derived originally from 5 lb. of peas is added 1232 g. of ammonium sulfate (220 g./l. of extract) to give a final concentration of 37% saturation. After standing for 30 minutes at room temperature, the precipitate is collected by centrifugation for 30 minutes at 3000 $\times$ g and discarded. The supernatant solution is treated with an additional 670 g. of ammonium sulfate (120 g./l. of original crude extract), and after 1½ hr. of standing at room temperature the precipitate (53% saturation) is collected by centrifugation as above. The precipitate is suspended in 250 ml. of 0.1 M phosphate buffer, pH 7.0, and dialyzed against 20 l. of 0.01 M phosphate buffer at 4° overnight. An appreciable precipitate of inert protein results which is removed by centrifugation and discarded. The supernatant solution (fraction 2) containing the active enzyme experiences a two- to threefold increase in volume as a result of dialysis.

All subsequent purification steps are carried out at 0 to 4°.

Step 3. Alumina C_γ gel treatment. To fraction 2 (680 ml.) is added 450 ml. of alumina C_γ gel,[5] aged 9 months or longer (123 mg. dry weight per milliliter). After 15 minutes, the precipitate is collected by centrifugation and eluted for 15-minute intervals with five portions of 50 ml. each of 0.1 M phosphate buffer, pH 7.9. The eluates are combined with the supernatant solution of the alumina gel treatment (fraction 3).

Step 4. Further Fractionation with Ammonium Sulfate. To fraction 3 (1083 ml.) is added 263 g. of ammonium sulfate to give 40% saturation. After 15 minutes the precipitate is collected by centrifugation, discarded, and the supernatant solution treated with 141 g. of ammonium sulfate (60% saturation). The precipitate is collected by centrifugation, suspended in 120 ml. of 0.1 M phosphate buffer, pH 7.0, and dialyzed overnight against 0.01 M phosphate buffer, pH 7.0. The supernatant solution which results after centrifugation to remove inert precipitated protein represents fraction 4.

Step 5. Calcium Phosphate Gel Treatment. To fraction 4 (253 ml.) is

[5] For the preparation of alumina C_γ gel, see Vol. I [11].

added 150 ml. of calcium phosphate gel[6] (11 mg. dry weight per milliliter), and after 15 minutes the supernatant solution (fraction 5) is collected by centrifugation.

Step 6. Further Treatment with Alumina C_γ Gel. To fraction 5 (340 ml.) is added 90 ml. of alumina C_γ gel. After 15 minutes the precipitate is collected by centrifugation and eluted for 15-minute intervals with five portions of 30 ml. each of 0.1 M phosphate buffer, pH 7.9. The combined eluates (fraction 6) represent about 10% of the units in the crude starting material and a purification of fiftyfold.

Step 7. Lead Acetate Treatment and 40 to 60% Ammonium Sulfate. This step, which has been found to give a further twofold to fourfold increase in enzyme purity, works best with small volumes of enzyme. To 20 ml. of fraction 6 is added 0.2 to 1.5 ml. of basic lead acetate suspension (100 mg./ml.) in increments of 0.1 ml. until a heavy white precipitate appears. The latter is collected by centrifugation and discarded. To the supernatant solution is added 4.8 g. of ammonium sulfate to give 40% saturation followed by the addition of 2.6 g. more of ammonium sulfate to the resulting centrifuged supernatant solution to give 60% saturation. The 60% ammonium sulfate precipitate is dissolved in 10 ml. of 0.1 M phosphate buffer, pH 7.0, and dialyzed against 0.01 M phosphate buffer, pH 7.0, overnight. The dialyzate (fraction 7) increases in volume about twofold (19 ml.) and has a specific activity of 512, representing an over-all purification of approximately 200-fold as compared to the crude starting material as shown in the purification summary. This fraction is free of diaphorase activity using methylene blue and cyanide.[7]

In terms of turnover number fraction 7 catalyzes the oxidation of 25 moles of formate per mole of protein per minute, assuming a molecular weight for the enzyme of 100,000.

Properties

Stability of Enzyme. The enzyme is quite stable, maintaining most of its activity in the various fractionation stages when stored at pH 7 to 8 in the deep-freeze. Fraction 7 has also been stored for several weeks at 0° without losing more than half its activity. A 10-minute exposure of the enzyme to pH 5.3 results in protein precipitation and complete loss of enzyme activity. Maintenance of the enzyme for 10 minutes at 60° also causes a complete loss of activity. The enzyme is more stable during dialysis against phosphate buffer as compared to water.

pH Optimum. The enzyme does not exhibit a sharp pH optimum but

---

[6] For the preparation of calcium phosphate gel, see Vol. I [11].

[7] For the diaphorase test, see H. R. Mahler, Vol. II [120].

is active over a pH range of 5.5 to 8.0. The best activity appears to be at pH 5.5 to 6.5.

Specificity, Activators, and Inhibitors. There is a marked specificity for DPN as the electron acceptor. Azide at 5×10^{-6} M and 5×10^{-7} M final concentrations inhibits the enzyme 100% and 65%, respectively. Cyanide at 5×10^{-4} M and 10^{-4} M final concentrations inhibits 100% and 80%, respectively. There is 100% inhibition by 8-hydroxyquinoline at 5×10^{-3} M final concentration.

There is no effect on enzyme activity by the salts of Mg, Zn, Cu, Co, Mn, Fe, Ca, B, or Mo (10^{-4} M final concentration). There is also no effect by Versene, salicylaldoxime, *o*-aminophenol-*p*-sulfonic acid, potassium ethyl xanthate, sodium diethyldithiocarbamate, thiourea, fluoride, cysteine, orthophenanthroline, α,α-dipyridyl, or hydroxylamine (5×10^{-3} M final concentration).

SUMMARY OF PURIFICATION OF FORMIC DEHYDROGENASE FROM PEAS

Fraction	Total units	Total protein, mg.	Specific activity	Recovery, %
1. Crude extract	863,000	318,000	2.7	—
2. 37–53% ammonium sulfate	204,000	21,400	9.5	24
3. Alumina C_γ gel treatment	174,000	4,980	35.0	20
4. 40–60% ammonium sulfate	147,000	2,220	66.2	17
5. Calcium phosphate gel treatment	123,000	1,460	84.1	14
6. Alumina C_γ gel treatment	98,000	680	144.1	11
7. Lead acetate treatment and 40–60% ammonium sulfate	—	—	512	—

[88] Formic Hydrogenlyase from *Escherichia coli*

$$HCOOH \rightarrow H_2 + CO_2$$
$$(HCOOH + OH^- \rightarrow HCO_3^- + H_2)$$

By MAX BOVARNICK

Assay Method

Principle. The method generally used is that described by Stephenson and Stickland[1] or slight modifications thereof, and is based on the manometric measurement of gaseous hydrogen liberated by the enzyme in the presence of formate.

[1] M. Stephenson and L. H. Stickland, *Biochem. J.* **26**, 712 (1932).

Reagents

Formate solution (0.1 M). Dissolve 6.8 g. of sodium formate in 1 l. of water.

0.05 M sodium and potassium phosphate buffer, pH 6.2.

20% KOH.

Nitrogen gas rendered O_2 free by passing over copper heated to glowing.

Enzyme. Dilute stock enzyme solution with phosphate buffer to obtain suitable degree of activity (see below).

Procedure. Warburg vessels are set up as follows: 2 ml. of the diluted enzyme in the main vessel, 0.5 ml. of the formate solution in the side arm, 0.2 ml. of 20% KOH in the center well plus a slip of filter paper. The air is replaced by N_2, and the system is equilibrated at 30°, after which the formate is tipped in. Readings are taken up to a period of 1 hour. Suitable controls for the enzyme preparation and the substrate are included.

Definition of Specific Activity. Activity is expressed as Q_{H2}, i.e., microliters of H_2 produced per hour per milligram of protein N.

Purification Procedure

The enzyme system has only recently been prepared in cell-free form by Hughes[2] and by Gest and Gibbs.[3] In both instances disintegration of the organisms was achieved by subjecting them to pressure in the presence of abrasives. Hughes used a specially constructed apparatus. Gest and Gibbs employed the simpler procedure of McIlwain,[4] and this method is recommended here.

Step 1. Preparation of Organisms. A strain of *E. coli* known to possess hydrogenlyase (Strain Crookes, American Type Culture Collection 8739 has been used by Gest[3]) is grown in deep stationary culture at 37° in a 1% tryptone broth medium containing 1% glucose and 1% yeast extract. The cells are harvested at 12 hours, washed once, packed by centrifugation, and stored in the deep-freeze.

Step 2. The moist cells are ground for 30 seconds by mortar and pestle, using maximum hand pressure, with Alumina A-301 (Aluminum Co. of America) or with c.p. Al_2O_3 ignited powder (J. T. Baker Chemical Co.). The ground mixtures are extracted with 1.5 ml. of water for each gram of "wet" cells used. Intact cells and debris are removed by centrifugation at 16,000 × g for 25 minutes. The turbid supernatant is decanted and centrifuged for 1 hour at 20,000 × g in the high-speed head of an Inter-

[2] D. E. Hughes, *Brit. J. Exptl. Pathol.* **32**, 97 (1951).

[3] H. Gest and M. Gibbs, *J. Bacteriol.* **63**, 661 (1952).

[4] H. McIlwain, *J. Gen. Microbiol.* **2**, 288 (1948).

national refrigerated centrifuge to yield an opalescent yellowish supernatant fluid. This fluid is dispensed in tubes which are flushed with helium, stoppered, and stored in the deep-freeze. All manipulations are carried out in the cold.

Preparation	Abrasive	Activity, μl. of H_2/ml./hr.
1. *B. coli* Extract[a]	Al_2O_3	300
2. *B. coli* Extract[b]	Polishing alumina	460
	Emery flower	700
	Pyrex glass	560
	Diamantine	0

[a] H. Gest and M. Gibbs, *J. Bacteriol.* **63**, 661 (1952).
[b] D. E. Hughes, *Brit. J. Exptl. Pathol.* **32**, 97 (1951).

Properties

Specificity. There is some question concerning the identity of formic hydrogenlyase as a single enzyme or as a combination of formic dehydrogenase and hydrogenase with or without intermediate carrier systems. Until this question is resolved the definition of the specificity of the enzyme involved must remain in abeyance.

Activators and Inhibitors. The partial inactivation which occurs after dialysis against oxygen-free water is reversed by addition of boiled extracts of yeast, pigeon liver, or anaerobically or aerobically cultivated *E. coli* (Gest[5]). Irregular activation by diacetyl, and by pyruvate plus Co-carboxylase has also been reported by Gest.[5]

A metal requirement has been observed in that the presence of Fe^{++} or Mn^{++} seems to be required as a cofactor.

The activity in 0.05 M sodium maleate buffer at pH 6 is one-sixth that observed in phosphate buffer of the same concentration and pH.

The enzyme is inhibited by cyanide, formaldehyde, and metal-binding agents such as α,α-dipyridyl and 8-hydroxylquinoline. Nitrate diminishes the activity of enzyme solutions. Shaking an enzyme solution with air for 2 hours causes the appearance of a lag period in activity. This is shortened by addition of a dye such as methyl viologen ($E_0' - 0.466$v) but is unaffected by addition of homocysteine. Hydrogenlyase activity is partially inhibited by an atmosphere of 100% H_2, presumably owing to reversibility of the reaction.

Effect of pH. The optimum pH for the soluble cell-free enzyme is 6.0 to 6.2. No activity is found at pH 7.2.

[5] H. Gest, *in* "Phosphorus Metabolism" (W. D. McElroy and B. Glass, eds.), Vol. II, p. 522, The Johns Hopkins Press, Baltimore, 1952.

Section III

Enzymes of Lipid Metabolism

[89] Fatty Acid Oxidation in Mitochondria

By ALBERT L. LEHNINGER

Assay Method

Principle. Suspensions of mitochondria, isolated from sucrose homogenates of liver or kidney by the differential centrifugation procedure of Hogeboom *et al.*,[1] oxidize fatty acids at the expense of molecular oxygen when supplemented with ATP, Mg^{++}, orthophosphate, and, under certain experimental conditions, a tricarboxylic acid cycle intermediate.[2-5]

In rat *liver* mitochondria, the oxidation of straight-chain, saturated, even-carbon fatty acids proceeds with formation of an equivalent amount of acetoacetate if the reaction medium does not contain significant concentrations of tricarboxylic acid cycle intermediates. Octanoate is the substrate giving maximal rate of oxidation and leads to quantitative formation of 2 molecules of acetoacetate. Shorter chain acids are attacked more slowly; acids of 10 to 18 carbon atoms are attacked, but their insolubility and the fact that they are also highly inhibitory because of surface effects make them less suitable as reproducible test substrates. However, palmitate or oleate at concentrations of 0.0002 M are oxidized as fast as octanoate if properly brought into solution.

If tricarboxylic acid cycle intermediates are added to rat liver mitochondria, the yield of acetoacetate from octanoate is decreased as some of the Ac~SCoA formed on oxidation of the fatty acid condenses with oxalacetate and enters the cycle. Free acetoacetate, once formed, is not oxidized via the cycle by rat liver mitochondria. The fraction of Ac~SCoA entering the cycle depends somewhat on the concentration of oxalacetate precursor; with rat liver mitochondria, complete diversion of Ac~SCoA into the cycle has not been achieved in any published reports, and a residue of acetoacetate formation always remains.

In suspensions of *kidney* mitochondria, on the other hand, acetoacetate formation does not occur, and in fact no oxidation of fatty acid occurs unless an oxalacetate precursor is present. Oxidation of the fatty acid thus proceeds entirely via the tricarboxylic acid cycle in kidney preparations, yielding CO_2 as end product.

[1] G. H. Hogeboom, W. C. Schneider, and G. E. Palade, *J. Biol. Chem.* **172,** 619 (1948).
[2] E. P. Kennedy and A. L. Lehninger, *in* "Phosphorus Metabolism" (McElroy and Glass, eds.), Vol. 2, p. 253, The Johns Hopkins Press, Baltimore, 1952.
[3] D. E. Green, *Biol. Revs.* **26,** 410 (1951).
[4] J. D. Judah and K. R. Rees, *Biochem. J.* **55,** 664 (1953).
[5] H. A. Lardy and H. Wellman, *J. Biol. Chem.* **195,** 215 (1952).

Rat liver mitochondria therefore represent the preparation of choice for studying the reaction sequence

$$C_7H_{15}COOH + 3O_2 \rightarrow 2CH_3COCH_2COOH + 2H_2O \tag{1}$$

which proceeds with little or no endogenous respiratory activity in the absence of added cycle intermediate. Rat or rabbit kidney mitochondria, on the other hand, represent the preparation of choice for studying the *complete* oxidation of fatty acid to CO_2 and H_2O via the tricarboxylic acid cycle:

$$C_7H_{15}COOH + 11O_2 \rightarrow 8CO_2 + 8H_2O \tag{2}$$

Within limits, some proportionality exists between the concentration of mitochondria in the system and the rate of oxygen uptake, and some investigators have used this fact as a basis for quantitative assay of the "fatty acid oxidase system."[6] Obviously, in tissue homogenates or tissue fractions such an assay measures only the rate-limiting step of this multi-enzyme system, and the use of the test system described below as an "assay" in this way is therefore not to be recommended unless it is known that the rate-limiting step is similar in all samples assayed. At least one factor which can be rate-limiting is discussed below.

Reagents

ATP (0.015 M), as the sodium or potassium salt, pH 7.4.

0.1 M Na-K phosphate buffer, pH 7.4.

0.01 M octanoate solution. Prepared by dilution of a 0.1 M stock solution. For 100 ml. of 0.1 M octanoate, add 10.0 ml. of 1.0 N KOH to 1.58 ml. of redistilled octanoic (caprylic) acid (Eastman Kodak Co.), and shake until the latter is dissolved. Dilute to 100 ml. with H_2O, adjust pH to 7.4, and filter through moistened filter paper. The clear filtrate is stable for months either at 0° or frozen, and 0.01 M solutions for use as substrate are prepared by dilution with H_2O as required.

0.1 M $MgCl_2$.

1 M KCl.

0.25 M sucrose brought to pH 7.4.

0.1 M potassium fumarate.

3×10^{-4} M cytochrome c.

Suspension of freshly isolated mitochondria, in 0.25 M sucrose, such that 1.0 ml. contains the mitochondria derived from 0.5 g. of wet weight liver or kidney. The preparation of such suspensions is described in Vol. I [3]. Ordinarily the mitochondria freshly sedi-

[6] W. C. Schneider, *J. Biol. Chem.* **176**, 259 (1948).

mented from the sucrose homogenate are mechanically freed of the so-called "fluffy layer" of microsomes and then washed twice with sucrose in the refrigerated centrifuge before the final suspension is prepared. It is most desirable to use them immediately after preparation, but suspensions may often be stored for hours at 0° without appreciable loss of activity.

Quantitative Oxidation of Octanoate to Acetoacetate by Rat Liver Mitochondria

Reaction medium. To the main compartment of a Warburg vessel of 15 to 18 ml. total volume is added the following:

Component	Final concentration
0.4 ml. phosphate buffer	0.02 M
0.2 ml. ATP-K	0.0015 M
0.15 ml. MgCl$_2$	0.0075 M
0.2 ml. octanoate	0.001 M
0.15 ml. cytochrome c	0.000015 M
0.1 ml. KCl	0.05 M
0.30 ml. H$_2$O	

The center well is equipped with alkali and filter paper roll. To each vessel is added 0.5 ml. of suspension of rat liver mitochondria in 0.25 M sucrose. This addition, which brings the total volume to 2.0 ml., starts the reaction.

The vessels are quickly attached to manometers and equilibrated for 5 minutes before taps are closed and readings taken. It is most convenient to run the reaction at 25° (equilibration period 5 minutes), since it is relatively fast and the limited amount of substrate is rapidly used up. Higher concentrations of substrate cannot be employed, since excessive concentrations of octanoate and other higher fatty acids are highly inhibitory to the oxidation. Tipping of the substrate from the side arm after the equilibration period is not satisfactory, since activity of the oxidase system declines rapidly in the absence of active oxidation and phosphorylation. In this system, oxygen uptake proceeds almost linearly until the octanoate is completely oxidized (equation 1) and then stops abruptly. Under the conditions described above, at 25°, about 2.0 to 3.0 microatoms of oxygen will be taken up per 10-minute interval near the beginning of the reaction. There is virtually no oxygen uptake if octanoate is omitted.

Oxidation of Octanoate to CO$_2$ and Water by Kidney Mitochondria

The test system described above is used, with the substitution of 0.1 ml. of 0.01 M fumarate (final concentration, 0.0005 M) for 0.1 ml. of water and with 0.50 ml. of rat kidney mitochondria in 0.25 M sucrose (1.0 ml. ≈ 0.5 g. of wet weight kidney) for the liver mitochondria. The reaction is carried out in the same way. Oxygen uptake proceeds at about the same rate as in the liver system described but continues for a considerably longer period. Complete oxidation of the octanoate added to CO_2 and H_2O may be observed in most cases, although the mitochondria may become inactivated before this point is reached. A vessel not containing octanoate may be run to indicate the magnitude of the oxygen uptake due to oxidation of fumarate alone. However, subtraction of this oxygen uptake from that of the complete system does not necessarily yield a more valid measure of the rate of octanoate oxidation than does the oxygen uptake of the complete system.

Other Properties of the "Fatty Acid Oxidase" System

The rate-limiting step in fatty acid oxidation in systems as described above is not known with certainty, and it is likely that it may vary with experimental conditions, age of the preparation, tissue, presence of inhibitors, etc. In systems similar to those described, Lardy and Wellman[5] have found that the rate-limiting factor in freshly prepared mitochondria is the availability of phosphate acceptors for the phosphorylations which accompany fatty acid oxidation, since the addition of hexokinase plus glucose increased the rate of oxidation of fatty acid of the order of 100% or more. Such considerations therefore make use of this test as an "assay" for the oxidase system rather questionable.

Preparations of liver mitochondria which have been aged or exhaustively washed or exposed to isotonic or hypotonic salt solutions may require the addition of a small amount of cycle intermediate to "prime" or "spark" the oxidation of fatty acid; with such preparations the addition of ATP alone does not suffice to start the oxidation. The mechanism of the "priming" effect is not known, but it is probable that it is concerned with the formation of fatty acid-CoA complex.

Since the individual enzymatic steps in fatty acid oxidation are now known, it is possible to isolate and study the individual enzymes separately.[7,8] Clear extracts of acetone-dried mitochondria also may be used to study fatty acid oxidation.[9]

[7] F. Lynen and S. Ochoa, *Biochim. et Biophys. Acta* **12**, 299 (1953).
[8] H. Beinert, R. M. Bock, D. S. Goldman, D. E. Green, H. R. Mahler, S. Mii, P. G. Stansly, and S. J. Wakil, *J. Am. Chem. Soc.* **75**, 411 (1953).
[9] G. R. Drysdale and H. A. Lardy, *J. Biol. Chem.* **202**, 119 (1953).

[90] Fatty Acid Oxidation in Higher Plants

By P. K. STUMPF

In animal cells the site of fatty acid oxidation is in the mitochondrion.[1] Studies with extracts of germinated peanut cotyledons indicate that, in part, oxidation of palmitic acid is catalyzed by two independent systems found localized in (1) the microsomal particles, and (2) the soluble cytoplasmic proteins.[2]

Preparation of Enzyme Systems

Peanuts were selected as the source of the enzymes because of their availability, their relative ease of growth, and their high lipid metabolism. Raw peanuts (*Arachis hypogea L., var.* Virginia Jumbo) were purchased from Manning's Inc., San Francisco, California. Shelled peanuts are germinated under greenhouse conditions in flats filled with moist, sterilized peat moss. Excessive watering must be avoided, since vigorous *Penicillium* and *Rhizopus* growth will occur. After 5 to 8 days of germination, the peanuts are harvested, and the cotyledons are separated from seed coat, embryo, root, and hypocotyl. Cotyledons, washed several times with distilled water, are homogenized at maximum speed in a chilled Waring blendor for 15 seconds with 2 vol. of 0.4 M sucrose–0.2 M Tris buffer at pH 7.3 in the cold. The homogenate is filtered through cheesecloth to remove debris, and the filtrate is centrifuged for 30 minutes at 13,000 $\times$ g at 4° (high-speed attachment, refrigerated International centrifuge). The resulting supernatant is then centrifuged for 1 hour at 100,000 $\times$ g at 4° (No. 40 head in a Spinco preparative ultracentrifuge). The translucent yellow pellet is resuspended in 20 ml. of 0.2 M KCl and recentrifuged for 30 minutes at 100,000 $\times$ g at 4°. This material (preparation A) is dispersed in a small volume of 0.2 M KCl and stored at −5° in small Lusteroid tubes. After 2 weeks no appreciable loss in activity is observed.

The supernatant obtained from the first Spinco centrifugation is made 80% saturated with ammonium sulfate. The precipitate is centrifuged down in the cold, resuspended in 80% saturated ammonium sulfate (pH 7.6), and recentrifuged. The sediment is dissolved in a small volume of 0.2 M Tris buffer at pH 7.5 (preparation B). This solution can be stored at −5° for a week with little loss in activity.

---

[1] See Vol. I [89].

[2] E. H. Newcomb and P. K. Stumpf *in* "Phosphorus Metabolism" (McElroy and Glass, eds.), Vol. II, p. 291, The Johns Hopkins Press, Baltimore, 1952.

Reagents

0.001 M palmitic acid. Sufficient radioactive palmitic acid (approximately 1000 c.p.m. per 0.1 micromole) is added to carrier palmitic acid to make a total of 2.56 mg. of substrate which is suspended in 5 ml. of distilled water. The suspension is heated to boiling, and 1 M NH_4OH is added dropwise until the fatty acid is dissolved. The solution is diluted to a final volume of 10 ml. and stored at $-5°$. The palmitate solution is always heated to 100° to bring the substrate into complete solution and then added to the Warburg flasks in required amounts.

0.1% DPN "90" Sigma.

0.2 M Tris buffer adjusted to pH 7.5 with 6 N HCl.

0.2 M phosphate-veronal buffer, pH 5.5.

0.2 M phosphate buffer, pH 7.5.

20% KOH. KOH is dissolved in boiled distilled water.

20% barium acetate. The salt is dissolved in boiled distilled water.

5 N sulfuric acid.

Reaction Mixture. OXIDATION BY MICROSOMAL PARTICLE.

$$\text{Palmitic-1-}C^{14} \rightarrow X + C^{14}O_2 \qquad (1)$$

Each Warburg vessel contains 0.1 ml. of preparation A, 0.1 micromole of palmitic-1-C^{14}, 0.5 ml. of 0.2 M phosphate-veronal buffer at pH 5.5, 0.2 ml. of 20% KOH in the center well, 0.3 ml. of 5 N sulfuric acid in the side arm, and water to a final volume of 2.0 ml. The incubation period is 1 hour at 30°. At the end of the incubation period, the acid is tipped from the side arm into the main compartment to stop the reaction and release the bound CO_2. Shaking is continued for 10 minutes, after which the contents of the center well are removed with washing, and the respiratory CO_2 is precipitated as $BaCO_3$ with 20% $BaAc_2$, washed three times with 15 ml. of 50% ethanol, plated on an aluminum disk, and counted with a thin-window Geiger-Muller tube.

$$\text{Palmitic-2-}C^{14} \rightarrow Y + C^{14}O_2 \qquad (2)$$

The reaction mixture and procedure is essentially the same as for (1) with the following exceptions: 0.1 micromole of palmitate-2-C^{14}, 0.5 ml. of 0.2 M phosphate buffer at pH 7.2, and 0.1 ml. of 0.1% DPN.

OXIDATION BY SOLUBLE CYTOPLASMIC PROTEIN

$$\text{Palmitic-1-}C^{14} \rightarrow Z + C^{14}O_2 \qquad (3)$$

The reaction mixture and procedure is the same as for (1) with the following exceptions: 0.5 ml. of preparation B, 0.5 ml. of 0.2 M phosphate buffer at pH 7.2, and 1 ml. of boiled crude supernatant. Final volume, 3 ml.

Properties

Unlike the cofactor requirements for the animal enzyme systems of fatty acid oxidation, no ATP, CoA, cysteine, or Krebs cycle intermediates is required in any of these systems. Moreover no acetoacetate or acetate accumulates in the reaction mixture. All attempts to identify the large fragments (X, Y, Z) by conventional chromatographic techniques have failed to reveal the nature of the intermediates of oxidation.

Activators and Inhibitors. Preparation B requires an unidentified cofactor for complete activity. This material has been isolated from peanut tissue by a series of treatments involving acid and base hydrolysis, ether extraction, nonadsorbability on Darco-activated charcoal, and paper chromatography.[3] It is stable to 0.1 M acidic or 0.1 M basic hydrolysis for 15 minutes at 100°, is extracted by ether under acid conditions, is adsorbed by Dowex-1 but not by Dowex-50, and is not adsorbed on charcoal. The activation effect of this material cannot be duplicated by ATP, CoA, cytochrome c, Krebs cycle intermediates, Mn^{++}, Mg^{++}, molybdate, lipoic acid, lipothiamide-pyrophosphate, biotin, cocarboxylase, glutathione, DPN, TPN, or FAD. It has been isolated from wheat germ, yeast extract, liver, and *Cl. kluyveri.*

The three reactions are inhibited by hydroxylamine (10^{-3} M), arsenite (10^{-4} M), imidazole (10^{-3} M), and hypophosphite (10^{-3} M). Fluoride, iodoacetate, DPN, cyanide, azide, and arsenate are inert.

[3] P. Castelfranco, P. K. Stumpf, R. Contopoulou, and T. E. Humphreys, *Federation Proc.* **13**, 189 (1954).

[91] Butyrate Enzymes of *Clostridium kluyveri*

$$CH_3CH_2CH_2COOH + Pi + O_2 \rightarrow \text{Acetyl Phosphate} + Ac$$

By H. A. BARKER

The above reaction is catalyzed by a multienzyme system that can be extracted from *Cl. kluyveri.* The individual enzymes participating in the reaction include phosphotransacetylase and CoA transphorase, described in Vol. I [**98, 99**], and several other enzymes that have not been isolated or studied in detail from this source. The object of this section is to describe methods that can be used to obtain cell-free extracts that carry out the above reaction and estimate their over-all activity.

Assay Method

Principle. The rate of oxygen uptake is measured manometrically as described by Stadtman and Barker.[1,2] Two components required in catalytic amounts, CoA and acetyl phosphate,[3] are added, although they are usually present in nonlimiting amounts in crude extracts. The assay gives no indication as to which of several component enzymes and coenzymes is rate-limiting.

Reagents

Potassium butyrate, 0.4 M.

KPO_4 buffer, pH 7.6, 0.5 M.

Dilithium acetyl phosphate, 0.01 M. This solution can be used for months if stored at $-18°$.

CoA, approximately 0.001 M. Dissolve 6 mg. of CoA[4] (65% pure) in 1 ml. of 0.05 M KPO_4 buffer, pH 6.8, containing 50 micromoles of reduced glutathione. Incubate for 60 minutes at room temperature, and then dilute to 5 ml. This solution is stable for weeks at $-18°$.

Enzyme. Prepare a crude extract containing approximately 25 mg. of protein per milliliter.

Procedure. Use a Warburg vessel with one side arm. To the main compartment add 0.1 ml. of 0.4 M butyrate, 0.2 ml. of 0.5 M KPO_4 buffer, 0.1 ml. of 0.01 M acetyl phosphate, 0.1 ml. of 0.001 M CoA, and 0.5 ml. of water. To the side arm add 1 ml. of the enzyme solution. After equilibration with air at 30°, mix the enzyme with the contents of the main compartment and measure the oxygen uptake manometrically at 5-minute intervals for 30 minutes.

Preparation of Cell-Free Extract

Cl. kluyveri is grown in an ethanol-acetate medium,[5] and the washed cells are dried in a thin layer over calcium chloride in an evacuate desiccator or are lyophilized. The thoroughly dried cells retain their activity for years when stored at $-15°$.

[1] E. R. Stadtman and H. A. Barker, *J. Biol. Chem.* **180**, 1085 (1949).

[2] E. R. Stadtman and H. A. Barker, *J. Biol. Chem.* **180**, 1095 (1949).

[3] In the presence of phosphotransacetylase and CoA, acetyl phosphate is equivalent to acetyl~SCoA, the compound actually required.

[4] CoA of this purity may be obtained commercially from Pabst Laboratories. For methods of isolation and purification of this compound, see Vol. III [131].

[5] See Vol. I [98] for composition of medium.

One gram of finely ground dried cells is homogenized in 15 ml. of 0.05 M KPO$_4$, pH 7.0, containing 0.001 M Na$_2$S, and incubated in an evacuated flask for 1 to 3 hours at room temperature with occasional shaking. The time required for maximal extraction of the butyrate enzymes varies with different batches of dried cells. The suspension is then centrifuged at 10,000 r.p.m. The enzyme in the clear supernatant retains its full activity for weeks at $-18°$. The cell-free extract contains approximately 50% of the protein of the dried cells.

Activity of Extracts. A satisfactory extract consumes oxygen at the rate of approximately 10 micromoles/hr./25 mg. of protein. Such extracts also catalyze several other reactions,[2,6] including the oxidation of ethanol to acetyl phosphate,[7] the reductive conversion of acetyl phosphate to butyrate and caproate, and the dismutation of vinyl acetate to butyrate, acetate, and acetyl phosphate, and associated intermediate reactions.

[6] E. R. Stadtman and H. A. Barker, *J. Biol. Chem.* **180**, 1117 (1949); **180**, 1169 (1949); **181**, 221 (1949); **184**, 769 (1950).

[7] R. M. Burton and E. R. Stadtman, *J. Biol. Chem.* **202**, 873 (1953).

[92] Butyryl Coenzyme A Dehydrogenase

$$RCH_2CH_2CO{\sim}SCoA \rightleftarrows RCH{=}CHCO{\sim}SCoA + 2H^+ + 2e^-$$

By Henry R. Mahler

Assay Methods

Principle. The methods to be described use dyes as electron acceptors.[1] Dye reduction is then followed spectrophotometrically. The indophenol assay is easy and convenient but is suitable only for purified preparations, whereas the tetrazolium method, although more laborious, can be used with crude preparations as well.[2]

[1] D. E. Green, S. Mii, and H. R. Mahler, *J. Biol. Chem.* **206**, 1 (1954).

[2] The reason for this apparent discrepancy is to be found in the presence of enzymes in crude preparations which liberate free CoASH from butyryl$\sim$SCoA. The sulfhydryl groups thus freed interact rapidly with indophenol but only relatively slowly with tetrazolium under the conditions described here. The following rationale underlies the use of the combination pyocyanine (or methylene blue) plus TTZ: The enzymatic reduction of TTZ by flavoproteins is very slow, much slower than the corresponding reduction of pyocyanine (or methylene blue). On the other hand, the nonenzymatic reduction of an excess of TTZ by the (enzymatically) reduced carrier pyocyanine is very rapid. Thus the combination of acceptors actually employed is capable of rapid reduction by the enzyme and at the same time still takes advantage of the solubility behavior of the reduced formazan.

Reagents

0.01 M butyryl$\sim$CoA.[3]
0.10 M Tris buffer, pH 7.0.
0.01% aqueous 2,6-dichlorophenolindophenol.
1.0 M glycylglycine buffer, pH 8.2.
0.2% aqueous pyocyanine.
1.0% aqueous triphenyl tetrazolium chloride (TTZ).
0.1% serum albumin.

Procedure. INDOPHENOL ASSAY. Into 1-ml. cuvettes place 0.1 ml. of Tris buffer, 0.2 ml. of indophenol, 0.01 ml. of butyryl CoA solution, and 0.53 ml. of H_2O. Read the optical density at 600 mμ (E_{600}); if any changes occur in the absence of enzyme, follow them for about 5 minutes. At the end of this period add 0.05 ml. of enzyme solution, and take readings at 30-second intervals against a blank containing all components (including dye and enzyme) but no butyryl CoA. An amount of enzyme catalyzing a ΔE_{600} of 0.010 to 0.050 per minute is to be used. Under these conditions the initial rates of dye reduction are strictly proportional to enzyme concentration over a wide range.

TTZ-PYOCYANINE ASSAY. Into 10-ml. test tubes place 0.10 ml. of butyryl CoA solution, enzyme (2 to 200 γ), 0.02 ml. of glycylglycine buffer, 0.04 ml. of pyocyanine, 0.12 ml. of TTZ, and 0.14 ml. of serum albumin solutions in a total volume of 0.38 ml. At the same time, set up two blanks, one containing no butyryl CoA and one containing no enzyme. Place the tubes inside a vacuum desiccator, which is then evacuated to ≤ 1 mm. of Hg pressure and incubated at 38°. After 1 hour admit air, and to each tube add 1 drop of 1 N HCl, 1.5 ml. of acetone, and 4.5 ml. of carbon tetrachloride. Extract the red formazan into the organic phase, the layers being separated by centrifugation. Then determine the E_{485}. Subtract the values for both blanks from that of the experimental. Under these conditions $0.5 \times E_{485}$ (corr.) = micromoles of formazan produced.

Definition of Units and Specific Activity. One indophenol unit of enzyme is defined as that amount which catalyzes a ΔE_{600} of 1.0 per minute at 22°. Specific activity equals units per milligram. One TTZ unit is defined as that amount which will catalyze the reduction of 1 micromole of TTZ per hour at 38°. Under the conditions of their respective procedures, one indophenol unit equals one-fifteenth TTZ unit.

[3] Butyryl CoA can be prepared by the interaction of thiolbutyric acid and coenzyme A [modification of the method of I. B. Wilson, *J. Am. Chem. Soc.* **74**, 3205 (1953)]. For methods of preparation of acyl CoA derivatives see Vol. III [137].

Purification Procedure

The procedure given here is that developed by the original investigators.[1,4]

Step 1. Preparation of Acetone-Dried Particles and Extraction. Fresh beef liver, chilled in ice immediately at the slaughterhouse, is freed of connective tissue and divided into 100-g. portions (all operations at 1 to 5°). Each portion is then cut up into smaller fragments and added to 300 ml. of a solution containing 85 g. of sucrose and 5 g. of K_2HPO_4 per liter. The mixture is blended for about 50 seconds in Waring blendors operating at half maximal speed, and the pH maintained at 7.0 by the addition of 6 M KOH. The homogenate derived from 1 kg. of liver is centrifuged at 1600 r.p.m. (2300 $\times$ g) for 6 minutes in the 4-l. International centrifuge. The supernatant fluid is passed through two layers of cheesecloth and added to 10 l. of 0.9% KCl with stirring. The suspension is then centrifuged in the Sharples supercentrifuge, and the particulate matter is scraped from the bowl, suspended in a total volume of 400 ml. by brief blending with 0.9% KCl, and stirred into 4 l. of acetone at $-15°$. The mixture is filtered with suction, and the insoluble material is washed with additional acetone and air-dried rapidly. 1000 g. of liver yields 40 to 50 g. of dry, light-tan powder; 150 to 200 g. can conveniently be obtained in 1 day's work. 100 g. of acetone powder is suspended with vigorous stirring in 1 l. of 0.02 M K_2HPO_4 and extracted for 1 hour at 0°. The insoluble residue is removed by centrifugation at 2300 $\times$ g.

Step 2. First Ammonium Sulfate Fractionation. To each 100 ml. of extract is added 28 g. of solid $(NH_4)_2SO_4$. The precipitate is removed by centrifugation, and an additional 22 g. of $(NH_4)_2SO_4$ is added to the supernatant. The precipitate is collected and dissolved in a minimal volume of 0.02 M $KHCO_3$.

Step 3. Second Ammonium Sulfate Fractionation. The above solution is made 40% with respect to saturated $(NH_4)_2SO_4$ at pH 7.0, and the precipitate formed is discarded. The saturation is then raised to 50%. The precipitate collected by centrifugation is dissolved in 0.02 M $KHCO_3$.

Step 4. Treatment with Zinc Hydroxide Gel. The solution is dialyzed for 2 hours against 100 vol. of 0.02 M $KHCO_3$. The protein concentration is then adjusted to 15 to 20 mg./ml., and an amount of $Zn(OH)_2$ gel[5] suffi-

---

[4] H. R. Mahler, S. J. Wakil, and R. M. Bock, *J. Biol. Chem.* **204**, 453 (1953).

[5] The gel used for enzyme purification is prepared as follows: 287.6 g. of $ZnSO_4$ is dissolved in 10 l. of deionized water, and 333 ml. of 1 N sodium hydroxide is then added with stirring (pH between 8 and 9). The gel is then washed by decantation three or four times with a large volume of water. This is followed by three washings with 0.02 M potassium bicarbonate. Finally the gel is concentrated by centrifuga-

cient to remove all red and brown proteins is then stirred in. The supernatant liquid should have a definite light-green color after treatment. Usually about 1.25 vol. of a gel having a dry weight of 20 to 30 mg./ml. will be required, and the final protein concentration in the supernatant solution will be between 2.0 and 3.0 mg./ml. 50 g. of solid ammonium sulfate is then added for each 100 ml. of supernatant solution, and the pH is kept constant at 7.0 by the dropwise addition of concentrated NH_4OH. The precipitate is dissolved in 0.02 M $KHCO_3$.

Step 5. First Ethanol Fractionation. The solution is dialyzed against 0.02 M $KHCO_3$ for 6 hours and is then fractionated with alcohol at low temperature. An equal volume of alcohol is stirred slowly into the cooled solution of enzyme, and the temperature is gradually lowered to -10 to 15° and then maintained at that point for 15 minutes. The precipitate is discarded. The alcohol concentration is then slowly raised, and fractions are removed whenever a significant amount of protein begins to separate. In this manner five fractions are collected between 50 and 75% (v/v) ethanol concentration. The precipitates are dissolved in 0.02 M $KHCO_3$, and the two fractions of highest specific activity (precipitating in the range 60 to 70% ethanol) are pooled. An alternative method for selecting the best fractions other than by enzymatic test consists in measuring the absorption at 430 mμ and determining the ratio E_{430} per protein (mg.). The fractions showing the highest ratio (≥ 0.1) are pooled.

Step 6. Adsorption on Alumina Cγ. The bright-green preparation is treated with alumina Cγ[6] until the supernatant solution is colorless. The gel is first washed with 0.01 M K_2HPO_4, the washings being discarded. The enzyme is eluted with several volumes of 0.04 M K_2HPO_4. The active protein is precipitated by the addition of 50 g. of $(NH_4)_2SO_4$ per 100 ml. of eluate and is then redissolved in 0.02 M $KHCO_3$.

Step 7. Second Ethanol Fractionation. The enzyme protein is then refractionated by means of ethanol with 0.1 M Tris chloride at pH 8.0 as supporting electrolyte. For this purpose the enzyme is first dialyzed for 2 hours against a buffer, 0.05 M in Cl, and neutralized to pH 8.0 with Tris hydroxide. The alcohol fractionation is then carried out essentially as described above. Four fractions are collected, and the two showing the highest activity (or alternatively a ratio ≥ 0.17–0.20) are pooled. At the end of this treatment the enzyme is about 30% pure. For most purposes the enzyme at the end of step 5 is quite satisfactory. If a completely homogeneous preparation is desired, step 8 has to be carried out.

---

tion and suspended in 0.02 M bicarbonate to a final concentration of about 20 mg. of dry solids per milliliter of suspension.

[6] See Vol. I [11] for the method of preparation of alumina Cγ.

Step 8. Further Purification. Further purification can be achieved by repeating steps 6 and 7. In this manner preparations showing purities of 0.75 to 0.80% can be obtained. An alternative method, which at the same time permits a determination of the purity of the enzyme, takes advantage of the fact that the protein of the green enzyme has by far the highest electrophoretic mobility (in Tris buffer of pH 8.0) of all the proteins contained in the preparation at the end of step 7. The enzyme solution is dialyzed for 6 to 12 hours against the buffer already described (see step 7) and is then subjected to analytical electrophoresis. The fast-moving peak containing from 20 to 60% of the total protein carries all the green color and all the enzymatic activity. If the specific activity is 2.0 or better, complete resolution can be achieved in about 3 to 4 hours, and the protein under the rapidly moving peak is removed by means of a syringe. The optical system of the Tiselius apparatus is used to determine the extent of separation and to assist in sampling.

SUMMARY OF PURIFICATION PROCEDURE

	Total protein, mg.	Specific activity, indophenol units/mg.	Recovery, %
1. Extract	100,000	0.016	—
2. First $(NH_4)_2SO_4$ fractionation	29,600	0.034	86
3. Second $(NH_4)_2SO_4$ fractionation	12,000	0.080	60
4. $Zn(OH)_2$ gel supernatant	5,350	0.180	42
5. First alcohol fractionation	336	1.00	21
6. Alumina Cγ eluate	152	1.60	15
7. Second alcohol fractionation	67	2.40	10
8. Electrophoresis	10	4.00	4.0

Properties[1,7]

Physical Properties. The mobility of butyryl CoA dehydrogenase at pH 8.0 in Tris chloride of 0.05 ionic strength is 5.0×10^{-5} cm.^{2} volt^{-1} sec.$^{-1}$. The enzyme after step 8 gave a single peak in the analytical ultracentrifuge. The sedimentation constant, corrected to 20° and water, is 7.4×10^{-13} second for a 0.2% protein solution. The molecular weight has not been determined accurately but is estimated to fall between 150,000 and 200,000.

Appearance and Absorption Spectrum. The enzyme is colored a deep, brilliant green. Its absorption spectrum shows maxima at 255, 355, 432.5, and 685 mμ. The ratios $E_{255}|E_{355}|E_{432.5}|E_{685}$ are 7.5:0.87:1.00:0.25. On

[7] H. R. Mahler, *J. Biol. Chem.* **205**, 16 (1954).

treatment with hydrosulfite or butyryl CoA, all three peaks in the visible range disappear. They reappear completely on prolonged exposure to oxygen and partially on incubation with crotonyl CoA.

Prosthetic Groups. Butyryl CoA dehydrogenase contains FAD as part of its prosthetic group. The nucleotide was identified by paper chromatography, absorption spectrum, fluorescence, stability, and the D-amino acid oxidase test. Riboflavin constitutes 1.2% of the weight of the purest enzyme. It also contains copper in the ratio of 2 gram atoms of copper per mole of flavin. Both FAD and Cu ions have been implicated as participating in the reversible oxido-reductions catalyzed by the enzyme. Enzyme-bound copper cannot be removed by boiling the enzyme, but it is, at least partially, split off by prolonged dialysis against 5×10^{-3} M KCN buffered at pH 7.5. Copper can be added back to an enzyme treated with cyanide, by incubating the protein briefly with 5×10^{-4} M CuSO$_4$. The FAD can be reversibly dissociated by exposure of the enzyme to pH 3.5 to 4.0 in strong salt solution. Full activity can be restored by treating an enzyme so split with FAD, but not with FMN. The dissociation constant for FAD determined in this manner equals 4.4×10^{-7} M.

Activators and Inhibitors. Although the enzyme contains copper, most metal-complexing agents such as ethylenediaminetetraacetate, 8-hydroxyquinoline, and pyrophosphate inhibit it only slightly, if at all. As just stated, copper can be removed by dialysis against CN$^-$. An enzyme so treated has not been impaired in its ability to reduce dyes. It is incapable of interacting with cytochrome c, however. It is this ability that is restored by incubation with CuSO$_4$. The enzyme is inactivated 40% with 10^{-4} M p-chloromercuribenzoate. This inhibition can be completely reversed by glutathione. Butyryl CoA and FAD exert a protective action against chloromercuribenzoate inhibition.

Specificity. The enzyme is active with short-chain acyl derivatives of coenzyme A. With the activity of C$_4$-CoA taken as 100, activities with homologs are: C$_2$, 0; C$_3$, 30; C$_5$, 54; C$_6$, 40; C$_7$, 20; C$_8$, 10.

Nature of Electron Acceptor. Various dyes can be used, as can ferricyanide and cytochrome c. The rate of reduction of the latter is probably too low to be of physiological importance.

Kinetics and Turnover. Under the conditions of the indophenol assay, the initial rate of dye reduction is of zero order and proportional to enzyme concentration. Under the assay conditions, the Michaelis constants for butyryl CoA and indophenol are 1.4×10^{-5} M and 3.3×10^{-4} M, respectively. Under the same experimental circumstances, and correcting for limiting velocities, the turnover number[8] of the enzyme has been cal-

---

[8] Moles of substrate oxidized per mole of enzyme.

culated to be 2000 at 22° and pH 7.0, assuming a molecular weight of 150,000 for the enzyme.

Variation with pH. In Tris or histidine buffer initial dye reduction rates are optimal at pH 7.0. The rates at pH 6.5, 7.5, and 8.0 are 42, 78, and 65% of the pH 7.0 rate, respectively. Rates in phosphate show a similar pattern but are somewhat lower.

[93] Crystalline Crotonase[1] from Ox Liver

$$R—CH_2—CH=CH—CO—S—CoA + H_2O$$
$$\rightleftarrows R—CH_2—CHOH—CH_2—CO—S—CoA$$

By JOSEPH R. STERN

Assay Method

Principle. The assay is based on the observation of Seubert and Lynen[2] (cf. Lynen and Ochoa[3]) that S-crotonyl-N-acetylthioethanolamine has a characteristic absorption band in the ultraviolet with a maximum at 263 mμ, which may be attributed to diene conjugation between the α,β double bond and the carbonyl double bond of the thio ester, which is lacking in the free acid. In the case of crotonyl-S-CoA,[3] the 263-mμ band is largely obscured by the absorption of the adenine moiety and becomes apparent only when this interference is eliminated by taking a difference spectrum between the intact and the hydrolyzed thio ester. The product of hydration, β-hydroxybutyryl-S-CoA, exhibits no specific absorption in the 263-mμ region other than that due to its adenine moiety. Vinylacetyl-S-CoA also does not exhibit a specific absorption at 263 mμ, since a conjugate double-bond system is lacking.

Procedure. Two quartz cuvettes ($d = 0.5$ cm.) are made up as follows:

	Blank cell	Experimental cell
Assay mixture	0.20 ml.	0.20 ml.
Adenylic acid, 1 mg./ml.	0.15 ml.	—
Crotonyl-S-CoA, 3.2 μM./ml.	—	0.05 ml.
Crotonase fraction	0.01 ml.	0.01 ml.
Water	1.14 ml.	1.24 ml.

[1] J. R. Stern, A. del Campillo, and I. Raw, to be published.
[2] W. Seubert and F. Lynen, *J. Am. Chem. Soc.* **75**, 2587 (1953).
[3] F. Lynen and S. Ochoa, *Biochim. et Biophys. Acta* **12**, 299 (1953).

The assay mixture is prepared by mixing M Tris buffer pH 7.5 (1.0 ml.), 0.1% egg albumin (1.0 ml.), 0.1 M potassium EDTA pH 7.4 (0.15 ml.), and water (0.85 ml.).

Adenylic acid is added to the blank cell to compensate for the absorption of the adenine moiety of crotonyl-S-CoA. The optical density of a solution of adenylic acid decreases slowly on standing at 3°. An amount should be added which gives an initial reading (E) of 0.4 to 0.5 for the experimental cell.

Crotonase has such a high turnover number that even in crude tissue extracts only 1 to 20 γ of protein need be used in the assay. For purposes of assay, the fraction is diluted with 0.02 M potassium phosphate buffer, pH 7.4, to 0.5 mg. protein per milliliter. Any necessary further dilution of the fraction is made with the same buffer containing 0.001 M potassium EDTA and 0.1% egg albumin. It is essential to insure adequate mixing in making high dilutions of enzyme.

The reaction is started by addition of 0.01 ml. of enzyme solution (containing 0.0025 to 2 γ, depending on the purity), and the decrease in optical density at 263 mμ is recorded at 0.5-minute intervals. The initial rate of $-\Delta E_{263}$ is proportional to enzyme concentration, provided that it does not exceed 0.03 per minute. One unit of crotonase is defined as the amount which causes a decrease in optical density of 0.01 per minute at 25° under the above conditions. The specific activity is expressed as units *per microgram* of protein. Protein is determined by the method of Warburg and Christian.[4] The molecular extinction coefficient of crotonyl-S-CoA at 263 mμ is taken to be 6700, as for S-crotonyl-N-acetyl thioethanolamine.[2,3] Therefore one unit corresponds to the hydration of 0.0045 μM. of crotonyl-S-CoA.

Preparation of Crotonyl-S-CoA. 20 mg. of CoA[5] (Pabst) is dissolved in 2.0 ml. of cold water. The total sulfhydryl, as determined[6] on a 0.02-ml. sample, is about 20 μM. 0.40 ml. of 1.0 M potassium bicarbonate is added, then 0.008 ml. of crotonic anhydride. The solution is mixed and stirred at 0° for 10 minutes. The anhydride dissolves as its reacts. Redetermination of sulfhydryl (0.04-ml. sample) shows complete acylation. The solution contains 18 to 20 μM. of thio ester based on sulfhydryl disappearance. By enzymatic assay, 45% of the total thio ester is crotonyl-S-CoA, and 15%

---

[4] O. Warburg and W. Christian, *Biochem. Z.* **310**, 384 (1941–42).

[5] This CoA assays 50 to 65% pure and contains up to 20% (molar basis) glutathione. Any oxidized CoA can be reduced with sodium borohydride. Excess borohydride should be destroyed before acylation, since high concentrations (0.1 M) split thio esters.

[6] By the method of R. R. Grunert and P. H. Phillips, *Arch. Biochem.* **30**, 217 (1951).

crotonyl-S-glutathione. Part of the remainder probably represents compounds resulting from addition of unreacted thiol across the ethylenic bond of the thio esters. Often it is desirable to remove the free crotonic acid and any residual anhydride by acidification to pH 3 to 4 and extraction with several volumes of peroxide-free ether, taking care to remove the latter completely, or by lyophilization.[7]

Crotonyl-S-glutathione is not hydrated by crotonase, nor is it inhibitory. Ordinarily, relatively crude preparations of liver deacylase[8] or glyoxylase II[9] may be used to eliminate thio esters of glutathione, since the CoA esters are scarcely attacked, if at all. However, in the case of the crotonyl thio esters, the contamination of these enzyme fractions with crotonase results in simultaneous hydration of most of the crotonyl-S-CoA.

Assay of Crotonyl-S-CoA

Direct Optical Method. Crotonyl thio esters ($E_{263} = 6700$) react with hydroxylamine to yield crotonohydroxamic acid which has no significant absorption at 263 mμ. Thus the $-\Delta E_{263}$ which results from addition of neutralized hydroxylamine (200 μM.) to a solution of crotonyl-S-CoA in 0.067 M Tris buffer, pH 7.5 ($d = 0.5$ cm., volume 1.50 ml.) is a measure of the total crotonyl thio ester. To determine crotonyl-S-CoA specifically, an ox liver fraction[10] containing crotonase and S-acyl glutathione deacylase is used. The $-\Delta E_{263}$ at equilibrium caused by this fraction is a measure of the sum of hydration of crotonyl-S-CoA and deacylation of crotonyl-S-glutathione. At this point hydroxylamine is added, and the resultant $-\Delta E_{263}$ measures the unhydrated crotonyl-S-CoA. The $-\Delta E_{263}$ due to both enzyme and hydroxylamine represents the total crotonyl thio ester content of the solution. The $-\Delta E_{263}$ at equilibrium is redetermined on a separate sample with either crystalline crotonase or crotonase fractions in which the deacylase, if not already absent, is effectively removed by dilution. The difference in the $-\Delta E_{263}$ at equilibrium between liver fraction and crotonase represents crotonyl-S-glutathione and is subtracted from the total crotonyl thio ester to obtain the crotonyl-S-CoA content.

Indirect Optical Method. Crotonase can be coupled with β-hydroxybutyryl-S-CoA dehydrogenase (β-keto reductase)[3] to effect the reduction

---

[7] Amounts of crotonyl-S-CoA solution greater than 0.05 ml. may become inhibitory to crotonase. The inhibitor can be largely removed by ether extraction.

[8] W. W. Kielley and L. B. Bradley, *J. Biol. Chem.* **206**, 327 (1954).

[9] E. Racker, *J. Biol. Chem.* **190**, 685 (1951).

[10] Prepared from the initial crude extract by precipitation with (NH₄)₂SO₄, 0.3 to 0.6 saturation, and heating at neutral pH and 50° for 10 minutes.

of DPN by crotonyl-S-CoA at alkaline pH.[11,12] At pH 10[13] the equilibrium is such that DPN reduction and acetoacetyl-S-CoA formation, measured at 340 and 305 to 310 mμ, respectively, are practically quantitative. By addition of thiolase and the malic dehydrogenase-condensing enzyme system, crotonyl-S-CoA can be converted to citrate[11] with concomitant reduction of two additional molecules of DPN, thereby increasing the sensitivity of the method. The indirect method does not distinguish among crotonyl-S-CoA, vinylacetyl-S-CoA, and β-hydroxybutyryl-S-CoA. The last can be measured independently with crotonase-free dehydrogenase.

Purification Procedure

Step 1. Preparation of Extract. Ox liver is removed immediately after slaughter, frozen in dry ice, and chopped into small pieces. All subsequent steps are performed at 0 to 3° unless otherwise stated. 150-g. portions of partially thawed liver are placed in a Waring blendor, and 100 ml. of 0.2 M potassium bicarbonate containing 0.005 M L-cysteine and adjusted to pH 8.2 is added. The suspension is homogenized until it is of uniform consistency (about 5 minutes); then another 200 ml. of the solution is added, and homogenization is continued for 5 minutes. The suspension is passed through two layers of cheesecloth and centrifuged in a Servall angle centrifuge at 13,000 r.p.m. for 20 minutes. The supernatant is passed through cheesecloth again to remove fat particles. Approximately 2400 ml. of an opalescent dark reddish-brown extract is obtained from 1 kg. of frozen liver.

Step 2. Acid and Heat Treatment. 1200 ml. of liver extract is diluted to 6000 ml. with 0.02 M potassium phosphate buffer, pH 7.4. The pH of the solution is now 7.6. To each liter of solution 43 ml. of 1.0 N acetic acid is added with efficient stirring at 0° to bring the pH to 5.5. 1-l. portions of the acidified suspension in a 2-l. glass beaker are placed in a 55° bath and stirred continuously. The temperature of the suspension rises to 49 to 50° over a period of 10 minutes and is maintained at 50° for 3 minutes, then placed in an ice bath. The pH is adjusted to 7.0 with 1.0 M potassium bicarbonate (37 ml.), and the precipitate is removed by centrifugation at 13,000 r.p.m. for 5 minutes.

Step 3. Acetone Precipitation. The clear red supernatant (pH 7.0) is divided into two equal portions. To each portion (2880 ml.) cold acetone (2360 ml.) is added slowly, with continuous stirring, to achieve a final concentration of 45 vol.%, the temperature of the suspension being

[11] J. R. Stern and A. del Campillo, *J. Am. Chem. Soc.* **75**, 2277 (1953).

[12] S. J. Wakil and H. R. Mahler, *J. Biol. Chem.* **207**, 125 (1954).

[13] Glycine buffer should be avoided because of apparent nonenzymatic acylation of glycine by crotonyl-S-CoA at this pH.

lowered gradually to $-5°$. Practically no precipitation of protein occurs below 20% acetone. The suspension is centrifuged at 2000 r.p.m. for 20 minutes, the temperature being maintained at $-5°$. The precipitate is dissolved in 150 ml. of 0.02 M potassium phosphate buffer, pH 7.4, containing 0.003 M potassium EDTA, and dialyzed overnight against 10 l. of the same buffer. A small precipitate may form on dialysis; it is removed by centrifugation and discarded. Although crystallization can be effected at this stage, the crystals are of low activity and admixed with other protein. It is best to crystallize crotonase after step 4.

Step 4. Fractionation with Ammonium Sulfate. The 0 to 45 acetone fraction (402 ml.) is diluted to 1000 ml. with 0.02 M potassium phosphate buffer, pH 7.4, containing 0.003 M potassium EDTA in order to bring the protein concentration to about 15 mg./ml. Powdered ammonium sulfate (280 g.) is added slowly over a period of 20 minutes, with mechanical stirring, bringing the solution to 40% saturation. After standing for 10 minutes the suspension is centrifuged at 13,000 r.p.m. for 20 minutes. The precipitate is discarded. The supernatant is brought to 65% saturation (based on the original volume) by addition of 180 g. of ammonium sulfate in the manner indicated. The mixture is centrifuged, and the precipitate dissolved in about 100 ml. of 0.02 M potassium phosphate buffer, pH 7.4, containing 0.003 M potassium EDTA and 0.001 M neutralized glutathione, and dialyzed against 4 l. of the same buffer with glutathione omitted.

Step 5. Crystallization. 0.1 vol. of cold 95% ethanol is added slowly to the dialyzed 40 to 65 ammonium sulfate fraction (152 ml.) at 0°. During this addition the solution becomes opalescent, and the onset of crystallization is signaled by the appearance of Schlieren patterns when the solution is stirred and viewed in transmitted light. The solution is kept at 0°. The crystals settle gradually on standing. After 1 hour 65% of the enzyme has crystallized. Within 16 to 18 hours essentially all the enzyme is recovered as crystals. At this stage, microscopic examination at 0° reveals the presence of the typical crotonase crystals and many fine, circular highly refractile bodies (crystals?). Since the crotonase crystals redissolve as the drop evaporates, one must work rapidly at 0° to observe them or suspend the drop in 65% ethanol and seal the coverslip. The crystals are harvested by centrifugation at 10,000 r.p.m. for 5 minutes. The clear red supernatant is discarded. The white residue is dissolved in an excess of 0.02 M potassium phosphate buffer, pH 7.4, containing 0.003 M potassium EDTA. Any residue is centrifuged, washed with buffer, and the wash combined with the main solution. The residue, which contains the highly refractile particles but no crotonase crystals, as determined microscopically, is discarded.

The enzyme is readily recrystallized from the phosphate-EDTA solution by adding 0.1 to 0.2 vol. of 95% ethanol and allowing crystallization to proceed for 16 to 18 hours at 0° before harvesting as above. Although not essential, seeding may be used to facilitate crystallization. Since the solubility of crotonase is only about 0.9% at 0° in 0.02 M potassium phosphate buffer, pH 7.4, containing 0.003 M potassium EDTA, care must be taken not to discard crystalline enzyme on redissolving. With these precautions the recovery of enzyme can be made almost complete at each recrystallization. The specific activity reaches a maximum value of 780 after one recrystallization and remains unchanged after three recrystallizations.

The crystalline enzyme is soluble in water as well as in dilute salt solution. It is quite stable in the presence of EDTA and can be kept frozen for many months without appreciable loss of activity. From the specific activity of the crystals, it can be calculated that about 0.12% of the protein in the initial liver extract is crotonase.

Crotonase crystallizes from ethanol as irregular hexagonal plates which have alternate long and short sides. The crystals can readily be observed under the microscope at room temperature if they are suspended in 65% ethanol in the manner indicated above.

Properties

Purity. The crystalline enzyme is homogeneous on ultracentrifugal analysis. Its estimated molecular weight is 210,000. Thus 1 mole of crystalline crotonase hydrates 730,000 moles of crotonyl-S-CoA per minute at pH 7.5 and 25°.

Specificity. The crystalline enzyme hydrates β-methyl-crotonyl-S-CoA, *trans*-2-pentenoyl-S-CoA, and *trans*-2-hexenoyl-S-CoA among the compounds so far tested, but less rapidly than crotonyl-S-CoA. It hydrates vinylacetyl-S-CoA to β-hydroxybutyryl-S-CoA, and *trans*-3-hexenoyl-S-CoA to β-hydroxyhexanoyl-S-CoA. Therefore, crotonase acts as an isomerase. The evidence suggests, but is not conclusive, that the isomerization is indirect and involves intermediate formation of the hydroxy derivative. The 2-ethylenic bond of sorbyl-S-CoA (2,4-hexadienoyl-S-CoA) is not hydrated by the crystalline enzyme, although an enzyme hydrating this bond has been found in crude ox liver extracts.[14] A partially purified crotonase from liver has been prepared by Wakil and Mahler.[12] This crotonase preparation, which is only 1% pure, acts on various α,β unsaturated or β-hydroxyacyl-CoA derivatives including $\dot{C}_{12}$. The product of

[14] J. R. Stern and A. del Campillo, unpublished experiments.

hydration of crotonyl-S-CoA is presumably d-β-hydroxybutyryl-S-CoA, since it reduces DPN in the presence of β-hydroxybutyryl-S-CoA dehydrogenase, which is strongly indicated to be specific for the d-isomer.[15]

Crystalline crotonase hydrates both the *cis*- and *trans*-isomers of unsaturated acyl CoA derivatives. Isocrotonyl-S-CoA (the *cis*-isomer) is hydrated at about one-third the rate of *trans*-crotonyl-S-CoA; *cis*-2-hexenoyl-S-CoA is hydrated rather less rapidly than *trans*-2-hexenoyl-S-CoA. The claim of Wakil and Mahler[12] that crotonase is specific for the *trans*-isomer is therefore erroneous.

Crystalline crotonase does not hydrate free crotonic acid, S-crotonyl-N-diethylthioethanolamine, crotonyl-S-glutathione, or crotonyl-S-thioglycollate.

Inhibitors. The crystalline enzyme is inhibited by various sulfhydryl reagents as described for crude preparations.[12] Incubation of crystalline crotonase for 10 minutes at 0°, with the concentration of p-chloromercuribenzoate indicated in brackets, gives the following maximal inhibitions: 25% (10^{-6} M), 51% (10^{-5} M), 74% (10^{-4} M), 100% (10^{-3} M). The p-chloromercuribenzoate inhibited enzyme is reactivated only 24% by prolonged incubation with 0.25 M glutathione. Preincubation of crotonase with crotonyl-S-CoA affords little protection against the inhibitory action of p-chloromercuribenzoate. o-Iodosobenzoate (10^{-3} M) inhibits 56%. Iodoacetate (5×10^{-2} M) inhibits about 10%. The sulfhydryl group of crotonase is resistant to oxidation by hydrogen peroxide or oxygenation in the presence of traces of iron.

Equilibrium Constant. Crotonase acts as both hydrase and isomerase catalyzing an equilibrium between crotonyl-S-CoA, vinylacetyl-S-CoA, and β-hydroxybutyryl-S-CoA. Its actions are analogous to aconitase where, however, the unsaturated, rather than the hydroxy, derivative is the intermediate. The ratio β-hydroxybutyryl-S-CoA/crotonyl-S-CoA is 3.5 at equilibrium as determined by the optical method. The ratio β-hydroxyhexanoyl-S-CoA/*trans*-2-hexenoyl-S-CoA is 1.5. The ratio is independent of H^+ ion concentration, as expected. The value of the equilibrium constant is greater for the *cis*-isomers.

Effect of pH. The hydration of crotonyl-S-CoA by crotonase is influenced greatly by H^+ ion concentration. Crotonase activity increases continuously over the pH range 5 to 10. At pH 5.8 its activity is one-half, at pH 9.4 twice, that at pH 7.5.

Effect of Substrate Concentration. The K_m for crotonyl-S-CoA is 2.0 $\times$ 10^{-5} M. The enzyme is saturated at 8.5 $\times$ 10^{-5} M.

[15] A. L. Lehninger and G. D. Greville, *Biochim. et Biophys. Acta* **12**, 188 (1953).

SUMMARY OF CRYSTALLIZATION PROCEDURE

Fraction	Volume, ml.	Units × 10⁻³	Protein, mg.	Specific Activity, units/γ protein	Yield, %
1. Bicarbonate extract[a]	6000[b]	140,000	146,000	0.96	100
2. Acid-heat supernatant	5760	124,000	53,900	2.3	89
3. Acetone ppt., 0–45	402	137,000	14,000	9.8	98
4. (NH₄)₂SO₄, 0.40–0.65	152	96,300	6,020	16.0	69
5. Crystals (ethanol)	25.8	95,800	223	429	68
6. Supernatant of crystals	150	5,210	5,210	1.0	4
7. First recrystallization	10.7	87,400	112	780	62
8. Second recrystallization	10.2	68,900	89	773	49

[a] From 500 g. of ox liver.
[b] After dilution 1:5.

[94] β-Ketoreductase

$$CH_3-CO-CH_2-CO-S-CoA + DPNH + H^+$$
$$\rightleftharpoons CH_3-CH(OH)-CH_2-CO-S-CoA + DPN^+$$

By FEODOR LYNEN and OTTO WIELAND

β-Ketoreductase (β-ketohydrogenase, β-hydroxyacyl CoA dehydrogenase) is the enzyme of the fatty acid cycle catalyzing the reversible oxidation-reduction between acetoacetyl CoA or homologous β-ketoacyl CoA derivatives and DPNH.[1-5]

Assay Method

Principle. In accordance with Warburg's optical test, the rate at which DPNH disappears in the above reaction is followed by measurement of the light absorption at 340 mμ or at 366 mμ. However, the substrate employed is not acetoacetyl CoA but S-acetoacetyl-N-acetylcysteamine (AAC), which is similar in structure and readily synthesized.[1]

[1] F. Lynen, L. Wessely, O. Wieland, and L. Rueff, *Angew. Chem.* **64**, 687 (1952).
[2] F. Lynen, *Federation Proc.* **12**, 683 (1953).
[3] A. L. Lehninger and G. D. Greville, *Biochim. et Biophys. Acta* **12**, 188 (1953).
[4] S. J. Wakil, D. E. Green, S. Mii, and H. R. Mahler, *J. Biol. Chem.* **207**, 631 (1954).
[5] F. Lynen and S. Ochoa, *Biochim. et Biophys. Acta* **12**, 299 (1953).

Preparation of AAC. This compound is prepared by reaction of N-acetylcysteamine with diketene:

$$HS—CH_2—CH_2—NH—COCH_3 + CH_2\!=\!C—CH_2—C\!=\!O$$
$$\underset{\big|\underline{\quad O \quad}\big|}{}$$
$$\rightarrow CH_3—CO—CH_2—CO—S—CH_2—CH_2—NH—COCH_3$$

To a solution of 10 g. of N-acetylcysteamine (prepared by the method of Kuhn and Quadbeck[6]) in 10 ml. of ether, cooled in an ice bath, is added slowly a solution of 7.05 g. of diketene in 10 ml. of ether. The reaction begins at low temperature but is not complete until the mixture is brought to room temperature and allowed to stand for several hours. Upon subsequent cooling to $-10°$ the oily reaction mixture solidifies to a crystalline paste. The solution is filtered, and the residue washed with cold ether. Several recrystallizations from ether lead to the pure product in the form of colorless leaves or needles, melting point 60°. Yield, 8 to 10 g.

Reagents

> AAC stock solution (0.1 M). 203 mg. of the crystalline compound is dissolved in 10 ml. of water. This solution keeps for months in the refrigerator.
> 0.025 M sodium pyrophosphate-HCl buffer, pH 7.8.
> 0.025 M DPNH[7] solution (sodium salt) in water, pH 8 to 9.
> Enzyme. The stock solution is diluted with 10^{-3} M potassium ethylenediaminetetraacetate solution in water to give 3000 to 6000 units/ml. (For definition, see below.) The dilution should be made shortly before each assay, because dilute enzyme solutions from the last steps of the purification rapidly lose their activity.

Procedure.[8] Into a glass cuvette having a 1-cm. light path are placed 1.93 ml. of pyrophosphate buffer, 0.01 ml. of DPNH solution, and 0.01 ml. of enzyme solution. The reaction is started by adding 0.05 ml. of AAC solution, resulting in a total volume of 2.0 ml. The absorption at 366 mμ is read after 15 seconds, then at 15 to 60-second intervals, depending on the velocity of the reaction. The reaction is linear for at least 2 minutes, provided that the enzyme concentration given in the preceding section is used.

---

[6] R. Kuhn and G. Quadbeck, *Chem. Ber.* **84**, 844 (1951).

[7] For methods of preparing DPNH, see Vol. III [126, 127].

[8] Below is described the optical measurement at 366 mμ, at which wavelength glass cuvettes may be used. When quartz cuvettes are employed, it is preferable to read at 340 mμ, the absorption maximum of DPNH.

Definition of Unit and Specific Activity. One unit of enzyme is defined as that amount which, under the given conditions, causes an initial rate of change in optical density (ΔE_{366}) of 0.001 per minute at 20°. Specific activity is expressed as units per milligram of protein. Protein is determined either according to the nephelometric method of Kunitz[9] or by the biuret method as modified by Bücher.[10] The content of nucleic acid is found by measurement of absorption at 280 mμ and 260 mμ, according to Warburg and Christian.[11]

Application of Assay Method to Tissue Extracts. With extracts or homogenates of animal tissues, DPNH is oxidized in the absence of AAC. In this case the endogenous rate, that is, the decrease in optical density in the absence of AAC, should be determined and subtracted from the rate in the presence of substrate.

Procedure for Purification of Enzyme from Sheep Liver

Note. In order to bind traces of heavy metals which might damage the enzyme, only distilled water containing 10^{-3} M potassium ethylenediaminetetraacetate will be used in the course of the purification to dilute enzyme solutions and to dissolve precipitates.

Step 1. Preparation of Crude Extract. 1 kg. of sheep liver which has been kept frozen more than 48 hours in a freezer at $-20°$ is cut with a knife into small pieces. Portions of this are homogenized in a Waring blendor or Star-Mix with 2 l. of ice-cold 0.1 M Na_2HPO_4 solution. Subsequently the homogenate is stirred for about 30 minutes at 0° with a glass stirrer having several blades. Coarse connective tissue fibers form a heavy coat on the stirrer blades and are thus removed from the extract. Finally the extract is brought to pH 7.7 by cautious addition of 2 N KOH. $\frac{1}{1000}$ vol. of M potassium ethylenediaminetetraacetate solution is added.

Step 2. Denaturation of Inactive Proteins by Heating to 58°. The homogenate is divided into two parts and with vigorous stirring heated to 57 to 58° in a 2-l. enamel basin which is immersed in a water bath at 80 to 90°. This process should take less than 5 minutes. Both parts are then combined and placed for 40 to 45 minutes in a constant-temperature water bath at 58°, again with constant stirring. During this time the temperature of the extract should not exceed 58°. The homogenate, which has become viscous owing to the presence of denatured protein, is placed

[9] M. Kunitz, *J. Gen. Physiol.* **35**, 423 (1952).

[10] G. Beisenherz, H. J. Boltze, T. Bücher, R. Czök, K.-H. Garbade, E. Meyer-Arendt, and G. Pfleiderer, *Z. Naturforsch.* **8b**, 576 (1953).

[11] O. Warburg and W. Christian, *Biochem. Z.* **310**, 284 (1941); see Vol. III [73].

in an ice bath at −18°. 1.5 l. of ice-cold water[12] is added, and the solution is cooled to 0° with agitation. 675 g. of kieselgur[13] (infusorial earth) (150 g./l. of liquid) is stirred in, and the heavy paste is distributed among four Büchner funnels (internal diameter 24 cm., Schleicher and Schüll filter paper No. 505) and filtered with suction in the cold room. The residue is not washed and should therefore be dried as much as possible by suction. It still contains 20 to 25% of total activity, but for technical reasons no attempt is made to recover this activity.

Step 3. Acetone Precipitation. To the clear, red filtrate having a volume of 2.6 l. and which is stirred and cooled in an ice-salt mixture, 1.4 l. of freshly distilled acetone is added slowly in a fine stream. The temperature during the addition should not exceed 0° and at the end should be −10°. The precipitate can be separated easily at 2000 × *g* and consists, in addition to protein, largely of secondary sodium phosphate. The supernatant is discarded, and the pellet is stirred with about 80 ml. of water.[12] This dissolves almost all the protein but only part of the phosphate. The phosphate residue is centrifuged down and washed three times with 25-ml. portions of water.[12] The combined solution, which is slightly turbid, medium to dark brown, is made up to 250 ml. with water.[12]

Step 4. First Alcohol-Acid Precipitation. To 250 ml. of solution from step 3 kept in an ice bath is added slowly 50 ml. of 96% ethanol at −15°. The precipitate is centrifuged at high speed at 0° and discarded. The light-brown, clear filtrate is brought to pH 6.5 at 0° by slow addition of *M* acetic acid. After 30 minutes the precipitate is centrifuged off at 2000 × *g*. If an assay of the filtrate indicates that it still contains more than 20% of initial activity, the rest can be precipitated by further addition of *M* acetic acid to pH 6.3.

The muddy precipitates are nucleic acid precipitates which dissolve in 70 ml. of water[12] only after the pH is raised to 7.2 by addition of a few drops of 2 *M* ammonia. The brownish, slightly opalescent solution is diluted to 100 ml. with water.[12]

Step 5. Adsorption on Gel and First Ammonium Sulfate Precipitation. 100 ml. of the solution from step 4 is stirred, and 80 ml. of calcium phosphate gel[14] is added slowly. After being stirred for 15 minutes, the gel is centrifuged. The green filtrate which should contain only 12 to 16% of the activity is discarded. If the filtrate contains more activity, additional gel should be used. The gel residue is washed with centrifugation first with 20 ml. of 0.1 *M* phosphate buffer, pH 7.5, and then with 6.5 ml. of

---

[12] See note p. 568.

[13] Kieselgur No. IA from Schenk Filterbau Co., Schwäbisch Gmünd, Germany. This material is used as a filter aid.

[14] Prepared according to M. Kunitz;[9] see Vol. I [11].

0.1 M tetrasodium pyrophosphate. When the residue is treated with 10 ml. of 0.1 M Na₄-pyrophosphate, the gel dissolves to a great extent and the activity goes into the solution. The turbid, yellow solution is brought to a volume of 30 ml. with 0.1 M Na₄-pyrophosphate. Without removing the turbidity, 20 ml. of ammonium sulfate solution saturated at 0°, pH 9.6, is added slowly with stirring [160 g. of crystalline ammonium sulfate + 190 ml. of 0.1 M Na₄-pyrophosphate + 0.19 ml. of M potassium ethylenediaminetetraacetate + 10 ml. of concentrated ammonia ($D = 0.91$) are mixed and kept tightly stoppered at 0°]. After standing for 30 minutes, the solution is centrifuged with refrigeration at 15,000 r.p.m. The residue is discarded, and the supernatant, which is clear, has little color, and contains over 80% of the activity eluted from the gel, is brought to pH 7.4 with 2 N H₂SO₄. Its volume is then 43 ml. The enzyme is precipitated by addition of 3.5 g. of crystalline ammonium sulfate. After standing for 30 minutes the precipitate is centrifuged at 15,000 r.p.m. and then dissolved in water[12] to a final volume of 15 ml.

In order to separate nucleic acids still present, salts must be removed from the solution. This is accomplished by dialysis at 0° with stirring for 12 hours against 0.05 M phosphate buffer, pH 8.0, then for an additional 12 hours against 0.005 M phosphate buffer of the same pH. Both buffers should contain 10⁻³ M potassium ethylenediaminetetraacetate. Since this dialysis occasionally causes up to 50% loss of activity, it is recommended that the separation be achieved by a second alcohol precipitation. However, this does not increase the specific activity.

Step 6. Second Alcohol-Acid Precipitation. 15 ml. of solution from step 5, cooled in an ice bath, is brought to pH 5.9 with 0.3 ml. of M acetic acid. With vigorous cooling 10 ml. of 96% alcohol chilled to −15° is added slowly. The mixture is allowed to stand for 1 to 2 hours at −14°. The dense, yellow-white precipitate is centrifuged at −10°, washed once with 2.5 ml. of 30% alcohol, and then suspended in 5 ml. of water.[12] Only when the pH is raised to 8.0 by addition of dilute ammonia does the residue form an almost clear solution. The latter is diluted with water[12] to 10 ml.

Step 7. Removal of Nucleic Acids by Means of Protamine Sulfate and Alumina Gel. 10 ml. of solution from step 6 is mixed with portions of a 2% solution of protamine sulfate, pH 6.45, until spot tests on the filtrate are negative to protamine and to nucleic acids. This requires 1.0 to 1.1 ml. of protamine solution. After the dense precipitate is centrifuged off at 0°, the clear, almost colorless filtrate, whose volume is 11 ml., is mixed with 1.1 ml. of 50 vol. % aluminum hydroxide C$_\gamma$. After standing for 5 minutes, the gel is centrifuged at high speed, leaving 10.4 ml. of almost water-clear filtrate.

Step 8. Second Ammonium Sulfate Precipitation. To 10.4 ml. of solution is added slowly 2.75 g. of crystalline ammonium sulfate and 235 mg. of $Na_2P_2O_7 \cdot 10H_2O$. This yields a solution of pH 7.2 to 7.4 which is 0.45 saturated in ammonium sulfate. After 1 hour the precipitate is centrifuged and suspended in water[12] to a total volume of 1 ml.

TABLE I
SUMMARY OF PURIFICATION PROCEDURE

Fraction	Total volume, ml.	Total units, thousands	Protein, mg.	Specific activity	Recovery, %
1. Homogenate	—	7600	160,000	40–60	—
2. Solution after heating	2400	3900	23,100	170	100
3. Acetone ppt.	250	2600	7,000	370	67
4. First EtOH ppt.	100	1660	2,630	630	43
5. Calcium phosphate gel eluate	30	1400	—	—	36
6. First (NH₄)₂SO₄ precipitation	15	900	250	3600	23
7. Second EtOH ppt.	10	675	200	3400	17.3
8. Filtrate after protamine pptn. and alumina C_γ adsorption	10.4	520	97	5650	13.4
9. Second (NH₄)₂SO₄ precipitation	1.0	270	35	7800	8.4

A method for the purification of β-Ketoreductase from beef liver mitochondria has been published recently by Wakil *et al.*[4]

Properties

Purity and Stability. The enzyme preparation from step 8 of the purification is practically free of the other enzymes of the fatty acid cycle: crotonase, β-ketothiolase, and butyryl CoA dehydrogenase. The considerable stability of the enzyme to alkali should be noted: No loss of activity is observed after keeping at pH 10.5 and 0° for 20 hours.

Kinetics of Enzyme Reaction. Figure 1 shows the dependence of DPNH oxidation on AAC concentration. The Michaelis constant is 1.1×10^{-2} M. In contrast to this, the affinity of the enzyme for the biological substrate, acetoacetyl CoA, is considerably higher. A Michaelis constant of 1.1×10^{-4} M was found.

Equilibrium and Specificity. According to Lehninger and Greville,[3] the enzyme is specific for D-β-hydroxybutyryl CoA. When racemic β-hydroxybutyryl CoA or S-β-hydroxybutyryl-N-acetylcysteamine is used, only half of the starting material reacts. Owing to the participation of H$^+$, the position of equilibrium of hydrogen transfer

D-β-Hydroxybutyryl CoA + DPN$^+$ $\rightleftarrows$ Acetoacetyl CoA + DPNH + H$^+$

is strongly pH-dependent. At 25° we found the constant to be

$$K' = \frac{[\text{D-}\beta\text{-hydroxybutyryl CoA}] \times [\text{DPN}^+]}{[\text{Acetoacetyl CoA}] \times [\text{DPNH}] \times [\text{H}^+]} = 5.25 \times 10^9 \ M$$

S. J. Wakil[4] gives a value of 1.6×10^{10} at 25°. Above pH 8 there is an additional effect owing to the dissociation of a proton from acetoacetyl

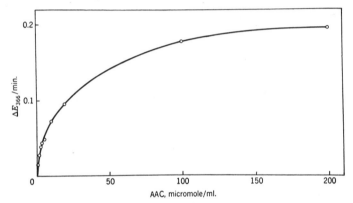

Fig. 1. Dependence of DPNH oxidation on AAC concentration. 3.9 γ of β-keto-reductase from step 8, containing 30 units, in 0.025 M pyrophosphate buffer, pH 7.8; 0.25 micromoles of DPNH; AAC as shown. Total volume, 2.00 ml.; $T = 23.5°$.

CoA, leading to the equilibrium equation

$$\overset{\displaystyle \text{OH}}{\underset{\displaystyle |}{\text{CH}_3\text{—CH—CH}_2\text{—CO—S—CoA}}} + \text{DPN}^+$$

$$\rightleftarrows \overset{\displaystyle \text{O}^-}{\underset{\displaystyle |}{\text{CH}_3\text{—C}}}\text{=CH—CO—S—CoA} + \text{DPNH} + 2\text{H}^+$$

At pH 9 and with an excess of DPN$^+$, this results in the almost complete dehydrogenation of D-β-hydroxybutyryl CoA or D-β-hydroxybutyryl-N-acetylcysteamine to the acetoacetyl compound. This circumstance can be used to advantage in the preparation of acetoacetyl CoA from β-hydroxy-butyryl CoA.[1,15]

Our results and those of Green and co-workers[4] indicate that the enzyme has little specificity with respect to chain length. All normal

[15] H. Beinert, *J. Biol. Chem.* **205**, 575 (1953).

β-hydroxyacyl CoA derivatives from C_4 to C_{12} are dehydrogenated by the enzyme at an equal rate. Higher S-β-hydroxyacyl-N-acetylcysteamine derivatives also react. However, the corresponding thio esters with glutathione or the ethyl esters or the β-hydroxy acids themselves are not acted upon by the enzyme.

Distribution. N. Hielschmann[16] has investigated the activity of β-ketoreductase in tissue extracts of human organs after preparation of a clear filtrate with kieselgur. The distribution found in man and animals is given in Table II.

TABLE II

DISTRIBUTION OF β-KETOREDUCTASE IN HUMAN AND ANIMAL TISSUES

Organ	Soluble protein, units/mg.					
	Man*	Rabbit	Rat	Calf	Pork	Sheep
Liver	31.4	38.8	8.3	27.9	33.1	92.6
Kidney	28.0	50.6	19.6	42.6	64.8	63.3
Adrenal	19.6	7.9	22.0	23.4	12.8	6.9
Spleen	13.6	—	4.1	—	—	—
Heart muscle	52.1	28.7	37.6	39.4	89.8	30.5
Skeletal muscle	32.2	—	3.0	—	—	—
Prostate	9.1	—	—	—	—	—
Thyroid	12.7	—	—	—	—	—
Hypophysis (pituitary)	14.8	—	—	—	—	—
Brain	——	4.7	1.2	8.1	3.3	0.8

* Mean values of 6 individuals.

[16] N. Hielschmann, Dissertation, Munich, 1954.

[95] Enzymes of Acetoacetate Formation and Breakdown

By JOSEPH R. STERN

It is now well established by the work of several laboratories (cf. reviews[1-3]) that the formation and breakdown of acetoacetate and higher β-keto acids proceeds via the β-ketoacyl-S-coenzyme A derivative as the active intermediate. Two enzymes catalyzing the conversion of acetoacetate to acetoacetyl-S-CoA have been described. One enzyme, an acetoacetate-activating enzyme present in liver,[4] kidney and yeast,[5] and, to a

[1] F. Lynen and S. Ochoa, *Biochim. et Biophys. Acta* **12**, 299 (1953).

[2] H. R. Mahler, *Federation Proc.* **12**, 694 (1953).

[3] F. Lynen, *Harvey Lectures* XLVIII, 210 (1952–53).

[4] J. R. Stern and S. Ochoa, *J. Biol. Chem.* **191**, 161 (1951).

[5] J. R. Stern, M. J. Coon, and A. del Campillo, *Nature* **171**, 28 (1953).

lesser extent, heart,[6] catalyzes the synthesis of acetoacetyl-S-CoA from acetoacetate, CoA-SH, and ATP, probably according to the mechanism indicated by reaction 1. The other enzyme, CoA transferase,[7] found in heart[6,7] and kidney[7] catalyzes a direct transfer of CoA from succinyl-S-CoA to acetoacetate, according to reaction 2.

$$\text{Acetoacetate} + \text{CoA-SH} + \text{ATP} \rightleftarrows \text{Acetoacetyl-S-CoA} + \text{AMP} + \text{PP} \tag{1}$$

$$\text{Acetoacetate} + \text{succinyl-S-CoA} \rightleftarrows \text{Acetoacetyl-S-CoA} + \text{succinate} \tag{2}$$

The acetoacetyl-S-CoA (or C_n β-ketoacyl-S-CoA) thus formed can be either split "thiolytically," in the presence of CoA-SH, to 2 molecules of acetyl-S-CoA (or C_{n-2} acyl-S-CoA + acetyl-S-CoA) by thiolase[1-3,5-8] (reaction 3) or hydrolyzed to acetoacetate and CoA-SH by a specific deacylase found only in liver[1,5] (reaction 4).

$$\text{Acetoacetyl-S-CoA} + \text{CoA-SH} \rightleftarrows 2 \text{ Acetyl-S-CoA} \tag{3}$$
$$\text{Acetoacetyl-S-CoA} + \text{H}_2\text{O} \rightarrow \text{Acetoacetate} + \text{CoA-SH} \tag{4}$$

The coupling of reaction 3 (from right to left) and reaction 4 accounts for the synthesis of acetoacetate from acetyl-S-CoA[9] in pigeon liver. In heart, which lacks acetoacetyl-S-CoA deacylase, acetoacetate synthesis from acetyl-S-CoA is effected by the coupling of reactions 3 and 2, in reverse, with a specific succinyl-S-CoA deacylase.[10-12]

Only the preparation of CoA transferase and thiolase will be described, since neither the activating enzyme nor the deacylase has been purified to any significant extent.

CoA Transferase[13]

Succinyl-β-ketoacyl-S-CoA Transferase from Heart

$$\text{Succinyl-S-CoA} + \text{Acetoacetate} \rightleftarrows \text{Acetoacetyl-S-CoA} + \text{Succinate}$$

Assay Method

Principle. The assay is based on the observation of Lynen *et al.*[8] that at alkaline pH acetoacetyl thio esters (e.g., S-acetoacetyl-N-acetyl thio-

---

[6] D. E. Green, D. S. Goldman, S. Mii, and H. Beinert, *J. Biol. Chem.* **202**, 137 (1953).

[7] J. R. Stern, M. J. Coon, and A. del Campillo, *J. Am. Chem. Soc.* **75**, 1517 (1953).

[8] F. Lynen, L. Wessely, O. Wieland, and L. Rueff, *Angew. Chem.* **64**, 687 (1952).

[9] E. R. Stadtman, M. Doudoroff, and F. Lipmann, *J. Biol. Chem.* **191**, 377 (1951).

[10] S. Kaufman, *in* "Phosphorus Metabolism" (McElroy and Glass, eds.), Vol. 1, p. 370, Johns Hopkins Press, Baltimore, 1951.

[11] S. Kaufman, C. Gilvarg, O. Cori, and S. Ochoa, *J. Biol. Chem.* **203**, 869 (1953).

[12] J. Gergely, P. Hele, and C. V. Ramakrishnan, *J. Biol. Chem.* **198**, 323 (1952).

[13] J. R. Stern, M. J. Coon, A. del Campillo, and M. C. Schneider, to be published.

ethanolamine and S-acetoacetyl CoA) exhibit a characteristic absorption band in the range 280 to 320 mμ, with a maximum at 303 mμ. This absorption is attributed to formation of an enolate ion and is increased by decrease in H+ ion concentration. Stern et al.[7] observed that this enolate absorption of acetoacetyl-S-CoA is markedly increased by Mg++ ions, presumably through formation of a chelate structure. The presence of Mg++ ions, which are not required for CoA transferase activity, increases the sensitivity of the optical assay. Succinyl-S-CoA does not absorb in the region 303 mμ. It is important to note that the enolate ion absorption is abolished by p-chloromercuribenzoate[14] (10^{-3} M) and by cysteine (6 $\times$ 10^{-3} M) but not by glutathione[15] or thioglycollate.[14]

Procedure. Two quartz cuvettes ($d = 0.5$ cm.) are made up as follows:

	Blank cell	Experimental cell
M Tris-HCl buffer, pH 8.1	0.10 ml.	0.10 ml.
0.8 M MgCl$_2$	0.01 ml.	0.01 ml.
M potassium acetoacetate	0.10 ml.	0.10 ml.
Succinyl-S-CoA, 1.8 μM./ml.	—	0.20 ml.
CoA transferase fraction	0.01 ml.	0.01 ml.
Water	To 1.50 ml.	

The succinyl-S-CoA is prepared by reaction of succinic anhydride and CoA-SH[16] (Pabst). The actual succinyl-S-CoA content of the solution is approximately 50% of the total thio ester formation calculated from the disappearance of sulfhydryl. This is partly accounted for by the presence of glutathione in the CoA. Succinyl-S-CoA is relatively unstable; its half-life at 25° is about 30 minutes. If kept frozen at neutral pH, the solution can be used over a period of at least 2 weeks.

The CoA transferase fraction is diluted with 0.02 M potassium phosphate buffer, pH 7.4, so that 0.01 ml. contains 3 to 40 γ depending on its activity.

The reaction is started by addition of enzyme, and the increase in optical density at 310 mμ is recorded at 0.5-minute intervals at 25°. The ΔE_{310} from 0 5 to 1.0 minute after the addition of enzyme is used to calculate its activity.

Definition of Unit and Specific Activity. One unit of enzyme is defined as the amount which causes an initial rate of increase in optical density of 0.01 per minute at 25°. The initial rate of increase in optical density is proportional to enzyme concentration, provided that the ΔE_{310} does not

[14] J. R. Stern and A. del Campillo, unpublished observations.
[15] F. Lynen, personal communication.
[16] E. J. Simon and D. Shemin, *J. Am. Chem. Soc.* **75**, 2520 (1953).

exceed 0.04 per minute. The initial rate is not maintained beyond 2 to 3 minutes, presumably because of the unfavorable equilibrium position when the reaction is measured in this direction. The specific activity is expressed as units per milligram of protein. Protein is estimated by the method of Warburg.[17] The molecular extinction coefficient (E_{310}) of acetoacetyl-S-CoA at pH 8.1 with Mg^{++} present[13,18] is 1.2×10^4. Therefore one unit corresponds to the formation of 0.0025 μM. of acetoacetyl-S-CoA.

Application of Assay Method to Crude Tissue Extracts. The direct optical method of assay can be employed only in fractions which are free of thiolase. This is accomplished by steps 3 and 4 of the purification procedure. In cruder fractions and tissue extracts, CoA transferase can be assayed optically by coupling with thiolase and trapping the acetyl CoA thus generated with oxalacetate, using the malic dehydrogenase-condensing enzyme system[19] (cf. ref. 6) or with *p*-nitroaniline,[20] using the aromatic amine acetylating enzyme of pigeon liver.[21] The sequence of reactions is as follows:

$$\text{Succinyl-S-CoA} + \text{acetoacetate} \rightleftarrows \text{Acetoacetyl-S-CoA} + \text{succinate} \tag{1}$$

$$\text{Acetoacetyl-S-CoA} + \text{CoA-SH} \rightleftarrows \text{2-Acetyl-S-CoA} \tag{2}$$

$$\text{2 Malate} + \text{2 DPN}^+ + \text{2 acetyl-S-CoA} \rightleftarrows \text{2 Citrate} + \text{2 CoA-SH} + \text{2 DPNH} + 2\text{H}^+ \tag{3a}$$

$$\text{2 Acetyl-S-CoA} + \text{2 } p\text{-nitroaniline} \rightarrow \text{2 } p\text{-Nitroacetanilide} + \text{2 CoA-SH} \tag{3b}$$

The reduction of DPN can be followed at 340 mμ, and the acetylation of *p*-nitroaniline at 420 mμ.[20]

The components of the DPN reduction assay are: Tris buffer pH 8.1 (100 μM.), $MgCl_2$ (8 μM.), potassium acetoacetate (100 μM.), CoA (0.2 μM.), neutralized glutathione (10 μM.), DPN (0.3 μM.), potassium L-malate (10 μM.), succinyl-S-CoA (0.35 μM.), purified thiolase (200 units, S.A. 15,000), purified malic dehydrogenase[22] (2 γ, S.A. 62,000), crystalline condensing enzyme[23] (16 γ, S.A. 330), and CoA transferase fraction (3 to 250 γ). The reaction is started by addition of succinyl-S-

[17] O. Warburg and W. Christian, *Biochem. Z.* **310**, 384 (1941–42).
[18] H. Beinert, *J. Biol. Chem.* **205**, 575 (1953).
[19] J. R. Stern, B. Shapiro, E. R. Stadtman, and S. Ochoa, *J. Biol. Chem.* **193**, 703 (1951).
[20] H. Tabor, A. H. Mehler, and E. R. Stadtman, *J. Biol. Chem.* **204**, 127 (1953).
[21] N. O. Kaplan and F. Lipmann, *J. Biol. Chem.* **174**, 37 (1948).
[22] F. B. Straub, *Biochem. J.* **34**, 483 (1940).
[23] S. Ochoa, J. R. Stern, and M. C. Schneider, *J. Biol. Chem.* **193**, 691 (1951).

CoA. One unit of CoA transferase is defined as the amount which causes an initial rate of increase in optical density at 340 mμ of 0.01 per minute at 25°. One unit corresponds to the formation of 0.0048 μM. of DPNH or 0.0024 μM. of acetoacetyl-S-CoA.

If the extract is contaminated with DPNH oxidase, although the use of cyanide to inhibit it may be tried,[6] the p-nitroaniline assay is indicated. The components of the latter are: Tris buffer pH 8.1 (100 μM.), MgCl₂ (8 μM.), potassium acetoacetate (100 μM.), CoA (0.2 μM.), neutralized glutathione (10 μM.), succinyl-S-CoA (0.35 μM.), p-nitroaniline (0.3 μM.), purified pigeon liver acetylating enzyme[20] (0.3 mg., S.A. 150 to 200), and CoA transferase fraction. Purified thiolase (200 units) must also be added after step 2 if not already supplied in excess by the acetylating enzyme fraction employed. The reaction is started by addition of succinyl-S-CoA. One unit of CoA transferase is defined as the amount which causes an initial rate of decrease in optical density at 420 mμ of 0.01 per minute at 25°. One unit corresponds to the formation of 0.0050 μM. of p-nitro-acetanilide or 0.0025 μM. of acetoacetyl-S-CoA.

Purification Procedure

Step 1. Preparation of Crude Extract. Pig hearts, removed immediately after death, are packed in ice. All subsequent steps are performed at 0 to 3°, unless otherwise indicated. Thirty hearts are trimmed of fat, blood clots, and connective tissue and passed twice through an electric mincer. The minced heart (about 4.5 kg.) is washed five times with 5 vol. of cold tap water in a tall glass cylinder, capacity 37 l. Cylindrical blocks of ice (volume 1 quart) are used to keep the suspension cold. The washed mince is then passed through a large table-type Büchner filter and sucked "dry." 170-g. portions of this preparation are placed in a Waring blendor, and 1.5 vol. (255 ml.) of 0.05 M potassium phosphate buffer, pH 7.4, containing 0.2 M KCl is added. The suspension is "blendorized" for 5 minutes at a rheostat setting of 70 volts; then another 1.5 vol. of buffer mixture is added, and "blendorization" is continued for another 5 minutes at a setting of 35 volts. The suspension is passed through two layers of cheesecloth and centrifuged in a Servall angle centrifuge at 13,000 r.p.m. for 7 minutes. The deep-pink supernatant fluid is passed through ten layers of cheesecloth to remove floating fat particles. About 10 l. of extract is obtained.

Step 2. Fractionation with Ammonium Sulfate. 5-l. portions of crude extract are brought to 35% saturation with powdered ammonium sulfate (245 g./l. of extract). The salt is added slowly over a period of 20 minutes with mechanical stirring. The mixture is stirred for 20 minutes and centrifuged at 13,000 r.p.m. for 15 minutes. The precipitate is discarded. The supernatant is brought to 65% saturation by adding 210 g. of ammonium sulfate for each liter of original solution over a period of 20 minutes. After another 20 minutes of stirring, the mixture is centrifuged as above. The precipitate is dissolved in 500 ml. of 0.017 M potassium phosphate buffer, pH 6.8, and dialyzed against 20 l. of the same buffer for 16 hours.

Step 3. Fractionation with Acetone. The 35 to 65 ammonium sulfate fraction is made 0.067 M with respect to potassium phosphate buffer, pH 6.8. Cold acetone ($-10°$) is then added to a concentration of 40.5% by volume with mechanical stirring, the temperature of the mixture being gradually lowered to $-5°$ during the 30 to 40-minute period required for the addition. Stirring is continued for another 15 minutes at $-5°$. The mixture is centrifuged at 3000 r.p.m. for 20 minutes at a temperature setting low enough to insure that the temperature of the supernatant will be -5 to $-6°$. The precipitate is dissolved in 150 ml. of 0.017 M potassium phosphate buffer, pH 7.4, and the solution dialyzed overnight against 20 l. of the same buffer. This 0 to 40.5 fraction is used for the purification of thiolase. The cloudy supernatant is brought to 57% acetone by volume at -9 to $-10°$, and after centrifugation at this temperature the precipitate is dissolved in 150 ml. of 0.017 M potassium phosphate buffer, pH 7.4, and the solution dialyzed overnight against 20 l. of the same buffer.

In practice, stocks of the 0 to 40.5 and 40.5 to 57 acetone fractions are accumulated as starting material for the further purification of thiolase and transferase, respectively, and kept frozen. Transferase is remarkably stable at all stages of purification and can be kept in the deep-freeze for many months with little or no loss of activity.

Step 4. Heat and Acid Treatment. The 40.5 to 57 acetone fraction is diluted with 0.017 M potassium phosphate buffer, pH 7.4, to bring the protein concentration to 10 mg./ml. 300-ml. portions contained in a 1-l. glass beaker are placed in a 55° bath and stirred continuously. The temperature of the solution is allowed to rise to 50° during an interval of 9 minutes, the bath temperature falling to about the same reading. The solution is maintained at 50° for another 6 minutes and then placed in an ice bath. The precipitate is removed by centrifugation at 13,000 r.p.m. for 20 minutes and discarded. To 300-ml. portions of supernatant, about 65 ml. of 0.1 N acetic acid are added rapidly, with mechanical stirring, until the pH falls to 5.8. The precipitate is removed by centrifugation at 13,000 r.p.m. for 5 minutes and discarded.

Step 5. Adsorption on Alumina Gel. The acid supernatant is quickly adjusted to pH 6.65 by addition of about 4 ml. of 1.0 M potassium bicarbonate. Alumina C_γ gel[24] (dry weight, 13 mg./ml.) is added to make a final concentration of 30% by volume, and the mixture is stirred for 10 minutes at 0°. The gel is separated by centrifugation, and the supernatant discarded. The gel is suspended in a volume of cold 0.1 M potassium phosphate buffer, pH 7.4, equal to that of the acid supernatant, and stirred continually for 10 minutes at 0°. After centrifugation, this eluate (718 ml. containing 1750 mg. of protein, S.A. 4) is discarded, and the gel is again eluted by being suspended in a similar volume of 0.2 M potassium phosphate buffer, pH 8.0 (20°), and stirring at 20° for 20 minutes. After removal of gel by centrifugation, the pH 8 eluate containing $\sim$1 mg. of protein per milliliter is concentrated by addition of solid ammonium sulfate to 85% saturation, centrifuging, and dissolving the precipitate in a volume of 0.017 M Tris buffer, pH 7.5, equal to 5% that of the eluate. This fraction is dialyzed overnight against the same buffer to remove residual inorganic phosphate as well as ammonium sulfate.

Step 6. Ethanol Fractionation in the Presence of Zinc. The dialyzed pH8 eluate is diluted to a protein concentration of about 5 mg./ml. with 0.017/M Tris buffer, pH 7.5, and 0.025 vol. of M potassium succinate, pH 6.2 is added. 0.25 vol. of 0.1 M zinc acetate is added slowly with stirring at 0°. After 5 minutes the mixture is centrifuged, and the precipitate (89.4 mg. of protein, S.A. 4) discarded. Cold absolute ethanol is added to the clear supernatant at 0° to a concentration of 15% by volume, and, after 10 minutes of stirring, the mixture is centrifuged, and the red-brown precipitate[25] (262 mg. of protein, S.A. 12) discarded. The supernatant is brought to 35 vol.% ethanol, stirred for 10 minutes at 0°, then centrifuged. The yellow-white precipitate is dissolved in 10 ml. of 0.1 M potassium phosphate buffer, pH 7.4, containing 0.01 M potassium EDTA and 0.1% glutathione, and the solution dialyzed overnight against 0.017 M potassium phosphate buffer, pH 7.4. Any turbidity which may appear is removed by centrifugation, leaving a clear yellow solution.

Step 7. Fractionation with Ammonium Sulfate. The procedure is that of step 2. The 15 to 35 Zn-ethanol fraction is diluted to a protein concentration of 5 mg./ml. with 0.017 M potassium phosphate buffer, pH 7.4, and then brought to 45% saturation with ammonium sulfate. The precipitate (5.6 mg. of protein, S.A. 168) is discarded, and the supernatant obtained after centrifugation is brought to 58% saturation. This precipitate is collected and dissolved in several milliliters of 0.017 M potassium

---

[24] For preparation, see Vol. I [11].
[25] This precipitate contains highly purified phosphorylating enzyme (cf. S. Kaufman, Vol. I [120]).

phosphate buffer, pH 7.4, to give a clear yellow solution which is dialyzed overnight against the same buffer.

SUMMARY OF PURIFICATION PROCEDURE

Fraction	Volume	Units	Protein, mg.	Specific[b] activity, units/mg. protein	Yield, %
1. Phosphate extract[a]	10,500	181,000	151,000	1.2[c]	100
2. (NH₄)₂SO₄, 0.35–0.65	1,240	156,000	25,900	6.0[d]	86
3. Acetone ppt., 40.5–57	474	86,700	8,250	10.5[d]	48
4. Acid supernatant	723	68,300	4,880	16.9[d] (14[e])	38
5. Alumina eluate[f]	46	63,400	576	110[e]	35
6. Zn-ethanol, 15–35	15.6	42,700	95.1	449[e]	24
7. (NH₄)₂SO₄, 0.45–0.58	1.8	34,300	34.3	1000[e]	18

[a] From 4.5 kg. of heart.
[b] Calculated on basis of 310 units.
[c] p-Nitroaniline assay.
[d] DPN reduction assay.
[e] Direct assay.
[f] After concentration.

Properties

Purity. The 45 to 58 ammonium sulfate fraction is practically the most active CoA transferase preparation which has been obtained. Ultracentrifugal analysis and paper electrophoresis show the presence in it of two protein components. The major component, comprising about 80% of the total protein, is the CoA transferase. The minor component, representing about 20% of the total protein, is a fluorescent flavoprotein which has tightly bound FAD as its prosthetic group and possesses very strong diaphorase activity. The flavoprotein can be largely, but not entirely, removed by adsorption on 0.12 vol. of alumina C$_\gamma$ gel at pH 6.0. The specific activity of the CoA transferase in the gel supernatant is increased about 20%.

Specificity. The purified enzyme catalyzes the transfer of CoA from succinyl CoA to acetoacetate, β-ketovalerate, β-ketoisocaproate, and β-ketocaproate, but not to β-ketoöctanoate, β-ketoadipate, saturated fatty acids, or crotonate. It catalyzes CoA transfer from acetoacetyl CoA to succinate, but not to malonate, butyrate, or crotonate.

Equilibrium Constant. The equilibrium constant $K_{equil.}$ = (succinate)(acetoacetyl-S-CoA)/(succinyl-S-CoA)(acetoacetate) is dependent on both pH and the presence or absence of Mg^{++}, as shown in the table.

EQUILIBRIUM CONSTANTS

pH	7.00	7.50	8.10	9.20
No Mg^{++}	2.83×10^{-3}	4.19×10^{-3}	4.34×10^{-3}	8.36×10^{-3}
With Mg^{++}	4.81×10^{-3}	5.94×10^{-3}	8.96×10^{-3}	1.50×10^{-2}

Activators and Inhibitors. No activators of CoA transferase are known. The enzyme is not inhibited by potassium EDTA (10^{-3} M) or by iodoacetate (10^{-2} M).

Turnover Number. The purest CoA transferase preparation catalyzes the transfer of 300 moles of CoA per minute from succinyl-S-CoA to acetoacetate per 100,000 g. of enzyme at 25°.

β-Keto Thiolase[26]

(Acetoacetate condensing enzyme, acetoacetate cleavage enzyme)

β-Keto Thiolase from Heart

Acetoacetyl-S-CoA + CoA-SH ⇌ 2 Acetyl-S-CoA

Assay Method

Principle. The principle is the same as for the direct assay of CoA transferase.[27] The CoA-dependent splitting of acetoacetyl-S-CoA to 2 molecules of acetyl-S-CoA is followed optically as a decrease in the absorption of light at 305 mμ. Mg^{++} ions are present to enhance the absorption of the enolate ion of the β-keto thio ester.

Preparation of Acetoacetyl-S-CoA. Acetoacetyl-S-CoA is made enzymatically from succinyl-S-CoA[28] and acetoacetate with CoA transferase which is free of thiolase.[29] Tris buffer pH 8.1 (500 μM.), MgCl$_2$ (40 μM.), potassium acetoacetate (500 μM.), succinyl-S-CoA (2μM.), and CoA transferase (50 to 100 units), in a volume of 4.0 ml., are incubated for 15 minutes at 25°. (The progress of the reaction can be followed optically on a 1.0-ml. sample as described below.) The whole solution (4.0 ml.) is brought to pH 5 with 1.0 N acetic acid, then incubated at 75° for 2 minutes, with stirring, and cooled rapidly to 0°. Alkali is added to bring the pH to 7.4. The precipitate is removed by centrifugation, washed with 0.5 ml. of cold water, and the original supernatant and wash combined.

---

[26] S. Ochoa, J. Harting, M. J. Coon, J. R. Stern, A. del Campillo, and M. C. Schneider, to be published.

[27] See previous section.

[28] E. J. Simon and D. Shemin, *J. Am. Chem. Soc.* **75**, 2520 (1953).

[29] J. R. Stern, M. J. Coon, and A. del Campillo, *J. Am. Chem. Soc.* **75**, 1517 (1953).

The amount of acetoacetyl-S-CoA in this solution is estimated by determining its optical density (E) at 310 mμ and pH 8.1, in the presence of 5.3×10^{-3} M MgCl, before and after addition of hydroxylamine (200 μM., pH 8). Under these conditions $(d = 0.5$ cm., volume 1.50 ml.) 0.10 μM. of acetoacetyl-S-CoA causes a ΔE_{310} of 0.398. The yield is 30 to 40% of the succinyl-S-CoA added. 0.5 ml. of solution containing 0.04 μM. of acetoacetyl-S-CoA is used in the assay.

Acetoacetyl-S-CoA can also be prepared chemically by reacting acetoacetyl-S-thiophenol or diketene and CoA-SH.[30] The yield is about 50%. Acetoacetyl-S-CoA can also be generated enzymatically from chemically prepared crotonyl-S-CoA by coupling crotonase and β-hydroxybutyryl CoA dehydrogenase, in the presence of DPN, at alkaline pH.[31,32]

Procedure. Two quartz cuvettes $(d = 0.5$ cm.) are made up as follows:

	Blank cell	Experimental cell
M Tris-HCl buffer, pH 8.1	0.20 ml.	0.20 ml.
0.8 M MgCl$_2$	0.01 ml.	0.01 ml.
CoA (3 mg./ml.)	0.05 ml.	0.05 ml.
0.1 M glutathione (neut.)	0.10 ml.	0.10 ml.
M K acetoacetate	0.05 ml.	—
Acetoacetyl-S-CoA (0.08 μM./ml.)	—	0.50 ml.
Heart thiolase fraction	0.01 ml.	0.01 ml.
Water	To 1.50 ml.	

Acetoacetate is present in the blank cell to compensate for the absorption of the free acetoacetate in the acetoacetyl-S-CoA solution. An amount should be added which gives a reading of 0.2 to 0.3 for the optical density of the experimental cell.

Dilutions of thiolase fractions for assay are made with 0.02 M potassium phosphate buffer, pH 7.4, containing 0.0005 M glutathione and 0.1% egg albumin. 0.1 to 20 γ of protein is used, depending on the activity of the fraction.

Because acetoacetyl-S-CoA is somewhat unstable at alkaline pH, the optical density at 305 mμ is recorded for 1 to 2 minutes, then the reaction started by addition of enzyme, and the decrease in optical density (E) followed at 0.5-minute intervals. The initial rate of $-\Delta E_{305}$ is proportional to enzyme concentration over a sixfold range, provided that it does not exceed 0.1 per minute.

Definition of Unit and Specific Activity. One unit of thiolase is defined as the amount which causes a decrease in optical density of 0.01 per

[30] T. Wieland and L. Rueff, *Angew. Chem.* **65**, 186 (1953).
[31] J. R. Stern and A. del Campillo, *J. Am. Chem. Soc.* **75**, 2277 (1953).
[32] H. Beinert, *J. Biol. Chem.* **205**, 575 (1953).

minute at 25° under the above conditions. The specific activity is expressed as units per milligram of protein. Protein is determined by the optical method of Warburg.[33]

Application of Assay to Crude Tissue Preparations. The only enzymes which may interfere with this assay are the specific acetoacetyl-S-CoA deacylase found in liver, but absent from kidney and heart extracts, and the CoA transferase of heart if large amounts of succinate are present as a contaminant. In practice, the activity of liver acetoacetyl-S-CoA deacylase is so low compared to thiolase that it is effectively and simply removed by dilution.

Purification Procedure

Steps 1, 2, and 3. The first three steps are identical with those outlined for the purification of CoA transferase.[27] The low (0 to 40.5) acetone fraction is used for the further purification of thiolase. After storage of this fraction in the frozen state, a precipitate may be found on thawing. This is not thiolase and can be discarded, resulting in an even more active fraction. All subsequent steps are carried out at 0 to 3° unless otherwise indicated.

Step 4. Acid Precipitation. The 0 to 40.5 acetone fraction (200 ml.) is diluted with ice-cold water to a protein concentration of 16 mg./ml. Approximately 4.5 ml. of cold 1.0 N acetic acid is added slowly, with mechanical stirring, at 0° until the pH falls to 5.3. The precipitate is removed by centrifugation at 13,000 r.p.m. and discarded.

Step 5. Fractionation with Ammonium Sulfate. The clear supernatant is adjusted carefully to pH 7.4 with 2 N KOH (about 1.7 ml.), 0.01 vol. of neutral 0.2 M L-cysteine is added, and the protein concentration adjusted to 6 mg./ml. with water. Powdered ammonium sulfate (38.8 g. per 100 ml. of solution) is added to bring the solution to 55% saturation. The precipitate is discarded, and the supernatant obtained after centrifugation is brought to 70% saturation by addition of 10.6 g. of ammonium sulfate per 100 ml. of original supernatant. The precipitate is collected, dissolved in 0.033 M potassium succinate buffer, pH 6.3, containing 10^{-3} M L-cysteine, and dialyzed overnight against the same buffer.

Step 6. Ethanol Fractionation in the Presence of Zinc. The dialyzed 55 to 70 fraction is diluted to 8 mg. of protein per milliliter with 0.033 M potassium succinate buffer containing 5×10^{-4} M L-cysteine. Cold absolute ethanol is added, with mechanical stirring, to a concentration of 20% by volume, the temperature of the mixture being gradually lowered to −3°. An amount of 0.1 M zinc acetate equal to one-tenth the volume

[33] O. Warburg and W. Christian, *Biochem. Z.* **310**, 384 (1941–42).

of the diluted fraction is added, the stirring continued for 10 minutes, and the mixture centrifuged at 10,000 r.p.m. at $-3°$ to $-4°$. The precipitate (445 mg. of protein, S.A. 2000) is discarded. This supernatant (18% ethanol) is brought to 25% ethanol at $-6°$, the mixture centrifuged at $-6°$, and the precipitate (21.4 mg. of protein, S.A. 3750) again discarded. This supernatant is brought to 40% ethanol, the temperature of the mixture being lowered gradually to $-15°$. The mixture is centrifuged at -15 to $-16°$, and the precipitate is dissolved in 5 ml. of a solution containing potassium phosphate buffer (pH 7.4, 0.1 M), glutathione (0.01 M), and potassium EDTA (0.03 M), and dialyzed overnight against 3 l. of potassium phosphate buffer (0.01 M, pH 7.4) containing 10^{-3} M L-cysteine.

SUMMARY OF PURIFICATION PROCEDURE

Fraction	Volume, ml.	Units	Protein, mg.	Specific activity, units/mg. protein	Yield, %
1. Phosphate extract[a]	4170	8,340,000	60,000	139	100
2. $(NH_4)_2SO_4$, 0.35–0.65	483	3,620,000	10,000	362	43
3. Acetone ppt., 0–40.5	184	3,545,000	3,420	1,050	42
4. Acid supernatant, pH 5.3	198	2,970,000	2,040	1,450	36
5. $(NH_4)_2SO_4$, 0.55–0.70	13	1,950,000	491	4,120	23
6. Zn-ethanol, 25–40	6.5	950,000	33.8	28,000	11

[a] From 1.9 kg. of heart.

Properties

Purity. The purified enzyme is homogeneous in the ultracentrifuge.

Specificity. The purified enzyme is highly specific for acetoacetyl-S-CoA. β-Ketovaleryl-S-CoA reacts at 20% of the rate of acetoacetyl-S-CoA; β-ketocaproyl-S-CoA and β-ketoisocaproyl-S-CoA practically not at all. This is in contrast to the broader specificity of crude heart enzyme fractions[34,35] and of crude[36] and purified[37] preparations of liver thiolase and indicates that there may be a family of thiolase enzymes acting on β-ketoacyl-S-CoA derivatives of varying chain length.

Inhibitors. Thiolase has been shown by Lynen and collaborators[38] to be a sulfhydryl enzyme. It is inhibited by iodoacetate and arsenoxide. 3×10^{-4} M iodoacetate results in complete inhibition.

[34] J. R. Stern, M. J. Coon, and A. del Campillo, *Nature* **171**, 28 (1953).
[35] D. E. Green, D. S. Goldman, S. Mii, and H. Beinert, *J. Biol. Chem.* **202**, 137 (1953).
[36] J. R. Stern and S. Ochoa, *J. Biol. Chem.* **191**, 161 (1951).
[37] D. S. Goldman, *J. Biol. Chem.* **208**, 345 (1954).
[38] F. Lynen, *Federation Proc.* **12**, 683 (1953).

Equilibrium Constant. The equilibrium constant $K = \text{(acetyl-S-CoA)}^2/$ (acetoacetyl-S-CoA)(CoA-SH) has been determined by the optical method to be about 5×10^4 at pH 8.1 [29] and 1×10^4 at pH 9.0. It is therefore not practical to use this reaction to synthesize acetoacetyl-S-CoA. However, the sensitivity of the optical method at pH 9 to 10 permits the synthesis of acetoacetyl-S-CoA from large amounts of acetyl-S-CoA to be observed as a small increase in optical density at 305 mμ. Thiolase can also be coupled with β-keto reductase, in the presence of DPNH, to effect the synthesis of β-hydroxybutyryl-S-CoA from 2 molecules of acetyl-S-CoA.[39]

Turnover Number. The best preparations of thiolase so far obtained catalyze the cleavage of 6500 moles of acetoacetyl-S-CoA per minute per 100,000 g. of enzyme at 25°.

[39] F. Lynen, L. Wessely, O. Wieland, and L. Rueff, *Angew. Chem.* **64**, 687 (1952).

[96] Aceto-CoA-Kinase

$$\text{ATP} + \text{CoA} + \text{Ac} \rightleftarrows \text{Ac}\sim\text{S CoA} + \text{5-AMP} + \text{PP}$$

By MARY ELLEN JONES and FRITZ LIPMANN

Principle of Enzymatic Reaction

The mechanism of the aceto-CoA-kinase reaction involves a pyrophosphoryl split of ATP which is coupled with the acetylation of CoA by free acetate. So far no evidence has been obtained that more than one enzyme is involved. However, isotope analysis indicates a three-step reaction with all intermediates being enzyme-bound.[1]

$$\text{E} + \text{ATP} \rightleftarrows \text{E-AMP} + \text{PP} \qquad (1)$$
$$\text{E-AMP} + \text{CoA} \rightleftarrows \text{E-CoA} + \text{AMP} \qquad (2)$$
$$\text{E-CoA} + \text{Ac} \rightleftarrows \text{Ac}\sim\text{S CoA} + \text{E} \qquad (3)$$

The activation of the higher fatty acids proceeds in a manner similar to the activation of acetate yielding analogous reaction products.[2,3]

$$\text{Fatty acid} + \text{CoA} + \text{ATP} \rightleftarrows \text{acyl CoA} + \text{5-AMP} + \text{PP}$$

Jencks[4] isolated a related enzyme from hog liver which forms acyl CoA as the product from ATP, CoA, and a four- to twelve-carbon fatty

[1] M. E. Jones, F. Lipmann, H. Hilz, and F. Lynen, *J. Am. Chem. Soc.* **75**, 3285 (1953).
[2] A. Kornberg and W. E. Pricer, Jr., *J. Biol. Chem.* **204**, 329 (1953).
[3] H. R. Mahler and S. J. Wakil, *J. Biol. Chem.* **204**, 453 (1953).
[4] W. P. Jencks, *Federation Proc.* **12**, 703 (1953).

acid. However, this enzyme splits ATP to AMP and PP also in the absence of CoA with hydroxylamine and catalytic amounts of fatty acid present.

Assay Methods

Principle. In the assay methods for this enzyme, CoA is usually added in catalytic amounts and recycles by a transfer of the acetyl group from Ac∼S·CoA, the product of the reaction, to various acceptors. This can be accomplished:

(1) Chemically, with hydroxylamine[5,6] which liberates CoA with hydroxamic acid formation. The hydroxamic acid is then measured colorimetrically with $FeCl_3$ by the well-known method.[7]

$$Ac\sim S\ CoA + NH_2OH \rightarrow AcNHOH + CoA$$

(2) Enzymatically, by coupling the kinase reaction with an acetyl acceptor enzyme such as the aceto-arylamine-kinase from pigeon liver.[5,8]

$$Ac\sim S\ CoA + NH_2R \rightarrow AcNHR + CoA$$

The aceto-CoA-kinase has also been coupled with the citrate condensing enzyme.[9,10]

The chemical method using hydroxylamine will be described in detail. Some additional comments will also be made concerning the enzymatic assays.

Procedure for Chemical Trapping with Hydroxylamine

Reagents

$8.1 \times 10^{-4}\ M$ or 250 units/ml.[11] CoA solution.
$0.1\ M$ ATP solution, pH 7.5.
$0.2\ M$ potassium acetate solution.
$1\ M$ potassium phosphate buffer, pH 7.5.
$1\ M$ tris(hydroxymethyl)aminomethane buffer, pH 8.2.

[5] F. Lipmann, *J. Biol. Chem.* **160**, 173 (1945).
[6] M. E. Jones, S. Black, R. M. Flynn, and F. Lipmann, *Biochim. et Biophys. Acta,* **12**, 141 (1953).
[7] F. Lipmann and L. C. Tuttle, *J. Biol. Chem.* **159**, 21 (1945).
[8] T. C. Chou and F. Lipmann, *J. Biol. Chem.* **174**, 37 (1952); see Vol. I [101].
[9] J. R. Stern, B. Shapiro, E. R. Stadtman, and S. Ochoa, *J. Biol. Chem.* **193**, 703 (1951).
[10] H. Beinert, D. E. Green, P. Hele, H. Hift, R. W. von Korff, and C. V. Ramakrishnan, *J. Biol. Chem.* **203**, 35 (1953).
[11] N. O. Kaplan, and F. Lipmann, *J. Biol. Chem.* **174**, 37 (1948).

1 M potassium fluoride.

0.2 M magnesium chloride.

0.2 M reduced glutathione, neutralized to pH 4.5 with KOH.

2 M hydroxylamine solution, pH 7.4. This reagent must be freshly prepared by mixing equal volumes of 4 N NH$_2$OH·HCl and 4 N KOH.

Aceto-CoA-kinase. Dilutions of the enzyme are used such that a suitable aliquot (0.05 to 0.35 ml.) will form 0.1 to 0.6 micromole of acethydroxamate in 20 minutes.

Procedure. To a small test tube add 0.1 ml. of the CoA solution (25 units); 0.1 ml. of the ATP solution (10 micromoles); 0.1 ml. of potassium acetate (20 micromoles); 0.1 ml. of potassium phosphate buffer, pH 7.5 (100 micromoles) for yeast extracts or 0.1 ml. of Tris buffer, pH 8.2 (100 micromoles), for extracts of animal tissues; 0.1 ml. of the hydroxylamine solution (200 micromoles); 0.05 ml. of potassium fluoride (50 micromoles); 0.05 ml. of magnesium chloride (10 micromoles); 0.05 ml. of glutathione solution (10 micromoles); and water if necessary to make a volume of 1 ml. after the addition of the enzyme. The tubes are placed in a 37° bath for 5 minutes before the enzyme is added.

After the addition of the enzyme, the tubes are incubated for 20 minutes at which time 1.5 ml. of a ferric chloride reagent is added which contains 10% FeCl$_3$·6H$_2$O and 3.3% trichloroacetic acid in 0.66 N HCl. The acethydroxamate formed is measured in the Klett-Summerson colorimeter with a No. 54 filter using 1 ml. of water and 1.5 ml. of the ferric chloride reagent to set the zero point. In all cases a blank tube which contains no CoA is incubated along with the tube containing CoA. The blank reading is subtracted from the reading obtained when CoA is present. With increasing purification of the enzyme the blank reading becomes negligible.

Definition of Unit and Specific Activity. One unit of activity is defined as that amount of enzyme which gives a reading of 100 Klett units (0.4 micromole of acethydroxamic acid) under these conditions. The specific activity is defined as units per milligram of protein. The protein concentration of the solution is determined turbidimetrically by the method of Stadtman *et al.*[12]

Precautions and Comments. Since it is difficult to find conditions where the system is truly saturated with CoA or NH$_2$OH, the enzyme concentration must be adjusted so that not more than 0.6 micromole of acethydroxamate is formed in 20 minutes. With crude extracts the blank tube containing no CoA is essential. Furthermore, with crude preparations, the

[12] E. R. Stadtman, G. D. Novelli, and F. Lipmann, *J. Biol. Chem.* **191**, 365 (1951).

acetylation of hydroxylamine appears to be, in part, enzymatic. Crude yeast extracts in particular may contain in addition to the enzyme under consideration an "acetohydroxylamine kinase" the activity of which is prominent, however, only at lower hydroxylamine concentrations.[5] Despite these limitations, the ease of the method recommends its use. Beinert et al.[10] have experienced difficulty in applying a modification of this method to pig heart extracts. However, the method has served well in the purification of yeast and pigeon liver extracts.[6,8,13]

   Coupling with the Aceto-Arylamine-Kinase or Citrate Condensing Enzyme. The acetylation of sulfanilamide is the classical method for assaying for CoA.[11] This system may be used in the presence of larger quantities of CoA (approximately 25 units) for the assay of aceto-CoA-kinase. This assay has greater sensitivity than the hydroxamate method, and it is linear over a wider range of concentrations.[6] Several arylamines have been used as acceptors, and they include sulfanilamide,[11,14] p-amino-benzoic acid,[5] aminoazobenzene,[15] 4-aminoazobenzene-4'-sulfonic acid,[16] and p-nitroaniline.[17]

   The coupling of aceto-CoA-kinase with the citrate condensing enzyme has been reported, and a convenient assay is detailed by Beinert et al.[10]

   Purification of Aceto-CoA-Kinase. The partial purification of aceto-CoA-kinase has been described in detail for either fresh baker's yeast or fresh beef heart mitochondria.[6,18] Crude extracts and acetone or ammonium sulfate fractions of pigeon liver rich in aceto-CoA-kinase have also been described.[8] Aceto-CoA-kinase has also been extracted from various plant sources, and a partial purification of the enzyme from spinach has been described.[19]

   The purification from yeast is described below. Most brands of commercial baker's yeast were found to be suitable; the best preparations were obtained, however, with National brand yeast.

   Step 1. Preparation of Crude Extract. (A) QUICK-FROZEN YEAST PREPARATION. One part of yeast is mixed with 1 part of ether and 1.5 parts of dry ice and stirred well. After 30 minutes the liquid is poured off, the yeast is spread out in a thin layer on a cloth, and air is blown over it by a fan under a well-ventilated hood. An additional 1.5 parts of dry ice is

[13] H. Hilz, Thesis, University of Munich (1953).
[14] F. Lynen, R. Reichert, and L. Rueff, *Ann.* **574**, 1 (1951).
[15] R. E. Handschumacher, G. C. Mueller, and F. M. Strong, *J. Biol. Chem.* **189**, 335 (1951).
[16] S. P. Bessman, in preparation.
[17] H. Tabor, A. H. Mehler, and E. R. Stadtman, *J. Biol. Chem.* **204**, 127 (1953).
[18] P. Hele, *J. Biol. Chem.* **206**, 671 (1954).
[19] A. Millerd and J. Bonner, *Arch. Biochem. and Biophys.* **49**, 343 (1954).

mixed with the yeast, and ventilation is continued until practically all the ether is removed. The preparation is then stored in a deep-freeze. On thawing, the yeast autolyzes rather quickly. To promote this further, 33 ml. of M dipotassium phosphate is mixed with 1000 g. of yeast, and the mixture is stirred overnight in the cold room at 0 to 5°. The yeast autolyzate is centrifuged in the cold. The supernatant solution does not store well, and it is desirable to carry it immediately through the first ammonium sulfate fractionation step.

(B) SONIC PREPARATIONS. One part of fresh baker's yeast is mixed to a paste with 1 part of cold 0.1 M dipotassium phosphate and exposed to vibration in a 10-kc magnetostrictive oscillator (Raytheon Manufacturing Company) for 40 minutes. The mixture is centrifuged in the cold in a Servall centrifuge at top speed (12,000 r.p.m.) for 30 minutes. The supernatant solution, amounting to slightly more than half the original volume, may be frozen and kept for a considerable length of time in the deep-freeze.

This extract has the same specific activity as preparation A, above, but is only half as concentrated. It behaves on fractionation like the quick-frozen preparation and, after fractionation step 3, the two preparations become very similar in all respects.

Step 2. Protamine Precipitation. The nucleic acid content of the slightly turbid supernatant solution is determined turbidimetrically in the following manner: to 0.05 ml. of the solution to be analyzed are added 4.5 ml. of 0.05 M phosphate buffer, pH 6.1, and 0.5 ml. of 2% protamine sulfate solution. The tubes are mixed, and the turbidity is measured on a Klett colorimeter using filter No. 54. The reading is compared with a standard curve prepared with 0.1 to 0.6 mg. of yeast nucleate adjusted to pH 6. For each milligram of nucleic acid thus determined, 0.025 ml. of a 2% protamine sulfate solution is added. The protamine precipitate is centrifuged down at 0° and is discarded.

Step 3. First Ammonium Sulfate Precipitation. To 1000 ml. of clear supernatant solution (pH about 6) 350 g. of solid ammonium sulfate is added (55% saturation). The active material is precipitated and centrifuged off; the supernatant solution is discarded. The precipitate is dissolved with 66 ml. of 0.05 M potassium bicarbonate solution. The resulting solution is 20% saturated with ammonium sulfate.

Step 4. Second Ammonium Sulfate Precipitation. Enough ammonium sulfate solution (saturated at 0 to 3°), is added to bring the above enzyme preparation from 20% to 35% saturation with respect to ammonium sulfate. The precipitate is discarded, and the supernatant solution is brought to 45% saturation with the saturated ammonium sulfate solution. The precipitate is dissolved in 0.05 M bicarbonate solution and may be used in this form without further treatment. When it is desired to remove the ammonium sulfate, the solution may be dialyzed with agitation at 0° against a solution of 0.05 M potassium bicarbonate in 0.5% potassium chloride for 2 to 4 hours. Typical results of the above fractionation procedure are illustrated in the table.

Properties

Stability. Although partially purified extracts of both yeast (fraction 4) and beef heart (fraction 3)[18] are stable almost indefinitely in the frozen state, particularly in the presence of ammonium sulfate, further purification has led to rather unstable preparations. Upon dialysis yeast preparations slowly lose activity even when frozen and after a month have lost all activity; however, if used immediately they can be dialyzed as described above. The yeast enzyme is unstable to acidities lower than pH 5.5 even in the presence of ammonium sulfate; however, it is relatively stable in the alkaline range and may be incubated at 37° for 20 minutes from pH 7 to 9 with only a 20% loss in activity.

In contrast to the yeast preparation, crude heart preparations can be kept at pH 4.3 in the presence of 20% ammonium sulfate for several hours.[18]

Specificity. The yeast enzyme will acetylate both CoA and dephospho-CoA; however, the rate of acetylation of dephospho-CoA is one-half that for CoA.[20] Of the acids which will serve as substrates for the yeast enzyme, acetate is the most effective; propionate and acrylate are activated at one-third the rate of the acetate. The enzyme from beef heart[18] has the highest affinity ($K_m \times 10^{-3} = 1.42\ M$) for acetate, but propionate and acrylate serve as equally good substrates on a rate basis. Marginal activations have been obtained with glycolate, glyoxalate, β-hydroxy-pyruvate, L-lactate, formate, butyrate, and valerate.[18,16]

Equilibrium. The reaction is reversible, and two estimates have been made for the equilibrium constant, which lies near unity. The values which have been obtained are 2.7[21] and 0.86,[18] where $K = (\mathrm{Ac}{\sim}\mathrm{S}\ \mathrm{CoA})(5\text{-}\mathrm{AMP})(\mathrm{PP})/(\mathrm{Ac})(\mathrm{ATP})(\mathrm{CoA})$. These values, in confirmation of earlier studies, indicate that the energy of the acyl-mercapto bond of $\mathrm{Ac}{\sim}\mathrm{S}$ CoA is nearly equivalent to the energy of the phosphoryl bonds of ATP. In addition it would appear that the energy of the second pyrophosphate bond of ATP is equivalent to the terminal pyrophosphate bond.

Effect of Various Ions. There is a requirement for magnesium ion which may be replaced by manganese ion. In addition, there is a specificity for monovalent ions; the best enzyme activity is observed in the

[20] M. E. Jones, unpublished observations.
[21] M. E. Jones, *Federation Proc.* **12**, 708 (1953).

presence of potassium, ammonium, or rubidium ions, and the enzyme is inhibited by sodium or lithium ions.[22] With certain yeast enzyme preparations, the addition of 30 to 100 micromoles of phosphate per milliliter increases the rate of the hydroxamate reaction considerably. The phosphate, however, is not indispensable, and the mechanism of its effect is not clear.

Effect of pH. The yeast enzyme is maximally active at pH 7.2 but has a rather broad optimum.[6] The activity falls sharply as the pH is lowered below 7. Extracts of animal tissues appear to have a slightly more basic pH optimum, and for this reason they are assayed at pH 8.2 rather than 7.5.

ACTIVITY OF VARIOUS YEAST FRACTIONS

Fraction	Total volume, ml.	Units/ ml.	Protein, mg./ml.	Specific activity, units/ mg.	Total units	Recovery, %
1. Original extract (quick-frozen preparation)	1020	66	66	1.0	67,400	100
2. Protamine supernate	1000	66	33	2.0	66,000	98
3. Ammonium sulfate precipitate, 0–55%	132	495	97	5.1	65,200	97
4. Ammonium sulfate precipitate, 35–45%	32	1063	95	11.2	34,000	51

[22] R. W. von Korff, *J. Biol. Chem.* **202**, 265 (1953).

[97] Acetate Kinase of Bacteria (Acetokinase)[1]

Acetate + ATP $\rightleftarrows$ Acetyl-P + ADP

By IRWIN A. ROSE

Assay Method

Principle. The reaction is carried out in the presence of excess hydroxylamine with acetate and ATP as reactants. The assay method makes use of the ability of acyl phosphates to form hydroxamic acids rapidly at neutrality and the subsequent formation of the colored ferric-hydroxamate complex in acid solution.[2]

[1] According to I. A. Rose, M. Grunberg-Manago, S. R. Korey, and S. Ochoa, to be published.
[2] F. Lipmann and L. C. Tuttle, *J. Biol. Chem.* **159**, 21 (1945).

Reagents

3.2 *M* potassium acetate.

1.0 *M* Tris-HCl buffer, pH 7.4.

1.0 *M* MgCl₂.

4.0 *M* KOH.

28% hydroxylamine hydrochloride. This reagent is stored in the cold.

0.1 *M* ATP. Sodium salt neutralized with HCl.

FeCl₃ reagent. 1.25% FeCl₃ in 1.0 *N* HCl.

10% trichloroacetic acid.

Phosphate-cysteine. 0.1 *M* potassium phosphate buffer, pH 7.4, containing 0.005 *M* cysteine.

Enzyme. Dilute stock enzyme with phosphate-cysteine to obtain 10 to 100 units/ml. (See definition below.)

Procedure. The stock of substrate is made up fresh daily using potassium acetate, Tris, and MgCl₂ in a volume ratio of 25:5:1. A neutral solution of hydroxylamine is made fresh by mixing equal volumes of the hydrochloride and KOH. Mix 0.3 ml. of the stock substrate, 0.35 ml. of neutral hydroxylamine, 0.1 ml. of ATP, and an amount of water to make the final volume with enzyme to 1.0 ml. Place the tube at 29°. Add one unit or less of enzyme at zero time, and stop the reaction after at least 2 minutes by the addition of 1.0 ml. of trichloroacetic acid. Centrifuge off any precipitate. Color is developed by adding 4.0 ml. of FeCl₃ reagent and is read immediately in the Klett colorimeter using the 540-mμ filter. A reading of 69 units above a reagent blank corresponds to 1.0 micromole of acetohydroxamic acid.

Definition of Unit and Specific Activity. One unit of enzyme is defined as the amount which produces 1 micromole of hydroxamic acid per minute under the standard assay conditions given above. The assay is linearly responsive to enzyme concentration below 2 units of enzyme per tube. Specific activity is defined as units per milligram of protein. Protein is determined by the biuret reaction.[3] Alternatively, protein may be determined by the method of Warburg and Christian[4] when the ratio of optical densities at 280 to 260 mμ is greater than 0.8.

Application of the Assay Method to Bacterial Extracts. Acetokinase has been found to be distributed widely among bacteria but has not been reported in mammalian tissues. The bacterial extracts are usually prepared by grinding with alumina and extracting by stirring with buffer

---

[3] A. G. Gornall, C. J. Bardawill, and M. M. David, *J. Biol. Chem.* **177**, 751 (1949).

[4] O. Warburg and W. Christian, *Biochem. Z.* **310**, 384 (1941–42).

(see below). Especially with crude extracts, a positive assay is not definite evidence for the presence of acetokinase in so far as the acetate-activating enzyme[5] will give rise to acetohydroxamic acid in the presence of catalytic concentrations of CoA. The most rigorous method for ruling out this latter enzyme is by the inability of AMP plus pyrophosphate to replace ADP in causing the disappearance of acetyl phosphate in the presence of added transacetylase.[5] An alternative procedure is to treat the enzyme solution with an anion exchange resin to remove CoA.

Purification Procedure from *E. coli*

Step 1. Preparation of Crude Extract. The starting material was a lypholyzed powder of *E. coli*, strain 4157 (for preparation, see Vol. I [114]). This powder had been stored in the cold for two years. Six-gram amounts of an equal weight mixture of alumina and bacterial powder are ground at room temperature with wetting amounts of potassium phosphate buffer, 0.1 M, pH 7.4. The gummy paste is stirred with 5 vol. of buffer for 30 minutes at 2°. The accumulated mixture is centrifuged at 18,000 × g for 1 hour in the cold. The residue is extracted again with 2.5 vol. of buffer, stirred, and centrifuged as before, and the clear supernatant solutions are combined. The solution usually contains 10 to 14 mg. of protein per milliliter with a specific activity of 2 to 3. It may be stored with 0.005 M cysteine with no loss of activity.

Step 2. First Acetone Fractionation. The crude extract, pH 7.3, is brought to 45% concentration of acetone (by volume) beginning at 0° and ending at −10° with rapid stirring and dropwise addition of acetone (held at −10°). The heavy precipitate is removed by centrifugation at −10° for 1 hour at 500 × g. The clear supernatant fluid is then brought to 55% acetone concentration and again centrifuged. The carefully drained precipitate is well suspended at 0° in 0.1 vol. of 0.5 saturated[6] ammonium sulfate containing 0.005 M cysteine. The enzyme assay may be carried out in the presence of ammonium sulfate in order to follow the progress of the purification. The enzyme cannot be stored at this point.

Step 3. First Ammonium Sulfate Fractionation. The above suspension is centrifuged in the cold to remove inactive protein; the supernatant should contain about 7 mg. of protein per milliliter. Solid ammonium sulfate is next added to the cold solution with stirring to obtain three fractions at 0.55, 0.60, and 0.65 ammonium sulfate saturation. The three precipitates are dissolved in small volumes of phosphate-cysteine and their specific activities determined. The best two fractions are combined. The

[5] F. Lipmann, M. E. Jones, S. Black, and R. Flynn, *J. Am. Chem. Soc.* **74**, 2384 (1952).
[6] Made up by adding 35.0 g. of ammonium sulfate to 100 ml. of water.

frozen solution may be stored indefinitely, and loss may be held to a minimum by the periodic addition of neutral cysteine.

Step 4. Second Acetone Fractionation. The solution from step 3 is diluted to contain 7 mg. of protein per milliliter and dialyzed for 2 hours against 10 vol. of phosphate-cysteine at 0°. Any insoluble material is removed, and the solution is fractionated as before with acetone at −10°. Three fractions are collected between 45 and 50, 50 and 55, and 55 and 60% acetone concentration. The major part of the activity usually appears in the 50 to 55% fraction. The precipitates, taken up in a little phosphate-cysteine, cannot be stored.

Steps 5 and 6. Second and Third Ammonium Sulfate Steps. The most active fraction of the previous step is diluted to 5.0 mg. of protein per milliliter and adjusted to pH 6.7. The protein precipitating between 0.0 and 0.45 ammonium sulfate saturation is discarded. Three further fractions are obtained between 0.45 and 0.50, 0.50 and 0.55, and 0.55 and 0.60% ammonium sulfate saturation, and the most active combined as before.

Another 1.5-fold purification is achieved by repeating this step in successive 0.02 saturation fractions between the limits of 0.45 and 0.55 ammonium sulfate saturation. The highest specific activity achieved was about 300. The preparation is free of adenylic kinase (myokinase) and transacetylase. It is stored in phosphate-cysteine at −20°.

Properties

Specificity. Of a number of acids tested, acetate and propionate are the only ones which serve as substrate, acetate being ten times as active as propionate. No measurable reaction is obtained with fluoracetate. ADP cannot replace ATP as the primary phosphorylating agent. The following Michaelis constants were found: acetate, 0.3 M; propionate, 0.47 M; Mg^{++}, 0.005 M; ATP, 0.002 M; ADP, 0.0015 M; acetyl phosphate, 0.005 M. The V_{max} in the back direction is five times that in the forward direction. (The rate in the back direction is determined by the disappearance of acetyl phosphate, the reaction being stopped with p-chloromercuribenzoate before the addition of hydroxylamine.) The apparent equilibrium constant at pH 7.0 is 0.006. Thus the equilibrium favors the formation of ATP from acetyl phosphate + ADP.

Activators and Inhibitors. Mg^{++} or Mn^{++} is required for the reaction. High concentrations of Versene may be used to stop the reaction.

The following sulfhydryl reagents produce 50% inhibition of the reaction: Hg^{++}, 10^{-7} M; p-chloromercuribenzoate, 3×10^{-7} M; iodosobenzoate, 3×10^{-4} M; and phenylarsine oxide, 10^{-3} M. The inhibitions caused by the first three compounds are largely reversed by cysteine. Iodoacetate and iodoacetamide do not inhibit at 0.01 M. Fluoroacetate is not an inhibitor.

Effect of pH. The enzyme activity is optimum at pH 7.4 in either direction, the rate falling moderately on both sides. Its stability is limited to the range 6.5 to 7.8.

Distribution. Assaying extracts from alumina ground cells, the following organisms are found to give solutions with the noted specific activities of acetokinase: *Proteus vulgaris*, 10.8; *Clostridium* strain H.F., 5.9; *E. coli*, 3.0; *Streptococcus faecalis*, 2.7; *Clostridium kluyverii*, 0.04; *Streptococcus haemolyticus*, 0.4; and *Azotobacter vinelandii*, 0.05.

Determination of Acetate with Acetokinase. Acetokinase may be used readily for the quantitative determination of micromole amounts of acetate. To a tube containing ATP (0.01 M), MgCl₂ (0.01 M), Tris (0.05 M), neutral hydroxylamine (0.75 M) and 6 to 12 units of acetokinase from *E. coli*, add an amount of the neutralized solution to be tested which would contain 0.5 to 2.5 micromoles of acetate and adjust the volume to 1.0 ml. Incubate at 29 to 38° for 1 hour, and terminate the reaction with 1.0 ml. of 10% TCA. Color is developed with 4.0 ml. of FeCl₃ reagent and read in the colorimeter as above. This method has been applied with better than 95% recoveries to crude biological mixtures. Acetokinase prepared through *step 3* of the purification procedure is satisfactory for use in the determination.

SUMMARY OF PURIFICATION PROCEDURE

Fraction	Total volume, ml.	Total units	Total protein, mg.	Specific activity	Recovery, %
1. Extract[a]	410	12,200	4060	3.1	100
2. Acetone, 45–55%	56	6,750	450	15	55
3. AmSO₄, 0.5–0.6	21	6,100	140	42	49
4. Acetone, 50–55%	6	2,900	30	98	24
5. AmSO₄, 0.50–0.6	3	2,400	13	190	20
6. AmSO₄, 0.49–0.53	2	1,800	6	300	15

[a] Extract from 36 g. of dry cells.

[98] Phosphotransacetylase from *Clostridium kluyveri*

$$Ac{\sim}P + CoA \rightleftarrows Ac{\sim}SCoA + Pi \qquad (1)$$

By E. R. STADTMAN

Assay Method

Principle. The method is based on the fact that, when the above reaction is coupled with the acetylation of arsenate (reaction 2), an unstable acetyl~arsenate derivative is formed which undergoes rapid spontaneous hydrolysis (reaction 3).[1]

$$Ac{\sim}SCoA + arsenate \rightarrow Ac{\sim}arsenate + CoA \qquad (2)$$
$$Ac{\sim}arsenate + H_2O \rightarrow Acetate + arsenate \qquad (3)$$

The over-all reaction is a hydrolysis of Ac~P (reaction 4).

$$Ac{\sim}P + H_2O \rightarrow Acetate + Pi \qquad (4)$$

Under the given experimental conditions, the rate of reaction 4, which is measured by the disappearance of Ac~P, is proportional to the enzyme concentration.[2]

Reagents

0.1 M Tris buffer, pH 8.0.
0.06 M dilithium acetyl phosphate.
1.6 $\times$ 10^{-4} M CoA, 50 units/ml.[3]
0.1 M cysteine·HCl.
0.5 M potassium arsenate, pH 8.0.
Enzyme. Dilute the stock solution with 0.2 M Tris buffer to obtain 50 units of enzyme per milliliter. (See definition below.)

Procedure. 0.4 ml. of water and 0.1-ml. aliquots each of the Tris buffer, Ac~P, CoA, and cysteine·HCl reagents and diluted enzyme solution are mixed in the order indicated. After incubation at 28° for 5 minutes, 0.1 ml. of the potassium arsenate reagent is added, and the incubation is continued for 15 minutes; then the residual Ac~P is determined by the

[1] E. R. Stadtman and H. A. Barker, *J. Biol. Chem.* **184**, 769 (1950); E. R. Stadtman, G. D. Novelli, and F. Lipmann, *J. Biol. Chem.* **191**, 365 (1951).
[2] E. R. Stadtman, *J. Biol. Chem.* **196**, 527 (1952).
[3] The method of standardization and assay of CoA was previously described by E. R. Stadtman and A. Kornberg, *J. Biol. Chem.* **203**, 47 (1953); see Vol. III [131].

hydroxamic acid procedure.[4] For reference, a sample without enzyme is also included.

Definition of Unit and Specific Activity. One unit of phosphotransacetylase is defined as the amount of enzyme required to catalyze the arseonolysis of 1 micromole of Ac∼P in 15 minutes under the above conditions. The specific activity refers to the units per milligram of protein. Protein is measured turbidimetrically after precipitation with sulfosalicylic acid.[5]

Purification Procedure

Methods for the growth of *Cl. kluyveri* and preparation of dried cells and cell-free extracts were previously described.[6] The purification procedure is a slight modification of the method presented earlier.[2]

Step 1. Ethanol Fractionation. About 100 ml. of cell-free extract of *Cl. kluyveri*, adjusted to 0.01 M potassium phosphate (pH 6.6) and 0.2 M potassium chloride, is cooled to 0° and, with stirring, cold absolute ethanol ($-10°$) is added until a concentration of 50 vol.% is reached (8 to 10 minutes).

The precipitate is removed by centrifuging for 5 minutes at 12,000 r.p.m. and is discarded. The supernatant solution is adjusted to an ethanol concentration of 58 vol.%, the precipitate obtained by centrifuging as above is suspended in 15 to 20 ml. of 0.05 M phosphate buffer (pH 7.2). The insoluble material is removed by centrifuging. All operations are carried out in a cold room at about 1°. The temperature after the first alcohol addition is $10 \pm 2°$ and remains at this level throughout the subsequent operations. The over-all time is 20 to 30 minutes. Since careful control of time and temperature is of utmost importance, it has been found impracticable to attempt fractionation of large volumes of extract in one operation. Therefore, 100 to 125 ml. of extract is taken, and the fractions from three such lots are pooled before proceeding to step 2.

Step 2. Acid Ammonium Sulfate Fractionation. Three lots of the protein fractions precipitating between 50 and 58% ethanol are pooled, and the pH is adjusted to 6.7 with 0.1 M KH_2PO_4. With stirring, a solution of ammonium sulfate saturated at 0° is added to the enzyme solution to give 0.2 ammonium sulfate saturation. The solution is cooled to $-2°$, and the concentration of ammonium sulfate is increased to 0.49 saturation by the careful addition of a saturated solution of acid ammonium sulfate (prepared by mixing 3 ml. of concentrated sulfuric acid with 1 l. of ammonium

[4] F. Lipmann and L. C. Tuttle, *J. Biol. Chem.* **158**, 505 (1945); see Vol. III [39].

[5] Klett-Summerson photoelectric colorimeter clinical manual, Klett Manufacturing Company, New York.

[6] E. R. Stadtman and H. A. Barker, *J. Biol. Chem.* **180**, 1085 (1949); see Vol. I [84].

sulfate solution, saturated at $-2°$). The solution is stirred for 5 minutes and centrifuged for 5 minutes at 12,000 r.p.m. in a Servall centrifuge in a cold room at $10°$. The precipitate is discarded. Saturated acid ammonium sulfate solution is added to the supernatant to give 0.7 ammonium sulfate saturation. After stirring for 5 minutes the faint precipitate is removed by centrifuging and is dissolved in 7 to 10 ml. of Tris buffer (pH 8.0). If stored at $-10°$, the enzyme solution is stable for several months. See the accompanying table for a summary of the purification procedure.

SUMMARY OF PURIFICATION PROCEDURE

Fraction	Total volume, ml.	Protein, mg./ml.	Enzyme activity[a]		Recovery, %
			Units/mg.	Total units	
Cell-free extract	339	25	34	289,000	—
Step 1, 50–58% alcohol fraction	55	8.7	282	134,000	46
Step 2, 0.49–0.70 ammonium sulfate fraction	7.5	2.0	2940	44,000	15[b]

[a] The specific activity of the crude cell-free extracts is much higher than that reported in ref. 2. This is due to a slight modification of the assay method and to the more recent availability of an absolute CoA standard which revealed that the CoA used earlier was only about two-thirds as active as it was believed.

[b] The yield from step 2 has in some instances been as high as 25%.

Properties

Specificity. The enzyme appears to be specific for CoA. It is without effect on various degradation products of CoA such as desamino-CoA[7] and various dephosphorylated derivatives.[8] Propionyl~CoA is attacked at a rate 0.1 to 0.5 as rapid as is Ac~SCoA, and butyryl~SCoA is attacked at a rate 0.01 or less as rapidly as Ac~SCoA. Acyl~SCoA derivatives of formate, lactate, and higher fatty acids are not attacked at significant rates.

Activators and Inhibitors. The enzyme is inactive in the absence of potassium or ammonium ions. Maximal activation by these cations occurs at about 0.02 M concentration. Sodium and, to a lesser extent, lithium ions inhibit the enzyme, but this inhibition can be overcome by the addition of higher concentrations of potassium or ammonium salt.

The enzyme is equally active in Tris, glycylglycine, histidine, and

[7] N. O. Kaplan, personal communication.
[8] J. D. Gregory, G. D. Novelli, and F. Lipmann, *J. Am. Chem. Soc.* **74**, 854 (1952); T. P. Wang, L. Shuster, and N. O. Kaplan, *J. Am. Chem. Soc.* **74**, 3204 (1952).

triethanolamine buffers (0.1 M, pH 8.0). At the same concentrations and pH, potassium diethylbarbiturate and citrate buffers cause 50% inhibition, and inhibition by potassium pyrophosphate is about 90%. Relatively high concentrations (0.01 M) of ATP and ADP inhibit the enzyme partially.

The enzyme activity is lost upon dialysis and cannot be restored by the addition of known cofactors or crude boiled extracts.

The rate of the arsenolysis reaction is directly proportional to the concentration of CoA over a very wide range (0 to 200 units/ml.). The concentration of CoA needed to saturate the enzyme has not been determined.

The rate of arsenolysis is a linear function of the arsenate concentration over the range of 0 to 0.1 M; further, nonlinear increases in rate are observed up to arsenate concentrations of 0.2 M.

Effect of pH. The enzyme is active over the pH range of 6.6 to 9.0 with a broad optimum range of 7.4 to 8.4.

Spectrophotometric Measurement of Enzyme. Reaction 1 may be followed directly by measuring the changes in optical density at 232 to 240 mμ that are associated with the formation or cleavage of the thiol ester bond of Ac$\sim$SCoA.[9] Since this method requires considerably more CoA (0.1 micromole), it is not recommended as a routine assay procedure.

Equilibrium Constant. The equilibrium constant[10] for reaction 1 is:

$$K = \frac{(\text{Ac}\sim\text{SCoA})(\text{HPO}_4{}^{--})}{(\text{Ac}\sim\text{P})(\text{CoA})} = 74$$

[9] E. R. Stadtman, *J. Biol. Chem.* **203**, 501 (1953).

[10] An equilibrium constant of 60 ± 20 was reported earlier [E. R. Stadtman, *J. Biol. Chem.* **196**, 535 (1952)], but more recent studies have shown it to be about 74 (E. R. Stadtman, unpublished results).

[99] Coenzyme A Transphorase from *Clostridium kluyveri*

Butyryl$\sim$SCoA + Ac $\rightleftarrows$ Butyrate + Ac$\sim$SCoA

By H. A. BARKER, E. R. STADTMAN, and ARTHUR KORNBERG

Assay Method

Principle. The method is based on the decrease in optical density in the 232- to 240-mμ region when the above reaction is coupled with the arsenolysis of Ac$\sim$SCoA under the influence of phosphotransacetylase.[1] The net reaction is an apparent hydrolysis of butyryl$\sim$SCoA.

[1] E. R. Stadtman, *J. Biol. Chem.* **203**, 501 (1953).

$$\text{Butyryl}{\sim}\text{SCoA} + \text{H}_2\text{O} \rightarrow \text{Butyrate} + \text{CoASH}$$

At 232 mμ, the molar extinction coefficient of CoASH is about 55% of that of acyl$\sim$SCoA. The difference in the molar extinction coefficients of butyryl$\sim$SCoA and its hydrolysis products (ΔE_{232}) is 4.5×10^3 cm.^{2}/mole.

Reagents

Butyryl$\sim$SCoA (0.001 M). When the solution is adjusted to pH 6 with KOH, the compound is stable for months at $-15°$.

0.15 M KAc–0.33 M KAsO$_4$, pH 7.0.

Phosphotransacetylase.[2] Dilute the stock solution to 40 units/ml. with 0.05 M Tris-HCl buffer, pH 8.0. The enzyme solution is stable for months at $-15°$.

Enzyme. Dilute the stock solution to a concentration of 0.05 to 0.15 unit/ml. (See definition below.)

Procedure. Add 0.2 ml. of KAc–KAsO$_4$ solution and 0.1 ml. each of butyryl$\sim$SCoA and phosphotransacetylase solutions to 1.0 ml. of water in a quartz cell having a 1-cm. light path and a 1.5-ml. capacity. Then add 0.1 ml. of enzyme solution, and take readings at 232 mμ at 1-minute intervals. The reference cell contains all reagents except butyryl$\sim$SCoA and phosphotransacetylase in order to compensate for the high ultraviolet absorption by the enzyme. Since the rate of the reaction declines with time, the enzyme activity is calculated from the optical density change between the first and fourth minutes.

Definition of Unit and Specific Activity. One unit of enzyme is defined as that amount which causes the decomposition of 1 micromole of butyryl$\sim$SCoA per 3 minutes. This corresponds to a rate of optical density change (ΔOD_{232}) of 1.0 per 3 minutes at 25° under the above conditions. Specific activity is expressed as units per milligram of protein, determined by the method of Lowry *et al.*[3]

Purification Procedure

Only small purification of CoA transphorase has been achieved by the methods applied until now.[4] However, the best preparations have a reduced nucleic acid content ($OD_{280m\mu}/OD_{260m\mu} = 1.23$) compared to the

---

[2] E. R. Stadtman, *J. Biol. Chem.* **196**, 527 (1952). The fraction obtained by alcohol precipitation (50 to 58%) is satisfactory; see Vol. I [98].

[3] O. H. Lowry, N. J. Rosebrough, A. L. Farr, and R. J. Randall, *J. Biol. Chem.* **193**, 265 (1951); see Vol. III [73].

[4] A. Kornberg, H. A. Barker, and E. R. Stadtman, unpublished results.

starting material (0.62) and are relatively free of light-absorbing components, in both the visible and ultraviolet regions.

Step 1. Preparation of Crude Extract. 10 g. of dried and ground cells of *Cl. kluyveri*[5] is homogenized with 100 ml. of 0.002 M KPO$_4$, pH 7.4, and incubated with occasional stirring at 30° for 4 hours. The solution is centrifuged at 10,000 r.p.m. for 10 minutes, the supernatant liquid decanted, and the precipitate washed twice with 40 ml. of buffer. The combined supernatant liquid (volume, 140 ml.) retains its activity for months when stored.

Step 2. Protamine Treatment. To 1 vol. of extract from step 1 is added 0.2 vol. of a 1% solution of protamine sulfate (salmine). After 5 minutes at 0°, the solution is centrifuged in the cold, and the precipitate is discarded. To the supernatant solution add 0.2 vol. of 1 M KAsO$_4$, pH 7.0, and 0.05 vol. of 1 M KAc, and dilute to twice the original volume.

Step 3. Fractionation with Ammonium Sulfate. To the cold enzyme solution of step 2 is added sufficient powdered ammonium sulfate to give 0.45 saturation. After stirring for 5 minutes at 0°, the precipitate is removed by centrifuging at 12,000 r.p.m. The supernatant liquid is adjusted to 0.65 saturation by addition of solid ammonium sulfate, and, after centrifugation, the 0.45 to 0.65 fraction is dissolved in one-fifth the original volume of 0.05 M KAsO$_4$, pH 7.0. This preparation loses approximately 25% of its activity in two weeks at −15°.

Step 4. Fractionation with Ethanol. KCl, KAc, and acetic acid are added to the solution derived from step 3 in such amounts as to give final concentrations of 0.2 M, 0.02 M, and pH 6.3, respectively, when the protein concentration is adjusted to 3.0 mg./ml. Absolute ethanol cooled to −10° is slowly added to the solution, which is stirred continuously, until the alcohol concentration is 31% (v/v). After 5 minutes the precip-

SUMMARY OF PURIFICATION PROCEDURE[a]

Fraction	Total volume, ml.	Units/ml.	Total units	Protein, mg./ml.	Specific activity, units/mg.	Recovery of activity, %
1. Crude extract	80	1.00	80	13.9	0.072	(100)
2. Protamine supernate	90	0.65	58	6.4	0.101	73
3. (NH$_4$)$_2$SO$_4$ fraction, 0.45–0.65 satn.	20	2.08	42	11.0	0.190	52
4. Ethanol fraction, 31–49%	20	1.26	25	4.0	0.310	31

[a] H. A. Barker, A. Kornberg, and E. R. Stadtman, unpublished experiments.

[5] E. R. Stadtman and H. A. Barker, *J. Biol. Chem.* **180**, 1085 (1949); see Vol. I [84].

itate is removed by centrifugation, and more alcohol is added to the supernatant solution to give a final concentration of 49%. The solution is centrifuged, and the precipitate is immediately dissolved in the original volume of 0.05 M $KAsO_4$, pH 7. All operations are carried out at 0 to $-6°$. The product of step 4 retains 90% of its initial activity during storage for two weeks at $-10°$. For some purposes orthophosphate or triethanolamine is preferable to arsenate as a buffer for the final enzyme preparation. See the accompanying table for a summary of the purification procedure.

Properties

Specificity. Crude extracts of *Cl. kluyveri* catalyze the transfer of the CoA moiety of acetyl CoA to formate, acetate, propionate, *n*-butyrate, *n*-valerate, *n*-caproate, *n*-caprylate, vinyl acetate, lactate, and possibly glycolate.[1,6] The reactions are generally reversible. The specificity of the purified enzyme has not been investigated.

Effect of pH. The enzyme shows a pH optimum for activity of pH 6.8 to 7.0. The pH range for activity is from approximately 5.6 to above 8.0.

Effect of Substrate Concentration. Incomplete data indicate that the Michaelis constants for butyryl CoA and acetate are approximately 5 $\times$ 10^{-5} M and 7 $\times$ 10^{-3} M, respectively.

Heat Stability. When crude extracts are heated for 5 minutes at 55, 60, and 65°, the enzyme loses approximately 40, 52, and 78%, respectively, of its activity. Since phosphotransacetylase is much more thermolabile, preparations heated for 15 minutes or longer at 60° have a low content of this contaminating enzyme.[1]

[6] I. Lieberman and H. A. Barker, unpublished results.

[100] Deacylases (Thiol Esterase)

Succinyl CoA Deacylase

$$\text{Succinyl} \sim \text{SCoA} + \text{H}_2\text{O} \rightarrow \text{Succinate} + \text{HS-CoA}$$

By JOHN GERGELY[1]

Assay Method

Principle. The method described below[2] is based on the following facts. The enzymatic reduction of DPN by α-ketoglutarate (KG),[3] ac-

[1] Established Investigator of the American Heart Association.
[2] J. Gergely, P. Hele, and C. V. Ramakrishnan, *J. Biol. Chem.* **198**, 323 (1952).
[3] KG will be used throughout as an abbreviation of α-ketoglutarate or the free acid.

companied by an increase in optical density at 340 mμ, in the presence of catalytic amounts of CoA occurs only if succinyl CoA deacylase is added. The hydrolysis by this enzyme of the intermediate succinyl CoA can be made the rate-determining step in the reduction of DPN.

$$KG + DPN^+ + HSCoA \xrightarrow[\text{dehydrogenase}]{} Succinyl{\sim}SCoA$$
$$+ CO_2 + DPNH + H^+$$
$$Succinyl{\sim}SCoA + H_2O \xrightarrow[\text{deacylase}]{} Succinate + HSCoA$$

Reagents

> KG solution,[4] 0.2 M. Free acid, twice recrystallized from benzene and acetone, is used. The pH is brought to 7.0 with 2 M KOH. The solution should be kept frozen.
> 1 M glycine, brought to pH 9.0 with 6 M KOH.
> 0.1 M cysteine HCl. The acid solution should be kept frozen. The required amount should be freshly neutralized with 2 M KOH for daily use.
> 2% crystalline plasma albumin solution.
> DPN solution,[4] 0.06 M,[5] pH 7.0, neutralized with 2 M KOH.
> CoA solution,[5a] 10 units/ml., pH 7.0, neutralized with 2 M KOH.
> KG-dehydrogenase, prepared according to Sanadi and Littlefield.[6] The solution should contain 1000 to 2500 units/ml.
> Succinyl CoA deacylase. The solution to be used should contain 10 to 40 units/ml.

Procedure. Mix the following amounts of the reagents in a test tube: KG, 0.1 ml., glycinate, 0.1 ml.; cysteine, 0.3 ml.; plasma albumin, 0.2 ml.; CoA, 0.2 ml.; KG-dehydrogenase, 24 units; deacylase, 0.1 ml. Add distilled water to a final volume of 2.9 ml. and mix. Incubate at 30° for 5 minutes. Transfer the mixture in a quartz cell having a 1-cm. light path. Take readings at 340 mμ before, and at 30-second intervals after, mixing with 0.1 ml. of the DPN solution. The change in optical density may be preceded by a lag of 2 to 3 minutes.

Definition of Unit and Specific Activity. One unit of enzyme is defined as the amount that causes an initial rate of change in optical density (ΔE_{340}) of 0.010 per minute under the above conditions. Specific activity is expressed as units per milligram of protein. Protein is determined by

---

[4] For the preparation and assay of this compound see Vol. III [124, 125, 128].
[5] The actual amount to be dissolved depends on the purity of the preparation used.
[5a] For the preparation and assay of this compound see Vol. III [131, 132].
[6] D. R. Sanadi, J. W. Littlefield, and R. M. Bock, *J. Biol. Chem.* **197**, 854 (1953); see also Vol. I [120].

the method of Warburg and Christian,[7] from optical densities at 280 and 260 mμ. Should much nucleic acid or denatured protein be present, dry weights should be determined.

Application of Assay Method to Crude Tissue Preparation. The presence of certain enzymes and substrates capable of reacting with DPNH would interfere with this method. Interfering systems can be detected by adding DPNH to the assay mixture: a decrease of E_{340} will indicate their presence.[8] Succinyl CoA deacylase can be demonstrated in a crude system either by chemically following the splitting of added succinyl CoA (according to Lipman and Tuttle)[9] or in a dismutation system by means of manometric measurements.[10,11] These methods have not been adapted for a quantitative assay.

Purification Procedure

The following procedure is taken from the work of Gergely *et al.*[2] Steps 1 and 2 are based on the observations of Green and Beinert.[10] Another procedure has been developed by Kaufman *et al.*[12] All steps should be carried out at 0 to 4°.

Step 1. Preparation of Crude Extract and pH Fractionation. Diced pig heart muscle, trimmed of fat and connective tissue, is homogenized with 3 vol. of 0.03 M potassium phosphate buffer, pH 7.2, for 2 minutes in the Waring blendor. The homogenate is centrifuged at 2000 $\times$ g for 30 minutes. The resulting cloudy supernatant is taken to pH 5.4 with 10% acetic acid, and centrifuged again. The supernatant is called fraction 1.[13]

Step 2. First Ammonium Sulfate Fractionation. 29 g. of ammonium sulfate is added, in small portions under constant stirring, to each 100 ml. of fraction 1. The precipitate is centrifuged for 1 hour at 2000 $\times$ g, and the precipitate is dissolved in 0.02 M potassium phosphate buffer, pH 7.6, 50 ml. being taken for each 1000 ml. of fraction 1. The solution is dialyzed for 6 hours against the buffer. The dialysis is followed by centrifugation at 40,000 $\times$ g (Spinco Model L preparative ultracentrifuge)

---

[7] O. Warburg and W. Christian, *Biochem. Z.* **310**, 384 (1942); see also Vol. III [73].

[8] Glutamic dehydrogenase is usually present in crude preparations. Care should, therefore, be taken to remove NH_4^+ ions to eliminate the following reaction:

$$DPNH + KG + NH_4^+ \rightarrow DPN^+ + glutamate + H_2O$$

[9] F. Lipmann and L. C. Tuttle, *J. Biol. Chem.* **159**, 21 (1945).

[10] D. E. Green and H. Beinert, *in* "Phosphorus Metabolism" (McElroy and Glass, eds.), Vol. 1, p. 330, The Johns Hopkins Press, Baltimore, 1951.

[11] S. Kaufman, *in* "Phosphorus Metabolism" (McElroy and Glass, eds.), Vol. 1, p. 370, The Johns Hopkins Press, Baltimore, 1951.

[12] S. Kaufman, C. Gilvarg, O. Cori, and S. Ochoa, *J. Biol. Chem.* **203**, 869 (1953).

[13] The precipitate contains the KG-dehydrogenase.

for 1 hour. The precipitate is discarded. The removal of insoluble material by high-speed centrifugation, in this and subsequent steps, is important because of its tendency to absorb the enzyme and inactivate it. This preparation (fraction 2), if kept frozen, retains its full activity for several months.

Step 3. Second Ammonium Sulfate Fractionation. To each 100 ml. of fraction 2, 10.6 g. of ammonium sulfate is added, and the solution is centrifuged for 20 minutes at $18,000 \times g$. The small precipitate is discarded, and 11.3 g. of ammonium sulfate is added to the supernatant. This material is centrifuged as before, and the precipitate is dissolved in buffer and dialyzed for 6 hours against it. After dialysis, insoluble material is removed by centrifuging for 1 hour at $40,000 \times g$ (fraction 3). This fraction loses some activity even by overnight frozen storage, so that step 4 should immediately follow.

Step 4. Adsorption of Deacylase by Calcium Triphosphate Gel.[14] Fraction 3 is diluted to a protein concentration of 6 to 10 mg./ml., and 75 ml. of the gel (15 mg. weight per milliliter) is added to each 100 ml. of the diluted fraction 3. The suspension is allowed to stand for 20 minutes, with occasional stirring, then centrifuged for 15 minutes at $18,000 \times g$. The enzyme is eluted with 0.1 M potassium phosphate buffer, pH 7.6, 100 ml. being taken for each 100 ml. of diluted fraction 3. The gel is removed by centrifugation after standing for 20 minutes. The eluate can be lyophilized and stored *in vacuo* at $-10°$, or it may be kept frozen in the presence of crystalline plasma albumin, 20 mg./ml. (fraction 4). Either way, the preparation retains its full activity for 3 to 4 weeks.

Properties

Specificity. Fractions 1, 2, and 3 catalyze the breakdown of both succinyl CoA and Ac~CoA, but the eluate from the Ca-triphosphate gel, if the absorption was carried out at a low enough protein concentration, has no Ac~CoA deacylase activity. Data on the activity of the enzyme toward other acyl CoA derivatives are not available.

Activators and Inhibitors. The activity of the enzyme, as studied in the system described above, is greater in glycine than in phosphate buffer. The optimal pH is 8.2. 0.00066 M pyrophosphate causes 50% inhibition with 10 units of deacylase; 0.00033 M ATP inhibits by 72%. The rate of DPN reduction depends markedly on the CoA concentration but is independent of DPN concentration over the range 3×10^{-5} to 4×10^{-4} M. Cysteine is required to keep CoA in the reduced form.

Physical Properties. The purified preparation (fraction 4) contains two electrophoretically distinct components. 70% of the total material mi-

[14] For the preparation of calcium triphosphate gel, see Vol. I [11].

grates under one peak, and this peak contains all the succinyl CoA deacylase activity. The mobility of this component, at pH 7.6 in 0.05 M phosphate buffer, is 4×10^{-4} cm.^{2} volt^{-1}; that of the smaller component is 6×10^{-6} cm.^{2} volt^{-1}. Ultracentrifugal examination (0.01 M potassium phosphate buffer, pH 7.6, 0.15 M NaCl) shows one well-defined peak containing 55% of the material, the remaining 45% being distributed evenly on either side of the main peak. The material in the main peak has a sedimentation constant of about 4.5 Svedberg units.

Acetyl CoA Deacylase

$$Ac{\sim}SCoA + H_2O \rightarrow HAc + HSCoA$$

Assay Method

No quantitative method has been described, but a semiquantitative test is described below, based on the following reactions:

$$Ac{\sim}P + HSCoA \xrightarrow{\text{phosphotransacetylase}} Ac{\sim}SCoA + P \qquad (1)$$

$$Ac{\sim}SCoA + H_2O \xrightarrow{\text{deacylase}} HAc + HSCoA \qquad (2)$$

The net result is a breakdown of Ac$\sim$P, and it is followed by the method of Lipmann and Tuttle.[9] If acetyl CoA is available, the direct breakdown by deacylase of this compound can be measured. In principle, an assay could be carried out by following the reduction of DPN in a similar system as that described for succinyl CoA deacylase but containing pyruvate and pyruvic dehydrogenase instead of KG and KG-dehydrogenase, respectively. Traces of lactic dehydrogenase would, however, interfere with the assay, since DPNH would react with pyruvate to yield DPN and lactate. A manometric assay, based on a dismutation reaction,[15] could also be used.

Reagents

0.1 M Li-acetyl phosphate.[16]

0.1 M cysteine hydrochloride solution, freshly neutralized for use.

CoA solution,[5a] 100 units/ml.

Phosphotransacetylase,[17] diluted to contain 300 units/ml.

Deacylase. The solution should contain per milliliter an amount of enzyme leading to the hydrolysis of 3 to 5 micromoles of Ac$\sim$P

---

[15] R. S. Schweet, M. Fuld, K. Cheslock, and M. H. Paul, *in* "Phosphorus Metabolism" (McElroy and Glass, eds.), Vol. 1, p. 246 The Johns Hopkins Press, Baltimore, 1951.

[16] See Vol. III [39].

[17] E. R. Stadtman, *J. Biol. Chem.* **196**, 527 (1952); see also Vol. I [98].

in 30 minutes under the conditions of the test. 2 to 3 mg. of protein is required for this in a purified preparation. All solutions should be neutralized with KOH, since sodium ions inhibit the phosphotransacetylase.

Procedure. The following amounts of the reagents are mixed: 0.1 ml. of acetylphosphate, 0.1 ml. of cysteine, 0.1 ml. of CoA solution, and 1 ml. of deacylase. Final volume, 1.9 ml.; pH adjusted to 8.5 with 2 M KOH. The reaction is started by the addition of 0.1 ml. of phosphotransacetylase. Samples are withdrawn at 10-minute intervals and Ac∼P is determined by the method of Lipmann and Tuttle.[9] Blanks without CoA and phosphotransacetylase should be run.

Preparation of the Enzyme[2]

If the gel adsorption step (step 4) in the preparation of succinyl CoA deacylase is carried out, as directed, at a protein concentration of 6 to 10 mg./ml., almost all the succinyl CoA deacylase activity is removed, and Ac∼SCoA deacylase is left in solution. Some preparations contain a phosphatase-like enzyme, which decomposes Ac∼P in the absence of CoA and phosphotransacetylase. At pH 8.5, however, this contaminant has a very low activity compared with that of Ac∼SCoA deacylase. Preparations almost free of this interfering enzyme, but in poorer yield, can be obtained in the following way: Fraction 1 of the succinyl CoA deacylase procedure is warmed to 25°, and 250 g. of anhydrous sodium sulfate is added per liter. The precipitate is filtered under gravity and dissolved in 0.02 M potassium phosphate buffer, pH 7.6, and after dialysis against it the solution is treated with calcium triphosphate gel as previously described.

PURIFICATION OF SUCCINYL CoA DEACYLASE[a]

Fraction	Specific activity	Total units	Recovery of units, %
1. Supernatant after pH 5.4 precipitation; 4800 g. of heart	2.1	242,000	—
2. First ammonium sulfate fraction (0–50% sat.)	21.8	257,000	106[b]
3. Second ammonium sulfate fraction (20–40% sat.)	51.0	197,000	81
4. Eluate from Ca triphosphate gel	153.0	186,000	77

[a] J. Gergely, P. Hele, and C. V. Ramakrishnan, *J. Biol. Chem.* **198,** 323 (1952).

[b] The slight apparent increase in the total number of units may be due to the removal of interfering substances in this step.

[101] Acetylation of Amines with Pigeon Liver Enzyme[1]

$$RNH_2 + Ac{\sim}SCoA \rightarrow RNHCOCH_3 + CoA$$

By HERBERT TABOR

Assay Method

Principle. Acetyl coenzyme A (Ac$\sim$SCoA) acetylates various amines in the presence of an enzyme prepared from pigeon liver. Although the Ac$\sim$SCoA can be added directly, in the method described below it is generated in the incubation mixture from acetyl phosphate and CoA in the presence of phosphotransacetylase.

The method originally used by Kaplan and Lipmann[2] depended on the acetylation of sulfanilamide. Free sulfanilamide was measured in aliquots of the incubation mixture by the Bratton-Marshall diazotization procedure. The modification described below is based on the spectral shift obtained when p-nitroaniline is acetylated to p-nitroacetanilide. At 420 mμ the molar extinction coefficient of p-nitroaniline is 6000, whereas that of p-nitroacetanilide is negligible at this wavelength. This permits direct observation of the course of the acetylation reaction in the spectrophotometer. Because of the sulfhydryl nature of the acetylating enzyme, a reducing agent (sodium thioglycolate) and a chelating agent (disodium ethylenediaminetetraacetate) are also added to the assay.

Reagents

p-Nitroaniline solution (0.001 *M*). Dissolve 138 mg. of commercial p-nitroaniline in 10 ml. of ethanol. Add water with stirring to a final volume of 1 l.

0.1 *M* potassium phosphate buffer, pH 6.8.

Enzyme. Dilute the enzyme with water to a concentration of approximately 100 to 600 units of enzyme per milliliter. (See definition below.)

0.25 *M* dilithium acetyl phosphate. Dissolve 330 mg. in 10 ml. of water. (See Vol. III [39] for synthesis of dilithium acetylphosphate.)

Transacetylase. Dilute with water to obtain 80 units/ml. (See Vol. I [98] for preparation of this enzyme.)

---

[1] The method described here has been reported by H. Tabor, A. H. Mehler, and E. R. Stadtman, *J. Biol. Chem.* **204**, 127 (1953).

[2] N. O. Kaplan and F. Lipmann, *J. Biol. Chem.* **174**, 37 (1948).

Coenzyme A solution containing 2.4 micromoles/ml. Commercial samples are available and are satisfactory. For methods of isolation and purification of CoA, see Vol. III [**131**].

0.1 M sodium thioglycolate. Dissolve 114 mg. of sodium thioglycolate in 10 ml. of water. Other reducing agents, such as cysteine, glutathione, and H_2S, are also satisfactory.

0.1 M disodium ethylenediaminetetraacetate (EDTA). 337 mg. are dissolved in 10 ml. of water.

Procedure.[3] Both the blank and the experimental cells (1-cm. light path) contain 0.1 ml. of the phosphate buffer, 0.05 ml. of sodium thioglycolate, 0.05 ml. of EDTA, 0.05 ml. of dilithium acetyl phosphate, 0.05 ml. of transacetylase, 0.05 ml. of CoA, pigeon liver enzyme (usually 0.05 ml.), and water (added to a total volume of 0.9 ml.). An additional 0.1 ml. of water is included in the blank cell. 0.1 ml. of the p-nitroaniline solution is added to the experimental cell; after thorough mixing, readings are taken at 420 mμ at 15-second intervals.

Definition of Unit and Specific Activity. One unit of enzyme is defined as that amount which causes an initial rate of decrease in optical density at 420 mμ of 0.001 per minute at 25° under the above conditions. Specific activity is expressed as units per milligram of protein. Protein concentrations can be determined from the optical density at 280 mμ, using the absorption at 260 mμ to correct for nucleic acid. (See Vol. III [**73**].)

Application of Assay Method to Crude Tissue Preparations. The assay method described is suitable for use with crude as well as with purified preparations.

Purification Procedure

The use of pigeon liver acetone powders for the preparation of active extracts for the acetylation of aromatic amines was first reported by Kaplan and Lipmann[2] and has been repeated by many laboratories. Chou and Lipmann[4] have used acetone fractionations of pigeon liver extracts in their studies on the acetylation enzyme.

Step 1. Preparation of Pigeon Liver Acetone Powder. Freshly removed pigeon livers are blended with 10 vol. of cold ($-10°$) acetone for 1 minute in a Waring blendor. After filtration on a Büchner funnel with suction, the semidry filter cake is blended with 10 vol. of cold acetone and again

---

[3] Crude extracts of *Cl. kluyveri* (0.05 ml.) can be substituted for the transacetylase preparation. The acetyl phosphate, CoA, and transacetylase additions can be replaced by 0.1 micromole of Ac$\sim$SCoA. (See Vol. III [137] for preparation of Ac$\sim$SCoA.)

[4] T. C. Chou and F. Lipmann, *J. Biol. Chem.* **196**, 89 (1952).

filtered. The filter cake is pulverized by hand, spread over a large surface, and allowed to dry at room temperature. It is then stored at 0°. Under these conditions the powder is stable for several weeks, but then it gradually deteriorates.

Step 2. Preparation of Crude Extract. 12 g. of the pigeon liver powder is ground in a mortar with 120 ml. of distilled water for 10 to 15 minutes at room temperature. The suspension is then centrifuged at 22,000 × *g* for 10 minutes.

Step 3. The supernatant (96 ml.) is cooled to 0° and treated with 76 ml. of cold (0°) acetone. The precipitate is collected by centrifugation and dissolved in 15 ml. of cold water.

Step 4. The enzyme solution is then treated with 90 ml. of aged alumina gel C_γ (11 mg./ml.) (see Vol. I [11] for preparation of this gel). After 5 minutes the mixture is centrifuged, and the precipitate collected. The precipitate is washed with 100 ml. of water and then eluted with 100 ml. of 0.01 *M* potassium phosphate buffer (pH 7.8) in three portions (eluate I). Further elution with 50 ml. of the buffer yields additional activity of slightly lower specific activity (eluate II).

Properties

Stability. The enzyme solutions can be stored at −15°; the activity is essentially unchanged for at least three months. More concentrated enzyme solutions can be obtained by lyophilization after the addition of 100 micromoles of EDTA per 100 ml.

Specificity. Numerous amines,[5] including *p*-nitroaniline, *m*-nitroaniline, *p*-aminobenzoic acid, *o*-phenylenediamine, *o*-toluidine, *m*-toluidine, *o*-bromoaniline, *p*-bromoaniline, *o*-anisidine, *p*-anisidine, histamine, phenethylamine, and glucosamine are acetylated.[6] No evidence for

---

[5] Pigeon liver extracts have been used in other laboratories for the acetylation of sulfanilamide and *p*-aminobenzoic acid,[2] 4-aminoazobenzene [R. E. Handschumacher, G. C. Mueller, and F. M. Strong, *J. Biol. Chem.* **189**, 335 (1951)], 4-aminoazobenzene-4'-sulfonic acid [S. P. Bessman and F. Lipmann, *Arch. Biochem. and Biophys.* **46**, 252 (1953)], histamine [R. C. Millican, S. M. Rosenthal, and H. Tabor, *J. Pharmacol. Exptl. Therap.* **97**, 4 (1949)], and glucosamine [T. C. Chou and M. Soodak, *J. Biol. Chem.* **196**, 105 (1952), and Vol. I [102].

[6] Acetylation of the aromatic amines can be studied with spectrophotometric techniques; the acetylation of aliphatic amines is followed by the disappearance of ninhydrin-reacting material, using the quantitative ninhydrin method of Moore and Stein (see Vol. III [76]). With histamine, the reaction may also be followed by measuring the formation of acetylhistamine (Vol. III [90]). Glucosamine acetylation can be studied with the method of Morgan and Elson (see Vol. I [102] and Vol. III [12]).

The incubation mixture used to study the acetylation of histamine contains 2

acetylation is found with *o*-aminobenzoic acid, *o*-nitroaniline, orthanilic acid, sulfanilic acid, or *p*-nitromethylaniline.

Acylation of *p*-nitroaniline also takes place when butyryl CoA is used as the acetylating agent instead of Ac~SCoA; however, the rate is only 4% of that observed with Ac~SCoA. No acylation is observed with palmityl CoA.

Effect of pH. With *p*-nitroaniline as the substrate, the activity is essentially constant from pH 6 to over 9.5. With histamine as the substrate, the reaction falls off sharply below pH 8.5.

Activators and Inhibitors. The acetylating enzyme is a sulfhydryl enzyme and is completely inhibited by 10^{-5} *M* *p*-chloromercuribenzoate. The enzyme is also spontaneously inactivated in the usual incubation mixtures (particularly at pH 8 and higher), presumably owing to the effect of heavy metal impurities on the sulfhydryl groups. This inactivation can be completely reversed by reducing agents, such as sodium thioglycolate, cysteine, or hydrogen sulfide. Since the enzyme shows no inactivation in the presence of EDTA, this is routinely added to all incubation mixtures.

The acetylating enzyme is inhibited by free CoA. 0.09 micromole of free CoA causes a 50% inhibition in an incubation mixture containing 0.05 micromole of Ac~SCoA. Marked inhibition is produced by palmityl CoA.

Stoichiometry. With limiting amounts of *p*-nitroaniline and excess Ac~SCoA (or with the phosphotransacetylase–acetyl-P system), the reaction proceeds until the amine is completely acetylated. With limiting amounts of Ac~SCoA and excess *p*-nitroaniline, the quantity of *p*-nitroaniline acetylated is exactly equivalent to the amount of Ac~SCoA present.[7] This affords a convenient assay for AC~SCoA (see Vol. III [**131**]).

Affinity of the Enzyme for CoA and for p-Nitroaniline. With varying amounts of CoA, approximately 50% of the maximal rate is observed with a CoA concentration of 0.015 micromole/ml.; CoA concentrations of

---

micromoles of histamine, 300 micromoles of potassium pyrophosphate buffer (pH 9.3), 10 micromoles of lithium acetyl phosphate, 4 units of transacetylase, 0.1 micromole of CoA, 10 micromoles of sodium thioglycolate, 5 micromoles of EDTA, and 160 units of acetylating enzyme in 1 ml. final volume. Approximately 1 micromole of histamine is acetylated in 30 minutes. Under similar conditions with phenethylamine as the substrate 0.2 micromole is acetylated.

[7] In experiments with stoichiometric quantities of Ac~SCoA, it is advisable to omit the sodium thioglycolate from the incubation mixture to avoid any loss due to the nonenzymatic acetylation of the thioglycolate [E. R. Stadtman, *J. Biol. Chem.* **196**, 535 (1952)].

over 0.05 micromole/ml. give maximal rates. With p-nitroaniline 50%
saturation of the enzyme is attained at approximately 0.1 micromole/ml.

SUMMARY OF PURIFICATION PROCEDURE

Fraction	Total volume, ml.	Units/ml.	Total units	Protein, mg./ml.	Specific activity, units/mg.	Recovery, %
1 and 2. Crude extract	96	406	39,000	38	10.6	—
3. Acetone fractionation	16	1620	26,000			67
4. C$_\gamma$						
Eluate I	100	150	15,000	0.55	275	38
Eluate II	50	70	3,500	0.38	184	9

[102] Acetylation of D-Glucosamine by Pigeon Liver Extracts

$$Ac{\sim}SCoA + \text{D-Glucosamine} \rightarrow \text{N-Acetyl-D-glucosamine} + CoA$$

By MORRIS SOODAK

Assay Method

Principle. Pigeon liver extracts contain an acetokinase which transfers
the acetyl group of Ac$\sim$SCoA to glucosamine and galactosamine.[1] The
appearance of the acetylated amino sugar is followed by a modification
of the Morgan and Elson method,[2] which is specific for N-acetylamino
sugars. The formation of N-acetyl-D-glucosamine is confirmed chromatographically on paper.

Reagents

    0.1 M N-acetylglucosamine.[3]
    0.1 M D-glucosamine·HCl.
    1.0 M KAc.
    0.1 M cysteine·HCl.
    0.1 M MgCl$_2$.
    0.1 M ATP (Na$^+$ or K$^+$).
    1.0 M KF.
    CoA solution, 200 units/ml.
    2.0 M Tris, pH 8.2.
    0.25 M lithium acetyl phosphate.

[1] T. C. Chou and M. Soodak, *J. Biol. Chem.* **196**, 105 (1952).
[2] W. T. J. Morgan and L. A. Elson, *Biochem. J.* **28**, 988 (1934).
[3] N-Acetylglucosamine was obtained from Krishell Laboratories, Portland, Oregon.

1.0 M NH$_4$Cl.

Enzymes. (1) Crude and aged extract of pigeon liver acetone powder.[4] (2a) Partially purified glucosamine acceptor enzyme from pigeon liver[5] (see below). (2b) Partially purified transacetylase from *Clostridium kluyveri*.[6]

5% TCA.

1.0 M Na$_2$CO$_3$.

95% ethanol.

Ehrlich's reagent. Dissolve 1.6 g. of p-dimethylaminobenzaldehyde[7] in a cooled mixture of 30 ml. of conc. HCl and 30 ml. of 95% ethanol.

Procedure 1. Acetylation with Aged Pigeon Liver Extract. The reaction mixture, final volume 1 ml., which is incubated for 2 hours at 37°, in 16 × 100-mm. test tubes, is as follows: 10 micromoles glucosamine·HCl (0.1 ml.), 50 micromoles acetate (0.05 ml.), 10 micromoles cysteine·HCl (0.1 ml.), 10 micromoles MgCl$_2$ (0.1 ml.), 10 micromoles ATP (0.1 ml.), 50 micromoles KF (0.05 ml.), 20 units CoA (0.1 ml.), 200 micromoles Tris (0.1 ml.), and 0.3 ml. aged pigeon liver extract (corresponding to 30 mg. acetone powder).

Procedure 2. Acetylation with Partially Purified Glucosamine Acceptor System from Pigeon Liver (Fraction A-60).[7] This preparation no longer contains the acetate-ATP reaction found in crude pigeon liver extracts. Ac∼SCoA is here supplied by the transacetylase system. The reaction mixture, which is incubated for 2 hours at 30°, is as follows: 10 micromoles glucosamine·HCl (0.1 ml.), 25 micromoles acetyl phosphate (0.1 ml.), 10 micromoles cysteine·HCl (0.1 ml.), 5 micromoles MgCl$_2$ (0.05 ml.), 100 micromoles NH$_4$Cl (0.1 ml.), 20 units CoA (0.1 ml.), 10 units transacetylase (0.05 ml.), 0.1 ml. water, and 0.3 ml. liver Fraction A-60.

Control tubes are run which contain no ATP or acetyl phosphate. To another set of control tubes is added 0.5 to 2 micromoles of N-acetylglucosamine. These latter tubes serve as internal standards.

The enzymatic reaction is terminated by the addition of 2 ml. of TCA. The protein precipitate is spun down. A 1-ml. aliquot is transferred to a 16 × 150-mm. test tube, followed by 0.5 ml. of water and 0.5 ml. of M Na$_2$CO$_3$. The tubes are heated in a boiling water bath for 10 minutes

---

[4] N. O. Kaplan and F. Lipmann, *J. Biol. Chem.* **174,** 37 (1948). The preparation of pigeon liver acetone powder and of the extracts therefrom is described in Vol. I [101]; and also see Vol. III [132].

[5] T. C. Chou and F. Lipmann, *J. Biol. Chem.* **196,** 89 (1952).

[6] See section on transacetylase (Vol. I [98]).

[7] A preparation of p-dimethylaminobenzaldehyde (highest purity) from Phanstiehl Chemical Co., Waukegan, Illinois, may be used without recrystallization.

and cooled in cold water. 2.5 ml. of 95% ethanol is added to each tube with shaking, followed by 0.5 ml. of Ehrlich's reagent. After shaking, the tubes are kept at 37° for 30 minutes for color development and then read on a Klett colorimeter with a 540-mμ filter.

In the experiments with transacetylase plus the partially purified glucosamine acceptor enzyme, sufficient N-acetylglucosamine is formed to confirm its appearance by paper chromatography. In this case, after the enzymatic incubation, a 0.5-ml. aliquot is removed and treated with 1 ml. of TCA for the colorimetric determination as above. The remainder of the incubation mixture (0.5 ml.) is treated in the following manner. 3 ml. of 95% ethanol is added, and the tubes are allowed to stand half an hour for protein precipitation. After spinning, the supernatants are evaporated to dryness in a bath at 40° or below by means of a stream of air. The residue is extracted with 2 ml. of 95% ethanol, and the extracts are brought to dryness again. The final residue is dissolved in 0.1 to 0.2 ml. of water, and appropriate aliquots are applied to Whatman No. 1 sheets. Ascending chromatograms are run for 16 hours with 80% propanol containing 0.8% ammonium acetate as the solvent. The spots are developed by the methods described by Partridge.[8] The R_f for glucosamine and galactosamine is 0.32. The R_f for acetylglucosamine is 0.46.

Preparation of the Partially Purified Glucosamine Acceptor Enzyme from Pigeon Liver Extracts.[5] The fraction precipitating between 40 and 60% acetone concentration contains the glucosamine acetokinase. All operations are carried out in the cold. For centrifugation a refrigerated centrifuge at −15° is used throughout. A 10% suspension of pigeon liver acetone powder in a 25% acetone solution containing 0.1 M Tris at pH 8.2 is stirred gently for 5 to 10 minutes at about −5°. The insoluble material is centrifuged at 4000 r.p.m. for 10 minutes and discarded. The supernatant is brought to 40% acetone concentration by the cautious addition of cold acetone (about −5°). The precipitate is centrifuged as above. The supernatant is further treated with acetone to attain a 60% concentration. This 40 to 60% fraction is spun down. The 25 to 40% (Fraction A-40) and the 40 to 60% (Fraction A-60) precipitates are separately dissolved in 1 ml./g. of original powder of M Tris at pH 8.2 and dialyzed for 12 hours with one change of fluid against 4.5 l. of a solution containing 0.001 M cysteine·HCl, 0.02 M KHCO$_3$, and 0.5% KCl (pH 8 adjusted with K$_2$CO$_3$). Fraction A-40 contains the acetate-ATP reaction, and the enzymes for the synthesis of citrate, acetoacetate, and glutamine. Fraction A-60 contains the arylamine acetylation acceptor enzyme as well as that for the amino sugars (see below). All the enzymes used are stored at −15 to −30° and are stable for many months.

[8] S. M. Partridge, *Biochem. J.* **42**, 238 (1948).

Results. Crude aged pigeon liver extracts (0.3 ml.) under the conditions stated above, or when combined with the transacetylase system as acetyl∼SCoA source, synthesize from 0.1 to 0.25 micromole of acetylglucosamine. With the transacetylase system plus Fraction A-60, from 1 to 2 micromoles is formed. This corresponds to about a seven-fold concentration of the enzyme in the A-60 fraction. In the latter case, the acetylglucosamine formation is confirmed by paper chromatography.

The synthesis of acetylgalactosamine via the Fraction A-60 system when followed by the Morgan and Elson colorimetric method using acetylglucosamine as a standard is about 30% of that of acetylglucosamine. Aminoff *et al.*,[9] however, have recently shown that the color yield of acetylgalactosamine is only 20% of that of acetylglucosamine. Taking this into account, similar amounts of acetylglucosamine and acetylgalactosamine are formed.

Comments. Any source of Ac∼SCoA, Ac∼SCoA itself, the transacetylase system, or an acetate-ATP-CoA system may be used in conjunction with Fraction A-60. Since the first two sources may not be available, a convenient source is the acetate-ATP-CoA reaction of yeast of Jones *et al.*[10]

The identity of the amino sugar acetokinase has not been definitely established. Tabor *et al.*[11] have purified the enzyme from pigeon liver which acetylates aromatic amines some 25-fold, by acetone fractionation and alumina C_γ adsorption. This preparation acetylates glucosamine at 4% of the rate of *p*-nitroaniline acetylation and also acetylates histamine and β-phenylethylamine. The enzyme contains an essential sulfhydryl group. Bessman and Lipmann[12] have found that their arylamine acetokinase preparation from pigeon liver, purified 38-fold by acetone fractionation, can also carry out a low-energy acetyl transfer between various aromatic amines. Some of their unpublished observations indicate that low-energy acetyl is transferred to glucosamine at a slow rate. Further work is necessary to prove the existence of a special acetokinase for the amino sugars.

Katz *et al.*[13] have shown that dried-cell preparations of *Clostridium kluyveri* in the presence of acetyl phosphate catalyze a cyanide-induced acetylation of glucosamine.

[9] D. Aminoff, W. T. J. Morgan, and W. M. Watkins, *Biochem. J.* **51**, 379 (1952).
[10] M. E. Jones, S. Black, R. M. Flynn, and F. Lipmann, *Biochim. et Biophys. Acta* **12**, 141 (1953); see also Vol. I [96].
[11] H. Tabor, A. H. Mehler, and E. R. Stadtman, *J. Biol. Chem.* **204**, 127 (1953).
[12] S. P. Bessman and F. Lipmann, *Arch. Biochem. and Biophys.* **46**, 252 (1953).
[13] J. Katz, I. Lieberman, and H. A. Barker, *J. Biol. Chem.* **200**, 417 (1953).

[103] Amino Acid Acetylase of *Clostridium kluyveri*

$$\text{Ac} \sim \text{SCoA} + \text{Glycine} \rightarrow \text{Acetyl Glycine} + \text{HSCoA}$$

By I. LIEBERMAN and H. A. BARKER

Assay Method

Principle. In the presence of a rather high concentration of cyanide, extracts of *Cl. kluyveri* catalyze the transfer of the acetyl group of acetyl phosphate to glycine and other amino acids.[1] The available evidence[2,3] indicates that this reaction requires two enzymes: phosphotransacetylase, which catalyzes the formation of Ac∼SCoA from acetyl phosphate and HSCoA; and the amino acid acetylase which in the presence of cyanide catalyzes the above reaction. The method described below was developed by Lieberman[3] and is based upon the observation that under certain conditions the disappearance of acetyl phosphate is a measure of amino acid acetylation.

Reagents

Tris-HCl buffer (2 M), pH 8.5.
Glycine (0.5 M), pH 8.5.
Dilithium acetyl phosphate (0.2 M). This solution is stable for weeks at $-15°$.
KCN-H$_2$SO$_4$ (2 M), pH 8.5. This solution must be prepared each day.
CoA solution, 400 Kaplan and Lipmann[4] units/ml. Pabst CoA ($\sim$250 units/mg.) is satisfactory. Dissolve the CoA in one-fifth the final volume of solution containing 0.05 M reduced glutathione and 0.02 M KPO$_4$ buffer, pH 6.8, and incubate at room temperature for 60 minutes to reduce the CoA; then dilute to volume.[5] This solution is stable for several weeks at $-15°$.
Phosphotransacetylase, 40 units/ml.[6]
Enzyme. The concentration of the stock enzyme solution was adjusted to contain 50 to 100 units/ml. (See definition below.)

[1] E. R. Stadtman, J. Katz, and H. A. Barker, *J. Biol. Chem.* **195,** 779 (1952).
[2] J. Katz, I. Lieberman, and H. A. Barker, *J. Biol. Chem.* **200,** 417 (1953).
[3] I. Lieberman, Ph.D. Dissertation, University of California, 1952.
[4] N. O. Kaplan and F. Lipmann, *J. Biol. Chem.* **174,** 37 (1948).
[5] R. M. Burton and E. R. Stadtman, *J. Biol. Chem.* **202,** 873 (1953).
[6] E. R. Stadtman, *J. Biol. Chem.* **196,** 527 (1952). The purification of the enzyme is described in Vol. I [98].

Procedure. Add 0.05 ml. of Tris buffer, 0.05 ml. of glycine solution, 0.1 ml. of acetyl phosphate solution, 0.1 ml. of CoA solution, 0.1 ml. of phosphotransacetylase, and 0.1 ml. of enzyme to sufficient water to give a total volume of 0.8 ml. At zero time add 0.2 ml. of KCN solution and incubate at 27°. After 30 minutes, estimate the residual acetyl phosphate by the method of Lipmann and Tuttle[7] on a 0.2-ml. aliquot. Incubate a solution containing all the constituents except the enzyme under the same conditions to determine the initial acetyl phosphate, corrected for nonenzymatic decomposition. Use a blank with enzyme but without acetyl phosphate to correct for blue color caused by interaction of cyanide with iron salts in the enzyme preparation.

Definition of Unit and Specific Activity. One unit of enzyme is defined as that amount which catalyzes the disappearance of 1.0 micromole of acetyl phosphate in 30 minutes under the conditions of the assay. Specific activity is expressed as units per milligram of protein. Protein was estimated by the turbidimetric method of Stadtman *et al.*[8]

Purification Procedure

Step 1. Preparation of Crude Extract. 1 g. of finely ground dried cells of *Cl. kluyveri*[9] is homogenized in 15 ml. of 0.05 M KPO_4, pH 7.0, containing 0.001 M Na_2S, and is incubated under vacuum for 60 minutes at room temperature with occasional shaking. The suspension is then centrifuged at 10,000 r.p.m. The enzyme in the clear supernatant retains full activity for weeks at $-18°$.

Step 2. Removal of Inactive Protein with Protamine. 1 ml. of a freshly prepared 3 % (w/v) solution of protamine sulfate is added to 10 ml. of cold crude extract containing 24 mg. of protein per milliliter. After 5 minutes, the supernatant which contains most of the activity is obtained by centrifugation.

Step 3. Fractionation with Ammonium Sulfate. 2 vol. of cold saturated ammonium sulfate is added to 1 vol. of cold protamine supernatant. The solution is centrifuged, and the precipitate discarded. The supernatant is saturated with solid ammonium sulfate, and, after centrifugation, the 0.66 to 1.0 saturation fraction is dissolved in half the original volume of 0.03 M KPO_4, pH 7.0. This preparation retains most of its activity for weeks when stored at $-18°$. See the accompanying table for a summary of the purification procedure.

[7] F. Lipmann and L. C. Tuttle, *J. Biol. Chem.* **158**, 505 (1945).
[8] E. R. Stadtman, G. D. Novelli, and F. Lipmann, *J. Biol. Chem.* **191**, 365 (1951).
[9] E. R. Stadtman and H. A. Barker, *J. Biol. Chem.* **180**, 1085 (1949); see Vol. I [84] for content of *Cl. kluyveri* medium.

Summary of Purification Procedure[a]

Fraction	Ml.	Activity, units/ml.	Total activity, units	Protein, mg.	Specific activity, units/mg.	Recovery of activity, %
1. Original extract	10	99	990	240	4.1	(100)
2. Protamine supernatant	10	91	910	95	9.6	92
3. (NH₄)₂SO₄ fraction, 0.66–1.0 satn.	5	99	495	22.5	22.0	50

[a] I. Lieberman, Ph. D. Dissertation, University of California (1952).

Properties

Specificity. Most studies on this enzyme have been done with crude extracts, and it is not certain whether the reactions observed are catalyzed by one or several enzymes. With crude extracts, all the amino acids tested appear to be acetylated, namely glycine, leucine, serine, lysine, and glutamate and/or aspartate.[1,2] Glycyl-L-leucine, DL-alanylglycine, glycine ethyl ester, and ethanolamine react at about the same rate as glycine. Glucosamine, methylamine, dimethylamine, ethylamine, and β-alanine react from 50 to 75% as fast as glycine. It appears that egg albumin also can be acetylated. Sulfanilamide and p-aminobenzoic acid are acetylated at a very slow rate. Choline and ammonia are unreactive.

With crude extracts containing phosphotransacetylase, propionyl phosphate can be substituted for acetyl phosphate as an acyl donor.[10] This indicates that propionyl~SCoA can serve as a substrate for the enzyme. The formation of n-butyryl, isobutyryl, n-valeryl, and n-caproyl derivatives of glycine in reaction mixtures containing acetyl phosphate, CoA, glycine, the appropriate fatty acid, and extracts containing phosphotransacetylase, CoA transphorase, and the amino acid-acetylating enzyme, suggests that a variety of acyl~SCoA derivatives can replace acetyl~SCoA in the acetylase reaction. The rate of acylation decreases with increasing chain length of the acyl derivative beyond the propionyl compound. It is probable, but not certain, that the rate-limiting step is the acetylase reaction rather than the CoA transphorase reaction.[11]

Activators. The acetylation of amino acids is dependent upon the presence of cyanide.[1,2] The normal function of the enzyme, in the absence of cyanide, is not known. The acetylation reaction is just detectable by the method described with 0.01 M KCN. The rate increases almost lin-

[10] J. Katz, I. Lieberman, and H. A. Barker, *J. Biol. Chem.* **200**, 431 (1953).
[11] E. R. Stadtman, *J. Biol. Chem.* **203**, 501 (1953).

early with cyanide concentration up to 0.2 M and increases more slowly at higher cyanide levels up to 0.8 M.

The only compound that has been found to replace cyanide as an activator is azide. At 0.025 M azide is as effective as cyanide in inducing acetylation, but at higher concentrations it is less effective. Sulfide, sulfite, or α,α'-dipyridyl cannot replace cyanide.

Effect of pH. The enzyme is active from pH 6.0 to above pH 10; the optimum lies between 7.7 and 8.3.[2]

Substrate Concentration. The Michaelis constant for glycine is of the order of 1×10^{-3} M.[2] Direct observations on the relation between acetyl CoA concentration and rate are not available, but the effect of CoA concentration in the presence of excess acetyl phosphate and phosphotransacetylase indicates that the Michaelis constant for acetyl CoA is of the order of 5×10^{-5} M.

Stability. The activity is not decreased by incubation at pH 5.0 or 9.9 for 4 hours. The enzyme is moderately heat-stable in crude extracts.[3] After 10 minutes' heating at 60° of a solution adjusted to pH 6.2, 84% of the initial activity was recovered; after 10 minutes at 65°, 44% of the activity remained. Crude extracts heated for 45 minutes at 60° are relatively free of phosphotransacetylase and require addition of this enzyme for maximal activity when acetyl phosphate and coenzyme A are used as substrates.

[104] Choline Acetylase

By DAVID NACHMANSOHN and IRWIN B. WILSON

Introduction: Definition and Biological Significance

In 1943 it was discovered that extracts of brain and electric tissue acetylate choline in presence of ATP (Nachmansohn and Machado[1]). This was the first enzymic acetylation obtained in a soluble system. The experiments were performed on the assumption that in the elementary process of conduction acetylcholine breakdown is the primary event and precedes that of ATP. The hydrolysis of the latter was, therefore, postulated to provide the energy for the acetylation of choline.[2] The detailed mechanism of the reaction remained obscure. Two years later Nachman-

[1] D. Nachmansohn and A. L. Machado, *J. Neurophysiol.* **6**, 397 (1943).

[2] D. Nachmansohn, R. T. Cox, C. W. Coates, and A. L. Machado, *J. Neurophysiol.* **6**, 383 (1943).

sohn and Berman[3,4] and simultaneously Lipton[5] in Barron's laboratory observed that the acetylation of choline requires a coenzyme. Concurrently Lipmann and Kaplan[6,7] found that ATP provides the energy source of acetylation of sulfanilamide by liver extracts and demonstrated the existence of a coenzyme for this acetylation. Subsequently it became obvious that the coenzyme is the same in both reactions and in fact in all acetylations and was therefore termed by Lipmann coenzyme A (for acetylation)—CoA. Today it is well established by the work of various laboratories that the first step in acetylation is the formation of acetyl CoA. The formation of this intermediate is catalyzed in animal tissue and in yeast from acetate and CoA by the enzyme acetylkinase, whereas in most bacteria the acetyl group is transferred from acetyl phosphate to CoA by the enzyme phosphotransacetylase. The acetyl group of acetyl CoA is then transferred to the compound acetylated in the reaction by an enzyme more or less specific for the acceptor. In the case of choline the enzyme has been termed choline acetylase.

I. Sources and Preparation of the Enzyme

Choline acetylase, like acetylcholinesterase, is present in all conducting tissue, nerve, and muscle, but in considerably lower concentrations. Very few tissues are therefore a suitable starting material for the preparation of the enzyme. Electric tissue either does not contain large amounts of the enzyme, or the extracting methods tried so far have not been adequate. Brains of small animals, such as rats, mice, guinea pigs, and rabbits, are usable; those of larger animals, such as sheep, oxen, or cat, are a very poor source. By far the best material proved to be the head ganglia of squid.[8]

The drawback of this material is obviously the difficulty of obtaining it and the special arrangements required. The squids should be frozen on dry ice, since otherwise the enzyme deteriorates very rapidly. It takes some experience to take the head ganglia out of the animal; on the other hand, shipment of the whole animal at long distances may be quite expensive. From about 250 lb. of squids about 700 to 800 g. of ganglia may be obtained, which yield about 70 to 80 g. of acetone-dried powder.

A. Squid Ganglia. Acetone-dried powder may be prepared in the following way: 150 g. of ganglia is homogenized in the Waring blendor

---

[3] D. Nachmansohn, *Ann. N. Y. Acad. Sc.* **47**, 395 (1946).
[4] D. Nachmansohn and M. Berman, *J. Biol. Chem.* **165**, 551 (1946).
[5] M. A. Lipton, *Federation Proc.* **5**, 145 (1946).
[6] F. Lipmann, *J. Biol. Chem.* **160**, 173 (1945).
[7] F. Lipmann and N. O. Kaplan, *J. Biol. Chem.* **162**, 743 (1946).
[8] D. Nachmansohn and M. S. Weiss, *J. Biol. Chem.* **172**, 677 (1948).

with about the same amount of H_2O and poured into 3 l. of acetone of -5 to $-10°$. After 10 minutes the suspension is rapidly filtered with suction through a Buchner funnel. The solid is suspended again in 2 l. of acetone, filtered after standing for 30 minutes in acetone, and then dried in the usual way. The yield is about 15 to 18 g. of powder per 150 g. of ganglia.

A procedure by which a ten- to twelve-fold purification of the enzyme starting with this powder may be obtained may be briefly described.[9] The powder is extracted with 0.05 M K_2HPO_4 buffer of pH 7.4 and centrifuged. The proteins of the supernatant solution are precipitated with an equal volume of 50% ammonium sulfate and centrifuged. The precipitate containing the enzyme is dissolved in 0.1 M K_2HPO_4 buffer of pH 7.8. The ammonium sulfate is removed by dialysis against 0.5 M phosphate buffer of pH 7.4. Protamine sulfate in a 2% solution is then added (1 mg. per 10 mg. of protein); the precipitate formed is removed by centrifugation. When large amounts of material are used, a second ammonium sulfate fractionation at this stage is advisable; the protein should be fractionated between 16 and 28% of ammonium sulfate. With smaller amounts of material the next step may be carried out directly. Treatment of the solution with 3.3 mg. of calcium PO_4 gel per milligram of protein at pH 6.2 leads to absorption of practically all the enzyme. The gel is centrifuged and eluted with 0.05 M phosphate buffer of pH 7.5, and the eluate is discarded. A second elution with 0.2 M phosphate buffer of pH 8.2 contains most of the enzyme. The eluate is again fractionated with ammonium sulfate. The fraction between 16 and 32% contains the enzyme.

The data of one preparation in which a tenfold purification was obtained are given in the table. The enzyme solution at this stage kept in the deep-freeze has a satisfactory stability for at least 4 to 6 weeks.

SUMMARY OF PURIFICATION PROCEDURE

	Specific activity[a]	Recovery, %
Crude extract	5.9	100
First ammonium sulfate	11	60
Protamine sulfate	10	60
Ca gel	40	60
Second ammonium sulfate	58	45

[a] Micromoles of ACh formed per milligram of protein per hour.

[9] R. Berman, I. B. Wilson and D. Nachmansohn, *Biochim. et Biophys. Acta* **12**, 315 (1953).

B. Mammalian Brain. Acetone-dried powder of brains of rats, guinea pigs, and rabbits yields an extract which forms 8.5 to 14 micromoles of acetylcholine per gram of powder per hour. Per gram of protein the amount of ester formed is about 35 micromoles per hour at 37°. With high-speed centrifugation (48,000 r.p.m. for 60 minutes) a slight separation may be achieved, increasing the amount formed per gram of protein per hour to about 55 in the bottom layer. At two- to three-fold purification may be obtained by fractionation with ammonium sulfate. The protein precipitating between 16 and 35% of ammonium sulfate contains most of the activity. Cysteine in 10^{-3} M concentration should be added to the buffer against which the enzyme is dialyzed. This procedure was used in 1949. At that time the system was not yet fully understood, and neither phosphotransacetylase nor acetylkinase was separated. It appears most likely that a better purification may be obtained at present when the assay is made with addition of all the components required for acetyl CoA formation which would then greatly facilitate the purification of choline acetylase from brain extracts.

C. Lactobacillus plantarum. Choline acetylase has been prepared by Girvin and Stevenson[10] from *Lactobacillus plantarum.* Lyophilized cells were disrupted manually with an equal weight of glass powder. The clumps were broken, and the powder was extracted with thirty times its weight of *n*-butyl alcohol at 5° for 16 hours, filtered, dried and extracted with twenty times its weight of acetone cooled with dry ice, dried in a vacuum dessicator, and finally lyophilized. The powder was stored at $-25°$ in evacuated ampoules. Each 18 mg. of powder was extracted with 1 ml. of 0.17 M KCl, 0.00125 M $MgCl_2$, and 0.006 M Na_2HPO_4 (pH 7.1) at 5°. The preparation has an activity at 28° of 120 to 350 micromoles of acetylcholine chloride produced per hour per gram of protein.

No acetyl CoA-producing system was added by the authors in testing this preparation. It is possible that much higher activity would be obtained if acetyl phosphate and transacetylase were added in the assay.

II. Methods of Assay

An assay of choline acetylase should have all the other components of the acetylcholine-synthesizing system added in excess. This includes the acetyl CoA-synthesizing system, which can readily be supplied by CoA (5×10^{-5} M), phosphotransacetylase, and acetyl phosphate (0.04 M),

---

[10] G. T. Girvin and J. W. Stevenson, *Can. J. Biochem. and Biophysiol.* **32**, 131 (1954).

or by CoA, acetylkinase, ATP, and acetate.[11] As an alternative acetyl CoA (preparation[12,13]) can be used as a substrate.

Besides choline chloride (0.05 M), the medium should also contain cysteine (0.05 M) and Mg^{++} (0.005 M) (unless acetyl CoA is used as a substrate), as well as 0.010 potassium phosphate buffer at pH 7. Tetraethylpyrophosphate (0.001 M) is added to inhibit esterases. A small amount of ethylenediaminetetraacetic acid (0.001 M) may be beneficial.

A. Colorimetric Hydroxamic Acid Method. After incubation the acetylcholine formed may be measured directly by Hestrin's acethydroxamic acid method, which has been described in the article on acetylcholinesterase (Vol. I [107]). If acetyl phosphate is present, it is first destroyed by boiling for 4 minutes at pH 4.5.

B. Bioassay with Frog's Rectus Abdominus Muscle. Another direct test for acetylcholine is the contracture of the rectus abdominus muscle of the frog.[14] A kymograph is not necessary for the procedure. Satisfactory measurements may be made with a light, but not too light, lever arm moving against a background of coordinate paper. The movement of the pointer in 2 minutes is calibrated with known amounts of acetylcholine (0.2 μg./ml.). If the muscle is well aerated, rested 5 minutes between measurements, and calibrated frequently, an accuracy of 5% can be attained. Large amounts of choline (50 μg./ml.) or K^+ (0.02 M) interfere. This method is far more sensitive than chemical methods and is more specific.

An unknown solution is diluted to about 0.2 μg. of acetylcholine per milliliter with frog's ringer containing 12 μg. of eserine per milliliter. Frog's ringer plus 12 μg. of eserine per milliliter bathing the resting muscle is drawn off, and the diluted unknown is added. At the same time a timer is started. The position of the pointer is read every half minute for 2 minutes. The solution is drained, and the muscle is washed with eserine ringer and bathed in eserine ringer for 5 minutes before another sample or calibrating solution is applied.

C. Nitroprusside Test for SH Groups. If acetyl CoA is used as substrate, the nitroprusside test for SH groups described by Grunert and Philips[15] may be applied. The following is an example of the use of this method.[9] 0.2 ml. of an unknown solution containing about 0.1 μM. of free SH is added to 0.1 ml. of 0.033 N HCl and 1.2 ml. of saturated NaCl

[11] S. Korkes, A. del Campillo, S. R. Korey, J. R. Stern, D. Nachmansohn, and S. Ochoa, *J. Biol. Chem.* **198**, 215 (1952).

[12] I. B. Wilson, *J. Am. Chem. Soc.* **74**, 3205 (1952).

[13] E. J. Simon and D. Shemin, *J. Am. Chem. Soc.* **75**, 2520 (1953).

[14] H. C. Chang and J. H. Gaddum, *J. Physiol.* **79**, 225 (1933).

[15] R. R. Grunert and P. H. Phillips, *Arch. Biochem.* **30**, 217 (1951).

plus 0.2 g. of solid NaCl. After the inactivation of the enzyme, 0.1 ml. of 0.033 N NaOH is added. The flocculated protein is removed by centrifugation—the mixture is stable at this stage. An aliquot (0.8 ml.) is used for spectrophotometric determination. To the aliquot is added 0.1 ml. of 0.067 M sodium nitroprusside and 0.1 ml. of 1.5 M Na_2CO_3 and 0.067 M NaCN. The optical density is determined at 520 mμ exactly 30 seconds after the addition of base. The optimum pH for color development is 10. Glutathione may be used as a standard for the procedure.

Acetyl CoA hydrolyzes slowly at pH 10. There is, therefore, a small increase in color between the time of addition of the alkali and the reading 30 seconds later which is proportional to the amount of acetyl CoA present at the time of the test. The optical density is corrected by subtracting the color so produced by the remaining acetyl CoA.

[105] Acetoacetate Decarboxylase

$$CH_3COCH_2COOH \rightarrow CH_3COCH_3 + CO_2$$

By H. W. Seeley

The paper by Davies[1] should be consulted as a valuable reference on some aspects of this topic.

Assay Method

Principle. The procedure is based on the manometric measurement of carbon dioxide liberated by the enzyme from acetoacetic acid.

Reagents

Acetoacetic acid. Dissolve 13 g. of redistilled ethyl acetoacetate in 100 ml. of N NaOH. Hold at room temperature for 10 hours. Acidify the ice-cooled solution with H_2SO_4 (25 ml. of 20% by volume). Extract twice with 150 ml. of ether; evaporate the ether at room temperature. Neutralize the residue with N NaOH; extract the unchanged ester with ether. Evaporate the ether *in vacuo*. Prepare an 0.5 M solution for use in manometer cups. Store at 4°. A slow, spontaneous decarboxylation of the stock solution occurs, which on prolonged storage is troublesome as a result of the release of carbon dioxide on tipping into acid buffer.

Buffer. 0.5 M acetic acid–Na acetate, pH 5.0.

Enzyme. Purified clostridial decarboyxlase.

[1] R. Davies, *Biochem. J.* **37**, 230 (1943).

Procedure. Employ standard manometric methods for the Warburg device. Place 1.0 ml. of 0.5 M acetate buffer, 0.1 to 0.5 ml. of purified enzyme, and distilled water (to total cup volume of 3.0 ml.) in a cup; place 0.5 ml. of 0.5 M Na acetoacetate in the side arm. Temperature 37.5°; gas phase, air.

Definition of Unit. One enzyme unit is defined as the amount that will liberate 100 μl. of CO_2 in 5 minutes at 37.5° under optimum conditions.

Effect of Extraneous Organic Materials on Decarboxylation. The results of experiments in which crude preparations or proteinaceous materials are added to the cup should be considered critically, since serum, organ extracts, peptones, and several amino acids (Pollack[2]), precipitated yeast protein (Mayer[3]), and commercial yeast extract (VanDemark and Seeley[4]) are known to stimulate the nonenzymatic decarboxylation of acetoacetic acid.

Purification Procedure

Step 1. Cell Production. The enzyme is present in variable degree in certain strains of *Clostridium acetobutylicum* and closely allied microorganisms.

In practice, a spore stock preparation maintained in sand is passed through two test-tube transfers and then to 8 l. (two 4-l. batches in Erlenmeyer flasks) of a medium of the following composition: pepticase 2.0%, yeast extract 0.1%, sucrose 2.0%, KH_2PO_4 0.5%, $(NH_4)_2HPO_4$ 1.0%, salts B 1.0%.[5] The pH is adjusted to 6.5 with 0.5 M K_2HPO_4. The inoculum is 50 to 60 ml. of clostridial culture. Light mineral oil or vaspar is layered ($\frac{1}{4}$ in.) over the medium. The culture is held at 37° for about 72 hours or until the fermentation has become quiescent. The cell crop may be harvested by a Sharples centrifuge.

Step 2. Acetone Preparation of Cells. The cells from step 1 are washed with 300 to 400 ml. of distilled water, resuspended in 50 ml. of H_2O, and blown from a pipet into 500 ml. of acetone, the acetone being constantly swirled during the process. After a few minutes, the preparation is filtered on a Büchner funnel, washed with acetone and with ether, and dried. The yield, 6 to 7 g. of active cell powder, is stable in dry form for several months at 4°.

Step 3. Concentration. 1 g. of acetone-dried cells is placed in 190 ml. of H_2O and dispersed with a high-speed mixer for 1 to 2 minutes. The dis-

[2] L. Pollack, *Beitr. chem. Physiol. Pathol.* **10**, 232 (1907).
[3] P. Mayer, *Biochem. Z.* **50**, 283 (1914).
[4] P. J. VanDemark and H. W. Seeley, unpublished data.
[5] Other media used for industrial production of acetone are suitable.

persed material is stored at 37° for 4 hours and refrigerated for 12 to 18 hours. It is then centrifuged, and the sediment discarded. Glacial acetic acid is added until pH 4.0 is reached. The preparation is allowed to stand for 15 minutes and then centrifuged. The precipitate is resuspended in 40 ml. of 0.01 M K_2HPO_4, and $(NH_4)_2SO_4$ is added to the 50% saturation point. After standing for 20 minutes, it is centrifuged and redissolved in 40 ml. of distilled water.

Step 4. Resolution.[6] A partial resolution of the enzyme may be achieved by the following procedure. 3 mg. of trypsin ($\frac{1}{250}$) is added to 5 ml. of enzyme from step 3. The preparation is allowed to stand for 24 hours at room temperature, then refrigerated for 2 to 3 days before using. This treatment inactivates both the apoenzyme and the coenzyme, the former at a slower rate, resulting in a partial resolution. Activity can be restored by adding Kochsaft (100 mg. of acetone preparation of cells boiled for 10 minutes in 10 ml. of H_2O). To prevent a more complete enzyme loss, the resolved enzyme may be freeze-dried in small vials. Activity is retained on storage. The presence of trypsin does not appear to affect subsequent manometric work.

SUMMARY OF PURIFICATION PROCEDURE

	Total volume, ml.	Units/ml.	Q_{CO_2} (dry wt.)	Recovery, %
Whole cells			330	
Acetone powder dispersed in H_2O	190	850	1930	
Supernatant, after centrifuging	190	717	4100	84
pH_4 with acetic acid, resuspended	40	2593	6470	79
$(NH_4)_2SO_4$ ppt, resuspended	40	1817	23000	75

Properties[7]

Specificity. The purified enzyme is specific for acetoacetic acid, having no action on ethyl acetoacetate, pyruvic acid, acetopyruvic acid, oxalacetic acid, α-ketoglutaric acid, or acetonedicarboxylic acid.

Effect of pH. The pH for optimum activity is 5.

Effect of Temperature. For the range 20 to 30°, the $Q_{10} = 1.82$; for 30 to 40°, the $Q_{10} = 2.16$. The optimum is about 37°. In 5 minutes at 80° almost complete destruction of the enzyme occurs.

[6] H. W. Seeley and P. J. VanDemark, *J. Bacteriol.* **59**, 381 (1950).

[7] These data were largely secured by R. Davies and reported in the previous reference[1] to his work.

Effect of Inhibitors. Silver and Hg ($10^{-5}\,M$) and KCN ($10^{-4}\,M$) inhibit enzyme activity 90%, 100%, and 82%, respectively; NaN$_3$ ($10^{-2}\,M$), and iodoacetate ($10^{-2}\,M$) inhibit activity 53% and 63%, respectively. α-Keto acids inhibit to varying degrees, acetopyruvic ($10^{-6}\,M$, 65% inhibition) being outstanding.

Effect of Oxygen. Oxygen appears to have little or no effect on the stability or activity of the enzyme.

[106] Lipases

$$RCOOR' + H_2O \rightarrow RCOOH + R'OH$$

By M. BIER

Within the wide class of enzymes catalyzing the hydrolysis of various esters, we differentiate between lipases and esterases on the basis of their relative preferential specificity. The natural substrates for lipases are oils and fats, i.e., triglycerides of long-chain fatty acids, whereas esterases act on simple esters of low molecular weight acids. Both groups are unusually nonspecific in their action, and the nature of either the fatty acid or the alcohol residue has mostly but a secondary effect, influencing only the rate of hydrolysis of the esters.

A review of the procedures for assay and purification has, therefore, to include a discussion of methods suitable to a variety of systems. The applicability of the methods to the different enzymes will depend on their particular properties. This report will, of course, not include the more specific esterases, such as cholinesterases, lecithinases, liver esterases, etc., which are dealt with in independent articles.

Assay Methods

A surprisingly large number of widely different procedures were reported for the determination of the activity of various lipase preparations. This only signifies that there is not one method which is satisfactory, and practically every worker in studying lipases has thought it best to start by developing a new one. The most frequently employed procedures are based on the titrimetric determination of the acids liberated by the action of the enzyme, with either water-soluble or water-insoluble esters as substrates. Colorimetric assay methods employing chromogenic substrates are also useful in certain instances.

A number of more indirect methods of lipase assay are also available, but they will be mentioned only briefly. Tributyrin is a very strongly surface-active substance, and its hydrolysis can be followed by measuring the increase in the surface tension of its aqueous solutions. On the basis of this principle, Rona and Michaelis[1] have developed the stalagmometric method using a drop pipet for the estimation of the surface tension. This procedure underwent numerous modifications[2-4] and was criticized by Sobotka and Glick,[5] who have observed that it gives too high initial hydrolysis rates, as compared to other methods. Rona and Lasnitzki[6] have proposed the use of the Warburg manometric method for the determination of lipase activity, by measuring the CO_2 displaced from a bicarbonate buffer by the acid liberated through enzymatic hydrolysis of esters. Whereas these authors used tributyrin as substrate, Nicolai[7] chose the more soluble monobutyrin, and Copenhaver et al.[8] have applied Tween 20.[9] The volume change accompanying the hydrolysis of esters is utilized in the dilatometry of lipase determinations.[10] But this procedure, as well as the turbidimetric method,[11] are only of limited interest.

TITRIMETRIC METHOD USING WATER-INSOLUBLE SUBSTRATES

Principle. The method consists in the titrimetric estimation of the fatty acids liberated from the esters. The natural substrates of lipases, i.e., the esters of long-chain fatty acids (C_{10}–C_{18}) are practically insoluble in water (with the exception of the Tweens). This compels us to work with emulsions of the substrates, and all proposed methods suffer, therefore, from the general inadequacies common to all measurements of reaction rates in heterogeneous systems. The main difficulty is, of course, the preparation of stable and reproducible emulsions, as the rate of hydrolysis will depend on the state of dispersion of the enzyme.[12] Any substance

[1] P. Rona and L. Michaelis, *Biochem. Z.* **31,** 345 (1911).
[2] E. Bamann and E. Ullmann, *Biochem. Z.* **312,** 9 (1942).
[3] H. O. Lagerlöf, *Acta Physiol. Scand.* **13,** 301 (1947).
[4] R. Ammon and M. Jaarma, *in* "The Enzymes" (Sumner and Myrbäck, eds.), Vol. I, p. 390, Academic Press, New York, 1950.
[5] H. Sobotka and D. Glick, *J. Biol. Chem.* **105,** 199 (1934).
[6] P. Rona and A. Lasnitzki, *Biochem. Z.* **152,** 504 (1924).
[7] H. W. Nicolai, *Biochem. Z.* **174,** 343 (1926).
[8] J. H. Copenhaver, Jr., R. O. Stafford, and W. H. McShan, *Arch. Biochem.* **26,** 260 (1950).
[9] Tween is the trade name for the water-soluble esters of fatty acids and polyalkylene derivative of sorbitol, containing 1 mole of fatty acid per sorbitol residue. (Atlas Powder Company, Wilmington, Del.) Tween 20 is the monolaurate ester.
[10] P. Rona and R. Ammon, *Biochem. Z.* **249,** 446 (1932).
[11] P. Rona and H. Kleinmann, *Biochem. Z.* **174,** 18 (1926).
[12] F. Schönheyder and K. Volqvartz, *Acta Physiol. Scand.* **9,** 57 (1945).

affecting the interfacial tension will have a direct effect on the apparent activity, and it is probable that many activators and inhibitors described in the literature were surface-active compounds acting only on the state of emulsification, and not on the enzyme itself.[13] Although numerous variants of this method were employed from the very beginning of the studies on lipases, the procedure of Willstätter et al.[14] is most frequently quoted in text and reference books.[14-16] Their method was, however, subject to many modifications and did not attain general acceptance. In the experience of this reviewer, the following procedure of Fiore and Nord[17] gave satisfactory results in the study of mold, castor bean, and also pancreas lipases.

Emulsification. 10 g. of polyvinyl alcohol (PVA)[18] is stirred mechanically into 1 l. of distilled water, and to the fine dispersion obtained 5 ml. of 0.1 N HCl is added. The mixture is heated at 75 to 85° until solution is completed (about 1 hour), then cooled, filtered, and brought to the desired pH with 0.1 N NaOH. Olive oil or any other desired substrate is then added to an aliquot of the above solution, to bring the concentration of the substrate to 0.1 M, and emulsified in a Waring blendor for 5 minutes.

Procedure. In a 125-ml. Erlenmeyer flask are mixed 10 ml. of the freshly prepared PVA-emulsified substrate, 5 ml. of buffer (McIlvaine), and 5 ml. of the enzyme preparation to be tested. The mixture is shaken gently and incubated for 4 hours at 37° with constant shaking. At the end of the incubation time, 30 ml. of a 1:1 alcohol-acetone solution is added to stop the reaction and break the emulsion. Phenolphthalein indicator is added, and the solution is titrated with 0.05 N NaOH. Controls are determined in the same way, with an enzyme preparation inactivated by a short boiling. The concentration of the enzyme should be chosen so as to give an activity corresponding to 2 to 5 ml. of 0.05 N NaOH (10 to 25% hydrolysis). With very active preparations, the incubation time can be conveniently shortened.

Modifications of the Procedure. EMULSIFYING AGENTS. In alternate procedures the following emulsifying agents were used: albumin and

[13] R. A. Boissonnas, *Helv. Chim. Acta* **31**, 1571 (1948).

[14] R. Willstätter, E. Waldschmidt-Leitz, and F. Memmen, *Z. physiol. Chem.* **125**, 93 (1923).

[15] E. Waldschmidt-Leitz and A. Schäffner, *in* "Die Methoden der Fermentforschung" (Bamann and Myrbäck, eds.), p. 1547, Thieme, Leipzig, 1941.

[16] R. Ammon, *in* "Handbuch der Enzymologie" (Nord and Weidenhagen, eds.), p. 350, Akademische Verlagsgesellschaft, Leipzig, 1940.

[17] J. V. Fiore and F. F. Nord, *Arch. Biochem.* **23**, 473 (1949).

[18] Elvanol, medium viscosity, grade 52–22, E. I. Du Pont de Nemours and Co., Wilmington, Del.

sodium oleate,[14,19] gum arabic,[20] gum acacia,[21] bentonite,[22] etc. Triglycerides of stearic and palmitic acids are particularly difficult to emulsify, and the best method appears to be to dissolve the substrate in glycerine and dry ox bile, to which mixture then water, buffer, and enzyme are added.[23] For the testing of the apparently insoluble castor bean lipase, water in oil emulsions, rather than oil in water, were also proposed, with only a minimum amount of water.[24,25]

SUBSTRATES. Whereas olive oil is the most commonly used natural substrate for lipases, many other esters and triglycerides, natural or synthetic, were utilized.[5,23,26-29] Benzyl stearate and benzyl butyrate were singled out for their very rapid hydrolysis.[30]

ACTIVATORS. Calcium salts, either chlorides or lactates, were frequently added as activators, their alleged role being to precipitate the stearic acid liberated.[14,23,31,32]

TITRIMETRIC METHODS USING WATER-SOLUBLE SUBSTRATES

Principle. This method is also based on the titrimetric determination of the fatty acids liberated from esters, but to avoid the difficulties encountered in the emulsification water-soluble esters are employed. The most frequently used esters were tributyrin and the more soluble esters such as methyl butyrate or triacetin. These low molecular weight esters are, however, unspecific for lipases, and the introduction of the Tweens in lipase studies by Gomori[33] and Archibald[34] was, therefore, welcome. Tween 20, a monolaurate ester, was chosen as the most appropriate, because the esters of fatty acids of C_6 to C_{14} chain length are split with the highest velocity by pancreatic lipase. Although not a chemically pure substance, different batches give reproducible results. Either the liberated

[19] L. Vogel and P. Laeverenz, *Z. physiol. Chem.* **234**, 176 (1935).
[20] P. J. Fodor, *Nature* **158**, 375 (1946).
[21] L. A. Crandall and I. S. Cherry, *Proc. Soc. Exptl. Biol. Med.* **28**, 570 (1931).
[22] C. W. Bauer and E. E. Wilson, *J. Am. Pharm. Assoc. Sci. Ed.* **36**, 109 (1947).
[23] A. K. Balls, M. B. Matlack, and I. W. Tucker, *J. Biol. Chem.* **122**, 125 (1937).
[24] H. E. Longenecker and D. E. Haley, *J. Am. Chem. Soc.* **59**, 2156 (1937).
[25] W. G. Rose, *J. Am. Oil Chem. Soc.* **28**, 47 (1951).
[26] H. Kraut, Ä. Weischer, and R. Hügel, *Biochem. Z.* **316**, 96 (1943).
[27] G. Kabelitz, *Biochem. Z.* **316**, 409 (1944).
[28] S. S. Weinstein and A. M. Wynne, *J. Biol. Chem.* **112**, 641, 649 (1935–1936).
[29] F. Schönheyder and K. Volqvartz, *Enzymologia* **11**, 178 (1944).
[30] A. K. Balls and M. B. Matlack, *J. Biol. Chem.* **123**, 679 (1938); **125**, 539 (1938).
[31] F. Schönheyder and K. Volqvartz, *Acta Physiol. Scand.* **10**, 62 (1945); **11**, 349 (1946).
[32] H. Steudel, *Biochem. Z.* **318**, 205 (1947).
[33] G. Gomori, *Proc. Soc. Exptl. Biol. Med.* **58**, 362 (1945).
[34] R. M. Archibald, *J. Biol. Chem.* **165**, 443 (1946).

lauric acid can be extracted with ether and titrated after the evaporation of the ether,[34] or the whole mixture can be titrated directly.[13] In the experience of this reviewer, the following adaptation of the two procedures seems to give satisfactory results:

Reagents

Tween 20. The commercial product can be used directly, but it is preferable first to extract the free fatty acids. 30 ml. of Tween 20 and 50 ml. of a mixture of one part of diethyl ether and two parts of petroleum ether are placed in a 100-ml. glass-stoppered centrifuge flask. The mixture is acidified with 0.25 ml. of phosphoric acid, shaken, and centrifuged. The ether layer is rejected, and the extraction repeated with 25 ml. of the mixture. Finally, the ethers dissolved in the Tween are removed *in vacuo*, and the residue neutralized with NaOH, then decanted from the crystals of sodium phosphate.

Sodium acetate, 0.2 N.

Aqueous solution of phenyl red, 0.02%.

NaOH, 0.02 N.

Decylic alcohol.

Procedure. The substrate is prepared by mixing 100 ml. of buffer with 50 ml. of Tween, 10 ml. of the indicator, and 90 ml. of distilled water. The pH of the mixture is about 7.2, and the substrate can be used for about a week, with a gradual decrease in pH and an increase in the control blanks. To 1 ml. of the enzyme preparation is added 5 ml. of the substrate, and the test tube is stoppered and placed in a thermostat at 20°. After 9 minutes a drop of decylic alcohol is added to prevent foaming, and the solution is titrated so as to reach the end point at exactly 10 minutes after the mixing of the substrate and the enzyme. The content of the test tube is kept stirred during the titration by a slow stream of nitrogen, which also prevents the absorption of CO_2. Blank values are obtained by titrating under identical conditions separately 5 ml. of the substrate and 1 ml. of the enzyme preparation. The difference between the alkali consumed in the test and the sum of the two blank values is taken as the measure of the enzyme activity, and there is a direct proportionality up to 3 ml. of 0.02 N NaOH. The indicator is yellow at the beginning, and its transition from rose to red-violet is taken as the end point (pH 8.3). By definition, 1 ml. of 0.02 N NaOH is taken as corresponding to 100 lipase units.

COLORIMETRIC METHOD

Principle. The development of the characteristic color of *p*-nitrophenol by the enzymatic hydrolysis of its colorless esters furnishes a delicate test of esterase activity.[35] It differs from the previous methods as it measures the amount of hydroxyl groups liberated, rather than carboxyl. Owing to the great coloring power of the p-nitrophenol, it is a very sensitive method and, therefore, requires only short digestion periods. It has definite advantages when used as a qualitative test for esterase activity or in the detection of very low enzymatic activities. Unfortunately, however, because of the extremely low solubility of esters of long-chain fatty acids, the method is applicable preferentially to esterases, and, in the experience of this reviewer, a typical lipase, such as the castor bean lipase, produced no hydrolysis.

Reagents

Synthesis of the substrate.[35] p-Nitrophenol acetate (PNPA) is synthesized by refluxing for 1 hour 0.1 mole of p-nitrophenol with 1.2 g. of magnesium turnings and 0.12 mole of acetyl chloride in 30 ml. of benzene. The resulting solution is taken up in 150 ml. of ether and extracted with water, aqueous sodium bicarbonate, and again water, then evaporated *in vacuo*. For further purification, PNPA is recrystallized three times from about 150 ml. of ether. Although this treatment still does not eliminate all the free nitrophenol, the product is sufficiently pure for use as substrate.

$M/15$ phosphate buffer, pH 7.0.

0.1 N NaOH.

Standard solutions of p-nitrophenol. A series of solutions is prepared, containing 0.1 to 0.7 micromole of p-nitrophenol per milliliter (mol. wt., 139.1).

Substrate. About 63 mg. of PNPA is dissolved in 10 ml. of methanol and stored in the refrigerator, where it can be kept for about a week, with only a small increase in free nitrophenol. 1 ml. of this solution is slowly added to 100 ml. of distilled water with strong agitation, to prevent precipitation. This solution has to be prepared fresh every day, and it can be used directly for qualitative or semiquantitative activity determinations. For exact quantitative assays, because of the very low substrate concentrations employed, it has to be standardized. The solution as prepared is slightly more concentrated than required and has to be adjusted before each series of measurements. To this effect, standard curves are constructed for solutions containing 1 ml. of the nitrophenol standards and (a) 9 ml. of $M/15$ phosphate, pH 7, or (b) 9 ml. of 0.1 N NaOH. The colorimetric readings performed immediately with a 400-mμ filter are plotted

[35] C. Huggins and J. Lapides, *J. Biol. Chem.* **170**, 467 (1947).

on the logarithmic scale of a semilogarithmic paper against the concentration of p-nitrophenol in micromoles per milliliter. Just before the enzyme assay is performed, the concentration of the substrate is determined by taking the colorimetric readings on 1 ml. of substrate, added again to (a) 9 ml. of the phosphate buffer or (b) 9 ml. of NaOH. The first reading gives the concentration of free p-nitrophenol in the substrate ($conc._1$), whereas the second reading gives the total concentration of p-nitrophenol in free or esterified state, due to the instantaneous hydrolysis of the ester ($conc._2$). A simple calculation permits the substrate adjustment to a concentration of exactly 0.333 micromole/ml.:

$$X = \frac{(conc._2 - conc._1)}{0.333} \times 98 - 98$$

The value X represents the volume to be added to the 98 ml. of dilute substrate (100 ml. − 2 ml. used for the assay) to bring it to the desired concentration.

Procedure. For the actual determination of activity, 1 ml. of the enzyme solution in appropriate concentration is mixed with 2 ml. of the phosphate buffer and 5 ml. of distilled water, directly in the colorimeter test tubes. The tubes are brought to 25° in a water bath, and at precise time intervals, usually 1 minute, 2 ml. of the adjusted substrate is added. The tubes are placed in the colorimeter, and the galvanometer set at 100. Also the air blank reading, with the test tube removed, is noted. After 20 minutes, the galvanometer is set to the same air blank reading, the tube is inserted, and the final measurement taken. These readings are corrected for the nonenzymatic hydrolysis, which is determined separately for the same time interval. From the standard curves of p-nitrophenol in phosphate buffer, the amount of p-nitrophenol liberated is determined, and this value is multiplied by the dilution factor of the enzyme preparation. In the assay of human rat serum, a 25-fold dilution was found to be most convenient, whereas for rabbit serum, a 100-fold dilution is better. One unit of esterase activity is defined as the amount of enzyme liberating 1 micromole of p-nitrophenol under above specified conditions.

Modifications. This method, although slightly complicated by the unstability of the substrate, was found by the reviewer to be of convenience where a large number of assays has to be run. As other chromogenic substrates, Nachlas and Seligman[36] have proposed esters of β-naphthol, and Hofstee[37] has suggested esters of salycilic acid. In the

[36] M. M. Nachlas and A. M. Seligman, *J. Biol. Chem.* **181**, 343 (1949).
[37] B. H. J. Hofstee, *Science* **114**, 128 (1951).

latter case, the free salycilic acid is determined by measuring the light adsorption at 290 to 300 mμ in a spectrophotometer.

Pancreas Lipase

Purification Procedures

Preparation of Dry Tissue Powders. 5 kg. of fresh hog pancreas glands (richest in lipase) are cooled to 0° within 4 to 6 hours of the animal's death. By maintaining this low temperature, the glands are freed mechanically of fatty and connective tissue and are passed several times through a meat grinder until a quite liquid paste is obtained. This paste (ca. 2.5 l.) is shaken for 4 to 6 hours with 2.5 l. of acetone and is then centrifuged. The residue is reextracted three more times with the same volume of acetone, twice with acetone-ether mixture (1:1), and twice with ether. The product is dried for 48 hours in vacuum, and a stable, friable powder is obtained. This method of Meyer et al.[38] is similar to that used by Willstätter and Waldschmidt-Leitz,[33] except for the much smaller volumes of solvent involved. Such preparations can be obtained from many other tissues, and soluble extracts are prepared by treating the powder with either 87% glycerol or dilute ammonia. Both the powder and glycerol solutions are quite stable.

Fractionation Scheme. Boissonnas[40] proposed the following purification scheme for pancreas:

Step 1. 60 g. of powdered dry hog pancreas is extracted for 36 hours at 4° on a shaker with 840 ml. of 0.1 M MgSO$_4$ added with 1 ml. of toluene and 2 drops of decylic alcohol. The insoluble material is eliminated by centrifugation in the cold, and the supernatant brought to pH 4.5 with 0.1 N HCl.

Step 2. To this solution, kept at 0°, 1000 ml. of a water-acetone mixture is added (60% by volume acetone), precooled to −20°. The mixture is stirred for 20 minutes, then centrifuged, and the supernatant discarded. A loss of about 28% of activity in this step could not be prevented.

Step 3. The precipitate is triturated slowly with 730 ml. of iced distilled water, stirred for 10 minutes, and then centrifuged. The supernatant is again rejected.

Step 4. The precipitate is treated with 40 ml. of 0.5 M SO$_4$Mg, and then to it are added 320 ml. of iced water and 12 ml. of 0.2 M acetic acid solution. The solution is stirred for 10 minutes and again centrifuged. This time the precipitate is discarded.

[38] K. H. Meyer, E. H. Fischer, and P. Bernfeld, *Helv. Chim. Acta* **30**, 75 (1947).
[39] R. Willstätter and E. Waldschmidt-Leitz, *Z. physiol. Chem.* **125**, 132 (1923).
[40] R. A. Boissonnas, *Helv. Chim. Acta* **31**, 1577 (1948).

Step 5. To the supernatant, an equal volume of distilled water is added, and the precipitated material is again eliminated by centrifugation.

Step 6. To the supernatant is added 1400 ml. of iced water, containing 12 ml. of 0.2 M acetic acid, and stirred for 15 minutes. The slight precipitate which occurs is isolated by centrifugation and is reported to contain 21.5% of the activity of the crude extract, with a 17.2-fold enrichment in terms of Kjeldahl nitrogen. Added to its weight in water, the product loses 15% of its activity in 2 days, at 0°. A preparation of a dry material from the precipitate was not reported. From an incomplete electrophoretic examination of the material in veronal buffer, 0.1 M pH 7.2, the presence of at least eight components, none of which constitutes more than 20% of the total protein material, was concluded.

SUMMARY OF PURIFICATION PROCEDURE FOR PANCREATIC LIPASE

Fraction	Activity[a] per mg. nitrogen	Enrichment factor	Recovery, %
1. Crude extract	145	1	100
2. Precipitate	568	3.9	63
3. Precipitate	810	5.6	51
4. Solution	1140	7.9	31
5. Solution	1160	8.0	22.5
6. Precipitate	2500	17.2	21.5

[a] The activity is expressed in Tween units, cf. p. 631.

Alternate Procedures. Glick and King[41] have reported to have obtained a 10-fold enrichment of the pancreas lipase by a salt fractionation scheme, involving two steps. Their method is used in certain commercial preparations of steapsin. The possibility of purification of the lipase by fractional adsorption methods was exhaustively investigated by Willstätter and Waldschmidt-Leitz.[39] Bamann and Laeverenz[42] have reported an accidental single crystallization of the lipase, which could not be repeated by Boissonnas.[40]

Properties of Pancreas Lipase

In this presentation, we are mainly concerned with the most important characteristics of the pancreas lipase and with its action on the more common substrates. Reference to much work will have to be omitted, owing to space limitations. This refers particularly to the complex question of

[41] D. Glick and C. G. King, *J. Am. Chem. Soc.* **55**, 2445 (1933).
[42] E. Bamann and P. Laeverenz, *Z. physiol Chem.* **223**, 1 (1934).

the stereochemical specificity and the synthetic action of this enzyme. These and other topics were recently reviewed in greater detail by Ammon and Jaarma.[4]

Unfortunately, there is, as yet, no crystalline lipase available, and most of the work was carried out with rather crude tissue homogenates or extracts. This fact probably accounts for the numerous contradictory findings, because the action of lipases depends to a very high degree on the extraneous substances present, particularly if surface active.

Specificity. Balls *et al.*[23] have found no products of partial hydrolysis in the digestion of tristearin, in distinction to Frazer and Sammons,[43] who isolated large amounts of mono- and diglycerides after both *in vivo* and *in vitro* digestion of olive oil. In a more detailed study, Desnuelle *et al.*[44] have observed that in the absence of calcium ions diglycerides accumulate, whereas the addition of these ions accelerates further hydrolysis. The presence of two lipases with different specificity in pancreas was, however, suggested by Fodor[45] and Boissonnas.[40]

It would appear that this contention could be settled from a study of the comparative hydrolysis rates of the tri-, di-, and monoglycerides. Such a study was carried out by Sobotka and Glick[5] on esters of butyric

	Tributyrin	Dibutyrin	Monobutyrin	Methyl butyrate
Relative velocities	7.12	4.00	2.96	1.00
Michaelis constant	6×10^{-4}	100×10^{-4}	750×10^{-4}	900×10^{-4}

acid using hog pancreas. However, it was repeatedly observed that the extent of hydrolysis of mixed substrates does not parallel the relative velocity rates of the isolated components, owing to competitive substrate inhibition.[46-49]

The relative velocity rates for various substrates are best summarized in the following table.

[43] A. C. Frazer and H. G. Sammons, *Biochem. J.* **39**, 122 (1945).

[44] P. Desnuelle, M. Naudet, and M. J. Constantin, *Biochim. et Biophys. Acta* **5**, 561 (1950); **7**, 251 (1951).

[45] P. J. Fodor, *Arch. Biochem.* **26**, 307 (1950); **28**, 274 (1950).

[46] R. Willstätter, R. Kuhn, O. Lind, and F. Memmen, *Z. physiol. Chem.* **167**, 303 (1927).

[47] E. Bamann and M. Schmeller, *Z. physiol. Chem.* **188**, 251 (1930).

[48] W. F. Shipe, *Arch. Biochem.* **30**, 165 (1951).

[49] P. J. Fodor, *Arch. Biochem. and Biophys.* **35**, 311 (1952).

COMPARISON OF RELATIVE INITIAL VELOCITIES OF HYDROLYSIS

Substrate	Hog pancreas lipase[a]	Hog pancreas lipase[b]	Hog pancreas lipase[c]	Bovine pancreas lipase[c]	Woman milk lipase[c]	Human gastric lipase[d]
Triacetin		53	0.33	1.9	2.4	
Tripropionin		100	20.7	18.9	18.9	27
Tributyrin	7.12	87	100	100	100	100
Trivalerin	2.80	40				
Tricaproin	3.40	71	36.6	36.8	58	39
Triheptylin	5.00					
Tricaprylin	4.54		36.6	40.6	57.5	
Tricaprin			31.9	37	51.5	13
Trilaurin			31.1	23.5	8.1	2
Trimyristin			11.0	12.7	0.49	
Tristearin	0.005		0.47	1.3	0.37	0.8
Triolein	0.17					Negligible

[a] H. Sobotka and D. Glick, *J. Biol. Chem.* **105**, 199 (1934).

[b] S. S. Weinstein and A. M. Wynne, *J. Biol. Chem.* **112**, 641, 649 (1935–1936).

[c] F. Schönheyder and K. Volqvartz, *Enzymologia* **11**, 178 (1944).

[d] F. Schönheyder and K. Volqvartz, *Acta Physiol. Scand.* **10**, 62 (1945); **11**, 349 (1946).

It should be noted that Balls *et al.*[23] have found that the hydrolysis of the higher, but not of the lower, saturated triglycerides is very dependent on the temperature. At low temperatures the pancreatic lipase would appear to be an esterase, and with increase in temperature the maximum rate of hydrolysis shifts toward longer chain fatty acids (C_7–C_{10}). Schönheyder and Volqvartz[29] caution that erroneous conclusions can be reached if the comparison is not drawn on the basis of initial velocities, as proportionality exists only up to 20% of hydrolysis. Besides, the optimum shifts from pH 7 to 8.8 with increase in chain length of the fatty acids.

There is general agreement that unsaturated fatty acid glycerides are more easily split than their saturated analogs by most lipases, with the exception of gastric human lipase, which is apparently more active toward tristearin than triolein.[31] The very fast hydrolysis of benzyl esters was already referred to (cf. p. 630), and the great influence of alcohols in general on the rate of hydrolysis of stearic acid esters was studied by Balls and Matlack.[30] Even or odd numbers of carbons in the fatty acids, obtained either by the Fischer-Tropsch synthesis[26] or by addition of one methylene group to the coconut fatty acids,[27] do not seem to influence the hydrolysis.

The Michaelis constant of hog pancreas and tributyrate was reported as 6×10^{-4} (see above) and was found to be influenced by the addition of

octyl alcohol as activator. For bovine and hog pancreas lipase and triacetin the constant (K_s = 0.25) was found, on the other hand, not to be
influenced by sodium oleate.[50] Weinstein and Wynne[28] could not fit their
experimental data to the Michaelis equation. Schönheyder and Volqvartz,[51] in limiting their investigation to the strict solubility range of
substrates, could not confirm the above data and found direct proportionality between substrate concentration and lipase activity. This relationship was not altered by the addition of activating $CaCl_2$, bile salts, or
egg albumin.

Activators. Calcium salts increase the activity of the enzyme in homogeneous as well as in heterogeneous systems on both sides of the optimum
pH, whereby on the acid side the range of hydrolytic activity was found
to be broadened by one pH unit.[31] This effect can be ascribed in part to
the precipitation of the calcium salts of inhibitory higher fatty acids, but
the same mechanism cannot account for the activation of the hydrolysis
of esters of lower fatty acids. It appears more probable that this is a case
of a specific action of calcium ions on the enzyme itself, similar to the
now well-established mechanism in the case of trypsin.[52]

A number of surface active substances, such as octyl alcohol,[5] bile
salts,[23] and albumin,[14] were found to increase the activity of the pancreas
lipase. Some of these substances may act as activators, at higher pH
values, or as inhibitors, at lower values, and no correlation between this
inversion point and their dissociation constants could be found.[50] Glick
and King[53] observed that a series of higher alcohols activate the pancreas
lipase but inhibit the liver esterase, there being a parallelism between the
activating, resp. inhibiting action and the surface activity of the compounds. Boissonnas[13] could not find, however, any influence of decylic
alcohol in homogeneous systems. Of considerable interest also is the
activating effect of glycyl and leucyl polypeptides.[54,55]

Inhibitors. The competitive substrate inhibition in mixtures of substrates was already mentioned (cf. p. 636). There is, however, a long list
of other substances which have an outright inhibiting effect on pancreas
lipase. A variety of ketones and aldehydes with free carbonyl groups are
inhibitory, the effect being proportional to their molecular volume.
Phenols and cresols, except dihydroxy phenols, are relatively inert.

[50] K. Krähling and H. H. Weber, *Biochem. Z.* **298**, 227 (1938).
[51] F. Schönheyder and K. Volqvartz, *Acta Physiol. Scand.* **7**, 376 (1944).
[52] M. Bier and F. F. Nord, *Arch. Biochem. and Biophys.* **31**, 335 (1951); **33**, 320 (1951);
     F. F. Nord and M. Bier, *Biochim. et Biophys. Acta* **12**, 56 (1953).
[53] D. Glick and C. G. King, *J. Biol. Chem.* **94**, 497 (1931–32); **97**, 675 (1932).
[54] R. Willstätter and F. Memmen, *Z. physiol. Chem.* **129**, 1 (1923).
[55] E. Abderhalden and W. Geidel, *Fermentforschung* **13**, 156 (1933).

Heavy metals and also halogen ions seem to inhibit in the sequence $F > I > Br > Cl$.[28] Inhibition is also caused by alkaloids,[56] alcohols,[57] fluorophosphate,[58] chloroform, bromoform and formaldehyde,[59] various lactones, but not lactides.[60] Atoxyl inhibits the liver esterase but not the pancreatic lipase, and quinine inhibits the pancreatic lipase but not the liver esterase or stomach lipase.[61]

The pancreatic lipase is activated by cyanides and thio compounds,[28] and inhibited by the usual —SH blocking reagents,[62] but not to such an extent as other —SH sensitive enzymes. Singer[63] studied in greater detail the similar behavior of wheat germ lipase and found that the apparent inhibition increased with the molecular volume of the substrate employed in the assay. Thus, it appears that the —SH groups are not directly indispensable for the hydrolytic activity, but that the —SH blocking reagents cause a steric interference to the approach of the substrate to the active sites on the enzyme.

Optimal Conditions. The optimal conditions of temperature and pH depend on the substrate, state of purity of the enzyme, buffers, method of assay, etc. In the case of pancreatic lipase the optimum temperature for most substrates is about 37° (cf. p. 637). With Tween as a substrate, in phosphate buffer, the optimum is, however, about 25°.[34]

The optimum pH for hydrolysis of lower triglycerides is around pH 7. For the higher triclycerides, it is shifted up to pH 8.8.[31] The optimum for the hydrolysis of Tweens is around pH 6.8 in phosphate buffer.[34] With glycine,[28] NH_4OH—NH_4Cl,[14] or borate buffer,[64] the optimum pH seems to be shifted toward more alkaline pH values than in phosphate or mixed phosphate-borate buffer.[28,64] However, for the hydrolysis of esters of dicarboxylic acids, such as succinic and malonic acid esters, hydrolysis takes place only in acid media, around pH 5.[65] The optima for the lipases of other animal tissues were listed by Ammon.[16]

Stearn[66] has calculated the energy and the entropy of thermal inac-

[56] P. Rona and R. Pavlovic, *Biochem. Z.* **130**, 225 (1922); P. Rona and R. Ammon, *ibid.* **181**, 49 (1928).
[57] D. R. P. Murray and C. G. King, *Biochem. J.* **23**, 292 (1929); **24**, 190, 1890 (1930).
[58] E. C. Webb, *Biochem. J.* **42**, 96 (1948).
[59] L. S. Palmer, *J. Am. Chem. Soc.* **44**, 1527 (1922).
[60] E. Bamann and M. Schmeller, *Z. physiol. Chem.* **194**, 14 (1931); E. Bamann, E. Schweizer, and M. Schmeller, *ibid.* **222**, 121 (1933).
[61] P. Rona and H. Petow, *Biochem. Z.* **146**, 144 (1924).
[62] T. P. Singer and E. S. G. Barron, *J. Biol. Chem.* **157**, 241 (1945).
[63] T. P. Singer, *J. Biol. Chem.* **174**, 11 (1948).
[64] B. S. Platt and E. R. Dawson, *Biochem. J.* **19**, 860 (1925).
[65] E. Bamann and E. Rendlen, *Z. physiol. Chem.* **238**, 133 (1936).
[66] A. E. Stearn, *Ergeb. Enzymforsch.* **7**, 1 (1938).

tivation at pH 6 of the pancreatic lipase as being 46,000 cal./mole and 68.2 cal./degree/mole, respectively. For the activation energy of the pancreatic hydrolysis of tributyrin at pH 8.1, Sizer[67] found a sharp break in the curve at about 0°, the value being 37,000 cal./mole in the temperature range −70 to 0°, and only 7600 in the range 0 to 50°. Such discontinuities in the curve can be expected in the light of the changes observed by Nord and co-workers[68,69] which take place on freezing in the physical and colloidal properties of various systems. The values themselves fall within the range of values encountered in other enzymatic systems and are in good agreement with the value of 8700 cal./mole found by Schwartz[70] for the activation energy of the hydrolysis of tributyrin. For the hydrolysis of trivalerin, tricaproin, triheptylin, and tricaprylin, the activation energies were reported to vary from 8500 to 23,700 cal./mole, as the substrate concentration was decreased from 0.1 M to 0.002 M.

Lipases from Fungi and Seeds

Preparation of Lipase from *Fusarium lini* Bolley

According to Fiore and Nord,[71] *Fusarium lini* Bolley, strain No. 5140, of the Biolog. Reichsanstalt, Berlin-Dahlem, is grown on a Raulin-Thom medium containing 2.5% glucose. After 21 days of growth the mycelium is separated, air dried, and then dried over sulfuric acid in a vacuum desiccator. Before extraction of the lipase, the mold is ground to fine powder, and exhaustively extracted with ethyl ether in a Soxhlet.

Step 1. 30 g. of the fat-free powder is homogenized for 5 minutes in a blendor with 400 ml. of distilled water and centrifuged. The cloudy supernatant is cooled to 0°, brought to pH 5 with 1 N HOAc, and centrifuged again; this procedure clarifies the supernatant without a significant loss in activity.

Step 2. The clear solution, cooled to 0° (essential), is then poured slowly with stirring into 3 vol. of 95% alcohol, precooled to −50° in a dry ice-acetone bath. The precipitate formed is rapidly separated by high-speed centrifugation, without allowing the temperature to rise above −10°, and is then dehydrated by washing with precooled acetone and again centrifuged. This residue contains most of the original activity; it is freed from acetone in high vacuum. It is obtained as a white powder, stable in

[67] I. W. Sizer, *Advances in Enzymol.* **3**, 35 (1943).
[68] F. F. Nord, *Ergeb. Enzymforsch.* **1**, 77 (1932).
[69] F. F. Nord and M. Bier, *in* "Handbuch der Kältetechnik" (Plank, ed.), Vol. IX, p. 84, Springer Verlag, Berlin, 1952.
[70] B. Schwartz, *J. Gen. Physiol.* **27**, 113 (1943).
[71] J. V. Fiore and F. F. Nord, *Arch. Biochem.* **26**, 382 (1950).

the dry state, and is soluble. If the dehydration is not carried out properly, however, a brown, inactive gum will be obtained.

Step 3. Considerable further purification can be obtained by repeating the alcohol precipitation. The alcohol precipitate is dissolved in distilled water at 0°, the solution filtered over a thin layer of amphibole with the aid of suction, dialyzed for 2 to 3 hours in a shaking dialyzer against cold distilled water, and then the alcohol precipitation is repeated as described in Step 2. The dried product is a white, friable powder, stable for long periods of time in the icebox and perfectly soluble. Electrophoretic analysis at pH 7.4 in a phosphate buffer showed that besides an immobile component (probably not proteinlike in nature) there are two major components present. Its specific activity was increased 15-fold over this of the starting material.

SUMMARY OF PURIFICATION PROCEDURE OF LIPASE FROM *Fusarium lini* B.

Fraction	Activity[a] per mg. nitrogen	Enrichment factor	Recovery, %
1. Starting product	0.4	1	100
2. Alcohol precipitate	2.4	6	20
3. Final product	6.4	16	10

[a] The lipase unit is defined as that amount of enzyme which liberates 0.01 millimole of acetic acid when allowed to act on 10 ml. of 5% triacetin and 0.5 ml. of McIlvaine buffer at pH 7, 37°, for 3 hours, the final volume being 12.5 ml.

Other Lipases

The presence of lipases was reported in *Mycotorula lipolytica*,[72] *Penicillium roqueforti* and *Aspergillus niger*,[48] and yeast.[73] The esterase component of takadiastase of *Aspergillus oryzae*[74] and various bacteria[75,76] was also studied.

The usual source of plant lipases are seeds, and the most active one is found in castor beans.[77] It is an unusual enzyme, in so far as it appears to be insoluble in the usual solvents.[24,25,78,79] The oat lipase can be also

[72] I. I. Peters and F. E. Nelson, *J. Bacteriol.* **55**, 581 (1948).

[73] G. Gorbach and H. Güntner, *Sitzber. Akad. Wiss. Wien. Math.-naturw. Kl.* Abt. IIb, 141, 415 (1932).

[74] P. Rona and R. Ammon, *Biochem. Z.* **181**, 49 (1927).

[75] M. Bayliss, D. Glick, and R. A. Siem, *J. Bacteriol.* **55**, 307 (1948).

[76] M. P. Starr and W. H. Burkholder, *Phytopathology* **32**, 598 (1942).

[77] W. Connstein, E. Hoyer, and H. von Wartenberg, *Ber.* **35**, 3988 (1902).

[78] E. Hoyer, *Z. physiol. Chem.* **50**, 414 (1907).

[79] R. Willstätter and E. Waldschmidt-Leitz, *Z. physiol. Chem.* **134**, 161 (1924).

isolated in an insoluble form by repeated freezing.[80] Germination of seeds seems to give rise to a modified enzyme, called blastolipase, as was observed in castor beans,[79] cotton,[81] and pine seeds.[82]

The lipase of wheat germs was particularly studied by Singer and Hofstee,[83] and a number of other sources were screened by Bamann et al.[84]

Properties of the Lipase Found in *Fusarium lini* Bolley

The specificity of the *Fusarium* lipase can be illustrated by the following series of esters, presented in decreasing order of their hydrolysis rate: triacetin, tributyrin, ethyl acetate, *n*-butyl butyrate, ethyl propionate, ethyl butyrate, *Fusarium* oil, olive oil, ethyl laurate, and ethyl stearate.[17] The lipase is very stable in the dry state and contains no —SH or —S—S— groups essential for its activity. The instability of the lipase in solution is probably due to the presence of a proteolytic enzyme. The activity curve presents a sharp optimum at pH 7 in McIlvaine buffer, and the temperature optimum is at 37°.[71]

[80] H. F. Martin and F. G. Peers, *Biochem. J.* **55**, 523 (1953).

[81] H. S. Olcott and T. D. Fontaine, *J. Am. Chem. Soc.* **63**, 825 (1941).

[82] H. W. Nicolai, *Biochem. Z.* **174**, 373 (1926).

[83] T. P. Singer and B. H. J. Hofstee, *Arch. Biochem.* **18**, 229 (1948).

[84] E. Bamann, E. Ullmann, and N. Tietz, *Biochem. Z.* **323**, 489 (1953); **324**, 134, 249 (1953).

[107] Acetylcholinesterase

By DAVID NACHMANSOHN and IRWIN B. WILSON

Introduction: Biological Significance

Acetylcholine is an ester of great biological significance. Its powerful pharmacological action, discovered in 1906 by Reid Hunt and Taveau, led some pharmacologists and physiologists to believe that the ester is a "neurohumoral transmitter" from the nerve endings to the effector organ or from the nerve ending to a second nerve cell. However, extensive investigations during the last fifteen years based on biochemical and physicochemical methods have shown that the action of acetylcholine is essential for the generation of bioelectric currents which propagate impulses along nerve and muscle fibers.[1-3] At present it appears that the

[1] D. Nachmansohn, *in* "Modern Trends in Physiology and Biochemistry" (E. S. G. Barron, ed.), p. 229, Academic Press, New York, 1952.

[2] D. Nachmansohn, *Harvey Lectures* (1953/54) in press.

[3] I. B. Wilson and D. Nachmansohn, *in* "Ion Transport across Membranes" (H. T. Clarke, ed.), p. 35, Academic Press, New York, 1954.

ester generates these currents by changing the permeability to sodium ion whereby ionic concentration gradients between the inside of the fiber and its outer environment, the potential sources of the electromotive force, become effective. Having produced its action, the ester is inactivated by the enzyme acetylcholinesterase, hydrolyzing it to its two inactive components: choline and acetic acid. The resting condition is thus restored, and the next impulse is able to pass. The action of the enzyme is thus essential for the elementary process of conduction. Owing to its vital function the enzyme has attracted the interest of many investigators, and its properties have been discussed in several reviews (see, e.g., footnotes 4, 5).

I. Special Properties and Distinction from Other Esterases

Esterases of various types are present in many tissues. The esterase whose physiological function is to hydrolyze acetylcholine has, however, a number of special properties by which it can be easily distinguished from other types of esterases. It has, therefore, been termed acetylcholinesterase.[6] The historical development by which these properties have been established may be found elsewhere.[4] Here are summarized a few of the main characteristics which determine whether or not an esterase may be classified as acetylcholinesterase.

The enzyme has a well-defined optimum substrate concentration for acetylcholine which is about 4 to 7 micromoles per milliliter, depending on the source. At higher concentrations the activity decreases. If the activity is plotted against pS, the negative logarithm of substrate concentration, the resulting curve has a "bell-shaped" form. Increasing the number of C atoms of the acyl chain from 2 to 3 does not markedly affect the activity; however, with butyrylcholine as substrate the activity is sharply decreased.[7] To obtain the same activity as with acetylcholine the enzyme concentration must be increased by a factor of 140. The hydrolytic power is still more decreased with benzoylcholine as substrate. Here the factor is about 2000 (I. B. Wilson, unpublished data). Acetyl-β-methyl choline, on the other hand, is hydrolyzed rapidly although at a considerably lower rate than acetylcholine itself. However, there is not such a sharp optimum of substrate concentration. No inhibition at high substrate concentration is observed with dimethylaminoethyl acetate

[4] D. Nachmansohn and I. B. Wilson, *Advances in Enzymol.* **12**, 259 (1951).

[5] K.-B. Augustinsson, *in* "The Enzymes" (J. B. Sumner and K. Myrbäck, eds.), Vol. I, Part 1, p. 443, Academic Press, New York, 1950.

[6] K.-B. Augustinsson and D. Nachmansohn, *Science* **110**, 98 (1949).

[7] D. Nachmansohn and M. A. Rothenberg, *J. Biol. Chem.* **158**, 653 (1945).

as substrate (Wilson[8]). Noncholine esters are split, but the concentrations required are much higher and the maximum rates are usually very much lower. The Michaelis constant of electric eel acetylcholinesterase, when the substrate is acetylcholine, is 4.6×10^{-4}.

For classifying an esterase as acetylcholinesterase it is sufficient in most cases to determine whether there is an optimum substrate concentration and whether the hydrolysis of butyrylcholine is very low compared to that of acetylcholine. Occasionally, of course, it may be desirable to test some of the other characteristics. It may be stressed that the properties of the enzyme have been discussed only as a means of determining the type of esterase. Much has been learned during the last few years about many other characteristics and properties and especially about the molecular forces acting between the enzyme protein and substrate. The reader interested in these aspects is referred to other reviews.[4,9]

There are other types of esterases which hydrolyze choline esters faster than noncholine esters but differ distinctly from acetylcholinesterase. These esterases can be readily distinguished by three features: (1) the rate of hydrolysis increases if the length of the acyl chain increases from 2 to 4 C atoms; (2) the activity substrate concentration relationship does not show a well-defined optimum; i.e., there is no substrate inhibition at higher concentrations; (3) it does not hydrolyze acetyl-β-methyl choline. Since the physiological substrate of these esterases, if any, is not yet established, a proper name cannot be assigned at present; following usual enzyme terminology they may be referred to as cholinesterases. Human and horse serum have, for instance, predominantly this type of esterase. It has been reported that other tissues also contain enzymes resembling the type present in serum.

II. Sources and Preparations

Since acetylcholinesterase is essential for conduction, the enzyme is found in all conductive tissue throughout the animal kingdom. The enzyme is localized exclusively in the active membrane, however; therefore the concentration per gram of tissue is generally low. There are, however, several sources which offer a material which is very rich in acetylcholinesterase. A very high concentration of the enzyme is found, for instance, in the nucleus caudatus in the brain of mammals.[10] Nucleus

[8] I. B. Wilson, *J. Biol. Chem.* **197**, 215 (1952).
[9] I. B. Wilson, *in* "The Mechanism of Enzyme Action" (W. D. McElroy and B. Glass, eds.), p. 642, The Johns Hopkins Press, Baltimore, 1954.
[10] D. Nachmansohn, *Bull. soc. chim. biol.* **21**, 761 (1939).

caudatus of ox brain can be readily obtained in sufficiently large amounts to provide a starting material for purification. Per gram fresh weight this tissue is able to hydrolyze 1.5 millimoles of acetylcholine per hour. On homogenization in the Waring blendor and after centrifugation of the suspension, about half of the enzyme is found in the precipitate, but the rest is in the supernatant solution. However, no satisfactory method of purification has been developed. Squid head ganglia are a material with a very high concentration of acetylcholinesterase.[11] Per gram of fresh weight 15 to 30 millimoles of acetylcholine are split per hour. The protein content of the ganglia is about 20%. However, the material is not too readily obtainable, and so far no method has been worked out for purification. A third source is various cobra snake venoms which per milligram weight may split 0.5 to 1.5 millimoles of acetylcholine per hour (Zeller[12]). The enzyme has all the characteristics of acetylcholinesterase (Augustinsson[13]). Here again the material is not too readily available.

A. Electric Organs. An unusually favorable source for obtaining highly active enzyme solutions and for purifying the enzyme is the electric organs of electric fish. The suitable species are the *Electrophorus electricus* and the various types of *Torpedo*. The *Electrophorus* is found in the Amazon River; its electric tissue hydrolyzes 10 to 20 millimoles of acetylcholine per gram fresh weight, although the tissue contains only 2% of proteins and 92% of water. Specimens of the different types of *Torpedo* are found in various parts of the world. The enzyme concentration and the protein content are similar to those in *Electrophorus*. The main drawback is obviously the difficulty encountered in the procurement of electric organs. It may be mentioned, however, that in Europe for instance there are two marine biological stations in which *Torpedos* may be obtained—in Naples, Italy, and in Arcachon, France.

Purification of acetylcholinesterase is favored by the great stability of the enzyme. It may be kept in the refrigerator for years without loss of activity. Freezing does not affect the enzyme. As a preliminary step to purification the mucin which, as was shown by Bailey,[14] is present in the tissue in considerable amounts must be removed, for it is a disturbing factor for the process of purification.[15] The mucin may be readily removed in the following manner. The tissue is cut into small pieces and allowed to remain in the refrigerator under toluene for about 1 month. A considerable amount of exudate separates, which has a high mucin content.

[11] D. Nachmansohn and B. Meyerhof, *J. Neurophysiol.* **4,** 348 (1941).

[12] E. A. Zeller, *Advances in Enzymol.* **8,** 459 (1948).

[13] K.-B. Augustinsson, *Arch. Biochem.* **23,** 111 (1949).

[14] K. Bailey, *Biochem. J.* **23,** 255 (1939).

[15] M. A. Rothenberg and D. Nachmansohn, *J. Biol. Chem.* **168,** 223 (1947).

The exudate is discarded periodically. At the end of 4 to 5 weeks the tissue contains very little mucin.[16]

Once the mucin has been removed, a relatively high degree of purity may be obtained by fractional ammonium sulfate precipitation. The tissue is homogenized in the Waring blendor with twice its weight of 5% ammonium sulfate. The suspension is centrifuged, and the precipitate discarded. The yields of a second extraction are relatively small. About 1 l. of extract is obtained per kilogram of tissue freed from mucin (about 40% of the fresh weight), with a total activity equivalent to 5 to 10 moles of acetylcholine hydrolyzed per hour, depending on the starting material. The smaller the specimen, the higher the concentration. The extract is successively fractionated between 15 and 40, 19 and 29, 19 and 28, and finally 20.5 and 27% ammonium sulfate. The last fraction is reprecipitated at 27% ammonium sulfate, and a product is obtained having a specific activity of 120 millimoles of acetylcholine hydrolyzed per hour per milligram of protein, representing a 75-fold purification over the homogenate.[15] The yield is about 15%. The fractionations are performed at pH 5.6 to 5.9. Higher acidities may result in loss of activity.

The fractional ammonium sulfate precipitation is a rather tedious procedure, and occasionally unexpected difficulties may be encountered. Although some promising attempts of improving the method have been made in this laboratory, a simple and really satisfactory procedure has not yet been worked out.

A further separation of inactive protein from the enzyme protein may be obtained by high-speed centrifugation. In this way solutions were obtained in which the specific activity was close to 400 millimoles. At this stage both analytical ultracentrifuge runs and electrophoretic studies showed only one component. On the basis of the sedimentation rate the molecular weight of the enzyme is close to $2\frac{1}{2}$ million.[15]

B. *Red Blood Cells.* A method has recently been reported for purifying acetylcholinesterase from ox red blood cells (Cohen and Warringa[17]) which produces a 250- to 400-fold purification of the esterase as compared with red cell hemolyzates. Freeze-dried stromata are extracted with cold, dry butanol, vacuum-dessicated over silica, treated in a Waring blendor with 0.01 M phosphate buffer, pH 8, and centrifuged at 14,000 r.p.m. The enzyme is then soluble in the supernatant.

[16] Recently Hargreaves and Lobo [*Arch. Biochem. and Biophys.* **10**, 195 (1953)] have reported in a preliminary note that fresh electric tissue may be treated in a way which, without removal of mucin, brings the degree of purity to a relatively high level within 2 days. An attempt to repeat this procedure in our laboratory failed, but this is probably due to the rather incomplete information given in the short preliminary note, and the evaluation of the method has to wait for full description.

[17] J. A. Cohen and M. G. P. J. Warringa, *Biochim. et Biophys. Acta* **10**, 195 (1953).

Ammonium sulfate fractionation, treatment with Lloyd's reagent, and precipitation with ammonium sulfate yields a preparation with an activity of 250 micromoles per hour per milligram of protein. The authors give no indication concerning the yield, however, and the method represents a purification of only 20-fold over the stromata.

A method for purifying human erythrocyte acetylcholinesterase (Zittle et al.[18]) uses polyoxymethylene sorbitan monolaurate (Tween 20) and toluene to solubilize the enzyme. The method involves hemolysis with 1 M acetic acid, extensive washing of stromata, precipitation with cadmium acetate, solubilization with Tween 20, ammonium sulfate precipitations, calcium phosphate gel, dialysis, and freeze drying. The dry powder is extracted successively with acetone, ethanol, and ether. The final product contains 10 to 20% of the original enzyme at a purification of 240-fold over red cells (11-fold over stroma) and hydrolyzes 440 micromoles of acetylcholine per hour per milligram of protein at 38°.

C. Serum Esterase. Purification of cholinesterase from horse serum was first described by Stedman and Stedman.[19] A much higher purification has been obtained by Strelitz.[20] The method utilizes ammonium sulfate and high acidity (pH 2.8), followed by repeated ammonium sulfate fractionations and Lloyd's reagent. The final dialyzed solution is dried *in vacuo* over $CaCl_2$. The preparation hydrolyzed 1.2×10^4 micromoles of acetylcholine (0.03 M) per hour per milligram of powder at 37.5°.

Human serum esterase has been purified (Surgenor et al.[21]) by alcohol fractionation starting with a preparation called Fraction IV by the group in E. J. Cohn's laboratory at Harvard. The preparation designated IV-6 had an activity of 45 micromoles of acetylcholine hydrolyzed per milligram per hour. Further purification yields a fraction IV-6-3 with an activity of about 360 micromoles per milligram per hour.

III. Measurement of Activity

Even though purified enzyme is very soluble, in many cases the enzyme is not readily separated from the tissue debris of a homogenate and brought into solution. It is therefore generally necessary to assay a suspension of the homogenate without centrifugation if full activity is to be obtained. The medium should have an ionic strength of about 0.1 to 0.2 and contain at least 0.01 M $MgCl_2$. Although purified enzyme does

[18] C. A. Zittle, E. S. Della Monica, and J. H. Custer, *Arch. Biochem. and Biophys.* **48,** 43 (1953).
[19] E. Stedman and E. Stedman, *Biochem. J.* **29,** 2563 (1935).
[20] F. Strelitz, *Biochem. J.* **38,** 86 (1944).
[21] D. M. Surgenor, L. E. Strong, H. L. Taylor, R. S. Gordon, Jr., and D. M. Gibson, *J. Am. Chem. Soc.* **71,** 1223 (1949).

not require Mg^{++}, impure preparations are frequently greatly activated by Mg^{++} ion. A suitable medium contains 0.1 M NaCl and 0.04 M MgCl$_2$, in addition to any buffer which the method may require. The optimum substrate concentration varies with the source of acetylcholinesterase, being, for example, 0.004 M for electric tissue esterase and 0.007 M for erythrocyte enzyme, but since the maxima are quite broad the selection of any substrate concentration in this range will usually be satisfactory. The contra ion to acetylcholine does not appear to be of importance— in general the common anions in low concentration do not affect the activity.

A. *Titration Methods.* The acetic acid formed by the hydrolysis of acetylcholine may be titrated with standardized alkali using indicators such as phenol red, phenolphthalein, or bromothymol blue (Stedman *et al.,*[22] Vahlquist,[23] Bovet and Santenoise,[24] Glick,[25] Sawyer[26]), but a pH meter is preferable (Alles and Hawes,[27] Glick[28]). The pH is maintained within reasonably narrow limits of a preselected value by the frequent addition of alkali. The addition of alkali may be performed automatically with an automatic titrator (Wilson[29]). Little or no buffer is required.

B. *Manometric Method.* The enzyme is assayed in a bicarbonate-carbon dioxide-buffered solution. The acetic acid produced by the hydrolysis converts bicarbonate to CO_2, which, since the volume of the reaction vessel is maintained constant, is reflected by an increase of pressure. The Warburg manometric method was first applied by Ammon[30] and because of its convenience has been widely used. The pH can be varied, but only over a relatively small range, by altering the concentration of bicarbonate and the partial pressure of CO_2. With 0.025 M NaHCO$_3$ and 5% CO_2 at 1 atmosphere and 25°, the pH is 7.4.

The Cartesian diver microchemical method was first used for esterase activity by Linderstrøm-Lang and Glick.[31] It is of advantage if only extremely small quantities of tissue are available, as, e.g., in studies by Boell and Nachmansohn[32] of enzyme localization in the nerve axon.

[22] E. Stedman, E. Stedman, and L. H. Easson, *Biochem. J.* **26,** 2056 (1932).
[23] B. Vahlquist, *Skand. Arch. Physiol.* **72,** 133 (1935).
[24] D. Bovet and D. Santenoise, *Compt. rend. soc. biol.* **135,** 844 (1941).
[25] D. Glick, *J. Gen. Physiol.* **21,** 289 (1938).
[26] C. H. Sawyer, *J. Exptl. Zoöl.* **94,** 1 (1943).
[27] G. A. Alles and R. C. Hawes, *J. Biol. Chem.* **133,** 375 (1940).
[28] D. Glick, *Biochem. J.* **31,** 521 (1937).
[29] I. B. Wilson, *J. Biol. Chem.* (1954) in press.
[30] R. Ammon, *Pflügers Arch. ges. Physiol.* **233,** 486 (1933).
[31] K. Linderstrøm-Lang and D. Glick, *Compt. rend. trav. lab. Carlsberg, sér. chim.* **22,** 300 (1938).
[32] E. J. Boell and D. Nachmansohn, *Science* **92,** 513 (1940).

C. Colorimetric Method. A quantitative colorimetric procedure for acetylcholine has been developed by Hestrin,[33] based on the reaction of O-acyl derivatives with alkaline hydroxylamine (see, for example, the textbook by Meyer and Jacobson[34] to yield acethydroxamic acids). All hydroxamic acids give a red-purple color with ferric ion in acid solution, ascribed to the formation of a soluble inner complex salt:

$$
\begin{array}{c}
\text{H} \\
\text{N—O} \\
\diagup \qquad | \\
\text{R—C} \qquad | \\
\diagdown \qquad | \\
\text{O—Fe/3}
\end{array}
$$

Feigl *et al.*[35,36] have described and used these reactions as spot tests for carboxylic acids, esters, and anhydrides. Lipmann and Tuttle have developed the method for acyl phosphates.[37] Anhydrides and S-acyl derivatives react at pH 6 or higher, but highly alkaline conditions are required for esters.

A solution of the ester is added to alkaline hydroxylamine, and after 1 minute the mixture is brought to pH 1.2 ± 0.2 with HCl. Finally $FeCl_3$ (10% in 0.1 N HCl) is added, and the optical absorption is measured at 520 to 540 mμ.

This method is very convenient and may be run at any pH, whereas the manometric method can be run only in a limited pH range. The method is more convenient than the titration method, but it is less accurate, since it measures the remaining ester and the activity is obtained by difference. Generally about 30% of the ester must be hydrolyzed. There is an advantage for some purposes in that the enzymatic hydrolysis can be run for as little as one-half minute. The incubation medium must be buffered; 0.01 M phosphate solution at pH 7.0 is a convenient buffer.

D. Determination of the Esterases in Blood. A simple and rapid method for determining the individual activities of serum esterase (cholinesterase) and erythrocyte esterase (acetylcholinesterase) in whole blood has recently been described (Augustinsson and Heimbuerger[38]). Two 0.05-ml. samples are removed from the fingertip with a blood pipette, transferred to filter paper, and dried in air. If desired, the dried papers may be stored

[33] S. Hestrin, *J. Biol. Chem.* **180**, 249 (1949).
[34] V. Meyer and P. Jacobson, "Lehrbuch der organischen Chemie," Vol. I, Berlin and Leipzig, 1922.
[35] F. Feigl, V. Anger, and O. Frehden, *Mikrochemie* **15**, 12 (1934).
[36] F. Feigl, V. Anger, and O. Frehden, *Mikrochemie* **15**, 18 (1934).
[37] F. Lipmann and L. C. Tuttle, *J. Biol. Chem.* **159**, 21 (1945).
[38] K.-B. Augustinsson and G. Heimbuerger, *Acta Physiol. Scand.* **30**, 45 (1953).

at 4° for months before assay. Each blood spot is cut into strips and assayed, paper and all, one with butyryl choline ($6.4 \times 10^{-3} M$) and the other with acetyl-β-methylcholine ($6.8 \times 10^{-3} M$) as substrate. In the above paper[38] the manometric method was used at 25°. The strips must soak for half an hour before the substrate is introduced. As already described, butyrylcholine assays only the serum esterase, and acetyl-β-methylcholine only the erythrocyte esterase.

The individual activities can be converted to acetylcholine ($7.4 \times 10^{-3} M$) values by multiplying by 2.05 and by 0.333 for the erythrocyte and serum esterases, respectively.

E. Histochemical Staining Technique. In view of its role in conduction of nerve impulses, the localization of acetylcholinesterase is of great biological interest. In addition to attempts to localize the enzyme by micro- and ultramicrobiochemical techniques, a histochemical method has been described by Koelle and Friedenwald.[39] The method consists in incubating the tissue preparation in a medium containing acetylthiocholine as a substrate in concentrations optimal in solutions ($4 \times 10^{-3} M$ for acetylcholinesterase of most tissues) and $2 \times 10^{-3} M$ copper glycinate saturated with copper thiocholine for 10 to 60 minutes. After this treatment the tissue sections are briefly rinsed with saturated copper thiocholine solution and then immersed in an ammonium sulfide solution. The latter converts the white precipitate of copper thiocholine to a dark-brown amorphous deposit of copper sulfide.

Whereas the first observations with this technique on the localization of acetylcholinesterase seemed to be in agreement with earlier biochemical results, the extension of the technique to various tissues ran into serious difficulties. Koelle became aware of these difficulties and improved the method initially described.[40] The technique is, however, in many respects still unsatisfactory. The results vary with the pH and are best in the acid range where the enzyme activity is known to be markedly poorer, the optimum of enzyme activity being between pH 8 and 9. The pH dependents vary, moreover, with the different types of tissue used. Permeability barriers play a disturbing role in many instances, especially since acetylthiocholine is a quaternary ammonium salt. Distinction from other esterases by use of inhibitors which have different affinities to the various types of esterases is a doubtful procedure, especially in the presence of structural barriers. No really quantitative evaluation supported by a study of enzyme kinetics has been so far attempted to substantiate the results of the histochemical technique. Variations of ionic strength, too long or too short exposure, and a series of other factors lead to artifacts

[39] G. B. Koelle and J. S. Friedenwald, *Proc. Soc. Exptl. Biol. Med.* **70**, 617 (1949).
[40] G. B. Koelle, *J. Pharmacol. Exptl. Therap.* **103**, 153 (1951).

and yield misleading pictures of localization. As in the case of other histochemical techniques, conclusions require confirmation by biochemical methods. A critical and extremely lucid evaluation of this technique may be found in an article by Couteaux and Taxi.[41] The authors discuss the pitfalls as well as the presently available results and suggest several valuable improvements as to the use of pH, ionic strength, and other factors.

[41] R. Couteaux and J. Taxi, *Arch. anat. microscop. et morphol. exptl.* **41**, 352 (1952).

[108] Dialkylphosphofluoridase[1]

$$(C_3H_7O)_2{=}P{\overset{O}{\underset{F}{\diagup\diagdown}}} + H_2O \rightarrow (C_3H_7O)_2{=}P{\overset{O}{\underset{OH}{\diagup\diagdown}}} + HF$$

By A. MAZUR

Assay Method

Principle. The method described is based on the fact that the hydrolysis of the phosphorus-fluorine bond in diisopropyl fluorophosphate (DFP) results in the liberation of H^+ as well as of F^-. The reaction is carried out in Warburg vessels with a bicarbonate-carbon dioxide buffer at pH 7.4 so that the resulting liberation of carbon dioxide can be taken as an index of H^+ formation. Analysis for F^- may be performed at the end of the reaction period.

Reagents

$NaHCO_3$, 0.025 M.

DFP, 0.035 millimole per 3 ml. dissolved in 0.025 M bicarbonate.

Enzyme. Plasma is used per se or diluted with 0.025 M bicarbonate.

Tissue extracts are prepared with 0.025 M bicarbonate (see below).

Preparation of Active Enzyme Extracts. Blood is collected with heparin to prevent clotting. Plasma is used per se or suitably diluted with 0.025 M

[1] Among the names used to describe this enzyme are dialkylfluorophosphatase and dialkylfluorophosphate esterase. Neither of these names is correct in the light of our knowledge concerning phosphatases and esterases. The enzyme does not split at an ester linkage, nor does it form inorganic phosphate and an alcohol. Instead it splits at a phosphorus-halogen bond and must tentatively be termed a phosphofluoridase until its natural substrate, if any exists, is determined.

bicarbonate. Red cells are washed twice with 5 to 10 vol. of saline and diluted four times with 0.025 M bicarbonate. Muscle and brain are freed of coagulated blood, homogenized in a Waring blendor with 9 vol. of bicarbonate, and filtered through coarse filter paper. Heart, lung, liver, and kidneys are perfused with saline in order to reduce the quantity of blood in these tissues, homogenized, and filtered as above. (For purification see below.)

Procedure. 3 ml. of DFP (0.035 millimole) is placed in the large well of a Warburg flask. 0.5 ml. of enzyme in 0.025 M bicarbonate is placed in the

TABLE I

EFFECT OF RABBIT PLASMA ON HYDROLYSIS OF DIISOPROPYL FLUOROPHOSPHATE[a]

The control vessels contained 0.035 millimole of diisopropyl fluorophosphate in bicarbonate buffer, pH 7.4. Others contained in addition 0.5 ml. of rabbit plasma. The data reported for plasma are not corrected for control hydrolysis in water. The values are expressed in millimoles produced per millimole of diisopropyl fluorophosphate.

Time, min.	CO_2 production in presence of		Fluoride production in presence of	
	Water	Plasma	Water	Plasma
20	0.01	0.34		
30	0.02	0.53	0.03	0.36
40	0.04	0.70		
60	0.05	1.01	0.08	0.54
90	0.08	1.31		
120	0.10	1.43	0.07	0.83
130	0.09	1.44		
140	0.10	1.46	0.07	0.99
150	0.11	1.48		
160	0.11	1.49	0.08	1.01

[a] A. Mazur, *J. Biol. Chem.* **164**, 271 (1946).

first side arm. 0.5 ml. of 0.025 M bicarbonate or a solution of inhibitor or activator dissolved in bicarbonate is placed in the second side arm. A control vessel contains 0.5 ml. of bicarbonate instead of the enzyme solution, in order to measure the extent of spontaneous hydrolysis of DFP. The solutions are equilibrated with shaking at 38° with a gas mixture of 95% N_2 and 5% CO_2. The reaction is started by tipping the contents of the side arms into the main chamber of the vessel. Readings are taken at 10-minute intervals, and activities are calculated from the initial reaction velocity of the straight-line portion of the reaction curve.

Table I lists the results obtained when the reaction is allowed to go to completion, using rabbit plasma as a source of enzyme. Table II gives the relative activities of crude extracts of various tissues from the rabbit and

TABLE II

HYDROLYSIS OF DIISOPROPYL FLUOROPHOSPHATE BY TISSUE EXTRACTS[a]

The velocity of hydrolysis of diisopropyl fluorophosphate was determined as microliters of CO_2 liberated in 30 minutes from a bicarbonate buffer, pH 7.4, by 0.5 ml. of a 1:10 tissue extract. The plasma and red blood cell activities were calculated from the activities determined experimentally on more concentrated solutions and also expressed as those of a 1:10 dilution. All data are corrected for the hydrolysis of the diisopropyl fluorophosphate in water.

	Velocity of hydrolysis by	
	---	---
Tissue	Rabbit, $\mu l.\ CO_2$	Human, $\mu l.\ CO_2$
Liver	274	457
Kidney	187	319
Small intestine	129	—
Plasma	127	19
Lung	55	50
Heart	30	—
Brain	20	29
Muscle	12	40
Red cells	9	13

[a] A. Mazur, *J. Biol. Chem.* **164**, 271 (1946).

TABLE III

RELATIONSHIP BETWEEN ENZYME ACTIVITY AND PROTEIN CONTENT OF RABBIT TISSUES[a]

Protein N is determined as the difference between total N and nonprotein N in terms of milligrams of N per 0.5 ml. of a 1:10 extract of tissue or dilution of plasma or red cells.

Tissue	Enzyme activity[b] per mg. of protein N
Kidney	370
Liver	294
Plasma	22
Lung	16
Heart	12
Brain	9
Red cells	2
Muscle	1

[a] A. Mazur, *J. Biol. Chem.* **164**, 271 (1946).
[b] Microliters CO_2 for 30 minutes.

man. Table III shows the relative activities of rabbit tissues on the basis of protein N content.

Purification Procedure

A partial purification of the enzyme from rabbit kidney has been reported by the author.[2] This procedure yielded a product which represented a purification of some 13 times. A more recent study by Mounter et al.[3] outlines a method which results in a purification of 65 to 100 times.

TABLE IV

DIISOPROPYLPHOSPHOFLUORIDASE ACTIVITIES IN HOG KIDNEY FRACTIONS[a]

Experiment	Fraction	pH	EtOH, %	Activity, μl. CO_2/mg. N/ 30 min.	Recovery, %	Purification factor
71	Kidney	6.8		230		
	Fraction A	5.8	60	1,600	98	7
	Fraction A-1	4.98	10	6,500	87	28
	Fraction A-2	4.81	12.5	23,000	41	100
73	Kidney	6.8		290		
	Fraction A	5.78	60	2,300	93	8
	Fraction A-1	5.02	10	4,300	85	15
	Fraction A-2	4.83	12.5	18,800	31	65
33	Kidney	6.8		300		
	Fraction A	5.8	60	1,800	95	6
	Fraction A-1	5.05	10	3,750	85	12.5
	Fraction A-1	4.7	13.5	9,650	40	32
	Fraction A-1	4.8	17.5	7,700	27	26
	Fraction A-1	4.8	12.5	11,000	70	37
	Fraction A-2	4.8	12.5	21,000	34	70

[a] L. A. Mounter, C. S. Floyd, and A. Chanutin, J. Biol. Chem. 204, 221 (1953).

The following is an outline of this method which utilizes frozen hog kidneys:

Step 1. 2500 g. of kidneys is ground in a meat grinder and homogenized in a Waring blendor for 3 minutes. The mixture is suspended in 10 l. of water, adjusted to pH 6.8 with acetate buffer, pH 4.0, and stirred for 30 minutes at 0°. The pH is reduced to 5.8 with buffer, and the mixture centrifuged for 20 minutes at 45,000 × g at 0°. The clear supernatant is filtered to remove lipid particles and adjusted to pH 5.8. Ethanol (60%) at 5° is added, and the mixture is allowed to stand overnight. The precip-

[2] A. Mazur, J. Biol. Chem. 164, 271 (1946).

[3] L. A. Mounter, C. S. Floyd, and A. Chanutin, J. Biol. Chem. 204, 221 (1953).

itate is recovered by centrifugation at 3000 $\times$ g. This represents fraction A, which is 6 to 8 times purified.

Step 2. The precipitate is suspended in 2 l. of 0.002 M NaHCO$_3$ and stirred for 1 hour at 0°. The insoluble material is discarded. The solution is adjusted to pH 5.0 with buffer, and a small precipitate removed by centrifugation. The solution is then adjusted to 10% ethanol at $-5°$ and allowed to stand for 30 minutes. The precipitate is removed and dissolved in 0.025 M NaHCO$_3$. This is fraction A-1 and represents a purification of 15 to 37 times.

Step 3. The solution is adjusted to pH 5.0 at room temperature and clarified by centrifugation. The supernatant is adjusted to 12.5% ethanol, and the mixture allowed to stand for 40 minutes at 25°. The precipitate which forms is removed, and the supernatant fluid allowed to stand at $-5°$ for 2 hours. The precipitate is recovered and brought into solution as before. This is fraction A-2 and represents an over-all purification of 65 to 100 times. Table IV illustrates the activities of the various fractions and the yields obtained.

Properties

Specificity. The enzyme is active in the splitting of the P-F linkage in a variety of fluorophosphates. Table V illustrates the activity of plasma and

TABLE V

Hydrolysis of Related Fluorophosphates[a]

The activities are expressed as microliters of CO$_2$ liberated per minute when 0.5 ml. of 1:1 rabbit plasma or 1:10 rabbit liver extract was used in the standard hydrolysis mixture containing 0.035 millimole of the dialkyl fluorophosphate in a bicarbonate buffer, pH 7.4. The hydrolysis of the fluorophosphate in water was determined by substituting bicarbonate solution for the enzyme.

| | | Enzyme activity of | |
| | Hydrolysis | | |
Fluorophosphate	in water	Plasma	Liver extract
Dimethyl fluorophosphate	10	114	73
Diethyl fluorophosphate	2	77	97
Diisopropyl fluorophosphate	1	15	22
Ethylmethyl fluorophosphate	2	14	51

[a] A. Mazur, *J. Biol. Chem.* **164**, 271 (1946).

of a liver extract on the dimethyl, diethyl, ethyl methyl, and diisopropyl fluorophosphates. However, no normal substrate was found which could be split by the enzyme. Table VI shows the lack of identity of the enzyme with acetylcholinesterase, glycerophosphatase, or lipase. When the enzyme splits DFP, no inorganic phosphate is liberated.

Activators. The partially purified enzyme from rabbit kidney as well as that from hog kidney is inhibited by dialysis. Addition to the dialyzed partially purified enzyme of the dialyzate, Ca^{++}, or Mg^{++} restored the enzyme activity. Mounter *et al.*[3] report a striking activation by Mn^{++} and to a slighter extent by Co^{++}. Histidine was also found to be a potent activator of the enzyme in the presence of Mn^{++}. Other amino acids which potentiated the activity of the enzyme are cysteine, thiolhistidine, and serine. The greatest enzyme activity was obtained by these workers in the presence of purified enzyme, Mn^{++}, and 2,2'-dipyridyl.

TABLE VI

RELATIONSHIP BETWEEN FLUOROPHOSPHATE-HYDROLYZING ENZYME AND PHOSPHATASE, CHOLINESTERASE, AND ESTERASE ACTIVITIES[a]

The phosphatase activities are measured by the production of inorganic phosphate from β-glycerophosphate in 1 hour. The activities of esterase and cholinesterase are expressed as microliters of CO_2 liberated by the action of 0.5 ml. of the enzyme solution in a bicarbonate buffer, pH 7.4, containing 0.015 mole per liter of acetylcholine, triacetin, tripropionin, or methyl butyrate.

	Enzyme activity in	
Substrate	Fraction A, Table IV	Fraction H, Table IV
Diisopropyl fluorophosphate, μl. CO_2	388	3640
Acetylcholine, μl. CO_2	0	0
Glycerophosphate, mg. PO_4	0.144[b]	0.080
Triacetin, μl. CO_2	288	14
Tripropionin, μl. CO_2	510	49
Methyl butyrate, μl. CO_2	277	7

[a] A. Mazur, *J. Biol. Chem.* **164**, 271 (1946).

[b] This quantity of inorganic phosphate accounts for but 4.6% of the available phosphate from glycerophosphate. In a similar experiment with rat intestinal phosphatase, 61% of the available phosphate from glycerophosphate was hydrolyzed in 1 hour.

Inhibitors. The enzyme is inhibited by heavy metal ions such as Hg^{++} and Cu^{++}. It is only slightly inhibited by iodoacetate, and not at all by iodosobenzoate or fluoride. Marked inhibition of the purified enzyme by *p*-chloromercuribenzoate and by phenyl mercuric nitrate is reported.

Effect of Temperature and pH. Exposure of the partially purified enzyme solution from liver to 60° for 5 minutes results in an almost complete inactivation (6% of its original activity). The enzyme is stable at icebox temperatures for 18 hours at a pH of 7.0 to 9.8. Below 7.0 its activity falls, and at pH 4.6 it is reduced to less than 2% of the activity of the enzyme at pH 7.0. The purified enzyme has its optimum activity from pH 75. to 8.0.

[109] Liver Esterase

By ELMER STOTZ

Assay Method

This enzyme, sometimes referred to as "nonspecific esterase" or "aliesterase," catalyzes the hydrolysis of lower fatty acid esters. Hence a variety of assay methods are available: (a) titrimetric, in which the acid liberated is measured by alkali titration; (b) manometric, in which the carbon dioxide liberated from bicarbonate by the acid formed is measured; and (c) colorimetric, in which esters give rise to a colored base, for example p-nitrophenyl acetate or propionate → p-nitrophenol,[1] or in which the base may be coupled with diazonium salts to form colored azo dyes, for example β-naphthyl acetate → β-naphthol → azo dyes.[2,3] Only the titrimetric and manometric methods will be described.

Titrimetric Assay.[4] The enzyme solution is added to a mixture containing 2.0 ml. of a saturated solution of redistilled ethyl butyrate, 0.25 ml. of 0.1% bromothymol blue, and 5.75 ml. of distilled water. During incubation over a 30-minute period at room temperature, standard alkali is drop-titrated into the mixture to adjust the reaction to the original pH. Esterase activity is expressed as milliliters of 0.01 N NaOH required to neutralize the butyric acid per milligram of dry weight of enzyme preparation.

The authors claim that the rate of hydrolysis under the specified conditions is proportional to the amount of enzyme preparation and is constant over the 30-minute period.

Manometric Assay.[5] This assay is carried out in standard-size Warburg flasks, employing a N_2-CO_2 atmosphere and the solution buffered with bicarbonate at pH 7.4.

The main compartments of the flasks contained 0.45 ml. of 0.15 M NaHCO$_3$, 0.038 ml. of 100% redistilled methyl butyrate (final concentration 0.10 M), and distilled water to make a final volume of 2.70 ml. The side arms contained 0.10 ml. of 0.15 M NaHCO$_3$, from 0.1 to 0.5 ml. of enzyme preparation, and sufficient distilled water to make a final volume of 0.60 ml. A gas mixture containing 95% N_2–5% CO_2 was passed through

[1] C. Huggins and J. Lapides, *J. Biol. Chem.* **170**, 467 (1947).
[2] M. M. Nachlas and A. M. Seligman, *J. Natl. Cancer Inst.* **9**, 415 (1949).
[3] A. M. Seligman and M. M. Nachlas, *J. Clin. Invest.* **29**, 31 (1950).
[4] C. J. Harrer and C. G. King, *J. Biol. Chem.* **138**, 111 (1941).
[5] W. M. Connors, A. Pihl, A. L. Dounce, and E. Stotz, *J. Biol. Chem.* **184**, 29 (1950).

the vessels for 4 minutes, the stopcocks closed, and temperature equilibrium (30°) established. After dumping the contents of the side arm, readings are taken at 4-minute intervals for a period of 30 minutes. Ideal rates are from 200 to 600 μl. of CO_2 per hour. Under the conditions described, a zero-order reaction is obtained, and the rate is proportional to the amount of enzyme solution employed.

Purification Procedure

Mohammed[6] has described the preparation of a crystalline esterase from horse liver, but Connors et al.[5] found this material very low in esterase activity, and its activity disappeared on recrystallization. The latter workers describe a liver esterase preparation from horse liver which is 270-fold purified over the acetone powder starting material.

Step 1. Fresh horse liver[7] is finely minced in an electric grinder. A 2000-g. portion is thoroughly stirred at room temperature with 8 l. of ice-cold acetone. The mixture is drained through a large Büchner funnel (without paper), and the residual acetone squeezed from the material by hand. Two further washings with 8 l. of acetone are done. The material is then spread out in a thin layer to dry at room temperature for 12 hours. The dry material is ground in a burr mill with care to avoid overheating. Yield, approximately 560 g. of acetone powder; stable for at least three months at 5°. 1 g. of powder contains about 1000 units of esterase; purity, 1.02.

Step 2. 100 g. of acetone powder is added to 1 l. of cold distilled water, and the material is extracted by stirring in the cold for 12 hours. After centrifuging (refrigerated), the supernatant liquid (840 ml.) contains 55,430 units; purity, 3.2.

Step 3. To the 840 ml. of solution is added an equal volume of saturated (at 5°) ammonium sulfate solution neutralized to pH 7.5 with ammonia. The mixture is allowed to stand for 12 hours in the cold, and the precipitate is removed by centrifugation.

For every 100 ml. of supernatant fluid, 10 g. of ammonium sulfate is added with stirring. After the mixture has stood in the cold for 12 hours, it is centrifuged, and the precipitate is dissolved in 0.145 M phosphate buffer, pH 7.5, and made up to 130 ml. Total units, 42,750; purity, 8.6.

Step 4. The 130 ml. of solution is brought to 63° within 1 minute in a water bath and held at that temperature for 5 minutes. The mixture is

---

[6] S. M. Mohammed, *Acta Chem. Scand.* **2**, 90 (1948); **3**, 56 (1949).

[7] Beef liver has a much lower esterase activity than horse liver, and the method is not applicable to the former.

then quickly cooled in an ice bath. The mixture is centrifuged, and the supernatant fluid (96 ml.) saved. Total units, 32,250; purity, 17.

Step 5. To 90 ml. of the above solution at 4° is added with stirring 30 ml. of ice-cold acetone. The mixture is allowed to stand in the cold for 12 hours, after which the precipitate is removed by centrifugation and discarded. To 106 ml. of the cold supernatant fluid is added 38.5 ml. of cold acetone. After standing in the cold for 8 hours, the mixture is centrifuged, and the supernatant fluid discarded. The precipitate is thoroughly suspended in about 30 ml. of distilled water and allowed to stand for 5 hours in the cold for extraction of the esterase. The mixture is then centrifuged, and the clear, slightly yellow supernatant fluid saved. Volume, 38 ml.; total units, 18,600; purity, 41.

Step 6. To the 38 ml. of solution from above is added 450 mg. of solid $Cu(CH_3COO)_2 \cdot H_2O$ with stirring. The solution is allowed to stand in the cold for 4 hours, after which it is centrifuged and the precipitate discarded.

The supernatant fluid (purity 103) is dialyzed exhaustively against distilled water in the cold, and the precipitated proteins discarded after centrifugation. The solution is reduced in volume to approximately 10 ml. by pressure filtration through Cellophane.[8] Purity of final solution, 277; contains approximately 2 mg. of solids per milliliter.

The solution can be stored at 5° for at least 1 month without loss of activity.

Properties

Esterase prepared in this manner shows fifty times as much activity toward methyl butyrate as toward acetylcholine. Its optimum pH in borate buffer is 8.0, it is quite heat sensitive (partial destruction about 30°), and it has a Michaelis constant of 0.022 M methyl butyrate. Energy of activation of the enzyme-catalyzed hydrolysis of methyl butyrate was calculated to be about 4,400 cal./mole.

The electrophoretic peak associated with the esterase component appeared to comprise about 70% of the total protein in the purified esterase preparation.

---

[8] A simple and suitable apparatus for this purpose consists of a Cellophane sausage casing contained within a strong muslin sleeve. By an adequately clamped connection with a nitrogen cylinder, 25 to 35 lb. of pressure can be exerted on the solution in the Cellophane bag, producing a rapid ultrafiltration.

[110] Phospholipases

Phospholipase A

Lecithin + H_2O → Lysolecithin + Unsaturated fatty acid

By OSAMU HAYAISHI

Assay Method

Principle. The activity of phospholipase A[1] can be measured either by the decrease of lecithin, the increase in free acid liberated, or the increase of lysolecithin. For the determination of lecithin, the method for choline determination (see below) can be applied at neutral pH. The sample, at pH 7.0, is treated directly with iodine reagent, and the insoluble lecithin iodine complex is collected by centrifugation, dissolved in ethylene dichloride, and measured in the spectrophotometer.[2,3] Lecithin can be spread out as a monomolecular film on the surface of a saline solution, and the enzyme activity can be measured by the decrease of the surface potential of the film when the enzyme is added on the film.[4]

The unsaturated fatty acid liberated by the enzyme can be isolated by extraction with organic solvents[5] or by filtration[6] and is determined by titration with alkali, phenolphthalein being used as an indicator.

The production of lysolecithin can be measured by its hemolytic action[7] or by the application of the iodine method (see below), since the lysolecithin iodine complex is soluble at neutral pH but is precipitated by acidifying the solution.[3] The hemolysis test is sensitive and simple, but it is not practical because quantitative results can be obtained only when

---

[1] There has been some confusion of terminology as to the nomenclature and the classification of a group of enzymes which acts upon lecithin and related compounds. As has been pointed out in a recent review [W. D. Celmer and H. E. Carter, *Physiol. Revs.* **32**, 167 (1952)], standardization of nomenclature is greatly needed in this field. Since other phospholipids are also suitable substrates as well as lecithin, the term "phospholipase" appears to be preferable to the old name "lecithinase." For further differentiation of the phospholipases, the nomenclature proposed by Zeller ["The Enzymes" (Sumner and Myrbäck, ed.), Vol. I, Part 2, p. 986. Academic Press, New York] is employed in the following discussion.

[2] O. Hayaishi, *Federation Proc.* **12**, 216 (1953).

[3] O. Hayaishi and A. Kornberg, *J. Biol. Chem.* **206**, 647 (1954).

[4] A. Hughes, *Biochem. J.* **29**, 437 (1935).

[5] D. Fairbairn, *J. Biol. Chem.* **157**, 633 (1945).

[6] M. Francioli and A. Ercoli, *in* "Die Methoden der Fermentforschung" (Bamann and Myrbäck, ed.), Thieme, Leipzig, 1941; Academic Press, New York, 1945.

[7] A. W. Bernheimer, *J. Gen. Physiol.* **30**, 337 (1947).

the amount of lysolecithin is adjusted within a certain range.[8] Also, bacterial and venom toxins may contain substances other than lysolecithin which are capable of causing hemolysis.

The method described below[2,3] is based on the rate of choline liberation from purified egg lecithin in the presence of potent lysolecithinase and purified GPC[9]-diesterase. In order to avoid the difficulty of emulsifying lecithin, Hanahan developed a method of performing the reaction in diethyl ether as a solvent.[10] With a partially purified enzyme prepared from Merck's pancreatin, he was able to study the degradation of egg lecithin in diethyl ether after the appearance of lysolecithin or the liberation of unsaturated fatty acids. This method appears to be simple and precise, but somehow it was not applicable when bacterial phospholipase A was used instead of partially purified pancreatin. Since the substrate used in each case is prepared by different methods, it is not quite clear whether this discrepancy is due to the different enzyme preparations or to some substances which might be present in the substrate and might affect the solubility of the enzyme, substrate, or enzyme-substrate complex in diethyl ether.

Reagents

0.01 M lecithin suspension. A purified preparation of egg yolk lecithin[11] is suspended in distilled water and treated in a sonic oscillator (Raytheon, Model 9 KC) for 10 to 15 minutes at approximately 20°.

Lysolecithinase preparation. Crude fraction S is used when fraction R is assayed, and crude fraction R is used when fraction S is assayed (see below). Other lysolecithinase preparations may be employed, provided that they do not contain phospholipase A activity.

GPC diesterase. Purified preparation containing approximately 500 units/ml.

0.2 M phosphate buffer, pH 7.0.

Iodine reagent. 12.5 g. of KI and 9.8 g. of I_2 are dissolved in water to a total volume of 250 ml.

Procedure. The test system contains 0.1 ml. of lecithin suspension, 0.05 ml. of buffer, 0.05 ml. of purified GPC diesterase, 0.1 ml. of lysolecithinase, and water to a total volume of 0.4 ml. Incubation is carried out

[8] H. B. Collier, *J. Gen. Physiol.* **35**, 617 (1952).
[9] GPC: glycerophosphorylcholine.
[10] D. J. Hanahan, *J. Biol. Chem.* **195**, 199 (1952).
[11] M. C. Pangborn, *J. Biol. Chem.* **188**, 471 (1951).

at 25° for 1 hour, and, at the end of the incubation, protein, remaining lecithin, and lysolecithin are removed by treatment with 5% cold perchloric acid. An aliquot containing 0.01 to 0.5 micromole of free choline is removed, and the amount of free choline is measured by the method of Appleton et al.[12] 0.5 ml. of a sample is treated with 0.2 ml. of iodine reagent and kept in an ice bath for 15 minutes. After centrifugation, the supernatant fluid is removed by aspiration, the precipitate is dissolved in ethylenedichloride (reagent grade) to a volume of 10 ml., and the optical density determined at 365 mμ in a Beckman Model DU spectrophotometer. The molar extinction coefficient, referred to choline, is approximately 2.7×10^4 cm.^{2}/mole.

When purified GPC diesterase is not available, acid-labile choline (100° in 1 N HCl for 20 minutes[13]) is regarded as GPC.

Definition of Unit and Specific Activity. One unit of enzyme is defined as that amount which causes an initial rate of hydrolysis of the substrate of 0.1 micromole per hour under the conditions described above. Specific activity is expressed as units per milligram of protein.

Purification Procedure

The enzyme has been found in a number of species of snake venom,[14,15] various animal tissues,[16–18] and bacteria.[3] In tissue extracts, phospholipase A is generally found together with other phospholipase activities. Gronchi reported that the former can be distinguished by its remarkable heat resistance and solubility in 50% ethanol and can be obtained free from phospholipase B, protease, amylase, and lipase.[19]

1. Crystalline phospholipase A from the venom of *Crotalus terrificus*.[14] 730 mg. of amorphous crotoxin (the venom of *Crotalus terrificus*) is dissolved in 8.5 ml. of 1% acetic acid at 55°, and 4.5 ml. of 1% pyridine is added with stirring. The solution, containing 0.6% pyridine acetate at pH 4.4, is left in a water bath and is gradually cooled. In about 30 minutes, crystallization slowly starts to occur. The next day crystals can be isolated and dried *in vacuo*. The crystalline enzyme has to be washed thoroughly with water in order to avoid denaturation due to trace

[12] H. D. Appleton, B. N. La Du, Jr., B. B. Levy, J. M. Steele, and B. B. Brodie, *J. Biol. Chem.* **205**, 803 (1953); see also footnote 3 for detail.

[13] G. Schmidt, L. Hecht, N. Strickler, and S. J. Thannhauser, *Federation Proc.* **9**, 146 (1950).

[14] K. H. Slotta and H. L. Fraenkel-Conrat, *Ber.* **71**, 1076 (1938).

[15] S. S. Pe, *Ann. Biochem. and Exptl. Med. (India)* **4**, 45 (1944).

[16] M. Francioli, *Fermentforschung* **14**, 241 (1934).

[17] G. J. Scheff and A. J. Awny, *Am. J. Clin. Pathol.* **19**, 615 (1949).

[18] Z. Nikuni, *Proc. Imp. Acad. (Tokyo)* **8**, 300 (1932).

[19] V. Gronchi, *Sperimentale* **90**, 223, 262 (1936).

amounts of acid and pyridine. (Solubility of crystalline crotoxin is about 2.3 mg. per 1 ml. of water.) Starting with 140 ml. of amorphous crotoxin, 74 mg. of crystalline material is obtained after five recrystallizations. This material was later shown to be homogeneous in the ultracentrifuge and by electrophoresis studies.[20,21]

2. Phospholipase A of *Serratia plymuthicum*.[2,3] A strain of *Serratia plymuthicum* was isolated from soil by enrichment culture technique on a medium containing 0.1% GPC, 0.1% Difco yeast extract, 0.15% K_2HPO_4, 0.05% KH_2PO_4, and 0.02% $MgSO_4 \cdot 7H_2O$, and was used as a source of enzymes. The formation of choline from egg yolk has been reported by a number of other strains belonging to the genus *Serratia*.[22]

Step 1. Cultivation of an Organism and Preparation of Crude Extracts. Cells are grown in a medium containing 0.2% lecithin (crude egg lecithin made by the Nutritional Biochemical Corporation), 0.1% Difco yeast extract, 0.15% K_2HPO_4, 0.05% KH_2PO_4, and 0.02% $MgSO_4 \cdot 7H_2O$. The lecithin is dissolved in ether, sterilized separately, and poured into the hot medium. Ether is removed by mechanical shaking while the whole medium is hot. The cells are cultivated in 20-l. glass carboys containing 10 l. of the medium at about 26° for 16 hours with constant, reciprocal shaking. The cells are harvested by centrifugation in a Sharples super-centrifuge, washed once with a 0.5% NaCl–0.5% KCl solution, and frozen. At −10°, cells retain their activity for at least six months. The yield of cells (wet weight) is approximately 4 g./l. of medium in a large-scale culture, and 6 to 10 g./l. in smaller scale cultures.

Cell-free extracts are prepared by grinding the wet cells in a mortar with four times their weight of alumina (Alcoa A-301) at 0° for 5 minutes. The cell paste is extracted with 10 parts of glycylglycine buffer (pH 9.0, 0.02 M); the whole slurry is centrifuged at 0° at about 6000 × g for 15 minutes. The supernatant fluid is a slightly turbid suspension of cell fragments and contains a negligible number of intact cells. This preparation, referred to as "bacterial homogenate," contains 3.5 to 4.0 mg. of protein per milliliter and usually has a specific activity of about 2.0.

Step 2. Separation of Soluble and Particulate Fractions. The homogenate is centrifuged in a Spinco centrifuge (Model L, rotor No. 40) at 110,000 × g, for 1 hour. The clear upper layer, representing about 90% of the fluid, is removed with a syringe; the turbid fluid layer overlying the residue is discarded. The residue is washed once with a volume of glycylglycine buffer equal to that of the original homogenate by centrifuging at 110,000 × g for 1 hour and resuspended in the same volume of buffer.

[20] C. H. Li and H. L. Fraenkel-Conrat, *J. Am. Chem. Soc.* **64**, 1588 (1942).
[21] N. Gralin and T. Svedberg, *Biochem. J.* **32**, 1375 (1938).
[22] V. Monsour and A. R. Colmer, *J. Bacteriol.* **63**, 597 (1952).

The clear supernatant fluid and the washed residue, referred to as fraction S and fraction R, contain approximately 2.2 and 1.3 mg. of protein per milliliter, respectively.

Step 3. Solubilization of Fraction R. Fraction R is made soluble by butanol treatment according to Morton.[23] To 5 ml. of the chilled fraction R is added 0.5 ml. of *n*-butanol with constant mechanical stirring. After 5 minutes at 0°, the mixture is centrifuged in a Servall angle centrifuge (SS-1) at 13,000 r.p.m. for 30 minutes. Almost 100% of the activity remains in the supernatant if Ca ion is added to the incubation mixture; prior to the butanol treatment, the activity is completely sedimented under these conditions. The supernatant fluid is dialyzed against 0.01 M K_2HPO_4 at 3° for 4 to 6 hours (until the odor of butanol is no longer detectable). Excess dialysis leads to inactivation of the enzyme. This preparation contains 80 to 100% of the original activity and only 25 to 30% of the protein. (Additional purification (two- to fourfold) can be obtained by fractionation between ammonium sulfate saturation of 30 to 50%; but this latter step is not included routinely because yields of less than 30% are obtained when the removal of butanol in the previous step is incomplete.) An almost obligatory requirement for Ca ion is observed at this stage. The saturation level is reached with 0.01 M Ca^{++}, and inhibitory effects are observed at concentrations above 0.02 M. Mg^{++} and Mn^{++} have no demonstrable effect.

Properties of Phospholipase A

Specificity. It has been generally accepted that phospholipase A acts not only on lecithin but also on cephalin, but no activity was shown toward cerebrosides, sphingomyelin, or acetal phosphatides.[5] Although synthetic lecithins, containing two saturated fatty acids, were reported not to be acted on,[24] more recent reports[25] indicate the presence of activity which attacks lecithins containing two identical fatty acids. *Serratia* phospholipase A metabolizes not only lecithin and cephalin but also dipalmytoleyllecithin, dimyristoyllecithin, and dimyristoylcephalin. Sphingomyelin[3] is not attacked. Hanahan[26] also reported in snake venom and commercial pancreatin the presence of activities which split off palmitoleic acid from dipalmitoleyl-L-α-lecithin.

Stability. Phospholipase A is outstanding for its resistance toward heat. At pH 5.9 snake venom enzyme can be boiled without any loss of

[23] R. K. Morton, *Nature* **166**, 1092 (1950).
[24] A. Contardi and P. Latzer, *Biochem. Z.* **197**, 222 (1928).
[25] K. Ogawa, *J. Biochem. (Japan)* **24**, 389 (1936).
[26] D. J. Hanahan, *Federation Proc.* **12**, 214 (1953).

activity.[4] The bacterial enzyme can be heated to 97° at pH 9.0 at least for 5 minutes without any significant change in activity.[3]

Activators and Inhibitors. Bacterial enzyme has been separated into two components, fractions R and S, both of which are required for the lecithin breakdown. The former consists of at least two components, namely Ca ion and another fraction which appears to be of protein nature. The fraction S can be substituted by ferrous ion to a limited extent. However, the action of Fe^{++} ceases after only 20 to 30% of the substrate is removed, although the rate of the reaction is comparable to or even greater than the rate with crude fraction S.

Activation by Ca ion has been reported with enzymes from other source material.[27,28] Sodium fluoride,[4] eserine,[29] and potassium cyanide[25] have no inhibitory effect.

Effect of pH. Optimum pH appears to be around 7.0.[3,4,10]

Phospholipase B

Lysolecithin + H_2O → GPC + Saturated fatty acid

Assay Method

Principle. The activity can be followed either by a decrease of lysolecithin, an increase in free fatty acid liberated, or the appearance of GPC.

The quantitative determination of lysophospholipids can be performed either by iodine methods or by measuring the hemolytic activity (cf. Phospholipase A). Titration of the fatty acid liberated has been done with KOH after the addition of ethanol to the incubation mixture.[30] Minute quantities of GPC can be determined accurately either by a chemical method or, more specifically, by the use of a specific enzyme, GPC diesterase. Free choline liberated from GPC either by acid hydrolysis (100° in 1 N HCl for 20 minutes)[13] or by enzymatic means is estimated by a micromethod already mentioned.[12] The method described below has been employed for the assay of bacterial phospholipase B and was found to be sufficiently accurate in a microscale experiment.[3]

Reagents

    0.01 M lysolecithin solution.
    0.2 M potassium phosphate buffer, pH 7.0.
    GPC diesterase. Purified preparation containing approximately 500 units/ml.

[27] C. Delezenne and E. Fourneau, *Bull. Soc. Chim. Biol.* **15**, 421 (1914).
[28] R. Kudicke and H. Sachs, *Biochem. Z.* **76**, 359 (1916).
[29] M. Francioli, *Enzymologia* **3**, 204 (1937).
[30] S. Belfanti, A. Contardi, and A. Ercoli, *Ergeb. Enzymforsch.* **5**, 213 (1936).

Procedure. The test system contains 0.1 ml. of lysolecithin solution, 0.05 ml. of the buffer, 0.05 ml. of GPC diesterase, lysolecithinase, and water to a total volume of 0.4 ml. The incubation is carried out at 25° for 30 minutes. At the end of the incubation, remaining lysolecithin and protein are removed by treatment with 5% ice-cold perchloric acid. An aliquot containing 0.01 to 0.5 micromole of free choline is removed, and the amount of free choline is determined by the method described above.[12]

Definition of Unit and Specific Activity. One unit of enzyme is defined as that amount which causes an initial rate of hydrolysis of the substrate of 0.1 micromole per hour under the conditions described above. Specific activity is expressed as units per milligram of protein.

Purification Procedure

Phospholipase B has been found in pancreas[31] and other animal tissues,[16] wasp venoms,[24] rice bran,[32] castor beans,[24] molds,[31,32] and bacteria,[3] but extensive purification has not been conducted. Phospholipase B of *Serratia plymuthicum* can be isolated as described in the preceding section. The activity is by far the most potent among the series of enzymes involved in lecithin degradation in this organism, its specific activity in the crude homogenate (about 30) being almost fifteen times as high as that of phospholipase A, and almost three times as high as that of GPC diesterase. After high-speed centrifugation, it is found in both fractions R and S, although the specific activity is about three times as high in fraction R as in fraction S. As stated above, fraction R is practically free of phospholipase A and devoid of GPC diesterase activity, as will be discussed below.

Phospholipase B from Pancreas.[33] *Step 1.* An acetone powder of ox pancreas is extracted twice with cold ether and dried in a desiccator. The dry powder is stored in a refrigerator. The activity remains unchanged for many weeks, although lecithinase A activity is lost after prolonged storage. Samples (10 g.) of this powder are extracted with 75 ml. of 50% (v/v) glycerol by grinding in a mortar for 15 minutes at room temperature; 75 ml. of 0.1% NaHCO₃ solution is added, and the extraction continued for another 15 minutes. The suspension is centrifuged at 10,000 r.p.m. for 15 minutes. The resulting glycerol extract keeps its activity for several weeks when stored in the cold. This extraction yields 118 ml. of solution with a total activity of 650 units and a specific activity of 0.15 unit/mg. of protein. (A unit of activity is defined as the

[31] D. Fairbairn, *J. Biol. Chem.* **173**, 705 (1948).
[32] A. Contardi and A. Ercoli, *Biochem. Z.* **261**, 275 (1933)
[33] B. Shapiro, *Biochem. J.* **53**, 663 (1953).

amount of enzyme catalyzing the hydrolysis of 1 micromole of lyso-lecithin in 1 hour at 37°, at pH 6.0.)

Step 2. Portions of the glycerol extract (110 ml.) are diluted with 440 ml. of water, and 1 *N* acetic acid is added with stirring to bring the pH to 5.0. The precipitate is collected by centrifugation and suspended in 100 ml. of 0.04 *M* phosphate buffer, pH 6.0.

Step 3. The suspension is stirred for 15 minutes, and the undissolved material, which possesses very low activity, is discarded. The solution is cooled in ice, and $(NH_4)_2SO_4$ (35 g. per 100 ml.) is added in small portions over a period of 30 minutes with constant stirring. The precipitate is centrifuged in the cold and dissolved in 30 ml. of water. At this stage the enzyme is unstable when kept in aqueous solution but is relatively stable in the presence of ammonium sulfate.

Step 4. The solution is dialyzed for 1 hour against ice-cold distilled water, with stirring. (Prolonged dialysis causes marked inactivation.) The dialyzed solution is diluted with water to 80 ml., and acetic acid is added to bring the pH to 5.0. After centrifugation, the clear supernatant is fractionated with ammonium sulfate by adding 20 g. per 100 ml. of the ice-cold solution with mechanical stirring. The resulting precipitate is discarded, and to the supernatant 5 g. of ammonium sulfate is added for each 100 ml. of solution. The precipitate contains 240 units with a specific activity of 1.4 units/mg. of protein. By increasing the ammonium sulfate concentration by another 3 g. per 100 ml., a highly active fraction is obtained containing 200 units with a specific activity of 4.0 units/mg. of protein.

Still higher activity can be obtained with low yield, by allowing the supernatant obtained after the addition of 25 g. of ammonium sulfate per 100 ml. to warm up to 25° and stand for 24 hours at that temperature. A crystalline material separates out, having a specific activity of 6.5 units/mg. of protein. The specific activity remains unchanged during several recrystallizations. The over-all purification through step 4 is approximately 25-fold, with a reasonable yield of 30 to 35%. A further purification to about 45-fold, but giving low yields, can be obtained by the crystallization procedure.

SUMMARY OF PURIFICATION PROCEDURE

Fraction	Total activity, units	Specific activity, units/mg. protein
Glycerol extract	610	0.15
Step 2	450	0.28
Step 3	580	1.4
Step 4	200	4.0

Properties of Phospholipase B

Specificity. It has long been a matter of dispute whether phospholipase B is an independent enzyme or a part of phospholipase A activity. However, recent studies[3,31] with relatively pure preparation of phospholipase B clearly have demonstrated the presence of a specific enzyme which attacks lysophosphatides but not phosphatides. Wasp venom lysolecithinase[32] was shown to have no lipase activity.

Activators and Inhibitors. Unlike phospholipase A, this enzyme does not show any requirement for Ca ion,[31] although some authors have claimed that Ca ions activate.[32] Penicillium phospholipase B is inhibited by 0.001 M cyanide to the extent of 87%.[31]

pH Optimum. Bacterial phospholipase B shows its maximum activity at about pH 6.0. Enzymes from other sources seem to have much lower optimum pH (3.8 to 5.5).[31,32]

Stability. Phospholipase B from animal origin[34] is rather heat-labile as compared with the accompanying phospholipase A. The enzyme from *Penicillium notatum* is most stable at pH 4.5 and is rapidly inactivated at a pH higher than 7.0.[35] *Serratia* phospholipase B, on the other hand, appears to be more heat-resistant than phospholipase A from the same origin, or at least is as stable as the latter. Heating of bacterial homogenate at 100° for 10 minutes at pH 7.0 causes only a 50 to 80% inactivation of both activities.[3]

GPC Diesterase[36]

$$\text{L-}(\alpha)\text{-GPC} + H_2O \rightarrow \text{L-}(\alpha)\text{-GP}^{37} + \text{Choline}$$

Assay Method

Principle. Determination of free choline by the iodine method[12] is the most accurate and satisfactory method, although L-(α)-GP can be measured by manometric means by the use of L-(α)-GP dehydrogenase.[38]

Reagents

GPC solution, 0.02 M, pH 7.0.
0.2 M glycylglycine buffer, pH 8.85.

---

[34] S. Noguchi, *J. Biochem.* (*Japan*) **36**, 113 (1944).
[35] P. G. Zamecnik, L. E. Brewster, and F. Lipmann, *J. Exptl. Med.* **85**, 381 (1947).
[36] Strictly speaking, GPC does not belong to the phospholipids, but GPC-diesterase is included in this chapter, since the enzyme is a very useful tool for the study of other phospholipases.
[37] GP: glycerophosphate.
[38] D. E. Green, *Biochem. J.* **30**, 629 (1936).

Procedure. The test system contains 0.2 ml. of GPC solution, 0.1 ml. of the buffer, and 0.1 ml. of enzyme. After 10 minutes at 25°, the reaction is stopped by the addition of 0.4 ml. of perchloric acid (6%), and free choline is determined in a 0.2-ml. aliquot.

Definition of Unit and Specific Activity. A unit of enzyme is defined as the amount hydrolyzing 1 micromole of GPC during a 1-hour interval under these conditions, and the specific activity is defined as units per milligram of protein. With about 1 unit of enzyme in this test system, the reaction is linear for more than 1 hour. With a crude, cell-free extract, proportionality between rate and the amount of enzyme is obtained between 0 and 4 units, whereas with purified enzyme preparations proportionality is observed between 0 and 10 units.

Purification Procedure

Step 1. The cells are grown, and the crude extracts are obtained according to the method described above (cf. Phospholipase A). To each 100 ml. of the crude cell-free extract is added 31.5 g. of $(NH_4)_2SO_4$. After 30 minutes at 0°, the precipitate is centrifuged, and 30 g. of $(NH_4)_2SO_4$ is added to the supernatant. The precipitate is collected by centrifugation and dissolved in 50 ml. of glycylglycine buffer (0.02 M, pH 9.0) (ammonium sulfate I). A twofold increase in the total activity at this stage is probably due to the removal of metallic inhibitors in the crude extract.

Step 2. To each 50 ml. of ammonium sulfate I is added 18.2 g. of $(NH_4)_2SO_4$. The precipitate is centrifuged, and to the supernatant fluid is added 6.5 g. of $(NH_4)_2SO_4$. The precipitate, collected by centrifugation, is dissolved in 10 ml. of glycylglycine buffer (0.04 M, pH 9.0) (ammonium sulfate II).

Step 3. To 9 ml. of ammonium sulfate II in a 0° bath are added 1 ml. of 1 M sodium acetate and then 6.5 ml. of acetone ($-10°$) dropwise with mechanical stirring. After standing for 5 minutes, the precipitate is removed by centrifugation, and 4 ml. of acetone is added to the supernatant fluid. After 5 minutes, the precipitate is collected by centrifugation and dissolved in 4 ml. of 0.01 M K_2HPO_4 (acetone fraction.)

SUMMARY OF PURIFICATION PROCEDURE

Fraction	Total activity, units	Specific activity, units/mg. protein
Crude extract	4545	24
Ammonium sulfate I	9800	438
Ammonium sulfate II	4488	1155
Acetone	3630	5706

Properties of GPC Diesterase

Specificity. Glycerophosphorylethanolamine (GPE) is the only substrate other than GPC among a group of phosphate diesters tested which is split by the enzyme preparation at a rate comparable with the splitting of GPC. Under test conditions, which resulted in the release of 1 micromole of choline from GPC in 60 minutes, there is no detectable choline release from purified egg lecithin, lysolecithin, sphingomyelin, dipalmitoleyllecithin, or dimyristoyllecithin in a 120-minute incubation period.

Noncholine-containing phosphate diesters, except GPE, did not appear to be attacked. Diphenyl phosphate, yeast ribonucleic acid, and polynucleotides resistant to the ribonuclease action are not metabolized. Very slight activity is observed with α-GP and adenosine-5-phosphate, probably owing to a slight contamination of these activities.

Effect of pH. Maximal activity for the splitting of both GPC and GPE is observed in the range of pH 8.0 to 9.0.

Activators and Inhibitors. At a concentration of 0.001 M, Mg and Mn ions inhibited GPC splitting by more than 90%, and Zn ion inhibited the rate by about 60%. Ethylenediaminetetraacetate (0.0001 M) inhibited splitting by 90%, but this inhibition cannot be overcome by any metals tested. Prolonged dialysis against water did not reduce the activity of the enzyme.

Influence of Substrate Concentration. The Michaelis constants are 1.2×10^{-3} and 2.5×10^{-3} for GPC and GPE, respectively. GPC and GPE are observed to be competitive inhibitors to each other. The dissociation constants of the inhibitor-enzyme complex are calculated to be 2.4×10^{-3} M for GPE and 1.8×10^{-3} M for GPC.

Stability of the Enzyme. The purified preparations are stored at $-10°$ for least six months without any appreciable loss of activity. Heating the enzyme preparation at pH 9.0 for 5 minutes at 50, 75, or 100° destroyed approximately 15, 95, or 100% of the activity, respectively.

Phospholipase C

$$\text{Lecithin} + H_2O \rightarrow \text{Diglyceride} + \text{Phosphorylcholine}$$

Assay Method

Principle. Since a phosphoryl choline is water-soluble, an increase of water-soluble phosphate can be a measure of activity.[39] A more recent method, which is described below, is based on the fact that an equivalent amount of CO_2 is liberated from bicarbonate buffer in the region of pH 7.0 upon the liberation of phosphorylcholine (pK_2 5.6) from lecithin.[35]

[39] M. G. MacFarlane and B. C. J. G. Knight, *Biochem. J.* **35,** 884 (1941).

Reagents

4% lecithin solution in water.

1 M NaHCO$_3$.

0.04 M CaCl$_2$.

Procedure. 0.7 ml. of lecithin solution, 0.1 ml. of NaHCO$_3$ solution, 0.1 ml. of distilled water, and 0.1 ml. of CaCl$_2$ solution are placed in the main compartment of a Warburg flask, and 0.1 ml. of enzyme solution, diluted in 1% albumin as stabilizer, and 0.004 M CaCl$_2$ are placed in the side arm. The vessels are filled with CO$_2$, and CO$_2$ evolution is measured after the enzyme is tipped into the main chamber. The method is reproducible to within ±5% down to approximately 1 M.L.D. of toxin.

Purification Procedure

Little attempt has been made to purify this enzyme. MacFarlane *et al.*[39] grew *Clostridium welchii* in a mixture of 800 ml. of 3.3% Evans peptone and 100 ml. of Na$_2$SO$_4$ muscle extract containing 4.5% glucose. The culture is either dried or dialyzed against glycerol at 3°. Under the conditions described above, 17 M.L.D. of toxin produces approximately 150 μl. of CO$_2$ during the first 10-minute interval.

Properties of Phospholipase C

Specificity. MacFarlane[39,40] found that this enzyme attacks lecithin and sphingomyelin, but not cephalin, phosphatidylserine, lysolecithin, cerebrosides, mono- and diphenylphosphate, β-glycerophosphate, or nucleic acids.

Activators and Inhibitors. Ca ion is required for the maximum activity. Magnesium, manganese, cobalt, and zinc are less effective.

Fluoride, citrate, and phosphate are inhibitory presumably because of the reduction of calcium ion concentration. Sodium dodecyl sulfate causes 80 to 100% inhibition at a concentration of 0.025%.[39] Cuprous, strontium, cadmium, aluminum, ferrous, ferric, and barium ions are inhibitory.[35]

Effect of pH. The pH optimum lies around 6.0 to 7.6, varying with the method employed. With the manometric method described above, pH optimum was found to be 6.7.

Stability. This enzyme, like other phospholipases, is rather heat-resistant. At pH 7.6, heating at 100° for 10 minutes destroys only about 50% of the original activity.

[40] M. G. MacFarlane, *Biochem. J.* **42**, 587 (1948).

Phospholipase D[41]

Lecithin $+ H_2O \rightarrow$ Phosphatidic Acid $+$ Choline

Assay Method

Principle. Although lecithin and phosphatidic acid are ether-soluble, free choline is not soluble in ether. Therefore, the decrease in ether-soluble nitrogen is taken as an index of enzymatic activity.

Reagents

An ether solution containing 100 mg. of soybean phospholipid in about 10 ml.

0.1 M phosphate buffer, pH 5.6.

Procedure. The substrate solution is transferred to a 50-ml. glass-stoppered volumetric flask, and the solvent removed by evaporation at 55 to 60°. The phospholipid is then mixed with 2.5 ml. of enzyme extract and 2.5 ml. of the buffer, and the mixture is vigorously shaken. One drop of chloroform is then added to the emulsified mixture as a preservative, and the mixture is incubated at 25° for 2 hours. At the end of the incubation period, ethyl ether is added to the reaction mixture, and the contents are diluted to volume with ethyl ether and vigorously shaken. About 45 minutes is allowed for complete extraction. The ether-soluble and aqueous phases are then analyzed for nitrogen, choline, and phosphorus. The loss of ether-soluble nitrogen is the index of enzymatic activity.

Purification Procedure

100 g. of fresh cabbage leaves is ground and homogenized with 75 ml. of distilled water in a Waring blendor, and the mixture is allowed to stand at 5° for 1 to 2 hours. It is then filtered through a Büchner funnel; the entire filtrate is then centrifuged. The supernatant is used within 12 hours after preparation. With the standard assay system, about 40% of the ether-soluble nitrogen is removed with 2.5 ml. of this enzyme preparation.

Properties of Phospholipase D

Effect of pH. Optimum activity is observed at a pH range between 5.1 and 5.9. Little activity is found below pH 2.0 or above 8.0.

Stability of the Enzyme. The enzyme retains 30 to 40% of its activity after being heated at 100° for 15 minutes. Nevertheless, the enzyme is more active at 25° than at 37°, presumably because the phospholipid emulsion is more easily broken at 37° than at 25°.

[41] D. J. Hanahan and I. L. Chaikoff, *J. Biol. Chem.* **172**, 191 (1948).

[111] Enzymatic Phospholipid Synthesis : Phosphatidic Acid

$$RCOOH + ATP + CoASH \rightarrow RCOSCoA + PP + A\text{-}5\text{-}P \qquad (1)$$

$$
\begin{array}{ccc}
CH_2OH & & CH_2O\cdot CO\cdot R \\
| & & | \\
CH\ OH & + 2RC{=}O \rightarrow & CHO\cdot CO\cdot R + 2CoASH \qquad (2) \\
| & | & | \\
CH_2OP^*O_3H_2 & SCoA & CH_2OP^*O_3H_2
\end{array}
$$

$* = P^{32}$ isotope present

$RCOOH = C_{14}$ to C_{18} fatty acid

By ARTHUR KORNBERG

Assay Method

Principle. The enzymes[1–3] responsible for phospholipid synthesis are as yet known only in the crude state and, therefore, justify no further description. The purpose of this section is to describe an easy and extremely sensitive method for determining the extent of conversion of a water-soluble substance to a water-insoluble, fat-soluble substance. An isotopically labeled substrate of high specific activity is used in the incubation mixture. Usually the residue obtained by the addition of acid (i.e., to yield a final concentration of 3.5% perchloric acid) can be washed completely free of radioactivity unless some of the isotope has been incorporated into a lipid or other water-insoluble substance. Lipids can be readily removed from the residue by extraction with an organic solvent.

Reagents

α-Glycerophosphate-P^{32} (0.025 M, 5 × 10⁵ c.p.m./micromole).[2]
ATP (0.04 M).
CoA (4 × 10⁻⁴ M).
Palmitate (0.02 M) (adjusted to pH of about 9 with NH₄OH and heated briefly in a steam bath to obtain almost complete solution).
MgCl₂ (0.3 M).
NaF (0.5 M).
Cysteine (0.2 M).
KH₂PO₄–K₂HPO₄ (0.5 M, pH 7.0).
Enzyme.[2]

[1] A. Kornberg and W. E. Pricer, Jr., *J. Biol. Chem.* **204**, 329 (1953).
[2] A. Kornberg and W. E. Pricer, Jr., *J. Biol. Chem.* **204**, 345 (1953).
[3] E. P. Kennedy, *J. Biol. Chem.* **201**, 399 (1953).

Procedure. The incubation mixture contained 0.1 ml. of ATP, 0.1 ml. of α-glycerophosphate-P^{32}, 0.1 ml. of CoA, 0.1 ml. of palmitate, 0.02 ml. of $MgCl_2$, 0.1 ml. of NaF, 0.05 ml. of cysteine, 0.1 ml. of phosphate buffer, enzyme, and water to a final volume of 1.2 ml. The mixture was incubated for 30 minutes at 23°.

The rate and extent of glycerophosphate esterification (phospholipid synthesis) were assayed as follows: the mixture, incubated in a 6-ml. round-bottom centrifuge tube (for use in the International multispeed centrifuge), was treated with an equal volume of 7% perchloric acid and centrifuged to collect the precipitate.[4] The latter was then washed three times with 2.5-ml. portions of 3.5% perchloric acid.[5] The precipitate was routinely extracted with 2 ml. of ethanol, but methanol and an alcohol: ether mixture (3:1) was used on occasion. The radioactivity of the ethanolic extract was taken to represent α-glycerophosphate converted to a lipid form, since no counts were obtained in control experiments in which heated enzyme was used or an essential component of the reaction, such as ATP or CoA, was omitted. Extraction of the fat-soluble P^{32} from the residue was essentially complete (>95%) with a single ethanol treatment at room temperature; similar results were obtained with extractions at 2 to 5° or at boiling temperature, and also with the other solvents mentioned. In view of the high specific activity of the starting α-glycerophosphate, the conversion of less than 1 millimicromole was readily measured. In experiments designed for the isolation of phospholipids, the ethanolic extracts were kept cold and also promptly neutralized.

[4] When the size of the precipitate appeared to be too small to handle conveniently, 0.1 ml. of crystalline bovine serum albumin (6% solution) was added.
[5] Quick and effective washing of the residue was effected by the use of a plunger made of a glass rod with a bulbous end (about 1 ml. in volume) closely fitted to the bore of the tube.

[112] Choline Oxidase

By J. H. QUASTEL

The fact that there is an enzyme, choline oxidase, in liver that accomplishes the oxidation of choline to betaine aldehyde was first demonstrated by Mann and Quastel,[1] who isolated the aldehyde as the reineckate and aurichloride. Bernheim and Bernheim[2] had already found that the presence of acetylcholine increases the rate of oxygen uptake of a

[1] P. J. G. Mann and J. H. Quastel, *Biochem. J.* **31**, 869 (1937).
[2] F. Bernheim and M. L. C. Bernheim, *Am. J. Physiol.* **104**, 438 (1933).

rat liver suspension, the extra amount of oxygen consumed being proportional to the amount of acetylcholine added. They[3] later showed that choline, itself, undergoes oxidation presumably to a mixture of betaine aldehyde and betaine, although these substances were not isolated. Trowell,[4] working with rat liver slices and with preparations of washed liver, had also come to the conclusion that choline was oxidized by this tissue. Mann et al.[5] found, by manometric studies, that the oxygen consumption due to the addition of choline to a rat liver preparation corresponded to the conversion of 1 mole of choline to 1 mole of betaine, but that in presence of semicarbazide the oxygen consumption was halved corresponding to the formation of 1 mole of betaine aldehyde. They further showed that the choline oxidase system involves choline dehydrogenase which brings about, in presence of choline, the anaerobic reduction of ferricyanide or of ferricytochrome. They concluded that choline oxidase contains the reaction chain: choline dehydrogenase, cytochrome c, and cytochrome oxidase. Rothschild, Cori and Barron[6] have found from spectrophotometric measurements of cytochrome C reduction by choline oxidase, before and after dialysis, that flavin adenine dinucleotide is also involved in the reaction chain (see also Kelley,[7] and Ebisuzaki and Williams[8]). The presence of magnesium ions seems to be necessary.

It is evident that the conversion of choline to betaine involves the activities of two enzymes, choline dehydrogenase and betaine aldehyde dehydrogenase, and that the choline oxidase system may be considered to involve both enzymes. Although DPN has been reported to serve as a cofactor both for choline oxidase (Strength et al.[9]) and for betaine aldehyde dehydrogenase (Klein and Handler[10]), it seems to be clear from the work of Williams[11] that only the latter dehydrogenase is DPN linked.

Occurrence

Choline oxidase occurs in the liver and kidney of rat[1,2] and of cat[2] but apparently not of guinea pig.[1,2] The enzyme system is located in the mitochondria (Kensler and Langemann,[12] Williams[13]). Exhaustive washing

[3] F. Bernheim and M. L. C. Bernheim, *Am. J. Physiol.* **121**, 55 (1938).

[4] O. A. Trowell, *J. Physiol.* **85**, 356 (1935).

[5] P. J. G. Mann, H. E. Woodward, and J. H. Quastel, *Biochem. J.* **32**, 1025 (1938).

[6] H. A. Rothschild, O. Cori, and E. S. G. Barron, *J. Biol. Chem.* **208**, 41 (1954).

[7] B. Kelley, *Federation Proc.* **11**, 238 (1952).

[8] K. Ebisuzaki and J. N. Williams, Jr., *J. Biol. Chem.* **200**, 297 (1953).

[9] D. R. Strength, J. R. Christensen, and L. J. Daniel, *J. Biol. Chem.* **203**, 63 (1953).

[10] J. R. Klein and P. Handler, *J. Biol. Chem.* **144**, 537 (1942).

[11] J. N. Williams, Jr., *J. Biol. Chem.* **206**, 191 (1954).

[12] C. J. Kensler and H. Langemann, *J. Biol. Chem.* **192**, 551 (1951).

[13] J. N. Williams, Jr., *J. Biol. Chem.* **194**, 139; **195**, 37 (1952).

of the mitochondria does not remove the choline dehydrogenase, and removes only a part of the more soluble betaine aldehyde dehydrogenase.[11,14]

Preparation

Colter and Quastel,[15] in investigations involving aerobic oxidation of choline, homogenized one part of rat liver tissue, freshly obtained, with two parts of 0.1 M sodium phosphate buffer, pH 7.4, and strained the homogenate through several thicknesses of muslin, the homogenate being used per se. For the study of the anaerobic oxidation of choline, these authors[15] prepared 3.5 ml. of liver homogenate, made as indicated above, diluted it to 10 ml. with distilled water, and centrifuged it for 5 minutes at 3000 r.p.m. After the supernatant liquid was removed, the solid residue was washed twice with 10-ml. portions of saline-sodium phosphate buffer (0.05 M, pH 6.6) by resuspension and centrifugation. It was finally suspended in 5 ml. of distilled water. Such preparations absorb no oxygen in the presence of choline but oxidize choline actively under anaerobic conditions.

Soluble choline dehydrogenase may be prepared by extraction of acetone-ether dried mitochondria with sodium choleate.[14] Williams[11] describes the following technique for the study of choline oxidase. Mitochondria are isolated from rat livers by differential centrifugation of 20% homogenates in 0.25 M sucrose.[16] The mitochondria from 4 g. of liver are diluted at the last stage to 25 ml. with ice-cold distilled water. Such a mitochondrial preparation may be used directly. It may also be frozen in a mixture of dry ice and acetone and rapidly thawed at 37°. The frozen and thawed mitochondria are then centrifuged at 25,000 × g for 30 minutes and resuspended in cold distilled water to the original volume.

Mitochondrial preparations, dried by washing with cold acetone and subsequently with ether, the ether being finally removed in vacuo, retain their choline dehydrogenase and betaine aldehyde dehydrogenase activities for months if the dried mitochondria are kept at −5°.[11]

Betaine aldehyde dehydrogenase has been prepared from rat liver[17] by homogenization and differential centrifugation at 0° and by fractionation of the clear supernatant with ammonium sulfate.

Assay

Choline oxidase and dehydrogenase activities may be conveniently estimated by means of the conventional Warburg manometric tech-

[14] J. N. Williams, Jr., and A. Sreenivasan, J. Biol. Chem. **203**, 899 (1953).
[15] J. S. Colter and J. H. Quastel, Arch. Biochem. and Biophys. **41**, 305 (1952).
[16] J. N. Williams, Jr., J. Biol. Chem. **197**, 709 (1952); see also Vol. I [3].
[17] H. A. Rothschild and E. S. G. Barron, J. Biol. Chem. **209**, 511 (1954).

niques,[1,3,5,11] ferricyanide being a useful substrate for anaerobic assays.[5] Manganese dioxide may also be used as an anaerobic terminal hydrogen acceptor, with a trace of ferricyanide as carrier, for choline oxidation studies.[18]

Spectrophotometric assays for choline and betaine aldehyde dehydrogenase may be carried out (Williams[11]) in the following manner: To a series of test tubes are added 1 ml. of phosphate buffer (pH 6.8 for choline dehydrogenase) and pH 7.8 for betaine aldehyde dehydrogenase), 0.2 ml. of 2% choline chloride or 2% betaine aldehyde, 0.5 ml. of 10 mg. % 2,6-dichlorophenolindophenol, and water to make 2.2 ml. The contents of the tubes are transferred to cuvettes of a model DU Beckmann spectrophotometer set at 607 mμ 1 ml. of the enzyme solutions are added, the cuvette contents mixed, and optical density readings taken periodically.

Properties

Choline dehydrogenase attacks both choline and arsenocholine[5] (the affinity of the latter for the enzyme being less than that of the former) but not ethanol.[3,5] It is inhibited by ammonium and trimethylammonium ions and by betaine.[5] It is also competitively inhibited by a variety of amines, e.g., benzedrine, ephedrine, tyramine, histamine, mono- and dimethylamines and mono-, di-, and triethanolamines.[15] The dissociation constant for the choline oxidase-choline complex is given as 0.0029 $\pm$ 0.0002; that for choline oxidase-benzedrine is 0.00025 $\pm$ 0.00003; and that for choline oxidase-ephedrine is 0.0014 $\pm$ 0.0002.[15] Barron et al.[19] have shown that choline oxidase is strongly inhibited by the nitrogen mustards. Protection against inactivation by nitrogen mustard may be secured by the presence of choline, or of the amines, described above, that have been termed anticholine oxidases.[15]

Choline dehydrogenase also attacks dimethylethylcholine, diethylmethylcholine, homocholine, β-methylcholine, α,α-dimethylcholine, α-methyl-α-hydroxymethylcholine, and is inhibited by a number of choline analogues.[20]

Choline oxidase of rat liver is inactivated by both crude and heated cobra venom (Braganca and Quastel[21]), the lecithinase present being held responsible.

The enzyme system is inhibited by HCN, complete inhibition taking place at 10^{-3} M, but choline dehydrogenase is comparatively cyanide

[18] R. M. Hochster and J. H. Quastel, Arch. Biochem. and Biophys. **36**, 132 (1952).
[19] E. S. G. Barron, G. R. Bartlett, and Z. B. Miller, J. Exptl. Med. **87**, 489 (1948).
[20] I. C. Wells, J. Biol. Chem. **207**, 575 (1954).
[21] B. M. Braganca and J. H. Quastel, Biochem. J. **53**, 88 (1953).

insensitive.[5] It is a thiolenzyme,[22,23,24] the system inducing aerobic phosphorylation during choline oxidation.[6]

Williams[11] claims that the presence of ATP stimulates the choline oxidase activity of frozen and thawed mitochondria but that ATP has no effect on the activity of choline dehydrogenase. There is an enhancement of activity due to the addition of DPN, whose effect, however, is confined to betaine aldehyde dehydrogenase. There is little or no effect of TPN.

[22] J. J. Gordon and J. H. Quastel, *Nature* **159**, 97 (1947).
[23] J. J. Gordon and J. H. Quastel, *Biochem. J.* **42**, 337 (1948).
[24] E. S. G. Barron and T. P. Singer, *Science* **97**, 356 (1943).

[113] Cholesterol Dehydrogenase from a *Mycobacterium*[1]

$$\text{Cholesterol} + \tfrac{1}{2}O_2 \rightarrow \Delta^4\text{-Cholestenene-3-one} + H_2O$$

By THRESSA C. STADTMAN

Assay Method

Principle. The method involves measurement of disappearance of cholesterol by the Liebermann-Burchard reaction. Alternatively, cholestenone formation can be followed in purified extracts either by measuring its carbonyl absorption at 240 mμ or by measuring the absorption of its 2,4-dinitrophenylhydrazone at 390 mμ.

Reagents

> Cholesterol stock solution (10 mg./ml.). Dissolve 1 g. of cholesterol (m.p. 149°) in acetone, and dilute to 100 ml. at refrigerator temperature. This solution may be stored for several months at 5 to 10°.
> Colloidal cholesterol suspension. (See Vol. III [63].)
> 1 *M* potassium phosphate buffer, pH 7.5.
> Enzyme. (a) *Cell suspensions.* Dilute suspensions of fresh or dried glycerol-grown cells to about 75 mg. (dry weight) per milliliter. Such preparations oxidize cholesterol only as far as cholestenone.[2]

[1] T. C. Stadtman, A. Cherkes, and C. B. Anfinsen, *J. Biol. Chem.* **206**, 511 (1954).
[2] An adaptive enzyme is involved in the further oxidation of cholesterol, and the organism must be grown on the sterol in order to demonstrate activity on chloes-

(b) *Soluble enzyme preparation.* Soluble enzyme preparations are adjusted so that 1 ml. oxidizes 0.5 to 1.0 mg. of cholesterol (1.25 to 2.5 micromoles) during a 3- to 4-hour incubation period.

Reagents for determination of cholesterol (see Vol. III [63]).

Absolute ethanol + acetone (1:1).

Procedure. Prepare 25-ml. Erlenmeyer flasks containing 0.1 ml. of 1.0 M potassium phosphate buffer (pH 7.5), 1.0 ml. of enzyme, and 0.9 ml. of colloidal cholesterol suspension (or 0.1 ml. of acetone solution of cholesterol plus 0.8 ml. of water).[3] Flush the flasks with oxygen, stopper, and incubate at 37° with shaking. One or more boiled controls are included in each experiment to test recovery of the added sterol by the extraction and analytical methods employed. After 3 to 4 hours' incubation, the reaction is stopped by the addition of absolute ethanol-acetone (1:1) to 80% by volume. This is conveniently carried out by transferring the reaction mixture to a centrifuge tube and rinsing the flask with aliquots of ethanol-acetone until 8 ml. has been added to the tube. This procedure serves both to extract the sterol and to deproteinize the reaction mixture. The supernatant and washings are evaporated to dryness and analyzed for cholesterol (see Vol. III [63]).

Specific Activity. Activity of the preparations is expressed as micromoles of sterol oxidized per hour per milligram of protein or per 10 mg. of dried cells. Protein is determined by the sulfosalicylic acid turbidimetric method.[4]

Alternate Assay Procedures for Purified Enzyme Preparations. Partial purification of crude extracts of the *Mycobacterium* removes, to a large extent, materials that absorb at 240 mμ and enables one to follow cholestenone formation by measuring increases in absorption at this wavelength due to the carbonyl band of the reaction product. This is conveniently done on aliquots removed from the reaction mixtures at appropriate intervals and fixed by the addition of absolute ethanol to 90%. Suitable dilutions of the supernatants are analyzed spectrophotometrically at 240 mμ (see Vol. III [63]).

Cholestenone may also be estimated as its 2,4-dinotrophenylhydrazone (see Vol. III [63]). Aliquots of reaction mixtures deproteinized by the

---

tenone. Glycerol-grown cells do not form detectable amounts of enzymes responsible for oxidation of cholestenone during incubation with this compound for periods as long as 12 hours.

[3] Acetone is not oxidized by the *Mycobacterium;* acetone solutions of steroids can be added to cell suspensions or soluble enzyme preparations provided that the final concentration of acetone does not exceed about 5% (v/v).

[4] Klett-Summerson photoelectric colorimeter clinical manual.

addition of ethanol to 80 or 90% (but *not* ethanol + acetone) are reacted with alcoholic dinitrophenyl hydrazine reagent, and after suitable dilution with chloroform are analyzed spectrophotometrically at 390 mμ. Both of these assays are less time-consuming than is the Liebermann-Burchard method for cholesterol.

Preparation of Cells

The cholesterol dehydrogenase is obtained from a soil microorganism[5] tentatively identified as a *Mycobacterium*. The medium employed for growth contains the following ingredients in grams per 100 ml.: $(NH_4)_2$-SO_4, 0.2; K_2HPO_4, 0.2; $MgSO_4 \cdot 7H_2O$, 0.02; $CaCl_2 \cdot 2H_2O$, 0.001; $FeSO_4 \cdot 7H_2O$, 0.001; and glycerol, 0.5. The addition of 0.01% tryptone or 0.05% acid-hydrolyzed "vitamin-free" casein allows more rapid growth of the organism. The final pH of the medium is 7.0 to 7.5. Liquid cultures started with a 5% inoculum are incubated 20 to 24 hours at 26 to 32° with aeration. Growth does not occur at 37°.

Cells are harvested in a Sharples supercentrifuge (a flow rate of about 250 ml. per minute at 27,000 r.p.m. removes all the cells). After removal from the Sharples rotor the cells are resuspended in 0.5% KCl + 0.5% NaCl and centrifuged at 6000 to 7000 r.p.m. in a Servall centrifuge. This is repeated one or two times. Addition of salt to the wash water facilitates sedimentation of the cells.

The cell paste is lyophilized or frozen to preserve activity. The average yield is slightly more than 1 g. of wet packed cells per liter of medium or 250 to 300 mg. of dry cells. Higher yields are obtained if the cultures are allowed to grow longer, but such material has much lower specific activity. Therefore, it is advantageous to harvest the cells early.

Enzyme Preparation

Cell pastes (freshly harvested or frozen) are ground in a mortar with dry ice for 10 to 20 minutes until a very uniform buff-colored powder is obtained. The powder is extracted with about 10 vol. of 0.05 M potassium phosphate buffer, pH 7.5, for $\frac{1}{2}$ to 1 hour in the cold. Solids are sedimented by centrifugation at 2° in a Servall centrifuge at 8000 r.p.m. The supernatant is decanted and centrifuged at 15,000 r.p.m. to obtain a clear extract. This extract can be stored at $-10°$ for at least 1 to 2 months.

Alternatively, frozen or dried cells may be ground with two and one-half times their weight (dry weight basis) of alumina (Aluminum Corporation of America, A-301), according to the method of McIlwain.[6] They

[5] This organism was isolated from cholesterol enrichment cultures by Schatz *et al.*: A. Schatz, K. Savard, and I. J. Pintner, *J. Bacteriol.* **58**, 117 (1949).

[6] H. McIlwain, *J. Gen. Microbiol.* **2**, 288 (1948); see Vol. I [7].

are then extracted and centrifuged as above. The yield of extracted protein (about 5 to 10% of the dry weight of the cells) is low by either of these procedures, but a variety of other cell-rupturing techniques gave even poorer results. The nitrogen content of the *Mycobacterium* cells (9.5% on a dry weight basis) is not markedly lower than other species of bacteria. Hence the low protein yield is presumably caused by the resistant nature of the cell wall.

The cholesterol dehydrogenase activity of the crude cell-free extracts (1 to 10 mg. of protein per milliliter) can be concentrated by the slow addition (with stirring, at 2°) of solid ammonium sulfate to 60% saturation. The precipitate containing the enzyme is collected by centrifugation at 2° and dissolved in a small volume of 0.01 M pH 7.4 potassium phosphate buffer.

Instead of ammonium sulfate, acetone can be used to precipitate the enzyme from the dilute extracts. This is accomplished by the slow addition of acetone to 60% (v/v) at −10° with stirring. The precipitate, after centrifugation, is then dissolved in dilute buffer as above. Preliminary attempts to fractionate the extracts resulted in a twofold purification of the enzyme in the 42 to 50% acetone fraction, but the yield was very poor. Ammonium sulfate, either at pH 2.0 or at neutrality, was generally unsatisfactory because every fraction contained the enzyme and no purification was achieved. However, selective adsorption of the enzyme on cholesterol appears to offer some hope as a purification technique.[7]

Properties

Specificity. A closely related compound, ergosterol, which is also a Δ^5-3β-OH steroid, is not oxidized by the enzyme.

General Properties. The enzyme is not sedimented by centrifugation at 110,000 $\times$ g for 30 minutes. It can be heated at 60° for 10 minutes with only 50% loss of activity. It withstands dialysis for 18 to 24 hours against distilled water at 2°. The slight loss of activity resulting from dialysis could not be restored by the addition of DPN, TPN, or boiled extracts.

The rate of oxidation of cholesterol is essentially constant over the range of pH 6.5 to 9.0 but falls off markedly above pH 10.0.

Cyanide (0.01 M) or arsenite (0.05 M) does not affect the activity of the enzyme.

[7] In preliminary experiments Dr. R. K. Brown found that the cholesterol dehydrogenase in *Mycobacterium* extracts could be adsorbed on colloidal cholesterol particles and precipitated by centrifugation.

Section IV

Enzymes of Citric Acid Cycle

[114] Crystalline Condensing Enzyme from Pig Heart[1]

$$\text{Acetyl-S-CoA} + \text{Oxalacetate}^{--} + H_2O \rightleftarrows \text{Citrate}^{---} + \text{HS-CoA} + H^+ \quad (1)$$

By Severo Ochoa

Assay Method

a. Chemical Assay. *Principle.* This assay is based on the conversion of a mixture of acetyl phosphate and oxalacetate to citrate and orthophosphate, in the presence of transacetylase,[2] CoA, and condensing enzyme. Under these conditions reaction 2 (catalyzed by transacetylase) couples with reaction 1 to give reaction 3. Although the reactions are

$$\text{Acetyl phosphate} + \text{HS-CoA} \rightleftarrows \text{Acetyl-S-CoA} + \text{orthophosphate} \quad (2)$$

$$\text{Acetyl phosphate} + \text{oxalacetate}^{--} + H_2O \xrightleftharpoons{\text{(CoA)}}$$
$$\text{Citrate}^{---} + \text{orthophosphate} + H^+ \quad (3)$$

reversible, the equilibrium position of reaction 3 is markedly in favor of citrate synthesis. Under the condition of the assay the rate of acetyl phosphate disappearance and/or citrate synthesis is proportional to the condensing enzyme concentration over a 25-fold range of enzyme dilutions.

Reagents

0.25 M potassium phosphate buffer, pH 7.4.
0.1 M potassium L-cysteine.[3]
0.2 M potassium oxalacetate.[3]
0.1 M dilithium acetyl phosphate.[4]
Escherichia coli (strain 4157) extract. This extract is used as a source of both transacetylase and CoA.

Preparation of E. coli Extract. E. coli, strain 4157 (American Type Culture Collection), is grown as described by Umbreit and Gunsalus.[5]

[1] S. Ochoa, J. R. Stern, and M. C. Schneider, *J. Biol. Chem.* **193**, 691 (1951).
[2] E. R. Stadtman, *J. Biol. Chem.* **196**, 527 (1952); see Vol. I [98].
[3] Freshly prepared solutions of L-cysteine and oxalacetic acid are neutralized with potassium hydroxide to pH 7.4 just before use. Sodium hydroxide is to be avoided, since sodium ions inhibit transacetylase. Cysteine is used to maintain CoA in the reduced state.
[4] E. R. Stadtman and F. Lipmann, *J. Biol. Chem.* **185**, 549 (1950); see Vol. III [39].
[5] W. W. Umbreit and I. C. Gunsalus, *J. Biol. Chem.* **159**, 333 (1945).

After 40 hours of growth at 30°, without aeration, the organisms are harvested by centrifugation and washed twice with 0.4% sodium chloride. The yield of fresh cells is about 1.0 g./l. of medium. The washed cells are made into a paste with distilled water and lyophilized. 4.0 g. of cells yields about 1.0 g. of dry powder. This powder keeps indefinitely when stored in closed containers in the cold.

First 0.5 g. of powder is ground in a mortar with 0.5 g. of Alumina A-301 (−325 mesh, Aluminum Company of America) and then extracted for 10 minutes at room temperature with 6.5 ml. of 0.02 M potassium phosphate buffer, pH 7.0, with continuous grinding. The suspension is centrifuged at 13,000 r.p.m. and 0° (high-speed head of International centrifuge) for 20 minutes, and the solution decanted for use. 1.0 ml. of this extract contains about 20 mg. of protein (as determined by the method of Lowry et al.[6]) and supplies 20 to 25 units of transacetylase[2] and 15 to 20 units of CoA;[7] it contains little or no condensing enzyme and is free of aconitase. The extract gives uniform results provided that it is used on the same day it is prepared.

Procedure. The reaction mixture contains 0.1 ml. of buffer (25 micromoles), 0.1 ml. of cysteine (10 micromoles), 0.1 ml. of oxalacetate (20 micromoles), 0.1 ml. of acetyl phosphate (10 micromoles), 0.2 ml. of freshly prepared *E. coli* extract (about 4 mg. of protein, 4 to 5 units of transacetylase, 3 to 4 units of CoA), approximately 3 to 6 units (see definition below) of condensing enzyme, and water to a final volume of 1.0 ml. The additions are made into a tube kept in ice. After addition of the enzyme, the tube is placed in a bath at 25°, shaken for 10 minutes, and then replaced in ice. A 0.5-ml. aliquot is immediately withdrawn for determination of acetyl phosphate as hydroxamic acid by the method of Lipmann and Tuttle.[8] 3.0 ml. of 12% trichloroacetic acid is added to the remainder. The mixture is centrifuged, and a suitable aliquot of the supernatant solution used for the determination of citric acid.[9] The appropriate volume of the aliquot is estimated from the disappearance of acetyl phosphate. Citric acid is determined by the method of Natelson et al.[10] A control tube without condensing enzyme is included; it supplies the initial value of acetyl phosphate and can be used to correct for any

[6] O. H. Lowry, N. J. Rosebrough, A. L. Farr, and R. J. Randall, *J. Biol. Chem.* **193'** 265 (1951); see Vol. III [73].

[7] N. O. Kaplan and F. Lipmann, *J. Biol. Chem.* **174,** 37 (1948).

[8] F. Lipmann and L. C. Tuttle, *J. Biol. Chem.* **159,** 21 (1945).

[9] Since the disappearance of acetyl phosphate is always identical to the formation of citric acid (except in the initial extracts where the former slightly exceeds the latter, probably because of the presence of weak acylphosphatase activity), the determination of citrate is not indispensable.

[10] S. Natelson, J. B. Pincus, and J. K. Lugovoy, *J. Biol. Chem.* **175,** 745 (1948).

citrate synthesis caused by traces of condensing enzyme in the *E. coli* extract.

Definition of Unit and Specific Activity. One unit of enzyme is defined as that amount which, under the condition of the above assay, causes the synthesis of 1.0 micromole of citrate (or the disappearance of 1.0 micromole of acetyl phosphate) in 10 minutes at 25°. Specific activity is expressed as units per milligram of protein. The specific activity of pure condensing enzyme is 330 ± 15.

The protein content of the enzyme solutions is determined by two methods. Prior to elution from calcium phosphate gel (see below), when the nucleic acid content is relatively high, protein is estimated by the method of Lowry *et al.*[6] At later stages protein is determined spectrophotometrically as described for the preparation of "malic" enzyme (see Vol. I [124]).

b. Optical Assay. *Principle.* In this assay acetyl CoA is used as one of the substrates, and a mixture of L-malate, DPN⁺, and malic dehydrogenase (see Vol. I [123]) is substituted for oxalacetate. Under the conditions reaction 4, catalyzed by malic dehydrogenase, couples with reaction 1 to give reaction 5.[11]

$$\text{L-Malate}^{--} + \text{DPN}^+ \rightleftarrows \text{Oxalacetate}^{--} + \text{DPNH} + \text{H}^+ \qquad (4)$$

$$\text{Acetyl-S-CoA} + \text{L-malate}^{--} + \text{DPN}^+ + \text{H}_2\text{O}$$
$$\rightleftarrows \text{Citrate}^{---} + \text{HS-CoA} + \text{DPNH} + 2\text{H}^+ \quad (5)$$

Under the conditions of the assay (see below) the equilibrium position of reaction 5 is in favor of citrate synthesis. The early rate of reduction of DPN (which is equal to that of citrate synthesis) is proportional to the concentration of condensing enzyme within a five- to sixfold range of enzyme dilutions. The measurement is carried out in the Beckman spectrophotometer at wavelength 340 mμ using Corex or silica cells of 0.5-cm. light path.

Reagents

0.1 *M* tris(hydroxymethyl)aminomethane-HCl buffer, pH 8.0.
0.05 *M* potassium L-malate.
0.007 *M* (approximately) acetyl CoA.
0.003 *M* DPN⁺.
Malic dehydrogenase (specific activity 50,000 to 70,000; see Vol. I [123]).

[11] J. R. Stern, B. Shapiro, E. R. Stadtman, and S. Ochoa, *J. Biol. Chem.* **193**, 703, (1951); J. R. Stern, S. Ochoa, and F. Lynen, *J. Biol. Chem.* **198**, 313 (1952).

Preparation of Acetyl CoA. This preparation is based on the procedure developed by Simon and Shemin[12] for the synthesis of succinyl CoA. 10 mg. (about 8 micromoles) of CoA (Pabst)[13] is dissolved in 1.0 ml. of water, and an aliquot (0.05 ml.) is taken for determination of the sulf-hydryl content;[14] to the remainder is added 0.1 ml. of 1.0 M KHCO₃, and the mixture is cooled to 0°. An amount of ice-cold acetic anhydride (freshly prepared 0.1 M aqueous solution[15]), containing 1.3 micromoles of anhydride per micromole of —SH in the CoA solution, is added (0.13 to 0.16 ml.), and, after mixing, the solution is allowed to stand for 4 minutes at 0°. After this time most of the —SH has disappeared. This is controlled by —SH determination on an aliquot. The solution, which contains about 6.7 micromoles of acetyl-S-CoA and 4.3 micromoles of acetyl-S-glutathione per milliliter, is used directly for the condensing enzyme assay. Since the equilibrium position of reaction 1 is very far in the direction of citrate synthesis, the amount of citrate synthesized from an unknown amount of acetyl CoA in the presence of condensing enzyme and excess oxalacetate provides an accurate method for the quantitative determination of acetyl CoA if required. Acetyl glutathione does not interfere with this determination when carried out at pH 7.4.

Procedure. The reaction mixture (in 1.5-ml. cells, $d = 0.5$ cm.) contains 0.75 ml. of buffer (75 micromoles), 0.2 ml. of malate (10 micromoles), 0.1 ml. of DPN⁺ (0.3 micromole), an amount of malic dehydrogenase solution containing about 3 γ of enzyme (150 to 200 units), 0.03 ml. of acetyl CoA (about 0.2 micromole), condensing enzyme (0.04 to 0.22 *chemical assay* units), and water (adjusted to a temperature of 20 to 23°) to a final volume of 1.5 ml. The reaction is carried out at room temperature and is started by the addition of either acetyl CoA or condensing enzyme as soon as the malic dehydrogenase reaction has reached equilibrium. This takes approximately 2 minutes and results in a ΔE of about +0.04. Readings of the optical density (wavelength 340 mμ) are made, against a blank containing buffer and water, at 30-second intervals for 4 to 5 minutes. Under these conditions the optical density increases linearly with time during the first 3 minutes. The rate of increase in optical density (ΔE/minute) between 30 seconds and 2 or 3 minutes after the start of the reaction is used to calculate the enzyme activity.[16] Within

[12] E. J. Simon and D. Shemin, *J. Am. Chem. Soc.* **75**, 2520 (1953).
[13] This preparation contains 60 to 70% CoA-SH and 15 to 20% reduced glutathione. Molecular weight of CoA, 767.
[14] By the method of R. R. Grunert and P. H. Phillips, *Arch. Biochem.* **30**, 217 (1951).
[15] The acetic anhydride content is determined by the hydroxamic acid reaction of Lipmann and Tuttle.[8]
[16] The optical assay is not reliable with the initial extracts because of the presence of DPNH-oxidase.

the range of condensing enzyme concentrations specified above, the ΔE/minute varies between 0.006 and 0.03.

Definition of Unit and Specific Activity. One unit of enzyme is defined as the amount which causes an increase in optical density of 0.01 per minute under the above conditions. Specific activity is expressed as units per milligram of protein. The specific activity of pure condensing enzyme in the optical assay is 4860. Thus 1.0 optical unit is equal to $330/4860 = 0.068$ chemical unit.

Purification Procedure

A description of the preparation below has been previously published.[1] Some improvements in a preparation obtained more recently[17] will be pointed out.

Pig hearts, removed immediately after death, are packed in ice. All operations are performed at about 0°.

Step 1. Extraction. 70 pig hearts are trimmed of fat and connective tissue and ground in a mechanical mincer. 10 kg. of mince is extracted with 20 l. of 0.02 M potassium phosphate buffer, pH 7.4, for 30 minutes with mechanical stirring. The suspension is squeezed through four layers of cheesecloth with the aid of a press, yielding a deep-red extract. The residue is extracted with another 20 l. of buffer in the same manner, and the second extract is combined with the first. The volume of combined extracts was 33 l.; pH 6.5; protein, 312 g.; specific activity, 0.91.

Step 2a. Isoelectric Precipitation. To the extract is added slowly, with mechanical stirring, 348 ml. of 1.0 M acetic acid over a period of about 1 hour until the pH falls to 5.5. The suspension is allowed to settle overnight. The clear supernatant fluid is siphoned, and the remainder centrifuged in the cold at 2000 r.p.m. for 1 hour. The precipitate is taken up in 800 ml. of 0.02 M potassium phosphate buffer, pH 7.4, and shaken mechanically for 30 minutes at 3 to 4°. The suspension is centrifuged in a Servall angle centrifuge at 13,000 r.p.m. for 30 minutes, and the insoluble residue discarded. The deep-pink supernatant fluid (1295 ml.) is dialyzed against 10 l. of 0.1 M potassium phosphate buffer, pH 7.9, for 48 hours. The solution of the isoelectric precipitate contains 8.77 g. of protein; specific activity, 7.3.

Step 2b. Adsorption on Calcium Phosphate Gel. The supernatant fluid from the above step (32.5 l.) contains 275 g. of protein; specific activity, 0.89. To this is added 3.6 l. of calcium phosphate gel (25 mg. (dry weight)/ml.), prepared according to Keilin and Hartree.[18] The mixture is stirred

[17] February 1954. In collaboration with D. O. Brummond and M. C. Schneider.
[18] D. Keilin and E. F. Hartree, *Proc. Roy. Soc. (London)* **B124**, 397 (1938); see Vol. I [11].

mechanically for 4 hours and allowed to settle overnight. The clear solution is siphoned, and the gel centrifuged at 2000 r.p.m. for 45 minutes. The solution (37 l.) containing 207 g. of protein (specific activity, 0.42) is discarded. The precipitate is taken up in 1.5 l. of 0.1 M potassium phosphate buffer, pH 7.9, and shaken mechanically for 1 hour in a 4-l. bottle containing a handful of glass beads. A few drops of capryl alcohol are added to minimize foaming. The suspension is then centrifuged at 2000 r.p.m. for 30 minutes. Four more elutions of the gel are carried out in the above manner. The volume of combined eluates was 7240 ml.; protein, 51 g.; specific activity, 2.7.

Step 3. Ethanol Precipitation. The dialyzed solution of the isoelectric precipitate and the calcium phosphate eluates are combined to give 8480 ml. of solution containing 43.5 g. of protein; specific activity, 5.5. The solution is brought to 0°, and 3640 ml. of ethanol cooled to $-50°$ is added dropwise, with mechanical stirring, at such a rate that about 6 hours is required for complete transfer. The final ethanol concentration was 30% by volume. As the ethanol runs in, the temperature of the solution is gradually lowered to $-12°$ or $-14°$. The mixture is transferred to a tall, narrow glass cylinder and stored at $-18°$ overnight. Because of the large volumes involved, the alcohol precipitation has been carried out in two lots, each half the total volume. The next morning as much of the clear fluid as possible is siphoned, and the remainder is centrifuged at $-20°$ and 2000 r.p.m. for 45 to 60 minutes. The temperature after the run was $-15°$. The supernatant solution is decanted, collected, and again stored at $-18°$. Additional active precipitates form over a period of 2 weeks. These have been collected and further fractionated in the manner outlined for the first ethanol precipitate. The first ethanol precipitate is dissolved in 140 ml. of 0.02 M potassium phosphate buffer, pH 7.4, and dialyzed against 4 l. of the same buffer for 7 hours to remove ethanol. After dialysis, a slight precipitate is centrifuged at 13,000 r.p.m. and 0° for 15 minutes; 440 ml. of deep-red, clear supernatant fluid is obtained containing 15.5 g. of protein; specific activity, 6.0. The solution of the second ethanol precipitate contained 11.76 g. of protein; specific activity, 5.0. The harvesting of several precipitates can be avoided by allowing the ethanol mixture to stand at $-18°$ for 3 days.

Step 4. First Ammonium Sulfate Fractionation. The solution of the first ethanol precipitate is diluted with 0.02 M potassium phosphate buffer, pH 7.4, to bring the protein concentration to 12 to 15 mg./ml., and is brought to 50% saturation with solid ammonium sulfate (35.3 g./100 ml.). The salt is added slowly over a period of 1 hour at 0° with mechanical stirring. The mixture is allowed to stand for 30 minutes, and the precipitate centrifuged at 13,000 r.p.m. for 20 minutes. The precipitate (4.79 g.

of protein; specific activity, 2.4) is discarded. The supernatant solution is brought to 60% saturation by adding 7.1 g. of ammonium sulfate for each 100 ml. of original solution over a period of 30 minutes. After standing for 20 to 30 minutes, the precipitate is centrifuged as above and dissolved in 15 ml. of 0.02 M potassium phosphate buffer, pH 7.4, to give 20 ml. of a pink solution, containing 428 mg. of protein; specific activity, 31. The supernatant solution is brought to 70% saturation by adding 7.1 g. of ammonium sulfate for each original 100 ml. in the same manner as above. The precipitate is dissolved in 15 ml. of the above buffer, giving 21 ml. of a clear, deep-red solution containing 877 mg. of protein; specific activity, 34. The supernatant solution is discarded.

Step 5. Second Ammonium Sulfate Fractionation. The two fractions from the previous step are refractionated individually without previous dialysis. The solution is diluted with 0.02 M potassium phosphate buffer, pH 7.4, to bring the protein concentration to 10 mg./ml. It is then brought to 40% saturation by adding 28.2 g. of ammonium sulfate per milliliter over a 20-minute period. After standing 20 minutes, the precipitate is centrifuged and discarded. The supernatant solution is brought to 60% saturation by adding 14.2 g. of ammonium sulfate per 100 ml. of original solution. The precipitate is centrifuged and dissolved in the minimum volume of 0.02 M potassium phosphate buffer, pH 7.4, required for solution. The solutions together (6.0 ml.) contain 721 mg. of protein; specific activity, 65.

Step 6. Crystallization. The solutions from the previous step, each containing about 12% protein in rather concentrated ammonium sulfate, are kept at 3 to 4°. Crystallization of the condensing enzyme begins within a few hours and is allowed to continue for 48 hours or more. The crystals are harvested by centrifugation at 15,000 r.p.m. for 45 to 60 minutes at 0°. Almost all the enzyme is present in the crystalline precipitate. The clear red supernatant solution, which contains only about 2% of the enzyme originally present in the ammonium sulfate fractions (specific activity, 2), is discarded. The two lots of crystals are combined, washed in the centrifuge at 0° with a few milliliters of 60% saturated ammonium sulfate, and dissolved in 0.02 M potassium phosphate buffer, pH 7.4, to give a clear, light-yellow solution containing 202 mg. of protein; specific activity, 200. Any small residue at this stage is centrifuged, washed with a small volume of buffer, and the wash combined with the main solution.

Step 7. Recrystallization. The enzyme is easily recrystallized from 3 to 4% solutions in 0.02 M potassium phosphate buffer, pH 7.4, by dropwise addition, with stirring, of saturated ammonium sulfate at 0° until crystallization begins. This occurs when the ammonium sulfate saturation is about 45%. Crystallization is allowed to proceed for 3 or more days in the cold. The crystals are harvested by centrifugation at 0°, washed once with a little 60% saturated ammonium sulfate, and dissolved in buffer to give a colorless water-clear solution. After one or two recrystallizations, the specific activity becomes 330 ± 15 and remains unchanged on further recrystallization. This activity corresponds to the synthesis of 3300 moles of citrate per minute per 100,000 g. of enzyme at 25°, under the conditions of the standard chemical assay. The turnover number increases to a value of about 5000 when the amount of CoA in the assay system is increased to 25 units. Increase of transacetylase to 10 units did not affect the rate of citrate synthesis. Data on the purification of the enzyme are summarized in the table. The procedure has been repeated several times with uniform results. From the specific activity of the pure enzyme, it can be calculated that about 0.3% of the protein in the initial pig heart extract is condensing enzyme.

The specific activity of recrystallized condensing enzyme in the optical assay (4860) corresponds to the reduction of 460 micromoles of DPN^+ (i.e., to the synthesis of 460 micromoles of citrate) per 10 minutes per milligram of enzyme (temperature, ~25°). Under these conditions the turnover number is 4600 moles of citrate per minute per 100,000 g. of enzyme.

SUMMARY OF PURIFICATION PROCEDURE[a]

Step	Volume of solution, ml.	Units[b]	Protein, mg.	Specific activity	Yield, %
1. Extract	33,000	284,000	312,000	0.9	100
2. Calcium phosphate gel eluate and isoelectric precipitate combined	8,480	241,000	43,500	5.5	85
3. First ethanol precipitate[c]	440	92,000	15,500	6.0	32
4. First AmSO₄ fractionation (50–70)	41	43,500	1,305	33	15
5. Second AmSO₄ fractionation (40–60)	6	46,210	721	65	16
6. Crystals	5	40,300	202	200	14
Mother liquor of crystals	4	840	392	2.0	
7. First recrystallization	2.8	26,800	82	327	9.5

[a] 10 kg. of pig heart.

[b] *Chemical assay* units.

[c] The second ethanol precipitate was treated in the same manner as the first. Specific activity 330 was reached in this case only after three recrystallizations; yield, 11,200 units. The total yield of crystals of specific activity 330 was, therefore, 26,800 + 11,200 = 38,000 units, or 13% of the enzyme in the original extract.

The condensing enzyme crystallizes as small slender needles under the conditions outlined above. The refractive index of the crystals differs very little from that of the ammonium sulfate solution. As in the case of the crystalline lactic dehydrogenase from heart muscle,[19] the crystals take up methylene blue and become much more easily visible. They can also be readily observed by means of phase contrast microscopy.

Notes to Purification Procedure. The procedure described above has been improved and simplified[17] by carrying out the isoelectric precipitate and gel adsorption steps (2a and 2b) in one single step and by obtaining a single ammonium sulfate fraction (between 50 and 70% saturation) in step 4. Larger yields of enzyme are obtained by repeating the initial extraction twice instead of only once and by carrying out eight elutions (instead of five) of the calcium phosphate gel.

Extraction of 10 kg. of minced pig heart yielded 460,000 (*chemical assay*) units of condensing enzyme (specific activity, 1 to 1.5). The units were quantitatively absorbed in the gel step, and the elution yielded 475,000 units (specific activity, 5.7). The yield of crystals in step 6 was 445 mg. containing 78,500 units (specific activity, 176); in step 7, 146 mg. containing 49,000 units (specific activity, 335). The 0 to 50 ammonium sulfate fraction discarded in step 4 was found to contain about 50% of the units present in the 50 to 70 fraction. Part of these units were recovered as crystalline enzyme on refractionation of the dialyzed 0 to 50 fraction between 0 to 40 and 40 to 70% ammonium sulfate saturation, discarding the 0 to 40 fraction. This suggests that ammonium sulfate fractionation of step 4 should be carried out between 0 and 40% and 40 and 70% saturation, discarding the first precipitate.

Properties

Stability. The condensing enzyme is very stable. Dilute solutions in phosphate buffer, pH 7.4, can be kept for several weeks in the refrigerator without loss of activity. The crystalline enzyme is usually kept as a suspension in 50 or 60% saturated ammonium sulfate at 3 to 4°. Under these conditions, its activity remains unchanged for many months. The enzyme withstands dialysis against dilute (0.02 M) phosphate buffer, pH 7.4, for several days. Activity is destroyed by heating at 100° for 2 to 3 minutes.

Purity and Specificity. Condensing enzyme recrystallized three times is homogenous on electrophoresis (1% in 0.05 M phosphate buffer, pH 7.4) and in the ultracentrifuge. Enzyme recrystallized four times is free of acylphosphatase, aconitase, fumarase, lactic dehydrogenase, and aldolase;

[19] F. B. Straub, *Biochem. J.* **34**, 483 (1940).

it contains traces of isocitric dehydrogenase (specific activity, 0.9; see Vol. I [116]).

There is a very slow reaction in the optical assay when propionyl CoA is substituted for acetyl CoA. The reaction with propionyl CoA would lead to homocitric acid, but the product has not been identified. No other acyl CoA derivatives have been investigated so far. When added to the components of the chemical assay in concentrations equal to that of oxalacetate, there is but slight inhibition by maleate, malate, and tartrate.

Other Properties. The affinity of the condensing enzyme for its two substrates[11] is very high; concentrations of each acetyl CoA and oxalacetate around 4×10^{-4} M are sufficient to saturate the enzyme. The apparent equilibrium constant[11] of reaction 1 ($K' = $ [citrate][CoA]/[acetyl][CoA][oxalacetate][H_2O]) is 8.38×10^3 (liters $\times$ mole^{-1}) at pH 7.2 and 22°. Thus the equilibrium position of the condensing reaction is far in the direction of citrate synthesis. The ΔF of the reaction is about -7000 cal./mole. Since the ΔF of synthesis of citrate from acetate and oxalacetate has been calculated[20] to be about 5000 cal./mole, the free energy of hydrolysis of acetyl-S-CoA (to acetate + HS-CoA) would be $7000 + 5000 = 12,000$ cal./mole. This value agrees with the values calculated from the equilibrium constants of the reactions catalyzed by transacetylase (reaction 2) and the CoA- and DPN-linked aldehyde dehydrogenase of *Clostridium kluyveri*.[21]

Distribution. The condensing enzyme is widely distributed in animal tissues, yeast, and other microorganisms (with the possible exception of obligate anaerobes)[1] and is undoubtedly present in the tissues of higher plants. As would be expected, the enzyme is present in highest concentrations in tissues which, like heart and pigeon breast muscle, have a high aerobic metabolism.

[20] N. O. Kaplan, *in* "The Enzymes" (Sumner and Myrbäck, eds.), Vol. II, p. 55. Academic Press, New York, 1951.
[21] E. R. Stadtman, *J. Biol. Chem.* **196**, 535 (1952); R. M. Burton and E. R. Stadtman, *J. Biol. Chem.* **202**, 873 (1953).

[115] Aconitase from Pig Heart Muscle

Citrate $\leftrightarrows$ cis-Aconitate $+ H_2O \rightleftarrows$ D-Isocitrate

By CHRISTIAN B. ANFINSEN

Assay Method

Principle. The method that has been most commonly used in studies on aconitase is based on the chemical determination of citrate formed from cis-aconitate. Such a method was employed in the purification studies[1] summarized in the table below, and sufficient detail is given in this table to permit the use of this assay method. Citrate is added as a stabilizer in this preparation, and therefore the chemical method should be used in following the purification steps, since endogenous citrate introduces difficulties into the optical method described below. The relatively simple and sensitive optical method,[2] based on the light absorption, at 240 mμ, of cis-aconitic acid and other α,β-unsaturated carboxylic acids, is recommended for general use and is described in detail in the next section.

Reagents

0.03 M citrate in 0.05 M phosphate buffer, pH 7.4, *or* 0.01 M D,L-isocitrate in 0.05 M phosphate buffer, pH 7.4.

Enzyme. Dilute the enzyme with 0.1 M phosphate buffer, pH 7.4, to give a solution containing approximately 500 units/ml. (See definition of unit below.)

Procedure. To 2.9 ml. of substrate-buffer solution, in a quartz cell with a 1-cm. light path, add 0.1 ml. of the suitably diluted enzyme. Measure extinction values at 240 mμ, at intervals of 15 seconds for a few minutes, at 25°.

Definition of Unit and Specific Activity. One unit of enzyme is defined as that amount which causes an initial rate of increase in optical density (ΔE_{240}) of 0.001 per minute under the above conditions. Specific activity is expressed as units per milligram of protein. Protein may be determined spectrophotometrically, after suitable dilution, by measurement of optical density at 280 mμ. The extinction coefficients (in a quartz cell having a 1-cm. light path) of the original extract and of the final product in the purification summarized in the table were identical and, for a protein concentration of 1 mg./ml. (based on Kjeldahl-N $\times$ 6.25), equaled 1.6.

[1] J. M. Buchanan and C. B. Anfinsen, *J. Biol. Chem.* **180**, 47 (1949).

[2] E. Racker, *Biochim. et Biophys. Acta* **4**, 211 (1950).

Application of Assay Method to Crude Tissue Preparations. There appear to be no significant competing reactions in crude tissue extracts which might lead to error in the assay of aconitase. The chemically measured production of citrate from *cis*-aconitate proceeds equally well under aerobic or anaerobic conditions, and significant removal of isocitrate through the action of isocitric dehydrogenase takes place only in the presence of added TPN.

Particularly turbid extracts should be clarified by centrifugation for assay in the optical test of Racker.[2]

Purification Procedure

The procedure described below has been carried out successfully in several laboratories. Experience has indicated the importance of maintaining low temperatures during the various steps as indicated. Aconitase is an especially unstable enzyme, particularly in solution, but is partially stabilized by the presence of its substrate, citrate. A further possibility in regard to its stabilization during fractionation is mentioned below. (See section on Properties.)

Step 1. Preparation of Crude Extract. 1 kg. of pig heart,[3] freed of fat and connective tissue, is diced and homogenized in portions in a Waring blendor with 3 l. of buffer (0.002 M sodium citrate, 0.002 M citric acid). The homogenate is centrifuged. (A 4-l. centrifuge was used for the purification study shown in the table, although the 1-l. angle head in the International centrifuge can also be used successfully.)

Step 2. First Ethanol Fractionation. The supernatant from the centrifuged homogenate is cooled rapidly in a dry ice-alcohol bath to 0°, and 90% alcohol is added to bring the concentration to 15% by volume. Care must be taken to keep the temperature as near as possible to the freezing point of the aqueous alcohol solution. Filter aid[4] is added, the solution filtered with suction, and the precipitate discarded. In order to assure precipitation of the enzyme by the further addition of alcohol, 1.25 ml. of saturated NaCl is added to each 100 ml. of 15% alcohol filtrate. The alcohol concentration is then slowly increased to 45%, the temperature of the solution being gradually lowered to about −10. The precipitate is allowed to settle overnight, and the supernatant is siphoned off. The precipitate is centrifuged compactly in a refrigerated centrifuge. The bulk of the enzyme activity is in the precipitate.

---

[3] The pig hearts may be used immediately after removal from the animal, but they may also be frozen and stored at −15° for as long as 2 weeks without significant loss of enzyme.

[4] Hyflo Super-Cel, Johns Manville Corporation.

Step 3. Second Ethanol Fractionation. The precipitate is taken up in sufficient ice-cold water to give an extinction of 20 at a wavelength of 280 mμ in a 1-cm. cell. This corresponds to a protein concentration of about 12.5 mg. of protein per milliliter. Disregarding the alcohol in the precipitate, sufficient 90% alcohol (8.5 ml./100 ml. of enzyme solution) is added at a gradually diminishing temperature to bring the concentration to 8 vol. %. The inactive precipitate is centrifuged off in a refrigerated centrifuge. Alcohol is then added to the supernatant to bring the alcohol concentration to 23%. The temperature is meanwhile lowered to -7 to $-8°$. The solution is centrifuged, and the supernatant discarded.

Step 4. Third Ethanol Fractionation. The precipitate is taken up in sufficient ice-cold water to give an extinction of 24 at 280 mμ in a 1-cm. cell (15 mg. protein per milliliter). Disregarding the alcohol content of the precipitate, sufficient alcohol is added to bring the concentration to 8%; the precipitate is centrifuged off and discarded. The pH of the supernatant, which is about 5.6, is then adjusted to 6.8 by the careful addition of 5% sodium carbonate. Alcohol is added to 23%, and the precipitate, containing the activity, is centrifuged off.

Step 5. Ammonium Sulfate Fractionation. The precipitate from step 4 is taken up in 0.1 M phosphate buffer, pH 7.4, containing 0.05 M sodium citrate (extinction of protein solution at 280 mμ = 16, equivalent to 10 mg. of protein per milliliter). Ammonium sulfate is added slowly until the solution is 66% saturated. During the addition, the temperature is lowered progressively to $-10°$. The mixture is centrifuged for 5 minutes at

SUMMARY OF PURIFICATION PROCEDURE[1]

Step	Volume of solution, ml.	Units[a]	Protein, mg.	Specific activity, units/mg. protein	Yield, %
Citrate extract	2360	310,000	31,000	10	100
First ethanol fractionation	600	133,000	7000	19	42
Second ethanol fractionation	110	102,000	1640	62	36
Third ethanol fractionation	50	63,000	480	131	20
Ammonium sulfate fractionation (0.66–0.72 sat.)	10	37,000	157	235	12

[a] The unit used in this table is the amount of enzyme which forms 1.0 μM. of citric acid (in 15 minutes at 38° in 0.2 M phosphate buffer, pH 7.4.) from *cis*-aconitate. An aliquot of stock 0.5 M *cis*-aconitate solution (see Vol. III [68]) is added to suitably diluted enzyme in phosphate buffer to give a final concentration of 0.025 M after final dilution to 5 ml. with buffer. The reaction is stopped by adding 1 ml. of 20% TCA. After centrifugation of precipitated protein an aliquot of the supernatant is analyzed for citric acid as described in Vol. III [69].

8000 r.p.m., and the precipitate is discarded. This centrifugation is carried out at −5 to −10°. Ammonium sulfate is then added to the supernatant until the solution is 73% saturated. The precipitate is centrifuged down as before, and the supernatant discarded. The precipitate contains the bulk of the enzyme, which has been purified about 23-fold. In the precipitated form it may be frozen with dry ice and stored for a relatively long time without loss in activity.

If further purification of the enzyme is attempted, it is probably inadvisable to use ammonium sulfate in any other than the last step, since salts cannot be removed by dialysis without loss of enzyme activity. The table summarizes the yields and degrees of purification obtained at each step in the preparation of the enzyme.

Properties of Enzyme

Specificity. The purified enzyme is contaminated with isocitric dehydrogenase but contains relatively little fumarase.

Activators and Inhibitors. Aconitase is strongly inhibited by cyanide and sulfide. This property may be due to the presence of Fe^{++} as a cofactor which has been shown to reactivate aged or dialyzed crude preparations.[5] A recent note,[6] unfortunately lacking details, reports the effect of cysteine and Fe^{++} on 24-fold purified aconitase.

The enzyme is inhibited by copper and mercury ions at low concentrations and competitively by *trans*-aconitate.

Equilibrium. At equilibrium, citrate, aconitate, and isocitrate are present in the proportions 89.5:4.3:6.2.[7] This equilibrium mixture is obtained irrespective of pH between 6.8 and the pH optimum of the enzyme, 7.4. Mg^{++} shifts the equilibrium toward citrate.

Although the existence of two aconitases, a and b, has been suggested, on the basis of this equilibrium, fractionation and stability studies indicate only one.[1]

Kinetic Properties. The initial rate of dehydration of isocitrate is nearly three times that of citrate. Racker[2] reports K_m (citrate) $= 1.1 \times 10^{-3}$ and K_m (D-isocitrate) $= 4 \times 10^{-4}$ mole $\times$ liter^{-1} at 25° and pH 7.4.

[5] S. R. Dickman and A. A. Cloutier, *Arch. Biochem.* **25**, 229 (1950).

[6] J. F. Morrison, *Biochem. J.* **55**, iv (1953).

[7] H. A. Krebs and L. V. Eggleston, *Biochem. J.* **37**, 334 (1943).

[116] Isocitric Dehydrogenase System (TPN) from Pig Heart[1-4]

$$d\text{-Isocitrate}^{---} + \text{TPN}^+ \underset{\xrightarrow{(\text{Mn}^{++})}}{\rightleftarrows} \alpha\text{-Ketoglutarate}^{--} + CO_2 + \text{TPNH} \quad (1)$$

$$\text{Oxalosuccinate} \xrightarrow{(\text{Mn}^{++})} \alpha\text{-Ketoglutarate} + CO_2 \quad (2)$$

By SEVERO OCHOA

Extracts of a number of animal and plant tissues catalyze reactions 1 and 2. Both reactions require Mn^{++}. In the absence of added Mn^{++} reaction 3 is obtained.[1]

$$d\text{-Isocitrate}^{---} + \text{TPN}^+ \rightleftarrows \text{Oxalosuccinate}^{---} + \text{TPNH} + H^+ \quad (3)$$

Evidence for reaction 3 is based on the reduction of TPN^+ by isocitrate and the oxidation of TPNH by oxalosuccinate as followed spectrophotometrically. Several observations suggest that reactions 1 and 2 may be catalyzed by an enzyme analogous to the "malic" enzyme (cf. Vol. I [124]). Such an enzyme might be referred to as "isocitric" enzyme. Reaction 3 may be catalyzed by a separate enzyme, an isocitric dehydrogenase proper. The relationship between these two enzymes would thus be similar to that between "malic" enzyme and malic dehydrogenase. If this view is correct, it should be possible to separate the two enzymes, as has been the case with the "malic" enzyme and malic dehydrogenase. However, such separation has not yet been accomplished.

Assay Method

a. Oxidative Decarboxylation of *d*-Isocitrate. *Principle.* The activity determination is based on reaction 1. The early rate of reduction of TPN^+ in the presence of enzyme, Mn^{++}, and an excess of isocitrate is proportional to the concentration of the enzyme within certain limits. The measurement is carried out in the Beckman spectrophotometer at wavelength 340 mμ using Corex or silica cells of 1.0-cm. light path.

[1] S. Ochoa, *J. Biol. Chem.* **174**, 133 (1948).

[2] Ochoa and E. Weisz-Tabori, *J. Biol. Chem.* **174**, 123 (1948).

[3] A. L. Grafflin and S. Ochoa, *Biochim. et Biophys. Acta* **4**, 205 (1950).

[4] S. Ochoa, *in* "The Enzymes" (Sumner and Myrbäck, eds.), Vol. II, Part 2, p. **929**. Academic Press, New York, 1951.

Reagents

0.25 M glycylglycine buffer, pH 7.4.[5]
0.018 M $MnCl_2$.
0.00135 M TPN^+.
0.006 M dl-isocitrate (or 0.003 M d-isocitrate).

Procedure. The reaction mixture contains 0.3 ml. of buffer, 0.1 ml. of $MnCl_2$ (1.8 micromoles), 0.1 ml. of TPN^+ (0.135 micromole), 0.1 ml. of isocitrate (0.3 micromole of d-isomer), enzyme, and water (adjusted to a temperature of 22 to 23°) to a final volume of 3.0 ml. The reaction is carried out at room temperature and is started by the addition of either isocitrate or enzyme. Readings of the optical density are made, against a blank containing all components except TPN, at intervals of 15 seconds for 1 or 2 minutes. The increase in optical density ($\Delta \log I_0/I$) between 30 and 45 seconds after the start of the reaction is used to calculate the enzyme activity. The amount of enzyme used in a test is adjusted so that the density change, for the period between 30 and 45 seconds, lies between +0.005 and +0.025. The presence of phosphate in concentrations higher than 0.0003 M should be avoided because turbidity, due to precipitation of manganous phosphate, may develop.

Definition of Unit and Specific Activity. One unit of enzyme is defined as that amount which causes an increase in optical density of 0.01 per minute under the above conditions. Specific activity is expressed as units per milligram of protein. Protein is determined as described for the preparation of "malic" enzyme (cf. Vol. I [124]).

b. Oxalosuccinic Carboxylase Activity. *Principle.* This activity (reaction 2) can be determined manometrically by measuring the rate of enzymatic liberation of CO_2 from oxalosuccinate under controlled conditions. It can also be determined by means of a more rapid and much more sensitive optical test. This test is based on the fact that, in the presence of Mn^{++} and oxalosuccinate, the enzyme causes a pronounced increase in the absorption of light at 240 mμ, presumably as a result of increased formation of an oxalosuccinate-Mn complex; this increase is followed by a rapid drop, indicating decarboxylation.[6] The early rate of increase of light absorption is, within certain limits, proportional to the concentration of enzyme. The measurement is carried out in the Beckman spectrophotometer using silica cells of 1.0-cm. light path.

---

[5] Veronal or tris(hydroxymethyl)aminomethane buffer has been used with identical results.
[6] A. Kornberg, S. Ochoa, and A. H. Mehler, *J. Biol. Chem.* **174,** 159 (1948).

Reagents

> 1.34 M KCl.
> 0.005 M MnCl$_2$.
> Approximately 0.005 M oxalosuccinate, pH 7.4.

Preparation of Oxalosuccinate Solution. Owing to the instability of oxalosuccinate, all operations are carried out at 0°. The required amount of barium oxalosuccinate is suspended in a little water, brought into solution with a few drops of 2.0 N HCl, and the Ba precipitated with the calculated amount of 1.0 N H$_2$SO$_4$. The precipitate of BaSO$_4$ is centrifuged off and discarded without washing; the supernatant is adjusted with NaOH to pH 7.4 and is brought with water to the desired volume. About 70% of the calculated oxalosuccinate is present in solution. For more accurate data, a determination with aniline citrate should be carried out.

Solutions of oxalosuccinate can conveniently be made up just before use but, in most cases, they can be used over a period of several hours if kept at 0°. The decrease in oxalosuccinate concentration, through spontaneous decarboxylation, is small under these conditions.

Procedure. The reaction mixture, in a final volume of 3.0 ml., contains 0.3 ml. of KCl (402 micromoles), enzyme, 0.1 ml. of MnCl$_2$ (0.5 micromole), and 0.1 ml. of oxalosuccinate (about 0.5 micromole). The volume is made up with water adjusted to a temperature of 15°. The blank cell contains no oxalosuccinate. The reaction is started by the addition of oxalosuccinate which is blown into the mixture from a Lang-Levy micropipet,[7] and readings of the optical density at 240 mμ are made at 15-second intervals thereafter for 1 or 2 minutes. The optical density of the oxalosuccinate is determined separately and furnishes the zero time value. The amount of enzyme is adjusted so as to obtain an increase in optical density between 0.07 and 0.20 in the first 15 seconds. The reason for the presence of KCl in the reaction mixture is that it has been found to increase the carboxylase activity of the enzyme. This appears to be a nonspecific effect of increased ionic strength.[6] The presence of phosphate in concentrations higher than 0.0003 M should be avoided for the reasons already stated.

Although the use of the $\Delta \log I_0/I$ for the first 15 seconds to calculate the enzyme activity is fairly satisfactory, better results are obtained by using the initial rate (V_0) calculated by plotting the rates for successive 15-second intervals against time and extrapolating to zero time.

Definition of Unit and Specific Activity. One unit is defined as the amount of enzyme causing an increase in optical density of 0.01 in 1 min-

---

[7] M. Levy, *Compt. rend. trav. Lab. Carlsberg, Sér. Chim.* **21,** 101 (1936).

ute calculated for the first 15-second period after the start of the reaction. Specific activity is expressed as units per milligram of protein.

c. Isocitric Dehydrogenase. No attempts have as yet been made to isolate a true isocitric dehydrogenase. An assay specifically designed to test for this activity should be based on the rate of reaction 3 from right to left in the absence of Mn^{++} or other divalent cations. This assay would utilize the rate of oxidation of TPNH in the presence of enzyme and oxalosuccinate and would be analogous to the assay for malic dehydrogenase described in Vol. I [123].

Purification Procedure

Partial purification of the activities corresponding to reactions 1 and 2 has been obtained through a procedure involving salt and ethanol fractionation.[3] The starting material is an acetone powder of pig heart prepared and extracted essentially according to the procedure described by Straub[8] for the isolation of malic dehydrogenase.

Preparation of Acetone Powder. Fresh pig hearts, previously cooled in ice, are freed from fat and connective tissue and minced in a meat grinder in the cold. The mince is washed three times with 10 vol. of ice-cold tap water with mechanical stirring. The washed mince is stirred with 3 vol. of cold 90% acetone for about 30 minutes, and the residue again extracted with the same volume of 100% acetone. The residue is dried by spreading on filter paper under a fan and ground to a fine powder in a mechanical mortar. If kept in the cold in closed containers the powder ratains its activity for a long time.

Step 1. Extraction. The extract is prepared by stirring the acetone powder mechanically for 30 minutes at room temperature with 10 vol. of 0.1 M phosphate buffer, pH 7.4. The mixture is centrifuged; the residue is re-extracted with 3 vol. of buffer, and the supernatants are combined.

Step 2. Ammonium Sulfate Fractionation. The extract is cooled to 0°, brought to 50% saturation with solid $(NH_4)_2SO_4$, and the mixture is filtered with suction in the cold using a filter aid (Hyflo Super-Cel) to facilitate filtration. The precipitate is discarded, and the supernatant is brought to 60% saturation with $(NH_4)_2SO_4$. The mixture is filtered. The supernatant is discarded, and the precipitate is dissolved in 0.04 M phosphate buffer, pH 7.4, to give a concentration of about 3% protein. The solution is clarified by filtration and dialyzed against 0.01 M phosphate buffer, pH 7.4, at 2 to 3° for 4 to 5 hours.

Step 3. Ethanol Fractionation. The dialyzed solution from the above step is fractionated with ethanol at −5°, and the fractions collected by centrifugation at the same temperature. The most active fractions are

[8] F. B. Straub, *Z. physiol. Chem.* **275**, 63 (1942).

usually obtained between 20 and 25% ethanol by volume. The fractions are dissolved in ice-cold 0.01 M phosphate buffer, pH 7.4, and dialyzed for a few hours in the cold against the same buffer to remove the alcohol. These preparations are very unstable and lose activity rather rapidly even when stored at 0°. If dried from the frozen state, 30 to 40% of the activity is lost, but the remainder persists unchanged for many months when the dry powder is stored in the cold over $CaCl_2$.

The procedure is summarized in the table. There is about a sixfold purification of both the over-all (reaction 1) and the oxalosuccinic carboxylase activity (reaction 2) of the extract. The malic dehydrogenase activity is purified only about two and one-half times so that the ratio of oxalosuccinic carboxylase to malic dehydrogenase increases two and one-half times. The preparations contain no aconitase and only traces of lactic dehydrogenase.

SUMMARY OF PURIFICATION PROCEDURE[a]

Step	Volume of solution, ml.	Protein, mg.	Reaction 1		Reaction 2		Yield, %
			Units	Specific activity	Units	Specific activity	
1. Extract	8300	48,200	7,968,000	165	22,244,000	462	100
2. AmSO₄ fractionation	134	7,210			12,542,400	1740	55
3. Ethanol fractionation	37	1,254	1,108,150	885	3,596,400	2860	16

[a] 800 g. of pig heart acetone powder.

Occasionally the purification after $(NH_4)_2SO_4$ and ethanol fractionation may be only about fourfold. The purity of these preparations can be increased about one and one-half times, with a yield of 60% or better, by adsorption on calcium phosphate gel prepared according to Keilin and Hartree.[9] For this the enzyme solution is diluted with 0.01 M phosphate buffer, pH 7.4, to give a protein concentration of about 1%. The adsorption is carried out successively with small amounts of the gel, and the precipitates are separately eluted with 0.1 M phosphate buffer, pH 7.4. The eluates are tested, and the best ones are combined.

Properties

The ratio (activity in catalysis of reaction 1)/(activity in catalysis of reaction 2) remains constant throughout the purification so far achieved.

[9] D. Keilin and E. F. Hartree, *Proc. Roy. Soc. (London)* **B124**, 397 (1938); see Vol. I [11].

Moreover, under optimum conditions for each reaction the rate of CO_2 evolution through oxidative decarboxylation of d-isocitrate or decarboxylation of oxalosuccinate is the same for a given amount of enzyme (about 400 moles of CO_2/min./100,000 g. of protein for the best preparations obtained). These properties are strongly reminiscent of those of the "malic" enzyme (cf. Vol. I [124]). Furthermore, just as the decarboxylation of oxalacetate by "malic" enzyme is strongly inhibited by malate, the decarboxylation of oxalosuccinate by the isocitric enzyme described here is strongly inhibited by isocitrate; citrate, and cis-aconitate are not inhibitory. Mn^{++} is also an essential requirement for both reactions 1 and 2; Mg^{++} is much less effective. The equilibrium constant of the reversible reaction 1 ($[d$-isocitrate$^{---}][TPN^+]/[\alpha$-ketoglutarate$^{--}][CO_2]$ $[TPNH]$) is 1.3 (liters $\times$ mole^{-1}) at pH 7.0 and 22°.

In the case of reaction 1 the enzyme is specific for d-isocitrate and TPN^+; l-isocitrate, citrate, cis-aconitate, and DPN^+ do not react. In the case of reaction 2, the enzyme is active on oxalosuccinate but not on oxalacetate or acetoacetate.

The affinity of the enzyme for d-isocitric acid is extremely high; under the conditions of the optical assay, reaction 1 still proceeds at optimal rates with concentrations of d-isocitrate as low as 1.2×10^{-5} M.[1,10] The optimum concentration of Mn^{++} has been given by Adler et $al.$[10] as 5×10^{-4}. The dissociation constant of the enzyme-TPN complex has not been determined. The pH optimum of the reaction is around 7.0 to 7.5.[10]

In the case of reaction 2 the dissociation constant of the enzyme-Mn complex has been found to be about 3×10^{-4} (g.-atom/l.), and that of the enzyme-oxalosuccinate complex, 2.6×10^{-2} (mole/l.). These values were obtained at pH 5.6 and 18° using the manometrically determined CO_2 evolution as the assay.[2] With the same technique the rate of oxalosuccinate decarboxylation by the pig heart enzyme was found to be about the same at pH 5.6, 6.3, and 7.4.[2]

Inhibitors. The inhibition of oxalosuccinate decarboxylation by isocitrate has already been mentioned. With pig heart extracts Lotspeich and Peters[11] have found that reaction 1 is inhibited by Cu and Hg compounds ($CuSO_4$, p-chloromercuribenzoate, phenylmercuric nitrate) and by certain disubstituted As derivatives (diphenylchloroarsine, phenarsazines). The inhibition is reversed by monothiols and BAL. Other compounds including arsenite, Lewisite, iodoacetate, iodoacetamide, and fluoroacetate were noninhibitory.[11] Inhibition by orthophosphate and pyrophosphate is due to removal of Mn^{++}.[10,11]

[10] E. Adler, H. von Euler, G. Günther, and M. Plass, *Biochem. J.* **33**, 1028 (1939).
[11] W. D. Lotspeich and R. A. Peters, *Biochem. J.* **49**, 704 (1951).

[117] Isocitric Dehydrogenase of Yeast (TPN)[1]

d-Isocitrate + TPN $\rightleftarrows$ Oxalosuccinate + TPNH

Oxalosuccinate $\rightleftarrows$ α-Ketoglutarate + CO_2

By ARTHUR KORNBERG

Assay Method

Principle. The method is essentially that of Ochoa[2] and depends on the spectrophotometric determination of TPN reduction. The molecular extinction coefficient of 6.22×10^6 cm.^{2}/mole is used.[3]

Reagents

KH_2PO_4–K_2HPO_4 buffer (0.5 M, pH 7.0).
$MgCl_2$ (0.1 M).
TPN (0.025 M).
dl-Isocitrate (0.005 M).

Procedure. The incubation mixture in a cuvette (1-cm. light path) contains 0.2 ml. of phosphate buffer, 0.1 ml. of $MgCl_2$, 0.02 ml. of TPN, 0.1 ml. of isocitrate, enzyme solution, and water to a final volume of 3.0 ml. The assay is run at room temperature. The blank cell lacked isocitrate, and with purified preparations the blank cell contained only water. The increase in optical density at 340 mμ resulting from the reduction of TPN was observed at 30-second intervals for 3 minutes. When the amount of enzyme added did not exceed 0.2 unit, the density increase remained linear for 3 minutes.

Definition of Unit and Specific Activity. One unit of enzyme is defined as that amount which causes the reduction of 1 μM. of TPN in 20 minutes. Specific activity is expressed as units per milligram of protein. Protein was determined by the method of Lowry *et al.*[4]

Purification Procedure

The enzyme was identified in both baker's yeast and brewer's yeast but purified only from the former.

Baker's Yeast. The starting material was the supernatant obtained on removal of the DPN isocitric dehydrogenase fraction, Ammonium Sul-

[1] A. Kornberg and W. E. Pricer, Jr., *J. Biol. Chem.* **189**, 123 (1951).

[2] S. Ochoa, *J. Biol. Chem.* **174**, 133 (1948).

[3] B. L. Horecker and A. Kornberg, *J. Biol. Chem.* **175**, 385 (1948).

[4] O. H. Lowry, N. J. Rosebrough, A. L. Farr, and R. J. Randall, *J. Biol. Chem.* **193**, 265 (1951).

fate A.[5] To this supernatant was added 15 g. of ammonium sulfate; the precipitate was separated by centrifugation and discarded. Ammonium sulfate (21 g.) was added to the supernatant, and the precipitate, collected by centrifugation, was dissolved in water to a volume of 25 ml. (Ammonium Sulfate I). This fraction was placed in a 60° water bath for 5 minutes, and the precipitate which developed was removed by centrifugation. The supernatant was dialyzed for 2 hours against running, cold distilled water and then adjusted with 0.1 N acetic acid to pH 4.7. Ethanol fractionation was carried out at $-2°$ to $-5°$. To 27.5 ml. was added 2.2 ml. of ethanol, and the precipitate was centrifuged and discarded. Ethanol (8.5 ml.) was added to the supernatant fluid, and the resulting precipitate dissolved in phosphate buffer (0.03 M, pH 7.0) to a volume of 15 ml. This fraction was fully active after 4 weeks at $-15°$.

Beer Yeast. Dried beer yeast was autolyzed in 3 vol. of 0.1 M NaHCO$_3$ for 4 hours at 34° and then centrifuged to remove the residue. This supernatant fluid was not purified further.

TPN ENZYME FROM BAKER'S YEAST

Step	Total units	Specific activity, units/mg. protein
Extract	825	0.6
Ammonium Sulfate I	625	2.1
Heating	612	2.5
Ethanol	386	12.4

TPN ENZYME FROM BEER YEAST

Autolyzate	1100	0.4

The lability of yeast DPN isocitric dehydrogenase may be used as a means of freeing a TPN isocitric dehydrogenase preparation of this activity, and separation of the DPN and TPN systems may be achieved by fractionation with ammonium sulfate.

Properties

Specificity. There was no detectable reaction with DPN in the case of the purified enzyme preparation, nor was there any significant inhibitory effect by DPN. The enzyme preparation catalyzed the decarboxylation of oxalosuccinate, the reductive carboxylation of α-ketoglutarate,

---

[5] A. Kornberg, Vol. I [118].

and the reduction of oxalosuccinate and thus resembles the TPN isocitric dehydrogenase system described by Ochoa in animal tissues.[2]

Activators and Inhibitors. Mg^{++} and Mn^{++} both gave about equal stimulation of the enzyme. NaCN (0.01 M) and NaN_3 (0.01 M) showed no inhibition.

[118] Isocitric Dehydrogenase of Yeast (DPN)[1]

d-Isocitrate + DPN → α-Ketoglutarate + DPNH + CO_2

By Arthur Kornberg

Assay Method

Principle. The method is the same as in the case of the TPN isocitric dehydrogenase of yeast.[2]

Reagents

KH_2PO_4–K_2HPO_4 (0.5 M, pH 7.0).
$MgCl_2$ (0.1 M).
Adenosine-5'-phosphate (0.025 M).
DPN (0.05 M).
dl-Isocitrate (0.005 M).

Procedure. The incubation mixture in a cuvette (1-cm. light path) contains 0.2 ml. of phosphate buffer, 0.1 ml. of $MgCl_2$, 0.02 ml. of adenosine-5'-phosphate, 0.02 ml. of DPN, 0.1 ml. of dl-isocitrate, enzyme solution, and water to a final volume of 3.0 ml. The assay is run at room temperature. The blank cell lacked isocitrate, and with purified preparations the blank cell contained only water. The increase in optical density at 340 mμ resulting from the reduction of DPN was observed at 30-second intervals for 3 minutes. When the amount of enzyme added did not exceed 0.2 unit, the density increase remained linear for 3 minutes.

Definition of Unit and Specific Activity. One unit of enzyme is defined as that amount which causes the reduction of 1 μM. of DPN in 20 minutes. Specific activity is expressed as units per milligram of protein. Protein was determined by the method of Lowry *et al.*[3]

[1] A. Kornberg and W. E. Pricer, Jr., *J. Biol. Chem.* **189**, 123 (1951).
[2] A. Kornberg, Vol. I [117].
[3] O. H. Lowry, N. J. Rosebrough, A. L. Farr, and R. J. Randall, *J. Biol. Chem.* **193**, 265 (1951).

Purification Procedure

The enzyme was purified from both baker's and brewer's yeast.

Baker's Yeast. Fresh, pressed baker's yeast (40 g.) was mechanically shaken with 200 g. of sea sand (40 to 80 mesh) and 200 ml. of 0.1 M sodium bicarbonate for 90 minutes at 3°, essentially as described by Curran and Evans for the disintegration of bacterial cells.[4] A slightly opalescent supernatant was obtained on centrifugation for 10 minutes at 10,000 r.p.m. in a Servall angle centrifuge. To 145 ml. of this extract (see the table) was added 35 g. of ammonium sulfate. The precipitate was separated by centrifugation and discarded. Ammonium sulfate (9.4 g.) was added to the supernatant, and the precipitate collected by centrifugation and dissolved in water to a volume of 50 ml. (Ammonium Sulfate A). (The supernatant was used for purification of TPN isocitric dehydrogenase.[2]) An equal volume of acetate buffer (0.1 M, pH 5.5) and then 26 g. of ammonium sulfate were added. The precipitate was removed by centrifugation and discarded. Ammonium sulfate (6.0 g.) was added to the supernatant, and the precipitate, collected by centrifugation, was dissolved to a volume of 12 ml. (Ammonium Sulfate B). Water (50 ml.) and then aluminum hydroxide gel C_γ[5] (6.0 ml. containing 93 mg. of dry weight) were added. After 5 minutes the gel was centrifuged, washed twice with 60-ml. portions of phosphate buffer (0.02 M, pH 7.2), and eluted twice with 50-ml. portions of phosphate buffer (0.1 M, pH 7.6) (C_γ eluate). This eluate lost about 40% of its activity after 24 hours at 3° and about 70% after 48 hours; at −15° only 10 to 20% was lost after 48 hours. All operations were carried out at 3° unless otherwise specified.

Ale Yeast. Dried ale yeast (30 g.) was autolyzed in 3 vol. of 0.1 M sodium bicarbonate for 40 hours at 30° and then centrifuged to remove the residue. To 60 ml. of supernatant were added an equal volume of cold water and 42 g. of ammonium sulfate. The precipitate collected by centrifugation was dissolved in water to a volume of 25 ml. (Ammonium Sulfate *a*). Salmine sulfate (50 mg. in 2.5 ml. of water) was added, the precipitate centrifuged and discarded, and the supernatant dialyzed for 60 minutes against running, cold distilled water. The dialyzate (26 ml.) was then diluted with 26 ml. of water and 13 ml. of acetate buffer (0.1 M, pH 5.0) and fractionated with ethanol at −2 to −4°. The precipitate formed by the addition of 5 ml. of ethanol was removed by centrifugation and discarded. To the supernatant was added 6.0 ml. of ethanol, and the precipitate was dissolved in water with the aid of a few drops of glycylglycine buffer (0.25 M, pH 7.4) to a volume of 7.6 ml. (ethanol fraction). Ammonium sulfate (2.5 g.) was added. The precipitate was

[4] H. R. Curran and F. R. Evans, *J. Bacteriol.* **43**, 125 (1942).
[5] R. Willstätter and H. Kraut, *Ber.* **56**, 1117 (1923).

removed by centrifugation, and 0.7 g. of ammonium sulfate was added to the supernatant. The precipitate was dissolved in 4 ml. of glycylglycine buffer (0.06 M, pH 7.4) (Ammonium Sulfate b). The stability of this fraction was similar to that obtained from baker's yeast.

DPN ENZYME FROM BAKER'S YEAST

Step	Total units	Specific activity, units/mg. protein
Extract	1370	0.9
Ammonium sulfate A	1425	5.0
Ammonium sulfate B	925	16.7
C$_\gamma$ eluate	777	46.2

DPN ENZYME FROM ALE YEAST

Autolyzate	205	0.2
Ammonium sulfate a	168	0.7
Protamine	163	1.7
Ethanol	93	7.3
Ammonium sulfate b	56	15.4

Properties

Specificity. There was no detectable reaction with TPN in the case of the purified enzyme preparation, nor was there any significant inhibitory effect of TPN. The enzyme preparation failed to catalyze the reductive carboxylation of α-ketoglutarate and does not appear to have oxalosuccinate as an intermediate.

Activators and Inhibitors. There is an almost complete dependence on adenosine-5'-phosphate. The Michaelis constant is approximately 9×10^{-6} M. The following were inactive in replacing adenosine-5'-phosphate: adenosine, adenosine-2'-phosphate, adenosine-3'-phosphate, inosine-5'-phosphate, ADP, and ATP.

Mn^{++} produced a higher stimulation of the enzyme than did Mg^{++} but was inhibitory at higher concentrations.

Inhibitions of about 50% were produced by NaCN (0.01 M), NaN$_3$ (0.01 M), molybdate (0.01 M), and H_2S-reduced molybdate (0.00025 M).

pH Optimum. The pH optimum of the enzyme was found to be approximately 7.5, with a decrease in activity of about 40% at pH 6.5 and 8.5. The rates obtained with phosphate and tris(hydroxymethyl)-aminomethane buffers at given pH values were in agreement. Although there is no evidence for a requirement of phosphate by this enzyme (Ammonium Sulfate B fraction), faint traces of orthophosphate were still present in the nucleotide additions.

[119] Diphosphopyridine Nucleotide Isocitric Dehydrogenase from Animal Tissues

d-Isocitric acid + $DPN^+ \rightarrow$ α-Ketoglutaric acid + CO_2 + $DPN{\cdot}H$ + H^+

By G. W. E. PLAUT* and S.-C. SUNG

Assay Method

Principle. The progress of the reaction is determined by the rate of appearance of reduced DPN, as measured at 340 mμ.

Procedure. The following reaction mixtures are used for the assay of DPN and TPN isocitric dehydrogenase activity, respectively. The *DPN system* contains 1.0 ml. of 0.1 M cacodylate buffer, pH 6.5, enzyme solution, 0.1 ml. of 0.01 M DPN, 0.1 ml. of 0.01 M AMP, 0.1 ml. of 0.02 M MnSO$_4$, and 0.1 ml. of 0.80 M dl-isocitrate; volume made up to 3 ml. with water. The *TPN system* contains 1.0 ml. of 0.1 M tris (hydroxymethyl)aminomethane buffer, pH 7.4, enzyme solution, 0.2 ml. of 0.0015 M TPN, 0.1 ml. of 0.02 M MnSO$_4$, and 0.05 ml. of 0.08 M dl-isocitrate; volume made up to 3 ml. with water. In all cases a control containing all reaction components except isocitrate accompanies the particular sample tested. All assays are performed at room temperature in a Beckman model DU spectrophotometer in cuvettes of 1-cm. light path.

Definition of Unit and Specific Activity. A unit of activity is defined as the amount of enzyme causing an increase in optical density of 0.01 per minute, under conditions for which the rate of density increase remains linear for at least 5 minutes. Specific activity is expressed as units per milligram of protein. Protein is determined by the method of Warburg and Christian.[1]

Purification Procedure

DPN isocitric dehydrogenase is much more labile than the corresponding TPN enzyme. All the operations described below are therefore conducted at 0 to 2° unless specified otherwise.

Conventional methods of enzyme fractionation are used except for the application of precipitation chromatography[2] in step 4. If simple ammo-

---

* Established Investigator of the American Heart Association. Supported in part by a grant (No. H-1279) from the National Heart Institute, United States Public Health Service.

[1] O. Warburg and W. Christian, *Biochem. Z.* **310**, 384 (1941).

[2] E. H. Fischer and H. M. Hilpert, *Experientia* **9**, 176 (1953).

nium sulfate fractionation is used at this stage, a preparation with about one-third of the specific activity of the chromatographed product can be obtained. Nevertheless, even in this enzyme preparation the specific activity of TPN isocitric dehydrogenase has been reduced to about 5% of that of the original extract. The alternative procedure is described in step 4a.

Step 1. Preparation of Acetone Powder. Fresh beef hearts are packed in ice at the packing plant and transported to the laboratory within ½ hour. The hearts are trimmed to remove fat, ligament, and the auricles. The ventricles are cut into approximately 1-cm. cubes. 2 kg. of this tissue is ground with 5.5 l. of a medium containing 0.25 M sucrose and 0.03 M K_2HPO_4 in a Waring blendor for 1 minute at "fast" and 1 minute at "slow" speed (about 150 g. of wet tissue and 400 ml. of suspending medium used per Waring blendor bowl). The resulting suspension is centrifuged at 600 × g for 10 minutes to remove large particles. The resulting supernatant is passed through a double layer of cheesecloth to remove floating pieces of fat. The filtrate is adjusted to pH 5.8 to 5.9 with dilute acetic acid and centrifuged at 1800 × g for 20 minutes. The residue is taken up in a minimum quantity of 0.25 M sucrose and centrifuged at 5000 × g for 30 minutes. The well-packed, washed residue preparation is then added to about 500 ml. of cold ($-20°$) acetone in a Waring blendor bowl. The suspension is mixed for 1 minute and centrifuged in 250-ml. glass cups at $-10°$. The supernatant is discarded, and the residue is suspended in acetone as before; this procedure is repeated two to three times. After the last centrifugation the residue is distributed onto the walls of the centrifuge bottles, and a single layer of cheesecloth is placed over the mouth of the bottles. The containers are placed sideways in a large vacuum desiccator which is evacuated. The vacuum drying process is most conveniently done at room temperature and takes 1 to 2 hours. About 25 g. of a light-tan powder is obtained. It can be stored in a tightly closed container at $-10°$. Under these conditions the powder loses about 25% of the DPN isocitric dehydrogenase but none of the TPN enzyme activity in one month.

Because of the lability of the enzyme, all the operations described below and the assays must be done on the same day.

Step 2. Preparation of Crude Extract. 5 g. of acetone powder is extracted with 100 ml. of 0.01 M phosphate, pH 6.5, for 15 to 30 minutes. The mixture is centrifuged at 18,000 × g for 20 minutes.

Step 3. Adsorption by Calcium Phosphate Gel. 80 ml. of supernatant solution is treated with 7 ml. of calcium phosphate gel (2.2 mg./mg. of protein), the latter prepared according to the directions of Swingle and Tiselius.[3] After a contact period of 10 minutes, the suspension is centri-

[3] S. Swingle and A. Tiselius, *Biochem. J.* **48**, 171 (1951).

fuged for 10 minutes at $18,000 \times g$. The supernatant solution should contain less than 5% of the activity and is discarded. The residue is suspended in 80 ml. of 0.1 M phosphate, pH 6.5, and centrifuged. The supernatant contains about 40% of the protein but little activity. The residue is subsequently washed in a similar manner with 60 ml. of 0.6 saturated ammonium sulfate,[4] which removes a considerable amount of TPN enzyme. The DPN isocitric dehydrogenase is eluted from the gel with 30 ml. of 0.3 saturated ammonium sulfate.

Step 4. $(NH_4)_2SO_4$ Fractionation and Precipitation Chromatography. The ammonium sulfate concentration of the centrifuged eluate is adjusted to 0.6 with saturated ammonium sulfate, centrifuged, and the residue taken up in 9 ml. of 0.01 M phosphate, pH 6.5. 6 ml. of saturated $(NH_4)_2$-SO_4 is added to this solution. Upon centrifugation the supernatant layer of the 0.4 saturated solution is placed on a starch-Celite (1:1) column (2.2 cm. in diameter, height 2.6 cm.), prepared according to the directions of Fischer *et al.*[2] The column is washed with 25 ml. of 0.4 saturated $(NH_4)_2SO_4$, which removes none of the activity but considerable protein. The eluant concentration is then changed to 0.3 saturation, and three 8-ml. fractions are collected. The second fraction usually contains most of

SUMMARY OF PURIFICATION PROCEDURE[a]

Step	Fraction	Specific activity, units/mg.	Recovery of original, %	Puri- fication, fold	TPN enzyme specific activity, units/mg.
2	Crude extract	31			600
3	Ca₃(PO₄)₂ gel:				
	Raffinate	0	0		
	0.1 M phosphate, pH 6.5, wash	5	8		
	0.3 satn. $(NH_4)_2SO_4$ eluate	138	70	4.5	
4	0.4 satn. $(NH_4)_2SO_4$ supernatant	261	60	1.9	
4	Starch column effluent	1410	34	5.4	10–50
4a	0.4 satn. $(NH_4)_2SO_4$ supernatant	220	65	1.6	
4a	0.4–0.5 satn. $(NH_4)_2SO_4$	400[b]	49	1.8	35
1–4	Over-all purification	1410	34	46	
1–4a	Over-all purification	400	49	13	

[a] Data in part from G. W. E. Plaut and S.-C. Sung, *J. Biol. Chem.* **207**, 305 (1954).
[b] Specific activities up to 600 have been obtained with 0.45 to 0.5 saturation ammonium sulfate. However, only about one-half of the enzyme activity is recovered when compared to the 0.4 to 0.5 cut.

[4] The saturation levels referred to here are in terms of saturation of $(NH_4)_2SO_4$ at 25°, although fractionations are done at 0 to 2°.

the activity; its ammonium sulfate concentration is adjusted to 0.5, and the resulting precipitate is collected by centrifugation and dissolved in 5 ml. of 0.01 M phosphate, pH 6.5, for assay.

Step 4a. Ammonium Sulfate Fractionation. The ammonium sulfate concentration of the centrifuged eluate from step 3 is adjusted to 0.4 and centrifuged. The precipitate is discarded, and the ammonium sulfate saturation of the supernatant is increased to 0.5. The resulting precipitate is collected by centrifugation and dissolved in 5 ml. of 0.01 M phosphate, pH 6.5, for assay. Information pertinent to this fractionation is summarized in the table, shown on page 712.

Properties

Specificity. The only substrate that has been found to be active for this enzyme, so far, has been d-isocitric acid. The following substances have been tested and found to be inactive for the purified enzyme: citrate, *cis*-aconitate, d- and l-α-hydroxyglutarate, malate, dl-β-hydroxybutyrate, and glutaconate. In contrast to the TPN enzyme, which has a very high affinity for isocitrate ($K_s < 1.2 \times 10^{-5}$ M),[5] the DPN enzyme has a much higher K_s (approximately 4.5×10^{-4} M); K_{DPN} is approximately 6×10^{-5} M.

Even though the purified DPN isocitric dehydrogenase still contains a small amount of TPN enzyme, no significant inhibitory or activating effect by TPN is observed. Pyridine nucleotide transdehydrogenase, which was found by Colowick *et al.*[6] in extracts of *Pseudomonas fluorescens* and by Kaplan *et al.* in heart muscle,[7] does not seem to be present in this preparation. Nicotinamide mononucleotide is not active in the crude or purified preparation either in the absence or in the presence of various concentrations of ATP. The DPN enzyme is almost completely inhibited by 0.002 M ATP.

Activators. DPN isocitric dehydrogenase is completely inactive in the absence of certain metal ions. Manganous ion is a more effective activator than magnesium. In contrast to DPN isocitric dehydrogenase from yeast,[8] the addition of AMP is not absolutely required for activity, although it has been found to stimulate the reaction at times and therefore has been routinely added to the assay medium. 0.01 M Cyanide inhibits the enzyme.

[5] S. Ochoa, *J. Biol. Chem.* **174**, 133 (1948).

[6] S. P. Colowick, N. O. Kaplan, E. F. Neufeld, and M. M. Ciotti, *J. Biol. Chem.* **195**, 95 (1952).

[7] N. O. Kaplan, S. P. Colowick, and E. F. Neufeld, *J. Biol. Chem.* **205**, 1 (1953).

[8] A. Kornberg and W. E. Pricer, Jr., *J. Biol. Chem.* **189**, 123 (1951).

Effect of pH. The DPN isocitric dehydrogenase of guinea pig and beef heart appears to be much more sensitive to the pH of the reaction medium than the corresponding TPN enzyme. DPN isocitric dehydrogenase exhibits a sharp optimum at pH 6.5, whereas the TPN enzyme has a broad range of reaction-pH dependency with a peak at about pH 7.8.

Reaction Products. In view of the studies with TPN-enzyme from pig heart[5] and DPN-isocitric dehydrogenase from yeast,[8] it was to be expected that the products of oxidation of the substrate by DPN would be α-ketoglutarate, bicarbonate, and DPN·H. The identity of DPN·H formed in this reaction was confirmed by its reoxidation to DPN by acetaldehyde and crystalline yeast alcohol dehydrogenase at pH 6.5. When the product of this reaction was converted to the 2,4-dinitrophenyl-hydrazone, it was found to migrate at the same rate as an authentic sample of the 2,4-dinitrophenylhydrazone of α-ketoglutarate on silica gel columns[9] using different solvent systems.

No reoxidation of DPN·H by either α-ketoglutarate or oxalosuccinate was obtained with purified isocitric dehydrogenase from beef heart or yeast[8] under the experimental conditions used.

Occurrence. DPN-isocitric dehydrogenase activity has been found in the extracts of acetone powders of either mitochondria or washed residues of the following tissues: guinea pig, beef, and pigeon heart, pigeon breast muscle, rat kidney, and rat liver. However, the DPN and TPN enzymes were not separated except with guinea pig and beef heart and pigeon breast muscle. In such extracts of beef or guinea pig heart from ten to thirty times as much TPN enzyme as DPN-isocitric dehydrogenase was found.

[9] D. O. Brummond and R. H. Burris, *Proc. Natl. Acad. Sci. U.S.* **39**, 754 (1953).

[120] α-Ketoglutaric Dehydrogenase System and Phosphorylating Enzyme from Heart Muscle[1]

By SEYMOUR KAUFMAN

Dehydrogenase

α-Ketoglutarate + DPN$^+$ + HSCoA
$$\to \text{Succinyl-SCoA} + CO_2 + DPNH + H^+$$

Assay Method

Principle. When α-ketoglutarate is oxidized in the presence of DPN$^+$ and CoA, the formation of DPNH can be used to follow the reaction by

[1] S. Kaufman, C. Gilvarg, O. Cori, and S. Ochoa, *J. Biol. Chem.* **203**, 869 (1953).

measuring the increase in optical density at 340 mμ in the Beckman spectrophotometer. With catalytic amounts of CoA, succinyl CoA deacylase is added to regenerate CoA.[2-4] The deacylase does not become limiting until it is separated from the dehydrogenase by the $Ca_3(PO_4)_2$ gel step, at which time it must be added back to the assay system.

Reagents

0.0013 M CoA.

1.0 M potassium phosphate buffer, pH 7.4.

0.0027 M DPN.

0.083 M L-cysteine.

0.25 M potassium α-ketoglutarate.

Purified deacylase (pH 7.4 calcium phosphate gel eluate with about 0.5 mg. of protein) when needed (see Vol. I [100]).

α-Ketoglutaric dehydrogenase solution to be assayed.

The CoA, DPN, and α-ketoglutarate solutions are stable for several weeks when kept frozen. The cysteine solution is prepared by neutralization of a stock solution of cysteine-HCl with KOH. Although the cysteine-HCl stock solution is stable for weeks when stored in the cold, the neutralized solution is unstable and is prepared immediately before the assay.

Procedure. 0.1 ml. of each solution is added to a silica cell having a 1-cm. light path. Water is added to make a final volume of 2.90 ml. The reaction is started by the addition of 0.1 ml. of the α-ketoglutarate solution. Final volume, 3.0 ml. Readings are taken at wavelength 340 mμ for at least four consecutive 15-second periods. The temperature of the reaction is maintained at 25 to 26° by circulating cold water through the cell compartment block.

Definition of Unit and Specific Activity. One unit is defined as the amount of enzyme catalyzing the oxidation of 1.0 micromole of α-ketoglutarate per minute under the standard conditions. The molecular extinction coefficient of DPNH at 340 mμ is taken as 6.22×10^6 (cm.^{2}/mole).[5] Thus, 1 micromole of DPNH in 3.0 ml. is equivalent to an optical density of 2.07. Specific activity is expressed as units per milligram of protein. Protein is determined spectrophotometrically by the absorption of light

[2] S. Kaufman, *in* "Phosphorus Metabolism" (McElroy and Glass, eds.), Vol. 1, p. 370, The Johns Hopkins Press, Baltimore, 1951.

[3] J. Gergely, P. Hele, and C. V. Ramakrishnan, *J. Biol. Chem.* **198**, 323 (1952).

[4] S. Kaufman, *Federation Proc.* **12**, 704 (1953).

[5] B. L. Horecker and A. Kornberg, *J. Biol. Chem.* **175**, 385 (1948).

at wavelengths 280 and 260 mμ, with a correction for the nucleic acid content by the data given by Warburg and Christian.[6]

Assay in Crude Tissue Preparations. Because of the presence of a DPNH oxidase, the initial extracts and the 0 to 50 ammonium sulfate fraction are assayed by measuring the rate of CO_2 evolution due to the dismutation,[7] 2α-ketoglutarate + NH_3 → Succinate + CO_2 + L-glutamate, in the presence of glutamic dehydrogenase. The assay system contains the following components (in micromoles): potassium phosphate buffer, pH 7.4, 100; DPN, 0.15; CoA, 0.13; L-cysteine, 10; potasssium α-ketoglutarate, 50; NH_4Cl, 50; excess glutamic dehydrogenase; and enzyme. Final volume, 2.0 ml.; temperature, 25°. The reaction is started by tipping in α-ketoglutarate from the side bulb of the Warburg vessels after temperature equilibration. One unit is the amount of enzyme catalyzing the evolution of 1 μM. of CO_2 per minute.

Purification Procedure

Step 1. Extraction. Pig hearts, removed immediately after death, are packed in ice. They can be used either fresh or after variable periods of frozen storage. All operations are performed at about 0° unless otherwise indicated. The hearts are trimmed of fat and connective tissue and ground in an electric mincer. The minced tissue is washed with cold tap water until the washings are almost colorless. The water is squeezed out through cheesecloth, and the residue is extracted in a Waring blendor with an equal volume of 0.05 M potassium phosphate buffer, pH 7.4, made 0.2 M with respect to potassium chloride. After 5 minutes, another 2 vol. of the buffer-KCl solution is added, and the blending is continued for a further 15 minutes. The mixture is centrifuged at 18,000 × g for 45 minutes, and the sediment is discarded.

Step 2. First Ammonium Sulfate Fractionation. The extract is brought to 50% saturation with solid ammonium sulfate. The salt is added slowly with mechanical stirring. Stirring is continued for 30 minutes, and the mixture is then centrifuged at 18,000 × g for 30 minutes. Gravity filtration through fluted filter paper (Reeve Angle No. 802) may be substituted for centrifugation. The precipitate is taken up in 0.033 M potassium phosphate buffer, pH 7.4, and the solution is dialyzed overnight against a large excess of the same buffer. Prior to the dialysis, iodoacetate is usually added to the enzyme solution to a final concentration of 0.001 M.

Step 3. Second Ammonium Sulfate Fractionation. After the insoluble matter is removed by centrifugation, the dialyzed solution from the previous step, containing 20 to 30 mg. of protein per milliliter, is diluted with

[6] O. Warburg and W. Christian, *Biochem. Z.* **310**, 384 (1941).

[7] F. E. Hunter, Jr., and W. S. Hixon, *J. Biol. Chem.* **181**, 67 (1949).

water to a protein concentration of 10 mg./ml. The fractionation is conducted as before, except that the pH of the mixture is maintained close to 7.4 by the dropwise addition of 1.0 N KOH when necessary. The solution is brought to 30% saturation with solid ammonium sulfate. The mixture is centrifuged at 18,000 × g for 20 minutes, and the precipitate dissolved in 0.033 M potassium phosphate buffer, pH 7.4 (0 to 30 fraction). Solid ammonium sulfate is added to the supernatant fluid to give 42% saturation, and the precipitate is discarded. The supernatant solution is brought to 70% saturation with ammonium sulfate; the precipitate is collected by centrifugation and dissolved in potassium phosphate buffer as above (42 to 70 fraction). The 0 to 30 and 42 to 70 fractions are dialyzed overnight against 0.033 M potassium phosphate buffer, pH 7.4. The 0 to 30 ammonium sulfate fraction, which contains both the dehydrogenase and deacylase activities, is usually kept frozen, at least for several days, before further fractionation. On thawing, a bulky precipitate of inactive protein is removed by centrifugation and discarded. If the protein so removed amounts to 50% or more of the protein originally present in the fraction, the ethanol fractionation described below can be omitted.

Step 4. Ethanol Fractionation. The ice-cold, clear enzyme solution is diluted with 0.033 M potassium phosphate buffer, pH 7.4, to a protein concentration of 10 mg./ml. Ethanol, chilled to −30°, is added dropwise with mechanical stirring to a concentration of 10% by volume; the temperature is allowed to drop to −4°. The precipitate is removed by centrifugation at −4° and discarded. The supernatant solution is gradually brought to −7°, and ethanol is added as above to a final concentration of 20% by volume. The precipitate is collected by centrifugation at −7°, dissolved in 0.033 M potassium phosphate buffer, pH 7.4, and dialyzed against the same buffer to remove ethanol.

Step 5. Adsorption on Calcium Phosphate Gel. The solution from the preceding step is adjusted, if required, to a protein concentration of 10 to 15 mg./ml. with the 0.033 M phosphate buffer. 0.1 vol. of calcium phosphate gel is added, and the mixture is stirred for 10 minutes. The gel is removed by centrifugation and discarded. 0.3 vol. of calcium phosphate gel is added to the supernatant solution. The mixture is stirred for 10 minutes, then centrifuged, and the supernatant solution, which is free of dehydrogenase, is discarded.

Step 6. Fractional Elution. The gel is eluted first by stirring for 10 minutes with a volume of 0.1 M potassium phosphate buffer, pH 7.4, equal to the original volume of enzyme solution. The deacylase is eluted in this fraction free of dehydrogenase. The gel is washed once more with the same volume of 0.1 M buffer, pH 7.4, and then eluted as above with 0.2 M potassium phosphate buffer, pH 8.0. This eluate contains the dehydrogenase with traces of deacylase. By repetition of the absorption step once or twice, the dehydrogenase can be obtained completely free of deacylase. The preparation is free of lactic dehydrogenase. An outline of the purification of the dehydrogenase is given in Table I. The over-all purification from the initial extract is about 140-fold.

TABLE I

SUMMARY OF PURIFICATION OF α-KETOGLUTARIC DEHYDROGENASE

(2140 g. of pig heart)

Step	Volume, ml.	Enzyme, units	Protein, mg.	Specific activity, units/mg. protein	Yield, %
Extract	5200	265	88,000	0.003	100
First ammonium sulfate fractionation (0–50)	890	215	16,550	0.013	81
Second ammonium sulfate fractionation (0–30)	196	165	7,850	0.021	62
Ethanol fractionation (10–20)	172	82	1,860	0.044	31
Calcium phosphate gel eluate, pH 8.0	60	45	108	0.42	17

Properties

Although the enzyme has been purified extensively from pig heart by two independent methods, there is no indication of a separation of enzymatic activities.[1,8] By ultracentrifugal and electrophoretic criteria, one of these preparations is 90% pure.[8] Its molecular weight has been estimated as 2,000,000, and it contains bound diphosphothiamine and α-lipoic acid.[8] Nevertheless, in analogy with the pyruvate oxidation system,[9] the α-ketoglutaric dehydrogenase unit probably consists of two or more individual enzymes. In fact, Gunsalus and his co-workers[10] have recently succeeded in resolving the α-ketoglutaric dehydrogenase of E. coli (Crooke's strain) into two enzyme components similar to the pyruvic oxidation fractions A and B.[9] For this reason, the enzyme has been referred to as the α-ketoglutaric dehydrogenase system.[1]

P Enzyme

$$\text{Succinyl-SCoA} + \text{ADP} + \text{P} \rightleftarrows \text{Succinate} + \text{ATP} + \text{HSCoA}$$

[8] D. R. Sanadi, J. W. Littlefield, and R. M. Bock, J. Biol. Chem. **197**, 851 (1952).

[9] S. Korkes, A. del Campillo, I. C. Gunsalus, and S. Ochoa, J. Biol. Chem. **193**, 721 (1951).

[10] I. C. Gunsalus, in "Mechanism of Enzyme Action" (McElroy and Glass, eds.), The Johns Hopkins Press, Baltimore, 1953.

Assay Method

Principle. The enzyme is most conveniently assayed from the succinate, ATP, CoA side.[1,11] The succinyl CoA generated reacts to form succinohydroxamic acid in the presence of hydroxylamine. Under these conditions, CoA is constantly regenerated. The succinohydroxamic acid is determined with the FeCl₃ reagent of Lipmann and Tuttle.[12]

Reagents

1 M Tris buffer, pH 7.4.
1 M potassium succinate.
0.1 M ATP.
0.0013 M CoA.
0.20 M GSH.
0.1 M MgCl₂.
2.4 M NH₂OH.
Enzyme solution to be assayed.
The NH₂OH solution must be freshly prepared before the assay. A stock solution of 4 M NH₂OH·HCl, which is stable for months, is neutralized with 7 N KOH. 3.0 ml. of the NH₂OH·HCl + 2 ml. of 7 N KOH gives a neutral solution which is 2.4 M. The GSH is weighed out and neutralized immediately before the experiment.

Procedure. To 0.4 ml. of the neutralized NH₂OH solution, add 0.1 ml. of each of the other reagents and water up to 2.0 ml. (allowing for the subsequent addition of the enzyme). The reaction is started by the addition of the enzyme. The mixture is incubated at 37° for 30 minutes. The reaction is stopped by the addition of a freshly prepared solution containing equal volumes of 5 % FeCl₃, 3.0 N HCl, and 12 % TCA.[12] The precipitated protein is removed by centrifugation, and the optical density of the supernatant solution is determined in a Klett photoelectric colorimeter (filter No. 54). It has been found that, under these conditions, 1 micromole of succinohydroxamic acid (prepared from succinic anhydride) has an optical density of 61. The amount of enzyme used in the assay must not give more than 1.5 micromole of succinohydroxamic acid. Beyond this point, the rate of the reaction is no longer proportional to enzyme concentration.

Definition of Unit and Specific Activity. One unit is defined as the amount of enzyme catalyzing the formation of 1.0 micromole of succinohydroxamic acid in 30 minutes at 37°. Specific activity is expressed as

[11] S. Kaufman, *Abstr. 122nd Meeting Am. Chem. Soc., Atlantic City* 32c (Sept., 1952).
[12] F. Lipmann and L. C. Tuttle, *J. Biol. Chem.* **158**, 505 (1945).

units per milligram of protein, the latter determined as described for the ketoglutaric dehydrogenase.

Application of Assay Method to Crude Tissue Preparations. The method is suitable for use in all stages of the purification procedure. In the early stages, it is advisable to run a blank containing the complete system, to which the FeCl₃ mixture is added at zero time. This serves as a correction for any color or turbidity due to the crude enzyme solution.

Purification Procedure

The P enzyme is present in the 42 to 70 ammonium sulfate fraction from the preparation of the dehydrogenase. This fraction is usually kept frozen until required and, if necessary, is clarified by centrifugation before further fractionation.

Acetone Fractionation. The clear 42 to 70 fraction is adjusted to a protein concentration of 10 mg./ml. and made 0.067 M with respect to potassium phosphate buffer, pH 6.8. Cold acetone is then added to a concentration of 45% by volume with mechanical stirring, the temperature being allowed to drop to −5 to −6°. The acetone is added dropwise over a period of 20 to 30 minutes, and the stirring is continued for another 15 minutes. The precipitate is removed by centrifugation at 600 × g and −5° for 20 minutes and discarded. The cloudy supernatant solution is brought to 57% acetone by volume at −9 to −10°, and, after centrifugation at the same temperature, the precipitate is dissolved in 0.017 M potassium phosphate buffer, pH 7.4. The solution is dialyzed overnight against the same buffer.

Adsorption on Alumina Gel. The solution from the previous step is adjusted to a protein concentration of 10 mg./ml. with the 0.017 M phosphate buffer and kept at 50° for 6 minutes. The solution is cooled, and the precipitate is removed by centrifugation at 10,000 × g for 10 minutes. The pH of the clear supernatant fluid is adjusted to 5.7 with 1.0 N acetic acid, and the precipitate is removed by centrifugation (5 minutes at 18,000 × g); the pH is then brought to 6.5 with 1.0 M potassium bicarbonate. It is necessary to work rapidly at this stage because the enzyme is labile at pH 5.7. Alumina gel C$_\gamma$ (dry weight, 13 mg./ml.) is added to make 30% by volume, and the mixture is stirred for 10 minutes at 0°. The gel is separated by centrifugation and washed once with a volume of 0.1 M potassium phosphate buffer, pH 7.4, equal to that of the solution before addition of the alumina gel. This eluate is discarded. The P enzyme is eluted by stirring at room temperature for 20 minutes with 1 vol. of 0.2 M potassium phosphate buffer, pH 8.0, as above. The eluate is dialyzed almost free of phosphate against 0.05 M potassium chloride. Removal of

phosphate is important because it inhibits the activity of the P enzyme
as determined by the hydroxamic acid assay.

TABLE II
SUMMARY OF PURIFICATION OF P ENZYME
[2200 g. of pig heart (frozen)]

Step	Volume, ml.	Enzyme, units	Protein, mg.	Specific activity, units/mg. protein	Yield, %
Extract	6000	1410	47,500	0.030	100
First ammonium sulfate fractionation (0–50)	990	1060	20,000	0.053	75
Second ammonium sulfate fractionation (42–70)	181	1242	4,980	0.25	88
Acetone fractionation (45–57)	110	520	946	0.55	37
Alumina gel eluate, pH 8.0	154	370	150	2.5	26
Zn-ethanol fractionation (20–30)	10	180	36	5.0	13

Zn-Ethanol Fractionation. The enzyme is concentrated by precipitation
with ammonium sulfate at 85% saturation. The protein is dissolved in
$M/60$ Tris, pH 7.4, and dialyzed against the same buffer overnight. This
procedure usually results in some loss of activity.

For each 10 ml. of the concentrated eluate, containing 5 to 6 mg. of
protein per milliliter, 0.38 ml. of 0.9 M succinate buffer, pH 6.3, and
2.6 ml. of 0.1 M zinc acetate solution are added. The temperature is maintained between -2 and $-1°$ while ethanol is added to 20% by volume.
The precipitate is removed by centrifugation at 10,000 $\times$ g for 10 minutes. Ethanol is then added to the supernatant to a final concentration of
30% by volume, and the precipitate is removed by centrifugation. The
precipitate, which contains the enzyme, is dissolved in a solution of 0.1 M
phosphate buffer, pH 7.4, which is 0.01 M with respect to Versene, and
dialyzed against $M/60$ phosphate buffer, pH 7.4, overnight.

Properties

The enzyme prepared by this procedure is still contaminated with
myokinase. It is free of ATP-ase, pyrophosphatase, the acetate-activating
enzyme,[13] CoA transferase,[14,15] succinyl CoA deacylase, and α-ketoglu-

[13] F. Lipmann, M. E. Jones, S. Black, and R. M. Flynn, *J. Am. Chem. Soc.* **74,** 2384 (1952).
[14] J. R. Stern, M. J. Coon, and A. del Campillo, *J. Am. Chem. Soc.* **75,** 1517 (1953).
[15] D. E. Green, D. S. Goldman, S. Mii, and H. Beinert, *J. Biol. Chem.* **202,** 137 (1953).

taric dehydrogenase.[1] The enzyme is inhibited 50% in 0.075 M KF. NH_2OH concentrations much higher than that used in the standard assay lead to a marked inhibition. DNP at 1×10^{-4} M is without effect. The enzyme can be incubated in 0.001 M iodoacetate without inhibition, provided that the excess iodoacetate is dialyzed out prior to assay.[1]

Recently, the enzyme has been prepared from spinach leaves by a procedure which is a modification of the one described here. The enzyme from spinach is completely free of myokinase.[16] The equilibrium constant of the reaction is 0.30 at pH 7.4, at 20° [K = (succinyl CoA)(ADP)(P)/ succinate (ATP)(CoA)[16]]. Pantethiene cannot replace CoA, and with the myokinase-free enzyme, ADP cannot replace ATP.[16] The K_m values for the various components are as follows: ATP, $5 \times 10^{-4}M$; succinate, 1.5×10^{-2} M; CoA, 1×10^{-4} M; Mg^{++}, 3.5×10^{-3} M.[16] The enzyme seems to be specific for succinate. The following compounds have been tested and found to be inactive: acetate, citrate, α-ketoglutarate, formate, glutamate, malate, maleate, fumarate, tartarate, butyrate, aspartate, malonate, lactate, glutarate, and oxalate.[16] The enzyme from heart muscle is also specific for succinate.[17]

[16] S. Kaufman and S. G. A. Alivisatos, unpublished experiments.
[17] H. Hift, L. Ouellet, J. W. Littlefield, and D. R. Sanadi, *J. Biol. Chem.* **204**, 565 (1953).

[121] Succinic Dehydrogenase

By WALTER D. BONNER

Assay Method

Principle. No simple or successful method for the separation of succinic dehydrogenase from cellular inclusions has been devised to date; hence, any assay for this enzyme has to be carried out in the presence of some or all components of the succinic dehydrogenase-cytochrome system. Cytochrome oxidase may be inactivated by the use of cyanide or by carrying out the estimation anaerobically. Under either of these conditions and with the addition of a hydrogen acceptor, the activity of the "succinic dehydrogenase system" may be estimated readily. The expression "succinic dehydrogenase system" refers to the system which catalyzes the anaerobic oxidation of succinate and which probably includes cytochrome b in addition to succinic dehydrogenase and the hydrogen acceptor.[1] The activity of this system may be estimated by one of three

[1] E. C. Slater, *Biochem. J.* **45**, 1 (1949).

methods, the choice of method depending on the relative purity and activity of the enzyme preparations; hence, each method will be described. It should be emphasized that these methods give comparative activities only; succinic dehydrogenase works at far below its full capacity in any test that has been devised as yet.

Thunberg Method. This method, introduced by Thunberg[2] and developed by Szent-Györgyi[3] and others,[4] is dependent on the rate of reduction of a suitable hydrogen acceptor under anaerobic conditions. Observations may be made either photometrically or visually, the time required for 90% reduction of the hydrogen acceptor being the end point. Methylene blue and 2,6-di-chlorophenolindophenol are commonly employed as hydrogen acceptors; the enzyme concentration should be adjusted so that 90% reduction is obtained in 5 to 10 minutes. The method is well described by Burris.[5]

Reagents. Concentrations of components inevitably vary with the application; the following are suggested for average conditions.

Suitable buffer, Phosphate buffer is commonly employed.
$2.67 \times 10^{-4} M$ methylene blue or $2.5 \times 10^{-3} M$ 2,6-dichlorophenol-indophenol.
$0.2 M$ sodium succinate.

Procedure. To the Thunberg tube are added 1 ml. each of the dye and Na-succinate and sufficient buffer to give a final concentration of $0.1 M$ in a total of 6 ml. The enzyme preparation is placed in the side cap, and the joint of the cap is greased with a good grade of high-vacuum grease and firmly placed on the tube. For ordinary routine work 3 minutes of evacuation with an efficient aspirator is sufficient. If the complete removal of oxygen is desired, the tube should be flushed out three times with O_2-free N_2. After 10 minutes of temperature equilibration the contents are mixed and the reduction of the dye observed, either colorimetrically or visually. In the latter event comparison should be made against a tube which contains one-tenth the dye concentration and no substrate; this tube represents 90% reduction of the dye in the assay.

Definition of Activity. The activity is most conveniently expressed

[2] T. Thunberg, *Skand. Arch. Physiol.* **22**, 430 (1909).
[3] A. Szent-Györgyi, *Biochem. Z.* **150**, 195 (1924).
[4] F. Battelli and L. Stern, *Biochem. Z.* **30**, 172 (1910).
[5] R. H. Burris, *in* "Manometric Techniques and Tissue Metabolism" (Umbreit, Burris, and Staufer, eds.), 2nd ed., p. 105. Burgess Publishing Co., Minneapolis, Minn., 1949.

as Q_{MB}[6] (or Q dichlorophenolindophenol). The Q_{MB} is defined as microliters of H_2 transferred to methylene blue per hour per milligram dry weight. 1 ml. of 10^{-3} M MB requires 22.4 μl. of H_2; hence, the 1 ml. of 2.67×10^{-4} M MB used in the assay requires 5.98 μl. of H_2 or, at 90% reduction, 5.44 μl. of H_2.

Manometric Method. The assay of the succinic dehydrogenase system by the manometric method depends on the measurement of oxygen consumption during succinate oxidation in the presence of cyanide and methylene blue. This method is described by Slater.[1]

Reagents

    0.01 M methylene blue.
    0.1 M potassium cyanide, neutralized.
    0.4 M sodium succinate.
    Suitable buffer (pH 7.2.) Phosphate buffer is generally employed.

Procedure. Sufficient buffer is used to give a final concentration in 3.0 ml. of 0.1 M. To this are added 0.3 ml. each of the methylene blue and KCN and 0.2 ml. of Na-succinate; the suitably diluted enzyme preparation is placed in the side arm. The oxygen uptake is followed in the usual manner. The enzyme preparation is kept separate from the other components during equilibration in order to prevent the inactivation of succinic dehydrogenase by cyanide which occurs only in the absence of succinate.[7]

Definition of Activity. The activity of the succinic dehydrogenase system when measured by this method is expressed as Q_{O_2}. One correction is required; namely, the observed oxygen uptake must be halved because the final reaction with oxygen in this assay is with leucomethylene blue and the product of this oxidation is H_2O_2. For further information on this detail, see Slater and Bonner.[8]

Spectrophotometric Method. This assay, described by Slater and Bonner,[8] depends on measuring the reduction rate of $K_3Fe(CN)_6$ in the presence of sufficient KCN to inhibit cytochrome oxidase. It is not known what components of the succinic oxidase system are involved in the reduction of ferricyanide, although succinic dehydrogenase is probably the most important. This method is particularly useful for partially purified and highly active preparations; it is further a reliable assay for succinic dehydrogenase from plants.

[6] H. S. Corran, J. G. Dewan, A. H. Gordon, and D. E. Green, *Biochem. J.* **33**, 1694 (1939).

[7] C. L. Tsou, *Biochem. J.* **49**, 512 (1951).

[8] E. C. Slater and W. D. Bonner, *Biochem. J.* **52**, 185 (1952).

Reagents

0.1 M potassium cyanide, neutralized.
0.01 M $K_3Fe(CN)_6$.
0.2 M sodium-succinate.
Suitable buffer (pH 7.2).

Procedure. To a 1-cm. cell are added 0.3 ml. each of the KCN and $K_3Fe(CN)_6$ and 0.2 ml. of Na-succinate. Buffer (phosphate buffer is usually employed) is added to give a final concentration of 0.1 M in a total of 3.0 ml. Water only is used in the reference cell. At zero time 0.2 ml. of the suitably diluted preparation is added to both cells, and the optical density at 400 mμ is followed as a function of time. At the close of each determination the temperature is measured.

Definition of Activity. The activity of the system which reduces $K_3Fe(CN)_6$ may be expressed as the decrease of optical density at 20°. The rate of change of optical density may be corrected to 20°, assuming the same temperature coefficient as was found for the complete succinic oxidase system by Slater.[9] The validity of this correction is not certain, but the temperature range encountered will be small; hence, any error introduced by the correction will be negligible. The energy of activation found by Slater is 7520 cal.; hence, the observed $V_{uncorr.}$ may be converted to $V_{corr.}$ by the expression $\log_{10} (V_{corr.}/V_{uncorr.}) = 7520(T_s - T)/2.303RT_sT$, where T is the observed absolute temperature, T_s is the absolute temperature chosen as the standard of comparison (here 293), and R is the gas constant.

Preparation. Succinic dehydrogenase is present in almost all organisms, homogenates, mitochondrial preparations, and the like. Many methods have been used for the preparation of this enzyme from various sources.[10,11] By the discreet application of the method described below, preparations which possess low endogenous respiratory activity and high succinic dehydrogenase activity may be made from most tissues. This method of preparation was developed by Keilin and Hartree[12] (also see Slater[1]) for the study of the succinic dehydrogenase-cytochrome system in heart muscle. The well-minced tissue is thoroughly washed with water. It should frequently be collected on muslin and given a hard squeeze in order to facilitate the removal of water-soluble substances. With some tissues the washing may be carried out more effectively in a centrifuge.

---

[9] E. C. Slater, *Biochem. J.* **46**, 484 (1950).
[10] W. H. McShan, *in* "Respiratory Enzymes" (Lardy, ed.), rev. ed., p. 116. Burgess Publishing Co., Minneapolis, Minn., 1949.
[11] C. A. Price and K. V. Thimann, *Plant Physiol.* **29**, 113 (1954).
[12] D. Keilin and E. F. Hartree, *Biochem. J.* **41**, 500 (1947).

The washed tissue is next gently ground in a mortar with sand and dilute phosphate buffer; 0.02 M buffer (pH 7.2) is an adequate extracting medium for most animal tissues; plant tissues require a stronger buffer (0.1 M). The ground material is next centrifuged for 20 minutes at 1500 × g, and the resulting cloudy supernatant fluid is subjected either to acid precipitation or to further centrifugation. For the acid precipitation the fluid is cooled to 0 to 4° and brought to pH 5.7 with N acetic acid. The precipitate is immediately collected by centrifuging for 15 minutes in the cold at 1500 × g and suspended in an equal volume of 0.1 M phosphate buffer (pH 7.2). If it is desirable to avoid the acid precipitation, the material may be collected by centrifuging for 20 minutes at 10,000 × g. Keilin and Hartree introduced the acid precipitation for a number of reasons, chiefly to concentrate the preparation and remove last traces of soluble substances; this step increases the activity.[1] The keeping qualities of the final suspension are improved if 0.1 M borate-boric acid buffer (pH 7.2) is substituted for phosphate.[13]

Specificity. Succinate is the only substrate catalytically oxidized by the succinic dehydrogenase system. It has been reported that mono-alkylated methyl and ethyl succinate are oxidized at a slow rate; succinate derivatives with a large alkyl group are inhibitory.[14-16]

Activators and Inhibitors. At the present time there are no known activators of the succinic dehydrogenase system. The system is inhibited by a variety of agents. Depending on the type of substance used, the inhibition will belong to one of three categories, viz., inhibition through combining with sulfhydryl groups, competitive inhibition, and non-specific inhibitions.

Hopkins and his co-workers[17,18] first demonstrated the existence and importance of SH groups in succinic dehydrogenase in their classical experiments with oxidized and reduced glutathione. Many studies of the action of SH-combining compounds on succinic dehydrogenase (and succinic oxidase) have appeared since the publication of the papers by Hopkins and his co-workers.[10,16,19-24] These experiments have utilized

[13] W. D. Bonner, *Biochem. J.* **56**, 274 (1954).
[14] T. Thunberg, *Biochem. Z.* **258**, 48 (1933).
[15] W. Franke and D. Siewerdt, *Z. physiol. Chem.* **280**, 76 (1944).
[16] F. Schlenk, *in* "The Enzymes" (Sumner and Myrbäck, eds.), Vol. II, p. 316. Academic Press, New York, 1951.
[17] F. G. Hopkins and E. J. Morgan, *Biochem. J.* **32**, 611 (1938).
[18] F. G. Hopkins, E. J. Morgan, and C. Lutwak-Mann, *Biochem. J.* **32**, 1829 (1938).
[19] E. J. Morgan and E. Friedmann, *Biochem. J.* **32**, 862 (1938).
[20] V. R. Potter and K. P. Dubois, *J. Gen. Physiol.* **26**, 391 (1943).
[21] E. S. G. Barron and T. P. Singer, *J. Biol. Chem.* **157**, 221 (1945).
[22] E. S. G. Barron and G. Kalnitsky, *Biochem. J.* **41**, 346 (1947).
[23] E. C. Slater, *Biochem. J.* **45**, 130 (1948).
[24] E. C. Slater, *Biochem. J.* **45**, 14 (1949).

the following types of compounds: alkylating agents, oxidizing agents, arsenicals, mercaptide-forming agents, maleic acid, and metal ions. Many of these investigations, although referring to succinic dehydrogenase, have dealt with the complete succinic oxidase system; hence, in some cases whether the point of attack is succinic dehydrogenase remains open to question.

Succinic dehydrogenase is competitively inhibited by a variety of substances. Malonate inhibition of succinic dehydrogenase, a classical example of a structural analogue acting as a competitive inhibitor[25–27] has been studied in detail by Thorn.[28,29] The ratio of affinities of succinic dehydrogenase for malonate and succinate was found to be about 3. This value is discussed by Thorn in relation to other values previously reported.[30–33,20] The inhibition by oxalacetate and its competitive nature has been studied by Das,[34] Pardee and Potter,[35] and Keilin and Hartree;[36] the latter workers showed that inhibition of succinic dehydrogenase by cozymase was due to the formation of oxalacetate. Pyrophosphate is a powerful competitive inhibitor of succinic dehydrogenase.[7,37]

Succinic dehydrogenase is inhibited by fluoride, but phosphate is required for this inhibition.[13,38,38a] Both fluoride and phosphate separately produce slight inhibitions at low succinate concentrations, but a rapid and much greater inhibition is obtained by fluoride and phosphate acting together; these inhibitions are competitive with respect to succinate.[8] Various workers have shown that compounds such as fumarate, oxalate, methyl succinate, glutarate, adipate, and suberate are weak competitive inhibitors of succinic dehydrogenase.[10]

There are a variety of substances which produce nonspecific inhibitions of succinic dehydrogenase. Free hematin in a concentration of $2 \times 10^{-4} M$ inhibits succinic dehydrogenase almost completely.[39] Glyoxalin and

[25] J. H. Quastel and W. R. Wooldridge, *Biochem. J.* **22**, 689 (1928).

[26] J. H. Quastel and A. H. M. Wheatly, *Biochem. J.* **25**, 117 (1931).

[27] P. J. G. Mann and J. H. Quastel, *Biochem. J.* **35**, 502 (1941).

[28] M. B. Thorn, *Biochem. J.* **53**, i (1953).

[29] M. B. Thorn, *Biochem. J.* **54**, 540 (1953).

[30] W. W. Ackermann and V. R. Potter, *Proc. Soc. Exptl. Biol. Med.* **72**, 1 (1944).

[31] H. A. Krebs and W. A. Johnson, *Tabul. biol.*, *Hague* **19**, part 3, 100 (1948).

[32] H. A. Krebs, S. Gurin, and L. V. Eggleston, *Biochem. J.* **51**, 614 (1952).

[33] D. Keilin and E. F. Hartree, *Biochem. J.* **44**, 205 (1949).

[34] N. B. Das, *Biochem. J.* **31**, 1124 (1937).

[35] A. B. Pardee and V. R. Potter, *J. Biol. Chem.* **176**, 1085 (1948).

[36] D. Keilin and E. F. Hartree, *Proc. Roy. Soc. (London)* **B129**, 277 (1940).

[37] L. F. Leloir and M. Dixon, *Enzymologia* **2**, 81 (1937).

[38] W. D. Bonner, *Biochem. J.* **49**, viii (1951).

[38a] E. C. Slater, *Biochem. J.* **58**, i (1954).

[39] D. Keilin and E. F. Hartree, *Biochem. J.* **41**, 503 (1947).

denatured globin give marked protection against hematin inhibition, and, because these substances form stable parahematin compounds, Keilin and Hartree[39] have concluded that the inhibition is probably due to parahematin formation between the hematin and an iron-binding group essential for the catalytic activity of succinic dehydrogenase.

Succinic dehydrogenase is slowly and irreversibly inactivated by cyanide.[7] The rate of inactivation is independent of pH between pH 6.5 and 8, an indication that the reaction involves hydrocyanic acid an ! not cyanide ion. Protection from inactivation is obtained with succinate and sodium dithionate. Tsou[7] has presented evidence showing that succinic dehydrogenase is the immediate electron acceptor from succinate and that only the oxidized form of the enzyme is susceptible to attack by cyanide.

Many narcotics inhibit succinic dehydrogenase, urethan and phenyl-urethan being most frequently used. Stoppani[40,41] found that narcotics inhibited the reduction of methylene blue but not the reduction of ferri-cyanide by succinate. Tsou[7] has suggested that Stoppani's findings may be explained by assuming that ferricyanide reacts directly with succinic dehydrogenase, whereas methylene blue can react only with cytochrome b. If this explanation is correct, narcotics should affect the interaction between succinic dehydrogenase and cytochrome b.

Effect of pH. The succinic dehydrogenase system exhibits a rather sharp pH optimum between pH 7 and 8.[5] In general, experiments are carried out at pH 7.3. Price and Thimann[11] have observed an optimum of 6.8 for preparations from seedlings.

Phosphate. There is no requirement for phosphate. Maximal activity is obtained in a variety of phosphate concentrations and in the almost complete absence of phosphate.[13,33]

Nature of Enzyme. No reliable method has yet been found to separate succinic dehydrogenase from the particulate succinic dehydrogenase-cytochrome-cytochrome oxidase system; hence, its nature remains obscure. The postulate that succinic dehydrogenase is a metal enzyme has been put forward from time to time. The supposition of Banga and Porges[42] that copper is an integral part of the enzyme is difficult to reconcile with the powerful inhibitory effects of this metal. Similarly, the possibility that manganese could be associated with the enzyme[43] is removed because of the competitive nature of pyrophosphate inhibition. There have been a considerable number of papers which have dealt with the

[40] A. O. M. Stoppani, *Nature* **160**, 52 (1947).
[41] A. O. M. Stoppani, *Enzymologia* **13**, 165 (1949).
[42] I. Banga and E. Porges, *Z. physiol. Chem.* **254**, 200 (1938).
[43] C. Massart, R. Dufait, and G. van Gremberger, *Z. physiol. Chem.* **262**, 270 (1940).

possibility that succinic dehydrogenase may be identical with cytochrome b.[44-46] However, Tsou[7] has presented strong evidence that these two components of the succinic dehydrogenase system are *not* identical. Finally, starting with the work of Fisher and co-workers, evidence has accumulated indicating the possibility that succinic dehydrogenase may be a flavin.[47-50] Keilin and Hartree[36] observed two flavin bands in their heart muscle preparations which disappeared on reduction with succinate. It has further been shown[51] that livers of rats grown on a riboflavin-deficient diet are low in succinic oxidase activity. Since the known components of the succinic oxidase system are not flavin compounds, there is the possibility that the low activity was due to the dehydrogenase. Against this possibility is the fact that the nature of Slater's "factor" is also not known. Further information must be obtained before a final judgment can be made.

[44] S. J. Bach, M. Dixon, and L. G. Zerfas, *Biochem. J.* **40,** 229 (1946).
[45] E. G. Ball, C. B. Anfinson, and O. Cooper, *J. Biol. Chem.* **168,** 257 (1947).
[46] A. M. Pappenheimer and E. D. Hendee, *J. Biol. Chem.* **180,** 597 (1949).
[47] F. G. Fisher and H. Eysenbach, *Ann.* **530,** 99 (1937).
[48] F. G. Fisher, A. Roedig, and K. Rauch, *Naturwissenschaften* **27,** 197 (1939).
[49] F. G. Fisher, A. Roedig, and K. Rauch, *Ann.* **552,** 203 (1942).
[50] E. G. Ball and O. Cooper, *J. Biol. Chem.* **180,** 113 (1949).
[51] A. E. Axelrod, V. R. Potter, and C. A. Elvehjem, *J. Biol. Chem.* **142,** 85 (1942).

[122] Fumarase

Fumarate + Water $\rightleftarrows$ l-Malate

By VINCENT MASSEY

Assay Method

Fumarase activity can be measured in many ways. Dakin,[1] Clutterbuck,[2] and other workers made use of the optical rotation of *l*-malic acid to follow the reaction. Laki and Laki[3] and Scott and Powell[4] followed the reaction by titration of fumaric acid with permanganate, according to the procedure of Straub.[5] By far the most satisfactory and sensitive method is that developed by Racker,[6] which is based on the high ultraviolet

[1] H. D. Dakin, *J. Biol. Chem.* **52,** 183 (1922).
[2] P. W. Clutterbuck, *Biochem. J.* **21,** 512 (1927); **22,** 1193 (1928).
[3] E. Laki and K. Laki, *Enzymologia* **9,** 139 (1941).
[4] E. M. Scott and R. Powell, *J. Am. Chem. Soc.* **70,** 1104 (1948).
[5] F. B. Straub, *Z. physiol. Chem.* **236,** 43 (1935).
[6] E. Racker, *Biochim. et Biophys. Acta* **4,** 20 (1950).

absorption of fumarate. Malate has an almost negligible ultraviolet absorption in comparison to fumarate, so the reaction may be conveniently followed spectrophotometrically in either direction. Racker[6] recommends using a wavelength of 240 mμ, but any wavelength from 210 up to 300 mμ may be used; in fact, higher wavelengths must be used in studying the hydration reaction at high fumarate concentrations owing to the high extinction coefficient of fumarate at lower wavelengths.

An alternative procedure used by the author for studying fumarase action under conditions where the spectrophotometric method is unsatisfactory depends on the fluorometric assay of malic acid.[7] Aliquots are withdrawn at various time intervals, and the malate produced or consumed is estimated by this very sensitive method. This method is not so convenient as the spectrophometric assay, since in the latter procedure the reaction may be followed continuously with time.

As the activity of fumarase is markedly dependent on the nature and concentration of anions present (see below), the assay procedure must be modified according to the requirements of the particular experiment. However, the following assay procedure has been found convenient for use in the isolation of the enzyme.[8]

Procedure. 0.017 M sodium fumarate, 0.033 M phosphate buffer, pH 7.3, and enzyme in a total volume of 3 ml. are mixed in a quartz cell having a 1-cm. light path. The decrease in optical density with time is followed at 300 mμ, readings being taken at frequent intervals. The reaction rate is at first linear and gradually falls off with time. The initial rate is calculated from the linear part of the curve.

Definition of Unit and Specific Activity. The unit of activity is defined as the amount of enzyme which causes an initial rate of change of optical density at 300 mμ of 0.01 per minute at 20° and pH 7.3. Protein was determined by optical density at 277 mμ in a quartz cell of 1-cm. light path. The specific activity is then defined by the activity per milliliter per D_{277}.

Application of Assay Method to Crude Tissue Preparations. The procedure outlined above is applicable also to crude extracts, provided that the fumarase content is reasonably high, as the wavelength used is higher than the absorption maxima of most of the protein constituents of such extracts.

Purification Procedure

The following procedure, which is very slightly modified from the method as originally published,[8] has given extremely reproducible results

[7] J. P. Hummel, *J. Biol. Chem.* **180**, 1225 (1949).

[8] V. Massey, *Biochem. J.* **51**, 490 (1952).

in the hands of the author in over thirty preparations. An alternative method of preparation, differing in several respects from the following, has recently been worked out by Frieden et al.[9] The enzyme prepared by this method has the same specific activity and properties as the enzyme isolated by the following procedure.

Step 1. Preparation of Crude Extract. About twelve pigs' hearts are freed of fat and connective tissue, minced, and washed (preferably accompanied by mechanical stirring) with large quantities of tap water until the washings are almost colorless. The washed mince is then squeezed through cheesecloth and homogenized with three times its weight of cold 0.01 M Na$_2$HPO$_4$ for 2 to 3 minutes. The homogenate is then centrifuged for 30 minutes at 1800 $\times$ g, and the residue rejected.

Step 2. Removal of Inert Protein by Lowering the pH. The supernatant from step 1 is adjusted to pH 5.2 with 1 M acetate buffer, pH 4.0, and the suspension again centrifuged; 10 minutes at 1800 $\times$ g is sufficient to produce a hard-packed precipitate and a clear, slightly reddish supernatant. This supernatant has almost the same activity as the supernatant of step 1, but the protein concentration is reduced to 10 to 14% of the amount in the first supernatant.

Step 3. Adsorption of Fumarase on Calcium Phosphate Gel. The fumarase is now adsorbed almost quantitatively on calcium phosphate gel. The amount of gel required depends on the condition of the gel and often on the individual enzyme preparation and should be determined by a pilot experiment. The amount needed is generally about 4 ml. of gel (30 mg. dry weight per milliliter) per 100 ml. of supernatant. The gel is then centrifuged (or allowed to settle—supernatant removed by suction and residue centrifuged). The fumarase is then eluted from the gel by several extractions with 0.1 M phosphate, pH 7.3, containing 50 g. of ammonium sulfate per liter (9% saturation). A 30- to 40-fold purification from the initial extract is thus obtained with little loss of enzyme, and the volume has been reduced about 10-fold.

Step 4. Fractionation with Ammonium Sulfate. The combined eluates from step 3 are next fractionated between 45 and 60% ammonium sulfate saturation by the addition of solid ammonium sulfate in two steps, each accompanied by high-speed centrifugation. The 45 to 60% fraction contains the bulk of the enzyme.

Step 5. Further Fractionation with Calcium Phosphate Gel.[9a] The 45 to 60% ammonium sulfate precipitate is dissolved in water and dialyzed with stirring against 4 l. of cold distilled water. Considerable purification

[9] C. Frieden, R. M. Bock, and R. A. Alberty, *J. Am. Chem. Soc.* **76,** 2482 (1954).

[9a] Crystallization may often be induced omitting step 5, especially if the preparation is seeded in step 7 with crystals from a previous isolation.

DETAILS OF A REPRESENTATIVE PREPARATION OF FUMARASE

Procedure	Activity, units/ml.	Protein concentration $(D_{277m\mu})^a$	Purity, units/ $D_{277m\mu}$	Volume, ml.	Total units	Times purified	Yield, %
Twelve pigs' hearts minced, washed, and homogenized with 0.01 M Na₂HPO₄, supernatant	9	19.6	0.46	4800	43,000		
Adjusted to pH 5.2, supernatant	8.5	1.35	6.3	4750	41,000	14	94
Combined eluates from adsorption on calcium phosphate	66	4.05	16.3	590	39,000	36	91
Fractionated between 45 and 60% saturation with (NH₄)₂SO₄, dialyzed against distilled water	750	11.7	64	42	31,500	140	73
Added 0.5 ml. calcium phosphate gel, supernatant	750	11.1	68	42	31,500	148	73
Added 1.0 ml. calcium phosphate gel, supernatant	750	9.45	79	42	31,500	174	73
Added 2.5 ml. calcium phosphate gel, supernatant	700	7.35	95	43	30,000	209	70
Added 5.0 ml. calcium phosphate gel, supernatant	680	5.55	120	45	30,000	265	70
Added 10.0 ml. calcium phosphate gel, supernatant	510	2.55	200	50	25,500	438	59
Crystallized, crystals dissolved in 0.01 M phosphate pH 7.3	1200	1.44	835	15.1	18,100	1830	42
Recrystallized	1800	1.86	965	9.5	17,100	2110	40
Recrystallized again	1560	1.62	965	10.5	16,400	2110	38

^{a} An optical density of 1.0 at 277 mμ is equivalent to a protein concentration of approximately 2.0 mg./ml.

is now achieved by the stepwise addition of small amounts of calcium phosphate gel to remove impurities. Details of a typical preparation are given in the accompanying table. If excessive amounts of fumarase are adsorbed by the addition of too much gel, the enzyme may be eluted from the fraction as in step 3.

Step 6. Refractionation with Ammonium Sulfate. This is the same as in step 4, except that often there is no precipitate produced on the addition of ammonium sulfate to 45% saturation.

Step 7. Crystallization and Recrystallization. The precipitate from 60% saturation with ammonium sulfate is dissolved in a small volume of 0.05 *M* phosphate buffer, pH 7.3, and saturated ammonium sulfate solution is carefully added until a very slight turbidity is produced. When left in the cold, fumarase gradually crystallizes in the form of rather solid rods. The crystals so produced are comparatively insoluble in water, and almost pure enzyme may be obtained by a single recrystallization. The crystals are centrifuged, washed twice with cold distilled water, and then dissolved in 0.05 *M* phosphate buffer. The fumarase is then precipitated by 60% saturated ammonium sulfate, and the precipitate dissolved in a small volume of 0.05 *M* phosphate buffer, pH 7.3, and recrystallized as in the first crystallization. Fumarase may be crystallized in several different forms, according to the conditions used. If the pH is raised to pH 8, the crystals produced are in the form of long, thin needles which are much more soluble in water than the more solid rods. Fumarase may also be obtained in the form of thin, flat plates without the addition of ammonium sulfate by lowering the pH toward the isoelectric point.[8] The pH of a concentrated solution of fumarase is gradually lowered by dialyzing against phosphate buffer of pH 6.3. The resulting flat crystals are only very slightly soluble in water and in phosphate solutions. The technique of gradually changing the composition of the crystallizing medium by dialysis can also be recommended in the crystallization with ammonium sulfate. This has been successfully employed by Frieden *et al.*[9]

Purity. The enzyme reaches a constant specific activity after two recrystallizations. The specific activity of the various crystalline forms is identical. Sedimentation and electrophoretic studies at various pH's show a high degree of homogeneity.[8–11] The molecular weight has been calculated[10] as 200,000. The rather scanty electrophoretic data so far available[8,9] indicate an isoelectric point in the neighborhood of pH 5.2. The turnover number of the purified enzyme is over 100,000 (molecules of fumarate converted to *l*-malate per minute per molecule of enzyme) at pH 7.3 and 20° in the presence of 0.033 *M* phosphate.[8]

[10] R. Cecil and A. G. Ogston, *Biochem. J.* **51**, 494 (1952).
[11] P. Johnson, V. Massey, and K. Shooter, unpublished.

Stability. The enzyme seems to be indefinitely stable in the crystalline state when kept at 4° in ammonium sulfate suspension. In solution the enzyme is more unstable, especially in dilute solution and at low pH's and low ionic strength. Dilute solutions are conveniently stabilized by the addition of alkaline phosphate buffer and by storing in the cold.

Properties

Specificity. The purified enzyme appears to be absolutely specific to fumarate and *l*-malate. A large number of structurally related compounds are not acted on by the enzyme.[12]

The Effect of pH and the Effect of Anions on Activity. The pH-activity curves are influenced to a very great degree by the nature and concentration of the anions present, including the substrates. Under a given set of conditions the pH optimum for fumarate is lower than for *l*-malate.[4,13,14] For both substrates phosphate[14-16] and a large number of other polyvalent anions[14] will cause extensive activation and produce a shift in the pH optimum toward alkaline values. Similarly, a number of monovalent anions such as Cl⁻, Br⁻, I⁻, and CNS⁻ will cause inhibition over part of the pH range with a shift of the pH optimum toward the alkaline.[14] These anions will also cause activation at higher pH values.[17] The effect on the pH-activity curves of these anions and by the substrates themselves has been shown to be due largely to the alteration of the ionization constants of groups at or close to the active center on the addition of these anions.[14,17]

The Effect of Temperature on Activity. For accurate activity estimations the reaction temperature must be carefully controlled, especially as the energies of activation with the two substrates are markedly different under certain conditions.[4,18] The effect of temperature on reaction rates is very complicated; it is found that in common with many other enzymes fumarase shows marked changes in activation energies under various conditions, especially in its reaction with fumarate as substrate.[18] Michaelis constants and inhibition constants are also markedly temperature-dependent.[12,18] It has been shown that many of these effects of temperature on catalytic properties are paralleled by changes in the physical properties of the enzyme.[19]

[12] V. Massey, *Biochem. J.* **55**, 172 (1953).

[13] C. N. Ionescu, N. Stanciu, and V. Radulescu, *Ber.* **72B**, 1949 (1939).

[14] V. Massey, *Biochem. J.* **53**, 67 (1953).

[15] P. J. G. Mann and B. Woolf, *Biochem. J.* **24**, 427 (1930).

[16] S. J. Davydova, *Biochemistry* (*U.S.S.R.*) **12**, 135 (1947).

[17] V. Massey and R. A. Alberty, *Biochim. et Biophys. Acta* **13**, 354 (1954).

[18] V. Massey, *Biochem. J.* **53**, 72 (1953).

[19] V. Massey, unpublished.

Effect of Phosphate Concentration on Michaelis Constants. As well as varying markedly with temperature and pH, Michaelis constants are dependent on phosphate concentration. It has been shown over a wide range of pH that the Michaelis constants for the two substrates are directly proportional to the phosphate concentration over the range investigated (0.005 to 0.133 M).[20] Inhibition constants for competitive inhibitors are also dependent on the phosphate concentration.[21]

The Equilibrium Constant between Fumarate and l-Malate. The equilibrium constant $\frac{[\text{Malate}]}{[\text{Fumarate}]} = K_{\text{equil.}}$ has been found to vary with temperature. The ΔH of the over-all reaction at neutral pH values has been fixed within the limits of 3300 to 4100 cal./mole by various workers using enzyme of various states of purity.[4,18,22-26] The equilibrium constant at 25° is between 4.0 and 4.5, except at pH values below 5.0, when the equilibrium constant rises owing to the ionization of *l*-malic acid in this pH region.[22,26]

A relation derived by Haldane[27] equating the equilibrium constant to the ratios of maximum initial velocities and Michaelis constants for the two substrates is found to be obeyed over a wide range of conditions.[20,22]

[20] R. A. Alberty, V. Massey, C. Frieden, and A. Fulbrigge, *J. Am. Chem. Soc.* **76**, 2485 (1954).

[21] R. A. Alberty, private communication.

[22] R. M. Bock and R. A. Alberty, *J. Am. Chem. Soc.* **75**, 1921 (1953).

[23] K. Jacobsohn, *Biochem. Z.* **224**, 167 (1934).

[24] H. A. Krebs, D. H. Smyth, and E. A. Evans, Jr., *Biochem. J.* **34**, 1041 (1940).

[25] P. Ohlmeyer, *Z. physiol. Chem.* **282**, 37 (1945).

[26] H. A. Krebs, *Biochem. J.* **54**, 78 (1953).

[27] J. B. S. Haldane, "The Enzymes," p. 81, Longmans, Green & Co., London, 1930.

[123] Malic Dehydrogenase from Pig Heart

l-Malate + DPN$^+$ $\rightleftarrows$ Oxalacetate + DPNH + H$^+$

By Severo Ochoa

Assay Method[1]

Principle. The equilibrium position of the reaction catalyzed by malic dehydrogenase markedly favors the reduction of oxalacetate to l-malate. The activity determination is, therefore, based on the measurement of the rate of oxidation of DPNH (decrease in optical density at 340 mμ) in the presence of the enzyme and excess oxalacetate. The early rate of DPNH oxidation is proportional to the concentration of the enzyme within certain limits. The measurement is carried out in the Beckman

[1] A. H. Mehler, A. Kornberg, S. Grisolia, and S. Ochoa, *J. Biol. Chem.* **174**, 961 (1948).

spectrophotometer at wavelength 340 mμ, using Corex or silica cells of 1.0-cm. light path.

Reagents

>0.25 M glycylglycine buffer, pH 7.4.[2]
>0.0015 M DPNH.[3]
>0.0076 M oxalacetate, pH 7.4. This solution should be freshly prepared, as the presence of pyruvate should be avoided.

Procedure. The reaction mixture contains 0.3 ml. of buffer, 0.1 ml. of DPNH (0.15 micromole), 0.1 ml. of oxalacetate (0.76 micromole), enzyme, and water (adjusted to a temperature of 22 to 23°) to a final volume of 3.0 ml. The reaction is carried out at room temperature and is started by the addition of either oxalacetate or enzyme. Readings of the optical density are made, against a blank containing all components except DPNH, at intervals of 15 seconds for 1 or 2 minutes. The decrease in optical density ($\Delta \log I_0/I$ at 340 mμ) between 30 and 45 seconds after the start of the reaction is used to calculate the enzyme activity. The amount of enzyme used in a test is adjusted so that the rate of decrease of optical density, for the period between 30 and 45 seconds, does not exceed -0.025.

Definition of Unit and Specific Activity. One unit of enzyme is defined as that amount which causes a decrease in optical density of 0.01 per minute under the above conditions. Specific activity is expressed as units per milligram of protein. Protein is determined as described for the preparation of "malic" enzyme (cf. Vol. I [124]).

Application of Assay Method to Crude Tissue Preparations. The method can be used with crude tissue extracts or dilute homogenates, provided that previous tests have shown that DPNH is not oxidized or destroyed. The presence of lactic dehydrogenase does not interfere if pyruvate is absent from the enzyme or oxalacetate solutions.

Purification Procedure[4]

The enzyme is extracted from acetone powder of washed minced pig heart. The purification involves fractionation with ammonium sulfate and with ethanol in the presence of ammonium sulfate.

[2] Phosphate, veronal, or tris(hydroxymethyl)aminomethane buffer has been used with identical results.

[3] Solutions of DPNH are kept at pH 8.5 to 9.0. For preparation of DPNH, see Vol. III [126 and 127].

[4] According to F. B. Straub, *Z. physiol. Chem.* **275,** 63 (1942). Different batches of malic dehydrogenase have been prepared at various times in this laboratory by Drs. A. H. Mehler, I. Harary, and M. Schwartz. The preparation described was carried out by Dr. I. Harary.

Preparation of Acetone Powder. Mincing and washing of the heart and the preparation of the acetone powder are carried out as described for the TPN-specific isocitric dehydrogenase system (cf. Vol. I [116]). Six pig hearts yield on the average 1.0 kg. of mince and 165 g. of acetone powder.

Step 1. Extraction. The extraction is also carried out as for the isocitric dehydrogenase system (cf. Vol. I [115]). 165 g. of powder yields about 1600 ml. of extract, containing 7 to 8 mg. of protein per milliliter; specific activity, 526.

Step 2. Treatment with Calcium Phosphate. 1500 ml. of the phosphate extract is cooled to 0°, and 226 ml. of ice-cold 15% $CaCl_2$ is added with mechanical stirring. The stirring is continued for 15 minutes; the heavy precipitate of calcium phosphate is centrifuged off at 0° and discarded. The rather strongly acid supernatant solution is immediately neutralized (to pH about 7.0) with 10% trisodium phosphate ($Na_3PO_4 \cdot 12H_2O$); about 470 ml. is required. This precipitates most of the remaining calcium as calcium phosphate. The precipitate is centrifuged off as above and discarded.

Step 3. Ammonium Sulfate Fractionation. The clear pink supernatant solution from the above step (1676 ml.) is cooled to 0° and brought to 50% saturation with ammonium sulfate by the slow addition of 590 g. of powdered ammonium sulfate with mechanical stirring. Stirring is continued for 30 to 45 minutes, and the mixture is filtered in the cold through fluted filter paper. The precipitate is discarded. The filtrate (2020 ml.) is brought to 65% saturation at 0° by addition of 214 g. of ammonium sulfate as above. The precipitate is centrifuged off in the cold at 13,000 r.p.m. in the Servall angle centrifuge, and the supernatant fluid is discarded. The precipitate is dissolved in cold distilled water to give 40 ml. of solution containing 12 mg. of protein per milliliter; specific activity, 3750.

Step 4. Ethanol Fractionation in the Presence of Ammonium Sulfate. To 40 ml. of the ice-cold enzyme solution from the previous step, which contains about 8% ammonium sulfate (specific gravity, 1.046), is added 6.89 g. of solid ammonium sulfate to make its total concentration approximately 20%. Volume, 45 ml.; specific gravity, 1.118.[5] To the cold (0°) solution is added slowly, with mechanical stirring, 45 ml. (1 vol.) of ethanol cooled to −15°. After being stirred for 20 to 30 minutes the mixture is allowed to stand at 0° overnight; the precipitate is then centrifuged off at 0° and discarded. To the yellow supernatant fluid (82 ml.) is added 41 ml. of ethanol (0.5 vol.) as above. The precipitate is centrifuged off at 0°, dissolved in cold distilled water (18 ml.), and dialyzed for 3 to 4

---

[5] An alternate method to adjust the salt concentration is to dialyze the enzyme solution against 20% ammonium sulfate in the cold.

hours at 0° against 3 to 4 l. of 0.02 M potassium phosphate buffer, pH 7.4. The dialyzed solution (20 ml.) contains 1.45 mg. of protein per milliliter; specific activity, 50,000.

Step 5. Refractionation with Ammonium Sulfate. The solution from the above step is refractionated with ammonium sulfate at 0° between 50 and 65% saturation as in step 3. The precipitate between 0 and 50% ammonium sulfate saturation is centrifuged off at 0° and discarded. The 50 to 65 fraction containing the enzyme (specific activity, 65,000 to 70,000) is harvested by centrifugation at 0°. The white precipitate dissolves in water to give a clear, colorless solution which is free of lactic dehydrogenase, fumarase, and "malic" enzyme activities. The enzyme is kept as a suspension in 70% saturated ammonium sulfate in the refrigerator. Under these conditions the activity of the enzyme remains unchanged for many months.

A summary of the purification procedure is given in the table. The activity of the supernatant fluid from the calcium phosphate step (step 2) was not assayed in the preparation described. In our experience this step gives about 1.5-fold purification with a recovery of 70 to 80% of the enzyme units.

SUMMARY OF PURIFICATION PROCEDURE[a]

Step	Volume of solution, ml.	Units	Protein, mg.	Specific activity	Yield, %
1. Extract	1620	6,500,000	12,460	526	100
2. Calcium phosphate supernatant fluid	1700				
3. AmSO₄ fractionation (50–65)	40	1,800,000	480	3750	28
4. Ethanol-AmSO₄ fractionation	20	1,452,000	29	50,000	22
5. Refractionation with AmSO₄ (50–65)	3.4	1,300,000	19.5	66,000	20

[a] 165 g. of pig heart acetone powder.

Properties

Only 0.05 to 0.07 γ of the purified enzyme is required for a standard optical assay. Under these conditions the enzyme catalyzes the oxidation of about 35,000 moles of DPNH per minute per 100,000 g. of protein. The enzyme is specific for L-malic acid and is completely inactive toward D-malic acid.[4] Under the conditions of the optical assay the purified enzyme catalyzes the oxidation of TPNH at about 3% of the rate of that of DPNH.[1,6] In the opposite direction,[6] i.e., reduction of the oxidized pyridine nucleotide by L-malate (0.22 M L-malate, 1.2 $\times$ 10^{-4} M DPN$^+$ or TPN$^+$, pH 9.2), the initial rate of reduction of TPN$^+$ is only about 0.0016% of that of DPN$^+$. It remains open whether the slow reaction with TPN is an intrinsic property of malic dehydrogenase or is due to contamination of the preparation of DPN-specific malic dehydrogenase with small amounts of a TPN-specific enzyme.

Fumarate (but not malonate) inhibits malic dehydrogenase.[4] Under the conditions of Straub's assay (O$_2$ uptake with malic dehydrogenase, diaphorase, malate, DPN$^+$, methylene blue, and cyanide to bind the oxalacetate) the degree of inhibition is 30% when fumarate and L-malate are present in equal concentrations.

The apparent equilibrium constant of the reaction catalyzed by malic dehydrogenase (K' = [oxalacetate][DPNH]/[L-malate][DPN$^+$]), which is strongly pH-dependent, has an average value[7] of 2.33 $\times$ 10^{-5} at pH 7.2 and 22° (cf. also ref. 6).

[6] K. Burton and T. H. Wilson, *Biochem. J.* **54**, 86 (1953).
[7] J. R. Stern, S. Ochoa, and F. Lynen, *J. Biol. Chem.* **198**, 313 (1952).

[124] "Malic"·Enzyme

By SEVERO OCHOA

A. "Malic" Enzyme from Pigeon Liver and Wheat Germ[1-4]

$$\text{L-Malate}^{--} + \text{TPN}^+ \underset{\longleftarrow}{\overset{(\text{Mn}^{++})}{\longrightarrow}} \text{Pyruvate}^- + CO_2 + \text{TPNH} \qquad (1)$$

$$\text{Oxalacetate} \xrightarrow{(\text{Mn}^{++})} \text{Pyruvate} + CO_2 \qquad (2)$$

Assay Method

a. "Malic" Enzyme Activity. *Principle.* Activity determinations are based on reaction 1. The early rate of reduction of TPN$^+$, in the presence of enzyme, Mn^{++}, and an excess of L-malate is, within certain limits, proportional to the enzyme concentration. The formation of TPNH is followed in the Beckman spectrophotometer at wavelength 340 mμ.

[1] Ochoa, A. H. Mehler, and A. Kornberg, *J. Biol. Chem.* **174**, 979 (1948).
[2] J. B. Veiga Salles and S. Ochoa, *J. Biol. Chem.* **187**, 849 (1950).
[3] I. Harary, S. R. Korey, and S. Ochoa, *J. Biol. Chem.* **203**, 595 (1953).
[4] S. Ochoa, *in* "The Enzymes" (Sumner and Myrbäck, eds.), Vol. II, Part 2, p. 929. Academic Press, New York, 1952.

Reagents

0.25 M glycylglycine buffer, pH 7.4.[5]
0.05 M $MnCl_2$.
0.000675 M TPN^+.
0.030 M L-malate, pH 7.4.
The buffer is prepared by dissolving 1.65 g. of glycylglycine in 30 ml. of water. This solution (pH 5.3) is adjusted to pH 7.4 by the dropwise addition of 2.0 N NaOH (about 1.15 ml.), and the volume is made up with water to 50 ml. L-Malic acid is dissolved and neutralized with KOH to pH 7.4.

Procedure. The reaction mixture, in Corex or silica cells (d = 1.0 cm.), consists of 0.3 ml. of glycylglycine buffer (75 micromoles), 0.06 ml. of $MnCl_2$ (3.0 micromoles), 0.2 ml. of TPN^+ (0.135 micromole), 0.05 ml. of L-malate (1.5 micromoles), enzyme, and water to a final volume of 3.0 ml. The assay is carried out at room temperature (23 to 25°). The reaction is started by the addition of either malate or enzyme, and readings of the optical density are made, against a blank containing all components except TPN, at intervals of 15 seconds for 1 or 2 minutes. The increase in optical density ($\Delta \log I_0/I$) between 30 and 45 seconds after the start of the reaction is used to calculate the enzyme activity. With crude tissue extracts care must be taken that there is no interference by other systems reducing TPN^+ or oxidizing TPNH. Enzymes destroying pyridine nucleotides will also interfere. This is particularly the case with extracts of plant tissues.[6] The destruction can be prevented by addition of adenosine-5′-phosphate.[6]

Definition of Unit and Specific Activity. One enzyme unit is defined as the amount of enzyme causing an increase in optical density of 0.01 per minute under the above conditions. The amount of enzyme used in a test is adjusted so that the density change, for the period between 30 and 45 seconds, lies between +0.006 and +0.015. Specific activity is expressed as units per milligram of protein. Protein is determined spectrophotometrically by measuring the absorption of light at wavelengths 280 and 260 mμ with a correction for the nucleic acid content from the data given by Warburg and Christian.[7] The determination is carried out in silica cells (d = 0.5 cm.). One aliquot of the protein solution is suitably

---

[5] Veronal or tris(hydroxymethyl)aminomethane buffers have been used with equal results. Phosphate buffer is to be avoided because of possible precipitation of Mn^{++}.

[6] E. E. Conn, B. Vennesland, and L. M. Kraemer, *Arch. Biochem.* **23**, 179 (1949).

[7] O. Warburg and W. Christian, *Biochem. Z.* **310**, 384 (1941); see Vol. III [73] for details of the optical method for protein determination.

diluted with water or buffer[8] to a volume of 1.5 to 2.0 ml., and the optical density is read at 280 and 260 mμ against a blank cell containing the solvent. The product of the optical density reading at 280 mμ by the factor corresponding to the observed ratio of optical densities at 280 and 260 mμ (E_{280}/E_{260}) gives the protein concentration of the solution in the experimental cell.[7] This method is used only when E_{280}/E_{260} is above 0.7, i.e., when the concentration of nucleic acid is less than 10%; otherwise the method of Lowry et al.[9] is used.

b. Oxalacetic Carboxylase Activity. *Principle.* Between pH 4.0 and 6.0 all preparations of "malic" enzyme thus far available catalyze the decarboxylation of oxalacetate in the presence of Mn^{++} (reaction 2). Within this pH range there is practically no catalysis of reaction 1. Although tne spontaneous decarboxylation is kinetically of first order, the enzymatic decarboxylation follows zero-order kinetics as long as the enzyme is saturated with substrate.[2] The assay is based on the rate of CO$_2$ evolution in the presence of Mn^{++} and an excess of oxalacetate. This is proportional to the enzyme concentration within certain limits.

Reagents

1.0 M acetate buffer, pH 4.5 or 5.2.
0.01 M MnCl$_2$.
0.000675 M TPN$^+$.
0.08 M oxalacetate. Oxalacetic acid dissolved before use and adjusted to required pH with KOH.

Procedure. The test is carried out in small Warburg vessels (capacity about 6 ml.). For the assay of the pigeon liver enzyme the reaction mixture contains 0.1 ml. of acetate buffer (100 micromoles), pH 4.5, 0.1 ml. of MnCl$_2$ (1.0 micromole), 0.05 ml. of TPN$^+$ (0.034 micromole), 0.25 ml. of oxalacetate (20 micromoles), enzyme, and water to a final volume of 1.0 ml. For the assay of the wheat germ enzyme, acetate buffer, pH 5.2, is used and TPN$^+$ is omitted. Although TPN$^+$ (but not DPN$^+$) stimulates the decarboxylation of oxalacetate by the pigeon liver enzyme,[1,10] it inhibits the reaction with the wheat germ enzyme.[11] The mechanism of these effects of TPN$^+$ is not understood. The gas is air, and the temperature is 25°. The reaction is started, after temperature equilibration, by tipping the enzyme from the side bulb. Readings are taken at 5-minute

[8] Care must be taken that no turbidity develops.
[9] O. H. Lowry, N. J. Rosebrough, A. L. Farr, and R. J. Randall, *J. Biol. Chem.* **193**, 265 (1951).
[10] B. Vennesland, E. A. Evans, Jr., and K. I. Altman, *J. Biol. Chem.* **171**, 675 (1947).
[11] L. M. Kraemer, E. E. Conn, and B. Vennesland, *J. Biol. Chem.* **188**, 583 (1951).

intervals. A blank without enzyme or TPN is run simultaneously to correct for the spontaneous decarboxylation of oxalacetate.

Definition of Unit and Specific Activity. One enzyme unit is defined as the amount causing an excess CO_2 evolution over the blank of 1 μl. in 10 minutes, calculated for the second 5 minutes after tipping of the enzyme. The amount of enzyme used in a test is adjusted so that the rate of CO_2 evolution lies between 10 and 60 μl. per 10 minutes. Specific activity is expressed as units per milligram of protein.

Purification Procedure

a. **Pigeon Liver Enzyme.** The procedure is essentially as described by Ochoa *et al.*,[1] with some modifications.[4,12] The starting material is acetone powder of pigeon liver. Preparation of the powder and extraction are carried out essentially according to Evans *et al.*[13] The enzyme is purified by fractionation with ethanol at low temperature and fractional absorption on alumina gel.

Preparation of Acetone Powder. Pigeon livers, removed as rapidly as possible after decapitation, are cooled in ice and freed from fat and connective tissue. The livers are extracted with acetone, cooled to $-5°$, in a chilled Waring blendor; the suspension is then poured into sufficient acetone (at $-5°$) to make a total of 10 vol. of acetone. The mixture is stirred for a few minutes and filtered with suction in the cold room. The residue is re-extracted once more with acetone, spread on filter paper, and allowed to dry at room temperature. The dry material is freed from the larger threads of connective tissue, ground to a powder in a mechanical mortar, and stored in stoppered bottles in the ice box. Its activity keeps unchanged for several months.

Step 1. Extraction. 100 g. of acetone powder is extracted for 10 minutes with 1 l. of distilled water at 40° with mechanical stirring; the insoluble residue is centrifuged off at room temperature and discarded. This yields about 800 ml. of reddish-brown turbid extract, the protein concentration of which is about 25 to 30 mg./ml. Its specific activity is 4 to 6.

Step 2. First Ethanol Fractionation. To 800 ml. of extract is added 200 ml. of 0.2 M phosphate buffer, pH 7.4, after both have been cooled to 0°. Absolute alcohol (cooled to $-30°$) is now added to the solution with mechanical stirring. The alcohol is added very slowly, and the temperature is not allowed to rise above 0°. As the alcohol concentration rises, the temperature[14] is allowed to fall eventually to $-5°$ and is main-

[12] J. B. Veiga Salles and I. Harary, unpublished observations.

[13] E. A. Evans, Jr., B. Vennesland, and L. Slotin, *J. Biol. Chem.* **147,** 771 (1943).

[14] The desired temperatures are maintained with alcohol-water mixtures which are kept as semifrozen slushes by means of a mechanically refrigerated bath.

tained until enough alcohol has been added to make a concentration of 30 to 32% by volume.

The precipitate is centrifuged off at $-5°$ in the refrigerated centrifuge and discarded. The supernatant is allowed to stand overnight at $-15°$, whereby a further precipitate is obtained. This precipitate is centrifuged off at $-10°$ and dissolved in ice-cold 0.02 M phosphate buffer, pH 7.4. It gives about 125 ml. of a clear red solution containing about 50 mg. of protein per milliliter; specific activity, about 20. Solutions from several 100-g. batches of acetone powder, each obtained as described above can be combined at this stage. However, better results are obtained when the entire procedure is carried out on the solution from one 100-g. batch of acetone powder.

Step 3. Refractionation with Ethanol. The enzyme solution from step 2 is fractionally precipitated with ethanol at low temperature. Three fractions are obtained between the limits of 10 and 17% ethanol by volume (10 to 13, 13 to 15, and 15 to 17), at a temperature of -5 to $-6°$. These fractions, which contain most of the activity, are dissolved in ice-cold 0.02 M phosphate buffer, pH 7.4, and combined to give about 75 ml. of solution containing about 60 mg. of protein per milliliter; specific activity, 35 to 50.

Step 4. Fractional Adsorption on Alumina Gel. Solution 3 (76 ml.) is cooled to 0° and diluted to about 370 ml. with ice-cold distilled water. This solution is brought to pH about 5.0 by the slow addition of ice-cold 2.0 M acetate buffer, pH 4.7 (35 to 40 ml.), with brisk mechanical stirring. A small precipitate which occasionally forms at this stage contains some enzyme and is not removed. The solution (about 405 ml.) contains between 9 and 10 mg. of protein per milliliter. To this solution is added 27 to 30 ml. of alumina gel C_γ containing 10 to 12 mg. of Al_2O_3 per milliliter.[15] The mixture is stirred mechanically for 10 minutes at 0° and centrifuged in the cold. The supernatant is treated as above five more times, so that six alumina precipitates are obtained. The last supernatant contains little or no enzyme and is discarded. Each alumina precipitate is separately eluted with about 20 ml. of ice-cold 0.1 M phosphate buffer, pH 7.4. A second elution with 10 ml. of buffer yields some more enzyme of lower specific activity. The first eluate of the fifth alumina precipitate yields generally the largest amount of enzyme as well as that of highest specific activity. In the run here described this eluate (21 ml.) was a colorless solution containing 2.9 mg. of protein per milliliter; specific activity, 700.

[15] The alumina gel is prepared as described by E. Bauer, *in* "Die Methoden der Fermentforschung" (Bamann and Myrbäck, eds.), p. 1449. Thieme, Leipzig, 1941; see also Vol. I [11].

The purification procedure is summarized in the table. The yield of purified enzyme is about 30% from the initial extract. The increase in units in steps 2 and 3 is only apparent and is mainly caused by removal of large amounts of lactic dehydrogenase which interferes with the enzyme assay. Although it is not possible to estimate the degree of purity attained, some of the enzymes present in large amounts in the pigeon liver extracts, such as malic, isocitric, and glutamic dehydrogenases, are almost completely removed by the purification procedure described. Removal of lactic dehydrogenase, on the other hand, is incomplete.

SUMMARY OF PURIFICATION PROCEDURE OF PIGEON LIVER ENZYME[a]

Step	Volume of solution, ml.	Units	Protein, mg.	Specific activity
1. Aqueous extract	760	140,000	22,800	6
2. First ethanol fractionation	127	155,000	6,350	24
3. Ethanol refractionation	76	167,000	4,550	37
4. Elution from alumina gel	21	42,000	60	700

[a] 100 g. of acetone powder.

The enzyme is relatively stable, especially in the presence of phosphate. Solutions in 0.1 M phosphate buffer, pH 7.4, lose activity only very gradually if kept at 0°.

The specific oxalacetic carboxylase activity is about 7 at step 1 and 750 to 800 at step 4. Thus the ratio of "malic" enzyme (reaction 1) to oxalacetic carboxylase (reaction 2) activity remains constant throughout purification.[16]

b. Wheat Germ Enzyme. The enzyme was first found by Kraemer et al.[11] The procedure described below is that of Harary et al.[3]

Preparation of Acetone Powder. 250 g. of raw wheat germ is suspended in 2 to 3 vol. of acetone, cooled to −8°, and blended for 1 minute in a chilled Waring blendor. The suspension is then poured into sufficient acetone at −8° to make a total of 10 vol. of acetone. The mixture is stirred and then filtered by suction in the cold room. The residue is blended in the same manner and spread on filter paper to dry at room temperature.

Step 1. Extraction. 200 g. of acetone powder is extracted for 30 minutes with 2 l. of demineralized water at 25° with mechanical stirring. The residue is removed by centrifugation at room temperature and discarded.

[16] For a discussion of this fact in connection with the nature of the "malic" enzyme see ref. 4, especially p. 979.

1500 to 1600 ml. of a yellow turbid supernatant fluid is obtained. Protein concentration, 16 to 20 mg./ml.; specific activity, 1.0 to 1.4.

Step 2. Adsorption and Elution from Calcium Phosphate Gel. After being cooled to 0°, the supernatant fluid is brought to pH 5.1 with 1.0 M acetic acid and allowed to stand for 3 to 4 hours at 0°. The suspension is centrifuged in the cold, the precipitate discarded, and 600 ml. of fresh calcium phosphate gel added to the supernatant fluid with continuous stirring. After further stirring for 1 hour, the mixture is allowed to settle overnight in the cold. Much of the supernatant fluid can then be siphoned, and the remainder centrifuged at 0°. The supernatant fluid is tested for complete adsorption of enzyme and discarded. The gel is washed with 600 ml. of ice-cold demineralized water and then eluted with 600 ml. of 0.1 M phosphate buffer, pH 7.4; 630 ml. of a clear yellow solution containing 5 mg. of protein per milliliter (specific activity, 8.0) is obtained.

Step 3. Ammonium Sulfate Fractionation. The eluate is brought to pH 6.3 with 1.0 M acetic acid, and 185 g. of ammonium sulfate is added (42% saturation). After centrifugation, 62 g. of ammonium sulfate is added to the supernatant fluid to bring the concentration to 56% saturation. The precipitate obtained between 42 and 56% saturation is dissolved in a small volume of 0.01 M phosphate buffer, pH 7.4, and dialyzed overnight at 1 to 3° against the same buffer; 23 ml. of a solution with 16 mg. of protein per milliliter (specific activity, 50) is obtained.

Step 4. Ethanol Fractionation. After dialysis, the protein concentration is adjusted to 10 mg./ml. with 0.01 M phosphate buffer, pH 7.4, at 0°, and the pH is adjusted to 5.1 with 1.0 M acetic acid. To this solution is added absolute ethanol (cooled to $-30°$) with mechanical stirring. The alcohol concentration is brought to 12% by volume, and the temperature to $-3.5°$. The mixture is then centrifuged at $-3.5°$, and the precipitate dissolved in 0.02 M phosphate buffer, pH 7.4. The clear, slightly yellowish solution (11 ml.) contains 9.6 mg. of protein per milliliter; specific activity, 100.

Step 5. Removal of Inactive Protein with Calcium Phosphate Gel.[17] The yield of enzyme from step 4 amounts to 10,500 units. This amount dropped to 7700 units after the solution remained frozen at $-15°$ overnight. The specific activity dropped from 100 to 74. To 11 ml. of the solution (containing 700 units/ml.) 2 ml. of fresh calcium phosphate gel is added. The gel is centrifuged and discarded. The supernatant fluid contains 600 units/ml.; specific activity, 98. After addition of a further 2 ml. of the gel, the supernatant fluid has 520 units/ml.; specific activity, 135. To this is added 1.8 ml. of gel, resulting after centrifugation in

[17] Calcium phosphate gel prepared by the method of D. Keilin and E. F. Hartree, *Proc. Roy. Soc. (London)* **B124,** 397 (1938); see Vol. I [11].

14 ml. of a colorless solution containing 2.1 mg. of protein per milliliter; specific activity, 230.

Step 6. Second Ammonium Sulfate Fractionation. To the above solution is added 0.3 ml. of 0.05 M $MnCl_2$, 0.6 ml. of 0.5 M malate, and 2 ml. of 0.1 M cysteine. The pH is adjusted to 6.5. The solution is fractionated with ammonium sulfate between 48 and 70% saturation. The precipitate is dissolved in 0.02 M phosphate buffer, pH 7.4, and 8.2 ml. of a solution, containing 1.41 mg. of protein per milliliter, is obtained; specific activity, 332. The purification is summarized in the table.

The preparation should be carried out with minimal delay between the successive steps. The enzyme is unstable, and interruption of its purification results in loss of activity. Step 6 was found to be the least reliable in reproducibility. The instability of the partially purified enzyme made it impractical to attempt further purification. The purified enzyme is free of lactic dehydrogenase.

SUMMARY OF PURIFICATION OF WHEAT GERM ENZYME[a]

Step	Volume, ml.	Units	Protein, mg.	Specific activity, units/mg. protein	Yield, %
1. Aqueous extract	1530	30,000	25,500	1.2	100
2. Elution from calcium phosphate gel	630	25,200	3,160	8.0	83
3. First ammonium sulfate fractionation	23	18,400	368	50	60
4. Ethanol fractionation	11	10,500	104	100	34
5. Calcium phosphate gel supernatant	14	6,700	29.4	230	22
6. Second ammonium sulfate fractionation	8.4	4,200	17.9	332	14

[a] 200 g. of acetone powder.

The ratio of "malic" enzyme and oxalacetic carboxylase activities of the purified wheat germ enzyme is the same as that of the pigeon liver enzyme.

Properties

a. Pigeon Liver Enzyme. The two reactions catalyzed by the enzyme have widely separated, sharp pH optima. The optimum for reaction 1 is at pH 7.4; that for reaction 2 is at pH 4.5. Each activity falls off abruptly on either side of the optimum. At the respective pH optima the rate of evolution of CO_2 from malate and oxaloacetate catalyzed by a given amount of enzyme is the same.

The purified enzyme is strictly specific for TPN and for L-malate. Neither fumarate nor D-malate can be substituted for L-malate, nor DPN for TPN. Phospho(enol)pyruvate cannot replace pyruvate for reversal of reaction 1. The carboxylase activity is specific for oxalacetate; this keto acid cannot be replaced by acetoacetate or oxalosuccinate. Mn^{++} is an absolute requirement for both reactions 1 and 2. Mg^{++} is much less effective and can replace Mn^{++} only at much higher concentrations.

The dissociation constant of the protein-Mn complex is the same for reaction 1 (pH 7.4, 23°) and reaction 2 (pH 4.5, 25°), namely, 5×10^{-5} g.-atom of Mn per liter. The dissociation constant of the protein-L-malate complex (reaction 1) under the conditions of the assay is also 5×10^{-5} (mole of malate per liter); that of the protein-oxalacetate complex in reaction 2, under the assay conditions, is about 1×10^{-3} mole of oxalacetate per liter.

Inhibitors of Oxalacetate Decarboxylation. The catalysis of oxalacetate decarboxylation by the pigeon liver enzyme is inhibited by malate and, even more effectively, by malonate, whereas succinate and fumarate are only weakly inhibitory; D- and L-malate inhibit about equally. Although inhibition by the above compounds is probably of a structural nature, there is no evidence that it is competitive, since it is not reversed by increasing the oxalacetate concentration. Neither D-malate nor malonate inhibits reaction 1.

b. Wheat Germ Enzyme. The properties of this enzyme have been less extensively studied than those of the pigeon liver enzyme. The pH optima for catalysis of reactions 1 and 2 are also very sharp; the values are pH 7.3 and 5.2, respectively. Under the conditions of the assay, the dissociation constant of the protein-Mn complex for reaction 1 is 2.5×10^{-5}, and for reaction 2 it is 1×10^{-4} g.-atom/l.; the dissociation constant of the L-malate-protein complex (reaction 1) is 7×10^{-4}, and that of the protein-oxalacetate complex (reaction 2) is 6.5×10^{-3}.

The absence of interfering lactic dehydrogenase has made it possible to determine the equilibrium constant of reaction 1 using the wheat germ enzyme.[3] The average value of K ($[L\text{-malate}^{--}][TPN^+]/[pyruvate^-][CO_2]$ $[TPNH]$) is 19.6 (liters $\times$ mole^{-1}) at pH 7.4 and 22 to 25°.

Distribution of "Malic" Enzyme

The enzyme is widely distributed in animal tissues and in the tissues of higher plants.[4] Among the former, avian liver (pigeon, chicken, turkey) is the richest source, but the enzyme has also been found in brain,[18] heart muscle,[18] retina,[19] and diaphragm.[19] In plant tissue the enzyme has been

---

[18] J. R. Stern, unpublished observations.
[19] R. K. Crane and E. G. Ball. *J. Biol. Chem.* **188**, 819; **189**, 269 (1951).

found,[6] besides wheat germ, in parsley root, parsnips, beets, carrots, spinach, and peas. Among green leaves the enzyme has been found, besides spinach,[6,20] in the leaves of sugar beet,[20,21] sunflower,[21] and parsley.[20]

B. "Malic" Enzyme from *Lactobacillus arabinosus*[4,22,23]

$$\text{L-Malate} \xrightleftharpoons{(Mn^{++},DPN)} \text{L-Lactate} + CO_2 \qquad (1)$$

$$\text{Oxalacetate} \xrightarrow{(Mn^{++})} \text{Pyruvate} + CO_2 \qquad (2)$$

Washed resting cells of *L. arabinosus* (strain 17-5) evolve little CO_2 from L-malate anaerobically when harvested from media containing salts, vitamins, amino acids, and glucose, but they produce considerable amounts when harvested from media containing malic acid in addition to the above components. L-Malic acid is almost quantitatively converted to lactic acid and CO_2. These results are due to the formation of an adaptive enzyme.[24] Evidence has been obtained[22] that reaction 1 is the net result of reactions 3 and 4 below:

$$\text{L-Malate}^{--} + DPN^+ \xrightleftharpoons{(Mn^{++})} \text{Pyruvate}^- + CO_2 + DPNH \qquad (3)$$

$$\text{Pyruvate} + DPNH + H^+ \rightleftarrows \text{L-Lactate} + DPN^+ \qquad (4)$$

Thus, a DPN-specific "malic" enzyme would appear to be formed on adaptation of *L. arabinosus* to malic acid. L-Lactic dehydrogenase, which catalyzes reaction 4, is a constitutive enzyme in *L. arabinosus* and is present in the purest preparations of the bacterial enzyme so far obtained. It should be pointed out that adaptation to malate develops a capacity for catalysis of both reactions 1 and 2.

Assay Method

a. "Malic" Enzyme Activity. *Principle.* Activity determinations are based on reaction 1. The rate of liberation of CO_2 from L-malate, in the presence of DPN^+ and Mn^{++}, is proportional to the enzyme concentration within certain limits. The reaction is kinetically of zero order as long as the enzyme is saturated with substrate.

[20] K. A. Clendenning, E. R. Waygood, and P. Weinberger, *Can. J. Botany* **30**, 395 (1952).

[21] D. I. Arnon, *Nature* **167**, 1008 (1951).

[22] S. Korkes, A. del Campillo, and S. Ochoa, *J. Biol. Chem.* **187**, 891 (1950).

[23] S. Kaufman, S. Korkes, and A. del Campillo, *J. Biol. Chem.* **192**, 301 (1951).

[24] M. L. Blanchard, S. Korkes, A. del Campillo, and S. Ochoa, *J. Biol. Chem.* **187**, 875 (1950).

Reagents

0.8 M potassium phosphate or succinate buffer, pH 6.0.
0.02 M $MnCl_2$.
0.0012 M DPN^+.
0.6 M potassium L-malate, pH 6.0.

Procedure. The reaction mixture consists of 0.1 ml. of buffer (80 micromoles), 0.1 ml. of $MnCl_2$ (2.0 micromoles), 0.1 ml. of DPN^+ (0.12 micromole), 0.1 ml. of L-malate (60 micromoles), enzyme, and water in a final volume of 1.0 ml. Gas, air; temperature, 25°. The reaction is carried out in small Warburg vessels of about 6-ml. capacity and is started after temperature equilibration by tipping the enzyme from the side bulb. Readings are taken at 5-minute intervals. The volume of CO_2 produced is calculated by correcting the vessel constant for the small CO_2 retention at pH 6.0.

Definition of Unit and Specific Activity. One unit is defined as the amount of enzyme catalyzing the liberation of 1 μl. of CO_2 in 10 minutes calculated for the second 5 minutes after the tipping of the enzyme. The amount of enzyme used in a test is adjusted so that the rate of CO_2 evolution lies between 20 and 120 μl. per 10 minutes. Specific activity is expressed as units per milligram of protein.

b. Oxalacetic Carboxylase Activity. This assay (based on reaction 2) is as described for the other preparations of "malic" enzyme. The pH of the acetate buffer is 4.5, and the amount of $MnCl_2$ is 2.0 micromoles. No pyridine nucleotide is present in the reaction mixture. The definition of units and specific activity is the same.

Purification Procedure

The method here described[23] is a modification of that of Korkes *et al.*[22]

L. arabinosus (strain 17-5) is the starting material.[24] The organism is carried on agar stabs containing 1% glucose, 1% yeast extract, and 1.5% agar. Subcultures are made from the stabs into tubes containing 5 ml. of a mixture of 1% yeast extract, 1% glucose, and 1% DL-malate, adjusted to pH 6.8. The tubes are incubated at 30° for 18 hours; then 1 ml. of the suspension of organisms so obtained is pipetted into 1 l. of final growth medium, containing 20 g. of DL-malic acid, 20 g. of glucose, 10 g. of yeast extract, 10 g. of nutrient broth, 10 g. of sodium acetate ($3H_2O$), 1 g. of K_2HPO_4, 5 ml. of the salt mixture B of Wright and Skeggs,[25] and 5 γ of biotin, the mixture having been previously autoclaved for 10 minutes

[25] L. O. Wright and H. R. Skeggs, *Proc. Soc. Exptl. Biol. Med.* **56**, 95 (1944).

at 15 pounds. After 24 hours, the organisms are harvested by centrifugation and suspended in distilled iced water to make a thin paste.

Preparation of Acetone Powder. The chilled bacterial paste is poured all at once into 10 vol. of acetone chilled to $-10°$ in a Waring blendor; the mixture is homogenized for 2 minutes and then filtered by suction in the cold room. The cake is washed twice with small portions of cold acetone, and the suction is continued until the cake is almost dry. The residue is then crumbled by hand, spread on filter paper, and allowed to dry in the air at room temperature. The dry material is ground to a fine powder in a mechanical mortar and stored in stoppered bottles in the ice box. The yield of dry powder is about 3 g./l. of growth medium. In this state, the powder is stable for 1 year. The following steps are carried out in the cold.

Step 1. Extraction. 9 g. of powder is extracted with 450 ml. of 0.02 M phosphate buffer, pH 7, by grinding the mixture in a mechanical mortar for 3 hours at 3 to 4°. The insoluble residue is centrifuged off at 0° and discarded. The viscous extract (about 375 ml.) contains about 3 mg. of protein per milliliter; specific activity, about 60.

Step 2. Heating. 3600 ml. of extract (from 90 g. of acetone powder) is mixed with 460 ml. of 1.0 M MnCl$_2$ and 200 ml. of 2.0 M L-malate, pH 6.8, and stirred for 30 minutes at 3 to 4°. A heavy, stringy precipitate forms which is probably a mixture of $(Mn)_3(PO_4)_2$, Mn-nucleic acid, and Mn-nucleoprotein. The mixture is placed in a water bath, previously heated to about 55°, and stirred mechanically. Stirring is continued for 8 minutes after the temperature of the extract reaches 38 to 40°. The mixture is then placed in ice and stirred rapidly until the temperature falls below 20°. It is then filtered in the cold room on 50-cm. fluted filter paper (Reeve Angel No. 802). The filtrate is dialyzed overnight against 20 l. of running distilled water and then against 11 l. of 0.02 M phosphate buffer, pH 7.0, for 24 hours. A heavy precipitate which forms on dialysis against phosphate is filtered to give a clear filtrate. All subsequent steps are carried out at 0 to 4°.

Step 3. First Ammonium Sulfate Fractionation. The filtrate is brought to pH 7.3 with a few drops of 1.0 N NaOH with stirring. Solid ammonium sulfate is added slowly with mechanical stirring to 72% saturation, and the mixture is allowed to stand overnight. After filtration, the precipitate is discarded, the filtrate is brought to 100% saturation with solid ammonium sulfate and stirred for 2 hours, and the mixture is filtered. The precipitate is dissolved in 0.02 M phosphate buffer, pH 7.0, and dialyzed against 4 l. of the same buffer overnight.

Step 4. Precipitation of Nucleic Acid with Protamine. The dialyzed solution of the 0.7 to 1.0 ammonium sulfate fraction is brought to pH 6.0 with 1.0 N acetic acid, and 3 ml. of a solution of protamine sulfate

(20 mg./ml., adjusted to pH 5.0) is added with stirring. The optimum amount of protamine sulfate is 0.09 to 0.12 mg./mg. of protein. The precipitate is centrifuged and discarded.

Step 5. Removal of Inactive Proteins with Calcium Phosphate Gel.[17] The supernatant from step 4 is adjusted to pH 5.5 with a few drops of 1.0 N acetic acid, and then 11 ml. of $Ca_3(PO_4)_2$ gel is added slowly with stirring (0.024 to 0.028 ml. of $Ca_3(PO_4)_2$ per milligram of protein). The mixture is centrifuged after 15 minutes, and the precipitate discarded.

Step 6. Second Ammonium Sulfate Fractionation. The supernatant from step 5 is taken to pH 7.3 by careful addition of 1.0 N NaOH with stirring. 4 ml. of 0.3 M L-malate, pH 6.8, and 1.6 ml. of 0.05 M MnCl$_2$ are added to the solution (containing 280 mg. of protein). Solid ammonium sulfate is added to 75% saturation, the pH being maintained between 7.0 and 7.3 with 1 N NaOH. The bulk of the activity is thus precipitated. The precipitate is centrifuged and dissolved in 20 ml. of 0.02 M phosphate buffer at pH 7.

Step 7. Third Ammonium Sulfate Fractionation. The solution from step 6 is diluted with 0.02 M phosphate buffer, pH 7.0, to 40 ml. (1:2), and 0.8 ml. of 0.3 M L-malate, pH 6.8, and 0.32 ml. of 0.05 M MnCl$_2$ are added. The pH is brought to 7.3 with 0.5 N NaOH. Solid ammonium sulfate is then added to 80% saturation, and the mixture centrifuged. The precipitate is discarded. Ammonium sulfate is added to 90% saturation, and the mixture is stirred for 1 hour and then centrifuged. The precipitate is dissolved in a small volume of 0.02 M phosphate buffer, pH 7.0, and stored in the cold. A summary of the purification data is given in the table.

SUMMARY OF PURIFICATION OF *L. arabinosus* ENZYME[a]

Step	Volume of solution, ml.	Units	Protein, mg.	Specific activity	Yield, %
1. Phosphate extract	3570	576,000	10,300	56	100
2. Heated supernatant	4030	425,000	2,020	210	74
3. First ammonium sulfate fractionation	220	304,000	702	431	53
4. Protamine supernatant	220	295,000	390	760	51
5. Calcium phosphate supernatant	222	274,000	292	940	48
6. Second ammonium sulfate fractionation	21	245,000	189	1300	43
7. Third ammonium sulfate fractionation	5	90,000	23	3900	16

[a] 90 g. of acetone powder.

The specific protection of the "malic" enzyme with Mn^{++} and malic acid appears to be of some general interest. In the absence of any protection, heating of the "malic" enzyme under the same conditions as those used in the purification procedure leads to a 90% loss in activity. If the initial phosphate extracts are dialyzed against demineralized water, no precipitate forms on addition of Mn^{++} and malate. Subsequent heating leads to a marked loss in activity. When the undialyzed phosphate extracts are merely treated with Mn^{++} and malate without employing heat, a considerable purification can also be effected. Mn^{++} is more important than malate in protecting the enzyme, since 70 to 80% of the protection is afforded by Mn^{++} alone. In view of the above, it seems likely that adsorption of inactive proteins on $Mn_3(PO_4)_2$, Mn-nucleic acid, or Mn-nucleoprotein complexes, as well as heat denaturation, is responsible for the purification resulting from this step.

A weak fumarase activity present in the extracts of malate-adapted *L. arabinosus* is removed on purification of the enzyme. Malic dehydrogenase is not present even initially. The last purification step removes a D-lactic dehydrogenase which is present in the extracts along with L-lactic dehydrogenase.[23]

Properties

Activity with L-malate (reaction 1) shows a sharp pH optimum at pH 6.0, whereas activity with oxalacetate (reaction 2) exhibits an equally sharp optimum at pH 4.5. At pH 6.0 the rate of decarboxylation of oxalacetate is only about one-eighth of the maximal rate. At the respective pH optima the rate of evolution of CO_2 from malate and oxalacetate catalyzed by a given amount of enzyme is the same.

The purified enzyme is strictly specific for DPN and for L-malate (reaction 1). Neither fumarate nor D-malate can be substituted for L-malate, nor TPN for DPN. Mn^{++} is an absolute requirement for both reactions 1 and 2; this cation can be replaced by Co^{++}, but Mg^{++} is considerably less effective. For full activity in the catalysis of reaction 1 the system requires K^+ in addition to the other components;[26] in the absence of this cation activity drops to about 15% of the maximum.[22] K^+ can be completely replaced by Rb^+ and partially so by Cs^+.[26] Similar effects of monovalent cations on an analogous system in *Moraxella* had been previously described.[27]

The dissociation constants of the complex between the protein and each of the other components of reaction 1 are as follows: L-malate, 9.8×10^{-3}; DPN^+, 1.96×10^{-5}; Mn^{++}, 2.9×10^{-4}; all expressed in moles

[26] P. M. Nossal, *Biochem. J.* **49**, 407 (1951).
[27] A. Lwoff and H. Ionesco, *Compt. rend.* **225**, 77 (1947).

or gram-atoms per liter. In the case of reaction 2, the dissociation constant of the enzyme-oxalacetate complex is about 1×10^{-2} mole/l., i.e., essentially the same as that of the enzyme-L-malate complex; the dissociation constant of the enzyme-Mn complex is about 3×10^{-4} g.-atom/l. and is thus identical to the corresponding value for reaction 1.

As with the oxalacetic carboxylase activity of the pigeon liver enzyme, L-malate (D-malate not tried) and malonate are strong inhibitors of reaction 2. However, the reaction is unaffected by the absence or presence of pyridine nucleotides. D-Malate has no effect on the dissimilation of L-malate via reaction 1, while malonate is only weakly inhibitory at concentrations producing full inhibition of reaction 2. The rate of evolution of CO_2 from L-malate is inhibited by pyruvate, especially in the presence of CO_2. This is probably due to the reversibility of the reaction which was demonstrated with $C^{14}O_2$.[22]

[125] Oxalacetic Carboxylase of *Micrococcus lysodeikticus*

$$HOOC \cdot CH_2 \cdot CO \cdot COOH \rightarrow CH_3 \cdot CO \cdot COOH + CO_2$$

By DENIS HERBERT

The bacterium *Micrococcus lysodeikticus* is so far the only source from which a *specific* oxalacetic carboxylase, free from "malic enzyme," has been obtained in purified form (Herbert[1]).

Assay Method

Principle. The CO_2 evolved when oxalacetate (OAA) is decarboxylated by the enzyme is measured manometrically. Mn^{++} ions are added to activate the enzyme. To minimize blank corrections for spontaneous decarboxylation of OAA (which is catalyzed by Mn^{++}), high enzyme concentrations and short reaction times are employed, and concentrations of OAA and Mn^{++} are kept as low as possible consistent with saturation of the enzyme.

Procedure. Conical Warburg manometer vessels contain in the main compartment 0.33 M acetate buffer, pH 5.4 (1.0 ml.), 0.01 M MnSO$_4$ (0.3 ml.), enzyme solution (0.5 to 2 units; see below), and water[2] to 2.8 ml.; the side bulb contains 0.15 M OAA (0.2 ml., 30 micromoles)

---

[1] D. Herbert, *Symposia Soc. Exptl. Biol.* **4**, 52 (1951).

[2] When preparations of the higher purity levels are being tested, the enzyme is diluted in 0.1% crystalline bovine serum albumin (Armour), and 3 mg. of albumin is added to the contents of the main vessel to avoid surface denaturation.

freshly prepared and neutralized to pH 5.4. Total volume, 3.0 ml.; final concentrations: acetate, 10^{-1} M; OAA, 10^{-2} M; Mn^{++}, 10^{-3} M; temperature 30°; gas phase, air. After equilibration the OAA is tipped into the main compartment; the manometers are read at 2 minutes and 5 minutes after tipping, and the CO_2 evolved in this 3-minute interval is noted. A control containing all the above constituents except enzyme is treated similarly, and this "blank" CO_2 is subtracted from the value given in the experimental vessel.

Definition of Unit and Specific Activity. One unit of enzyme is defined as that amount which decarboxylates 1 micromole of OAA in 1 minute under the above conditions, i.e.:

$$\text{Enzyme units} = \frac{\text{Micromoles } CO_2 \text{ evolved between 2 and 5 min. in test}}{3}$$

Specific activity is expressed as units per milligram dry weight or, alternatively, as Q_{CO_2}, the conversion factor being

$$\text{Specific activity} \left(\frac{\text{micromoles } CO_2}{\text{mg. min.}}\right) \times 1344 = Q_{CO_2} \left(\frac{\mu l.\ CO_2}{\text{mg. hr.}}\right)$$

Purification Procedure

An essential preliminary to purification is the extraction of the enzyme from the bacterial cells, which is effected by lysing them with lysozyme.[3] The most important subsequent step involves precipitation of the enzyme at pH 3.8, which causes no loss of activity; other steps are conventional salt and solvent fractionations.

Step 1. Preparation and Lysis of Bacterial Suspension. Micrococcus lysodeikticus (strain N.C.T.C. No. 2665) is grown on agar in large enameled trays[3] for 24 hours at 35°, and the growth is suspended in 0.5% NaCl[3] to a density of 5% (bacterial dry weight). Crystalline lysozyme[4] is added (1 mg./g. bacteria), and lysis is allowed to take place for 2 hours at 30°.

Step 2. Acetone Precipitation. The lysed suspension is a translucent jelly, owing to the liberation of intracellular deoxyribonucleoprotein, which must be removed before any further purification can proceed. It is stirred vigorously, 0.1 vol. of M acetate buffer (pH 5.9) added, and cooled to 0°. Acetone, precooled to $-10°$, is added slowly to a concentration of 33% v/v. The gelatinous nucleoprotein precipitate is centrifuged off at $-10°$, washed with an equal volume of 0.1 M acetate pH 5.9, and the

---

[3] For details of growth procedure and optimum conditions for lysozyme action, see D. Herbert and J. Pinsent, *Biochem. J.* **43**, 193 (1948).

[4] Prepared according to G. Alderton and H. L. Fevold, *J. Biol. Chem.* **164**, 1 (1946).

washings combined with the supernatant. The combined supernatants are poured rapidly into 2.5 vol. of acetone at $-10°$, and the precipitate is collected by decantation followed by filtration on large Büchner funnels, washed with acetone and ether (all at $-10°$), and dried. The acetone powder retains its activity for at least one month if stored in a vacuum desiccator.

Step 3. Acid Precipitation. The acetone powder is dissolved in 10 parts of water, and $(NH_4)_2SO_4$ is added to a concentration of 10% w/v. $2 N$ Acetic acid is added with stirring until the pH (glass electrode) is 3.8. The bulky precipitate contains $> 90\%$ of the enzyme and much denatured protein. It is centrifuged off and suspended in an equal volume of water, and the pH is adjusted to 6.0 (glass electrode) by the addition of M K_2HPO_4; at this pH the enzyme all dissolves and most of the denatured protein remains insoluble. The insoluble protein is centrifuged off and washed once with an equal volume of water, the washings being added to the first supernatant. The entire process (precipitation at pH 3.8 in the presence of 10% $(NH_4)_2SO_4$ and extraction of enzyme from the precipitate at pH 6) is then repeated on the combined supernatants.

Step 4. Acetone Fractionation. The enzyme solution from step 3 is fractionated with acetone at $-10°$, the solvent being added slowly with careful stirring. Exact standardization of the procedure is not attempted, the aim being rather to divide the material into a series of roughly equal fractions. The precipitates are dissolved in 0.1 M acetate, pH 5.9, and analyzed for enzyme and total protein. In a typical experiment, fractions obtained between the limits 0 to 40%, 40 to 50%, 50 to 55%, and 55 to 60% (v/v) acetone contained, respectively, 6%, 71%, 13%, and 1.5% of the original enzyme activity.

The fractions containing most of the enzyme (the fractions at 40 to 50% and 50 to 55% acetone in the above case) are combined and the acetone fractionation repeated, this time fractions being taken off at smaller intervals of acetone concentration. The bulk of the enzyme is again usually found in the fractions precipitated between the limits 40 to 50% (v/v) acetone.

Step 5. $(NH_4)_2SO_4$ Fractionation. The purest fractions from the preceding step, dissolved in 0.1 M acetate, pH 5.9, to a protein concentration of ca. 4%, are fractionated with $(NH_4)_2SO_4$ solution (50% w/v). The same procedure is adopted as in the acetone fractions, a number of fractions being taken at gradually increasing $(NH_4)_2SO_4$ concentrations, and the precipitates being separately dissolved (0.1 M acetate, pH 5.9) and analyzed for enzyme and total protein. The purest fractions are combined and submitted to a second fractionation. The bulk of the enzyme is precipitated between the limits 14 to 18% (w/v) $(NH_4)_2SO_4$ concentration.

Step 6. Crystallization. The purest material from the preceding step is precipitated with $(NH_4)_2SO_4$ (20% w/v), centrifuged down, and redissolved in the smallest possible volume of 0.1 M acetate, pH 5.9, any trace of undissolved residue being spun off. The clear solution is treated with 50% $(NH_4)_2SO_4$ solution added drop by drop until the first appearance of permanent opalescence. It is then transferred to a small beaker and allowed to stand in a desiccator containing a dish of silica gel. The opalescence increases, and after some hours the solution shows a sheen on stirring, like a suspension of bacteria. It is allowed to stand for 24 hours over the silica gel, when ca. 75% of the total protein has precipitated. Microscopic examination shows the precipitate to consist largely of very fine, needle-shaped crystals, admixed with some amorphous material.

The enzyme at this stage has a specific activity of 890 units/mg. (Q_{CO_2} = 1,200,000) and is the purest material so far obtained. Nevertheless, examination in the ultracentrifuge discloses three components, and it is believed that the material, although mainly crystalline, is not more than about 50% pure. Further purification is limited by the small amount of material available; 100 g. (dry weight) of bacteria contain only about 16 mg. of enzyme of this purity level.

TABLE I

SUMMARY OF PURIFICATION PROCEDURE

Stage	Total volume, ml.	Total protein, g.	Total units	Specific activity, units/mg.	Yield, %
1. Lysed bacteria	11,800	600	89,000	0.15	100
2. Acetone powder	—	203	83,000	0.38	93
3. pH 3.8 pptn.	234	7.6	80,000	10.5	89
4. Two acetone fractionations	23	0.92	61,000	66	69
5. Two $(NH_4)_2SO_4$ fractionations	11	0.068	38,300	564	43
6. Crystalline preparation	—	0.030	26,800	890	30

Properties

General. The purified enzyme is a colorless protein and in the ultraviolet shows only the usual absorption peak at 275 mμ due to tyrosine and tryptophan. There have been suggestions that biotin is concerned in the decarboxylation of OAA, but it was not possible to detect biotin in the purified enzyme, nor did addition of biotin affect its activity. No other prosthetic group has been found. The enzyme is remarkably stable at low pH values, a fact utilized in the purification procedure (step 3).

Specificity. OAA is the only substance known to be attacked by the enzyme. The following keto acids and related compounds are not decarboxylated under the standard enzyme assay conditions: pyruvic, α-keto-glutaric, acetoacetic, acetone dicarboxylic, oxalosuccinic, dihydroxy-maleic, and dihydroxytartaric acids.

Activation by Metallic Cations. After step 2 of the purification procedure, the enzyme is completely inactive toward OAA unless certain divalent cations are added; the effect is shown in Table II. Some of the

TABLE II
EFFECT OF CATIONS

	μl. CO_2 in 3 min.		
Cation added	Cation	Cation + enzyme	Difference
Nil	4	4	0
Ba^{++}	5	5	0
Ca^{++}	15	35	20
Fe^{++}	14	37	23
Zn^{++}	59	101	42
Ni^{++}	72	134	62
Mg^{++}	5	83	78
Co^{++}	48	254	206
Cd^{++}	18	289	271
Mn^{++}	11	297	286

Standard test conditions; 10^{-2} M OAA, 4.3 enzyme units, pH 5.4, Ba^{++} and Ca^{++}, 10^{-2} M; other cations, 10^{-3} M.

cations are themselves quite efficient catalysts of OAA decarboxylation, but this effect does not run parallel with ability to activate the enzyme. Trivalent cations (Fe^{+++}, Al^{+++}, Cr^{+++}, La^{+++}) do not activate the enzyme.

Kinetics. Typical hyperbolic curves are obtained when OAA concentration is varied with constant Mn^{++}, or Mn^{++} concentration is varied with constant OAA, indicating that both substrate (S) and metal (M) combine reversibly with the enzyme to form a ternary E-M-S complex. Thus the rate equation for the metal-activated enzyme is:

$$- \frac{dS}{dt} = kE \left(\frac{S}{K_S + S} \right) \left(\frac{M}{K_M + M} \right)$$

where K_S is the dissociation constant for OAA ($= 2 \times 10^{-3}$ M) and K_M is the dissociation constant for the metal ($= 2 \times 10^{-4}$ M for Mn^{++}).

Effect of pH. This has only been measured over the range pH 3 to 6, since the manometric method fails at higher pH values, owing to retention of CO_2. In this range the activity is flat from pH 6 to 5.4 and then falls steeply to 50% of the maximum at pH 4.8 and 10% of the maximum at pH 4.1.

[126] Oxalacetate Synthesizing Enzyme

$$PEP^1 + IDP + CO_2 \leftrightarrows OAA^1 + ITP$$

By Merton F. Utter and Kiyoshi Kurahashi

Assay Method

Principle. The method is based on the incorporation of $C^{14}O_2$ into a pool of OAA in an exchange reaction. The radioactivity of the residual OAA is measured, and from a knowledge of the specific activity of the $C^{14}O_2$ of the reaction medium it is possible to calculate the amount of CO_2 incorporated or OAA synthesized.

Reagents

OAA, 0.3 M. A fresh solution is prepared for each assay by dissolving the free acid[2] in a little water and neutralizing to pH 6.2 to 6.4 with 0.5 M Na$_3$PO$_4$ and diluting to the proper volume (39.6 mg. of OAA per milliliter).

ITP, 0.02 M. Sodium form at pH 7 prepared from the barium salt[3] by dissolving in a minimal amount of acid and passing through a small column of IRC-50 cation exchange resin (sodium form).

MnCl$_2$, 0.04 M.

GSH,[1] 0.025 M. A fresh solution is prepared for each assay by dissolving in water and neutralizing to pH 7 with NaOH.

NaHC^{14}O$_3$, 0.33 M, containing approximately 6 μc. of C^{14} per milliliter. Prepared from BaC^{14}O$_3$ by liberation of $C^{14}O_2$ with perchloric acid in an evacuated system with collection of the $C^{14}O_2$ in an appropriate amount of NaOH.

Procedure. The assays are performed in standard Warburg vessels with two side arms and 9-ml. capacity. The main chamber receives 0.2 ml. of OAA (added last), 0.05 ml. of MnCl$_2$, 0.1 ml. of GSH, 0.1 ml. of ITP, and water sufficient to make the total volume of liquid in the vessel 1.0 ml. One side arm contains tissue extract with 1 to 5 units of the enzyme (see definition below), and the other side arm 0.15 ml. of NaHC^{14}O$_3$. The atmosphere is replaced with N$_2$ while the vessel is still at room temperature, then equilibrated at 38° for 7 minutes. The contents of the vessel are

---

[1] PEP, phosphoenolpyruvate; OAA, oxalacetate; GSH, reduced glutathione.

[2] For preparation of this compound see Vol. III.

[3] Some samples of the commercial barium salt of ITP contain substantial amounts of IDP which may be separated chromatographically, but this purification is probably not necessary for the purposes of this assay.

mixed, yielding a pH of approximately 7. After 5 minutes of incubation the reaction is stopped by the addition of 0.1 ml. of 10 N H_2SO_4. Accurate timing for the reaction period is essential, since the reaction is approximately linear with time. The contents of the vessels are removed, centrifuged to remove precipitated protein, and 0.1 ml. of the supernatant accurately pipetted into a counting dish containing a filter paper disk. The counting dishes are dried overnight at 5° *in vacuo* over $CaCl_2$ and counted immediately after removal from the desiccator. The latter point is important, since activity is gradually lost from the dishes on standing at room temperature.

The above procedure gives reasonably reproducible results when the assays are carried out at three protein concentrations for each sample. The $C^{14}O_2$ fixed is found almost entirely in OAA, with pigeon and chicken liver preparations at least, and the OAA is stable under the above procedure to about a 90% extent.

A more specific and accurate procedure for determination of the $C^{14}O_2$ content of the OAA may be used alternatively, although the process is more time consuming. In the alternative method, the entire contents of the reaction vessel are removed quantitatively after the reaction has been stopped by the addition of 0.1 ml. of 4 N H_2SO_4. After the precipitated protein has been centrifuged out and washed with a little water (0.5 ml.), the combined supernatants are gassed with tank CO_2 to remove any residual $C^{14}O_2$, and the CO_2 is removed by a subsequent 10 minutes of aeration. The entire solution is then transferred to the main chamber of a 20-ml. Warburg vessel with the aid of a little water (0.5 ml.). The vessel contains 1.0 ml. of 33% $Al_2(SO_4)_3 \cdot 18H_2O$ in one side arm, 1.0 ml. of 0.75 N potassium acid phthalate buffer in the other,[4] and 0.4 ml. of 2 N NaOH, (CO_3^- free) in the center well. After equilibration at 30 to 38° the contents of the vessel are incubated for 40 minutes to permit decarboxylation of the OAA and absorption of the CO_2. The NaOH from the center well is washed out with aid of a little NaOH and plated and counted as $BaCO_3$. In this procedure the entire contents of the vessel are utilized, so it is possible to reduce the C^{14} content of the $NaHC^{14}O_3$ to about one-tenth that mentioned above.

Although ATP has little or no direct activity in the reaction it can ordinarily be substituted for ITP with crude preparations, since ATP contains catalytic amounts of IDP, and all enzyme preparations obtained thus far are contaminated with an excess of nucleoside diphosphokinase which catalyzes the reaction[5]

$$ATP + IDP \rightleftarrows ITP + ADP$$

[4] H. A. Krebs and L. V. Eggleston, *Biochem. J.* **39**, 408 (1945).
[5] P. Berg and W. K. Joklik, *Nature* **172**, 1008 (1953).

With ATP, however, only about one-third as much OAA is synthesized per unit of enzyme as with ITP, and it is necessary to increase the C^{14} content of the $NaHC^{14}O_3$ accordingly. The results shown in the Summary of Purification Processes were obtained with ATP.

Definition of Unit and Specific Activity. As used in the accompanying table, one unit of activity is defined as that amount which causes the exchange incorporation of 0.01 micromole of $C^{14}O_2$ per minute under the conditions of the assay with ATP substituted for ITP. The values for ITP are approximately three times as large. Specific activity is defined as units per milligram of protein as determined by the colorimetric biuret method.[6]

Application of Assay Method. The method as outlined above has been used primarily with extracts of acetone powders of avian liver. A similar exchange reaction of $C^{14}O_2$ with OAA is shown by OAA decarboxylases of the type found in *Micrococcus lysodeikticus*.[7] However the bacterial enzyme does not require ATP or ITP and thus can be differentiated by omitting this component of the assay system.

Spectrophotometric Assay. After the initial steps of purification, it seems likely that a spectrophotometric test may be used for this enzyme by coupling the synthesis of OAA by this enzyme with malic dehydrogenase and DPNH. This system has been shown to be rapid and linear with time,[8] but it is not described in detail here, since it has not yet been applied to routine assays.

Purification Procedure

Step 1. Preparation of Crude Extract. Fresh livers are obtained from chickens weighing 2.5 to 3 pounds and treated immediately with 5 vol. (milliliters per gram wet weight) of cold acetone in a Waring blendor for about 1 minute. The suspension is filtered, the powder is washed with 2.5 vol. of acetone, and the procedure is repeated. The resulting powder is dried in a rather thin layer *in vacuo* over concentrated H_2SO_4. Ordinarily about 500 g. of liver from twenty chickens is treated in one series and yields about 100 g. of powder after removal of connective tissue. The powder may be stored at $-10°$ for several weeks.

For extraction 90 g. of the powder is triturated with 720 ml. of $M \times 10^{-3}$ potassium phosphate buffer (pH 7.8) and agitated at 38° for 15 minutes. After cooling, the suspension is centrifuged in a Spinco preparative Model L centrifuge for 10 minutes at 17,000 $\times$ g at 0°. The clear, dark-red liquid is decanted and used as the starting material.

[6] H. W. Robinson and C. G. Hogden, *J. Biol. Chem.* **135**, 707 (1940).
[7] L. O. Krampitz, H. G. Wood, and C. H. Werkman, *J. Biol. Chem.* **147**, 243 (1943).
[8] M. F. Utter and K. Kurahashi, *J. Biol. Chem.* **207**, 821 (1954).

Step 2. Heat Inactivation and Fractionation with $(NH_4)_2SO_4$. To the supernatant from step 1 are added 0.1 vol. of 0.5 M potassium phosphate buffer (pH 6.5), 0.1 vol. of 0.3 M OAA (neutralized to pH 6.5), and 0.02 vol. of sodium ATP (0.05 M). The temperature is brought to 46 to 48° within 2 to 3 minutes by swirling in a 75° bath and then transferred to a second bath at 49° for 5 minutes. The suspension is cooled rapidly to 2°, and the volume measured. An amount of $(NH_4)_2SO_4$ solution (saturated at 2°) calculated to yield 42.5% saturation is added slowly with stirring, maintaining the temperature at 2°. After 30 minutes of additional stirring at the same temperature, the suspension is centrifuged as in step 1, and the precipitate discarded. The volume of the supernatant is measured, and enough additional saturated $(NH_4)_2SO_4$ is added to yield 55% saturation under the same conditions as described above. The precipitate obtained after centrifugation is dissolved in 50 to 60 ml. of 0.9% KCl and dialyzed overnight in a rocker apparatus at 5° against a total volume of 40 l. of 5×10^{-3} M potassium phosphate buffer (pH 5.9).

Step 3. Treatment with $Ca_3(PO_4)_2$ *Gel.* The volume and protein content[6] of the dialyzed solution are measured. After cooling to 2°, enough cold water is added so that the final protein concentration will be approximately 13 mg./ml. after all additions. $Ca_3(PO_4)_2$ gel[9] is added with stirring at 2° equivalent to 0.75 g. (dry weight) of gel per gram of protein and enough KH_2PO_4 (0.5 M) to bring the pH to 5.8 to 5.9. After stirring for 15 minutes at 2° the suspension is centrifuged at 0° in an International refrigerated centrifuge for 10 minutes at 3000 $\times$ g. The supernatant is decanted, and its volume and protein content are measured.

Step 4. Treatment with Alumina C_γ *Gel.* After cooling to 2°, alumina C_γ gel[10] equivalent to 0.5 g. (dry weight) of gel per gram of protein is added. After 15 minutes of stirring, the gel is centrifuged out as in step 3, and the precipitate is eluted successively with 80 ml. of potassium phosphate buffer (0.04 M, pH 6.4), 40 ml. of pH 6.8 buffer, and 40 ml. of pH 7.4 buffer. For elution the gel is stirred for 10 minutes at 2° and centrifuged. The eluted fractions are termed 4a, 4b, and 4c, respectively. The supernatant from the first centrifugation with the gel is treated with another portion of the gel equivalent to the first and eluted in the same manner as previously to yield fractions 4d, 4e, and 4f.

Fractions 4a, 4d, and 4e ordinarily contain most of the activity and are usually combined. The protein solution is quite unstable but may be lyophilized after the addition of GSH (2.5 micromoles per milliliter). The resulting powder is stable for a period of months when stored dry and cold.

---

[9] See Vol. I [11]. The gel should be aged in the cold for no less than 6 and no more than 10 days before use. The pH of the adsorption operation is also critical.

[10] See Vol. I [11] for preparation.

Step 5. Repetition of Gel Steps. Further limited purification may be accomplished by a repetition of the gel steps. The lyophilized powder obtained from step 4 is dissolved in enough cold water to yield a protein concentration of 2.25 mg./ml., the pH carefully adjusted to 5.8 by the addition of M acetate buffer (pH 4), and the solution treated twice with alumina C_γ gel equivalent to 0.5 g. (dry weight) per gram of protein. The elutions are carried out as described in step 4 with 0.14 ml. of pH 6.4 buffer per milligram of protein and 0.10 ml. of pH 6.8 or 7.4 buffer per milligram of protein. The fractions corresponding to 4a, 4d, and 4e again ordinarily contain most of the activity and are combined, the protein concentration determined, and the pH adjusted to 5.8 with M acetate buffer (pH 4). $Ca_3(PO_4)_2$ gel equivalent to 1 g. (dry weight) per gram of protein is added, and the pH again adjusted to 5.8. After stirring and centrifuging, the supernatant is lyophilized after the addition of GSH (2.5 micromoles per milliliter).

SUMMARY OF PURIFICATION PROCESS

Fraction	Total volume, ml.	Units/ ml.	Total units, thou- sands	Pro- tein, mg./ ml.	Specific activity, units/ mg.	Recovery, %
1. Crude extract[a]	670	36.0	24.16	43.4	0.83	—
2. $(NH_4)_2SO_4$ fraction,[b] 0.425–0.55	80	109.2	8.73	40.6	2.69	36
3. $Ca_3(PO_4)_2$ gel supernatant	210	53.3	11.19	5.17	10.3	46
4. Alumina C_γ eluates a[c]	70	19.5	1.37	1.09	17.9	
d	70	29.0	2.03	1.42	20.4	19 (combined)
e	35	31.7	1.11	1.37	23.2	
5. Repetition of gel steps Supernatant	80	18.4	1.47	0.44	41.4	6

[a] 90 g. of acetone powder.

[b] Assays at this stage show low activity, probably because of presence of $(NH_4)^+$ ions.

[c] See text.

Properties

Metal Ion Requirements. The enzyme requires Mn^{++} for activity, and Mg^{++} cannot be substituted. The influence of other divalent metal ions has not been studied.

Inhibitors. The partially purified enzyme is often stimulated by GSH, and the enzyme is 80% inhibited by 1×10^{-4} M p-chloromercuribenzoate and almost completely inhibited by a 5×10^{-4} M concentration of the same inhibitor.[11]

Presence of Other Enzymes. The most highly purified fractions yet obtained are free from "malic" enzyme, ATP-ase, and pyruvic phosphokinase, but are contaminated with substantial amounts of myokinase, malic and lactic dehydrogenases, nucleoside diphosphokinase, and small amounts of enolase.[11,12]

[11] M. F. Utter, K. Kurahashi, and I. A. Rose, *J. Biol. Chem.* **207**, 803 (1954).
[12] M. F. Utter and K. Kurahashi, *J. Biol. Chem.* **207**, 787 (1954).

Author Index

A

Abderhalden, E., 638
Abul-Fadl, M. A. M., 33(21)
Ackermann, W. W., 727
Acree, S. F., 174
Adams, M., 255, 256
Adler, E., 391, 704
Aizawa, K., 234
Åkerlof, G., 82
Albers, E., 38
Albers, H., 38
Alberty, R. A., 731, 733(9), 734, 735
Albon, N., 258
Alburn, H. E., 169, 170(18)
Alderton, G., 124, 754
Alfin-Slater, R. B., 16(6)
Alivisatos, S. G. A., 722
Allfrey, V. G., 18(19)
Allen, M. C., 165
Allen, R. J. L., 193
Alles, G. A., 648
Altermatt, H., 158, 159(2), 162(2), 163(2), 164(2), 166(2)
Altman, K. I., 267, 741
Aminoff, D., 615
Ammon, R., 628, 629, 636, 639(16), 641, 648
Anderson, B., 467
Anderson, N. G., 28(10)
Anfinsen, C. B., 678, 695, 697(1), 698(1), 728
Anger, V., 649
Anyas-Weisz, L., 165
Appleby, C. A., 50, 445, 447(3), 448(3)
Appleton, H. D., 662, 665(12), 666(12), 668(12)
Archibald, R. M., 630, 631(34), 639(34)
Arnon, D. I., 25(19), 411, 748
Aschner, M., 186, 187(23), 189(23)
Ashton, G. C., 232
Ashworth, J. N., 90
Askonas, B. A., 40(49), 43, 44, 50
Asnis, R. E., 397
Astrup, T., 32(18), 86
Atkin, L., 54
Aubert, J. P., 474
Augustinsson, K.-B., 643, 645, 649, 650(38)

Averill, W., 194
Avineri-Shapiro, S., 186, 187(23), 188(22), 189(22, 23)
Awny, A. J., 662
Axelrod, A. E., 729
Axelrod, B., 20(2), 363
Axmacher, F., 461
Ayres, A., 162

B

Bach, H., 240
Bach, S. J., 44, 444, 445(1), 447(1), 448(1), 449(1), 728
Bachtold, J. G., 168, 170(12)
Bacila, M., 292
Bacon, J. S. D., 231, 255, 258, 259(2), 260(3), 261
Baer, E., 401, 412
Bailey, J. M., 193, 196, 197(6)
Bailey, K., 46, 50, 270, 271(7), 275(7), 645
Bain, J. A., 16(7)
Baker, G. L., 158
Balist, M., 250
Ball, C. D., 198
Ball, E. G., 33, 39(42), 728, 729, 747
Ballou, G. A., 159, 160(4), 161(4), 162(4)
Balls, A. K., 154, 156(13), 157(13), 630, 636(23), 637(23, 30), 638(23)
Bamann, E., 628, 635, 636, 639, 642
Bandurski, R. S., 20(2)
Banga, L., 728
Banks, J., 350
Baranowski, T., 307, 311, 312(15), 314(15), 320, 391, 396
Bard, R. C., 227, 328, 330(1), 331, 332(3), 334(1)
Bardawill, C. J., 524, 592
Barker, H. A., 53, 54(11), 132, 228, 229, 481, 519, 521(5), 552, 553(2), 596, 597, 600, 601, 602, 615, 616, 617, 618(1, 2), 619(2)
Barker, S. A., 191, 192(34), 194, 197(17)
Barlow, J. L., 40(46)
Barner, H., 366
Barnes, B. A., 32(19), 33(19)
Barnett, S. R., 18(19), 311, 313(10), 314(10)
Barnum, C. P., 18(18)

Barratt, R. W., 162, 165(25)
Barron, E. S. G., 464, 502, 503(8), 639, 675, 676, 677, 678(5a), 726
Bársony, J., 178
Bartlett, G. R., 677
Bate-Smith, E. C., 260
Bates, F. J., 232
Battelli, F., 723
Bauer, C. W., 630
Bauer, E., 97, 743
Bauer, W., 169
Baum, H., 194, 195
Bayliss, M., 641
Bealing, F. J., 258, 259(2)
Beaufays, H., 286, 287(4)
Beck, L. V., 311
Behrens, M., 18(19), 464
Beinert, H., 548, 572, 574, 576, 577(6), 582, 584, 586, 588(10), 604, 721
Beisenherz, G., 322, 387, 388(5), 391, 392(4), 394(4), 419, 436, 439(2), 440(2), 443, 568
Belfanti, S., 665
Bell, D. J., 261
Bellamy, W. D., 52, 54(5)
Bentley, R., 340, 343(7)
Berg, P., 759
Berger, J., 56
Berger, L., 270, 273(6), 274(6), 275(6), 276(6)
Bergold, G., 434
Bergstermann, H., 461
Berl, S., 472
Berman, M., 620
Berman, R., 621, 623(9)
Bernfeld, H., 267, 268(9)
Bernfeld, P., 149, 150, 154, 157, 199, 264, 267, 268(9), 634
Bernheim, F., 674, 675(2), 676(3)
Bernheim, M. L. C., 674, 675(2), 676(3)
Bernheimer, A. W., 660
Bernt, E., 248
Berthet, J., 281
Berthet, L., 281
Bessman, S. P., 588, 590(16), 610, 615
Bever, A. T., 194, 197(15)
Beyer, G. T., 311, 313(10), 314(9, 10)
Bhagvat, K., 22
Biale, J. B., 20(3)
Bier, M., 638, 640

Birch-Anderson, A., 86
Birkinshaw, J. H., 189, 242, 340, 341(2), 343(2b), 344(2)
Birmingham, M. K., 9(17)
Birnbaum, S. M., 42(56), 43(56), 47(68a), 49(56), 50
Björnesjö, K. B., 86
Black, S., 509, 510 (2, see 4), 586, 588(6), 591(6), 593, 615, 721
Blakley, R. L., 348
Blanchard, M. H., 74
Blanchard, M. L., 748, 749(24)
Blanchard, P. H., 258
Bliss, A. F., 498
Bliss, L., 194
Blix, G., 167, 291
Bloomfield, G., 259
Boardman, N. K., 118
Bock, R. M., 29(13), 34(13), 35(29a), 489, 526, 548, 555, 603, 718, 731, 733(9), 735
Boell, E. J., 648
Börnstein, R., 84
Bogard, M. O., 179, 181(3)
Boissonnas, R. A., 629, 631(13), 634, 635, 636, 638
Boltze, H. J., 322, 391, 392(4), 394(4), 419, 436, 439(2), 440(2), 443, 568
Bone, A. D., 275, 285
Bonner, D. M., 135
Bonner, J., 20(2), 21(4), 588
Bonner, W. D., 724, 726, 727(8, 13), 728(13)
Bonnichsen, R. K., 495, 496, 498(1), 499(1)
Booth, B. H., 59
Borenfreund, E., 363, 367, 377
Bornstein, B. T., 521
Bourne, E. J., 191, 192(34), 194, 195, 196, 197(17), 222
Bovet, D., 648
Boyd, W. C., 359
Boyer, P. D., 436, 439(4)
Bradley, L. B., 561
Braganca, B. M., 293, 677
Brand, F. C., 6
Braswell, E., 181
Brederech, H., 238
Brewster, L. E., 668, 670(35), 671(35)
Brink, N. G., 338, 339(4)

Brodie, A. F., 397
Brodie, B. B., 662, 665(12), 666(12), 668(12)
Brodie, T. M., 16(7)
Brown, A. E., 197
Brown, D. H., 274, 275(18)
Brown, R. K., 32(19), 33(19), 681
Brumm, A., 14(16)
Brummond, D. O., 689, 693(17), 714
Brunetti, R., 263
Bryant, G., 179, 181(3)
Buchanan, J. M., 695, 697(1), 698(1)
Bucher, N. L. R., 15(20)
Buchner, E., 55, 57
Buddle, H. L., 28
Bücher, T., 137, 322, 324, 362, 372, 382, 387, 388(5), 391, 392(4), 394(4), 415, 419, 421(1), 426, 434, 435, 436, 439(2), 440(2), 441, 443, 512, 568
Bueding, E., 472
Bullock, K., 47(68)
Bunting, A. H., 467
Burk, D., 332, 334(6), 508
Burkholder, W. H., 641
Burnet, F. M., 170
Burris, R. H., 61, 528, 529(1), 531(1), 532(1), 714, 723, 728(5)
Burton, K., 421, 739
Burton, R. M., 397, 518, 519(1), 520(1), 553, 616, 694
Butler, G. C., 96

C

Cabib, E., 294
Cain, C. K., 340, 341(4), 344(4b), 345(4b)
Campbell, D. H., 55
Caputto, R., 232, 250, 291, 292(5), 294, 296, 354, 355, 356(1)
Cardini, C. E., 291, 294, 296, 322, 354. 355, 356(1)
Carlson, W. W., 182, 184(8)
Carter, H. E., 660
Castelfranco, P., 551
Cecil, R., 343, 733
Celmer, W. D., 660
Chaikoff, I. L., 672
Chance, B., 454, 495, 498(1), 499(1)
Chang, H. C., 623
Chanutin, A., 654, 656(3)

Chargaff, E., 38(39)
Cherkes, A., 678
Cherry, I. S., 630
Cheslock, K., 489, 606
Chou, T. C., 194, 197(15), 586, 588(8), 609, 610, 612, 613, 614(5)
Christensen, J. R., 675
Christian, W., 85, 86, 299, 311, 312(12), 314(12), 316, 317(4), 323, 335, 371, 376, 387, 398, 401, 407, 408(1, see 6), 410(1, see 6), 413, 419, 425, 428, 429, 433(2), 434(2), 455, 484, 491, 501, 515, 529, 533(3), 560, 568, 576, 583, 592, 604, 710, 740, 741(7)
Ciotti, M. M., 334, 499, 503, 713
Clagett, C. O., 528, 529(1), 531(1), 532(1)
Clark, W. M., 138, 139(1), 142(1), 146(1)
Cleland, K. W., 16(4), 28(7), 282
Clendenning, K. A., 22(3), 23(3), 24(13, 16), 25(16), 748
Cloutier, A. A., 698
Clutterbuck, P. W., 729
Coates, C. W., 619
Cohen, J. A., 47(67), 50, 646
Cohen, P. P., 467
Cohen, S. S., 350, 353, 366, 367(1), 368, 369(1), 370(1)
Cohen-Bazire, G., 190, 247
Cohn, D. J., 259
Cohn, E. J., 32(19), 33(19), 70, 71, 74, 90, 647
Cohn, M., 132, 241, 242, 243(35), 247, 248, 295
Collier, H. B., 661
Colmer, A. R., 663
Colowick, S. P., 38(37a), 198, 269, 270, 273(6), 274(6), 275(6), 276(6), 277, 280, 284(3), 285(3), 288, 290, 291, 294, 295, 298, 355, 404, 440, 713
Colter, J. S., 676, 677(12)
Conn, E. E., 740, 741, 744(11), 748(6)
Connors, W. M., 657, 658
Connstein, W., 641
Constantin, M. J., 636
Contardi, A., 664, 665, 666(24), 668(32)
Contopoulou, R., 551
Coon, M. J., 573, 574(5), 575(7), 576(13), 581, 584, 585(29), 721
Cooper, G. R., 86
Cooper, O., 33, 728, 729

Cooper, P. D., 58, 59(30)
Cooperstein, S. J., 36(33)
Copenhaver, J. H., Jr., 628
Cori, C. F., 196, 198, 200, 205(1), 206,
    207(1), 215, 216, 270, 273(6), 274(6),
    275(6), 276(6), 277, 284(1), 285(1),
    289, 290(12), 294, 295, 298, 307, 371,
    401, 403(1), 406(1), 423, 424(2),
    425(2), 442
Cori, G. T., 71, 96, 198, 200, 201, 202,
    205(1), 206, 207(1, 2), 211, 212,
    214(1, 4), 215, 216, 224, 274, 277,
    280, 284(1), 285(1), 289, 290(12), 291,
    294, 295, 298, 307, 310, 312(2),
    313(2), 314(2), 315(2), 371, 381, 401,
    403(1), 406(1), 425, 427(7), 442
Cori, O., 574, 604, 675, 678(5a), 714,
    718(1), 719(1), 722(1)
Corman, J., 179, 181(3), 182(15), 184(15)
Corran, H. S., 724
Coulthard, C. E., 189, 242, 340, 341(2),
    343(2b), 344(2)
Courtois, J., 513
Couteaux, R., 651
Cowles, P. V. B., 51
Cox, R. T., 619
Cragg, J. M., 63
Craig, L. C., 99
Cramer, F. B., 274
Crandall, L. A., 630
Crane, R. K., 275, 277, 279(2), 280(2),
    281, 284, 285(2, 19, 20), 747
Crawford, E. J., 311
Curran, H. R., 708
Custer, J. H., 37(36a), 647
Czök, R., 322, 391, 392(4), 394(4), 419,
    436, 439(2), 440(2), 443, 568

                      D

Dagley, S., 398
Dakin, H. D., 729
Daniel, L. J., 675
Das, N. B., 727
David, M. M., 524, 592
Davies, R., 252, 624, 626(1, see 7)
Davison, D. C., 536
Davison, F. R., 158
Davydova, S. J., 734
Dawes, E. A., 398

Dawson, C. R., 96, 97(27)
Dawson, E. R., 639
Day, H. G., 196
DeBusk, B. G., 494
Debye, P., 70
Dedonder, R., 187, 188(27)
de Duve, C., 281, 286, 287(4)
Deffner, M., 342
de la Haba, G., 375, 380
Delavier-Klutchko, C., 50
del Campillo, A., 61, 137, 244, 486, 490(1),
    559, 562, 564, 573, 574(5), 575(7),
    576(13), 581, 582, 584, 585(29),
    623, 718, 721, 748, 749(22, 23, 24),
    752(22, 23), 753(22)
Delezenne, C., 665
Della Monica, E. S., 37(36a), 39(44) 50,
    647
Delory, G. E., 146
Deming, N. P., 27(3a), 29(3a)
de Montmollin, R., 154
DeMoss, R. D., 227, 328, 330(1), 331,
    332(3), 334(1), 504
Derouaux, G., 32(19), 33(19)
Desjobert, A., 513
Desnuelle, P., 636
de Stevens, G., 173, 177
Deuel, H., 158, 159(2), 161, 162(2),
    163(2), 164(2), 165, 166(2)
Deutsch, W., 4(7)
Dewan, J. G., 724
Dickman, S. R., 698
Dieu, H. A., 256, 257
Dingle, J., 162, 163
Dirscherl, W., 460
Dische, Z., 363, 367, 377
Dixon, M., 35, 44(63), 77, 78, 91, 95(4),
    275, 276(23), 444, 445(1), 447(1),
    448(1), 449(1), 727
Doisy, E. A., 340, 341(4), 344(4b),
    345(4b)
Dolin, M. I., 55, 61, 474, 475(16)
Dorfman, A., 168, 170, 172(25), 173(25)
Doty, P., 435
Doudoroff, M., 187, 190, 191(33), 192(33),
    197, 226, 228, 229, 230, 574
Dounce, A. L., 18(19), 55, 311, 313(10),
    314(9, 10), 657, 658(5)
Drake, B. B., 166
Drysdale, G. R., 35(29), 548

Dubois, K. P., 726, 727(20)
Duckert, F., 154
Dufait, R., 434, 728
Duran-Reynals, F., 168
Durrans, I. H., 42
Dustman, R. B., 159

E

Easson, L. H., 648
Ebisuzaki, K., 37(36b), 675
Edelhoch, H., 34(25), 435
Edelman, J., 255, 258, 260
Edsall, J. T., 90, 435
Effront, J., 434
Eggleston, L. V., 698, 727, 759
Ehrensvärd, G., 39(43)
Eichel, B., 36(33)
Elliott, K. A. C., 3(3), 6(9), 8(13), 9(16, 17)
Elser, W. J., 55
Elson, L. A., 610, 612
Elvehjem, C. A., 10, 12(1), 16(10), 322, 729
Emmens, C. W., 169
Enders, C., 178
England, S., 156, 157(14), 158(16)
Epstein, J. H., 286
Ercoli, A., 660, 665, 666, 668(32)
Erickson, J. O., 86
Erlandsen, R. F., 163
Euler, H. von, 174, 391, 453, 704
Evans, D. G., 170
Evans, E. A., Jr., 735, 741, 742
Evans, F. R., 708
Evans, H. J., 63, 95
Eysenbach, H., 729

F

Faber, V., 170
Fairbairn, D., 660, 664(5), 666, 668(31)
Falconer, J. S., 453
Fantes, K. H., 232
Farr, A. L., 150, 156(5), 207, 264, 270, 278, 296, 300, 312, 347, 441, 491, 536, 600, 686, 687(6), 705, 707, 741
Feger, V. H., 179, 181(3)
Feigelson, M., 16(4)
Feigl, F., 121, 649

Fellenberg, T., 159
Fellig, J., 255
Felsher, R. Z., 467
Fevold, H. L., 124, 754
Field, J., 2nd, 3(4), 6(11)
Fiore, J. V., 629, 640, 642(17, 71)
Fischer, E., 251, 252, 254, 255
Fischer, E. H., 150, 154, 158, 194, 197(18), 262, 634, 710, 712
Fischer, H. O. L., 513
Fish, V. B., 159
Fisher, F. G., 729
Fisher, H. F., 454
Fishman, W. H., 263, 264, 267, 268(9), 269(7)
Fiske, C. H., 206, 207(3), 215, 295, 311, 346
Fitting, C., 230, 231
Fleury, P., 513
Flosdorf, E. W., 55
Floyd, C. S., 654, 656(3)
Flynn, R. M., 586, 588(6), 591(6), 593, 615, 721
Fodor, P. J., 42(57), 50, 630, 636
Fontaine, T. D., 642
Forsyth, W. G. C., 180, 183(5)
Fourneau, E., 665
Fraenkel-Conrat, H. L., 662, 663
Francioli, M., 660, 662, 665, 666(16)
Franke, W., 340, 342(8), 343, 726
Fraser, D., 57
Frazer, A. C., 636
Fredericksen, E., 240
Freer, R. M., 18(19)
Frehden, O., 649
French, C. S., 23(9), 24(11), 64, 65(1), 67(1)
French, D., 196
Freudenberg, K., 177, 178
Frieden, C., 731, 733(9), 735
Friedenwald, J. S., 650
Friedkin, M., 361
Friedmann, B., 533
Friedmann, E., 726
Fromageot, C., 113
Fürth, P., 178
Fuhrman, F. A., 3(4), 6(11)
Fulbrigge, A., 735
Fuld, M., 154, 489, 606
Furness, G., 58, 59(30)

G

Gaby, W. L., 340, 341(4), 344(4b), 345(4b)
Gaddum, J. H., 623
Gaisford, W., 170
Gale, E. F., 55, 130, 136(8), 137
Garbade, K.-H., 322, 387, 388(5), 391, 392(4), 394(4), 419, 436, 439(2), 440(2), 443, 568
Garrod, A. E., 129
Gavard, R., 474
Gebhardt, L. P., 168, 170(12)
Geidel, W., 638
Geller, D. M., 483, 486(4)
Gergely, J., 574, 602, 604(2), 607(2), 715
Gest, H., 540, 541
Gibbs, M., 62, 411, 414(1), 540, 541
Gibian, H., 168
Gibson, D. M., 647
Gilbert, G. A., 194, 195
Gillespie, J. M., 32(19), 33(19)
Gilvarg, C., 574, 604, 714, 718(1), 719(1), 722(1)
Giri, K. V., 194, 197(16)
Girvin, G. T., 622
Glaid, A. J., 453, 454(12)
Glatthaar, C., 368
Glick, D., 628, 635, 636, 637, 638(5), 641, 648
Goldblum, J., 256
Goldman, D. S., 548, 574, 577(6), 584, 721
Goldstein, N. P., 286
Gomori, G., 82, 139, 141(3), 143, 145, 630
Gootz, R., 236
Gorbach, G., 641
Gordon, A. H., 310, 313(4), 314(4), 316, 523, 524(1, see 5), 525(1), 526(1), 527(1), 528(1), 724
Gordon, J. J., 678
Gordon, R. S., Jr., 647
Gorham, P. R., 22(3), 23(3), 24(13, 16), 25(16)
Gornall, A. G., 524, 592
Gottlieb, S., 232
Gottschalk, A., 274
Grafflin, A. L., 699, 702(3)
Graham, A. F., 269
Graham, W. D., 517

Gralin, N., 663
Granick, S., 22(1), 24(1)
Grant, P. T., 274
Grassmann, W., 174, 175, 177, 178(11)
Grauer, A., 32(20)
Gray, P. P., 54, 249
Green, A. A., 71, 73, 74, 78, 79, 82, 200, 202, 205(1), 216, 307, 310, 312(2), 313(2), 314(2), 315(2), 381, 425, 427(7), 442
Green, D. E., 11, 35(29a), 58, 59, 192, 194(1), 196(1), 197(1), 198(1), 199(1), 310, 313(4), 314(4), 316, 445, 460, 461(4), 463(4), 464(4), 472, 523, 524(1, see 5), 525(1, 2), 526(1, 2), 527(1, 2), 528(1), 545, 548, 553, 555(1), 557(1), 566, 571(4), 572(4), 574, 577(6), 584, 586, 588(10), 604, 668, 721, 724
Greenstein, J. P., 42(56), 43(56), 47(68a), 49(56), 50
Gregory, J. D., 598
Greif, R. L., 170
Greville, G. D., 439, 565, 566, 571
Griese, A., 323
Griffiths, M., 285
Grisolia, S., 443, 735, 739(1)
Gronchi, V., 662
Grunberg-Manago, M., 50, 591
Grunert, R. R., 560, 623, 688
Guarino, A. J., 351, 352, 353(2), 354(2)
Günther, E., 236
Günther, G., 391, 704
Güntner, H., 641
Gürtler, P., 157
Gunsalus, C. F., 61
Gunsalus, I. C., 51, 52, 54(5), 61, 96, 137, 227, 328, 330(1), 331, 332(3), 334(1), 474, 475(16), 490, 491(2), 494(2), 685, 718
Gunter, G. S., 169
Gurd, F. R. N., 32(19), 33(19)
Gurin, S., 727

H

Haas, E., 29, 95, 96, 325
Haehn, H., 38, 57
Hager, L. P., 483, 486(4), 490, 491(2), 494(2)

McClean, D., 168, 169
McColloch, R. J., 158, 159(2), 160, 162(2, 9), 163(2), 164(2, 22), 165(22), 166(2)
McCready, M. H., 170
McCready, R. M., 63, 163, 192, 222
McCullagh, D. R., 170
McDonald, M. R., 121, 269, 270(2), 274 (2), 275(2), 276, 277, 290, 409
MacDonnell, L. R., 161, 165
Macey, A., 195
McFarlane, A. S., 34(26), 41
MacFarlane, M. G., 670, 671(39)
McGilvery, R. W., 95
McGovern, N. H., 15(20)
Machado, A. L., 619
McIlvaine, T. C., 141
McIlwain, H., 28, 56, 60(23), 329, 506, 540, 680
MacKay, M. E., 84
Mackler, B., 523, 525(2), 526(2), 527(2)
Maclay, W. D., 163
McShan, W. H., 628, 725(10)
Madinaveitia, J., 169
Maehly, A. C., 97
Magee, R. J., 96, 97(27)
Mager, J., 199
Maher, J., 258
Mahler, H. R., 35(29a), 91, 96, 523, 525 (2), 526(2), 527(2), 538, 548, 553, 555(1), 557(1), 562, 564, 565, 566, 571(4), 572(4), 573, 574(2), 585
Mallette, M. F., 59
Malmgren, H., 167
Mandl, I., 32(20)
Mann, K. M., 285
Mann, P. F. E., 198
Mann, P. J. G., 674, 675(1), 676(1, 5), 677(5), 678(5), 727, 734
Marsh, P. B., 232
Martin, A. J. P., 99, 100, 107
Martin, H. F., 642
Maruo, B., 193, 194(8)
Mascré, M., 240
Maskell, E. J., 198
Massart, C., 728
Massey, V., 730, 733(8), 734, 735(18)
Mathews, M. B., 168, 171, 536
Mathies, J. C., 36(35), 39(35)
Matlack, M. B., 630, 636(23), 637(23, 30), 638(23)

Matus, J., 165, 166(34)
Mayer, P., 625
Mazur, A., 652, 653, 654, 655, 656
Meagher, W. R., 197
Meek, J. S., 480, 481(6)
Meeuse, B. J. D., 194
Mehler, A. H., 443, 576, 577(20), 588, 608, 615, 700, 701(6), 735, 736, 738 (4), 739(1, 4), 741(1), 742(1)
Mehlitz, A., 166
Mehta, M. L., 177
Meister, A., 94, 97, 442, 443, 451, 452, 453(7), 454(7)
Melin, M., 90
Mellanby, J., 69, 70(3)
Melnick, J. L., 461
Memmen, F., 629, 630(14), 636, 638(14), 639(14)
Meyer, K., 167, 168(2), 170(2)
Meyer, K. H., 150, 154, 158, 197, 199, 634
Meyer, V., 649
Meyer-Arendt, E., 322, 387, 388(5), 391, 392(4), 394(4), 419, 436, 439(2), 440(2), 443, 568
Meyerhof, B., 645
Meyerhof, O., 49(69), 271, 273(8), 308, 310, 311, 313, 316, 387, 390(2, 4), 391, 392, 403, 423, 425, 427, 433(1), 434(1), 439
Michaelis, L., 144
Michaelis, R., 189, 242, 340, 341(2), 343(2b), 344(2)
Mickle, H., 58
Miers, J. C., 163
Mii, S., 35(29a), 548, 553, 555(1), 557(1), 566, 571(4), 572(4), 574, 577(6), 584, 721
Milhaud, G., 474
Miller, A., 42(57), 50
Miller, E. C., 12(7), 15(7), 17(13), 19(13)
Miller, G. E., 262
Miller, J. A., 12(7), 15(7), 17(13), 166
Miller, Z. B., 677
Millerd, A., 20(2), 21(4), 22(6), 588
Millican, R. C., 610
Mills, G. T., 269
Milner, H. W., 23(9), 24(11), 64, 65(1), 67(1)
Mirick, G. S., 134
Mirsky, A. E., 18(19)

Mitsuhashi, S., 366, 370(2)
Mittelman, D., 32(19), 33(19)
Mittelman, N., 291, 292(5)
Miwa, T., 234
Möller, E. F., 503
Moelwyn-Hughes, E. A., 256
Mohammed, S. M., 658
Monod, J., 132, 189, 190(30), 191(30), 192(30), 241, 242, 243(35), 247, 248
Monsour, V., 663
Moore, S., 113, 116(10), 117(5), 351, 610
Morawetz, H., 86
Morgan, E. J., 28, 726
Morgan, W. T. J., 610, 612, 615
Morris, D. L., 223
Morrison, G. A., 398
Morrison, J. F., 698
Morrison, P. R., 435
Morton, R. K., 28(5, 6), 34, 36(5), 39, 40(47, 48), 41, 42(47, 48, 54), 43(48), 44(47, 54), 45, 47, 48(47, 48), 49(54), 50, 56, 445, 447(3), 448(3), 664
Moses, W., 54
Mottern, H. H., 158
Mounter, L. A., 654, 656
Mouton, R. F., 32(19), 33(19)
Moyer, J. C., 160
Müller, D., 340
Mueller, G. C., 588, 610
Muir, R. D., 340, 341(4), 344(4b), 345 (4b)
Mulford, D. J., 90
Munch-Petersen, A., 293
Munk, L., 467
Muntz, J., 439
Murray, D. R. P., 639
Myers, R. J., 113, 118(7)
Myrbäck, K., 256, 257

N

Nachlas, M. M., 633, 657
Nachmansohn, D., 619, 620, 621, 623(9), 642, 643, 644(4), 645, 646(15), 648
Nadel, H., 182
Najjar, V. A., 294, 296, 298(4), 299, 361, 362
Nakamura, M., 193, 194(8), 196, 197
Nason, A., 63, 95
Natelson, S., 686

Nath, R. L., 234, 237, 238, 239
Naudet, M., 636
Naylor, N. M., 194
Needham, D. M., 275, 276(23)
Neely, W. B., 198
Negelein, E., 325, 435, 436, 439(1), 440(1), 503
Neilands, J. B., 113, 117(2), 452, 453(8), 454
Neill, J. M., 180, 183(5)
Nelson, F. E., 641
Nelson, J. M., 256, 259
Nelson, N., 211, 232, 235, 262, 296, 357
Neuberg, C., 32(20), 53, 161, 257, 292, 310, 460, 464
Neuberger, A., 340, 343(7)
Neufeld, E. F., 38(37a), 713
Neurath, H., 86
Newcomb, E. H., 549
Nicolai, H. W., 628, 642
Niederland, T. R., 307, 311 312(15), 314(15)
Nikuni, Z., 662
Nisselbaum, J. S., 264, 267, 268(14), 269(14)
Noblesse, Mme., 187, 188(27)
Noe, E., 18(17)
Noelting, G., 149
Noguchi, S., 668
Nord, F. F., 177, 629, 638, 640, 642(17, 71)
Northrop, J. H., 122
Nossal, P. M., 752
Novelli, G. D., 484, 587, 596, 598, 617
Novikoff, A. B., 18(17)
Nygaard, A. P., 517

O

Ochoa, S., 61, 94, 95, 137, 280, 289, 443, 486, 490(1), 514, 528, 529(2), 532, 548, 559, 560(3), 561(3), 566, 573, 574(1), 576, 581, 584, 586, 591, 604, 623, 685, 687, 689(1), 694(1, 11), 699, 700, 701(6), 702(3), 704(1, 2), 705, 707, 713, 714, 718(1), 719(1), 722(1), 735, 739(1), 741(1, 2), 742 (1, 4), 744(3; 4, see 16), 747(3, 4), 748(4), 749(22, 24), 752(22), 753(22)
Oda, Y., 132, 133(16)

Oesper, P., 308, 387, 390(4), 391, 425, 439
Ofner, P., 454
Ogawa, K., 664, 665(25)
Ogston, A. G., 343, 733
Ohlmeyer, P., 434, 735
Okamoto, K., 162
O'Kane, D. J., 97, 257, 262, 479, 480, 481(1), 482(1)
Olcott, H. S., 642
Oldewurtel, H. A., 63
Olsen, A. G., 158
Olsen, N. S., 216
O'Neal, R., 187
Onodera, K., 194
Ord, M. G., 36, 37
Osborne, T. B., 69, 71, 530, 534(6)
Ostendorf, C., 161
Oster, G., 181
Otey, M. C., 47(68a), 50
Ott, M. L., 170, 172(25), 173(25)
Ott, P., 86, 435, 437, 439(1, 5), 440(1, 5), 441, 442(1)
Otto, R., 166
Oullet, L., 722
Owens, H. S., 163
Ozawa, J., 162

P

Paech, K., 194
Palade, G. E., 10(5), 15(5, 22), 16(1), 28(8), 545
Paladini, A. C., 294, 296, 354, 355, 356(1)
Paleus, S., 113, 117(2)
Palmer, L. S., 639
Pangborn, M. C., 661
Pappenheimer, A. M., 728
Pardee, A. B., 43, 727
Paris, A., 240
Partridge, S. M., 118, 260, 614
Paul, M. H., 489, 606
Pavlovic, R., 639
Pavlovina, O., 194
Pe, S. S., 662
Pearce, R. H., 168
Peat, S., 194, 195, 196, 197(17), 222
Peers, F. G., 642
Penrose, M., 51, 56(1), 61(1)
Pensky, J., 464, 465(15), 467, 469(21), 470(15, 21)

Perkins, F. T., 170
Perlman, P. L., 170
Perry, S. V., 16(5)
Peterjohn, H. R., 367
Petermann, M. L., 16(6), 18(20)
Peters, I. I., 641
Peters, O., 236
Peters, R. A., 704
Petersen, S. R., 240
Peterson, W. H., 56
Petow, H., 639
Pfleiderer, G., 322, 391, 392(4), 394(4), 419, 436, 439(2), 440(2), 443, 568
Phaff, H. J., 158, 159(2), 162(2), 163(2), 164(2), 166(2)
Phillips, P. H., 439, 560, 623, 688
Phillips, T. G., 194
Phipps, A., 162
Pigman, W. W., 237, 251
Piguet, A., 158
Pihl, A., 657, 658(5)
Pillemer, L., 84
Pincus, J. B., 686
Pinsent, J., 754
Pintner, I. J., 680
Pippen, E. L., 163
Plass, M., 704
Platt, B. S., 639
Plaut, G. W. E., 16(4), 55, 95, 712
Plaut, K. A., 16(4)
Ploetz, T., 177, 178
Plumel, M., 142
Podber, E., 18(17)
Podolsky, B., 279
Pollis, B. D., 40(50), 50, 97
Pollack, L., 625
Pollard, E., 256
Pollock, M. R., 57, 132
Porges, E., 728
Porter, H. K., 193, 194(7)
Porter, R. R., 96, 97(33), 107, 108, 109
Porter, V. S., 27, 29(3a)
Posternak, T., 295, 423, 424(2), 425(2)
Potter, A. L., 190, 191(33), 192(33), 197
Potter, V. R., 10(2, 3, 4), 11(4), 12(1), 13(3, 8, 9, 10, 11, 12, 13), 14(1, 14, 15, 16, 17, 18, 19), 15(19), 16(2, 10), 322, 726, 727(20), 729
Powell, R., 729, 734(4), 735(4)
Powell, W., 256

Prentiss, A. M., 70
Pressmann, D., 55
Price, C. A., 725, 728
Price, J. M., 12, 15(7), 17(13), 19(13)
Pricer, W. E., Jr., 436, 441, 512, 585, 673, 705, 707, 713, 714(8)
Pringsheim, H., 173, 176(2), 177, 178(2)
Proehl, E. C., 196
Propst, L. M., 63
Provasoli, L., 130
Puck, T. T., 62
Pullman, M. E., 294, 298(4), 404
Putman, E. W., 190, 191(33), 192(33), 229, 231
Putnam, F. W., 35, 36(32)

Q

Quadbeck, G., 567
Quastel, J. H., 51, 56(1), 61(1), 257, 674, 675(1), 676(1, 5), 677(5, 12), 678(5), 727
Quibell, T. H. H., 169

R

Racker, E., 277, 286(7b), 307, 314, 371, 375, 380, 384, 385(1), 387, 442, 454, 455(1), 457(1), 458(1, 4), 500, 503, 504, 514, 515(1), 517(1), 518, 561, 695, 696, 698, 729, 730
Radulescu, V., 734
Rafter, G. W., 272, 407, 408(2), 409(2), 411(2)
Raistrick, H., 189, 242, 340, 341(2), 343(2b), 344(2)
Ram, J. S., 194, 197(16)
Ramakrishnan, C. V., 574, 586, 588(10), 602, 604(2), 607(2), 715
Randall, M., 70, 600
Randall, R. J., 150, 156(5), 207, 264, 270, 278, 296, 300, 312, 346, 441, 491, 536, 686, 687(6), 705, 707, 741
Randles, C. I., 182
Rapoport, S., 425, 426
Rapp, R., 55
Rapport, M. M., 167, 168(2), 170(2)
Ratner, S., 96
Rauch, K., 729

Raw, I., 559
Reaume, S., 256
Recknagel, R. O., 13(8), 14(17, 19), 15(19)
Record, B. R., 84
Redfern, S., 262
Reed, L. J., 494
Rees, K. H., 545
Reichert, R., 588
Reichstein, T., 368
Reid, W. W., 162, 163
Reinfurth, E., 464
Reithel, F. J., 340, 341(4), 344(4b), 345(4b)
Rendlen, E., 639
Reuter, F., 33(22), 44(64)
Richardson, H. B., 5(8)
Richtmeyer, N. K., 256
Riggs, B. C., 3(6), 8(14)
Roberts, E. C., 340, 341(4), 344(4b), 345(4b)
Roberts, I. S., 257
Roberts, N. R., 311
Robertson, W. van B., 169
Robinson, D. S., 42(56), 43(56), 49(56), 50
Robinson, H. W., 223, 312, 358, 376, 384, 410, 455, 501, 515, 760, 761(6)
Roboz, E., 162, 165(25)
Roe, J. H., 286, 300, 320, 346
Roedig, A., 729
Rogers, L. A., 54
Rona, P., 628, 639, 641
Ronzoni, E., 315
Ropes, M. W., 169
Rosano, C. L., 182, 184(8)
Rose, I. A., 591, 763
Rose, W. G., 630, 641(25)
Roseborough, N. J., 150, 156(5), 207, 264, 270, 278, 296, 300, 312, 347, 441, 491, 536, 600, 686, 687(6), 705, 707, 741
Roseman, S., 168
Rosenthal, S. M., 610
Rothchild, H., 249
Rothenberg, M. A., 310, 643, 645, 646(15)
Rothschild, H. A., 675, 676, 678(5a)
Rowen, J. W., 361
Rueff, L., 566, 572(1), 574, 582, 585, 588
Runnström, J., 434

Ryan, J., 18(17)
Rydon, H. N., 234, 237, 238, 239

S

Sable, H. Z., 351, 352, 353(2), 354(2), 358, 359(7), 360(7), 504
Sachs, H., 665
Saetren, H., 18(19)
Saji, T., 178
Sakuma, S., 318
Salton, M. R. J., 59
Sammons, H. G., 636
Sanadi, D. R., 29(13), 34(13), 603, 718, 722
Santenoise, D., 648
Sarker, N. K., 269
Savard, K., 680
Sawyer, C. H., 648
Scatchard, G., 70, 71
Schäfer, O., 461
Schäffner, A., 200, 629
Schatz, A., 130, 680
Scheff, G. J., 662
Scherp, H. W., 229
Schlenk, F., 726
Schlockermann, W., 461
Schmeller, M., 636, 639
Schmid, K., 32(19), 33(19)
Schmidt, G., 39(43), 95, 662, 665(13)
Schmith, K., 170
Schmitz, H., 14(15)
Schmitz-Hillebrecht, E., 238, 240
Schneider, K., 252, 262
Schneider, M. C., 514, 574, 576(13), 581, 685, 689(1), 693(17), 694(1)
Schneider, W. C., 10(5), 13(9), 15(5, 21), 16(1, 2, 3), 17(14), 18(17, 19, 20), 27, 29(15), 545, 546
Schneidmüller, A., 238
Schoch, T. J., 149
Schönheyder, F., 628, 630, 637(29, 31), 638(31), 639(31)
Schubert, M. P., 256, 259
Schubert, P., 174, 176(4), 177
Schudel, G., 174
Schultz, T. H., 163
Schulz, W., 403, 423, 434
Schuster, P., 313
Schwartz, B., 640

Schwartz, M., 736, 738(4), 739(4)
Schweet, R. S., 29(12), 34(12), 488, 489 (6), 606
Schweizer, E., 639
Schwenk, E., 170
Schwert, G. W., 453, 454(12)
Schwertz, F. A., 25(19)
Schwimmer, S., 43, 154
Scott, D. A., 85
Scott, D. B. M., 353
Scott, E. M., 27(3a), 29(3a), 729, 734(4), 735(4)
Scott, R. L., 69
Seastone, C. V., 169
Seegmiller, C. G., 162, 165(21)
Seegmiller, J. E., 323, 366, 510(1), 513
Seeley, H. W., 625, 626
Seidman, M., 241
Seligman, A. M., 633, 657
Sell, H. M., 198
Senti, F. R., 182(15), 184(15)
Seubert, W., 559, 560(2)
Shaffer, P. A., 211
Shapiro, B., 576, 586, 666, 687, 694(11)
Sharpe, E. S., 180, 182(6), 183(6)
Shemin, D., 575, 581, 623, 688
Shepherd, A. D., 163
Shipe, W. F., 636, 641(48)
Shmukler, H. W., 40(50), 50, 97
Shooter, K., 733
Shorr, E., 5(8)
Short, A., 82
Short, W. F., 189, 242, 340, 341(2), 343 (2b), 344(2)
Shuster, L., 95, 96, 332, 598
Sibley, J. A., 311, 314(8), 321
Siekevitz, P., 13(13), 14(14, 16, 18)
Siem, R. A., 641
Siewerdt, D., 726
Silverman, M., 472, 475(8)
Simms, H. S., 273
Simon, E. J., 575, 581, 623, 688
Simonson, H. C., 13(13)
Singer, T. P., 156, 157(14), 158(16), 464, 465(15), 467, 469(21), 470(15, 21), 471, 639, 642, 678, 726
Singh, B. M., 177
Sitch, D. A., 196
Sizer, I. W., 256, 257, 640
Skeggs, H. R., 749

Skrimshire, G. E. H., 189, 242, 340, 341 (2), 343(2b), 344(2)
Slater, E. C., 16(4), 28(7), 277, 307, 722, 724, 725, 726, 727(8)
Sleeper, B. P., 53, 54(12)
Slein, M. W., 95, 270, 273(6), 274(6), 275(6), 276(6), 277, 280, 284(1), 285(1), 289, 290(12), 291, 300, 301 (1), 304(1), 307, 371, 401, 403(1), 406(1), 425, 442
Slotin, L., 742
Slotta, K. H., 662
Smith, C. S., 197
Smith, E. E. B., 269
Smith, F. W., 58
Smith, L., 36
Smith, V. M., 178
Smits, G., 291
Smolelis, A. N., 124
Smyrniotis, P. Z., 53, 94, 326, 353, 366, 372, 375(5)
Smyth, D. H., 735
Smythe, C. V., 95, 166
Snellman, O., 38(38), 167
Sobotka, H., 628, 636, 637, 638(5)
Sørensen, S. P. L., 139, 143(2), 145(2)
Solms, J., 158, 159(2), 162(2), 163(2), 164(2), 165, 166(2)
Solomons, G. L., 162, 163
Sols, A., 275, 277, 279(2), 280(2), 281, 284, 285(2, 19, 20)
Somogyi, M., 180, 211, 234, 262, 277, 296, 357
Soodak, M., 610, 612
Sorof, S., 157, 158(16)
Sowden, J. C., 197
Specht, H., 200
Spencer, B., 267
Sperber, E., 434
Sperry, W. M., 6
Spiegelman, S., 132, 133
Springer, B., 263, 267
Sreenivasan, A., 37(36b), 675, 676(11)
Stadie, W. C., 3(6), 8(14)
Stadler, R., 174, 175, 177, 178(11)
Stadtman, E. R., 53, 54(11), 481, 482(9), 483, 484, 518, 519(1), 520(1), 521(5), 552, 553(2), 574, 576, 577(20), 586, 587, 588, 596, 597(2), 599, 600, 601,

602(1), 606, 608, 610, 615, 616, 617, 618(1), 685, 687, 694(11)
Stadtman, T. C., 678
Stafford, H. A., 22, 504
Stafford, R. O., 628
Stage, A., 32(18)
Stahly, S. L., 182
Stanciu, N., 734
Standfast, A. F., 189, 242, 340, 341(2), 343(2b), 344(2)
Stanier, R. Y., 53, 54(12), 56, 60(24), 61, 128, 132, 133, 134
Stansly, P. G., 548
Starr, M. P., 641
Staub, A., 150, 288
Stauffer, J. F., 61
Stearn, A. E., 639
Stedman, E., 647, 648
Stedman, Ellen, 647, 648
Steele, J. M., 662, 665(12), 666(12), 668(12)
Steffen, E. I., 55
Stein, S. S., 255
Stein, W. H., 113, 116(10), 117(3, 5), 124, 125, 610
Steiner, R. F., 435
Stephenson, M., 539
Stern, H., 18(19)
Stern, J. R., 514, 559, 562, 564, 573, 574(5), 575(7), 576(13), 581, 582, 584, 585(29), 586, 623, 685, 687, 689 (1), 694(1, 11), 721, 739, 747
Stern, K. G., 181, 461
Stern, L., 723
Steudel, H., 630
Stevenson, J. W., 622
Stewart, H. B., 44
Stickland, L. H., 56, 61(22), 539
Stockhausen, F., 126
Stocking, C. R., 22, 24(2)
Stodola, F. H., 183
Stolzenbach, F., 397
Stoppani, A. O. M., 728
Stotz, E., 36, 657, 658(5)
Straub, F. B., 442, 450, 576, 693, 702, 729, 736, 738(4), 739(4)
Strecker, H. J., 338, 476, 478
Strelitz, F., 647
Strength, D. R., 675
Strickler, N., 662, 665(13)

Striebich, M. J., 15(21), 18(19)
Stringer, C. S., 182(15), 184(15)
Strong, F. M., 588, 610
Strong, L. E., 90, 647
Stumpf, P. K., 58, 192, 194(1), 196(1), 197(1), 198(1), 199(1), 411, 412(3), 414(3), 549
SubbaRow, Y., 206, 207(3), 215, 295, 311, 346
Subrahmanyan, V., 310, 313(4), 314(4), 316, 460, 461(4), 463(4), 464(4), 523, 524(1, see 5), 525(1), 526(1), 527(1), 528(1)
Suda, M., 132, 133(16)
Sugg, J. Y., 181, 184(7), 187(7)
Sumner, J. B., 55, 96, 97(29), 149, 194, 197(15), 257, 262, 269, 517
Sun, C. N., 194
Sung, S.-C., 95, 712
Surgenor, D. M., 32(19), 33(19, 23), 647
Sussman, M., 133
Sutherland, E. W., 204, 215, 216, 295, 298, 423, 424(2), 425(2)
Svedberg, T., 663
Swanson, M. A., 196
Swenson, H. A., 163
Swingle, S., 97, 245, 711
Swyer, G. I. M., 169
Sykes, G., 189, 242, 340, 341(2), 343(2b), 344(2)
Sylvén, B., 38(38)
Synge, R. L. M., 99
Szent-Györgyi, A., 27, 32(2), 723
Szulmajster, J., 50

T

Tabor, H., 517, 576, 577(20), 588, 608, 610, 615
Takagi, Y., 14(16)
Takai, Y., 177
Takeda, A., 162
Talalay, P., 263
Tallan, H., 113, 117(3), 124, 125
Tamura, E., 177
Tankó, B., 467
Tapano, K., 234
Tarr, H. L. A., 182
Tatum, E. L., 162, 165(25)
Taube, C., 513

Tauber, H., 232, 249
Taxi, J., 651
Taylor, D. B., 133, 453
Taylor, H. L., 90, 647
Taylor, J. F., 26, 90, 307, 310, 312(2), 313(2), 314(2), 315(2), 381, 406, 410, 425, 427(7), 442
Temple, J. W., 142
Teorell, T., 86
Teply, L. J., 34(25)
Testa, E., 289, 322
Thannhauser, S. J., 39(43), 662, 665(13)
Theorell, H., 454, 495, 498(1), 499(1)
Thimann, K. V., 725, 728
Thomas, G. J., 193, 197(6)
Thomas, R. A., 55
Thompson, R. H. S., 36, 37
Thompson, R. R., 154, 156(13), 157(13)
Thorn, M. B., 727
Thunberg, T., 723, 726
Thurlow, S., 35
Tietz, N., 642
Tiselius, A., 97, 193, 194(4), 196(4), 199(4), 245, 711
Tishkoff, G. H., 18(19)
Tolbert, N. E., 528, 529(1), 531(1), 532(1)
Tolksdorf, S., 170
Torriani, A. M., 189, 190(30), 191(30), 192(30)
Tóth, G., 174, 175, 177, 178(11), 250
Tracey, M. V., 177
Trager, W., 174, 176(6), 177
Trevelyan, W. E., 198
Trowell, O. A., 675
Trucco, R. E., 96, 232, 250, 291, 292(5), 294, 296, 354, 355, 356(1)
Tsao, T.-C., 43, 45, 46, 50
Tsou, C. L., 724, 727(7), 728
Tsuchida, M., 53, 54(12)
Tsuchihashi, M., 497
Tsuchiya, H. M., 179, 181(3), 182(6, 15), 183(2, 6), 184(2, 15)
Tucker, I. W., 630, 636(23), 637(23), 638(23)
Turvey, J. R., 194
Tuttle, L. C., 457, 458(5), 480, 482(2), 483, 487, 519, 586, 591, 597, 604, 606(9), 607(9), 617, 649, 686, 688(8, see 15), 719

U

Udenfriend, S., 409
Ullmann, E., 628, 642
Umbreit, W. W., 52, 54(5), 61, 685
Uroma, E., 32(19), 33(19)
Utter, M. F., 51, 56(1), 60, 61(1), 473, 476, 478, 760, 763

V

Vahlquist, B., 648
van Bruggen, J. T., 340, 341(4), 344(4b), 345(4b)
VanDemark, P. J., 625, 626
van Gremberger, G., 728
van Heyningen, R., 197, 274
Vas, K., 163
Veibel. S., 234, 236, 237, 238, 240, 251
Veiga Salles, J. B., 739, 741(2), 742
Velick, S. F., 315, 402, 403, 404, 405(4), 406(9), 409, 411, 503
Velluz, L., 463, 464(13)
Vennesland, B., 454, 467, 472, 504, 536, 740, 741, 742, 744(11), 748(6)
Vestling, C. S., 288
Vishniac, W., 275
Vogel, L., 630
Volkin, E., 95, 96
Volqvartz, K., 628, 630, 637(29, 31), 638(31), 639(31)
Vorsatz, F., 240

W

Waelsch, H., 42(57), 50
Wagner-Jauregg, T., 503
Wainio, W. W., 36(33)
Wajzer, J., 269, 274(27, see 22), 275, 277
Wakil, S. J., 35(29a), 548, 555, 562, 564, 565, 566, 571(4), 572(4), 585
Waksman, S. A., 165
Walden, M. K., 154, 156(13), 157(13)
Waldschmidt-Leitz, E., 281, 629, 630(14), 634, 635, 638(14), 639(13), 641, 642(79)
Walker, P. G., 37(37), 38(37), 267, 269 (13)
Wallenfels, K., 248, 503

Walpole, G. S., 140
Walseth, C., 176, 177
Walter, I. O., 99
Wang, T. P., 598
Warburg, O., 3, 9(1), 24(12), 85, 86, 299, 311, 312(12), 314(12), 315, 316, 317(1, 4), 318, 323, 335, 371, 376, 387, 391, 392(3), 398, 401, 407, 408 (1, see 6), 410(1, see 6), 412, 413, 419, 425, 428, 429, 433(2), 434(2), 455, 484, 491, 501, 515, 529, 533(3), 560, 568, 576, 583, 592, 604, 710, 716, 740, 741(7)
Warren, C. O., 3(2)
Warringa, M. G. P. J., 47(67), 50, 646
Wartenberg, H. von, 641
Watkins, W. M., 615
Watson, R. W., 366, 367, 370(2)
Watt, D., 474
Waygood, E. R., 748
Webb, E. C., 197, 270, 271(7), 275(7), 639
Weber, F., 165
Weber, G. M., 12(7), 15(7)
Weber, H. H., 638
Webley, D. M., 180, 183(5)
Wehrli, W., 174, 176(4), 177
Weibull, C., 193, 194(4), 196(4), 199(4)
Weichselbaum, T. E., 216, 300
Weidenhagen, R., 233, 257, 262
Weier, T. E., 22, 24(2)
Weil, R. M., 197
Weil-Malherbe, H., 275, 282, 285
Weinberger, P., 748
Weinhouse, S., 533
Weinstein, S. S., 630, 637, 638, 639(28)
Weischer, Ä., 630
Weiss, M. S., 620
Weisz-Tabori, E., 699, 704(2)
Wellman, H., 545, 548
Wells, I. C., 677
Werkman, C. H., 51, 56(1), 60, 61(1), 472, 473, 475(8), 476, 477(1), 478(1), 480, 481(5), 482(5), 760
Wessely, L., 566, 572(1), 574, 585
Westerfeld, W. W., 471, 472
Westerkamp, H., 461
Westheimer, F. H., 454
Wheatly, A. H. M., 727
Whelan, W. J., 193, 194, 196, 197(6)
Whitaker, D. R., 177

White, J. W., Jr., 258
Whitehead, B. K., 232
Whiteside-Carlson, V., 182, 184(8)
Whitley, R. W., 169, 170(18)
Wiebelhaus, V. D., 280, 285
Wieland, O., 566, 572(1), 574, 585
Wieland, T., 458, 582
Wigglesworth, V. B., 259
Wild, G. M., 196
Wilkinson, I. A., 194, 197(17)
Wilkinson, J. F., 292
Willaman, J. J., 158, 159, 160
Williams, J. N., Jr., 37(36b), 675, 676(8, 11), 677(8), 678(8)
Williams, R. T., 267
Williams-Ashman, H. G., 350
Wills, E. D., 275
Willstaedt, E., 256
Willstätter, R., 90, 97(2), 174, 252, 281, 513, 629, 630(14), 634, 635, 636, 638(14), 639(14), 641, 642(79), 708
Wilson, E. E., 630
Wilson, I. B., 554, 621, 623(9), 642, 643, 644(4), 648
Wilson, J., 480, 481(5), 482(5)
Wilson, T. H., 739
Winkler, G., 166
Winkler, S., 236, 240
Winogradsky, S., 131
Wolf, H. P., 320, 322, 397
Wolfe, R. S., 479, 480, 481(1), 482(1)
Wolff, J. B., 346
Wolfrom, M. L., 197
Wolken, J. J., 25(19)
Wolochow, H., 229
Wood, A. J., 51
Wood, H. G., 476, 760
Wood, W. A., 323

Woodward, G. E., 274
Woodward, H. E., 675, 676(5), 677(5), 678(5)
Wooldridge, W. R., 727
Woolf, B., 734
Wormall, A., 275
Wosilait, W. D., 215, 221
Wrede, H., 461
Wright, L. O., 749
Wright, R. C., 27(3a), 29(3a)
Wu, M.-L., 311
Wulff, H. J., 503
Wynne, A. M., 630, 637, 638, 639(28)

Y

Yamazaki, K., 193, 194(8)
Yates, E. D., 257
Yin, H. C., 194
Young, F. G., 44
Young, R. E., 20(3)

Z

Zamecnik, P. G., 668, 670(35), 671(35)
Zechmeister, L., 174, 175, 177, 178(11), 250
Zeile, K., 33(22), 44(64)
Zelitch, I., 528, 529(2), 531(4). 532
Zeller, E. A., 645, 660
Zerfas, L. G., 44(63), 444, 445(1), 447(1), 448(1), 449(1), 728
Ziegler, J., 439
Ziese, W., 174, 176(7), 177
Zittle, C. A., 37(36a), 39(44), 42(55), 50, 647
Zscheile, F. P., 25(19)

Subject Index

Each enzyme is indexed not only under its name, but also under the names of its substrates, and under the names of the tissues or organisms in which it occurs.

Enzymes marked by asterisk (*) are covered in detail in Volume II.

A

AAC, see S-Acetoacetyl-N-acetylcysteamine

Acetaldehyde,
acyloin formation from, 464
assay of with aldehyde dehydrogenase, 514
dismutation of in *Cl. kluyveri*, 518
in DR-aldolase reaction, 384–386
inhibition of wheat germ carboxylase by, 470
inhibition of yeast carboxylase by, 464
as substrate for alcohol dehydrogenase (DPN), 501
for alcohol dehydrogenase (TPN), 507
for aldehyde dehydrogenases, 510, 513, 517, 520
for glyceraldehyde-3-phosphate dehydrogenase, 404
for liver aldehyde oxidase, 527

Acetate,
conversion to acetyl CoA,
by aceto-CoA-kinase, 585–591
by CoA transphorase, 599–602
phosphorylation of by ATP, 481, 591–595
quantitative determination of with acetokinase, 595

Acetate-activating enzyme, see Aceto-CoA-kinase

Acetate buffer, 80, 140
table for various concentrations of, 80

Acetate kinase (acetokinase),
from *E. coli*, 591–595
from other sources, 595

Acetic acid,
as solvent for serum albumin, 83

Acetoacetate,
as acceptor in CoA transferase reaction, 580
conversion to acetoacetyl CoA, 573–581
by reaction with CoA and ATP, 573–574
by reaction with succinyl CoA, 574–581
formation and breakdown of, 573–585
formation from octanoate in rat liver mitochondria, 545–548
synthesis from acetyl CoA,
in heart, 574
in pigeon liver, 574

Acetoacetate cleavage enzyme, see Thiolase

Acetoacetate condensing enzyme, see Thiolase

Acetoacetate decarboxylase,
from *Clostridium acetobutylicum*, 624–627
coenzyme for, 626
resolution of, 626

Acetoacetic acid, preparation of, 624

S-Acetoacetyl-N-acetylcysteamine (AAC) (S-Acetoacetyl-N-acetylthioethanolamine),
absorption band of, 574–575
effect of p-chloromercuribenzoate and cysteine on, 575
of magnesium ion on, 575
preparation of, 567
as substrate for β-ketoreductase, 566–573

S-Acetoacetyl coenzyme A,
absorption band of,
effect of p-chloromercuribenzoate and cysteine on, 575
of magnesium ion on, 575
cleavage by deacylase, 574
by thiolase, 574, 581–585
determination with hydroxylamine, 582
formation of, 562
preparation of, 572, 581–582
chemically, 582
enzymatically, 581–582
from crotonyl CoA, 582
from succinyl CoA, 581
reduction of, 566–573

Acetoacetyl-S-thiophenol,
in synthesis of acetoacetyl CoA, 582

Aceto-arylamine-kinase, see also Acetylation of amines
in assay of aceto-CoA-kinase, 586, 588
Aceto-CoA-kinase,
in assay of choline acetylase, 623
from baker's yeast, 585–591
from heart, 588
from pigeon liver, 588
simulation of acetate kinase in crude extracts by, 593
Acetoin, see also α-Acetolactic acid and Acyloin condensation
colorimetric determination of, 471
synthesis by bacterial enzymes, 471–475, 486
from *A. aerogenes*, 471–475
from *B. subtilis*, 475
from *L. delbrückii*, 486
from *S. faecalis*, 474
synthesis of by fraction A of pyruvic oxidase, 493
by pyruvic oxidase of pigeon breast muscle, 489
by wheat germ carboxylase, 470, 471
α-Acetolactic acid,
decarboxylase for,
from *A. aerogenes*, 471–475
specificity for dextrorotatory form, 474
from *S. aureus*, 474
enzyme for formation of, 471–475
Acetone,
formation from acetoacetate, 624–627
phase diagrams for, 101, 106
as precipitant for human carboxyhemoglobin, 84
Acetone powders,
preparation from animal tissues, 34, 35, 609–610
from bacteria, 55
from higher plants, 63
Acetopyruvic acid,
inhibition of acetoacetate decarboxylase by, 627
Acetylation, see also Aceto-arylamine-kinase
of amines, 608–612
of amino acids, 616–619
of D-glucosamine, 612–615
with aged pigeon liver extract, 613

with purified glucosamine acceptor system, 613
Acetylcholine,
bioassay for, 623
determination with alkaline hydroxylamine, 623, 649
in generation of bioelectric currents, 642–643
inhibition of acetylcholinesterase by, 643
Acetylcholinesterase, 642–651, see also Cholinesterase
biological significance of, 642–643
criteria for classification as, 644
distinction from other esterases, 643–644
measurement of activity of, 647–651
colorimetric method, 649
histochemical staining technique, 650–651
manometric method, 648
application to blood, 649
titration methods, 648
sources and preparation of, 644–647
brain, 644–645
electric organs, 645
red blood cells, 646–647
snake venoms, 645
squid ganglia, 645
stability of, 645
Acetyl coenzyme A,
aceto-CoA-kinase as convenient source of, 615
in acetylation of amines, 608–612
of amino acids, 616–619
of choline, 620
of glucosamine, 612–615
assay by amine acetylation, 611
in citrate formation, 685–694
in CoA transphorase reaction, 599–602
formation of by aceto-CoA-kinase, 585–591
by aldehyde dehydrogenase, 518–523
in bacterial pyruvic oxidase system, 490–495
in pig heart pyruvic oxidase system, 486–490
free energy of hydrolysis of, 694
optical density change on cleavage of thioester bond of, 599

preparation of, 688
in thiolase reaction, 581
in transacetylase reaction, 596
trapping methods for, 576
Acetyl coenzyme A deacylase,
from pig heart, 606–607
N-Acetylcysteamine,
in synthesis of S-acetoacetyl-N-
acetylcysteamine, 567
N-Acetylgalactosamine,
chromatography of, 614
N-Acetylglucosamine,
chromatography of, 614
formation of, 612–615
as inhibitor for mammalian hexokinase, 284
Acetyl kinase, see Aceto-CoA-kinase
Acetyl-β-methylcholine,
for assay of erythrocyte acetylcholinesterase, 650
as substrate for identification of acetylcholinesterase, 643
Acetyl phosphate,
from acetate and ATP, 591–595
in assay of acetyl CoA deacylase, 606
cleavage by phosphatase-like enzyme,
607
determination of with hydroxylamine,
487, 686
formation from pyruvate in bacteria,
476–486
as phosphate donor to butyrate, 481
as product of butyrate oxidation, 551–
553
of dismutation of vinyl acetate, 553
of ethanol oxidation, 553
requirement for, in butyrate oxidation,
551
as substrate for formation of butyrate
and caproate, 553
in transacetylase reaction, 596
Acetylthiocholine,
as substrate for acetylcholinesterase,
650
Aconitase,
from heart muscle, 31, 695–698
optical method for, 695
Aconitate buffer, 139
cis-Aconitate,
light absorption by, 695

as substrate for aconitase, 695–698
trans-Aconitate,
inhibition of aconitase by, 698
Acrylate,
conversion to acrylyl CoA, 590
Ac~SCoA, see Acetyl coenzyme A
Actin, 50
Actomyosin, from heart muscle, 32
Acyl coenzyme A,
acylation of amino acids by, 618
extinction coefficient of, 600
Acyl-mercapto bond,
energy of, 590
Acyloin condensation,
catalysis by yeast carboxylase, 464
Acyl phosphates,
exchange with labeled orthophosphate,
405
Adaptation, 132–137
with resting cells, 135–136
Adaptive enzymes,
D-arabinose isomerase, 366
*barbiturase, 135, 136
cholestenone oxidizing system from
Mycobacterium, 678
β-galactosidase from E. coli, 248
gluconokinase from E. coli, 350–354
L-glycerophosphorylcholine diesterase,
135
kynureninase from Neurospora, 135
kynureninase from Pseudomonas, 133,
134
"malic" enzyme (DPN) from L.
arabinosus, 748–753
pyrocatecase, 134
*tryptophan peroxidase-oxidase sys-
tem, 134
*uracil-thymine oxidase, 135, 136
xylose isomerase, 366
Adenine,
inhibition of pentose phosphate isomerase by, 365
Adenosine,
inhibition of pentose phosphate isomerase by, 365
Adenosine diphosphate (ADP),
inhibition of mammalian hexokinase
by, 279, 285
phosphorylation by ITP, 759

Adenosine monophosphate, see 3'- and 5'-Adenylic acids

Adenosine triphosphate (ATP),
in acetylation of coenzyme A, 585–591
in assay of mammalian hexokinase, 278
as component of fatty acid oxidation system, 546
in conversion of succinate to succinyl-CoA, 718–722
in galactokinase reaction, 290–291
inhibition of beef heart isocitric dehydrogenase (DPN) by, 713
of glucose dehydrogenase by, 339
of G-1-P kinase by, 356
of succinyl CoA deacylase by, 605
in phosphatidic acid synthesis, 673
in phosphorylation of acetate, 591–595
of fructose-6-phosphate, 306
presence of IDP in, 759
reaction with IDP, 759
stimulation of choline oxidase by, 678
substitution for ITP in oxalacetate synthesis, 759
in yeast hexokinase reaction, 269–270

3'-Adenylic acid (adenosine-3'-phosphate) (3'-AMP),
inhibition of pentose phosphate isomerase by, 365

5'-Adenylic acid (adenosine-5'-phosphate) (5'-AMP),
activation of beef heart isocitric dehydrogenase (DPN) by, 713
of yeast isocitric dehydrogenase (DPN) by, 709
in assay of phosphorylase, 201, 215, 221
effect on muscle phosphorylase, 197, 205, 206, 208
inhibition of pentose phosphate isomerase by, 365
prevention of pyridine nucleotide destruction by, 740
as product of aceto-CoA-kinase reaction, 585–591

Adipate,
competitive inhibition of succinic dehydrogenase by, 727

ADP, see Adenosine diphosphate

Adrenal,
β-ketoreductase in, 573

Adsorbents,
for invertase, 257
for separation of proteins, 90–98

Adsorption,
agents for, 96

Aerobacter aerogenes,
α-acetolactic acid decarboxylase from, 471–475
α-acetolactic acid-forming enzyme from, 471–475
glycerol dehydrogenase from, 397–400
growth medium for, 398, 472
levansucrase from, 186–189

Agar,
as carbon source for microbes, 131

β-Alanine, acetylation of, 618

L-Alanine deaminase,
formation in E. coli, 136

DL-Alanylglycine,
acetylation of, 618

Albumin,
activation of β-glucuronidase by, 267
of lipase by, 638
egg,
acetylation of, 618
solubility of, 74
effect of pH on, 79
in thiolase assay, 582
serum,
precipitation by zinc ions, 85
as trapping agent for acetaldehyde, 466

Alcohol dehydrogenase (DPN),
in assay of aldehyde dehydrogenase, 518
from baker's yeast, 500–503
from brewer's yeast, 503
in DR-aldolase assay, 384
from liver, 495–500
in L. mesenteroides, 506

Alcohol dehydrogenase (TPN),
in baker's yeast, 503, 504
in higher plants, 504
from L. mesenteroides, 504–508

Alcohols, higher,
activation of pancreas lipase by, 638
inhibition of liver esterase by, 638

Aldehyde dehydrogenase (CoA-linked),
from Cl. kluyveri, 518–523

Aldehyde dehydrogenase (DPN),
  from liver, 514–517
Aldehyde dehydrogenase (DPN, TPN),
  from baker's yeast, 508–511
  activation of by potassium, 510
Aldehyde dehydrogenase (TPN),
  from baker's yeast, 511–514
Aldehyde oxidase (flavin-linked),
  from pig liver, 523–528
Aldehydes,
  aromatic,
    acyloin formation from, 464
    assayed with liver aldehyde dehydro-
      genase, 517
    inhibition of lipase by, 638
Aldolase,
  absence of from *L. mesenteroides*, 330
  in assay of diphosphoglyceric acid
    mutase, 425
  of phosphohexokinase, 307
  of transaldolase, 381
  crystallization of, 313, 394–395
  from muscle, 310–315, 316, 394–395
    action on F-1-P, 322
  from yeast, 315–320
Alfalfa,
  pectinesterase from, 159
  pentose phosphate isomerase from,
    363–366
Algae, unicellular,
  source of disintegrated chloroplasts, 23
Alginic acid,
  as adsorbent for polygalacturonidase,
    164
Alkaloids,
  inhibition of lipase by, 639
Alkylating agents,
  inhibition of succinic dehydrogenase
    by, 727
Alkyl glucosides,
  hydrolysis by β-glucosidase, 238
Allose-6-phosphate,
  as inhibitor of mammalian hexokinase,
    285
Alloxan,
  as inhibitor of mammalian hexokinase,
    285
Allyl alcohol,
  as substrate for yeast alcohol dehy-
    drogenase, 502

Altrose,
  as substrate for glucose aerodehydro-
    genase, 345
Alumina gel, Cγ,
  preparation of, 97, 98
  for removal of cytochrome *c*, 91
Alumina powder, see also Bauxite
  as adsorbent, 96
  for grinding bacterial cells, 60
  for grinding plant tissues, 63
Aluminum ion,
  activation of plant carboxylases by,
    463
  inhibition of wheat germ carboxylase
    by, 470
Amines,
  acetylation of, 608–612, 612–615
  aliphatic, ninhydrin method for, 610
  low energy acetyl transfer between, 615
Amino acid acetylase,
  from *Cl. kluyveri*, 616–619
*Amino acid decarboxylase,
  in *E. coli*, 137
*D-Amino acid oxidase,
  from liver, 33
Amino acids,
  activation of dialkylphosphofluoridase
    by, 656
Aminoazobenzene,
  acetylation of, 588, 610
4-Aminoazobenzene-4'-sulfonic acid,
  acetylation of, 588, 610
p-Aminobenzoic acid,
  acetylation of, 588, 610, 618
  effect on phosphorylase content of
    plants, 198
2-Amino-2-methyl-1,3-propanediol
  buffer (Ammediol), 145
*Aminopeptidase,
  in kidney, 50
  MnSO₄ in chromatography of, 107
  solubilization by butanol of, 42
Ammonium ion,
  activation of aceto-CoA-kinase by, 591
    of glycerol dehydrogenase by, 400
    of pyruvate kinase by, 439
    of transacetylase by, 598
    of yeast aldehyde dehydrogenase
      (DPN, TPN) by, 510

inhibition of choline dehydrogenase by,
    677
  of glycerol dehydrogenase by, 400
  of phosphomannose isomerase by,
    304
Ammonium sulfate,
  mechanical addition of, 394
  nomogram for adjustment of concen-
    tration of, 77
  recrystallization of, 393
  table for adjustment of concentration
    of, 76
  use of in protein fractionation, 75–78
5'-AMP, see 5'-Adenylic acid
n-Amyl alcohol,
  as substrate for yeast alcohol dehy-
    drogenase, 502
Amylases, α and β, 149–158
  α-Amylase,
    from A. oryzae, 154
    from B. subtilis, 154
    effect on plant phosphorylase, 193
    human pancreatic, 154
    inhibition of amylosucrase by, 185
    from malt, 154
    products of action of, 152
    removal from branching enzyme by
      starch adsorption, 223, 224
    from saliva, 150–154
    substrates for, 152
  β-Amylase,
    inhibition of plant phosphorylase by,
      193
    from malt, 158
    from sweet potato, 154–158
Amylo-1,6-glucosidase, 211–215
Amylomaltase,
  from E. coli, 189–192
Amylopectin,
  arsenolysis of, 197
  as substrate for α-amylase, 152
    for β-amylase, 156, 157
    for branching enzyme, 222
    for phosphorylase, 197, 205
Amylose,
  as product of phosphorylase action, 192
  as substrate for α-amylase, 152
    for β-amylase, 156
    for phosphorylase, 205

Amylosucrase,
  from Neisseria perflava, 184–185
Aniline,
  inactivation of invertase by, 257
Animal tissues,
  extraction of enzymes from,
    by acetone powder method, 34–35
    with aqueous solutions, 29–33
    use of ATP for, 32
    by autolysis, 38–39
    by butyl alcohol method, 34, 40–51
    by use of detergents, 35–38
    differential, 33
    disintegration of tissue for, 28, 29
      by high frequency oscillations, 29
      by freezing and thawing, 34
      by use of glycine, 32
      of glycylglycine, 32
    use of potassium thiocyanate for, 32
    using purified enzymes, 39, 40
Anions, see also under names of individ-
    ual anions
  activation of β-D-galactosidase by, 248
Anions, monovalent,
  activation of α-amylase by, 154
    of fumarase by, 734
    of β-glucosidase by, 239
  inhibition of fumarase by, 734
Anions, polyvalent,
  activation of fumarase by, 734
o-Anisidine,
  acetylation of, 610
p-Anisidine,
  acetylation of, 610
Antabuse, see Tetraethylthiuram disul-
    fide
Anthranilic acid mutant,
  isolation of, 128
Apoenzyme,
  acid ammonium sulfate for prepara-
    tion of, 486
  of yeast carboxylase, 464
D-Arabinose,
  conversion to D-ribulose, 366–370
D-Arabinose isomerase,
  from E. coli, 366–370
L-Arabinose,
  as glycosyl acceptor for sucrose phos-
    phorylase, 228

β-L-Arabinosides,
  as substrates for β-glucosidase, 237
*Arginase from liver, 33
Arsenate,
  activation of glyceraldehyde-3-phos-
      phate dehydrogenase (DPN), 414
  in assay of phosphotransacetylase, 596
  effect on fluoride inhibition of enolase,
      434
  inhibition of formate fixation by, 478
  of phosphoglucomutase by, 294
Arsenicals,
  inhibition of isocitric dehydrogenase
      by, 704
  of succinic dehydrogenase by, 727
  of yeast carboxylase by, 464
Arsenite,
  inhibition of acetoin-forming system
      by, 475
  of aldehyde oxidase by, 527
      protection against by substrate,
      527
Arsenocholine,
  as substrate for choline dehydrogen-
      ase, 677
Arsenolysis,
  of acetyl phosphate, 596–597
  of amylopectin, 197
  of sucrose and α-glucose-1-phosphate
      by sucrose phosphorylase, 228
Arsenomolybdate color reagent, Nelson's,
  preparation of, 235
  use of in β-glucosidase assay, 235
Arsenoxide,
  inhibition of thiolase by, 584
Aspartate, acetylation of, 618
Aspergillus niger,
  cellulase from, 176
  lipase from, 641
Aspergillus oryzae,
  cellulase from, 177
  cellobiase from, 178
  crystalline α-amylase from, 154
  lipase from, 641
  pH optimum of invertase from, 259
Atabrine, see Quinacrine
Atoxyl, inhibition of esterase by, 639
ATP, see Adenosine triphosphate
*ATP-ase, see Myosin ATP-ase

Avocado,
  pentose phosphate isomerase from, 366
Azide,
  activation of amino acid acetylase by,
      619
  inhibition of acetoacetate decarboxy-
      lase by, 627
  of acetone-forming system by, 475
  of aldehyde oxidase by, 527
  of formic dehydrogenase by, 539
  of yeast isocitric dehydrogenase
      (DPN) by, 709
Azotobacter vinelandii,
  acetate kinase from, 595

B

Bacillus subtilis,
  acetoin-forming system from, 475
  crystalline α-amylase from, 154
  extraction of by grinding, 59
  levan-splitting enzymes from, 187
Bacteria, see also under names of individ-
      ual species
  disintegration of by high-pressure ex-
      trusion, 64–67
  extraction of enzymes from, 51–62
  lyophilization of, 54–55
  preparation and extraction of acetone-
      dried, 55
  preparation and extraction of vacuum-
      dried, 53–54
  special techniques for enzyme studies
      with, 126–137
Baker's yeast, see Yeast, baker's
BAL (British Anti-Lewisite),
  inhibition of potato phosphorylase by,
      197
  reversal of metal inhibition of isocitric
      dehydrogenase by, 704
Barbital buffer (Veronal buffer), 144
*Barbiturase,
  formation of in Mycobacterium, 135,
      136
Barium ion,
  activation of yeast aldehyde dehydro-
      genase (TPN) by, 514
  inhibition of phospholipase C by, 671
Barium hydroxide-zinc sulfate reagents,
  in pentokinase assay, 357

preparation of, 235, 249
use of in α-galactosidase assay, 249
in β-glucosidase assay, 235
Barley,
malt,
cellulase from, 177
phosphorylase from, 194
shoots, etiolated,
extraction of, 62
Bauxite, see also Alumina powder
use in preparation of melibiase, 250
Bee, invertase from, 259
Beets,
"malic" enzyme (TPN) in, 748
Bentonite, as adsorbent, 96, 257
Benzaldehyde,
inhibition of liver aldehyde dehydrogenase (DPN) by, 517
as substrate for aldehyde dehydrogenase, 510, 517
for liver aldehyde oxidase, 527
Benzedrine,
inhibition of choline dehydrogenase by, 677
Benzene,
use of in removing lipid material, 45
Benzoylcholine,
as substrate for acetylcholinesterase, 643
Benzyl alcohol,
as acceptor for β-fructofuranosidase, 255
Benzyl butyrate,
as substrate for lipase, 630
4,6-Benzylidene glucose,
as substrate for glucose aerodehydrogenase, 345
Benzyl stearate,
as substrate for lipase, 630
Betaine,
formation by choline oxidase, 675
inhibition of choline dehydrogenase by, 677
Betaine aldehyde,
formation by choline oxidase, 674–678
as substrate for aldehyde dehydrogenase, 517
Betaine aldehyde dehydrogenase,
as component of choline oxidase system, 675

Bicarbonate, see also Carbon dioxide
inhibition of formate-fixation by, 478
Bile salts, see also Cholate and Deoxycholate
activation of lipases by, 638
Biotin,
and oxalacetic carboxylase, 756
Bisulfite (sodium hydrogen sulfite),
as activator of β-glucosidase, 240
inhibition of mannitol phosphate dehydrogenase by, 348
Blattella germanica (cockroach),
invertase from, 259
Borate,
effect on pentose phosphate isomerases, 370
inhibition of sorbitol dehydrogenase, 350
Borax-sodium hydroxide buffer, 146
Boric acid-borax buffer, 145
Brain,
acetylcholinesterase from, 644–645
choline acetylase from, 622
cholinesterase from, 36–37, 50
dialkylphosphofluoridase in, 653
hexokinase from, 281–282
β-ketoreductase in, 573
"malic" enzyme (TPN) in, 747
solubilization of cholinesterase from, 47
triosephosphate isomerase from, 391
Branching enzyme from liver, 222–225
Brilliant Alizarin Blue,
inhibition of sugar cane phosphorylase by, 198
Broad beans,
phosphorylase from, 194
Bromide ion,
activation and inhibition of fumarase by, 734
o-Bromoaniline, acetylation of, 610
p-Bromoaniline, acetylation of, 610
Bromoform,
inhibition of lipase by, 639
α-Bromoproprionic acid,
inhibition of acetoin-forming-system by, 475
Buffers, 80, 81, 138–146
acetate, 80, 140
table of various concentrations of, 80

aconitate, 139
2-amino-2-methyl-1,3-propanediol
(Ammediol), 145
barbital, 144
borax-sodium hydroxide, 146
boric acid-borax, 145
cacodylate, 142
carbonate-bicarbonate, 146
citrate, 140
citrate-phosphate, 141
glycine-hydrochloric acid, 139
glycine-sodium hydroxide, 146
hydrochloric acid-potassium chloride,
138
maleate, 142
phosphate, 81, 143
table of various concentrations of,
81
phthalate-hydrochloric acid, 139
phthalate-sodium hydroxide, 142
succinate, 141
tris(hydroxymethyl)aminomethane
(Tris), 144
tris-maleate, 143
veronal, see Barbital
1,4-Butandiol,
as substrate for glycerol dehydrogenase, 400
2,3-Butandiol,
as substrate for glycerol dehydrogenase, 400
Butyl alcohol,
for extraction of enzymes, 40–51
list of proteins purified by use of, 50
solubilization of fraction R of phospholipase A with, 664
as substrate for *Leuconostoc* alcohol
dehydrogenase (TPN), 507
as substrate for yeast alcohol dehydrogenase (DPN), 502
sec-Butyl alcohol,
as substrate for yeast alcohol dehydrogenase, 502
Butyl Carbitol,
phase diagram for, 106
Butyl Cellosolve,
phase diagrams for, 104, 105
n-Butyl-β-D-galactoside,
as substrate for β-galactosidase, 248

Butyraldehyde,
as substrate for alcohol dehydrogenase
(TPN), 507
for aldehyde dehydrogenase, 510,
517, 520
for glyceraldehyde-3-phosphate dehydrogenase, 404
for liver aldehyde oxidase, 527
Butyrate,
in coenzyme A transphorase reaction,
599–602
oxidation of in *Cl. kluyveri*, 551–553
phosphorylation by acetyl phosphate,
481
phosphorylation by ATP, 481
Butyric acid,
phase diagram for, 102
Butyrylcholine,
for assay of serum esterase, 650
as substrate for acetylcholinesterase,
643
Butyryl coenzyme A,
acylation of *p*-nitroaniline with, 611
in CoA transphorase reaction, 599–602
homologs of, 558
as substrate for transacetylase, 598,
Butyryl coenzyme A dehydrogenase,
from beef liver, 553–559
effect of adsorbents on, 96

C

Cabbage,
phospholipase D from, 672
Cacodylate buffer, 142
Cadmium ion,
activation of oxalacetic carboxylase by,
757
of plant carboxylases by, 463, 469
inhibition of phospholipase C by, 671
Calcium chloride,
activation and inhibition of alcohol dehydrogenase (TPN) by, 507
Calcium ion(s),
activation of α-amylase (malt) by, 157
of *Cl. welchii* lecithinase by, 299
of dialkylphosphofluoridase by, 656
of lipases by, 630, 638
of oxalacetic carboxylase by, 757
of phospholipase A by, 665

of phospholipase C by, 671
of plant carboxylases by, 463
of trypsin by, 638
of yeast aldehyde dehydrogenase
(TPN) by, 514
inhibition of enolase by, 434
of phosphomannose isomerase by,
304
of pyruvate kinase by, 439
for precipitation of casein, 85
Calcium phosphate gel,
preparation of, 98
n-Caproate,
as CoA acceptor in *Cl. kluyveri*, 602
n-Caprylate,
as CoA acceptor in *Cl. kluyveri*, 602
Carbazole reaction,
for keto sugars, 377
Carboligase, 464, see also Carboxylases,
plant
Carbonate-bicarbonate buffer, 146
Carbon dioxide, see also Bicarbonate
exchange into carboxyl of pyruvate,
catalysis by fraction A of pyruvic
oxidase, 493
fixation into d-isocitrate, 704
into malate, 753
into oxalacetate, 758–763
Carbon dioxide, C^{14} labeled,
in assay of oxalacetate synthesizing enzyme, 758–760
Carboxyhemoglobin,
selection of solvent for precipitation of,
84
solubility of,
in ammonium sulfate, 74
effect of ionic strength on, 73
from horse, 70–71
species variability of, 74
Carboxylases, plant, 460–471
α-carboxylase from wheat germ, 464–
471
from yeast, 460–464
Carrot,
extraction of, 62
"malic" enzyme (TPN) in, 748
Cartesian diver,
in measurement of acetylcholinesterase, 648

Casein,
partition of between solvents, 99
precipitation of by calcium ions, 85
purification by butyl alcohol, 50
Castor bean,
blastolipase from, 642
lipase from, 641
phospholipase from, 666
*Catalase,
as contaminant of glucose aerodehydrogenase, 341
formation in *Pseudomonas*, 136
from liver, 33
use in galactosidase assay, 242–243
in glucose aerodehydrogenase assay,
341
Catechol,
adaptation of *Pseudomonas* to, 135
Cations, see also under names of individual cations
activation of β-glucosidase by, 239
inhibition of β-galactosidase by, 248
Cations, divalent,
activation of oxalacetic carboxylase by,
757
of pectinesterase by, 162
stimulation of non-enzymatic decarboxylation of oxalacetate by, 757
Cations, monovalent,
activation of "malic" enzyme (DPN)
by, 752
of pectinesterase by, 162
of β-galactosidase by, 248
Celite as adsorbent, 96
Cell components, see also Chloroplasts,
Microsomes, Mitochondria, Nuclear
fraction, Nuclei
fractionation of in animal tissues,
16–19
Cellobiase,
from *Aspergillus oryzae*, 178
separation from cellulase, 175
from lichenase, 178
Cellobiose,
cleavage by cellobiase, 175
Cellodextrin,
cleavage by cellulase, 175
Cellotriose,
cleavage by cellobiase, 175

Cellulase,
  from *Helix pomatia* (snail), 173–178
  from other sources, 176–178
Cellulose,
  as adsorbent, 96
  cleavage by cellulase, 175
Cephalin, synthetic (saturated),
  as substrate for phospholipase A, 664
Cephalin, natural,
  as substrate for phospholipase A, 664
Cesium ion,
  activation of "malic" enzyme (DPN)
    by, 752
  inhibition of aldehyde dehydrogenase
    by, 510
Charcoal,
  as adsorbent for proteins, 96
  removal of bound DPN with, 406
Chitinase,
  separation of from melibiase, 250–251
Chloral hydrate,
  inhibition of DR-aldolase by, 386
  of liver aldehyde dehydrogenase by,
    517
Chlorella,
  disintegration of, 64
Chloride ions,
  activation of α-amylase by, 154
  activation and inhibition of fumarase
    by, 734
Chloroform,
  and cell permeability, 55
  inhibition of lipase by, 639
p-Chloromercuribenzoate,
  inhibition of butyryl coenzyme A de-
    hydrogenase by, 558
  protection against, by substrate and
    FAD, 558
  inhibition of other enzymes by, 157,
    285, 292, 334, 348, 405, 411, 453,
    464, 470, 507, 510, 527, 535, 558,
    565, 594, 611, 656, 704, 762–763
Chloroplasts,
  assay of chlorophyll content of, 25
  cofactors for activity of, 24
  disintegrated,
    particle size of, 24
    preparation of, 22–25
    from leaves, 23
    from unicellular algae, 23

lecithinase activity of, 24
oxidase system in, 24
  cytoplasmic activator for, 24
  particle size of, 24
  photochemical system in, 24
  chloride as activator for, 24
  preparation of, 22–25
    from leaves, 23
  purity of isolated, 24
  stability of enzyme systems of, 24
Cholate,
  for solubilization of enzymes, 36
Δ^4-Cholestenene-3-one,
  formation by cholesterol dehydro-
    genase, 678–681
Cholestenone,
  carbonyl absorption of, 678, 679
  2,4-dinitrophenylhydrazone of, 678,
    679–680
  oxidation by adaptive enzyme, 678
Cholesterol,
  measurement by Liebermann-Burchard
    method, 678
Cholesterol dehydrogenase,
  from a *Mycobacterium*, 678–681
Choline,
  formation from L-(α)-glycerophos-
    phoryl choline, 668–670
  formation by phospholipase D, 672
  interference with bioassay of acetyl-
    choline, 623
  liberation from GPC by acid, 662
    from lecithin, 661
  measurement of, 665, 666
    as iodine complex, 662
  phosphorylation during oxidation of,
    678
Choline acetylase, 619–624
  from *Lactobacillus plantarum*, 622
  from mammalian brain, 622
  from squid ganglia, 620–621
Choline analogues,
  inhibition of choline dehydrogenase by,
    677
Choline dehydrogenase,
  as component of choline oxidase sys-
    tem, 675
  extraction from liver, 37
  reduction of ferricyanide or ferricyto-
    chrome by, 675

Choline oxidase,
    from liver, 674–678
    other sources of, 675
Cholinesterase(s), see also Acetyl-
    cholinesterase
    brain,
        extraction of, 36–37
        lability of, 43
        purification with butyl alcohol, 50
        solubilization of, 47
    definition of, 644
    from intestine,
        purification with butyl alcohol, 50
    from red cell stroma, 47
        extraction of, 37
        purification with butyl alcohol, 50
    in serum, 644, 647
Chondroitinsulfuric acid,
    degradation of, 167
Chromatography,
    of sugars, 260–261
Chromatography of enzymes,
    ion exchange, 113–125
        application to chymotrypsinogen,
            122–124
        to cytochrome c, 113
        to β-glucuronidase, 265–266
        to lysozyme, 124–125
        to ribonuclease, 119–122
        pressure regulator for, 116–117
    partition, 98–112
        chymotrypsinogen and chymotryp-
            sin, 109–110
        insulin, 108–109
        penicillinase, 110–112
        ribonuclease, 107–108
        trypsin, 110, 111
*Chymotrypsin,
    partition chromatography of, 109–110
*Chymotrypsinogen,
    assay of, 122
    ion exchange chromatography of, 113,
        117, 122–124
    partition chromatography of, 109–110
Cinchonidine,
    as substrate for liver aldehyde oxidase,
        527
Citrate,
    ΔF for synthesis of, 694

formation by condensing enzyme, 685–
    694
    from crotonyl CoA, 562
    inhibition of aldehyde oxidase by, 527
    of phospholipase C by, 671
    of transacetylase by, 599
    as substrate for aconitase, 695–698
Citrate buffer, 140
Citrate-phosphate buffer, 141
Citric acid, determination of, 686
Citrus fruit,
    pectinesterase from, 160
Clostridium acetobutylicum,
    acetoacetate decarboxylase from, 624–
        627
    growth medium for, 625
Clostridium butylicum,
    growth medium for, 480
    phosphoroclastic system from, 479–482
Clostridium kluyveri,
    acetate kinase from, 595
    aldehyde dehydrogenase (CoA-linked)
        from, 518–523
    amino acid acetylase from, 616–619
    butyric acid oxidation in, 551–553
    CoA transphorase from, 599–602
    growth of, 521–523
    phosphotransacetylase from, 596–599
    preparation of cell-free extracts of,
        552–553
    preparation of dried cells of, 523
    reactions of acetyl phosphate in, 553
Clostridium strain H. F.,
    acetate kinase from, 595
Clostridium welchii,
    lecithinase (phospholipase C) from,
        299, 671
CoA (CoASH, HSCoA), see Coenzyme A
Cobaltous ion,
    activation of acetoin-forming system
        by, 474
    of dialkylphosphofluoridase by, 656
    of L. delbruckii pyruvic acid oxidase
        by, 482
    of "malic" enzyme (DPN) by, 752
    of oxalacetic carboxylase by, 757
    of phospholipase C by, 671
    of phosphoroclastic system from Cl.
        butylicum by, 482
    of plant carboxylases by, 463, 469

of yeast aldolase by, 319
inhibition of phosphoglucose isomerase by, 306
of phosphomannose isomerase by,
  304
Cobra venom,
  inactivation of choline oxidase by leci-
    thinase from, 677
Cocarboxylase, see Thiamine pyrophosphate
Cockroach, see *Blattella germanica*
Coenzyme A (CoA, CoASH, HSCoA),
  acetylation of, 585–591
  in acetylation of amines, 609
    of choline, 620
  acyl derivatives of, 558
    activities in butyryl CoA dehydro-
      genase system, 558
  in aldehyde dehydrogenase reaction,
    518–523
  in bacterial pyruvic oxidase system,
    490–495
  in butyrate oxidation in *Cl. kluyveri*,
    551
  chemical acetylation of, 688
  chemical formation of crotonyl deriva-
    tive, 560
  extinction coefficient of, 600
  in formation of succinyl-coenzyme A,
    718–722
  inhibition of amine acetylation by, 611
  in α-ketoglutaric dehydrogenase sys-
    tem, 714–718
  in phosphatidic acid synthesis, 673
  in phosphoroclastic splitting of pyru-
    vate, 478, 481
  in pyruvic acid oxidase system (ani-
    mal), 486–490
  in thiolase reaction, 581
  in transacetylase reaction, 596
Coenzyme A transferase (for reaction
    of succinyl CoA with acetoacetate),
  from pig heart, 574–581
Coenzyme A transphorase (for reaction
    of butyryl CoA with acetate),
  from *Cl. kluyveri*, 599–602
  role in acylation of amino acids, 618
  role in butyrate oxidation in *Cl.
    kluyveri*, 551

2,4,6-Collidine,
  inhibition of formate fixation by, 478
Concanavalin A,
  as precipitant for polysaccharides and
    invertase, 257
Condensing enzyme (citrate-forming),
  in assay of aceto-CoA-kinase, 586, 588
    of crotonyl CoA, 562
  distribution of, 694
  from pig heart, 685–694
Congo red,
  as inactivator of invertase, 257
Copper, see also Cupric ion, Cuprous
    ion
  as constituent of butyryl coenzyme A
    dehydrogenase, 558
  succinic dehydrogenase and, 728
Copper reagent(s),
  in assay of α-galactosidase, 249
    of β-glucosidase, 235
    of hexosidases, 232, 249
    of pentokinase, 357
  determination of glucose-6-phosphate
    with, 296
  preparation of, 234–235
Copper reduction method,
  interference with, by manganous ion,
    360
Cornstarch,
  as adsorbent, 96
Corynebacteria,
  rupture of by sonic oscillator, 58
Cotton seeds,
  blastolipase from, 642
Cozymase, see Diphosphopyridine
    nucleotide
*Creatine phosphokinase,
  from heart muscle, 31
  purification with butyl alcohol, 50
Crotalus terrificus venom,
  crystalline phospholipase A from, 662
Crotonaldehyde,
  inhibition of yeast aldehyde dehydro-
    genase (TPN) by, 514
  as substrate for aldehyde dehydro-
    genase, 510
  for liver aldehyde oxidase, 527
Crotonase,
  in determination of crotonyl CoA, 561
  from ox liver, 559–566

Crotonic anhydride,
  in preparation of crotonyl coenzyme A,
    560–561
S-Crotonyl-N-acetylthioethanolamine,
  as substrate for crotonase, 559–566
Crotonyl coenzyme A,
  conversion to citrate, 562
  effect on butyryl CoA dehydrogenase,
    558
  preparation of, 560–561
  reaction with hydroxylamine, 561
  as substrate for crotonase, 559–566
    comparison of *cis* and *trans* forms,
      565
Crotonyl glutathione,
  as by-product of crotonyl CoA syn-
    thesis, 561
Crotoxin, see *Crotalus terrificus* venom
Crystallization,
  of alcohol dehydrogenase from liver,
    497–498
  of alcohol dehydrogenase from yeast,
    502
  of aldolase from muscle, 313, 394–395
  of α-amylase from saliva, 152
  of β-amylase from sweet potato, 156
  of condensing enzyme (citrate-form-
    ing), 691–692
  of glyceraldehyde-3-phosphate dehy-
    drogenase from muscle, 404
    from yeast, 409–410
  of glycerophosphate dehydrogenase
    from muscle, 395–396
  of hexokinase from yeast, 274
  of lactic dehydrogenase from heart
    muscle, 451–452
    from rabbit muscle, 396
  of oxalacetic carboxylase from *M.
    lysodeikticus*, 756
  of phosphoglucomutase from muscle,
    296–298
  of phosphoglycerate kinase from yeast,
    420–421
  of phospholipase A from venom, 662–
    663
  of phosphorylase *a* from muscle, 203–
    204
  of phosphorylase *b* from muscle, 205
  of phosphorylase from potato, 195–196
    of pyruvate kinase from human muscle,
      438–439
    from rabbit muscle, 437
  of transketolase from baker's yeast,
    379–380
    modified procedure for, 380
  of triosephosphate isomerase from
    muscle, 389–390
  of "Yeast Protein No. 2," 272
  of "Yeast Protein No. 3," 272
  of "Yeast Yellow Protein," 272–273
CTP, see Cytidine triphosphate
Cupric chloride,
  activation of alcohol dehydrogenase
    (TPN) by, 507
Cupric ion,
  activation of butyryl CoA dehydro-
    genase by, 558
  of yeast aldolase by, 319
  as inhibitor of enzymes, 157, 176, 197,
    240, 248, 257, 298, 313, 414, 464,
    510, 656, 698, 704
Cuprous ion,
  inhibition of phospholipase C by, 671
Cyanide,
  activation of amino acid acetylase by,
    618
  of glycolic oxidase by, 531–532
  of lipase by, 639
  of 6-phosphogluconic dehydrogenase
    by, 327
  inhibition of acetoacetate decarboxyl-
    ase by, 627
  of acetoin-forming system by, 475
  of aconitase by, 698
  of aldehyde oxidase by, 527
  of beef heart isocitric dehydrogenase
    (DPN) by, 713
  of cellulase by, 176
    copper requirement for, 176
  of choline oxidase by, 677–678
  of formic dehydrogenase by, 539
  of formic hydrogenlyase by, 541
  of mannitol phosphate dehydrogen-
    ase by, 348
  of phospholipase B by, 668
  of sorbitol dehydrogenase by, 350
  of succinic dehydrogenase by, 727–
    728

of yeast isocitric dehydrogenase (DPN) by, 709

for oxalacetate binding in malic dehydrogenase assay, 739

in removal of copper from butyryl CoA dehydrogenase, 558

in succinic dehydrogenase assay, 724–725

Cysteine, see also Monothiols, Thiols

activation of acetate kinase by, 595

of aldehyde dehydrogenase (liver) by, 517

of aldehyde dehydrogenase (yeast) by, 510

of amine acetylation by, 611

of glyceraldehyde-3-phosphate dehydrogenases by, 405, 410–411, 414

of hexokinase (mammalian) by, 285

of lactic dehydrogenase by, 453

of phosphoglucomutase by, 298

of pyruvic oxidase system (animal) by, 487

of pyruvic oxidase (bacterial) by, 490

as inhibitor of cellulase, 176

of yeast aldolase, 318–319

reaction with glyceraldehyde-3-phosphate, 415

Cytidine triphosphate,

as substrate for phosphohexokinase, 310

*Cytochrome b,

as mediator of dye reduction in succinic dehydrogenase system, 728

possible identity with succinic dehydrogenase, 728

Cytochrome b₂, see Lactic dehydrogenase of yeast

*Cytochrome c,

as component of the fatty acid oxidation system, 546

as electron acceptor for butyryl CoA dehydrogenase, 558

copper requirement for, 558

elution of, 33

from heart muscle, 30, 31

ion exchange chromatography of, 113

reduction by aldehyde oxidase, 527

molybdenum requirement for, 527

phosphate requirement for, 527

solubility of oxidized and reduced, 44

*Cytochrome c reductase (DPN),

from breast muscle, 34

from kidney, 39

from liver, 50

from rat liver mitochondria, 29

*Cytochrome c reductase (TPN),

extraction from liver, 35

solubilization of by trypsin, 39

*Cytochrome oxidase,

from heart muscle, 29, 36

solubilization of by high frequency oscillations, 29

from rat liver mitochondria, 29

*Cytochrome peroxidase,

formation in Pseudomonas, 136

Cytoplasmic particles,

from peanut cotyledons, 550–551

fatty acid oxidation in, 550–551

unidentified cofactor for, 551

D

DAP, see Dihydroxyacetone phosphate

Deacylase(s) (Thiolesterase), 602–607

for acetoacetyl CoA, 574

for acetyl CoA, 606–607

in determination of crotonyl CoA, 561

for succinyl CoA, 574, 602–606

Deaminodiphosphopyridine nucleotide (deamino-DPN),

reduction of, by glyceraldehyde-3-phosphate dehydrogenase, 404

Debye-Hückel theory, 68–71

Dehydrogenases,

DPN-specific, 332, 346, 348, 391, 397, 401, 407, 411, 449, 486, 490, 495, 500, 506, 514, 518, 532, 536, 566, 675, 678, 707, 710, 714, 735, 748

TPN-specific, 323, 325, 411, 506, 511, 535, 699, 705, 739

unspecific toward DPN and TPN, 330, 339, 441, 508, 739

Deoxycholate,

in extraction of brain hexokinase, 282

as inactivator of ATP-ase, 280

for solubilization of enzymes, 36

6-Deoxy-D-glucose,

as inhibitor of mammalian hexokinase, 284

2-Deoxy-D-glucose,
   as substrate for hexokinases, 274, 284
3-Deoxyglucose-6-phosphate,
   as inhibitor of mammalian hexokinase,
      285
*Deoxyribonuclease,
   purification of virus protein by, 40
Deoxyribonucleic acid,
   as activator of β-glucuronidase, 267
Deoxyribonucleoproteins,
   from liver, 33
Deoxyribose-5-phosphate,
   cleavage by DR-aldolase, 384–386
   in transketolase reaction, 380
Deoxyribose phosphate aldolase (DR-
      aldolase),
   from E. coli, 384–386
Dephospho coenzyme A,
   acetylation of, 590
Detergents,
   for extraction of enzymes, 35–38
   inhibition of pectinesterase by, 162
   for protein fractionation, 86
   for separation of proteins from nucleic
      acids, 137
Dextran(s),
   as acceptor substrate for dextransu-
      crase, 182
   as donor substrate for dextransucrase,
      182
   as primers for phosphorylase, 196
Dextransucrase,
   from L. mesenteroides, 179–184
Dextrins,
   as substrates for α-amylase, 152
   for β-amylase, 156
DFP, see Diisopropyl fluorophosphate
Diacetyl,
   activation of formic hydrogenlyase by,
      541
Dialkylphosphofluoridase (dialkylfluoro-
      phosphate esterase, dialkylfluoro-
      phosphatase),
   action on diisopropyl fluorophosphate
      (DFP), 651–656
   action on other fluorophosphates, 655
   from hog kidney, 654–655
   other sources of, 653
Diamines,
   as activators of β-glucuronidase, 267

*Diaphorase,
   as contaminant of CoA transferase
      from heart, 580
   formation in Pseudomonas, 137
   from heart muscle, 32, 49, 50
   in malic dehydrogenase assay, 739
Diaphragm,
   "malic" enzyme (TPN) in, 747
2,6-Dichlorophenolindophenol,
   in assay of butyryl CoA dehydrogenase,
      554
      of aldehyde oxidase, 527
      of choline oxidase system, 677
      of glucose aerodehydrogenase, 345
      of glycolic acid oxidase, 528–532
      of pyruvic oxidase system, 493
      of succinic dehydrogenase, 723–724
2,4-Dichlorophenoxyacetic acid,
   effect on phosphorylase content of
      plants, 198
Dielectric constant,
   control by organic solvents, 82
Diethanolamine,
   inhibition of choline dehydrogenase
      by, 677
Diethylbarbiturate (Veronal), see also
      Barbital
   inhibition of transacetylase by, 599
Diethyl Carbitol,
   phase diagram for, 104
Diethylmethylcholine,
   as substrate for choline dehydrogenase,
      677
Diglyceride(s),
   formation by phospholipase C, 670–
      671
   as products of lipase action, 636
Diglycerol,
   as substrate for glycerol dehydro-
      genase, 400
Dihydroxyacetone,
   as substrate for glycerol dehydro-
      genase, 400
Dihydroxyacetone phosphate,
   conversion to D-Glyceraldehyde-3-
      phosphate, 387–391
   preparation of, 392
   reduction to L-α-glycerophosphate,
      391–397

Diisopropyl fluorophosphate (DFP),
  as substrate for dialkylphosphofluori-
    dase, 651–656
Diketene,
  in synthesis of S-acetoacetyl-N-
    acetylcysteamine, 567
  of acetoacetyl CoA, 582
α,γ-Diketo acids,
  as substrates for lactic dehydrogenase,
    453
Dimedon,
  as trapping agent for acetaldehyde, 466
Dimethoxytetraglycol,
  phase diagram for, 101
Dimethylamine,
  acetylation of, 618
  inhibition of choline dehydrogenase by,
    677
Dimethylaminoethyl acetate,
  as substrate for acetylcholinesterase,
    643–644
Dimethylaminoethyl acryllate polymer,
  in chemical lysis of cells, 62
Dimethylethylcholine,
  as substrate for choline dehydrogenase,
    677
α,α-Dimethylcholine,
  as substrate for choline dehydrogenase,
    677
Dimethyl formamide,
  phase diagram for, 103
Dimyristoylcephalin,
  as substrate for phospholipase A, 664
Dimyristoyllecithin,
  as substrate for phospholipase A, 664
Dioxane,
  phase diagram for, 102
  as protein precipitant, 84
Dipalmitoleyllecithin,
  as substrate for phospholipase A, 664
Diphenylamine,
  in measurement of deoxyribose phos-
    phate, 385
Diphenylchloroarsine,
  inhibition of isocitric dehydrogenase
    by, 704
1,3-Diphosphoglyceric acid,
  conversion to 2,3-diphosphoglyceric
    acid, 425–427

enzymatic formation of,
  from ATP and 3-PGA, 426
  by oxidation of GAP, 401–406, 407–
    411, 411–415
phosphorylation of ADP by, 415–422
2,3-Diphosphoglyceric acid,
  as coenzyme for phosphoglyceric
    mutase, 423, 425
  inhibition of diphosphoglyceric acid
    mutase by, 427
Diphosphoglyceric acid mutase,
  from rabbit erythrocytes, 425–427
2,3-Diphosphoglyceric acid phosphatase,
    426
Diphosphopyridine nucleotide (DPN),
  see also Dehydrogenases
  assay of with liver aldehyde dehydro-
    genase, 517
  binding by glyceraldehyde-3-phosphate
    dehydrogenase, 405–406
  by yeast alcohol dehydrogenase, 503
  inhibition of succinic dehydrogenase
    by, 727
Diphosphopyridine nucleotide, reduced
    (DPNH),
  binding by lactic dehydrogenase, 454
  complex with liver alcohol dehydrogen-
    ase, 498–499
  spectrum of, 498
  titration curve for, 499
  preparation of, 392
  as substrate for liver aldehyde oxidase,
    527
Diphosphothiamine, see Thiamine pyro-
    phosphate
α,α′-Dipyridyl,
  activation of dialkylphosphofluoridase
    by, 656
  inhibition of formic hydrogenlyase by,
    541
  of yeast aldolase by, 318
Disaccharide phosphorylases, 225–231
Disaccharides,
  polysaccharide synthesis from, 178–192
Disulfides,
  inhibition of glyceraldehyde-3-phos-
    phate dehydrogenase by, 405
1,2-Dithiopropanol,
  as activator of β-amylase, 158
DNA, see Deoxyribonucleic acid

DPN, see Diphosphopyridine nucleotide

DPNH, see Diphosphopyridine nucleotide, reduced

DR-aldolase, see Deoxyribose phosphate aldolase

Dyes,
  as electron acceptors for aldehyde oxidase, 527
  for butyryl CoA dehydrogenase, 553–554

E

Earthworms,
  cellulase from, 177
Edestin,
  crystallization of, 71
  solubility of, 69
EDTA, see Ethylenediaminetetraacetic acid
Egg albumin, see Albumin
Ehrlich's reagent,
  in measurement of acetylation of amino sugars, 613–614
Electrophorus electricus,
  acetylcholinesterase from, 645
Elution, agents for, 95
Emulsification, agents for, 629–630
Energy of activation, see Heat of activation
Enolase,
  from brewer's yeast, 427–435
  crystallization of by mercuric ions, 85, 432–433
  kinetics of reaction, 428–430
  molecular weight of, 434
Enrichment culture, 126–131
Enzymes, see also Proteins
  extraction of,
    from animal tissues, 25–51
    from bacteria, 51–62
    by freezing and thawing, 480–481
    from plants, 62, 63
  ion exchange chromatography of, 113–125
  partition chromatography of, 98–112
Ephedrine,
  inhibition of choline dehydrogenase by, 677

Equilibrium constant,
  relationship between maximum initial velocities, Michaelis constants, and, 735
Erythrocytes, see Red cell(s)
Erythrose,
  action of aldolase on, 322
L-Erythrulose,
  in transketolase reaction, 375, 380
Escherichia coli,
  acetate kinase from, 591–595
  *amino acid decarboxylase in, 137
  amylomaltase from, 189–192
  D-arabinose isomerase from, 366–370
  deoxyribose phosphate aldolase from, 384–386
  extraction by grinding, 59, 61
  formic hydrogenlyase from, 539–541
  β-galactosidase from, 241–249
  gluconokinase from, 350–354
  glucose-1,6-diphosphate formation in, 354–355
  *glutamic oxidase from, 64
  growth medium for, 350, 477, 491, 540
  mannitol-1-phosphate dehydrogenase from, 346–348
  phosphoroclastic split of pyruvate in extracts from, 476–478
  pyruvic oxidase (Coenzyme A-linked) from, 490–495
  rupture of cells by pressure release, 57
  transacetylase and CoA from, 685–686
  xylose isomerase from, 366–367
Esterase ("nonspecific esterase," "aliesterase"),
  from bovine liver, 33
  from horse liver, 657–659
  from insects, 50
  solubilization of from liver, 48
Ethanol, see also Alcohol dehydrogenase
  denaturing action on plasma proteins, 84
  for fractionation of human plasma, 89
  as solvent for zein, 83
Ethanolamine,
  acetylation of, 618
Ethanol-water mixtures,
  physical chemical constants of, 84
Ethylamine,
  acetylation of, 618

Ethyl carbamate, see Urethan
Ethyl Cellosolve,
    phase diagram for, 101, 105
Ethylenediaminetetraacetate (EDTA)
    (Versene),
    activation of glyceraldehyde-3-phos-
        phate dehydrogenases by, 405,
        414
    effect on yield of glyceraldehyde-3-
        phosphate dehydrogenase, 404
    inhibition of acetate kinase by, 594
    of aldehyde oxidase by, 527
    of GPC diesterase by, 670
    stabilization of liver aldehyde dehy-
        drogenase (DPN) by, 517
    use of, during purification of enzymes,
        568
Ethylene glycol,
    as substrate for glycerol dehydro-
        genase, 400
    for yeast alcohol dehydrogenase,
        502
Ethyl ether,
    as protein precipitant, 84
Ethylgluconate,
    phosphorylation by ATP, 334
Ethylgluconate-6-phosphate,
    as substrate for phosphogluconic de-
        hydrogenase, 334
Extraction of enzymes, see Animal tis-
    sues, Microorganisms, and Plant
    tissues

                    F

FAD, see Flavin adenine dinucleotide
Fatty acid oxidase,
    extraction from liver, 35
Fatty acids,
    conversion to acyl CoA derivatives, 585
    hog liver enzyme for, 585
    oxidation of,
        in animal tissues, 545–549
        by extracts of acetone dried mito-
            chondria, 548
        by rat kidney mitochondria, 547–
            548
        by rat liver mitochondria, 547
        rate-limiting step in, 548
        in Clostridium kluyveri, 551–553

    in higher plants, 549–551
        by microsomal particles, 549–550
        by soluble cytoplasmic proteins,
            550–551
    phosphorylation of, role of ATP and
        CoA in, 482
FDP, see Fructose-1,6-diphosphate
Ferric chloride reagent,
    for hydroxamic acid determinations,
        592
Ferricyanide,
    for anaerobic assay of choline oxidase,
        676
    as electron acceptor for butyryl coen-
        zyme A dehydrogenase, 558
    as oxidant in pyruvic oxidase system,
        493
    in spectrophotometric method for suc-
        cinic dehydrogenase, 724
*Fibrinogen,
    solubility of in ammonium sulfate, 74
Fibrinokinase, from pig heart, 32
Flavianic acid,
    in purification of invertase, 257
Flavin adenine dinucleotide (FAD),
    action in glycolic oxidase system, 531
    as cofactor for choline dehydrogenase,
        675
    for pyruvic acid oxidase in L. del-
        brückii, 482, 483, 486
    as component of butyryl coenzyme A
        dehydrogenase, 558
    of glucose aerodehydrogenase, 345
    of liver aldehyde oxidase, 526
    as hydrogen acceptor in phosphoro-
        clastic system, 482
    presence in jack bean phosphorylase,
        197
Flavin mononucleotide (FMN),
    as component of glycolic oxidase, 531
    of yeast lactic dehydrogenase, 448
    fluorescence of free and bound forms,
        449
    as hydrogen acceptor in phosphoro-
        clastic system, 482
"Flavocytochrome,"
    yeast lactic dehydrogenase as, 448
Flavoprotein,
    as contaminant of coenzyme A trans-
        ferase from pig heart, 580

Fluoride,
  inhibition of acetoin-forming system
    by, 475
  of *Cl. welchii* lecithinase by, 299
  of enolase by, 434
    phosphate requirement for, 434
  of phosphoglucomutase by, 298
  of phospholipase C by, 671
  of phosphorylating enzyme (succinic
    activating) by, 722
  of potato phosphorylase by, 197
  of succinic dehydrogenase by, 727
    phosphate requirement for, 727
Fluoroacetate,
  lack of reaction with acetate kinase,
    594
Fluoroethanol,
  inhibition of yeast alcohol dehydrogen-
    ase by, 502
Fluorophosphate,
  inhibition of lipase by, 639
FMN, see Flavin mononucleotide
Formaldehyde,
  inhibition of formic hydrogenlyase by,
    541
  of β-glucosidase by, 240
  of lipase by, 639
  of polygalacturonidase by, 166
  as substrate for alcohol dehydrogenase
    (TPN), 507
    for aldehyde dehydrogenases, 510,
      513, 517
Formate,
  as CoA acceptor in *Cl. kluyveri*, 602
  as product of glycolic acid oxidase, 528–
    532
    of phosphorolysis of pyruvate, 476–
    478
Formate, C^{14}-labeled,
  fixation during phosphoroclastic split-
    ting of pyruvate, 478
Formic dehydrogenase,
  from peas, 536–539
Formic hydrogenlyase,
  from *E. coli*, 539–541
F-1-P, see Fructose-1-phosphate
F-6-P, see Fructose-6-phosphate
β-Fructofuranosidase (invertase),
  pH optimum, variation with source,
    259

transglycosylase activity of, 258–262
  from yeast, 251–257
Fructokinase (ketohexokinase), 286–290
  from liver, 286–289
  from muscle, 289–290
D-Fructose,
  as acceptor for dextransucrase, 183
    for sucrose phosphorylase, 228
  inhibition of aldolase by, 313
  standard solution for fructokinase
    assay, 287
  as substrate for mammalian hexo-
    kinase, 284
  for yeast hexokinase, 274
Fructose-1,6-diphosphate (FDP),
  conversion to triose phosphates,
    by muscle aldolose, 310
    by yeast aldolase, 315–320
  inhibition of glucose dehydrogenase by,
    339
Fructose-1-phosphate,
  cleavage of by liver aldolase, 320–322
  as product of fructokinase reaction, 289
Fructose-1-phosphate aldolase,
  difference from fructose-1,6-diphos-
    phate aldolase, 321
  in jack bean, 322
  from liver, 320–322
  in myogen A, 322
Fructose-6-phosphate,
  colorimetric determination by Roe
    method, 300, 304
  conversion to fructose-1,6-diphosphate,
    306–310
    to glucose-6-phosphate, 304–306
  inhibition of aldolase by, 313
    of glucose dehydrogenase by, 339
  reduction to mannitol-l-phosphate,
    346–348
  as substrate for aldolase, 313
  in transaldolase reaction, 381
  in transketolase reaction, 375, 380
Fuchsin,
  as inactivator of invertase, 257
L-Fucose,
  as substrate for D-arabinose isomerase,
    370
Fuller's earth,
  as adsorbent, 96

Fumarase,
  from pig heart, 729–735
    crystalline forms of, 733
Fumarate,
  as component of fatty acid oxidation
    system, 546
  conversion to *l*-malate, 729–735
  inhibition of malic dehydrogenase by,
    739
    of succinic dehydrogenase by, 727
  permanganate titration of, 729
  ultra violet absorption by, 729–730
Fusarium lini Bolley,
  lipase from, 640–642

G

GA, see Glyceraldehyde
Galactobiose,
  formation by transglycosidation, 248
Galactokinase, 290–293
  from galactose-adapted yeast (*Saccharomyces fragilis*), 292–293
Galactosamine,
  acetylation of, 612
Galactose,
  as acceptor substrate for dextransucrase, 183
  as substrate for glucose aerodehydrogenase, 345
Galactose-1-phosphate,
  as product of galactokinase action, 290
  as substrate for galactowaldenase, 293
α-Galactosidase (melibiase),
  from sweet almond emulsin, 249–251
β-Galactosidase (lactase),
  from *Escherichia coli*, 241–248
  inducers for, 248
  substrates for, 248
  transglycosidation by, 248
α-D-Galactosides,
  as substrates for β-glucosidase, 237
Galactowaldenase, 293–294
  from *Saccharomyces fragilis*, 294
Gal-l-P, see Galactose-l-phosphate
GAP, see Glyceraldehyde-3-phosphate
Globin, denatured,
  protection against hematin inhibition
    of succinic dehydrogenase by, 727

γ-Globulin, rabbit,
  adsorption by kieselguhr, 100
Globulins,
  definition of, 69
  solubility of in salt solutions, 69, 70
    Debye equation for, 70
Gluconate,
  phosphorylation of, 350–354
Gluconokinase,
  from baker's yeast, 351
  from brewer's yeast, 351, 353–354
  from *E. coli*, 350–354
  from *L. mesenteroides*, 334
D-Gluconolactone,
  formation of, by glucose dehydrogenase, 335
δ-D-Gluconolactone,
  primary product of glucose aerodehydrogenase, 340
α-D-Glucopyranose,
  as substrate for glucose aerodehydrogenase, 345
β-D-Glucopyranose,
  as substrate for glucose aerodehydrogenase, 340
D-Glucosamine,
  acetylation of, 610, 612–615, 618
  enzyme for acetylation of, 612–615
    comparison with arylamine acetylating system, 615
  as substrate for mammalian hexokinase, 284
    for yeast hexokinase, 274
Glucose,
  as acceptor for amylomaltase, 191
    for dextran sucrase, 183
    for β-fructofuranosidase, 255
    for levansucrase, 188
  effect on fatty acid oxidation, 548
  inhibition of aldolase by, 313
    of muscle phosphorylase by, 205
  as product of α-amylase action, 152
  as substrate for mammalian hexokinase, 284
    for yeast hexokinase, 274
β-D-Glucose,
  as substrate for glucose dehydrogenase, 335
Glucose aerodehydrogenase (glucose oxidase),

from *Penicillium notatum* Westling,
340–345
Glucose dehydrogenase,
from liver, 335–339
Glucose-1,6-diphosphate,
as coenzyme for phosphoglucomutase,
294
for phosphoribomutase, 362
as contaminant of glucose-1-phos-
phate, 295
enzymatic synthesis of, 354–356
by dismutation of G-1-P, 355
by reaction of ATP with G-1-P,
354–356
formation by alkaline phosphatase, 355
in galactowaldenase assay, 293
as inhibitor of mammalian hexokinase,
285
Glucose oxidase (notatin), see also Glucose
aerodehydrogenase
of *Penicillium notatum*,
in assay of glucosidases, 233, 242–243
α-D-Glucose-1-phosphate,
role in fluoride inhibition of phosphoglucomutase, 298
as substrate for galactowaldenase, 293
for phosphoglucomutase, 294–299
for phosphorylase, 192, 193, 200, 215
for sucrose phosphorylase, 225
β-D-Glucose-1-phosphate,
as product of maltose phosphorylase,
229
Glucose-1-phosphate, P^{32}-labeled,
synthesis by sucrose phosphorylase,
229
Glucose-1-phosphate kinase,
from brewer's yeast, 354–356
from rabbit muscle, 354–356
Glucose-6-phosphate,
conversion to fructose-6-phosphate,
304–306
determination by copper reduction,
296
effect of on yeast hexokinase, 275
inhibition of glucose dehydrogenase
by, 339
of mammalian hexokinase by, 279,
285
of pentose phosphate isomerase by,
365

as substrate for phosphoglucomutase,
294–299
Glucose-6-phosphate dehydrogenase
(TPN) ("Zwischenferment"),
in assay of phospohexoisomerases, 299,
300, 305
of transaldolase, 381
from brewer's yeast, 323–326
Glucose-6-phosphate dehydrogenase
(DPN-TPN),
from *L. mesenteroides*, 328–332
β-Glucosidase,
separation from melibiase (α-galacto-
sidase), 250
from sweet almond emulsin, 234–240
Glucuronic acid,
as inhibitor of β-glucuronidase, 267
β-Glucuronidase,
from liver, 37, 262–269
role in degradation of hyaluronic acid,
167
from spleen, 269
Glucuronides,
as substrates for β-glucuronidase, 267
Glutamate,
acetylation of, 618
*Glutamic dehydrogenase,
in assay of α-ketoglutaric dehydrogenase system, 716
from liver, 33
*Glutamic oxidase,
from *E. coli*, 64
*Glutaminase I,
from kidney, 50
solubilization of, 47
γ-Glutamyl transpeptidase,
solubilization of by butanol, 42, 50
Glutarate,
competitive inhibition of succinic dehydrogenase by, 727
Glutathionase, see γ-Glutamyl transpeptidase
Glutathione (GSH), see also Monothiols
and Thiols
activation of alcohol dehydrogenase
(yeast) by, 503
of aldehyde dehydrogenase (*Cl.
kluyveri*) by, 520
of aldehyde dehydrogenase (yeast)
by, 510

of β-amylase (sweet potato) by, 158
of butyryl coenzyme A dehydrogen-
    ase by, 558
of crotonase by, 565
of galactokinase by, 292
of glyceraldehyde-3-phosphate de-
    hydrogenase (TPN) by, 414
of mannitol phosphate dehydrogenase
    by, 348
of oxalacetate synthesizing enzyme
    by, 762
of pyruvic oxidase (bacterial) by, 491
in assay of aceto-CoA-kinase, 587
inhibition of aldolase (yeast) by, 318
of cellulase by, 176
reaction with ketoaldehydes, 454
as substrate for glyoxalase I, 454–458
Glutathione, oxidized (GSSG),
inhibition of glyceraldehyde-3-phos-
    phate dehydrogenase by, 405
Glyceraldehyde,
as substrate for aldehyde dehydrogen-
    ase, 510
    for aldolase, 322
    for glyceraldehyde-3-phosphate de-
        hydrogenase, 404, 410
    for glycerol dehydrogenase, 400
in transketolase reaction, 375, 380
DL-Glyceraldehyde-1-bromide 3-phos-
    phoric acid dioxane addition com-
    pound, 401, 407, 412
D-Glyceraldehyde-3-phosphate,
conversion to dihydroxyacetone phos-
    phate, 387–391
in DR-aldolase reaction, 384–386
preparation of solution of, 401, 407,
    412
reaction with cysteine, 415
in transaldolase reaction, 381
in transketolase reaction, 371–375,
    375–380
Glyceraldehyde-3-phosphate dehydro-
    genase (DPN),
acyl enzyme derivative of, 404–405
in assay of aldolase, 315–316
    of diphosphoglyceric acid mutase,
        425
    of phosphoglycerate kinase, 415
    of phosphohexokinase, 307
from baker's yeast, 407–411

differentiation from aldehyde dehy-
    drogenase, 517
from muscle, 401–406
as "Yeast Protein No. 2," 272
Glyceraldehyde-3-phosphate dehydro-
    genase (TPN),
from plant tissue, 411–415
Glycerol,
as acceptor for β-fructofuranosidase,
    255
in extraction of phospholipase B, 666
as substrate for yeast alcohol dehy-
    drogenase, 502
Glycerol-α-chlorohydrin,
as substrate for glycerol dehydro-
    genase, 400
Glycerol dehydrogenase,
from A. aerogenes, 397–400
from liver, 320
L-(α)-Glycerophosphate,
formation from L-(α)-glycerophos-
    phoryl choline, 668–670
as substrate for glycerol dehydro-
    genase, 400
α-Glycerophosphate-P³²,
conversion to phosphatidic acid, 673–
    674
L-(α)-Glycerophosphate dehydrogenase,
in assay of DR-aldolase, 384
    of glycerophosphoryl choline dies-
        terase, 668
    of liver aldolase, 320
    of transketolase, 371, 375
    of triose phosphate isomerase, 387
DPN in crystallization of, 395
prosthetic group of, 397
from rabbit muscle, 391–397
L-(α)-Glycerophosphoryl choline (GPC),
conversion to L-(α)-glycerophosphate
    and choline, 668–670
determination of, 665
formation by phospholipase B, 665–668
inhibition of GPC diesterase action on
    GPE by, 670
organism utilizing, 129
Glycerophosphoryl choline diesterase,
in assay of phospholipase A, 661
from Serratia, 135, 668–670
Glycerophosphoryl ethanolamine (GPE),
inhibition of GPC diesterase by, 670

as substrate for GPC diesterase, 670
Glycine,
    acetylation of, 618
    inhibition of polygalacturonidase by,
      166
    lysis of cells by, 61, 62
Glycine ethyl ester,
    acetylation of, 618
Glycine-hydrochloric acid buffer, 139
Glycine-sodium hydroxide buffer, 146
Glycogen,
    as substrate for α-amylase, 152
      for β-amylase, 156
      for phosphorylase, 205
Glycolaldehyde,
    as substrate for alcohol dehydrogenase
      (TPN), 507
      for aldehyde dehydrogenase, 513,
        517, 520
      for aldolase, 322
      for liver aldehyde oxidase, 527
    in transketolase reaction, 375, 380
Glycolate,
    as CoA acceptor in Cl. kluyveri, 602
    as substrate for glyoxylic reductase,
      535
Glycolic oxidase,
    from liver, 529
    from spinach leaves, 528–532
    from tobacco leaves, 529
Glycylglycine,
    activation of 6-phosphogluconic de-
      hydrogenase by, 327
Glycyl-L-leucine,
    acetylation of, 618
Glyoxal,
    as substrate for alcohol dehydro-
      genase (TPN), 507
      for glyoxalase I, 457
Glyoxalases, 454–460
Glyoxalase I from bakers' yeast, 454–
    458
Glyoxalase II from beef liver, 458–460
Glyoxalin, see Imidazole
Glyoxylate,
    as product of glycolic acid oxidase, 528–
      532
Glyoxylic acid reductase,
    from spinach leaves, 532
    from tobacco leaves, 532–535

G-6-P, see Glucose-6-phosphate
GPC, see Glycerophosphorylcholine
GPE, see Glycerophosphorylethanolamine
Grana, 24
Green gram,
    phosphorylase from, 194
Growth, bacterial,
    measurement by turbidimetry, 136
GSH, see Glutathione
GSSG, see Glutathione, oxidized

H

Halogen ions,
    inhibition of lipase by, 639
Heart,
    acetoacetate-activating enzyme in,
      573–574
    aceto-CoA-kinase from, 588
    acetyl CoA deacylase from, 606–607
    acetyl phosphate cleavage in, 607
    aconitase from, 31, 695–698
    actomyosin from, 32
    coenzyme A transferase from, 574–581
    *creatine phosphokinase from, 31
    crystalline condensing enzyme (citrate
      forming) from, 685–694
    *cytochrome c from, 31
    *cytochrome oxidase from, 36
    deacylase for succinyl coenzyme A in,
      574, 602–606
    dialkylphosphofluoridase in, 653
    *diaphorase from, 32, 49, 50
    differential extraction of enzymes
      from, 31–32
    fibrinokinase from, 32
    fumarase from, 729–735
    hexokinase from, 282, 283
    isocitric dehydrogenase (DPN) from,
      711–714
    isocitric dehydrogenase (TPN) from,
      699–704
    α-keto acid oxidases from, 34
    α-ketoglutaric dehydrogenase system
      from, 714–718
    β-ketoreductase in, 573
    lactic dehydrogenase from, 449–454
    malic dehydrogenase from, 735–739
    "malic" enzyme (TPN) in, 747

"phosphorylating enzyme" (succinate activating) in, 718–722
pyruvic oxidase from, 487
succinic dehydrogenase from, 48–49
thiolase from, 581–585
Heart muscle mitochondria,
extraction of cytochrome c from, 30
Heat of activation,
for fumarase, 734
for β-glucosidase, 239–240
for β-glucuronidase, 268
Heats of reaction,
for β-glucosidase action on alkyl glucosides, 240
Heavy metal ions, see also under names of individual metals
inhibition of aldehyde dehydrogenase by, 510
of glyceraldehyde-3-phosphate dehydrogenases by, 405, 414
of 6-phosphogluconic dehydrogenase by, 327
of succinic dehydrogenase by, 727
Helix pomatia (snail),
cellulase from, 173–178
Hematin,
inhibition of succinic dehydrogenase by, 727
Hemoglobin,
removal of in enzyme purification, 497
solubility of,
effect of oxygen on, 69
of pH on, 73, 79
of temperature on, 73
Hemolysis,
by lysolecithin, 660–661, 665
Heparin,
inhibition of hyaluronidase by, 168
solubilization of lipoproteins by, 38
2,4-Hexadienoyl coenzyme A,
enzyme for hydration of, 564
trans-2-Hexenoyl coenzyme A,
as substrate for crotonase, 564, 565
comparison with cis form, 565
trans-3-Hexenoyl coenzyme A,
as substrate for crotonase, 564
Hexokinase(s),
from animal tissues, 277–286
from baker's yeast, 269–277
from brain, 281–282

effect on fatty acid oxidation, 548
from heart muscle, 282–283
from liver, 283
from skeletal muscle, 283–284
Hexose phosphate isomerase, see Phosphohexoisomerase
Hexose phosphate reductase, see Mannitol-l-phosphate dehydrogenase
Hexose reductase, see Sorbitol dehydrogenase
Hexosidases (Hexoside hydrolases), 231–257
Histamine,
acetylation of, 610–611
inhibition of choline dehydrogenase by, 677
Histidine,
as activator for phosphoglucomutase, 296
Histochemical staining technique,
for localization of acetylcholinesterase, 650
limitations of, 650–651
Holoenzyme,
of plant carboxylases, 464, 470
Homocholine,
as substrate for choline dehydrogenase, 677
Homocitric acid,
formation by condensing enzyme, 694
Homogenates, see Tissue homogenates
Hyaluronic acid,
purity of, 167
solution, 172
Hyaluronidase, 166–173
from bovine testes, 167
clinical use of, 168
from hemolytic streptococci, 168
from pneumococci, 168
Hydrochloric acid-potassium chloride buffer, 138
Hydrogen gas,
inhibition of formic hydrogenlyase by, 541
of phosphoroclastic system of Cl. butylicum by, 482
production of, by formic hydrogenlyase, 539–541
by phosphoroclastic splitting of pyruvate, 479–482

in reversal of phosphoroclastic system, 482

Hydrogen peroxide,
reaction with pyruvate, nonenzymatically, 483

Hydrogen sulfide,
activation of amine acetylation by, 611

Hydrosulfite,
reduction of flavins by, 526

S-β-Hydroxyacyl-N-acetylcysteamine,
derivatives of, as substrates for β-ketoreductase, 573

β-Hydroxyacyl coenzyme A,
derivatives of, as substrates for β-ketoreductase, 573

D-β-Hydroxybutyryl-N-acetyl cysteamine,
as substrate for β-ketoreductase, 572

β-Hydroxybutyryl coenzyme A,
measurement of, 562
as product of crotonase reaction, 559, 565
identification as D isomer, 565
as substrate for β-ketoreductase, 566, 571
specificity for D form, 571–572
synthesis from acetyl coenzyme A, 585

β-Hydroxybutyryl coenzyme A dehydrogenase, see β-Ketoreductase

β-Hydroxyhexanoyl coenzyme A,
as product of crotonase action, 564, 565

Hydroxylamine,
in assay of acetokinase, 591–595
of aceto-CoA-kinase, 586–588
of glyoxalase I, 457
inhibition of alcohol dehydrogenase (Leuconostoc) by, 507
of alcohol dehydrogenase (liver) by, 499
of alcohol dehydrogenase (yeast) by, 503
reversal by excess alcohol, 503
of fatty acid oxidation in higher plants, 551
of glycolic oxidase by, 532
reaction with crotonyl thioesters, 561
solution, preparation of, 719

Hydroxy-2-propanone acetate,
as substrate for glycerol dehydrogenase, 400

Hydroxypyruvate,
in transketolase reaction, 375, 380

Hydroxypyruvic aldehyde,
as substrate for glyoxalase I, 457

8-Hydroxyquinoline,
as activator for phosphoglucomutase, 298
inhibition of aldehyde oxidase by, 527
of formic dehydrogenase by, 539
of formic hydrogenlyase by, 541
of phosphoribomutase by, 362

Hypophosphite,
inhibition of fatty acid oxidation in higher plants by, 551

Hypophysis (pituitary),
β-ketoreductase in, 573

I

IAA, see Iodoacetate

L-Iditol,
oxidation to L-sorbose, 349

IDP, see Inosine diphosphate

Imidazole,
inhibition of fatty acid oxidation in higher plants by, 551
of formate fixation by, 478
protection against hematin inhibition of succinic dehydrogenase by, 727

5'-IMP, see 5'-Inosinic acid

Induced enzymes, see Adaptive enzymes

Inducers,
for β-galactosidase, 248

Induction, see Adaptation

Inosine diphosphate (IDP),
phosphorylation by ATP, 759
role in oxalacetate synthesis, 758–763

Inosine triphosphate (ITP),
phosphorylation of ADP by, 759
in reaction with oxalacetate, 758–763
as substrate for phosphohexokinase, 310

5'-Inosinic acid (IMP),
effect on liver phosphorylase, 221

i-Inositol,
as substrate for glycerol dehydrogenase, 400

Insect esterase, 50

Insulin,
crystallization with zinc, 85
partition of between solvents, 99

partition chromatography of, 108, 109
Intestine,
　*as source of alkaline phosphatase, 50
　　of cholinesterase, 50
　　of dialkylphosphofluoridase, 653
Inulins,
　action of invertase on, 256
Invertase, see β-Fructofuranosidase
Iodide ion,
　activation and inhibition of fumarase
　　by, 734
Iodine,
　complex with choline, 662
　complexes with lecithin and lysoleci-
　　thin, 660
　inactivation of aldolase by, 313
　　of lactic dehydrogenase by, 453
　use in assay of branching enzyme, 222–
　　223
Iodoacetamide,
　inhibition of galactokinase by, 292
　　of glyceraldehyde-3-phosphate de-
　　　hydrogenase (TPN) by, 414
Iodoacetate,
　inhibition of acetoacetate decarboxy-
　　lase by, 627
　　of acetoin-forming system by, 475
　　of alcohol dehydrogenases by, 503, 507
　　　protection against by substrate and
　　　　DPN, 503
　　of aldehyde oxidase by, 527
　　　protection against by substrate, 527
　　of glyceraldehyde-3-phosphate dehy-
　　　drogenases by, 405, 411, 503
　　　protection against by substrate, 503
　　of glycolic oxidase by, 532
　　of 6-phosphogluconic dehydrogenase
　　　by, 334
　　of thiolase by, 584
o-Iodosobenzoate,
　inhibition of enzymes by, 157, 285,
　　292, 405, 565, 594
Ion exchange chromatography of en-
　　zymes, 113–125, see also under Chro-
　　matography of enzymes
　application to α-amylase, 151
Ipomoea batatas, see Sweet potato
Iron,
　binding of, by cysteine, 318
　　by α,α'-dipyridyl, 318

Iron (ferric),
　activation of plant carboxylases by,
　　463
　inhibition of phospholipase C by, 671
　　of wheat germ carboxylase by, 470
Iron (ferrous),
　activation of aconitase by, 698
　　of formic hydrogenlyase by, 541
　　of oxalacetic carboxylase by, 757
　　of phospholipase A by, 665
　　of phosphoroclastic system from Cl.
　　　butylicum, 481–482
　　of plant carboxylases by, 463, 469
　　of yeast aldolase by, 319
　inhibition of phospholipase C by, 671
Iron-porphyrin,
　as component of liver aldehyde oxi-
　　dase, 526
Isobutyl alcohol,
　as substrate for yeast alcohol dehy-
　　drogenase, 502
　use of in extracting enzymes, 44
Isocitrate,
　conversion to α-ketoglutarate, 699–
　　704, 707–709, 710–714
　　to oxalosuccinate, 699–704, 705–707
　inhibition of oxalosuccinate decar-
　　boxylation by, 704
　as substrate for aconitase, 695–698
Isocitric dehydrogenase (DPN),
　from beef heart, 711–714
　　irreversibility of, 714
　　purification by precipitation chroma-
　　　tography, 712–713
　from yeast, 707–709
　　irreversibility of, 709
Isocitric dehydrogenase (TPN),
　from pig heart, 699–704
　　analogy to "malic" enzyme, 699,
　　　703–704
　　oxalosuccinic carboxylase activity
　　　of, 700–702
　　oxidative decarboxylation of d-iso-
　　　citrate by, 699–700
　　true isocitric dehydrogenase activity
　　　of, 699–702
　from yeast, 705–707
Isomaltose,
　as acceptor substrate for dextransu-
　　crase, 183

as donor substrate for dextransucrase, 182

Isomaltulose,
 formation by dextransucrase, 183

Isopropyl alcohol,
 phase diagram for, 106
 as substrate for glycerol dehydrogenase, 400
 for yeast alcohol dehydrogenase, 502

Isovaleraldehyde,
 as substrate for aldehyde dehydrogenase, 517

ITP, see Inosine triphosphate

J

Jack beans,
 fructose-1-phosphate aldolase in, 322
 phosphorylase from, 194

K

Kaolin, as adsorbent, 96, 257

α-Keto acid oxidases, see also α-Ketoglutaric dehydrogenase and Pyruvic oxidase
 from heart muscle, 34

α-Keto acids,
 inhibition of acetoacetate decarboxylase by, 627
 as substrates for lactic dehydrogenase, 453

β-Ketoacyl coenzyme A,
 reduction of, 566–573

Ketoaldehydes,
 reaction with glutathione, 454

α-Ketobutyrate,
 as substrate for lactic dehydrogenase, 443
 for wheat germ carboxylase, 470

α-Ketocaproate,
 as substrate for lactic dehydrogenase, 443

β-Ketocaproate,
 as acceptor in CoA transferase reaction, 580

Ketogenic factor,
 of calf intestinal mucosa, 44

α-Ketoglutarate,
 conversion to succinyl coenzyme A, 714–718

dismutation to succinate and glutamate, 716
 formation by isocitric dehydrogenase, 699–704
 from d-isocitrate, 707–709, 710–714
 from oxalosuccinate, 705–707
 reductive carboxylation to isocitrate, 699–704, 705–707
 as substrate for wheat germ carboxylase, 470

α-Ketoglutaric dehydrogenase system,
 analogy with pyruvate oxidation system, 718
 in assay of succinyl CoA deacylase, 603
 from E. coli, 718
 separation into components, 718
 from heart, 714–718

Ketoheptose-1-phosphate,
 formation by aldolase, 322

Ketohexose-1-phosphate,
 formation by aldolase, 322

β-Ketohydrogenase, see β-Ketoreductase

β-Ketoisocaproate,
 as acceptor in CoA transferase reaction, 580

Ketones,
 inhibition of lipase by, 638

Ketopentose-1-phosphate,
 formation by aldolase, 322

β-Ketoreductase (β-hydroxybutyryl coenzyme A dehydrogenase),
 in assay of crotonyl CoA, 561–562
 distribution of, 573
 from sheep liver, 566–573

Ketose,
 colorimetric determination of, 261

β-Ketothiolase, see Thiolase

α-Ketovalerate,
 as substrate for lactic dehydrogenase, 443

β-Ketovalerate,
 as acceptor in CoA transferase reaction, 580

β-Ketovaleryl coenzyme A,
 as substrate for thiolase, 584

KG, see α-Ketoglutarate

Kidney,
 acetoacetate-activating enzyme in, 573–574
 alkaline phosphatase from, 38, 50

*aminopeptidase from, 50
choline oxidase from, 675
CoA transferase from, 574
dialkylphosphofluoridase from, 654–655
DPN-cytochrome *c* reductase from, 39
fatty acid oxidation in mitochondria from, 545–548
*glutaminase I from, 50
isocitric dehydrogenase (DPN) from, 714
β-ketoreductase in, 573
Kieselguhr,
    partition chromatography with, 100
    silane treated, 100
*Kynureninase,
    as adaptive enzyme in *Neurospora*, 135
    in *Pseudomonas*, 133–134
Kynurenine,
    as energy source for microorganisms, 129

L

Lactase, see β-Galactosidase
Lactate,
    as CoA acceptor in *Cl. kluyveri*, 602
    formation by "malic" enzyme, 748–753
    specific stabilizing effect on lactic dehydrogenase, 445
    as substrate for glycolic oxidase, 531
Lactic acid bacteria,
    rupture of by sonic oscillator, 58
Lactic dehydrogenase,
    in assay of pyruvate kinase, 435–436
        of pyruvic oxidase, 487
    bacterial flavoprotein, 50
    from heart muscle, 449–454
    from muscle, 396, 441–443
    from yeast, 50, 444–449
        identity of with cytochrome b₂, 448
        presence of flavin mononucleotide in, 448
Lactobacillus arabinosus,
    growth medium for production of "malic" enzyme from, 749
    "malic" enzyme from, 748–753
Lactobacillus delbrückii,
    acetoin-forming system in, 486

growth medium for, 484
pyruvic acid oxidase from, 482–486
Lactobacillus pentosus,
    xylose isomerase from, 366–367
Lactobacillus plantarum,
    choline acetylase from, 622
Lactobionate,
    as substrate for β-galactosidase, 248
Lactobiose,
    formation by transglycosidation, 248
Lactones, inhibition of lipase by, 639
Lactose,
    optical rotation of, 232
    as substrate for β-galactosidase, 242
Lactositol,
    as substrate for β-galactosidase, 248
Lactotriose,
    formation by transglycosidation, 248
Lead acetate,
    in protein fractionation, 525, 538
    purification of β-amylase with, 155
Lead ion,
    as inhibitor for phosphoglucomutase, 298
Lead phosphate,
    as adsorbent for invertase, 257
Leaves, extraction of, 63
Lebedew juice, see also Yeast, extraction of
    preparation of, 317, 419, 446
Lecithin,
    iodine complex with, 660
    monomolecular film as substrate for phospholipase, 660
    as substrate for phospholipase A, 660–665
        for phospholipase C, 670–671
        for phospholipase D, 672
    suspension of, 661
Lecithinase, see also Phospholipases
    from bacteria, 50, 299
    from *Cl. welchii*, 299
        calcium or magnesium as activator for, 299
        fluoride as inhibitor for, 299
    inactivation of choline oxidase by, 677
Lecithins, synthetic (saturated),
    as substrates for phospholipase A, 664
Leucine, acetylation of, 618

Leuconostoc mesenteroides,
  absence of aldolase from extracts of, 330
    of transhydrogenase from extracts of, 331
  alcohol dehydrogenase (DPN) in, 506
  alcohol dehydrogenase (TPN) from, 504–508
  culture medium for, 181, 328
  dextran sucrase from, 179–184
  gluconokinase from, 334
  glucose-6-phosphate dehydrogenase (DPN-TPN) from, 328–332
  6-phosphogluconic dehydrogenase (DPN) from, 328–329, 332–334
  sucrose phosphorylase from, 225–229
    hydrolytic activity of, 229
Leucrose,
  in dextransucrase reactions, 182–183
Levans, action of invertase on, 256
Levan-splitting enzymes,
  from *Bacillus subtilis*, 187
Levansucrase,
  from *Aerobacter levanicum*, 186–189
Lichenase,
  separation from cellobiase, 178
Liebermann-Burchard reaction,
  for cholesterol, 678
Lima beans,
  phosphorylase from, 194
Limit dextrin,
  as product of action of phosphorylase, 205
    of α-amylase, 152
    of β-amylase, 157
  as substrate for amylo-1,6-glucosidase, 211
Lipase(s), 627–642
  assay methods for, 627–634
    colorimetric method, 631–634
    dilatometric method, 628
    manometric method, 628
    stalagmometric method (surface tension), 628
    titrimetric method, water-insoluble substrates, 628–630
      water-soluble substrates, 630–631
    turbidimetric method, 628
  from bacteria, 641
  chromogenic substrates for, 631–634

from fungi, 640–642
  gastric, 637
  from milk, 637
  from pancreas, 634–640
    activation energy of, 640
    energy and entropy of thermal inactivation of, 639–640
    properties of, 635–640
      activators, 638
      inhibitors, 638–639
      optimal conditions, 639–640
      specificity, 636–638
    purification of, 634–635
  from seeds, 641–642
  solubilization of enzymes by, 39
  use in extraction of brain hexokinase, 281
  from wheat germ, 639–642
Lipoic acid (protogen, pyruvate oxidation factor),
  acetylation of, by acetyl CoA, 494
  as component of α-ketoglutaric dehydrogenase system, 718
  of pyruvic oxidase, 489
  of pyruvic oxidase fraction A, 494
Lipoic dehydrogenase,
  in bacterial pyruvic oxidase fraction B, 494
Lipoic transacetylase,
  in bacterial pyruvic oxidase fraction A, 494
Lipoproteins,
  dissociation of by butanol, 42
  solubilization of by heparin, 38
Lipothiamide pyrophosphate,
  role in pyruvic oxidase system, 494
Lithium ion,
  inhibition of aceto-CoA-kinase by, 591
    of aldehyde dehydrogenase by, 510
    of pyruvate kinase by, 439
    of transacetylase by, 598
Liver,
  acetoacetate-activating enzyme in, 573–574
  aceto-CoA-kinase from, 588
  acetone powder, preparation from, 609–610
  alcohol dehydrogenase from, 495–500
  aldehyde dehydrogenase (DPN) from, 514–517

aldehyde oxidase (flavin-linked) from pig, 523–528
*D-amino acid oxidase from, 33
*arginase from, 33
branching enzyme from, 222–225
butyryl coenzyme A dehydrogenase from, 553–559
*catalase from, 33
choline oxidase from, 37, 674–678
crotonase from, 559–566
*cytochrome c reductase (DPN) from, 50
*cytochrome c reductase (TPN) from, 35
deacylase for acetoacetyl CoA in, 574
deoxyribonucleoproteins from, 33
dialkyl phosphofluoridase in, 653
enzyme for acetylation of amines from, 608–612
    of D-glucosamine from, 612, 615
esterase from, 33, 48, 657–659
fatty acid activating enzyme from, 585
fatty acid oxidase from, 35
fatty acid oxidation in mitochondria from, 545–548
    effect of aging, etc. on, 548
fatty acid oxidation in mitochondrial extracts from, 548
fructokinase from, 286–289
fructose-1-phosphate aldolase from, 320–322
glucose dehydrogenase from, 335–339
β-glucuronidase from, 37, 264–269
*glutamic dehydrogenase from, 33
glycerol dehydrogenase from, 320
glycolic oxidase from, 529
glyoxalase II from, 458–468
hexokinase from, 283
hydration of sorbyl CoA in, 564
isocitric dehydrogenase (DPN) from, 714
isolation of cell components from, 16–19
β-ketoreductase from, 566–573
"malic" enzyme from, 739–748
nuclei from, 18–19
nucleoside diphosphokinase in, 759
oxalacetate synthesizing enzyme from, 758–763
*peroxidase from, 33

phosphatase, acid from, 33
    alkaline from, 33, 50
phosphorylase from, 215–222
*proteinases from, 33
ribonucleoproteins from, 33
sorbitol dehydrogenase from, 348–350
succinic dehydrogenase from, 35, 37
transketolase fiom, 371–373, 375
*xanthine oxidase from, 33
LTPP, see Lipothiamide pyrophosphate
Lung,
    dialkylphosphofluoridase in, 653
Lyophilization of bacteria, 54–55
Lysine, acetylation of, 618
Lysolecithin,
    formation by phospholipase A, 660–665
    hemolytic action of, 660–661, 665
    iodine complex with, 660
    as substrate for phospholipase B, 665–668
Lysolecithinase,
    in phospholipase A assay, 661
Lysozyme,
    action on M. lysodeikticus, 754
    assay of, 124
    ion exchange chromatography of, 113, 117, 118, 124, 125
    for preparation of bacterial extracts, 161
D-Lyxose,
    as inhibitor for mammalian hexokinase, 284

M

Magnesium-ATP complex,
    evidence for in fructokinase reaction, 289
Magnesium chloride,
    activation of alcohol dehydrogenase (TPN) by, 507
Magnesium-fluoro-phosphate complex,
    inhibition of enolase by, 434
    of phosphoglucomutase by, 299
Magnesium ion,
    activation of acetate kinase by, 594
    of aceto-CoA-kinase by, 590
    of acetoin-forming system by, 474
    of aldehyde dehydrogenase (TPN) by, 514
    of choline oxidase by, 675

of dialkylphosphofluoridase by, 656
of enolase by, 433–434
of fructokinase by, 289
of galactokinase by, 292
of G-1-P kinase by, 356
of gluconokinase by, 354
of glucose-6-phosphate dehydrogenases by, 325, 332
of hexokinase by, 275
of isocitric dehydrogenases by, 704,
   707, 709, 713
of "malic" enzyme (DPN) by, 752
of "malic" enzyme (TPN) by, 747
of oxalacetic carboxylase by, 757
of pentokinase by, 360
of phosphoglucomutase by, 288
of 6-phosphogluconic dehydrogenase
   by, 327, 334
of phospholipase C (Cl. welchii) by,
   299, 671
of phosphoribomutase by, 362
of plant carboxylases by, 463, 469
of pyruvate kinase by, 439
of pyruvic oxidases (bacterial) by,
   482, 491
of transketolase by, 380
of xylose isomerase by, 370
as constituent of yeast carboxylase,
   463
effect on aconitase equilibrium, 698
inhibition of gluconokinase by, 354
   of GPC diesterase by, 670
   of phosphoglucose isomerase by,
      306
   of phosphomannose isomerase by,
      304
role in fluoride inhibition of phosphoglucomutase, 298
Malate,
   conversion to fumarate, 729–735
      to l-lactate and CO_2, 748–753
      to oxalacetate, 735–739
   fluorometric assay of, 730
   inhibition of oxalacetic carboxylase
      activity of "malic" enzyme by,
      747
   optical rotation of, 729
   oxidation to pyruvate and carbon
      dioxide, 739–748

protection of "malic" enzyme (DPN)
   by, 752
Maleate buffer, 142
Maleic acid,
   inhibition of succinic dehydrogenase
      by, 727
Malic dehydrogenase,
   in assay of condensing enzyme, 687–
      689
      of crotonyl CoA, 562
   from pig heart, 735–739
"Malic" enzyme (DPN),
   assay by CO_2 liberation from l-malate,
      748–749
   assay of oxalacetic carboxylase activ-
      ity of, 749
   from Lactobacillus arabinosus, 748–753
   in Moraxella, 752
      activation by monovalent cations,
         752
   reversibility of reaction, 753
   specific protection by malate and by
      manganous ion, 752
"Malic" enzyme (TPN),
   assay of dehydrogenase activity of,
      739–741
      of oxalacetic carboxylase activity
         of, 741–742
   distribution of, 747–748
   from pigeon liver, 742–744
   from wheat germ, 744–746
Malonate,
   inhibition of glycolic oxidase by, 532
      of oxalacetic decarboxylase activ-
         ity of "malic" enzyme by, 747
      of succinic dehydrogenase by, 727
Malt,
   β-amylase from, 158
   cellulase from, 176, 177
   crystalline α-amylase from, 154
Maltose,
   as acceptor substrate for amylomal-
      tase, 191
      for dextran sucrase, 183
   as donor substrate for amylomaltase,
      191
   optical rotation of, 232
   as product of α-amylase action, 152
β-Maltose,
   as product of β-amylase action, 157

Maltose, labeled,
  synthesis with maltose phosphorylase, 231
Maltose phosphorylase,
  from *Neisseria meningitidis*, 229–231
Maltotetraose,
  as primer for phosphorylase, 196
Maltotriose,
  as primer for phosphorylase, 196
Mammary gland,
  alkaline phosphatase from, 50
Manganese dioxide,
  for anaerobic assay of choline oxidase, 676–677
Manganous ion,
  activation of acetate kinase by, 594
    of aceto-CoA-kinase by, 590
    of acetoin forming system by, 474
    of α-acetolactic acid decarboxylase by, 474
    of aldehyde dehydrogenase (TPN) by, 514
    of dialkyl phosphofluoridase by, 656
    of enolase by, 433–434
    of formic hydrogenlyase by, 541
    of fructokinase by, 289
    of galactokinase by, 292
    of gluconokinase by, 354
    of G-1-P kinase by, 356
    of isocitric dehydrogenases by, 699–704, 707, 709, 713
    of "malic" enzyme (DPN), 752
    of "malic" enzyme (TPN), 747
    of oxalacetate synthesizing enzyme, 762
    of oxalacetic carboxylase by, 757
    of phospholipase C by, 671
    of phosphoroclastic system from *E. coli*, 478
    of plant carboxylases by, 461, 463, 469
    of pyruvic oxidases (bacterial) by, 482, 490
    of pyruvic oxidase system (animal) by, 487
    of xylose isomerase by, 370
  for chromatography of aminopeptidase, 107
  inhibition of GPC diesterase by, 670
    of phosphoglucose isomerase by, 306

  of phosphomannose isomerase by, 304
  interference with copper reduction method, 360
  protection of "malic" enzyme (DPN) by, 757
  reaction with Tris buffer, 504
  removal of by orthophosphate and pyrophosphate, 704
  succinic dehydrogenase and, 728
  use in separating nucleic acids from proteins, 137, 244, 329
Mannitol-1-phosphate dehydrogenase,
  from *E. coli*, 346–348
Mannose,
  as substrate for glucose aerodehydrogenase, 345
    for hexokinases, 274, 284
Mannose-6-phosphate,
  conversion to fructose-6-phosphate, 299–304
α-D-Mannosides,
  as substrates for β-glucosidase, 237
MB, see Methylene blue
Melezitose,
  optical rotation of, 232
Melibiase, see α-Galactosidase
Melibiose,
  as inducer for β-galactosidase, 248
  optical rotation of, 232
  as substrate for dextransucrase, 183
    for α-galactosidase (melibiase), 249
Mercaptide-forming agents,
  inhibition of succinic dehydrogenase by, 727
2-Mercaptoethanol,
  activation of yeast aldehyde dehydrogenase (DPN, TPN) by, 510
Mercuric ion,
  in crystallization of enolase, 85, 432
  as inhibitor of enzymes, 157, 197, 240, 248, 257, 267, 298, 464, 475, 507, 594, 627, 656, 698
Mercury-enolase complex,
  recovery of enolase from, 433
Meristems, extraction of, 63
Merulius lachrymans,
  cellobiase from, 178
  cellulase from, 177

Methanol,
    as acceptor for β-fructofuranosidase,
        255
    action on plasma proteins, 84
    in purification of β-glucuronidase, 266
        of transketolase, 372
    as substrate for yeast alcohol dehy-
        drogenase, 502
Methylamine, acetylation of, 618
Methyl butyrate,
    comparison with acetylcholine in liver
        esterase system, 659
    as substrate for lipases, 630
Methyl Cellosolve acetal,
    phase diagram for, 104
Methyl Cellosolve acetate,
    phase diagram for, 103
β-Methylcholine,
    as substrate for choline dehydrogenase,
        677
β-Methyl-crotonyl coenzyme A,
    as substrate for crotonase, 564
Methylene blue,
    as hydrogen acceptor for aldehyde
        oxidase, 527
    for yeast lactic dehydrogenase, 444
    in malic dehydrogenase assay, 739
    in manometric method for succinic
        dehydrogenase, 724
    in Thunberg method for succinic de-
        hydrogenase, 723–724
β-Methylfructofuranoside,
    as donor of β-fructofuranosidic group,
        255
Methyl-β-D-galactoside,
    as substrate for β-galactosidase, 248
N-Methyl-D-glucosamine,
    as inhibitor for mammalian hexo-
        kinase, 284
6-Methyl glucose,
    as substrate for glucose aerodehydro-
        genase, 345
α-Methyl glucoside,
    as acceptor substrate for dextran,
        183
Methylglyoxal,
    as substrate for glycerol dehydro-
        genase, 400
    for glyoxalase I, 454–458

α-Methyl-α-hydroxymethylcholine,
    as substrate for choline dehydrogenase,
        677
Methyl succinate,
    competitive inhibition of succinic
        dehydrogenase by, 727
Methyl viologen,
    effect on formic hydrogenlyase, 541
Michaelis constants,
    relationship between maximum initial
        velocities, equilibrium constant,
        and, 735
Micrococcus lysodeikticus,
    action of lysozyme on, 754
    assay of lysozyme with, 124
    oxalacetic carboxylase from, 753–757
Microorganisms, see also Bacteria and
        Yeast
    extraction of enzymes from, 51–62
        by chemical agents, 61, 62
        by enzymatic methods, 61
        by grinding, 59–61
        by mechanical pressure, 57
        by mechanical shaking with
            abrasives, 58, 59
        by pressure release, 57
        by sonic waves, 57, 58
Microsomal particles,
    from peanut cotyledons, 549–550
        fatty acid oxidation in, 549–550
    from rat liver, 18
Milk,
    *alkaline phosphatase from, 50
    lipase from, 637
    red protein of, 50
    *xanthine oxidase from, 35, 50
Minerals,
    as components of enrichment media,
        130
Mitochondria,
    animal tissue,
        betaine aldehyde dehydrogenase
            from, 714
        choline oxidase system in, 675–676
        *cytochrome oxidase from, 29
        disruption by high frequency oscil-
            lation, 29
        *DPN-cytochrome c reductase from,
            29

extraction of enzymes from, 48–49
extracts of acetone-dried, fatty acid oxidation in, 548
intact, fatty acid oxidation in, 545–549
isocitric dehydrogenase (DPN) from, 714
isolation technique, 17
soluble choline dehydrogenase from, 676
succinic oxidase from, 29
*transaminase from, 50
plant,
  isolation of, 19–22
  from avocado, 20
  from mung bean, 20
  from other sources, 22
  measurement of activity, 21
Molds,
  pectinesterase from, 160
  phospholipase B from, 666
  polygalacturonidase from, 164
Molybdate,
  inhibition of yeast isocitric dehydrogenase (DPN) by, 709
  replacement of, by tungstate in aldehyde oxidase system, 527
Molybdate, H₂S-reduced,
  inhibition of isocitric dehydrogenase (DPN) by, 709
Molybdenum,
  activation of aldehyde oxidase by, 527
  as component of liver aldehyde oxidase, 526
Monobutyrin,
  as substrate for lipase, 628
Monoethanolamine,
  inhibition of choline dehydrogenase by, 677
Monoethyl succinate,
  as substrate for succinic dehydrogenase, 726
Monoglycerides,
  as products of lipase action, 636
Monomethylamine,
  inhibition of choline dehydrogenase by, 677
Monomethyl succinate,
  as substrate for succinic dehydrogenase, 726

Monothiols, see also Cysteine, Glutathione,
  reversal of metal inhibition of isocitric dehydrogenase by, 704
Moraxella,
  "malic" enzyme in, 752
  activation by monovalent cations, 752
M-6-P, see Mannose-6-phosphate
Mucopolysaccharidases, 166–173
Muscle,
  aldolase from, 310–315, 316
  dialkyl phosphofluoridase in, 653
  fructokinase from, 289–290
  glucose-1-phosphate kinase from, 354–356
  glyceraldehyde-3-phosphate dehydrogenase from, 401–406
  α-glycerophosphate dehydrogenase from, 391–397
  hexokinase from, 283–284
  β-ketoreductase in, 573
  lactic dehydrogenase from, 396, 441–443
  phosphoglucomutase from, 294–299
  phosphoglucose isomerase from, 304–306
  phosphoglycerate kinase from, 418
  3-phosphoglyceric mutase from, 423–425
  phosphohexoisomerases from, 299–306
  phosphohexokinase from, 306–310
  phosphomannose isomerase from, 299–304
  phosphoribomutase from, 361–363
  phosphorylase from, 200–205
  PR enzyme from, 206–210
  pyruvate kinase from, 435–440
  triosephosphate isomerase from, 387–391
Muscle, heart, see Heart
Muscle, pigeon breast,
  condensing enzyme in, 694
  cytochrome c reductase (DPN) from, 34
  isocitric dehydrogenase (DPN) from, 714
  pyruvic oxidase from, 488, 490
Mustard gas (β,β-dichloroethylsulfide),
  as inactivator of yeast hexokinase, 275

Mutarotase,
  as contaminant of glucose aerodehydro-
    genase, 341, 342
Mycobacterium,
  *barbiturase formation in, 135, 136
  formation of uracil-thymine oxidase in,
    135, 136
Mycobacterium, cholesterol oxidizing,
  cholestenone oxidation by adaptive
    enzyme from, 678
  cholesterol dehydrogenase from, 678–
    681
  growth medium for, 680
Mycotorula lipolytica,
  lipase from, 641
Myogen A,
  fructose-1-phosphate aldolase in, 322
  in phosphohexokinase assay, 307
  as source of α-glycerophosphate dehy-
    drogenase, 308
*Myosin ATP-ase, 50
  protection of by ATP and actin, 43–44
  solubilization by butanol, 45

N

β-Naphthol esters,
  as substrates for lipases, 633
β-Naphthyl acetate,
  for colorimetric assay of esterases, 657
Narcotics,
  inhibition of succinic dehydrogenase
    by, 728
Neisseria perflava,
  amylosucrase from, 184–185
  phosphorylase from, 185
Nelson reagent, see Arsenomolybdate
    color reagent
Neurospora,
  kynureninase from, 135
Neurospora crassa,
  pectolytic enzymes from, 165
Neurospora sitophila,
  cellulase from, 177
Nickel ion,
  activation of oxalacetic carboxylase
    by, 757
  of plant carboxylases by, 469
Ninhydrin reagent,
  modification of, 117

β-Niphegal, see o-Nitrophenyl-β-D-
    galactoside
Nitrate,
  as electron acceptor for aldehyde
    oxidase, 527
  inhibition of formic hydrogenlyase by,
    541
*Nitrate reductase,
  from soybean leaves, 63
m-Nitroaniline,
  acetylation of, 610
p-Nitroaniline,
  acetylation of, 588, 608, 610
Nitrofuraldehyde semicarbazones,
  as hydrogen acceptors in phosphoro-
    clastic system, 482
"Nitrogen mustards,"
  inhibition of choline oxidase by, 677
    protection against, by choline or
      amines, 677
  inhibition of hexokinase by, 275
Nitrophenol,
  colorimetric determination, 232, 241
p-Nitrophenol acetate (PNPA),
  as substrate for lipase, 632–634
  synthesis of, 632
o-Nitrophenyl-α-L-arabinoside,
  as substrate for β-galactosidase, 248
p-Nitrophenyl esters,
  for colorimetric assay of esterases, 657
o-Nitrophenyl-β-D-galactoside,
  as substrate for β-galactosidase (lac-
    tase), 241, 248
p-Nitrophenyl-β-D-galactoside,
  as substrate for β-galactosidase, 248
Nitroprusside test,
  for sulfhydryl groups, 623–624
Nitrous acid,
  inactivation of invertase by, 257
Norit,
  adsorption of nucleotides by, for hexo-
    kinase assay, 279
Notatin, see Glucose aerodehydrogenase
    (Glucose oxidase)
Nuclear fraction,
  isolation from rat liver, 17
Nucleate, see Nucleic acid
Nuclei,
  isolation from rat liver, 18–19

Nucleic acids,
  precipitation by manganese sulfate,
    244
  protein fractionation with, 86, 137,
    378, 431
  removal of by Norit, 210
  separation of proteins from, 137, 138
  stabilizing effect on aldehyde dehy-
    drogenase, 511
Nucleoside diphosphokinase,
  in chicken liver, 759
3'-Nucleotidase,
  effect of adsorbents on, 95, 96
Nucleotides,
  adsorption of by Norit, 279

O

OAA, see Oxalacetate
Oats, lipase from, 641
Octanoate,
  oxidation of, in rat liver mitochondria,
    545–548
Octyl alcohol,
  activation of lipase by, 638
  inhibition of DR-aldolase by, 386
  as substrate for yeast alcohol dehy-
    drogenase, 502
Oleate,
  oxidation of, in rat liver mitochondria,
    545
Olive oil,
  as substrate for lipase, 630
Optical rotation,
  of oligosaccharides, before and after
    hydrolysis, 232
Orange flavedo,
  pentose phosphate isomerase from, 366
Organic solvents,
  control of dielectric constant by, 82
  protein fractionation with, 82–84, 88–
    90
Orotic acid, adaptation to, 135
Orthophosphate, see Phosphate
Oscillator, magnetostricture,
  use of for rupturing cells, 57, 58
Osmium tetroxide,
  inactivation of β-glucosidase by, 240
Oxalacetate (OAA),
  competitive inhibition of succinic de-
    hydrogenase by, 727

condensation with acetyl CoA,
  by crystalline condensing enzyme,
    685–694
  by rat liver mitochondria, 545
decarboxylation of by "malic" enzyme (DPN), 748–753
  inhibition by L-malate and malonate,
    753
decarboxylation of by "malic" enzyme (TPN), 739–748
  effect of manganous ion on, 741
  of TPN on, 741
  inhibition by malate and malonate,
    747
determination of $C^{14}O_2$ content of, 759
in malic dehydrogenase reaction, 735–
  739
Oxalacetate synthesizing enzyme,
  from chicken liver, 758–763
Oxalacetic carboxylase,
  "malic" enzymes as, 741, 749
  from M. lysodeikticus, 753–757
Oxalate,
  competitive inhibition of succinic de-
    hydrogenase by, 727
  inhibition of lactic dehydrogenase by,
    453
Oxalosuccinate,
  as intermediate in isocitric dehydro-
    genase reaction, 699–704, 705–707
  preparation of solution of, 701
Oxalosuccinate-manganese complex,
  light absorption by, 700
Oxamic acid,
  inhibition of lactic dehydrogenase by,
    453
Oxidation-reduction potential,
  of glucose-gluconate system, 339
Oxidizing agents,
  inhibition of succinic dehydrogenase
    by, 727
Oxygen,
  effect of on cysteine inhibition of yeast
    aldolase, 318–319
  as electron acceptor for aldehyde oxi-
    dase, 527
Oxyhemoglobin,
  application of Debye theory to solu-
    bility of, 70

Oxythiamine triphosphate,
  inhibition of yeast carboxylase by, 464
Ozone,
  inactivation of β-glucosidase by, 240

P

Palmitic acid,
  conversion to phosphatidic acid, 673
  oxidation of in rat liver mitochondria,
    545
Palmitic acid (C14-labeled),
  oxidation in higher plants, 550
Palmityl coenzyme A,
  inhibition of amine acetylation by, 611
Pancreas,
  α-amylase from human, 154
    from pig, 154
  lipase from, 634–640
  phospholipase B from, 666–668
  *ribonuclease from, 30, 107, 108, 119–
    122
Pancreatin,
  phospholipase A from, 661, 664
Parahematin compounds,
  formation from hematin, 727
Parsley (leaves and roots),
  "malic" enzyme (TPN) in, 748
Parsnips,
  "malic" enzyme (TPN) in, 748
Partition chromatography of enzymes,
  98–112, see also under Chromatog-
    raphy of enzymes
Pasteur effect,
  and aldolase action, 319
  and phosphate concentration, 320
PDA (DAP), see Dihydroxyacetone phos-
    phate
PE, see Pectinesterase
Pea leaves,
  glyceraldehyde-3-phosphate dehydro-
    genase (TPN) from, 413–415
  pentose phosphate isomerase from, 366
Peanut cotyledons,
  fatty acid oxidation systems from,
    549–551
Peas,
  formic dehydrogenase from, 536–539
  "malic" enzyme (TPN) in, 748
  phosphorylase from, 194

Pectase, see Pectinesterase
Pectic acids, definition of, 158
Pectic enzymes, 158–166
Pectinases, see Polygalacturonidases
Pectinesterase, 159–162
  sources of, 159–160
  from tomato, 159–161
Pectinic acids, definition of, 158
Pectin-methylesterase, see Pectinesterase
Pectins, definition of, 158
Pectolytic enzymes, see Polygalacturo-
    nidases
Penatin (Penicillin A, Penicillin B), see
    Glucose aerodehydrogenase (glucose
    oxidase)
*Penicillinase,
  partition chromatography of, 110–112
Penicillium,
  phospholipase B from, 668
Penicillium notatum,
  glucose aerodehydrogenase from, 340–
    345
  growth medium for, 343
Penicillium roqueforti,
  lipase from, 641
trans-2-Pentenoyl coenzyme A,
  as substrate for crotonase, 564
Pentokinase,
  from yeast, 357–360
Pentose isomerases, 366–370
Pentose phosphate isomerase,
  from alfalfa, 363–366
  from baker's yeast, 366, 376–377
  sources of, 366
P enzyme, see Phosphorylating enzyme
PEP, see Phosphoenolpyruvic acid
*Peroxidase,
  from liver, 33
PG, see Polygalacturonidases
PGA, see Phosphoglyceric acid
PGA-P, see 1,3-Diphosphoglyceric acid
Phase diagrams,
  preparation of, 101, 102
o-Phenanthroline,
  inhibition of aldehyde oxidase by, 527
Phenarsazine,
  inhibition of isocitric dehydrogenase
    by, 704
Phenethylamine,
  acetylation of, 610

Phenol,
  colorimetric determination, 232
Phenolphthalein β-glucuronide,
  as substrate for β-glucuronidase, 263
Phenylarsine oxide,
  inhibition of acetate kinase by, 594
o-Phenylenediamine,
  acetylation of, 610
Phenyl-β-D-galactoside,
  as substrate for β-galactosidase, 248
Phenyl-β-D-glucoside(s),
  influence of substituents on hydrolysis
    of, 239
  as substrate for β-glucosidase, 234
Phenylglyoxal,
  as substrate for glyoxalase I, 457
Phenylhydrazine,
  inhibition of glyoxylic reductase by,
    535
  of invertase by, 257
Phenyl mercuric nitrate,
  inhibition of dialkylphosphofluoridase
    by, 656
  of isocitric dehydrogenases by, 704
Phenylpyruvate,
  as substrate for lactic dehydrogenase,
    443
β-Phenylthiogalactoside,
  as inhibitor of β-galactosidase, 247
Phenylurethan,
  inhibition of succinic dehydrogenase
    by, 728
Phloretin,
  inhibition of muscle phosphorylase by,
    205
Phloridzin (Phlorhizin),
  inhibition of muscle phosphorylase by,
    205
  of plant phosphorylase by, 197
*Phosphatase, acid,
  from liver, 33
*Phosphatase, alkaline,
  adsorption of, effect of substrate on, 32
  assay of, in tissues, 49
  ethanol for extraction of, 33
  glucose-1,6-diphosphate formation by,
    355
  solubilization of by autolysis, 38
    by butanol, 42
    by lipase, 39

by trypsin, 39
  sources for purification by butyl alco-
    hol method, 50
Phosphate,
  activation of fumarase by, 734
  effect of on formic hydrogenlyase, 541
    on Michaelis constants of fumarase,
      734
  as glycosyl acceptor for sucrose phos-
    phorylase, 228
  inhibition of glucose-6-phosphate de-
    hydrogenase (L. mesenteroides)
    by, 332
    of glucose-6-phosphate dehydrogen-
      ase (yeast) by, 325
    of pentose phosphate isomerase by,
      365
    of phosphoglucose isomerase by, 306
    of phospholipase C by, 671
    of phosphomannose isomerase by,
      304
    of succinic dehydrogenase by, 727
    fluoride requirement for, 727
    of triosephosphate isomerase by, 390
  removal of manganous ion by, 704
  requirement for, in butyrate oxidation,
    551–553
    in liver aldehyde oxidase system, 527
    replacement of by arsenate, 527
    by pyruvic acid oxidase of L. del-
      brückii, 482
  role in fluoride inhibition of phospho-
    glucomutase, 299
  in transacetylase reaction, 596
Phosphate acceptors,
  effect on fatty acid oxidation, 548
Phosphate buffers, 81, 143
  table for various concentrations of, 81
Phosphate-transferring enzyme II, see
  Pyruvate kinase
Phosphatides, see Phospholipids
Phosphatidic acid,
  enzymatic synthesis of, 673–674
  formation by phospholipase D, 672
Phosphoenolpyruvic acid (PEP),
  extinction coefficient of, 428
  formation of from 2-phosphoglyceric
    acid, 427–435
  lability of phosphate group, 427
  oxaloacetate synthesis from, 758–763

phosphorylation of ADP by, 435–440
Phosphoglucomutase, see also Phosphoribomutase
from muscle, 294–299
6-Phosphogluconate,
   identification of, 353
6-Phosphogluconic dehydrogenase
   (DPN),
   from *L. mesenteroides*, 328–329, 332–
      334
6-Phosphogluconic dehydrogenase (TPN),
   from brewer's yeast, 323–324, 326–327
Phosphoglucose isomerase,
   from muscle, 304–306
Phosphoglycerate kinase,
   from brewer's yeast, 415–422
   from other sources, 418
2-Phosphoglyceric acid,
   conversion to phosphoenolpyruvic
      acid, 427–435
   conversion to 3-phosphoglyceric acid,
      423–425
3-Phosphoglyceric acid,
   as coenzyme for diphosphoglyceric
      acid mutase, 427
   conversion to 2-phosphoglyceric acid,
      423–425
   formation of, from 3-phosphoglycer-
      aldehyde, 401
   phosphorylation of, by ATP, 415–422
Phosphoglyceric acid mutases, 423–427
   diphosphoglyceric mutase from rabbit
      erythrocytes, 425–427
   3-phosphoglyceric mutase from rabbit
      muscle, 423–425
Phosphoglyceric acids,
   optical rotation in HCl solution, 424
      in molybdate solution, 423
Phosphohexoisomerase(s),
   in assay of transaldolase, 381
   from muscle, 299–306
Phosphohexokinase,
   from muscle, 306–310
1-Phosphoketoses,
   formation by aldolase, 322
Phospholipases, 660–672, see also
   Lecithinase
   glycerophosphorylcholine diesterase,
      668–670
   phospholipase A, 660–665

distribution of, 662–663
   from pancreatin, 661
   from *Serratia plymuthicum*, 663
   from venom of *Crotalus terrificus*,
      662
phospholipase B, 665–668
   distribution of, 666
   from pancreas, 666–668
   from *Penicillium*, 668
   from *Serratia plymuthicum*, 666
phospholipase C, 670–671
   from *Cl. welchii*, 671
phospholipase D from cabbage, 672
thermostability of, 664–665, 668, 671,
      672
Phospholipids,
   enzymatic cleavage of, 660–672
   enzymatic synthesis of, 663–674
Phosphomannose isomerase,
   from muscle, 299–304
Phosphopyruvate transphosphorylase,
   see Pyruvate kinase
Phosphoribomutase, see also Phosphoglucomutase
   from muscle, 361–363
   from yeast, 362
5-Phosphoribonic acid,
   inhibition of pentose phosphate iso-
      merase by, 365
Phosphoroclastic split of pyruvate,
   yielding formate, 476–478
   yielding hydrogen, 479–482
   ascorbate effect on, 481
Phosphorylase(s),
   from liver, 215–222
   from muscle, 71, 200–205
      form *a*, 202–204, 205
      form *b*, 204–205
   from *Neisseria perflava*, 185
   from plants, 192–200
      potato, 62, 194–199
      yeasts, 199–200
"Phosphorylating enzyme" (succinate
   activating),
   from heart, 718–722
   from spinach leaves, 722
*Phosphorylation, aerobic,
   during choline oxidation, 678
Phosphorylcholine,
   formation by phospholipase C, 670–671

Phosphotransacetylase, see Transacetylase
Phosphotungstic acid,
as protein precipitant, 85
Phthalate-hydrochloric acid buffer, 139
Phthalate-sodium hydroxide buffer, 142
Picric acid,
use in purification of invertase, 252, 257
Pine seed,
blastolipase from, 642
Plants, see also under names of individual
plants
extraction of enzymes from, 62, 63
glyceraldehyde-3-phosphate dehydrogenase (DPN) from, 411
glyceraldehyde-3-phosphate dehydrogenase (TPN) from, 411–415
isolation of mitochondria from, 19–22
Plasma,
dialkylphosphofluoridase in, 652–653
fractionation of by ethanol, 89
Plasmolysis of yeast, 271
Pneumococci,
hyaluronidase from, 168
PNPA, see p-Nitrophenol acetate
Polarimetry,
in assay of invertase, 259
Polyelectrolytes, see Nucleic acids and
Protamines
synthetic, use of in protein precipitation, 86
Polygalacturonic acid, 162
Polygalacturonidases (polygalacturon-
ases), 162–166
sources of, 164
Polymethacrylate,
precipitation by barium ions, 87
Polypeptides, glycyl and leucyl,
activation of lipase by, 638
Polysaccharide,
Pneumococcus type III,
as primer for phosphorylase, 196
precipitation by concanavalin A, 257
synthesis from disaccharides, 178–192
by amylomaltase, 189–192
by amylosucrase, 184–185
by dextransucrase, 179–184
by levansucrase, 186–189
synthesis from glucose-1-phosphate,
192–222

by liver phosphorylase, 215–222
by muscle phosphorylase, 200–215
by plant phosphorylases, 192–200
Polyvinyl alcohol (PVA),
for emulsification of lipids, 629
Potassium iodide,
activation of alcohol dehydrogenase
(TPN) by, 507
Potassium ion,
activation of aceto-CoA-kinase by, 591
of fructokinase by, 289
of "malic" enzyme (DPN) by, 752
of pyruvate kinase by, 439
of transacetylase by, 598
of yeast aldehyde dehydrogenase
(DPN, TPN) by, 508–511
Potato,
extraction of, 62
phosphorylase from, 194–199
Potato sprouts,
cellulase from, 177
Precipitation chromatography,
in purification of isocitric dehydrogenase (DPN) from beef heart,
712–713
PR enzyme,
action on phosphorylase, 202
preparation from muscle, 206–210
"Priming effect" (sparking effect),
on fatty acid oxidation in liver mitochondria, 548
n-Propanol,
as substrate for alcohol dehydrogenase
(TPN), 507
Propandiols,
as substrates for glycerol dehydrogenase, 400
Propionaldehyde,
acyloin formation from, 464
inhibition of DR aldolase by, 386
of wheat germ carboxylase by, 470
as substrate for aldehyde dehydrogenases, 510, 513, 517, 520
for glyceraldehyde-3-phosphate
dehydrogenase, 404
for liver aldehyde oxidase, 527
Propionate,
conversion to propionyl CoA,
by aceto-CoA-kinase of yeast, 590

by CoA transphorase of *Cl. kluy-veri*, 602
as substrate for acetate kinase, 594
Propionic acid,
phase diagram for, 103
Propionyl coenzyme A,
in acylation of amino acids, 618
as substrate for condensing enzyme, 694
for transacetylase, 598
n-Propyl alcohol,
as substrate for yeast alcohol dehydrogenase, 502
Propyl Cellosolve,
phase diagrams for, 102, 105
Prostate,
β-ketoreductase in, 573
*Prostatic phosphatase,
effect of adsorbents on, 96
Protamine sulfate,
use in protein precipitation, 86, 326, 534, 617
use in removal of nucleic acids, 138, 227, 378, 420, 431–432, 485, 492, 506, 570, 589, 601, 750–751
*Proteinases from liver, 33
Proteins, see also Enzymes
denaturation of by octyl alcohol-chloroform, 137
factors influencing solubility of, 68, 69
fractionation by solubility, 67–90
precipitation by acids, 85, 86
by detergents, 86
by metal ions, 85
by polyelectrolytes, 86
separation of by adsorbents, 90–98
separation of from nucleic acids, 137–138
solubility of,
effect of ionic strength on, 72
of pH on, 73, 78–82
of temperature on, 73
spectrophotometric determination of, 740
Proteus vulgaris,
acetate kinase from, 595
Protogen, see Lipoic acid
Protopectin, 158
Protopectinase, 158

Pseudoglobulin,
solubility of in ammonium sulfate, 74
Pseudomonas,
adaptation to catechol, 135
to tryptophane, 135
catalase formation in, 136
diaphorase formation in, 137
kynureninase from, 133, 134
peroxidase formation in, 136
pyrocatecase from, 134
tryptophan peroxidase-oxidase system from, 134
Pseudomonas hydrophila,
growth medium for, 367
xylose isomerase from, 366–370
Pseudomonas saccharophila,
sucrose phosphorylase from, 225–229
Pumpkin,
phosphorylase from, 194
PVA, see Polyvinyl alcohol
Pyocyanine,
in assay of butyryl CoA dehydrogenase, 553–554
Pyridine nucleotides, see also Diphosphopyridine nucleotide, Triphosphopyridine nucleotide
destruction in plant extracts, 740
Pyridine-3-sulfonate,
inhibition of lactic dehydrogenase by, 453
4-Pyridoxic acid,
inhibition of glucose dehydrogenase by, 339
Pyrocatecase,
as adaptive enzyme from *Pseudomonas*, 134
*Pyrophosphatase, inorganic,
elution of by pyrophosphate, 95
Pyrophosphate, inorganic,
inhibition of aldolase (yeast) by, 317–318
of phosphoglucose isomerase by, 306
of phosphomannose isomerase by, 304
of succinic dehydrogenase by, 727
of succinyl CoA deacylase by, 605
of transacetylase by, 599
as product of aceto-CoA-kinase reaction, 585–591
removal of manganous ions by, 704

Pyruvate,
    conversion to lactate, 441–443
    effect of on formic hydrogenlyase, 541
    homologs of, as substrates for yeast
        carboxylase, 464
    phosphoroclastic split of, yielding
        formate, 476–478
    phosphoroclastic split of, yielding hy-
        drogen, 479–482
    reductive carboxylation to L-malate,
        739–748
    as substrate for carboxylases, 460–471
Pyruvate kinase (pyruvate phospho-
        kinase, pyruvic phosphoferase),
    from muscle, 435–440
        human, 437–439
        rabbit, 436–437
    in yeast, 439
Pyruvate oxidation factor, see Lipoic
        acid
Pyruvic acid oxidase (coenzyme A-
        linked),
    from animal tissues, 486–490
        heart muscle, 487
        pigeon breast muscle, 488–490
    from bacteria, 490–495
Pyruvic acid oxidase (phosphate-linked),
    from *Lactobacillus delbrückii*, 482–486

Q

Q-enzyme,
    effect on plant phosphorylase, 193
Quinacrine (atabrine),
    inhibition of aldehyde oxidase by, 527
        reversal by FAD, 527
Quinine,
    inhibition of lipase by, 639

R

Raffinose,
    as donor substrate for levansucrase, 187
    as donor of β-fructofuranosidic group,
        255
    optical rotation of, 232
Red blood cells,
    acetylcholinesterase from, 646–647
    dialkylphosphofluoridase in, 653
    diphosphoglyceric acid mutase from
        rabbit, 425–427

Red cell stroma,
    cholinesterase from, 37, 47, 50
Red cloves,
    pectinesterase from, 159
Red protein from milk, 50
Resins,
    preparation of for chromatography,
        113–115
Resorcinol-thiourea reagent,
    for fructokinase assay, 287
Respiration,
    rates of, in rat tissues, 9
Retina,
    "malic" enzyme (TPN) in, 747
Rhodospirillum rubrum,
    disintegration of, 64
Riboflavin,
    as hydrogen acceptor in phosphoro-
        clastic system, 482
*Ribonuclease,
    assay of, 119
    extraction from pancreas, 30
    ion exchange chromatography of, 113,
        117, 118, 119–122
    partition chromatography of, 107, 108
Ribonucleoproteins, liver, 33
Ribose,
    phosphorylation of, 357–360
Ribose-1,5-diphosphate,
    formation by phosphoglucomutase, 362
Ribose-1-phosphate,
    conversion to ribose-5-phosphate, 361–
        363
Ribose-5-phosphate,
    conversion to ribulose-5-phosphate,
        363–366, 377
    inhibition of glucose dehydrogenase by,
        339
    as product of pentokinase reaction, 360
    in transketolase reaction, 371–375,
        375–380
D-Ribulose,
    conversion to D-arabinose, 366–370
    preparation of, 368
Ribulose-5-phosphate,
    equilibrium with ribose-5-phosphate,
        377
    formation from 6-phosphogluconate,
        323–324, 326–327

in transketolase reaction, 371–375, .375–380
Rice bran,
  phospholipase B from, 666
Roach, cellulase from, 176
Roots, extraction of, 63
Rosinduline GG,
  inhibition of sugar cane phosphorylase by, 198
R-1-P, see Ribose-1-phosphate
R-5-P, see Ribose-5-phosphate
Rubidium ion,
  activation of aceto-CoA-kinase by, 591
    of "malic" enzyme (DPN) by, 752
    of pyruvate kinase by, 439
    of yeast aldehyde dehydrogenase (DPN, TPN) by, 510

S

Saccharo-1,4-lactone,
  as inhibitor of β-glucuronidase, 267
Saccharomyces cerevisiae, see Yeast, baker's
Saccharomyces fragilis,
  galactokinase from galactose-adapted, 292–293
  galactowaldenase from, 294
Safranin,
  inactivation of invertase by, 257
Salicylaldehyde,
  as substrate for aldehyde dehydrogenase, 517
  for liver aldehyde oxidase, 527
Salicylic esters,
  as substrates for lipases, 633–634
Saliva,
  α-amylase from, 150–154
Salts,
  inhibition of phosphoribomutase by, 362
  use of in protein fractionation, 67–82
Sand,
  extraction of fresh yeast cells with, 708
Sarcina lutea,
  extraction of by grinding, 59
Sarcoma, rat,
  triosephosphate isomerase in, 391
Schardinger α-dextrin,
  as primer for phosphorylase, 196

Schizosaccharomyces,
  pH optimum of invertase from, 259
Sedoheptulose-7-phosphate,
  as substrate for phosphohexokinase, 310
  in transaldolase reaction, 381
  in transketolase reaction, 371–375, 375–380
Seeds, germinated,
  blastolipase from, 642
Semicarbazide,
  inhibition of glyoxylic reductase by, 535
Sequential induction, see Simultaneous adaptation
Serine, acetylation of, 618
Serratia,
  glycerophosphorylcholine diesterase from, 135, 668–670
  media for isolation and growth of, 663
  phospholipase A from, 663
  phospholipase B from, 666
Serum,
  cholinesterase from, 644, 647
  phosphoglycerate kinase in, 418
Serum proteins,
  separation by use of acid precipitants, 86
SH groups, see Sulfhydryl groups
SH-inhibiting reagents, see also under names of individual reagents
  action on liver alcohol dehydrogenase, 499
Silica jelly,
  as substitute for agar, 131
Silver,
  use of in purification of β-glucosidase, 237
Silver ion,
  as inhibitor of enzymes, 157, 197, 240, 257, 298, 313, 464, 507, 627
Silver oxide,
  as inactivator of α-galactosidase, 251
Simultaneous adaptation, 133
Snail (Helix pomatia),
  cellulase from, 173–178
Snake venoms,
  acetylcholinesterase from, 645
  phospholipase A from, 662–663, 664

Sodium chloride,
activation of alcohol dehydrogenase (TPN) by, 507
in protein precipitation, 509
Sodium dodecyl sulfate,
inhibition of phospholipase C by, 671
Sodium ion,
inhibition of aceto-CoA-kinase by, 591
of aldehyde dehydrogenase by, 510
of pyruvate kinase by, 439
of transacetylase by, 598, 607
Sodium lauryl sulfate,
as inhibitor of pectinesterase, 162
Somogyi reagent, see Copper reagents
Sonic oscillator,
in extraction of *L. mesenteroides*, 505
1,5-Sorbitan,
as substrate for mammalian hexokinase, 284
1,5-Sorbitan-6-phosphate,
as inhibitor of mammalian hexokinase, 285
Sorbitol dehydrogenase,
from liver, 348–350
L-Sorbose,
as glycosyl acceptor for sucrose phosphorylase, 228
as substrate for fructokinase, 289
L-Sorbose-1-phosphate,
as inhibitor of mammalian hexokinase, 285
of yeast hexokinase, 275
Sorbyl coenzyme A, see 2,4-Hexadienoyl coenzyme A
Soybean leaves,
*nitrate reductase from, 63
Sphingomyelin,
as substrate for phospholipase C, 671
Spinach leaves,
glyceraldehyde-3-phosphate dehydrogenase (TPN) from, 413
glycolic acid oxidase from, 528–532
glyoxylic acid reductase from, 532
"malic" enzyme (TPN) in, 748
pentose phosphate isomerase from, 366
"phosphorylating enzyme" (succinate activating) from, 722
transketolase from, 371–375
Spleen,
β-glucuronidase from, 269

β-ketoreductase in, 573
Spores, bacterial,
resistance to rupture by sonic oscillator, 58
Spores, fungus,
disintegration of, 64
Squash, phosphorylase from, 194
Squid ganglia,
acetylcholinesterase from, 645
choline acetylase from, 620–621
Staphylococci,
rupture of by sonic oscillator, 58
Staphylococcus aureus,
β-acetolactic acid decarboxylase from, 474
Starch-Celite column,
for precipitation chromatography of proteins, 712
Steapsin, see Lipase, pancreatic
Streptococci, hemolytic,
acetate kinase from, 595
hyaluronidase from, 168
Streptococcus faecalis,
acetate kinase from, 595
acetoin-forming system from, 474
extraction of by grinding, 61
Strontium hydroxide,
as adsorbent for invertase, 257
Strontium ion,
inhibition of enolase by, 434
of phospholipase C by, 671
Suberate,
competitive inhibition of succinic dehydrogenase by, 727
Succinate,
conversion to succinyl-CoA, 718–722
Succinate buffer, 141
Succinate derivatives,
inhibition of succinic dehydrogenase by, 726
Succinic dehydrogenase, 722–729
assay methods for, 722–725
manometric method, 724
spectrophotometric method, 724–725
Thunberg method, 723–724
extraction from liver, 35, 37
flavin as component of, 728
from heart muscle, 48–49, 725–726
lability of, 43

occurrence of, 725
oxidized form of, 728
  susceptibility to cyanide, 728
from plants,
  assay of, 724
  preparation of, 725–726
protection of by succinate and pyrophosphate, 44
purification by butyl alcohol method,
  50
relationship to cytochrome b, 728
Succinic oxidase,
  from rat liver mitochondria, 29
Succinic semialdehyde,
  as substrate for aldehyde dehydrogenase, 510
Succinohydroxamic acid,
  determination of, 719
Succinyl coenzyme A,
  formation by ketoglutaric dehydrogenase, 714–718
  preparation of, 575, 688
  reaction with hydroxylamine, 719
  role in ATP synthesis by "P enzyme,"
  718–722
Succinyl CoA deacylase,
  from pig heart, 602–606
  use in ketoglutaric dehydrogenase
  assay, 715
Succinyl-β-ketoacyl coenzyme A transferase, see Coenzyme A transferase
Sucrose,
  as acceptor for β-fructofuranosidase,
  255
  as donor of β-fructofuranoside group,
  255
  as donor substrate for amylosucrase,
  185
  as donor substrate for levansucrase,
  187
  labeled, synthesis by sucrose phosphorylase, 229
  optical rotation of, 232
  as substrate for dextransucrase, 182
  as substrate for invertase, 251
Sucrose phosphorylase (α-transglucosi-
  dase),
  from $Leuconostoc\ mesenteroides$, 225–
  229
  hydrolytic activity of, 229

  from $Pseudomonas\ saccharophila$, 225–
  229
Sugar beet,
  phosphorylase from, 194
Sugar beet leaves,
  "malic" enzyme (TPN) in, 748
Sugars,
  chromatography of, 260–261
Sulfanilamide,
  acetylation of, 608, 610, 618
  as coenzyme A assay, 588
Sulfhydryl compounds,
  activation of lipase by, 639
Sulfhydryl (SH) groups,
  nitroprusside test for, 623–624
  in succinic dehydrogenase, 726
Sulfide,
  inhibition of aconitase by, 698
Sulfonamides,
  effect on phosphorylase content of
  plants, 198
Sulfosalicylic acid,
  as protein precipitant, 85
Sunflower leaves,
  "malic" enzyme (TPN) in, 748
Suramin,
  as inhibitor of yeast hexokinase, 275
Sweet almond emulsin,
  α-galactosidase (melibiase) from, 249–
  251
  β-glucosidase from, 234–240
Sweet corn,
  phosphorylase from, 194
Sweet potato,
  β-amylase in, 154–158
  phosphorylase from, 194

T

Tagatose,
  as substrate for fructokinase, 289
Tagatose-1,6-diphosphate,
  as substrate for aldolase, 313
D-Tagatose-6-phosphate,
  as substrate for phosphohexokinase,
  310
Tannic acid,
  in purification of glucose aerodehydrogenase, 343

Tannin,
  in fractionation of glucosidase, 236
  as inactivator of α-galactosidase, 251
Temperature,
  effect on reaction rate of fumarase, 734
Testes, bovine,
  hyaluronidase from, 167
Tetraethylthiuram disulfide (antabuse),
    inhibition of liver aldehyde dehydro-
    genase (DPN) by, 517
Tetrahydrofurane,
  as protein precipitant, 84
Tetrazolium, neo-,
  as hydrogen acceptor in phosphoro-
    clastic system, 482
Tetrazolium, triphenyl (TTZ),
  in assay of butyryl CoA dehydro-
    genase, 553–554
  as hydrogen acceptor in phosphoro-
    clastic system, 482
Tetrose phosphate,
  in transaldolase reaction, 381
Thermostability,
  of phospholipases A, B, C, and D,
    664–665, 668, 671, 672
Thiamine,
  effect on phosphorylase content of
    plants, 198
Thiamine pyrophosphate (cocarbox-
    ylase),
  as coenzyme for acetoin formation,
    474, 493
  for α-acetolactic acid forming sys-
    tem, 474
  for carboxylases, 461, 463, 465, 469,
    470
  for formic hydrogenlyase, 541
  for α-ketoglutaric dehydrogenase
    system, 718
  in phosphoroclastic splitting of
    pyruvate, 478, 481
  for pyruvic oxidase, animal, 487
  for pyruvic oxidase of bacteria,
    Fraction A, 490, 493
  for pyruvic oxidase in L. delbrückii,
    482, 483, 486
Thiamine triphosphate,
  replacement of thiamine pyrophos-
    phate by, 463

Thiocyanate ion,
  activation and inhibition of fumarase
    by, 734
Thioglycolate,
  activation of amine acetylation by, 611
  as reducing agent, 609
Thiolase (β-ketothiolase, acetoacetate
    condensing enzyme, acetoacetate
    cleavage enzyme),
  in assay of crotonyl CoA, 562
  evidence for family of enzymes, 584
  from heart, 574, 581–585
Thiol compounds, see Sulfhydryl com-
    pounds
Thiol ester bond, see Acyl-mercapto bond
Thiol esterase, see Deacylase
Thiol esters,
  formation of, from ketoaldehydes and
    glutathione, 454
  reaction with hydroxylamine, 561
  as substrates for glyoxalase II, 458
Thiols,
  as acetyl acceptors, 405
  activation of glyceraldehyde-3-phos-
    phate dehydrogenase by, 405
  of yeast aldehyde dehydrogenase
    (DPN, TPN) by, 510
Thunberg method,
  for dehydrogenase measurement, 444
Thymine,
  in aerobic enrichment culture, 127
Thyroid gland,
  β-ketoreductase in, 573
Tissue homogenates, 10–15
  apparatus for, 11–13, 16
  definition of, 10, 11
  extent of disruption in, 15
  isolation of microsomes from, 18
  of mitochondria from, 17
  of nuclear fraction from, 17
  of nuclei from, 18
  media for, 13, 14
  preparation of, 12, 13, 17
  uses of, 14
Tissue slices,
  basis of calculation, 8
  limitations in use of, 9
  media for, 7, 8
  preparation of, humid chamber for, 6
  razor for, 4

Stadie-Riggs microtome for, 3–5
principle for use of, 3
respiration rates of, 9
technique, 3–9
Tobacco,
   glycolic acid oxidase, 529
   glyoxylic acid reductase from, 532–535
   pectinesterase from, 159
Toluene,
   as activator of β-glucosidase, 240
   and cell permeability, 55
   use of in extraction of enzymes, 55
m-Toluidine, acetylation of, 610
o-Toluidine, acetylation of, 610
p-Toluidine,
   as inactivator of invertase, 257
Tomato,
   pectinesterase from, 159
   polygalacturonidase from, 164
Torpedo,
   acetylcholinesterase from, 645
Torulopsis rotundata,
   phosphorylase from, 199
TPN, see Triphosphopyridine nucleotide
3'-TPN, see 3'-Triphosphopyridine nucleotide
TPP, see Thiamine pyrophosphate
Transacetylase (Phosphotransacetylase),
   in acetylation of amines, 608
     of glucosamine, 613
   in assay of acetyl CoA deacylase, 606
     of aldehyde dehydrogenase, 518
     of amino acid acetylase, 616
     of choline acetylase, 622
     of CoA transphorase, 599–600
     of condensing enzyme, 685
     of pyruvic oxidase, 487
   from Cl. kluyveri, 596–599
   heat inactivation of, 492, 602
   inhibition by sodium ion, 607
   role in acylation of amino acids, 618
     in butyrate oxidation Cl. kluyveri,
     551
Transaldolase,
   from brewer's yeast, 381–383
*Transaminase from mitochondria, 50
Transfructosidase, invertase as, 255
α-Transglucosidase, see Sucrose phosphorylase

Transglycosidation,
   by β-galactosidase, 248
Transglycosylase activity,
   of invertases, see also Transfructo-
     sidase, 255, 258–262
*Transhydrogenase (pyridine nucleotide),
   absence of from L. mesenteroides, 331
   extraction from animal tissues, 38
Transketolase,
   from baker's yeast, 375–380
   from liver, 371–373, 375
   from spinach, 371–375
Trehalose,
   optical rotation of, 232
Triacetin,
   as substrate for lipases, 630
Tributyrin,
   as substrate for lipase, 628, 630
Tricarboxylic acid cycle (citric acid
   cycle),
   role in fatty acid oxidation in mito-
     chondria, 545–548
Trichloroacetic acid,
   as protein precipitant, 85
Triethanolamine,
   inhibition of choline dehydrogenase
     by, 677
Triethanolamine buffer, 388, 392
Triglycerides,
   effect of chain length on rates of
     hydrolysis, 637
Trimethylammonium ion,
   inhibition of choline dehydrogenase
     by, 677
Triose phosphate,
   determination of,
     by alkali lability, 311
     by colorimetric methods, 311
     by enzymatic methods, 311
Triose phosphate dehydrogenase, see
   Glyceraldehyde-3-phosphate dehydrogenase
Triose phosphate isomerase,
   in assay of DR-aldolase, 384
   from calf muscle, 387–391
   from other sources, 391
Triphosphopyridine nucleotide (TPN),
   see also Dehydrogenases
   assay of, with aldehyde dehydro-
     genase, 514

effect of, on oxalacetic carboxylase
  activity of "malic" enzyme, 741
3'-Triphosphopyridine nucleotide,
  action with *Leuconostoc* glucose-6-
    phosphate dehydrogenase, 332
Tripolyphosphate, sodium,
  as inhibitor of hexokinase, 275
Tris(hydroxymethyl)aminomethane
  (Tris) buffer, 144
  reaction of with manganous ion, 504
Tris-maleate buffer, 143
*Trypsin,
  as inactivator of hexokinase, 275
  partition chromatography of, 110–111
  solubilization of enzymes by, 37
Tryptophan,
  adaptation of *Pseudomonas* to, 135
  in aerobic enrichment culture, 128
*Tryptophan peroxidase-oxidase system,
  as adaptive enzyme from *Pseudo-
    monas*, 134
TTZ, see Tetrazolium, triphenyl
Tumor cells,
  phosphoglycerate kinase from, 418
Tumor, mouse,
  triosephosphate isomerase in, 391
Tyramine,
  inhibition of choline dehydrogenase by,
    677
*Tyrosinase,
  as inactivator of invertase, 257
Tween 20 (polyoxymethylene sorbitan
    monolaureate),
  in solubilization of erythrocyte cholin-
    esterase, 647
  as substrate for lipase, 628, 630–631

U

UDPG, see Uridinediphosphoglucose
UDPGal, see Uridinediphosphogalactose
Uracil,
  in aerobic enrichment culture, 127
*Uracil-thymine oxidase,
  formation of in *Mycobacterium*, 135,
    136
Uranyl acetate,
  as adsorbent for invertase, 257
  in purification of glucose aerodehy-
    drogenase, 343

Urea,
  specific inactivation of polygalactur-
    onidase by, 166
Urethan (ethyl carbamate),
  inhibition of acetoin-forming system
    by, 475
  of succinic dehydrogenase by, 728
*Uricase,
  solubilization of by butanol, 42, 45, 50
Uridinediphosphogalactose (UDPGal),
  role in galactowaldenase reaction, 293
Uridinediphosphoglucose (UDPG),
  as coenzyme for galactowaldenase, 293
Uridine triphosphate,
  as substrate for phosphohexokinase,
    310
Uridyl-transferase,
  role in formation of UDPGal, 293
UTP, see Uridine triphosphate

V

n-Valerate,
  as CoA acceptor in *Cl. kluyveri*, 602
Veronal, see Barbital, diethylbarbiturate
Versene, see Ethylenediaminetetraace-
    tate (EDTA)
Vinyl acetate,
  as CoA acceptor in *Cl. kluyveri*, 602
Vinylacetyl coenzyme A,
  as substrate for crotonase, 559, 562,
    564, 565
Vitamin A,
  as substrate for liver alcohol dehydro-
    genase, 498

W

Wasp venoms,
  phospholipase B from, 666
Waxy maize,
  phosphorylase from, 194
Wheat germ,
  α-carboxylase from, 465–471
  lipase from, 639, 642
  "malic" enzyme from, 739–748
Wood-destroying fungi,
  cellulase from, 177

X

*Xanthine oxidase,
  from liver, 33

from milk, 35, 50

solubilization of by lipase or trypsin, 39

D-Xylose,

   beta form as substrate for glucose dehydrogenase, 338

   conversion to glucosido-xylose by maltose phosphorylase, 230

   conversion to D-xylulose, 366–370

   inhibition of mammalian hexokinase by, 284

Xylose isomerase,

   from *E. coli*, 366–367

   from *L. pentosus*, 366–367

   from *Ps. hydrophila*, 366–370

β-D-Xylosides,

   as substrates for β-glucosidase, 237

D-Xylulose,

   conversion to D-xylose, 366–370

Y

Yeast,

   acetoacetate-activating enzyme in, 573–574

   alcohol dehydrogenase (TPN) from, 503

   aldolase from, 315–320

   condensing enzyme (citrate-forming) in, 694

   cytochrome b_2 from, 50

   disintegration of, 64

   extraction of enzymes from, 51–62

   isocitric dehydrogenase (DPN) from, 707–709

   isocitric dehydrogenase (TPN) from, 705–707

   lipase from, 641

   pectinesterase from, 160

   pentokinase from, 357–360

   phosphoribomutase from, 362

   phosphorylase from, 199–200

   plasmolysis of, 271

   polygalacturonidase from, 164

   preparation and extraction of air-dried, 53

   pyruvate kinase in, 439

   quick-frozen, autolysis of, 588–589

   sonic extract of, 58, 589

   triosephosphate isomerase from, 391

Yeast, ale,

   α-carboxylase from, 460–464

Yeast, baker's,

   aceto-CoA-kinase from, 585–591

   alcohol dehydrogenase (DPN) from, 500–503

   alcohol dehydrogenase (TPN) in, 504

   aldehyde dehydrogenase (DPN, TPN) from, 508–511

   aldehyde dehydrogenase (TPN) from, 511–514

   extraction of fresh cells with sand, 708

   β-fructofuranosidase (invertase) from, 251–257, 259

   gluconokinase from, 351

   glyceraldehyde-3-phosphate dehydrogenase ("Protein 2") from, 272, 407–411

   glyoxalase I from, 454–458

   hexokinase from, 269–277

   invertase, see β-fructofuranosidase

   lactic dehydrogenase from, 444–449

   pentose phosphate isomerase from, 376–377

   phosphoglycerate kinase from, 418

   phosphorylase from, 199–200

   plasmolysis of, 253

   preparation of dried, 501

   transketolase from, 375–380

Yeast, brewer's (bottom yeast),

   alcohol dehydrogenase from, 503

   enolase from, 427–435

   gluconokinase from, 351, 353–354

   glucose-6-phosphate dehydrogenase (TPN) (Zwischenferment) from, 323–326

   glucose-1-phosphate kinase from, 354–356

   6-phosphogluconic dehydrogenase (TPN) from, 323–324, 326–327

   phosphoglycerate kinase from, 415–422

   transaldolase from, 381–383

"Yeast Protein No. 2,"

   crystallization of, 272

   identity with glyceraldehyde-3-phosphate dehydrogenase, 409

"Yeast Protein No. 3,"

   crystallization of, 272

"Yeast Yellow Protein,"

   crystallization of, 272–273

Z

Zein,
  organic solvents for, 83
Zinc,
  for crystallization of insulin, 85
  in ethanol fractionation of proteins,
    579, 720
  for precipitation of serum albumin, 85
Zinc hydroxide gel,
  as adsorbent, 91, 96
  preparation of, 555–556
Zinc ion,
  activation of aldolase (yeast) by, 319
    of enolase by, 433–434

    of oxalacetic carboxylase by, 757
    of phospholipase C by, 671
    of plant carboxylases by, 463, 469
  inhibition of β-galactosidase by, 248
    of GPC diesterase by, 670
    of phosphoglucomutase by, 298
    of potato phosphorylase by, 197
Zinc sulfate,
  as adsorbent for invertase, 257
Zinc sulfide,
  as adsorbent for invertase, 257
"Zwischenferment," see Glucose-6-phosphate dehydrogenase
Zymobacterium oroticum,
  adaptation to orotic acid, 135